ANNALS OF
THE NEW YORK ACADEMY
OF SCIENCES

Volume 936

EDITORIAL STAFF

Executive Editor
BARBARA M. GOLDMAN

Managing Editor
JUSTINE CULLINAN

Associate Editors
JOYCE HITCHCOCK
JOHN W. KENNEDY

The New York Academy of Sciences
2 East 63rd Street
New York, New York 10021

THE NEW YORK ACADEMY OF SCIENCES
(Founded in 1817)

BOARD OF GOVERNORS, September 2000–September 2001

BILL GREEN, *Chairman of the Board*
TORSTEN WIESEL, *Vice Chairman of the Board*
RODNEY W. NICHOLS, *President and CEO* [ex officio]

Honorary Life Governors
WILLIAM T. GOLDEN   JOSHUA LEDERBERG

JOHN T. MORGAN, *Treasurer*

Governors

ELEANOR BAUM	D. ALLAN BROMLEY	KAREN BURKE
LAWRENCE B. BUTTENWIESER		PRAVEEN CHAUDHARI
JOHN H. GIBBONS	MICHAEL GOLDEN	RONALD L. GRAHAM
ROBERT G. LAHITA	JACQUELINE LEO	WILLIAM J. McDONOUGH
JOHN F. NIBLACK	SANDRA PANEM	RICHARD RAVITCH
RICHARD A. RIFKIND	SARA LEE SCHUPF	JAMES H. SIMONS

HELENE L. KAPLAN, *Counsel* [ex officio]   NANCY B. EISENBERG, *Interim Secretary* [ex officio]

FIBRINOGEN

XVIth INTERNATIONAL FIBRINOGEN WORKSHOP

ANNALS OF THE NEW YORK ACADEMY OF SCIENCES
Volume 936

FIBRINOGEN

XVIth INTERNATIONAL FIBRINOGEN WORKSHOP

Edited by
Willem Nieuwenhuizen, Michael W. Mosesson,
and Moniek P.M. de Maat

The New York Academy of Sciences
New York, New York
2001

Copyright © 2001 by the New York Academy of Sciences. All rights reserved. Under the provisions of the United States Copyright Act of 1976, individual readers of the Annals *are permitted to make fair use of the material in them for teaching and research. Permission is granted to quote from the* Annals *provided that the customary acknowledgment is made of the source. Material in the* Annals *may be republished only by permission of the Academy. Address inquiries to the Permissions Department (editorial@nyas.org) at the New York Academy of Sciences.*

Copying fees: *For each copy of an article made beyond the free copying permitted under Section 107 or 108 of the 1976 Copyright Act, a fee should be paid through the Copyright Clearance Center, Inc., 222 Rosewood Drive, Danvers, MA 01923 (www.copyright.com).*

∞ *The paper used in this publication meets the minimum requirements of American National Standard for Information Sciences—Permanence of Paper for Printed Library Materials. ANSI Z39.48-1984.*

Library of Congress Cataloging-in-Publication Data

International Fibrinogen Workshop (16th : 2000 : Noordwijkerhout, Netherlands)
  Fibrinogen / XVIth International Fibrinogen Workshop ; edited by Willem
Nieuwenhuizen, Michael W. Mosesson, and Moniek P.M. de Maat.
      p. ; cm. — (Annals of the New York Academy of Sciences, ISSN 0077-8923 ; v. 936)
  Includes bibliographical references and index.
    ISBN 1-57331-314-9 (cloth : alk. paper) — ISBN 1-57331-315-7 (paper : alk. paper)
      1. Fibrinogen—Congresses. I. Nieuwenhuizen, Willem, 1942- II. Mosesson, Michael
W., 1934- III. De Maat, Moniek P. M. IV. Title. V. Series.
    [DNLM: 1. Fibrinogen—Congresses. WH 310 I564f 2001]

Q11 .N5 vol. 936
[QP93.5]
500 s—dc21
[612.1'15]

                                        2001030730
                                                CIP

K-M Research/PCP
Printed in the United States of America
ISBN 1-57331-314-9 (cloth)
ISBN 1-57331-315-7 (paper)
ISSN 0077-8923

ANNALS OF THE NEW YORK ACADEMY OF SCIENCES

Volume 936

FIBRINOGEN

XVIth INTERNATIONAL FIBRINOGEN WORKSHOP[a]

Editors
WILLEM NIEUWENHUIZEN, MICHAEL W. MOSESSON,
AND MONIEK P.M. DE MAAT

Officers and Counselors of the International Fibrinogen Research Society
DAVID L. AMRANI, PAUL BISHOP, STEPHEN O. BRENNAN, FRANK BROSSTAD,
CARL-ERIK DEMPFLE, PATRICK. J. GAFFNEY, GERD GRIENINGER,
PATRICIA J. SIMPSON-HAIDARIS, AGNES HENSCHEN-EDMAN, ROGER LIJNEN,
SUSAN T. LORD, GORDON D.O. LOWE, MICHIO MATSUDA, LEONID MEDVED,
MICHAEL W. MOSESSON, C. HARI NAIR, WILLEM NIEUWENHUIZEN,
DENNIS B. RYLATT, JOHN R. SHAINOFF, CLAUDINE SORIA,
JEANNETTE SORIA, AND ESTÉ H.H. VORSTER

CONTENTS

Preface. *By* MICHAEL W. MOSESSON FOR THE EDITORS. xi

Part I. Fibrinogen Structure and Function

Fibrinogen: Evolution of the Structure–Function Concept: Keynote Address
at Fibrinogen 2000 Congress. *By* BIRGER BLOMBÄCK 1

The Structure and Biological Features of Fibrinogen and Fibrin.
By MICHAEL W. MOSESSON, KEVIN R. SIEBENLIST, AND DAVID A. MEH . . . 11

Crystal Structure Studies on Fibrinogen and Fibrin. *By* RUSSELL F. DOOLITTLE,
ZHE YANG, AND IGOR MOCHALKIN 31

Contribution of the α_EC Domain to the Structure and Function of
Fibrinogen-420. *By* GERD GRIENINGER 44

Hereditary Disorders of Fibrinogen. *By* MICHIO MATSUDA AND TERUKO SUGO . . 65

A Database for Human Fibrinogen Variants. *By* M. HANSS AND F. BIOT 89

[a]This volume is the proceedings of a New York Academy of Sciences Conference entitled **Fribrinogen 2000: XVIth International Fibrinogen Workshop**, held 23–26 August, 2000, in Leiden, the Netherlands.

Molecular Mechanisms of Hypo- and Afibrinogenemia.
  By STEPHEN O. BRENNAN, ANDREW P. FELLOWES, AND PETER M. GEORGE . 91

Insight from Studies with Recombinant Fibrinogens.
  By SUSAN T. LORD AND OLEG V. GORKUN 101

Synthesis of a Mouse Model of the Dysfibrinogen Vlissingen/Frankfurt IV.
  By KELLY A. HOGAN, NOBUYO MAEDA, KIMBERLY D. KLUCKMAN,
  AND SUSAN T. LORD.. 117

Structural and Functional Role of the β-Strand Insert (γ381–390) in the
  Fibrinogen γ-Module: A "Pull Out" Hypothesis. *By* SERGEI YAKOVLEV,
  DMITRY LOUKINOV, AND LEONID MEDVED 122

Structure and Properties of Clots from Fibrinogen Bicêtre II (γ308 Asn→Lys):
  Increased Permeability due to Larger Pores, Thicker Fibers, and
  Decreased Rigidity. *By* RITA MARCHI, STÉPHANE LOYAU,
  EDUARDO ANGLÉS-CANO, AND JOHN W. WEISEL 125

Fibrinogen Longmont: A Heterozygous Abnormal Fibrinogen with Bβ Arg-166 to
  Cys Substitution Associated with Defective Fibrin Polymerization.
  By KARIM C. LOUNES, JERRY B. LEFKOWITZ, ANDREW I. COATES,
  ROY R. HANTGAN, AGNES HENSCHEN-EDMAN, AND SUSAN T. LORD 129

Part II. Blood Coagulation and Fibrin Formation

Determinants of Thrombin Specificity. *By* ENRICO DI CERA AND
  ANGELENE M. CANTWELL 133

The Fibrin Intermediate, Its Place in the Fibrinogen–Fibrin Transformation.
  By JOHN R. SHAINOFF, GARY B. SMEJKAL, PATRICIA M. DIBELLO,
  BARBARA CHASE, OLGA V. MITKEVICH, AND HELMUT LILL 147

Early Events in the Polymerization of Fibrin. *By* MATTIA ROCCO,
  SIMONETTA BERNOCCO, MARCO TURCI, ALDO PROFUMO,
  CARLA CUNIBERTI, AND FABIO FERRI 167

Conformational Changes upon Conversion of Fibrinogen into Fibrin:
  The Mechanisms of Exposure of Cryptic Sites. *By* LEONID MEDVED,
  GALINA TSURUPA, AND SERGEI YAKOVLEV 185

Polymerization Site *a* Function Dependence on Structural Integrity of Its
  Nearby Calcium Binding Site. *By* KARIM C. LOUNES, NOBUO OKUMURA,
  KELLY A. HOGAN, LIFANG PING, AND SUSAN T. LORD 205

Fibrin Formation and Proteolysis during Ancrod Treatment: Evidence for
  Des-A-Profibrin Formation and Thrombin Independent Factor XIII
  Activity. *By* CARL-ERIK DEMPFLE, SOTIRIA ARGIRIOU,
  SONJA ALESCI, KLAUS KUCHER, H. MÜLLER-PELTZER,
  KLAUS RÜBSAMEN, AND DIETER L. HEENE 210

Characterization of Crosslinking Sites in Fibrinogen for Plasminogen
  Activator Inhibitor 2 (PAI-2). *By* HELEN RITCHIE, LAURA C. LAWRIE,
  MICHAEL W. MOSESSON, AND NUALA A. BOOTH 215

The Formation of β Fibrin Requires a Functional *a* Site. *By* KELLY A. HOGAN,
  BETTINA BOLLIGER, NOBUO OKUMURA, AND SUSAN T. LORD 219

Mode of Perturbation of Asahi Fibrin Assembly by the Extra Oligosaccharides.
By TERUKO SUGO, OSAMU SEKINE, CHIZUKO NAKAMIKAWA,
HITOSHI ENDO, CARMEN L. AROCHA-PIÑANGO, AND MICHIO MATSUDA ... 223

Part III. Fibrinolysis and Related Subjects

Elements of the Fibrinolytic System. *By* H.R. LIJNEN 226

Fibrin-Mediated Plasminogen Activation. *By* WILLEM NIEUWENHUIZEN 237

Modulation of Fibrin Cofactor Activity in Plasminogen Activation.
By MICHAEL NESHEIM, JOHN WALKER, WEI WANG, MICHAEL BOFFA,
ANTON HORREVOETS, AND LASZLO BAJZAR......................... 247

Inhibition of Fibrinolysis by Lipoprotein(a). *By* EDUARDO ANGLÉS-CANO,
AURORA DE LA PEÑA DÍAZ, AND STÉPHANE LOYAU................... 261

Genetic Manipulation of Fibrinogen and Fibrinolysis in Mice.
By JAY L. DEGEN, ANGELA F. DREW, JOSEPH S. PALUMBO,
KEITH W. KOMBRINCK, JORGE A. BEZERRA, MARY JO S. DANTON,
KENN HOLMBÄCK, AND THEODORE T. SUH 276

Factor XIII: Structure, Activation, and Interactions with Fibrinogen and Fibrin.
By LASZLO LORAND 291

The Structure and Function of the αC Domains of Fibrinogen.
By JOHN W. WEISEL AND LEONID MEDVED......................... 312

Fibrinogen αC Domains Contain Cryptic Plasminogen and tPA Binding Sites.
By GALINA TSURUPA AND LEONID MEDVED 328

Clot Lysis of Variant Recombinant Fibrinogens Confirms that Fiber Diameter
is a Major Determinant of Lysis Rate. *By* JENNIFER L. MULLIN,
SUSAN E. NORFOLK, JOHN W. WEISEL, AND SUSAN T. LORD 331

Crosslinking of α_2-Antiplasmin to Fibrin. *By* KYUNG N. LEE,
CHUNG S. LEE, WEON-CHAN TAE, KENNETH W. JACKSON,
VICTORIA J. CHRISTIANSEN, AND PATRICK A. MCKEE................. 335

Part IV. Fibrinogen Interactions with Cell and Tissue Matrix Elements

Platelet–Fibrinogen Interactions. *By* JOEL S. BENNETT 340

Fibrin and Wound Healing. *By* RICHARD A.F. CLARK..................... 355

Recognition of Fibrinogen by Leukocyte Integrins.
By TATIANA P. UGAROVA AND VALENTIN P. YAKUBENKO 368

Interaction of Fibrin with VE-Cadherin. *By* JOSÉ MARTINEZ, ANDRÉS FERBER,
TAMI L. BACH, AND CHRISTOPHER H. YAEN 386

Tumors and Fibrinogen: The Role of Fibrinogen as an Extracellular Matrix
Protein. *By* P.J. SIMPSON-HAIDARIS AND BRIAN RYBARCZYK 406

Role of Fibrin Matrix in Angiogenesis. *By* VICTOR W.M. VAN HINSBERGH,
ANNEMIE COLLEN, AND PIETER KOOLWIJK......................... 426

Fibrinogen Modulates Gene Expression in Wounded Fibroblasts.
By MARIAN PEREIRA AND PATRICIA J. SIMPSON-HAIDARIS 438

Mutations on Fibrinogen (γ316–322) Are Associated with Reduction in Platelet Adhesion Under Flow Conditions. *By* JASPER A. REMIJN, KARIM C. LOUNES, KELLY A. HOGAN, SUSAN T. LORD, DENNIS K. GALANAKIS, JAN J. SIXMA, AND PHILIP G. DE GROOT 444

A New Model for *in Vitro* Clot Formation that Considers the Mode of the Fibrin(ogen) Contacts to Platelets and the Arrangement of the Platelet Cytoskeleton. *By* EBERHARD MORGENSTERN, MATTHIAS DAUB, AND ROLF DIERICHS 449

Mutation of W215 Compromises Thrombin Cleavage of Fibrinogen, but Not of PAR1 or Protein C. *By* YOUHNA M. AYALA, DANIELE AROSIO, AND ENRICO DI CERA 456

Platelets in Suspension Require Preactivation to Adhere to Immobilized Fibrinogen. *By* ARNAUD BONNEFOY, QINGDE LIU, W. GRAY JEROME, CHANTAL LEGRAND, AND MONY M. FROJMOVIC 459

Platelet Adhesion to Fibrinogen Coated at Various Densities. *By* MARKÉTA JIROUŠKOVÁ AND BARRY S. COLLER 464

Attenuation of Neointima Formation Following Arterial Injury in PAI-1-Deficient Mice. *By* VICTORIA A. PLOPLIS AND FRANCIS J. CASTELLINO 466

Part V. Biosynthesis and Metabolism

Transcriptional Control Mechanism of Fibrinogen Gene Expression. *By* GERALD M. FULLER AND ZHIXIN ZHANG 469

Fibrinogen Biosynthesis: Assembly, Intracellular Degradation, and Association with Lipid Synthesis and Secretion. *By* COLVIN M. REDMAN AND HUI XIA . 480

Fibrinogen Gene Mutations Accounting for Congenital Afibrinogenemia. *By* MARGUERITE NEERMAN-ARBEZ 496

Effects of Diet, Drugs, and Genes on Plasma Fibrinogen Levels. *By* MONIEK P.M. DE MAAT 509

Genetic and Immunological Characterization of Fibrinogen Inclusion Bodies in Patients with Hepatic Fibrinogen Storage and Liver Disease. *By* DANIELA MEDICINA, GIOVANNA FABBRETTI, STEPHEN O. BRENNAN, PETER M. GEORGE, BO KUDRYK, AND FRANCESCO CALLEA 522

Hypofibrinogenemia Associated with a Heterozygous C→T Nucleotide Substitution at Position −1138 BP of the 5′-Flanking Region of the Fibrinogen Aα-Chain Gene. *By* NOBUO OKUMURA, FUMIKO TERASAWA, OSAMU YONEKAWA, ETSUKO HAMADA, AND HIROSHI KANEKO 526

Modified Clotting Properties of Fibrinogen in the Presence of Acetylsalicylic Acid in a Purified System. *By* SHU HE, MARGARETA BLOMBÄCK, GINA YOO, RAKHI SINHA, AND AGNES H. HENSCHEN-EDMAN 531

Identification and Characterization of Five New Fibrinogen Gene Polymorphisms. *By* ANDREW P. FELLOWES, STEPHEN O. BRENNAN, AND PETER M. GEORGE 536

Development of Pulmonary Fibrosis in Fibrinogen-Deficient Mice.
  By JULIE A. WILBERDING, VICTORIA A. PLOPLIS, LAURA MCLENNAN,
  ZHONG LIANG, IVO CORNELISSEN, MATTHEW FELDMAN,
  MELANIE E. DEFORD, ELLIOT D. ROSEN, AND FRANCIS J. CASTELLINO 542

Part VI. Epidemiological and Clinical Subjects

Fibrinogen Polymorphisms and Atherothrombotic Disease. *By* F.R. GREEN 549

Fibrinogen and Its Degradation Products as Thrombotic Risk Factors.
  By GORDON D.O. LOWE AND ANN RUMLEY 560

Wound Healing: Role of Commercial Fibrin Sealants. *By* DAVID L. AMRANI,
  JAMES P. DIORIO, AND YVES DELMOTTE 566

Fibrinogen Non-Inherited Heterogeneity and Its Relationship to Function
  in Health and Disease. *By* AGNES H. HENSCHEN-EDMAN 580

Fibrin Degradation Products: A Review of Structures Found
  in Vitro and *in Vivo*. *By* PATRICK J. GAFFNEY 594

Antifibrinogen IgG, Fibrinogen, and Clq Complexes Circulating in a
  Hypodysfibrinogenemic Proband: Isolation, Stoichiometry, and
  Partial Characterization. *By* DENNIS K. GALANAKIS,
  AGNES HENSCHEN-EDMAN, JOHN WEISEL, AND SILVIA SPITZER 611

Factor XIII Prevents Development of Myocardial Edema in Children
  Undergoing Surgery for Congenital Heart Disease. *By* GERNOLD WOZNIAK,
  THOMAS NOLL, HAKAN AKINTÜRK, JOSEF THUL, AND MATTHIAS MÜLLER . 617

Fibrinogen: A Vascular Risk Factor, Why? Contributing Effect of Oncostatin M
  on Both Fibrinogen Biosynthesis by Hepatocytes and Participation in
  Atherothrombotic Risk Related to Modifications of Endothelial Cells.
  By F. MIRSHAHI, M. VASSE, L. VINCENT, V. TROCHON, J. POURTAU,
  J.P. VANNIER, H. LI, J. SORIA, AND C. SORIA 621

Purification of Fibrinogen and Virus Removal Using Preparative Electrophoresis.
  By ANDREW GILBERT, MICHAEL EVTUSHENKO, AND HARI NAIR 625

Effect of Moderate Alcohol Consumption on Fibrinogen Levels in
  Healthy Volunteers Is Discordant with Effects on C-Reactive Protein.
  By AAFJE SIERKSMA, MARTIJN S. VAN DER GAAG, CORNELIS KLUFT,
  AND HENK F.J. HENDRIKS 630

Fibrin Network Structure: Changes in Characteristics in Response to
  Physical Activity in Combination with a Pre-exercise Meal.
  By S.J. MOSS, M.M. MALAN, L.I. DREYER, AND H.H. VORSTER 634

Influence of Plasma Triglyceride and Plasma Cholesterol Levels on the
  Clearance Rate of Fibrinogen. *By* MAARTJE VERSCHUUR, MARIAN BEKKERS,
  MONIQUE G.M. VAN ERCK, JEF J. EMEIS, AND MONIEK P.M. DE MAAT 639

Index of Contributors 643

Financial assistance was received from:
- AMERICAN DIAGNOSTICA
- AMERICAN HEART ASSOCIATION
- AVENTIS
- BRISTOL-MYERS SQUIBB PHARMACEUTICAL RESEARCH INSTITUTE
- GLOBAL HEMOSTASIS INSTITUTE (GH)
- GRADIPORE
- INSTRUMENTATION LABORATORY
- JOHNSON & JOHNSON
- KORDIA (THE NETHERLANDS)
- MERCK & CO., INC. – USA
- MINISTRY OF HEALTH WELFARE AND SPORT (THE NETHERLANDS)
- NATIONAL HEART, LUNG AND BLOOD INSTITUTE, NIH
- ORGANON TEKNIKA
- PARKE DAVIS
- PHARMING
- ROCHE DIAGNOSTICS NETHERLANDS
- SIGMA DIAGNOSTICS
- UNILEVER HEALTH INSTITUTE
- ZYMOGENETICS INC.

The New York Academy of Sciences believes it has a responsibility to provide an open forum for discussion of scientific questions. The positions taken by the participants in the reported conferences are their own and not necessarily those of the Academy. The Academy has no intent to influence legislation by providing such forums.

Preface

Fibrinogen is a multifunctional blood protein of hepatic origin which, upon proteolytic conversion to fibrin by thrombin, polymerizes to form a fibrin clot, the main structural element of the intravascular thrombus. The fibrinogen research field encompasses multiple disciplines and subject areas including hemostasis and thrombosis, synthesis, assembly and secretion, platelets, transglutamination (crosslinking), fibrinolysis, wound healing and inflammation, cell migration, angiogenesis, tumor growth and metastasis, epidemiology, and genetic disorders. This volume results from the second Fibrinogen conference sponsored by New York Academy of Sciences, the first having taken place in 1982 in New York City. The volume containing the first conference proceedings (Volume 408 of the Annals, 1983) has served as a veritable "bible" for workers in the field, and continues to serve in this capacity from both a historical and informational perspective.

Since the 1982 conference, many details have been added to our knowledge of the structure and functions of fibrinogen, and new areas of inquiry that were barely perceptible at that time have emerged. Several controversies have been settled and new ones have taken their place. The primary sequence of structural variants and genetic or metabolic polymorphisms are now understood in finer detail, and we are well on our way to understanding fibrinogen gene regulation and expression. There is new information on fibrin and fibrinogen polymerization, structure and crosslinking, fibrin(ogen) intra- and intermolecular association sites, domain folding and conformation, and functional interactions with other proteins, growth factors, cells, lipids, and other surfaces. There have been significant advances concerned with the initiation and regulation of fibrinolysis and the fibrinolytic process itself. By now there is a clear understanding of its intracellular synthesis, assembly, and secretion, and a remarkable increase in the scope of our knowledge on congenital fibrinogen disorders, including afibrinogenemia and hypofibrinogenemia. In many cases, there is a detailed understanding of structure–function relationships, and some of these have led to useful correlations between clinical manifestations and management. New pathophysiological features of dysfibrinogenemic molecules have been reported, and this has broadened the categories of clinical interest to include conditions such as amyloidosis, renal failure, and hepatic cirrhosis. Many clinical and epidemiological studies on the relationship between fibrinogen levels and thromboembolic disease have been reported, and no end to this line of study is yet in sight. Scientists and pharmaceutical companies have introduced new therapeutic formulations of fibrinogen and fibrin, and several versatile drug therapies are now available for resolving, preventing, or managing fibrin-based vascular occlusive diseases, including myocardial infarction and stroke.

The second New York Academy of Sciences Fibrinogen conference, which was organized with the above overview in mind, took place under the chairmanship of Willem Nieuwenhuizen together with Michael W. Mosesson and Moniek P.M. de Maat, between August 24 and August 26, 2000 at the Leeuwenhorst Conference Centre in Noordwijkerhout, the Netherlands. Our first goal was to seed the conference agenda with invited plenary presentations by acknowledged experts that covered all the major subjects mentioned above, to which we added somewhat shorter oral presentations topically selected from among the submitted abstracts. We provided time at each session for discussion, which we anticipated would extend beyond the sessions among self-selected groups and individuals. To round out interactions and presentations by the participants we also included two extended luncheon/poster sessions, and we were delighted to hear from one of our most experienced and revered investigators, Birger Blombäck, who summarized his historical perspectives and accomplishments at our banquet dinner. Our second, and equally important, goal was to publish another New York Academy of Sciences Fibrinogen book that would serve as a long-lived and authoritative supplement to the first book. We believe we have achieved that goal with this present volume.

MICHAEL W. MOSESSON
For the editors

Fibrinogen: Evolution of the Structure–Function Concept

Keynote Address at Fibrinogen 2000 Congress

BIRGER BLOMBÄCK

Coagulation Laboratory, Karolinska Institutet, Stockholm, Sweden

ABSTRACT: Coagulation of blood is such an evident phenomenon that its observation can be traced back to earliest historical times. The great philosophers and physicians of antiquity discussed and provided interesting explanations. However, it was not until the end of the seventeenth century that the structural component of the blood clot was described by Malpighi as a white fibrous substance. In the middle of the nineteenth century this was identified as a constituent of pathological thrombi and given the name fibrin. At about that time its precursor in blood, fibrinogen, was isolated in a highly purified form by Hammarsten who suggested that, preceding fibrin formation, activation of fibrinogen by thrombin occurred by limited proteolysis. The activation mechanism was eventually clarified in the 1950s. It was shown to proceed in two discrete steps, by removal of low molecular weight activation peptides. Ferry postulated, based on physicochemical observations, that the activated molecules aligned in a half-staggered fashion to form polymers. The rapid post-war development of biochemical technology permitted evaluation of the primary structure of fibrinogen. With that followed identification of molecular domains in the activated fibrinogen molecules that participate in polymer formation, crosslinking of polymeric structures, and domains for cellular attachment. Crystallization of fragments and, recently, of the entire molecule has confirmed and extended this knowledge. Lately, it has also been possible to obtain detailed information on the architecture of the fiber network in the fibrin gel. The gel structure is primarily determined by the initial rate of fibrinogen activation, but without infringement of this primary rule, several factors in blood may modulate the structure. Fibrinogen and fibrin play important roles in normal hemostasis, wound healing, and pathological processes, such as thrombosis and atherosclerosis.

KEYWORDS: Fibrin; Fibrinogen; Structure–function.

Coagulation of blood is such evident phenomena that its observation can be traced back to early historical times. My presentation of the development of ideas on this phenomenon is focused on thoughts in the Western cultural sphere. Blood clotting certainly has attracted attention in Eastern philosophy as well, but is not addressed here. Much red blood flows in Homer's Iliad and Odyssey but clotting of blood was obviously of secondary interest and rarely, if ever, mentioned. On the other hand, the

---

Address for correspondence: Birger Blombäck, Coagulation Laboratory, Karolinska Institutet, Nobels väg 12A, SE-171 77 Stockholm, Sweden. Voice/fax: +46 (8) 33 83 95.
Birger.Blombäck@mbb.ki.se

astute philosophers took notice. Aristotle (cited by Beck in Ref. 12) in about 300 BC, declared that coagulation of blood was similar to freezing of water. Blood clotted when the energy provided by the heart dissipated. Clotting was connected with death and decay. This view was more or less shared by Galen (cited by Beck in Ref. 12) and others in antiquity.

It is of interest to notice that Galen, the teacher of humoral pathology, saw a phenomenon in blood (rediscovered 2,000 years later by Robert Fåhreus[1]) that has a bearing on fibrinogen. In many states of disease he saw that clotted blood had a whitish cap, *crusta phlogistica,* on top of the red cell mass. This is due, as we now know, to the increased sedimentation rate of blood cells caused by increased fibrinogen level in blood. Sedimentation occurred before clotting froze the process. Galen thought that the *crusta phlogistica* was due to an excessive and dangerous secretion of one of the four humors, the phlegm.

Another view we find in the Koran where we read about the sacred role of blood clots: "Recite you in the name of thy Lord who created:—Created man from clots of blood".

However, it was not until the end of the seventeenth century that the structural basis of the blood clot was described by Malpighi[2] as a white fibrous substance obtained from the red blood clot through washing with water. Although at that time it was widely believed that coagulation occurred through aggregation of the red blood cells, Hewson,[3] in about 1770, showed that coagulation takes place in plasma (or lymph as it was called). He found that neutral salts could inhibit coagulation and after sedimentation of the red cells the plasma could be decanted. On dilution with water it clotted spontaneously. Half a century later Müller[4] confirmed that clotting occurred in plasma, when he filtered off the large blood cells from frog's blood before coagulation occurred. However, what gave rise to this fibrous substance in blood was still not clear. Hewson[3] and later Virchow[5] believed that it was present as a particulate solute in plasma prior to clotting. Until the middle of the nineteenth century this view was upheld in one form or another. Nevertheless, through the studies of Virchow, it became clear that fibrin was an important factor in pathology, and that view is still with us today. It was at that time that our concepts of fibrinogen and fibrin dramatically changed.

Three names are in the forefront; Deni de Commercy,[6] in 1859, proposed that there existed a precursor of fibrin that he called fibrinogen and which could be salted out from plasma. Fibrinogen transformed spontaneously into fibrous fibrin by a process he called denaturation. At about this time it became clear through the work of, above all, Alexander Schmidt[7] that a "thrombic" enzyme was involved in fibrin formation. Then, on the scene came Olof Hammarsten.[8-11] He prepared a highly purified fibrinogen from horse plasma by precipitation with salt. The solubility of fibrinogen was dependent on the presence of physiological salt concentrations—it had low solubility in water and at high salt concentrations. Exposure to temperatures of 52–55°C caused irreversible denaturation. Hammarsten's preparation did not clot spontaneously but required the addition of "thrombic enzyme" in order to clot. Hammarsten was, like Buchanan before him, struck by the similarities between curdling of milk and clotting of fibrinogen. In his studies on casein he had observed a proteolytic cleavage of casein by rennin and suggested that there might also be a hydrolytic cleavage of fibrinogen during its transformation to fibrin, since he found that a small amount of protein like material was left in clot supernatants. The quantity of this material nevertheless varied appreciably in different preparations and Hammarsten refrained from taking any definite stand on the matter.

Then came a lull in fibrinogen research that, with few exceptions, lasted into mid twentieth century. It was the development of new technology in biochemistry and biophysics that spurred the hectic activity that then commenced. More extensive references to the early history of fibrinogen research than are cited here can be found in reviews by Beck[12] and by Blombäck.[13-14]

In the 1940s and 1950s numerous measurements of electrophoretic mobility, sedimentation and diffusion constants, osmotic pressure, intrinsic viscosity, flow birefringence, and light scattering were made on fibrinogen, especially by American investigators. From the different physicochemical measurements a picture of the fibrinogen molecule as a prolate ellipsoid of revolution emerged.[15,16] This model had a length of 500–700 Å and axial ratio of 5–20 Å. The molecular weight derived at by these measurements varied considerably but the most reliable was considered to be 330–340 kDa.[17,18] The electron microscopic picture by Hall and Slayter[19,20] of dehydrated specimens of fibrinogen was in general agreement with a hydrated prolate ellipsoid with a length of about 450 Å. However, Stryer[21] and associates, in low angle X-ray diffraction studies, found that hydrated specimens of fibrin and fibrinogen had axial repeats of 226 Å, which was about the same distance between dark and light bands found by Hawn and Porter[22] in fibrin fibers using the electron microscope. Since this was also the length of fibrinogen observed at the isoelectric point by Hall, the latter assumed that shrinkage of the molecule occurred during clotting and that fibrin was built up by end-to-end and side-to-side alignment, of such shrunken molecules. However, Astbury's group,[23] in their wide angle X-ray diffraction studies, found no fundamental internal rearrangement of the fibrinogen molecule during clotting.

Interest in the process of fibrin formation in presence of thrombin led to experiments by Ferry and coworkers in which clotting was inhibited by addition of different compounds (e.g., polyethylene glycol) before measurement. It was obvious that intermediary polymeric products precede visual clotting. Light scattering studies of partial polymers, stabilized in hexamethylene glycol solution, showed polymers of about 15 fibrinogen units and with mass/length ratio twice that of fibrinogen.[24,25] This geometry led, as recounted by Ferry in 1983, to the idea of assembly with staggered lateral overlapping, based partly on "the teleological argument [never mentioned] that such a design was good biological engineering" and furthermore no shrinkage of the molecule had to be assumed.[26] Nevertheless, electron microscopic pictures of early polymers by Bang[27] later confirmed this arrangement of monomers in polymers. It was also established that the physical properties of the clot were profoundly influenced by the milieu of its formation, that is, pH and ionic strength. Ferry and Morrison[28] showed that the clots formed under different experimental conditions could be classified as ranging in properties between two extreme types: the *fine* clot, which is transparent, elastic, friable, and non-synerizing and the *coarse* clot which is opaque, plastic, non-friable and synerizes very readily. The fine clot was favored by increasing pH, increasing ionic strength, and increasing temperature and was pictured as fibrinogen units joined end-to-end. In the coarse type, formed at low pH and low ionic strength, the chains were proposed to aggregate laterally in large bundles.

In the 1940s, Laki and Mommaerts[29] suggested that a two-stage process was involved in fibrinogen to fibrin transition. They found that, although no coagulation occurred at slightly acidic pH values, the fibrinogen was changed in some way by thrombin, since the coagulation time after neutralization was shorter for a fibrinogen solution that had been preincubated with thrombin. In fact Mihalyi[30] noticed a small difference in electrophoretic mobility of fibrinogen and fibrin in urea solution. Although prior to this time several investigators had found indirect proof of proteolytic action of thrombin on fibrinogen, there was no direct evidence. It was still, with few exceptions, believed that thrombin "denatured" fibrinogen in some way and caused its precipitation. There was a lack of suitable analytical technique to unequivocally answer this question. Direct evidence of a limited proteolytic activity of thrombin was not given until 1951, when Bailey, Bettelheim, Lorand, and Middlebrook,[31] by means of N-terminal analysis, showed a difference between the N-terminal amino acids of fibrinogen and fibrin and that two low molecular weight peptides, named A and B, were released,. The following scheme could then be put forward; F → f+P; nf → fn; fn → clot. (In this reaction, F denotes fibrinogen; f, fibrin monomer; P, fibrinopeptides; and fn, the intermediary polymer).

The rough outlines of the fibrinogen molecule started to emerge but little was known about its structure in terms of its constituent protein chains. Analysis with Sanger's FDNB method had given variable results.[32] With the use of Edman's PTH method, more reproducible results were obtained.[33] The latter studies suggested that fibrinogen was composed of three pairs of polypeptide chains, later named Aα, Bβ, and γ, and this applied to many animal species. It was also obvious that two different peptides were released and that the molecule was probably dimeric. The fibrinopeptides had acidic net charges and it was suggested that in the fibrin polymer the noncovalent bonding between fibrin monomers was caused by electrostatic forces as a result of the release of the acidic fibrinopeptides. The amino acid sequence of fibrinopeptides from a large number of species showed remarkable homologies. On the basis of these sequences and the nucleotide code words for amino acids, an evolutionary tree for animal species was suggested.[34] The preservation of a nonapeptide sequence close to the bond split by thrombin was remarkable, and especially a phenylalanyl residue that was always located at a fixed position from that bond. The idea was born that thrombin used this preserved sequence for binding to its active site. In fact, synthetic peptides modeled on the fibrinopeptide sequence were found to be inhibitors and substrates of thrombin;[35] and much later X-ray crystallographic studies showed how that part of the fibrinopeptide neatly fits into the active site of thrombin.[36] In addition to the sequences in the fibrinopeptide A region, a sequence of about 20 amino acid residues in the Aα-chain, beyond the fibrinopeptide A sequence, appeared to participate in binding and guiding thrombin.[37]

Three different chains with molecular weights between 48 and 70 kDa were isolated from reduced fibrinogen[38] and their amino acid sequence, and eventually the complete disulfide bonded structure, established by several investigators (Henschen,[39] Doolittle,[40] and our group[41]). The molecular weight of the chains suggested that the molecular weight of 340,000 for fibrinogen was for a dimer with pairs of each chain. Fragmentation of the molecule with CNBr proved to be important for establishing the covalent structure of fibrinogen since one of the fragments (N-DSK) contained N-terminal portions of all three chains. Thus, in the molecule of 340-kDa two three-chain monomers were linked together—but how? Hydrodynamic studies had indicated that fibrinopeptides were located at both ends of the molecule. These findings led to presentation of the molecule as two elongated bodies with fibrinopeptides at both ends.[41] However, this picture was soon to be reversed when it was found that symmetrical disulfides were linking two γ chains and two Aα chains at the very N-terminal end of the molecule. The N-terminal parts of the two monomers were linked at the N-terminal end.[42] Thus, it was probable that Hall and Slayter's middle nodule contained the N-terminal ends of the molecule. It was by that time known that fragmentation of fibrinogen gives rise to two main fragments called E and D.[43,45] Fragment E was shown to be analogues to N-DSK[44] and was, therefore, located in the middle nodule. Fragment D was then supposedly the distal nodule in the electron microscopic picture.[44] The E and D nodules seen in the electron microscope were held together by thin filamentous structures. Based on the amino acid sequences of the portions of the molecule that linked fragment E and D, Doolittle proposed for this part of the molecule a coiled-coil helical structure.[46] The picture of a three nodule structure for fibrinogen was later modified when it was shown that the end nodule consisted of two domains.[47] A fourth nodule (αC domain) formed by an extended Aα chain was shown in electron microscope pictures.[48–50] This nodule appeared to interact with the middle nodule and seemed to be attached by a thin filamentous structure with the D domain. In respect to fibrin, the electron microscopic studies verified Ferry's notion of staggered overlapping fibrinmonomers.

By the end of the 1970s the major architectural features as well as the complete covalent structure of fibrinogen was known. Recent crystallographic studies in the laboratories of Davie,[51] Doolittle,[52,53] and Cohen[54] have not only confirmed the early interpretations of fibrinogen structure and function, but also given exciting additional information at the atomic level.

The investigations by Robbins,[55] in 1944, had shown that fibrin can be classified into two types according to its solubility in alkali and acids. Fibrin formed in presence of calcium and a serum factor was insoluble. Solubility in other solvents like urea also turned out to depend on a factor found in serum and also contaminating the most pure fibrinogen preparations. Hydrodynamic measurement showed that whereas non-crosslinked fibrin dissociated in the presence of chaotropic reagents the crosslinked did not. Duckert et al.[56] brought to clinical relevance this factor when they described a patient with bleeding symptoms and deficiency of the crosslinking factor. Studies by Loewy[57] and Lorand[58] in the 1960s identified the factor as an enzyme that introduced isopeptide bonds (crosslinks) between the γ carboxy group of glutamine and an amino group in a different part of a neighboring molecule. Pisano[59] and coworkers identified the bond as ε-(γ-glutamyl) lysine. The process of crosslinking was further enlightened by the finding by Doolittle et al.,[60] who showed that specific residues in the C-terminal parts in the γ-chains took part in linking them in an antiparallel fashion. Their three-dimensional location was recently shown in the crystal structure of crosslinked fragment D by Doolittle.[52] Other crosslinks involving several residues in the α-chain were shown to be a final step in crosslinking of fibrin.[61]

The domains involved in polymerization were largely *terra incognito* in the 1960s. At that time an abnormal fibrinogen, fibrinogen Detroit, was discovered.[62] In this fibrinogen an amino acid substitution had occurred at a position three amino acid residues away on the amino side of the thrombin susceptible bond in the Aα chain. The clotting time with thrombin was much prolonged and homozygous patients had a severe bleeding disorder. Surprisingly the rate of cleavage of fibrinopeptide A was normal. At that time it was found by Heene and Matthias[63] that fibrinogen did adsorb to insolubilized fibrinogen provided that it had been treated with thrombin. It was subsequently found that N-DSK treated with thrombin adsorbed to insolubilized fibrinogen and also to fragment D, the plasmic degradation product of fibrinogen.[64] N-DSK from fibrinogen of the patient with fibrinogen Detroit did not absorb after treatment with thrombin.[65] A polymerization site in normal fibrinogen, detected in fibrinogen Detroit, was apparently unfolded by release of fibrinopeptide A in the N-terminal part of the molecule, and this site interacted with a site present in fragment D as had also been suggested previously.[66] This also made sense of the proposal by Ferry, 20 years previously, of a staggered overlapping of fibrin monomers in initial polymerization. Further support of this notion came from studies by Doolittle *et al.*,[67] who showed that a tetrapeptide with a wildtype amino acid sequence, FPRV, at the Detroit site did inhibit fibrin monomer polymerization, whereas a peptide containing this mutation did not. Most likely other structures in N-DSK are also required for interaction, but those structures have so far not been identified. The structures responsible for the interaction between the tetrapeptide and the fragment D domain were provided in recent crystallographic studies by Davie[51] and Doolittle.[52] These studies also identified in fragment D two of the three calcium atoms bound to fibrinogen and that are of importance for fibrinogen conformation and in fibrin formation.

The electron microscopic studies had confirmed Ferry's original concept of formation of linear or branched early polymers.[27,48–50,68,69] Nevertheless, the nature of the final network, the gel, was poorly understood. Studies employing turbidity, liquid permeation and three-dimensional confocal microscopy provided interesting observations on this structure and what governed its formation.[70] Thrombin, or rather the rate of activation of fibrinogen, was found to determine the network architecture. Fast activation yields a tight network with small pores (fine gel), whereas slow activation gives a porous network with thick fiber strands (coarse gel). More important, as studies with fibrin formation in plasma showed, the network established at the gel point remained essentially unchanged with time, despite large excess of thrombin and substrate. All fibrinogen molecules activated after the gel point are incorporated into the existing framework resulting in widening of strands but only minor decrease in meshwork area. Thus, it appears that fibrin gel formation is the result of a nucleation reaction. When the concentration of fibrin monomers have reached a critical concentration, the partial polymers formed acquire a higher functionality and non-linear structures develop with a higher order of complexity, according to the theory of Flory,[71] and gelation occurs. What can be the cause of this sudden higher functionality? One possibility would be branching of early polymers as a result of non-linear factor XIII crosslinking that was demonstrated by Mosesson.[69] Brosstad and associates[72] showed that thrombin formed in whole blood rapidly activated FXIII and crosslinked fibrinogen species were present at or even before the gel point. This has recently been confirmed by Mann.[73] Another source of higher functionality might be effected by release of fibrinopeptide B by thrombin, followed by interactions with αC domains on different protofibers or fibers.[50]

A polymerization site appears to be unfolded on release of FPB in the N-terminal part of the Bβ chain.[74–76] The complementary site may be located in the C-terminal part of the Bβ or Aα chain. It was early recognized that release of FPA was the leading event in fibrinogen activation and runs hand in hand with polymerization.[77] The release of FPB is much slower and is not correlated to the amount of fibrin formed. It was proposed in the 1950s, on basis of light scattering studies, that its release promoted side-to-side aggregation of linear polymers formed as a result of FPA release.[78] This view needed to be modified when it was shown by Godal and associates[79,80] and by our group[70] that when both FPA and FPB were released gelation occurred faster and at a smaller degree of activation than when only FPA was released. In fact, Shainoff[81] has shown that the association between monomers deficient in both FPA and FPB is greater than between monomers lacking only FPA. Since smaller activation is required a relatively coarser network is formed at the gelpoint and further release of FPB does not much affect the architecture of the network except for widening of strands.

Most studies on fibrin formation have been carried out on isolated fibrinogen thrombin systems. However, what is the nature of fibrin formed in whole blood? When the contact activation system (Factor XII, high molecular weight kininogen and prekallikrein) is inhibited in whole blood only FPA is released at a time when all fibrinogen has been converted to fibrin, as was shown recently by Mann et al.[73] and suggested previously.[74] Without inhibition of contact activation, and provided platelets are present, both peptides are released at a fast rate although FPA always is slightly preceding FPB.[82] This means that fast clotting and efficient hemostasis only occur at a lesion in the vasculature where the contact system is activated and platelets aggregate. Furthermore, at such sites Factor XIII is activated and fibronectin is incorporated into the polymer and, thereby, the fibrin is anchored to the vessel matrix.[83] Numerous *in vivo* experiments have been performed with certain snake venom enzymes that release FPA only from fibrinogen. Interestingly, no thrombotic complications occur. Activated fibrin monomers and dimers are present in blood but are rapidly taken up by various organs and degraded.[84] The reason why activation of fibrinogen follows a different pattern in native blood than when activated *in vitro* is an enigma. One explanation may be that during isolation of fibrinogen from blood conformational changes inevitably takes place and this makes the fibrinopeptide B site available to thrombin. In its native state in blood, the contact system and platelets are required to unfold this site.[85]

The role of fibrin as the structural element in pathological thrombi had been evident since Virchow's time. However, fibrin can also act as an "anticoagulant" by binding thrombin. This was first observed by Seegers and coworkers.[86] Trombin binds to both high and low affinity sites in fibrin.[87] The high affinity site appears to be located exclusively in the γ^1 chain variant.[88,89] In contrast, in fibrinogen, the fibrinopeptide residues (Aα9–16) bind to the active site in thrombin and binding to a succeeding sequence (Aα30–51) guides thrombin to the active site.[90–92]

Reports on the interaction of fibrinogen or fibrin with cellular elements started to emerge in the 1980s. It was shown that activated platelets expressed a receptor, GpIIb+IIIa, that bound to ligands in fibrinogen and fibrin.[93] One ligand was the ubiquitous RGDS sequence present in many proteins.[94] The other was specific and may physiologically be more important. It is present in the very C-terminal part of the γ chain but not in the $γ^l$ chain.[88,95] The interaction between platelets and fibrinogen directs fibrinogen to lesions in the vasculature, where fibrin formation is foremost needed.

Calcium plays an important role for fibrinogen structure and function. Pekelharing had in 1991 stated that calcium was necessary for both thrombin formation and fibrin deposition.[96] Hammarsten, however, did in 1896 claim that calcium ions had no influence on fibrin formation,[97] but acted only in the first phase of blood clotting. In mid twentieth century investigators observed that calcium chelating agents greatly retarded the clotting of fibrinogen by thrombin.[98] It later became well established that three calcium ions bind to fibrinogen and this binding is indispensable for its integrity in solution and during fibrin formation.[99] As mentioned previously, in the crystal structure two of the binding sites of calcium are clearly visible.[51,52]

Epidemiological studies have shown that increased fibrinogen level is a risk factor for ischaemic heart disease (for a review see Ref. 100). High fibrinogen levels cause tight network structures during clotting. Such structures are also formed, as I mentioned, at high thrombin concentrations. These structures are elastic, with thin fiber strands, and therefore brittle. Such structures we believe are thrombogenic in the sense that they are occluding and have propensity for embolization. Low fibrinogen and low thrombin concentrations on the other hand give rise to plastic structures that deform in flow. They can be considered nonthrombogenic.

In this review I have focused on the evolution of the structure–function concept for a protein of 340 kDa molecular weight. A new fibrinogen, with a molecular weight of 420 kDa, has been discovered.[101] This interesting protein is fully clottable but is present in very low concentrations (about 1–4% of total fibrinogen) and its role in physiology is so far enigmatic.

ACKNOWLEDGMENT

The technical assistance and editing by Birgitta Gruvfält is gratefully acknowledged.

REFERENCES

1. FÅHREUS, R. 1958. Acta Med. Scand. **161:** 151.
2. MALPIGHI, M. 1686. De polypo cordis dissertatio. *In* Opera Omnia, Londini.
3. GULLIVER, G. 1846. The works of William Hewson. Sydenham Soc., London.
4. MÜLLER, J. 1832. Poggendorfs Annalen, Vol. 25: 513.
5. VIRCHOW, R. 1847. Arch. Pathol. Anat. u. Physiol. (Virchow's, Berlin) **1:** 547.
6. DENIS DE COMMERCY, P.-S. 1859. Mémoire sur le sang. J.B. Baillière et fils, Paris.
7. SCHMIDT, A. 1872. Pflüg. Arch. ges Physiol. VI: 413.
8. HAMMARSTEN, O. 1876. Upsala Läkareförenings förhandlingar 1875–1876. **11:** 538.
9. HAMMARSTEN, O. 1876. Nova Acta Reg. Soc. Scient. Ups. Ser III. **10:** 1.
10. HAMMARSTEN, O. 1879. Pflüg. Arch ges Physiol. **19:** 563.
11. HAMMARSTEN, O. 1880. Pflüg Arch ges Physiol. **22:** 431.
12. BECK, E.A. 1968. The discovery of fibrinogen and its conversion to fibrin. *In* Fibrinogen. K. Laki, Ed. Marcel Dekker, Inc., New York.
13. BLOMBÄCK, B. 1958. Acta Physiol. Scand. **43**(Suppl.): 148.

14. BLOMBÄCK, B. 1967. Fibrinogen to fibrin transformation. In Blood Clotting Enzymology. W.H. Seegers, Ed. Academic Press Inc., New York.
15. EDSALL, J.T. 1954. J. Polymer. Sci. **12:** 253.
16. SCHERAGA, H.A. & M. LASKOWSKI, JR. 1957. Adv. Prot. Chem. **12:** 1.
17. CASPARY, E.A. & R.A. KEKWICK. 1957. Biochem. J. **67:** 41.
18. SHULMAN, S. 1953. J. Am. Chem. Soc. **75:** 5846.
19. HALL, C.E. 1949. J. Biol. Chem. **179:** 857.
20. HALL, C.E. & H.S. SLAYTER. 1959. J. Biophys. Biochem. Cytol. **5:** 11.
21. STRYER, L., C. COHEN & R. LANGRIDGE. 1963. Nature **197:** 793.
22. HAWN, C.V. & K.R. PORTER. 1947. J. Exptl. Med. **86:** 286.
23. BAILEY, K., W.T. ASTBURY & K.M. RUDALL. 1943. Nature **151:** 716.
24. FERRY, J.D. & S. SHULMAN. 1949. J. Am. Chem. Soc. **71:** 3198.
25. SHULMAN, S. & J.D. FERRY. 1951. J. Phys. Colloid. Chem. **55:** 135.
26. FERRY, J.D. 1983. Ann. N.Y. Acad. Sci. **408:** 1.
27. BANG, N.U. 1964. Thromb. Diath. Haemorrhag. **73**(Suppl. 13): 131.
28. FERRY, J.D. & P.R. MORRISON. 1947. J. Am. Chem. Soc. **69:** 388.
29. LAKI, K. & W.F.H.M. MOMMAERTS. 1945. Nature **156:** 664.
30. MIHALYI, E. 1950. Acta Chem. Scand. **4:** 317.
31. BAILEY, K., F.R. BETTELHEIM, L. LORAND, et al. 1951. Nature **167:** 233.
32. BAILEY, K. & F.R. BETTELHEIM. 1955. Brit. Med. Bull. **11:** 50.
33. BLOMBÄCK, B. & I. YAMASHINA. 1958. Arkiv Kemi. **12:** 299.
34. DOOLITTLE, R.F. & B. BLOMBÄCK. 1964. Nature **202:** 147.
35. BLOMBÄCK, B., M. BLOMBÄCK, P. OLSSON, et al. 1969. Scand. J. Clin. Lab. Inv. **24**(Suppl. 107): 59.
36. STUBBS, M.T. & W. BODE. 1993. Thromb. Res. **69:** 1.
37. HOGG, D.H. & B. BLOMBÄCK. 1978. Thromb. Res. **12:** 953.
38. HENSCHEN, A. 1964. Arkiv Kemi **22:** 397.
39. HENSCHEN, A., F. LOTTSPEICH, M. KEHL, et al. 1983. Ann. N.Y. Acad. Sci. **408:** 28.
40. DOOLITTLE, R.F. 1983. Ann. N.Y. Acad Sci. **408:** 13.
41. BLOMBÄCK, B., M. BLOMBÄCK, A. HENSCHEN, et al. 1968. Nature **218:** 130.
42. BLOMBÄCK, B. 1969. Br. J. Hæmatol. **17:** 145.
43. NUSSENZWEIG, V., M. SELIGMANN, J. PELMONT, et al. 1961. Ann. Inst. Pasteur. **100:** 377.
44. KOWALSKA-LOTH, B., B. GÅRDLUND, N. EGBERG, et al. 1973. Thromb. Res. **2:** 423.
45. MARDER, V.J., N. SHULMAN & W.R. CARROLL. 1969. J. Biol. Chem. **244:** 2111.
46. DOOLITTLE, R.F., D.M. GOLDBAUM & L.R. DOOLITTLE. 1978. J. Mol. Biol. **120:** 311.
47. NORTON, P. & H. SLAYTER. 1981. Proc. Natl. Acad. Sci. USA. **78:** 1661.
48. MOSESSON, M.W., J.F. HAINFELD, R.H. HASCHEMEYER, et al. 1981. J. Mol. Biol. **153:** 695.
49. ERICKSON, H.P. & W.E. FOWLER. 1983. Ann. N.Y. Acad. Sci. **408:** 146.
50. GORKUN, O.V., Y.I. VEKLICH, L.V. MEDVED, et al. 1994 Biochemistry **33:** 6986.
51. PRATT, K.P., H.C.F. CÔTÉ, D.W. CHUNG, et al. 1997. Proc. Natl. Acad. Sci. U.S.A. **94:** 7176.
52. DOOLITTLE, R.F. 1999. Thromb. Hæmostas. **82:** 271.
53. DOOLITTLE, R.F. 2000. XVIth International Workshop. Leiden, August 23–26.
54. BROWN, J.H., N. VOLKMANN, G. JUN, et al. 2000. Proc. Natl. Acad. Sci. U.S.A. **97:** 85.
55. ROBBINS, K.C. 1944. Am. J. Physiol. **142:** 581.
56. DUCKERT, F., E. JUNG & D.H. SCHMERLING. 1960. Thromb. Diath. Hæmorrhag. **5:** 179.
57. LOEWY, A.G. 1972. Ann. N.Y. Acad. Sci. **202:** 41.
58. LORAND, L. 1972. Ann. N.Y. Acad. Sci. **202:** 6.
59. PISANO, J.J., T.J. BRONZERT, M.P. PEYTON, et al. 1972. Ann. N.Y. Acad. Sci. **202:** 98.
60. DOOLITTLE, R.F., K.G. CASSMAN, R. CHEN, et al. 1972. Ann. N.Y. Acad. Sci. **202:** 114.
61. MCKEE, P.A., M.L. SCHWARTZ, S.V. PIZZO, et al. 1972. Ann. N.Y. Acad. Sci. **202:** 127.
62. BLOMBÄCK, M., B. BLOMBÄCK, E.F. MAMMEN, et al. 1968. Nature **218:** 134.
63. HEENE, D.L. & F.R. MATTHIAS. 1973. Thromb. Res. **2:** 137.
64. KUDRYK, B.J., D. COLLEN, K.R. WOODS, et al. 1974. J. Biol. Chem. **249:** 3322.
65. KUDRYK, B.J., B. BLOMBÄCK & M. BLOMBÄCK. 1976. Thromb. Res. **9:** 25.

66. BLOMBÄCK, B. & M. BLOMBÄCK. 1972. Ann. N.Y. Acad. Sci. **202:** 77.
67. LAUDANO, A.P. & F.R. DOOLITTLE. 1978. Biochemistry **19:** 1013.
68. WILLIAMS, R.C. 1983. Ann. N.Y. Acad. Sci. **408:** 180.
69. MOSESSON, M.W., K.R. SIEBENLIST, D.L. AMRANI, et al. 1989. Proc. Natl. Acad. Sci. U.S.A. **86:** 1113.
70. BLOMBÄCK, B., K. CARLSSON, K. FATAH, et al. 1994. Thromb. Res. **75:** 521.
71. FLORY, P. 1953. Principles of Polymer Chemistry. 347–398. Cornell University Press, Ithaca–London.
72. GRØN, B., C. FILION-MYKLEBUST, A. BENNICK, et al. 1992. Blood Coag. Fibrinol. **3:** 731.
73. BRUMMEL, K.E., S. BUTENAS & K.G. MANN. 1999. J. Biol. Chem. **274:** 22862.
74. BLOMBÄCK, B., B. HESSEL, D. HOGG, et al. 1978. Nature **257:** 501.
75. LAUDANO, A.P., B.A. COTTRELL & R.F. DOOLITTLE. 1983. Ann. N.Y. Acad. Sci. **408:** 315.
76. HASEGAWA, N. & S. SASAKI. 1990. Thromb. Res. **57:** 183.
77. BLOMBÄCK, B. & A. VESTERMARK. 1958. Arkiv. Kemi **12:** 173.
78. LAURENT, T.C. & B. BLOMBÄCK. 1958. Acta Chem. Scand. **12:** 1875.
79. HOLM, B., P. KIERULF & H.C. GODAL. 1986. Thromb. Res. **42:** 517.
80. HOLM, B., F. BROSSTAD, P. KIERULF, et al. 1985. Thromb. Res. **39:** 595.
81. SHAINOFF, J.R. & B.N. DARDIK. 1983. Ann. N.Y. Acad. Sci. **408:** 254.
82. BLOMBÄCK, B., B. HESSEL, M. OKADA, et al. 1981. Ann. N.Y. Acad. Sci. **370:** 536.
83. OKADA, M., B. BLOMBÄCK, CHANG MING-DER, et al. 1985. J. Biol. Chem. **260:** 1811.
84. EGBERG, N. 1973. Acta Med. Scand. **194:** 291.
85. BLOMBÄCK, B. 2000. Thromb. Res. **99:** 307.
86. SEEGERS, W.H., M. NIEFT & E.C. LOOMIS. 1945. Science **101:** 520.
87. LIU, C.Y., H. L.NOSSEL & K.L. KAPLAN. 1979. J. Biol. Chem. **254:** 10421.
88. WOLFENSTEIN-TODEL, C. & M.W. MOSESSON. 1981. Biochemistry **20:** 6146.
89. MEH, D.A., K.R. SIEBENLIST & M.W. MOSESSON. 1996. J. Biol. Chem. **271:** 23121.
90. HOGG, D.H. & B. BLOMBACK. 1978. Thromb. Res. **12:** 953.
91. VALI, Z. & H.A. SCHERAGA. 1988. Biochemistry **27:** 1956.
92. KACZMAREK, E. & J. MCDONAGH. 1988. J. Biol. Chem. **263:** 13896.
93. MARGUERIE, G.A., E.F. PLOW & T.S. EDGINGTON. 1979. J. Biol. Chem. **254:** 5357.
94. PLOW, E.F., M.D. PIERSCHBACHER, E. RAUSLAHTI, et al. 1985. Proc. Natl. Acad. Sci. U.S.A. **82:** 8057.
95. KLOCZEWIAK, M.S. TIMMONS, M.A. BEDNAREK, et al. 1989. Biochemistry **28:** 2915.
96. PEKELHARING, C.A. 1891. Virchow Festschrift **1:** 433.
97. HAMMARSTEN, O. 1896. Z. Physiol. Chem. Hoppe-Seylers **22:** 333.
98. GODAL, H.C. 1960. Scand. J. Clin. Lab. Invest. **12**(Suppl. 53): 1.
99. NIEUWENHUIZEN, W., I.A.M. VAN RUIJEN-VERMEER, W.J. NOOIJEN, et al. 1981. Thromb. Res. **22:** 653.
100. BLOMBÄCK, B. 1996. Thromb. Res. **83:** 1. 1.
101. FU, Y. & G. GRIENINGER. 1994. Proc. Natl. Acad. Sci. U.S.A. **91:** 2625.

The Structure and Biological Features of Fibrinogen and Fibrin

MICHAEL W. MOSESSON,[a] KEVIN R. SIEBENLIST,[b] AND DAVID A. MEH[a]

[a]*Blood Research Institute, The Blood Center of Southeastern Wisconsin, P.O. Box 2178, Milwaukee, Wisconsin 53201, USA*

[b]*Department of Biomedical Sciences, College of Health Sciences, Marquette University, P.O. Box 1881, Milwaukee, Wisconsin 53233, USA*

ABSTRACT: Fibrinogen and fibrin play important, overlapping roles in blood clotting, fibrinolysis, cellular and matrix interactions, inflammation, wound healing, and neoplasia. These events are regulated to a large extent by fibrin formation itself and by complementary interactions between specific binding sites on fibrin(ogen) and extrinsic molecules including proenzymes, clotting factors, enzyme inhibitors, and cell receptors. Fibrinogen is comprised of two sets of three polypeptide chains termed Aα, Bβ, and γ, that are joined by disulfide bridging within the N-terminal E domain. The molecules are elongated 45-nm structures consisting of two outer D domains, each connected to a central E domain by a coiled-coil segment. These domains contain constitutive binding sites that participate in fibrinogen conversion to fibrin, fibrin assembly, crosslinking, and platelet interactions (e.g., thrombin substrate, Da, Db, γXL, D:D, αC, γA chain platelet receptor) as well as sites that are available after fibrinopeptide cleavage (e.g., E domain low affinity non-substrate thrombin binding site); or that become exposed as a consequence of the polymerization process (e.g., tPA-dependent plasminogen activation). A constitutive plasma factor XIII binding site and a high affinity non-substrate thrombin binding site are located on variant γ' chains that comprise a minor proportion of the γ chain population. Initiation of fibrin assembly by thrombin-mediated cleavage of fibrinopeptide A from Aα chains exposes two E_A polymerization sites, and subsequent fibrinopeptide B cleavage exposes two E_B polymerization sites that can also interact with platelets, fibroblasts, and endothelial cells. Fibrin generation leads to end-to-middle intermolecular Da to E_A associations, resulting in linear double-stranded fibrils and equilaterally branched *trimolecular* fibril junctions. Side-to-side fibril convergence results in *bilateral* network branches and multistranded thick fiber cables. Concomitantly, factor XIII or thrombin-activated factor XIIIa introduce intermolecular covalent ϵ-(γ glutamyl)lysine bonds into these polymers, first creating γ dimers between properly aligned C-terminal γXL sites, which are positioned *transversely* between the two strands of each fibrin fibril. Later, crosslinks form mainly between complementary sites on α chains (forming α-polymers), and even more slowly among γ dimers to create higher order crosslinked γ trimers and tetramers, to complete the mature network structure.

KEYWORDS: Fibrin; Fibrinogen; Fibrinolysis; Factor XIII; Crosslinking; Endothelial cells; Leukocytes; Platelets; Integrins.

Address for correspondence: Michael W. Mosesson, Blood Research Institute, The Blood Center of Southeastern Wisconsin, P.O. Box 2178, Milwaukee, WI 53201-2178, USA. Voice: 414-937-3811; fax: 414-937-6284.
mwmosesson@bcsew.edu

INTRODUCTION

Fibrinogen and fibrin play important, overlapping roles in blood clotting, fibrinolysis, cellular and matrix interactions, inflammation, wound healing, and neoplasia. These events are regulated to a large extent by the fibrin formation process and by interactions between specific sites on fibrinogen or fibrin and extrinsic molecules

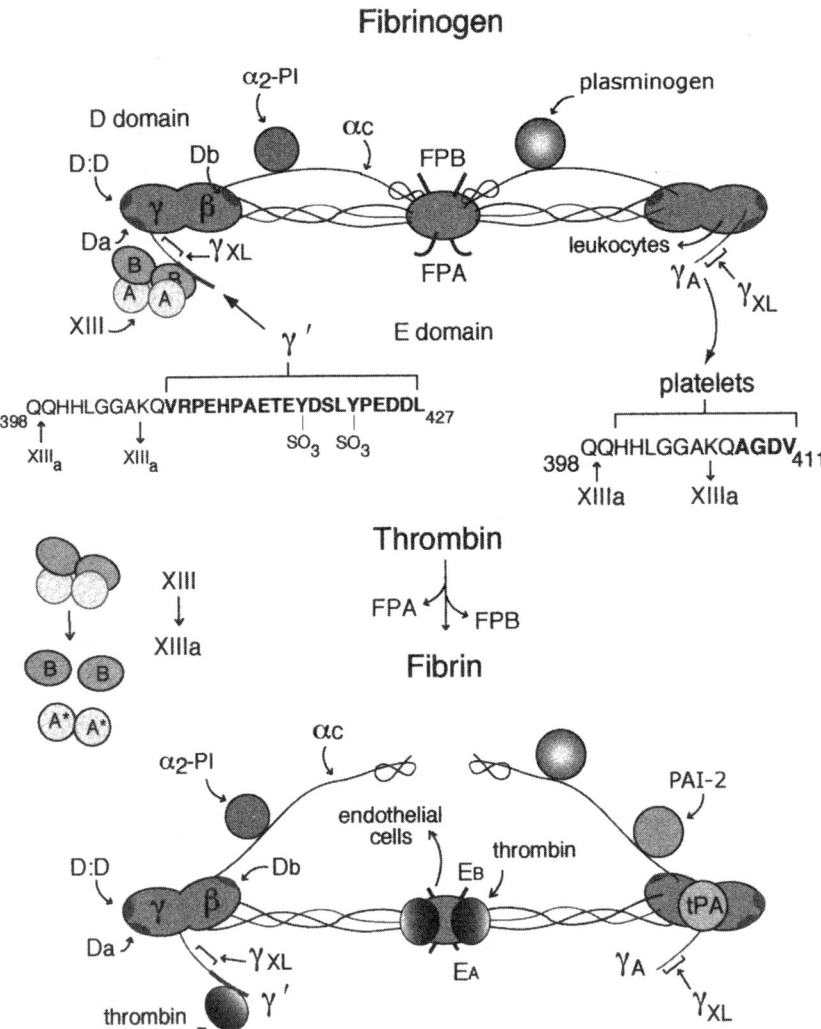

FIGURE 1. Schematic diagram of fibrinogen and fibrin showing the major structural domains, the association sites that participate in fibrin polymerization and crosslinking, and other molecular and cellular binding interactions.

such as proenzymes, clotting factors, growth factors, enzymes inhibitors, and cell receptors. Fibrinogen molecules are elongated 45-nm structures consisting of two outer D domains, each connected by a coiled-coil segment to a central E domain (see FIGURE 1). They are comprised of two sets of three polypeptide chains termed Aα, Bβ, and γ,[1] which are joined together within its N-terminal E domain by five symmetrical disulfide bridges, one pair at position Aα 28, two pairs between Aα36 and Bβ65, and a third set between the γ8 and γ9 positions.[2-4] Because these γ chain bridges are reciprocal (i.e., γ8 to γ9),[5] they orient the chains in an antiparallel manner, and this may contribute to the twofold axis of symmetry perpendicular to the long axis that fibrinogen displays.[6-8] Other non-symmetrical interchain disulfide bridges in this region form a so-called *disulfide ring*.[2,4] The Aα chain consists of 610, the Bβ chain 461, and the major form of γ chain, γA, 411 residues.[1] A minor γ chain variant termed γ', arises through alternative processing of the primary mRNA transcript,[9] and amounts to about 8% of the total γ chain population.[10] γ' chains consist of 427 residues and differ from the platelet-binding γA chains in that the four C-terminal γA residues, AGDV, are replaced by an anionic 20-amino acid sequence that includes two sulfated trosines.[11,12] Fibrinogen γ' chains bind plasma factor XIII B subunits and serve thereby as a carrier protein for the catalytic A subunits,[13] but they do not bind to the platelet fibrinogen receptor, $\alpha_{IIb}\beta_3$.[14] The C-terminal γA chain sequence, γ400–411, however, plays a critical role in mediating platelet aggregation via the platelet fibrinogen receptor.[15-17]

FIBRINOGEN CONVERSION TO FIBRIN AND FIBRIN ASSEMBLY

The N-terminal region of each Aα chain contains the FPA sequence, cleavage of which at Aα16R-17G by thrombin initiates fibrin assembly[18-20] by exposing a polymerization site, E_A. One portion of the E_A site is located at the N-terminus of the fibrin α chain comprising residues 17 to 20 (GPRV);[21] another portion is located in the fibrin β chain between residues 15 and 42.[22-25]

Each E_A site subsequently combines with a constitutive complementary binding pocket (Da) in the D domain of neighboring molecules that is located between γ337 and γ379.[26-28] These initial E_A:Da associations result in formation of double-stranded twisting fibrils in which fibrin molecules become aligned in an end-to-middle staggered overlapping domain arrangement (see FIGURE 2).[29-31] Fibrils undergo lateral associations and form branches that result in a complex fiber network.[32,33] Two types of branching account for fibrin network structures,[34] and they both play important roles in defining clot network structure. The first type occurs when double-stranded fibrils converge side-to-side to form the most widely appreciated fibril junction, a *tetramolecular* or *bilateral* branch point. More extensive lateral fibril associations result in large fiber bundles consisting of multiple fibrils and more condensed bilateral branch structures. This type of structure confers strength and rigidity to the network fibers.

The second type of branch junction, termed *trimolecular* or *equilateral,* forms by the coalescence of three fibrin molecules that connect three double-stranded fibrils of equal widths (FIG. 2). Equilateral branches probably form with greater frequency when fibrinopeptide cleavage is relatively slow. Under these conditions, especially

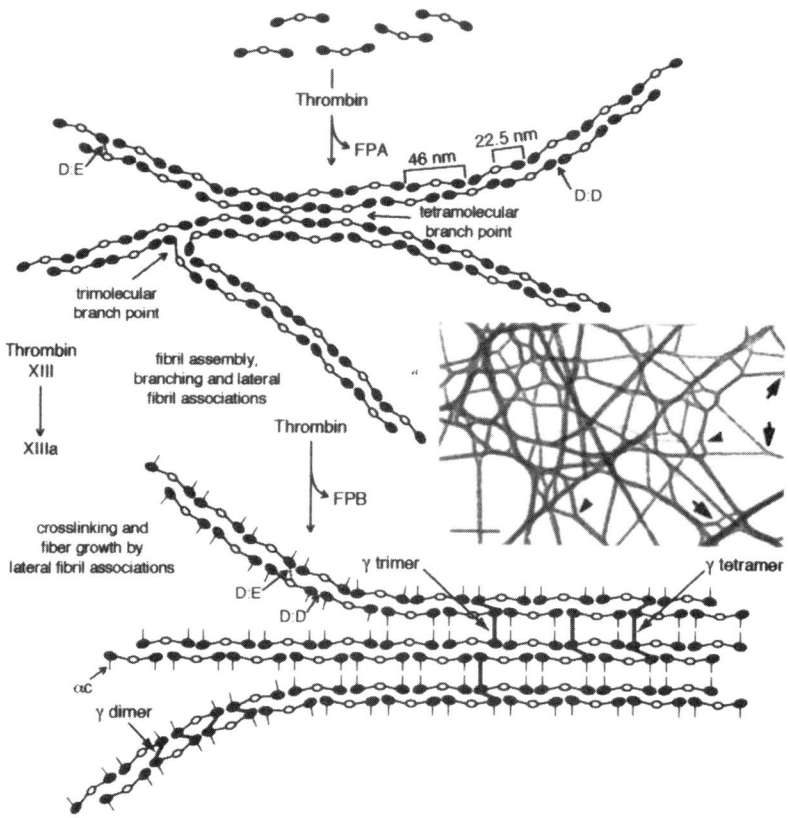

FIGURE 2. Schematic diagram of fibrin assembly and crosslinking. Assembly of fibrin begins with non-covalent interactions (D:E) between the EA and Da sites (*dotted lines*) to form end-to-middle staggered overlapping double-stranded fibrils (*upper*). Fibrils also branch and undergo lateral associations to form wider fibrils and fibers. [**Inset**: critical point dried thin fibril matrix containing equilateral (*arrows*) and bilateral (*arrowheads*) branch junctions; bar, 100 nm]. After cleavage of FPB (*lower*), αC domains become available for self-association with other αC domains, thereby promoting lateral fibril associations and fiber assembly. Factor XIII or XIIIa introduces ε-(γ-glutamyl) lysine isopeptide bonds between C-terminal γ_{XL} sites (*thick lines* between D domains) to rapidly form γ dimers. γ Trimers and γ tetramers form more slowly by interfibril γ chain crosslinking, and increase the resistance of the clot to fibrinolysis.

during the early phases of fibrin assembly, relatively large amounts of fibrin intermediates lacking a single FPA (des-A fibrin)[35,36] are formed. The availability and persistence of intermediate fibrin units very likely promotes higher degrees of equilateral branching as partially converted des-A fibrin intermediates search for complementary partners with only partial success. Blombäck *et al.*[37] have shown that the architecture of fibrin networks formed at low levels of thrombin is more branched and therefore tighter (i.e., less porous) than networks formed at high levels of thrombin. The type of network architecture that characterizes the tight fibrin network is entirely consistent with the higher degree of equilateral fibril branching that occurs at low thrombin levels.[38] Such branches enhance clot elasticity.

There are two constitutive self-association sites, γ_{XL} and D:D, that participate in the fibrin or fibrinogen assembly process.[39,40] The γ_{XL} site overlaps the crosslinking site in the C-terminal region of each γ chain. Intermolecular association between two γ_{XL} sites promotes alignment of crosslinking sites for subsequent factor XIII- or XIIIa-mediated transglutamination. The D:D sites are situated at the outer portion of each fibrin(ogen) D domain between residues 275 and 300 of the γ module.[41] They are necessary for proper end-to-end alignment of fibrinogen or fibrin molecules in assembling polymer structures, as exemplified by investigation of a dysfibrinogenemia, Tokyo II (γR275C),[40] which had defective D:D contacts. Crosslinked Tokyo II fibrinogen formed disorganized fibril structures despite the fact that γ chain crosslinking at the γ_{XL} site had proceeded at a normal rate. Tokyo II fibrin networks were characterized by increased fiber widths, tapering fibers, and an increased level of fiber branching, which evidently resulted from slower fibrin assembly and inaccurate end-to-end positioning of assembling fibrin monomers.

MOLECULAR AND CELLULAR INTERACTIONS FOLLOWING FIBRINOPEPTIDE B CLEAVAGE

FPB (Bβ1–14) release occurs more slowly than the release of FPA[18–20] and exposes an independent polymerization site, E_B,[42] beginning with β15–18 (GHRP).[21] E_B is utilized through interactions with a constitutive complementary D_b site located in the C-terminal β chain segment of the D domain.[28,42,43] The E_B:D_b interaction is not absolutely required for lateral fibril and fiber associations, but it contributes to this process through cooperative interactions resulting from alignment of D domains in the assembling polymer.[42] Polymerization of des BB-fibrin results in the same type of fibril structure as occurs with des AA-fibrin,[38] but the clot strength is lower than that of des AA-fibrin.[42]

The αC domain originates in the D domain at residue 111 and terminates at its C-terminus, Aα610.[1] Fibrin clots formed from circulating fibrinogen molecules such as fibrinogen *catabolite* fractions I-6 to I-9, which lack C-terminal portions of the αC domain, display a prolonged thrombin time, reduced turbidity, and produce thinner fibers.[44–46] In fibrinogen, the αC domains tend to be non-covalently tethered to the E domain,[6,47,48] but dissociate from it following FPB cleavage.[47,48] This event evidently makes αC domains available for interaction with other αC domains, thereby promoting lateral fibril associations and network assembly (FIG. 2).

In addition to its role in mediating fibrin assembly, the β15–42 sequence binds heparin[49,50] and participates in cell-matrix interactions. β15–42 mediates platelet spreading,[51] fibroblast proliferation,[52] endothelial cell spreading, proliferation and capillary tube formation,[50,52–54] and release of von Willebrand factor.[55,56] Binding to endothelial cells is mediated by the endothelial cell receptor, VE-cadherin.[54]

CROSSLINKING OF FIBRINOGEN AND FIBRIN

The C-terminal region of each fibrinogen or fibrin γ chain contains a single crosslinking site at which factor XIII or XIIIa catalyzes the formation of γ dimers[39,57–59] by incorporating intermolecular reciprocal ε-(γ-glutamyl)lysine bridges between a donor γ406 lysine of one chain and a glutamine acceptor at γ398/399 of another.[60–62] The same type of intermolecular cosslinking occurs more slowly among several amine donor and lysine acceptor sites in Aα or α chains,[32,65] and small amounts of internally crosslinked α–γ chain heterodimers are found in plasma fibrinogen molecules.[66] In addition to γ dimers, higher order forms of crosslinked γ chains, namely γ trimers and γ tetramers, evolve slowly.[32,59,65,67] Because there is only a single donor lysine residue at γ406,[60,62,63,68] it is safe to assume that trimeric and tetrameric structures form through utilization or reutilization of that same residue.

THE ALIGNMENT OF γ CHAIN PAIRS IN CROSSLINKED FIBRINOGEN OR FIBRIN POLYMERS

Interactions between D and E domains in an assembling fibrin polymer facilitate the antiparallel intermolecular alignment of γ chain pairs at the γ_{XL} site, thereby accelerating the rate of XIIIa-mediated crosslinking.[58,39] The orientation of these C-terminal γ chain pairs within an assembled and crosslinked fibrin or fibrinogen polymer has been at issue for some time. On the basis of evidence to be recited below, we believe that crosslinked γ chains become situated *transversely* between the D domains of opposing strands of a fibrin or fibrinogen fibril (see FIGURE 3). Despite the strength of this evidence, many others hold to the belief that these crosslinked regions are positioned across the distal ends of two linearly aligned molecules within a fibril strand in a so-called *DD-long* or end-to-end arrangement. It is valuable to summarize the evidence for these differing views, beginning with the end-to-end argument. Soon after the report by Fowler *et al.*,[69] which involved the demonstration of XIIIa-crosslinked fibrinogen dimers disposed end-to-end, it became almost axiomatic that crosslinked γ chains in assembled fibrin fibers were oriented in the same way. The DD-long model later drew support from observations on crosslinked fibrin degradation products that had been unfolded in acetic acid[70] and from the appearance of end-to-end crosslinked fibrinogen molecules that had been dissociated in acetic acid or that had been assembled on a fibrin fragment E template.[71] The conclusions drawn in both studies were ambiguous because they were based upon the appearance of acetic acid dissociated polymer structures that are known to become unfolded under such conditions.[72] In addition, side-to-side fibril associations, as well as end-to-end associations occurred in the fibrin E-fibrinogen study,[71] making it possible for either end-to-end or transverse crosslinking to occur. At most, these studies show that end-to-end positioning of crosslinked γ chains occurs under circumstances that are not closely related to those in assembled fibrin.

The idea of a DD-long arrangement has also gained inferential support from evaluation of D domain crystal structures,[41,73] even though in the studies cited the γ chain region involved in crosslinking had not been visualized. Nevertheless, from the

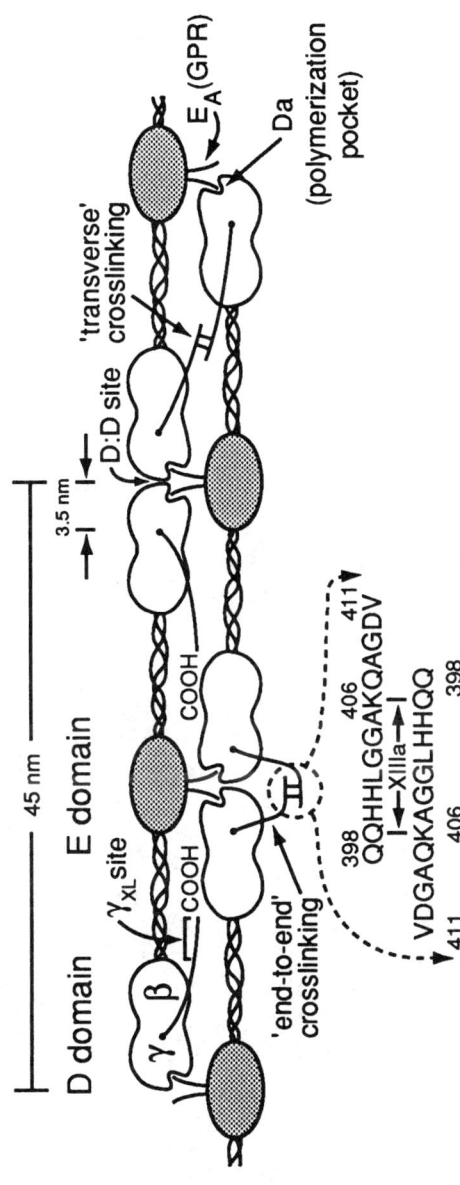

FIGURE 3. Schematic diagram of assembled fibrin molecules illustrating the intermolecular arrangements in a fibril. The positioning of the γ_{XL} sites in transverse or end-to-end orientations is shown, as is the C-terminal sequence of the critical residues of the γ_A chain crosslinking site. The emergence of the C-terminal portion of the γ chain from the middle of the γ module is consistent with recent crystallographic reports.[41,73] Reproduced from Mosesson et al.[78]

observed folding of structures comprising the D domain crystals and from calculation of the distances that each γ chain would presumably be required to traverse in order to become engaged with another γ chain, these groups concluded that transverse crosslinking could not occur in fibrin. These calculated discrepancies do not amount to a cohesive argument against the transverse crosslinking arrangement since they do not directly contradict the experimental evidence that provides proof for such structures in fibrin (*vide infra*). They should serve instead as a stimulus to reconcile potential differences between structures found in D domain crystals and those that enable transverse crosslinking to take place in fibrin. In this connection, Medved *et al.*[74] recently reported that the middle strand (i.e., γ381 to γ390) of a five-stranded β sheet structure in the γ module of D domain crystals, immediately preceding the sequence containing the $γ_{XL}$ site, can be displaced without disrupting the compact structure of the γ module. This suggests that the $γ_{XL}$ region in the fibrin polymer can be *pulled out* and thus become more extended, a behavior that might explain how transverse crosslinking takes place.

Several investigations relating to fibrin crosslinking, beginning with one by Selmayr *et al.*,[75] point directly to transverse positioning of γ chains in crosslinked fibrin fibrils. These investigators found that in the presence of XIIIa, fibrinogen molecules

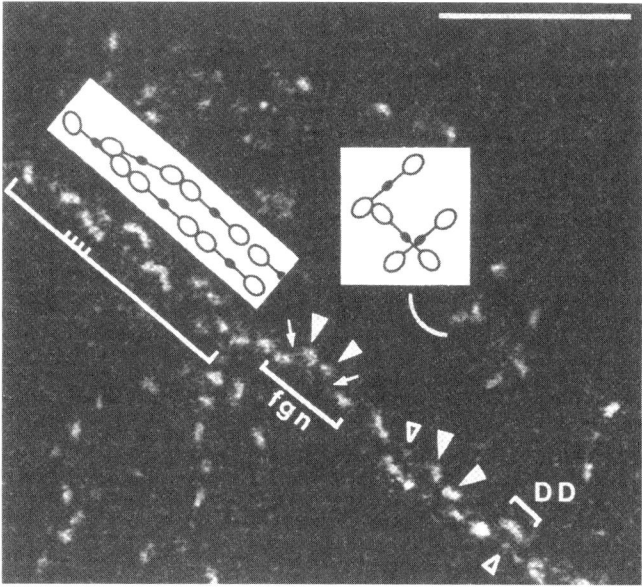

FIGURE 4. Scanning transmission electron microscope (STEM) image obtained from a sample of a XIIIa-fibrinogen crosslinking mixture. The micrograph shows a double-stranded γ chain-crosslinked fibrinogen fibril. The arrangement of the molecules comprising the fibril is drawn schematically. Filamentous structures connecting the fibril strands (*arrows*), probably represent transversely crosslinked γ chains. There are numerous non-crosslinked fibrinogen molecules and occasional oligomers in the field. *Bar*, 100 nm. Reproduced from Mosesson *et al.*[39]

formed a γ chain-crosslinked complex with immobilized fibrin, an arrangement that was interpreted as having arisen through transverse γ chain orientation. In later studies,[76] this group showed that crosslinked double-stranded fibrin fibrils were not depolymerized by 3M urea, a solvent that can otherwise disrupt D:E interactions. The observation reinforced the conclusion that crosslinked γ chains were positioned transversely in fibrin networks. In follow up studies triggered by the Selmayr reports,[a] electron microscopic analyses of crosslinked fibrinogen polymers demonstrated double stranded fibrils that had formed via crosslinked γ chains traversing the fibril strands[39] (see FIGURE 4). More focused evidence for transverse crosslinking came from a study demonstrating that mixtures of fibrin plus plasmic fragment D produced XIIIa-crosslinked D:fibrin:D complexes that must have come about through transverse γ chain positioning.[72] Moreover, EM images of D:fibrin:D complexes in that same study showed folding of the crosslinked molecular complex into transverse positions that could readily be converted to end-to-end positioning in the presence of dilute acetic acid, a solvent that disrupts non-covalent intermolecular fibrin associations. More recently, high resolution electron microscopy of gold-cadaverine-labeled γ chains in fibrinogen molecules or fibrin fibrils demonstrated that C-terminal γ chain regions are most often oriented toward the center of the molecule, suitably aligned for transverse crosslinking.[78] The most recent and perhaps most convincing evidence for transverse γ chain crosslinking, was obtained from analyses of the type and amount of radioactive γ dimers produced in crosslinking mixtures of fibrinogen 1 (γA, γA) and ^{125}I-labeled fibrin 2 (γA, γ′) (see FIGURE 5) or the converse, fibrin 1 and ^{125}I-fibrinogen 2 (not shown).[77] In the example shown, had DD-long crosslinking occurred, radioactive γ dimers would not have formed, whereas transverse crosslinking would have resulted in a 1:1 mixture of radioactive γA-γA and γA-γ′ dimers. In fact, a mixture of radioactive γ dimers in a 1:1 ratio was found as predicted for transverse crosslinking. In summary, although the possibility remains that some degree of end-to-end crosslinking may occur in fibrin, in sifting through all available information, we can find no durable evidence for this type of crosslinking in assembled fibrin networks. Instead, the evidence points to transverse crosslinking as the predominant, if not the only, crosslinking arrangement in fibrin.

FIBRINOGEN AND FIBRINOLYSIS

Tissue-type plasminogen activator (tPA)[79] circulates in blood and is synthesized by vascular endothelial cells.[80] tPA-mediated plasminogen activation is markedly accelerated in the presence of fibrin, whereas there is little effect in the presence of fibrinogen.[81,82] The stimulatory event is promoted not only by fibrin polymers, but also by XIIIa-crosslinked fibrinogen fibrils.[83] Plasminogen activation occurs through tPA binding to fibrin followed by the addition of plasminogen to form a ternary complex.[81] Two sites in fibrin are involved in enhancement of plasminogen activation by tPA, Aα148–160 and γ312–324.[84,85] These sites are cryptic in fibrinogen

---

[a]M.W.M. became interested in this problem as a member of the Scientific Review Committee for Professor Gert Müller-Berghaus' research group in Giessen, Germany, during which he had the opportunity of reviewing and discussing Eberhard Selmayr's experiments. These interactions stimulated our later interests in this subject.

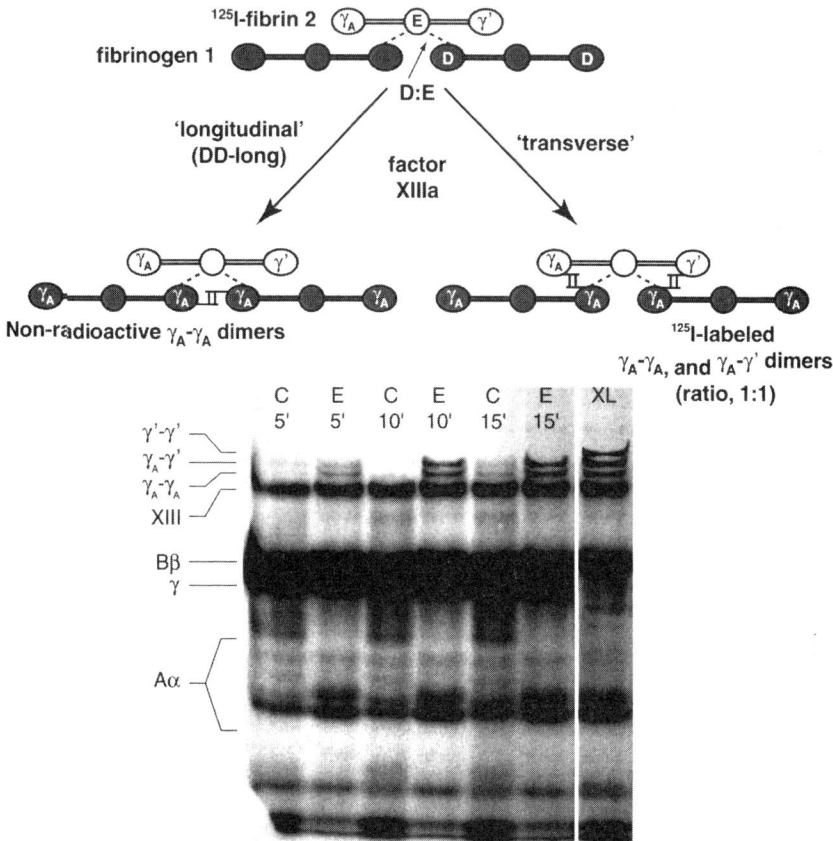

FIGURE 5. One of two protocols that were used to evaluate end-to-end versus transverse XIIIa-mediated crosslinking in a mixture of fibrinogen 1, ^{125}I-fibrin 2, and XIIIa (*upper*) and the results of autoradiography of a SDS-PAGE gel (*lower*). Modified from Siebenlist *et al.*[77]

but become exposed as a consequence of non-covalent D:E interactions in assembling fibrin fibrils. These associations induce conformational changes in the D region that result in exposure of tPA- and plasminogen binding sites, and the exposure is reversed after the complex dissociates.[86] There is a single high affinity plasminogen binding site in fibrinogen or fibrin,[87] and it seems to be located in the distal portion of the αC domain.[88] Proteolytic cleavage by plasmin creates numerous additional lysine binding sites,[89,90] thereby enhancing fibrinolysis by increasing the accumulation of plasminogen.

Other plasma proteins, notably α_2-antiplasmin, are crosslinked to fibrin α chains[91] at Aα303.[92] Although it has been recognized for many years that there is a potent antiplasmin activity that is bound to fibrinogen,[93] the specific entity in fibrinogen accounting for that activity had not been identified. Recently, however,

Siebenlist *et al.*, using an immunochemical technique, directly demonstrated the presence of covalently bound α_2-antiplasmin in normal fibrinogen as well as in a dysfibrinogenemic fibrinogen, Cedar Rapids (γR275C).[94] The transglutaminase activity responsible for the effect is probably plasma factor XIII, which has considerable constitutive fibrinogen crosslinking potential.[59] The presence of α_2-antiplasmin on fibrin(ogen) enhances resistance to fibrinolysis,[95] but some other protein bound to fibrin(ogen) might also contribute to this effect. That is, in addition to α_2-antiplasmin, PAI-2, an inhibitor of plasmin generation by urokinase or tPA, can also be crosslinked to fibrin at sites in the Aα chain that are remote from the Aα303 site for α_2-antiplasmin,[96,97] thus amplifying fibrin resistance to lysis.

Lipoprotein(a) is a highly atherogenic lipoprotein complex formed from apolipoprotein(a) that is disulfide bound to the apolipoprotein B-100 moiety of LDL. LP(a) binds to plasmin-degraded fibrin and fibrinogen,[98] competes with plasminogen for binding sites in fibrin(ogen),[99,100] and becomes crosslinked to fibrinogen in the presence of XIIIa,[101] effectively increasing the local concentration of LP(a) within the clot. The presence of this lipoprotein complex on fibrin(ogen) has an important negative effect on fibrinolysis.[99,100] (Also see HRGP.)

THROMBIN BINDING TO FIBRIN AND FIBRINOGEN

Thrombin binding to its substrate fibrinogen is mediated through an anion–binding fibrinogen recognition exosite in thrombin.[102,103] Binding at the substrate recognition sites in the E domain leads to proteolytic cleavage and release of FPA and FPB. In addition to fibrinogen–substrate binding, fibrin also has a significant thrombin binding potential that was highlighted more than 55 years ago by Walter Seegers, and which he termed antithrombin I.[104,105] Little more has been written on this topic from the standpoint of its relevance to thrombin metabolism and the pathophysiology of thrombophilia, but we do know that dysfibrinogenemic subjects with defective thrombin binding to fibrin have strikingly thrombophilic phenotypes.[106]

Antithrombin I activity is defined by two classes of non-substrate thrombin binding sites in fibrin,[107,108] one of low affinity in the E domain, and the other of high affinity in D domains of fibrin(ogen) molecules containing a γ chain variant termed γ' (γ'1–427L).[108] γ' chains contain a unique 20 residue highly anionic, tyrosine-sulfated C-terminal sequence (FIG. 1),[11,12] and differ from γA chains, which instead contain a C-terminal platelet-binding sequence (γA400–411V).[109] Low affinity thrombin binding evidently reflects residual aspects of fibrinogen substrate binding, and both N-terminal portions of α chains (α27–50) and β chains (β15–42) contribute to their formation.[24,108,110,111] The high affinity thrombin binding site in fibrin(ogen) is situated exclusively in the C-terminal aspect of γ' chains (γ'1–427L) between residues 414 and 427. Sulfated tyrosine residues at γ'418 and γ'422 significantly increase the thrombin binding affinity.[12]

There is good reason to believe that antithrombin I activity plays an important role in modulating thrombin activity. Dysfibrinogenemic subjects manifesting defective non-substrate thrombin binding develop severe thromboembolic disease,[106] most notably fibrinogens New York I (des Bβ9–72)[22,112] and Naples I (Bβ A68T).[113–115] Fibrinogen New York I was associated with recurrent deep venous thrombosis and fatal pulmonary embolism.[112] Two homozygous members of the Naples I kindred experienced occlusive arterial strokes at an early age, whereas a third homozygous family member suffered DVT following abdominal surgery.[116] In addition to the reports cited above, recent *in vitro* experiments indicate that thrombin generation in recalcified afibrinogenemic plasma[117,118] and in reptilase-defibrinated plasma[118] is significantly higher than in normal fibrinogen-containing plasma. Repletion of afibrinogenemic plasma with fibrinogen restores thrombin generation to normal. Fibrinogen 2, which contains both high and low affinity thrombin-binding sites has a more profound normalizing effect than fibrinogen 1 (low affinity site only).[117] The connection between antithrombin I activity and thrombotic disease remains to be more fully explored.

INTEGRIN BINDING SITES

Fibrinogen contains two integrin binding sites at Aα95–98 (RGDF) and at Aα572–575 (RGDS).[1] Many cellular interactions with fibrinogen and fibrin occur through binding to one or both of these recognition sequences. In addition to the ability of RGD sites to bind to platelet $\alpha_{IIb}\beta_3$ and compete for binding with the γA400–411 sequence,[120,121] $\alpha_V\beta_3$ integrins on endothelial cells,[122] melanoma cells,[123] and fibroblasts[124] also mediate RGD-dependent fibrinogen binding. There are other integrins on endothelial cells or fibroblasts that bind to fibrin(ogen) RGD sites.[125,126]

The leukocyte integrin *CD 11b/CD18 ($\alpha_M\beta_2$, Mac-1)* is a high affinity receptor for fibrinogen on stimulated monocytes and neutrophils that has been implicated in the inflammatory response. $\alpha_M\beta_2$ has been localized within the fibrinogen D domain at a site corresponding to the confluence of γ190–202 and γ377–395, which form two antiparallel β strands[127] (Fig. 1). It is valuable to recall that the γ381 to γ390 segment may undergo significant conformational rearrangements.[74]

OTHER BINDING INTERACTIONS

There are numerous examples of proteins that interact with fibrinogen or fibrin. The listing below is not intended to be exhaustive, but coupled with binding interactions covered elsewhere in this article, it reinforces the notion that fibrin(ogen) helps to orchestrate its own physiological destiny. *Plasma fibronectin* binds to the Aα chain of fibrinogen in its C-terminal region, since fibrinogen molecules lacking this part of the molecule do not interact with fibronectin.[128] Crosslinking of fibronectin to fibrin is mediated by factor XIIIa[91,129] between several potential lys residues in the fibrin α chain[130] and mainly gln-3 of fibronectin.[131] *Fibroblast growth factor-2* (FGF-2, bFGF) binds to fibrinogen[132] and is able to potentiate endothelial cell proliferation.[133] Furthermore, binding to fibrinogen protects it from proteolytic degradation.[134] Another growth factor, *insulin-like growth factor-1* (IGF-1), becomes bound to *insulin-like growth factor-binding protein-3* (IGFBP-3) which itself binds directly to fibrinogen or fibrin. IGF-1 participates in the wound healing process by stimulating stromal cell function and proliferation.[135]

A vitamin K-dependent clotting factor, *Xa,* has been shown to bind to fibrin and to fibrin/fibrinogen degradation products.[136] Xa bound specifically to a peptide corresponding to Aα 82D to 123K, a site that is located between the E and D domains, and quite distinct from the thrombin binding sites. *Tissue factor pathway inhibitor* binds to fibrin via a C-terminal region that is rich in Arg and Lys residues, but the location of the binding site in fibrin has not been determined.[137] *Gelsolin* is an actin-binding plasma protein that severs actin filaments, and that has been shown to bind to the fibrin clot.[138]

Histidine-Rich Glycoprotein is a plasma and platelet protein that binds specifically to fibrinogen and fibrin.[139] A high proportion of HRGP circulates as a complex bound to the lysine binding site of plasminogen and this serves to reduce the effective plasminogen concentration, inhibit binding of plasminogen to fibrin, and retard fibrinolysis *in vitro.*[140] The physiological relevance of this effect has been questioned.[141] Nevertheless, both high plasma levels of HRGP,[142,143] or HRGP deficiencies,[144,145] are associated with venous thromboembolism.

ACKNOWLEDGMENTS

The authors wish to express their gratitude to Pam Ried, Irene Hernandez, Kevin Thompson, and Diane Bartley for able technical assistance. Much of the work cited in this article from the authors' laboratory is supported by NIH R01 HL59507.

REFERENCES

1. HENSCHEN, A., F. LOTTSPEICH, M. KEHL & C. SOUTHAN. 1983. Covalent structure of fibrinogen. Ann. N.Y. Acad. Sci. **408:** 28–43.
2. BLOMBÄCK, B., B. HESSEL & D. HOGG. 1976. Disulfide bridges in NH_2-terminal part of human fibrinogen. Thromb. Res. **8:** 639–658.
3. HUANG, S., Z. CAO & E.W. DAVIE. 1993. The role of amino-terminal disulfide bonds in the structure and assembly of human fibrinogen. Biochem. Biophys. Res. Commun. **190:** 488–495.
4. ZHANG, J.-Z. & C.M. REDMAN. 1992. Identification of Bβ chain domains involved in human fibrinogen assembly. J. Biol. Chem. **267:** 21727–21732.
5. HOEPRICH, P.D., JR. & R.F. DOOLITTLE. 1983. Dimeric half-molecules of human fibrinogen are joined through disulfide bonds in an antiparallel orientation. Biochemistry **22:** 2049–2055.
6. MOSESSON, M.W., J.F. HAINFELD, R.H. HASCHEMEYER & J.S. WALL. 1981. Identification and mass analysis of human fibrinogen molecules and their domains by scanning transmission electron microscopy. J. Mol. Biol. **153:** 695–718.
7. WILLIAMS, R.C. 1983. Morphology of fibrinogen monomers and fibrin polymers. Ann. N.Y. Acad. Sci. **408:** 180–193.
8. MÉNDEZ, J.A., M.V. ALVAREZ, J.A. AZNÁREZ & J. GONZÁLEZ-RODRÍGUEZ. 1996. Two fold symmetry of human fibrinogen proved at the β chain distal domains by monoclonal-immunoelectron microscopy and image analysis. Biochemistry **35:** 634–637.
9. CHUNG, D.W. & E.W. DAVIE. 1984. γ and γ' chains of human fibrinogen are produced by alternative mRNA processing. Biochemistry **23:** 4232–4236.
10. MOSESSON, M.W., J.S. FINLAYSON & R.A. UMFLEET. 1972. Human fibrinogen heterogeneities. III. Identification of γ chain variants. J. Biol. Chem. **247:** 5223–5227.
11. WOLFENSTEIN-TODEL, C. & M.W. MOSESSON. 1981. Carboxy-terminal amino acid sequence of a human fibrinogen γ chain variant (γ'). Biochemistry **20:** 6146–6149.

12. MEH, D.A., K.R. SIEBENLIST, S.O. BRENNAN, *et al.* 2001. The amino acid sequences in fibrin responsible for high affinity thrombin binding. Thromb. Hæmostas. In press.
13. SIEBENLIST, K.R., D.A. MEH & M.W. MOSESSON. 1996. Plasma factor XIII binds specifically to fibrinogen molecules containing γ' chains. Biochemistry **35:** 10448–10453.
14. KIRSCHBAUM, N.E., M.W. MOSESSON & D.L. AMRANI. 1992. Characterization of the γ chain platelet binding site on fibrinogen fragment D. Blood **79:** 2643–2648.
15. BENNETT, J.S. & G. VILAIRE. 1979. Exposure of platelet fibrinogen receptors by ADP and epinephrine. J. Clin. Invest. **64:** 1393–1401.
16. KLOCZEWIAK, M., S. TIMMONS & J. HAWIGER. 1983. Recognition site for the platelet receptor is present on the 15-residue carboxy-terminal fragment of the γ-chain of human fibrinogen and is not involved in the fibrin polymerization reaction. Thromb. Res. **29:** 249–255.
17. MARGUERIE, G.A., E.F. PLOW & T.S. EDGINGTON. 1979. Human platelets possess an inducible and saturable receptor specific for fibrinogen. J. Biol. Chem. **254:** 5357–5363.
18. SCHERAGA, H.A. & M. LASKOWSKI, JR. 1957. The fibrinogen-fibrin conversion. Adv. Prot. Chem. **12:** 1–19.
19. BLOMBÄCK, B. 1958. Studies on the action of thrombotic enzymes on bovine fibrinogen as measured by N-terminal analysis. Arkiv Kemi **12:** 321–335.
20. BLOMBÄCK, B., B. HESSEL, D. HOGG & L. THERKILDSEN. 1978. A two-step fibrinogen-fibrin transition in blood coagulation. Nature **275:** 501–505.
21. LAUDANO, A.P. & R.F. DOOLITTLE. 1978. Studies on synthetic peptides that bind to fibrinogen and prevent fibrin polymerization. Proc. Natl. Acad. Sci. U.S.A. **75:** 3085–3089.
22. LIU, C.Y., J.A. KOEHN & F.J. MORGAN. 1985. Characterization of fibrinogen New York 1. J. Biol. Chem. **260:** 4390–4396.
23. PANDYA, B.V., C.S. CIERNIEWSKI & A.Z. BUDZYNSKI. 1985. Conservation of human fibrinogen conformation after cleavage of the Bβ chain NH$_2$-terminus. J. Biol. Chem. **260:** 2994–3000.
24. SIEBENLIST, K.R., J.P. DIORIO, A.Z. BUDZYNSKI & M.W. MOSESSON. 1990. The polymerization and thrombin-binding properties of des-(Bβ1-42)-fibrin. J. Biol. Chem. **265:** 18650–18655.
25. SHIMIZU, A., Y. SAITO & Y. INADA. 1986. Distinctive role of histidine-16 of the Bβ chain of fibrinogen in the end-to-end association of fibrin. Proc. Natl. Acad. Sci. U.S.A. **83:** 591–593.
26. SHIMIZU, A., G.M. NAGEL & R.F. DOOLITTLE. 1992. Photoaffinity labeling of the primary fibrin polymerization site: Isolation of a CNBr fragment corresponding to γ 337–379. Proc. Natl. Acad. Sci. U.S.A. **89:** 2888–2892.
27. PRATT, K.P., H.C.F. CÔTÉ, D.W. CHUNG, *et al.* 1997. The primary fibrin polymerization pocket: three-dimensional structure of a 30-kDa C-terminal γ chain fragment complexed with the peptide gly-pro-arg-pro. Proc. Natl. Acad. Sci. U.S.A. **94:** 7176–7181.
28. EVERSE, S.J., G. SPRAGGON, L. VEERAPANDIAN, *et al.* 1998. Crystal structure of fragment double-D from human fibrin with two different bound ligands. Biochemistry **37:** 8637–8642.
29. FERRY, J.D. 1952. The mechanism of polymerization of fibrinogen. Proc. Natl. Acad. Sci. U.S.A. **38:** 566–569.
30. KRAKOW, W., G.F. ENDRES, B.M. SIEGEL & H.A. SCHERAGA. 1972. An electron microscopic investigation of the polymerization of bovine fibrin monomer. J. Mol. Biol. **71:** 95–103.
31. FOWLER, W.E., R.R. HANTGAN, J. HERMANS, *et al.* 1981. Structure of the fibrin protofibril. Proc. Natl. Acad. Sci. U.S.A. **78:** 4872–4876.
32. MOSESSON, M.W., K.R. SIEBENLIST, D.L. AMRANI & J.P. DIORIO. 1989. Identification of covalently linked trimeric and tetrameric D domains in crosslinked fibrin. Proc. Natl. Acad. Sci. U.S.A. **86:** 1113–1117.

33. HEWAT, E.A., L. TRANQUI & R.H. WADE. 1983. Electron microscope structural study of modified fibrin and a related modified fibrinogen aggregate. J. Mol. Biol. **170:** 203–222.
34. MOSESSON, M.W., J.P. DIORIO, K.R. SIEBENLIST, et al. 1993. Evidence for a second type of fibril branch point in fibrin polymer networks, the trimolecular junction. Blood **82:** 1517–1521.
35. MEH, D.A., K.R. SIEBENLIST, G. BERGTROM & M.W. MOSESSON. 1995. Sequence of release of fibrinopeptide A from fibrinogen molecules by thrombin or Atroxin. J. Lab. Clin. Med. **125:** 384–391.
36. SHAINOFF, J.R., G.B. SMEJKAL, P.M. DIBELLO, et al. 1996. Isolation and characterization of the fibrin intermediate arising from cleavage of one fibrinopeptide A from fibrinogen. J. Biol. Chem. **271:** 24129–24137.
37. BLOMBÄCK, B., K. CARLSSON, K. FATAH, et al. 1994. Fibrin in human plasma: gel architectures governed by rate and nature of fibrinogen activation. Thromb. Res. **75:** 521–538.
38. MOSESSON, M.W., J.P. DIORIO, M.F. MULLER, et al. 1987. Studies on the ultrastructure of fibrin lacking fibrinopeptide B (β-fibrin). Blood **69:** 1073–1081.
39. MOSESSON, M.W., K.R. SIEBENLIST, J.F. HAINFELD & J.S. WALL. 1995. The covalent structure of factor XIIIa crosslinked fibrinogen fibrils. J. Struct. Biol. **115:** 88–101.
40. MOSESSON, M.W., K.R. SIEBENLIST, J.P. DIORIO, et al. 1995. The role of fibrinogen D domain intermolecular association sites in the polymerization of fibrin and fibrinogen Tokyo II (γ275 Arg$\rightarrow$Cys). J. Clin. Invest. **96:** 1053–1058.
41. SPRAGGON, G., S.J. EVERSE & R.F. DOOLITTLE. 1997. Crystal structures of fragment D from human fibrinogen and its crosslinked counterpart from fibrin. Nature **389:** 455–462.
42. SHAINOFF, J.R. & B.N. DARDIK. 1983. Fibrinopeptide B in fibrin assembly and metabolism: physiologic significance in delayed release of the peptide. Ann. N.Y. Acad. Sci. **408:** 254–267.
43. MEDVED, L.V., S.V. LITVINOVICH, T.P. UGAROVA, et al. 1993. Localization of a fibrin polymerization site complimentary to Gly-His-Arg sequence. FEBS Lett. **320:** 239–242.
44. HASEGAWA, N. & S. SASAKI. 1990. Location of the binding site "b" for lateral polymerization of fibrin. Thromb. Res. **57:** 183–195.
45. MOSESSON, M.W. & S. SHERRY. 1966. The preparation and properties of human fibrinogen of relatively high solubility. Biochemistry **5:** 2829–2835.
46. MOSESSON, M.W. 1983. Fibrinogen heterogeneity. Ann. N.Y. Acad. Sci. **408:** 97–113.
47. VEKLICH, Y.I., O.V. GORKUN, L.V. MEDVED, et al. 1993. Carboxyl-terminal portions of the α chains of fibrinogen and fibrin. J. Biol. Chem. **268:** 13577–13585.
48. GORKUN, O.V., Y.I. VEKLICH, L.V. MEDVED, et al. 1994. Role of the αC domains of fibrin in clot formation. Biochemistry **33:** 6986–6997.
49. ODRLJIN, T.M., J.R. SHAINOFF, S.O. LAWRENCE, et al. 1996. Thrombin cleavage enhances exposure of a heparin binding domain in the N-terminus of the fibrin β chain. Blood **88:** 2050–2961.
50. ODRLJIN, T.M., C.W. FRANCIS, L.A. SPORN, et al. 1996. Heparin-binding domain of fibrin mediates its binding to endothelial cells. Arterioscl. Thromb. Vasc. Biol. **16:** 1544–1551.
51. HAMAGUCHI, M., L.A. BUNCE, L.A. SPORN & C.W. FRANCIS. 1993. Spreading of platelets on fibrin is mediated by the amino terminus of the β chain including peptide β 15–42. Blood **81:** 2348–2356.
52. SPORN, L.A., L.A. BUNCE & C.W. FRANCIS. 1995. Cell proliferation on fibrin: modulation by fibrinopeptide cleavage. Blood **86:** 1801–1810.
53. CHALUPOWICZ, D.G., Z.A. CHOWDHURY, T.L. BACH, et al. 1995. Fibrin II induces endothelial cell capillary tube formation. J. Cell Bio. **130:** 207–215.
54. BACH, T.L., C. BARSIGIAN, C.H. YAEN & J. MARTINEZ. 1998. Endothelial cell VE-cadherin functions as a receptor for the β15–42 sequence of fibrin. J. Biol. Chem. **273:** 30719–30728.

55. RIBES, J.A., L.A. BUNCE & C.W. FRANCIS. 1989. Mediation of fibrin-induced release of von Willebrand factor from cultured endothelial cells by the fibrin β chain. J. Clin. Invest. **84:** 435–441.
56. FRANCIS, C.W., L.A. BUNCE & L.A. SPORN. 1993. Endothelial cell responses to fibrin mediated by FPB cleavage and the amino terminus of the β chain. Blood Cells **19:** 291–307.
57. MCKEE, P.A., P. MATTOCK & R.L. HILL. 1971. Proc. Natl. Acad. Sci. U.S.A. **66:** 738–744.
58. KANAIDE, H. & J.R. SHAINOFF. 1975. Cross-linking of fibrinogen and fibrin by fibrin-stabilizing factor (factor XIIIa). J. Lab. Clin. Med. **85:** 574–597.
59. SIEBENLIST, K.R., D. MEH & M.W. MOSESSON. 2001. Protransglutaminase (factor XIII) mediated crosslinking of fibrinogen and fibrin. Submitted.
60. CHEN, R. & R.F. DOOLITTLE. 1971. γ-γ Cross-linking sites in human and bovine fibrin. Biochemistry **10:** 4486–4491.
61. DOOLITTLE, R.F., R. CHEN & F. LAU. 1971. Hybrid fibrin: Proof of the intermolecular nature of γ-γ crosslinking units. Biochem. Biophys. Res. Commun. **44:** 94–100.
62. PURVES, L.R., M. PURVES & W. BRANDT. 1987. Cleavage of fibrin-derived D-dimer into monomers by endopeptidase from puff ader venom (bitis arietans) acting at cross-linked sites of the γ chain. Sequence of carboxy-terminal cyanogen bromide γ-chain fragments. Biochemistry **26:** 4640–4646.
63. SOBEL, J.H. & M.A. GAWINOWICZ. 1996. Identification of the α chain lysine donor sites involved in factor XIII_a fibrin cross-linking. J. Biol. Chem. **271:** 19288–19297.
64. MATSUKA, Y.V., L.V. MEDVED, M.M. MIGLIORINI & K.C. INGHAM. 1996. Factor XIIIa-catalyzed cross-linking of recombinant αC fragments of human fibrinogen. Biochemistry **35:** 5810–5816.
65. SHAINOFF, J.R., D.A. URBANIC & P.M. DIBELLO. 1991. Immunoelectrophoretic characterizations of the cross-linking of fibrinogen and fibrin by factor XIIIa and tissue transglutaminase. J. Biol. Chem. **266:** 6429–6437.
66. SIEBENLIST, K.R. & M.W. MOSESSON. 1996. Evidence for intramolecular cross-linked Aα γ chain heterodimers in plasma fibrinogen. Biochemistry **35:** 5817–5821.
67. SIEBENLIST, K.R. & M.W. MOSESSON. 1992. Factors affecting γ-chain multimer formation in cross-linked fibrin. Biochemistry **31:** 936–941.
68. SAMOKHIN, G.P. & L. LORAND. 1995. Contact with the N termini in the central E domain enhances the reactivities of the distal D domains of fibrin to factor XIIIa. J. Biol. Chem. **270:** 21827–21832.
69. FOWLER, W.E., H.P. ERICKSON, R.R. HANTGAN, *et al.* 1981. Crosslinked fibrinogen dimers demonstrate a feature of the molecular packing in fibrin fibers. Science **211:** 287–289.
70. WEISEL, J.W., C.W. FRANCIS, C. NAGASWANI & V.J. MARDER. 1993. Determination of the topology of factor XIIIa-induced fibrin γ-chain cross-links by electron microscopy of ligated fragments. J. Biol. Chem. **268:** 26618–26624.
71. VEKLICH, Y.I., E.K. ANG, L. LORAND & J.W. WEISEL. 1998. The complementary aggregation sites of fibrin investigated through examination of polymers of fibrinogen with fragment E. Proc. Natl. Acad. Sci. U.S.A. **95:** 1438–1442.
72. SIEBENLIST, K.R., D.A. MEH, J.S. WALL, *et al.* 1995. Orientation of the carboxy-terminal regions of fibrin γ chain dimers determined from the crosslinked products formed in mixtures of fibrin, fragment D, and factor XIIIa. Thromb. Hæmostas. **74:** 1113–1119.
73. YEE, V.C., K.P. PRATT, H.C.F. CÔTÉ, *et al.* 1997. Crystal structure of a 30 kDa C-terminal fragment from the γ chain of human fibrinogen. Structure **5:** 125–138.
74. MEDVED, L.V., S. YAKOVLEV, D. LOUKINOV & S.V. LITVINOVICH. 1999. The role of the β-strand insert in the central domain of the fibrinogen γ-module. Prot. Sci. **8** (Suppl. 2): 83.
75. SELMAYR, E., W. THIEL & G. MÜLLER-BERGHAUS. 1985. Crosslinking of fibrinogen to immobilized des AA-fibrin. Thromb. Res. **39:** 459–465.
76. SELMAYR, E. 1988. Chromatography and electron microscopy of cross-linked fibrin polymers-A new model describing the cross-linking at the DD-trans contact of the fibrin molecules. Biopolymers **27:** 1733–1748.

77. SIEBENLIST, K.R., D.A. MEH & M.W. MOSESSON. 2000. Position of γ chain carboxy-terminal regions in fibrinogen:fibrin crosslinking mixtures. Biochemistry **39**: 14171–14175.
78. MOSESSON, M.W., K.R. SIEBENLIST, D.A. MEH, *et al.* 1998. The location of the carboxy-terminal region of γ chains in fibrinogen and fibrin D domains. Proc. Natl. Acad. Sci. U.S.A. **95**: 10511–10516.
79. COLLEN, D. 1980. On the regulation and control of fibrinolysis. Thromb. Hæmostas. **43**: 77–89.
80. LEVIN, E. 1983. Latent tissue plasminogen activator produced by human endothelial cells in culture: evidence for an enzyme-inhibitor complex. Proc. Natl. Acad. Sci. U.S.A. **80**: 6804–6808.
81. HOYLAERTS, M., D.C. RIJKEN, H.R. LIJNEN & D. COLLEN. 1982. Kinetics of the activation of plasminogen by human tissue plasminogen. J. Biol. Chem. **257**: 2912–2919.
82. RÅNBY, M. 1982. Studies of the kinetics of plasminogen activation by tissue plasminogen activator. Biochim. Biophys. Acta **704**: 461–469.
83. MOSESSON, M.W., K.R. SIEBENLIST, M. VOSKUILEN & W. NIEUWENHUIZEN. 1998. Evaluation of the factors contributing to fibrin-dependent plasminogen activation. Thromb. Hæmostas. **79**: 796–801.
84. SCHIELEN, W.J.G., H.P.H.M. ADAMS, M. VOSKUILEN, *et al.* 1991. The sequence Aα-(154–159) of fibrinogen is capable of accelerating the tPA catalyzed activation of plasminogen. Blood Coag. Fibrinol. **2**: 465–470.
85. SCHIELEN, W.J.G., H.P.H.M. ADAMS, K. VAN LEUVEN, *et al.* 1991. The sequence γ-(312–324) is a fibrin-specific epitope. Blood **77**: 2169–2173.
86. YAKOVLEV, S., E.M. MAKOGONENKO, N. KUROCHKINA, *et al.* 2000. Conversion of fibrinogen to fibrin: the mechanism of exposure of tPA- and plasminogen-binding sites. Biochemistry **39**: 15730–15741.
87. BOK, R.A. & W.F. MANGEL. 1985. Quantitative characterization of the binding of plasminogen to intact fibrin clots, lysine-sepharose, and fibrin cleaved by plasmin. Biochemistry **24**: 3279–3286.
88. TSURUPA, G. & L. MEDVED. 2001. Identification and characterization of novel tPA- and plasminogen-binding sites within fibrin(ogen) αC-domains. Biochemistry **40**: 801–808.
89. SUENSON, E., O. LÜTZEN & S. THORSEN. 1984. Initial plasmin-degradation of fibrin as the basis of a positive feed-back mechanism in fibrinolysis. Eur. J. Biochem. **140**: 513–522.
90. HARPEL, P.C., T.S. CHANG & E. VERDERBER. 1985. Tissue plasminogen activator and urokinase mediate the binding of glu-plasminogen to plasma fibrin I: evidence for new binding sites in plasmin-degraded fibrin. J. Biol. Chem. **260**: 4432–4440.
91. TAMAKI, T. & H. AOKI. 1981. Cross-linking of α_2-plasmin inhibitor and fibronectin to fibrin by fibrin-stabilizing factor. Biochim. Biophys. Acta **661**: 280–286.
92. KIMURA, S. & N. AOKI. 1986. Cross-linking site in fibrinogen for alpha 2-plasmin inhibitor. J. Biol. Chem. **261**: 15591–15595.
93. MOSESSON, M.W. & J.S. FINLAYSON. 1963. Biochemical and chromatographic studies of certain activities associated with human fibrinogen preparations. J. Clin. Invest. **42**: 747–755.
94. SIEBENLIST, K.R., M.W. MOSESSON, D.A. MEH, *et al.* 2000. Coexisting dysfibrinogenemia (γR275C) and factor V Leiden deficiency associated with thromboembolic disease (fibrinogen Cedar Rapids). Blood Coag. Fibrinol. **11**: 293–304.
95. SAKATA, Y. & H. AOKI. 1982. Significance of cross-linking of α_2-plasmin inhibitor to fibrin in inhibition of fibrinolysis and in hemostasis. J. Clin. Invest. **69**: 536–542.
96. RITCHIE, H., L.C. LAWRIE, P.W. CROMBIE, *et al.* 2000. Cross-linking of plasminogen activator inhibitor 2 and α_2-antiplasmin to fibrin(ogen). J. Biol. Chem. **275**: 24915–24920.
97. RITCHIE, H., L.A. ROBBIE, S. KINGHORN, *et al.* 1999. Monocyte plasminogen activator inhibitor 2 (PAI-2) inhibits u-PA-mediated fibrin clot lysis and is cross-linked to fibrin. Thromb. Hæmostas. **81**: 96–103.

98. HARPEL, P.C., B.R. GORDON & T.S. PARKER. 1989. Plasmin catalyzes binding of lipoprotein(a) to immobilized fibrinogen and fibrin. Proc. Natl. Acad. Sci. U.S.A. **86:** 3847–3851.
99. LOSCALZO, J., M. WEINFELD, G.M. FLESS & A.M. SCANU. 1990. Lipoprotein(a), fibrin binding, and plasminogen activation. Arteriosclerosis **10:** 240–245.
100. HERVIO, L., V. DURLACH, A. GIRARD-GLOBA & E. ANGLÉS-CANO. 1995. Multiple binding with identical linkage: a mechanism that explains the effect of lipoprotein(a) on fibrinolysis. Biochemistry **34:** 13353–13358.
101. ROMANIC, A.M., A.J. ARLETH, R.N. WILLETTE & E.H. OHLSTEIN. 1998. Factor XIIIa cross-links lipoprotein(a) with fibrinogen and is present in human atherosclerotic lesions. Circ. Res. **83:** 264–269.
102. FENTON, J.W., II, T.A. OLSON, M.P. ZABINSKI & G.D. WILNER. 1988. Anion-binding exosite of human α-thrombin and fibrin(ogen) recognition. Biochemistry **27:** 7106–7112.
103. STUBBS, M.T. & W. BODE. 1993. A player of many parts: the spotlight falls on thrombin's structure. Thromb. Res. **69:** 1–58.
104. SEEGERS, W.H., M. NIEFT & E.C. LOOMIS. 1945. Note on the adsorption of thrombin on fibrin. Science **101:** 520–521.
105. SEEGERS, W.H., J.F. JOHNSON & C. FELL. 1954. An antithrombin reaction related to prothrombin activation. Amer. J. Physiol. **176:** 97–103.
106. MOSESSON, M.W. 1999. Dysfibrinogenemia and thrombosis. Semin. Thromb. Hæmost. **25:** 311–319.
107. LIU, C.Y., H.L. NOSSEL & K.L. KAPLAN. 1979. The binding of thrombin by fibrin. J. Biol. Chem. **254:** 10421–10425.
108. MEH, D.A., K.R. SIEBENLIST & M.W. MOSESSON. 1996. Identification and characterization of the thrombin binding sites on fibrin. J. Biol. Chem. **271:** 23121–23125.
109. KLOCZEWIAK, M., S. TIMMONS, T.J. LUKAS & J. HAWIGER. 1984. Platelet receptor recognition site on human fibrinogen. Synthesis and sturcture-function relationship of peptides corresponding to the C-terminal segment of the γ chain. Biochemistry **23:** 1767–1774.
110. VALI, Z. & H.A. SCHERAGA. 1988. Localization of the binding site on fibrin for the secondary binding site of thrombin. Biochemistry **27:** 1956–1963.
111. BINNIE, C.G. & S.T. LORD. 1991. A synthetic analog of fibrinogen α27–50 is an inhibitor of thrombin. Thromb. Hæmostas. **65:** 165–168.
112. LIU, C.Y., P. WALLEN & D.A. HANDLEY. 1986. Fibrinogen New York I: the structural, functional, and genetic defects and an hypothesis of the role of fibrin in the regulation of coagulation and fibrinolysis. *In* Fibrinogen, Fibrin Formation and Fibrinolysis. D.A. Lane, A. Henschen & M.K. Jasani, Eds.: 79–90. W. de Gruyter, Berlin.
113. KOOPMAN, J., F. HAVERKATE, S.T. LORD, *et al.* 1992. Molecular basis of fibrinogen Naples associated with defective thrombin binding and thrombophilia. J. Clin. Invest. **90:** 238–244.
114. LORD, S.T., E. STRICKLAND & E. JAYJOCK. 1996. Strategy for recombinant multichain protein synthesis: fibrinogen Bβ-chain variants as thrombin substrates. Biochemistry **35:** 2342–2348.
115. MEH, D., M.W. MOSESSON, K.R. SIEBEALIST, *et al.* 2001. Fibrinogen Naples I (BβA68T) non-substrate thrombin binding capacities. Thromb. Res. In press.
116. DI MINNO, G., J. MARTINEZ, F. CIRILLO, *et al.* 1991. A role for platelets and thrombin in the juvenile stroke of two siblings with defective thrombin-absorbing capacity of fibrin(ogen). Arterioscl. Thromb. **11:** 785–796.
117. DE BOSCH, N.B., M.W. MOSESSON, A. RUIZ-SAEZ, *et al.* 1999. Effects of γ'/γA chain fibrin on the plasma thrombin generation in afibrinogenemia. Thromb. Hæmostas. **82:** 320–320.
118. DUPUY, E., C. SORIA, P. MOLHO, *et al.* 2001. Embolized ischemic lesions of toes in an afibrinogenemic patient: possible relevance to *in vivo* circulating thrombin. Thromb. Res. In press.

119. ANDRIEUX, A., G. HUDRY-CLERGEON & J.-J. RYCKWAERT. 1989. Amino acid sequences in fibrinogen mediating its interaction with its platelet receptor, GP IIbIIIa. J. Biol. Chem. **264:** 9258–9265.
120. LAM, S.C.T., E.F. PLOW, M.A. SMITH, et al. 1987. Evidence that arginyl-glycyl-aspartate peptides and fibrinogen γ chain peptides share a common binding site on platelets. J. Biol. Chem. **262:** 947–950.
121. SANTORO, S.A. & W.J. LAWING, JR. 1987. Competition for related but nonidentical binding sites on the glycoprotein IIb-IIIa complex by peptides derived from platelet adhesive proteins. Cell **48:** 867–873.
122. CHERESH, D.A. 1987. Human endothelial cells synthesize and express an arg-gly-asp–directed adhesion receptor involved in attachment to fibrinogen and von Willebrand factor. Proc. Natl. Acad. Sci. U.S.A. **84:** 6471–6475.
123. FELDING-HABERMANN, B., Z.M. RUGGERI & D.A. CHERESH. 1992. Distinct biological consequences of integrin alpha v beta 3-mediated melanoma cell adhesion to fibrinogen and its plasmic fragments. J. Biol. Chem. **267:** 5070–5077.
124. GAILIT, J., C. CLARKE, D. NEWMAN, et al. 1997. Human fibroblasts bind directly to fibrinogen at RGD sites through integrin $\alpha_V\beta_3$. Exp. Cell Res. **232:** 118–126.
125. ASAKURA, S., K. NIWA, T. TOMOZAWA, et al. 1997. Fibroblasts spread on immobilized fibrin monomerby mobilizing a β_1-class integrin, together with a vitronectin receptor $\alpha_V\beta_3$ on their surface. J. Biol. Chem. **272:** 8824–8829.
126. SUEHIRO, K., J. GAILIT & E.F. PLOW. 1997. Fibrinogen is a ligand for integrin $\alpha 5\beta 1$ on endothelial cells. J. Biol. Chem. **272:** 5360–5366.
127. ALTIERI, D.C., J. PLESCIA & E.F. PLOW. 1993. The structural motif glycine 190-valine 202 of the fibrinogen γ chain interacts with CD11b/CD18 integrin ($\alpha_M\beta_2$, Mac-1) and promotes leukocyte adhesion. J. Biol. Chem. **268:** 1847–1853.
128. STATHAKIS, N.E. & M.W. MOSESSON. 1977. Interactions among heparin, cold-insoluble globulin, and fibrinogen in formation of the heparin precipitable fraction of plasma. J. Clin. Invest. **60:** 855–865.
129. MOSHER, D.F. 1975. Cross-linking of cold-insoluble globulin by fibrin-stabilizing factor. J. Biol. Chem. **250:** 6614–6621.
130. MATSUKA, Y.V., M.M. MIGLIORINI & K.C. INGHAM. 1997. Cross-linking of fibronectin to C-terminal fragments of the fibrinogen alpha-chain by factor XIIIa. J. Prot. Chem. **16:** 739–745.
131. MCDONAGH, R.P., J. MCDONAGH, T.E. PETERSEN, et al. 1981. Amino acid sequence of the factor XIIIa acceptor site in bovine plasma fibronectin. FEBS Lett. **127:** 174–178.
132. SAHNI, A., T. ODRLJIN & C.W. FRANCIS. 1998. Binding of basic fibroblast growth factors to fibrinogen and fibrin. J. Biol. Chem. **273:** 7554–7559.
133. SAHNI, A., L.A. SPORN & C.W. FRANCIS. 1999. Potentiation of endothelial cell proliferation by fibrin(ogen)-bound fibroblast growth factor-2. J. Biol. Chem. **274:** 14936–14941.
134. SAHNI, A., C.A. BAKER, L.A. SPORN & C.W. FRANCIS. 2000. Fibrinogen and fibrin protect fibroblast growth factor-2 from proteolytic degradation. Thromb. Hæmostas. **83:** 736–741.
135. CAMPBELL, P.G., S.K. DURHAM, J.D. HAYES, et al. 1999. Insulin-like growth factor-binding protein-3 binds fibrinogen and fibrin. J. Biol. Chem. **274:** 30215–30221.
136. IINO, M., H. TAKEYA, T. TAKEMITSU, et al. 1995. Characterization of the binding of factor Xa to fibrinogen/fibrin derivatives and localization of the factor Xa binding site on fibrinogen. Eur. J. Biochem. **232:** 90–97.
137. OHKURA, N., K. ENJYOJI, Y. KAMIKUBO & H. KATO. 1997. A novel degradation pathway of tissue factor pathway inhibitor: incorporation into fibrin clot and degradation by thrombin. Blood **90:** 1883–1892.
138. SMITH, D.B., P.A. JANMEY, T.-J. HERBERT & S.E. LIND. 1987. Quantitative measurement of plasma gelsolin and its incorporation into fibrin clots. J. Lab. Clin. Med. **110:** 189–195.
139. LEUNG, L.L.K. 1986. Interaction of histidine-rich glycoprotein with fibrinogen and fibrin. J. Clin. Invest. **77:** 1305–1311.

140. LIJNEN, H.R., M. HOYLAERTS & D. COLLEN. 1980. Isolation and characterization of a human plasma protein with affinity for the lysine binding sites in plasminogen. Role in the regulation of fibrinolysis and identification as hisitidine-rich glyocoprotein. J. Biol. Chem. **255:** 10214–10222.
141. ANGLÉS-CANO, E., D. ROUY & H.R. LIJNEN. 1992. Plasminogen binding by alpha 2-antiplasmin and histidine-rich glycoprotein does not inhibit plasminogen activation at the surface of fibrin. Biochim. Biophys. Acta **1156:** 34–42.
142. ANGLÉS-CANO, E., J.C. GRIS & J.F. SCHVED. 1992. Familial association of high levels of histidine-rich glycoprotein and plasminogen activator inhibitor-1 with venous thromboembolism. J. Lab. Clin. Med. **121:** 646–653.
143. CASTAMAN, G., M. RUGGERI, F. BUREI & F. RODEGHIERO. 1993. High levels of histidine-rich glycoprotein and thrombotic diathesis. Thromb. Res. **69:** 297–305.
144. SHIGEKIYO, T., T. OHSHIMA, H. OKA, et al. 1993. Congenital histidine-rich glycoprotein deficiency. Thromb. Hæmostas. **70:** 263–265.
145. SOUTO, J.C., M. GARÍ, L. FALKON & J. FONTCUBERTA. 1996. A new case of hereditary histidine-rich glycoprotein deficiency with familial thrombophila. Thromb. Hæmostas. **75:** 374–375.

Crystal Structure Studies on Fibrinogen and Fibrin

RUSSELL F. DOOLITTLE, ZHE YANG, AND IGOR MOCHALKIN

Center for Molecular Genetics, University California, San Diego, La Jolla, California 92093, USA

ABSTRACT: X-ray crystallography studies on fragments D and double-D from human fibrinogen and fibrin have revealed the details of knob-hole interactions between fibrin units, as well as the nature of the association at their ends. More recently, a lower-resolution structure of native chicken fibrinogen has provided details about the structure of the central domain, and particularly the arrangement of disulfide bonds. Parts of the fibrinogen molecule are so flexible that they have not been visualized in electron density maps. The elusive regions include the αC domain, the amino-terminal segments of the α and β chains, and the carboxyl-terminal segments of the γ chains. Nonetheless, when all the structural data are considered together, it is possible to construct a realistic model not only of a fibrinogen molecule but also of a fibrin protofibril.

KEYWORDS: Fibrinogen; Fibrin; Crystal structure; Synthetic peptide knobs; Protofibril model.

INTRODUCTION

The publication of several high resolution crystal structures of major fragments from fibrinogen and fibrin has introduced a new phase in research for fibrinogen and fibrin.[1-6] Moreover, the recent publication of lower resolution structures for a modified bovine fibrinogen[7] and a native chicken fibrinogen[8] make it possible for the first time to put these structural details in an overall context. Nowhere have these structures been more helpful than in providing a better understanding of the fibrinogen–fibrin transition.

In this regard, it has long been accepted that the removal of the fibrinopeptide A leads to the formation of intermediate polymers called protofibrils. It is also a long standing belief that the subsequent release of the fibrinopeptide B from the units in these oligomers allows the lateral association needed to form the thick fibers that constitute natural clots.[9] These conclusions were reached well before there was any structural information available about the relative locations of the sites involved. With time, it became clear that it was the exposure of centrally located Gly-Pro-Arg "knobs" fitting into "holes" near the ends of the fibrin units that gives rise to the formation of the intermediate protofibrils. More definitively, the holes for this set of knobs (*A knobs*) are situated on the γ-chain carboxyl terminal domains.

Address for correspondence: R.F. Doolittle, Ph.D., Center for Molecular Genetics, Rm. 206, University of California, San Diego, La Jolla, CA 92093-0634, USA. Voice: 858-534-4417; fax: 858-534-4985.

rdoolittle@ucsd.edu

A considerable body of biochemical evidence was accumulated to this end, including the identification of many variant human fibrinogens with amino acid substitutions in the critical areas,[10] as well as ligand-binding studies with synthetic peptides[11,12] and the photo-affinity labeling of γ-chain residue Tyr363 by Gly-Pro-Arg derivatives.[13,14]

Evidence for the role of the "B knobs," which begin with the sequence Gly-His-Arg, has been much harder to come by. The locations—or even the existence—of a corresponding "hole" for B knobs have never been correlated with variant fibrinogens, nor was it possible to demonstrate such a site by photo-affinity labeling. Furthermore, although synthetic peptides beginning with the sequence Gly-Pro-Arg are effective inhibitors of fibrin polymerization, Gly-His-Arg peptides are not.[11,12] Remarkably, Gly-His-Arg-Pro peptides actually accelerate fibrin formation.[11] In a similar vein, Gly-Pro-Arg-Pro affinity columns are very useful for preparing fibrinogen or fragment D, but Gly-His-Arg-Pro columns are wholly ineffective. It is also well known that fibrin can be formed by the removal of fibrinopeptide A only, as occurs with various snake venom enzymes.[15] Nonetheless, such fibrin can be distinguished from fibrin generated by the thrombin-catalyzed removal of both fibrinopeptides A and B, and there must be a physiological basis for removing the B peptide.

The crystal structures of fragment D and double-D from fibrin have shed light on these sometimes puzzling observations, first by uncovering the exact locations of the two kinds of hole, and then by demonstrating a conformational change involved with the formation of the β-chain holes.

STRUCTURES OF FRAGMENTS D AND DOUBLE-D

As expected, the crystal structure of the 86-kilodalton fragment D revealed two homologous globular domains attached to a residual stub of the coiled-coil connector (see FIGURE 1). All told, 707 of the 734 amino acid residues in the fragment were positioned.[3] The missing residues were all from the terminal regions of chains, including residues γ397–406.

The fold found in the globular domains corresponding to the carboxyl-terminal regions of the β and γ chains (βC and γC) had not been observed in other proteins, although amino acid sequence comparisons had previously determined that the carboxyl-terminal domains of numerous other extracellular proteins in animals are homologous.[16] The foundational core of these βC and γC domains is a centrally disposed five-stranded β sheet. There are also smaller subdomains at both the amino- and carboxyl-terminal ends. Remarkably, the central strand of the main β sheet is contributed by a stretch of residues separated from the main core by the carboxyl-terminal subdomain (FIG. 1). The γC domain is situated at what would be the very tip of the fibrinogen molecule, whereas the βC domain is folded back against the coiled coil. The angle between them is approximately 130°, and the two obvious holes are oriented in not quite opposite directions (FIG. 1).

The biggest surprise in the fragment D structure was that the α chain makes an abrupt reversal of direction at a position in the middle of the disulfide ring connection with the other chains. The retreating chain lies in a groove between the α and β chains of the coiled coil, masking a region of the α chain (residues 151–159) that has

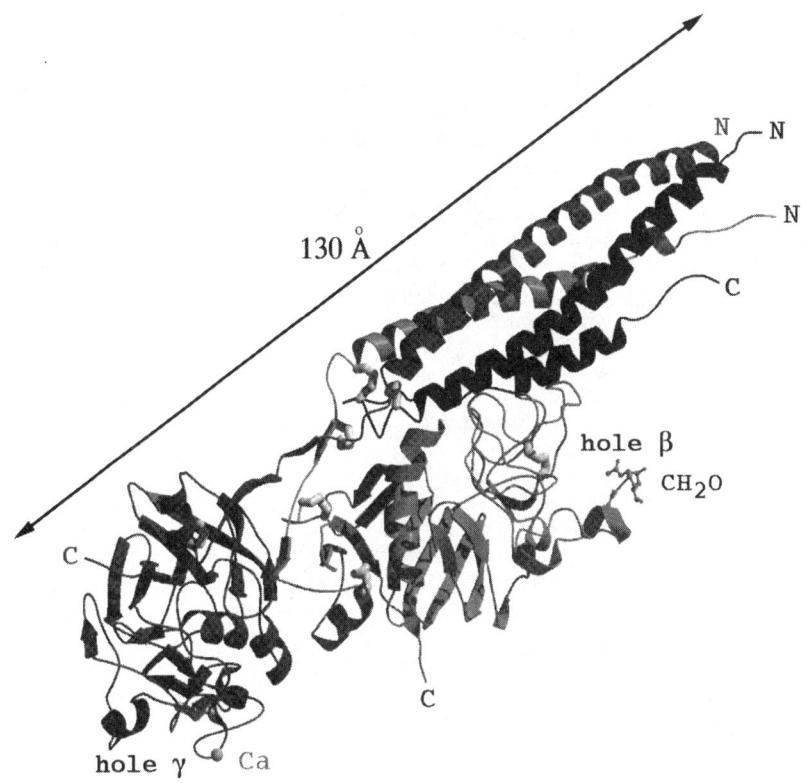

FIGURE 1. Ribbon representation of fragment D from human fibrinogen showing region of coiled coils and globular βC and γC domains, as well as holes for binding knobs (from Ref. 3, with permission).

been implicated in the activation of tPA.[17] The crystal structures show that this is the only region of the coiled coil not easily accessible from without.

The structure of the factor XIII-crosslinked double-D amounts to two molecules of fragment D butted end-to-end, any intrinsic conformational changes being too modest to detect. The two holes corresponding to the binding sites for Gly-Pro-Arg are situated on the same face of the dimer, opposite to that where the γ–γ crosslinks must be. With regard to the latter, it was one of the big disappointments of this work that the flexible linkages containing the crosslinks themselves do not appear in electron density maps (see FIGURE 2).

The interface between abutting D units has an offset nature that leads to the two associated units being stereochemically different. Thus, Argγ275 from one contributant is directed toward Tyrγ280 on the other side, whereas its opposite member on the other molecule is directed toward Serγ300. Although this condition is obligatory in the crystal, it is likely that under physiological conditions any given abutment may adopt either offset conformation, such that molecule A becomes B and B becomes

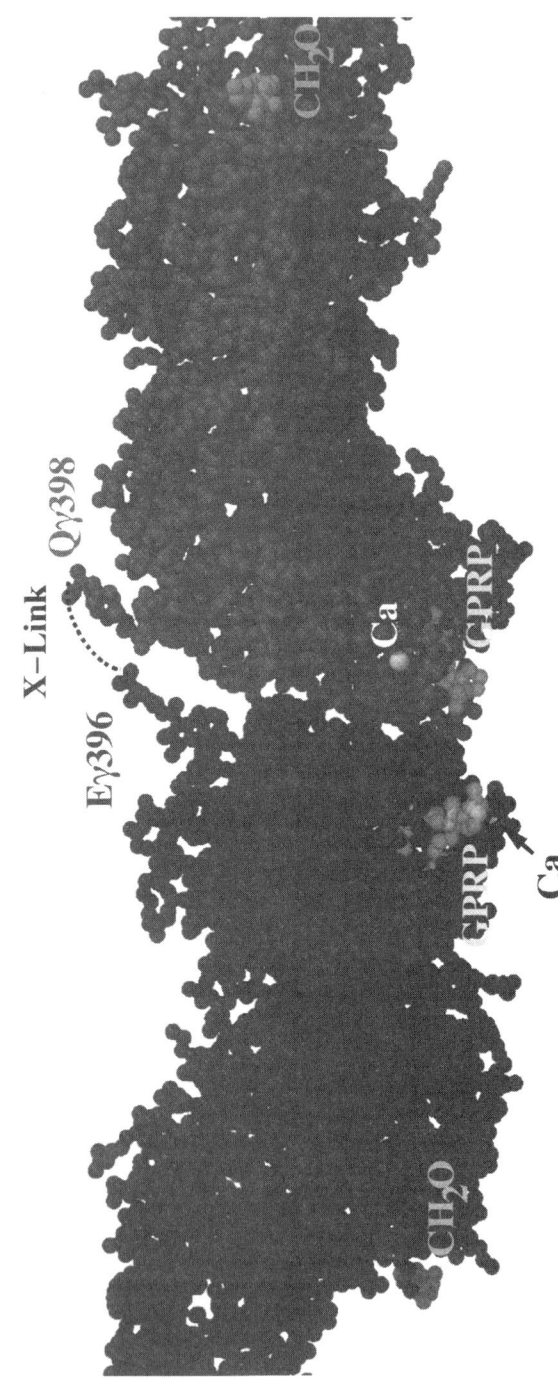

FIGURE 2. Close-up view of D–D interface showing positions of bound Gly-Pro-Arg-Pro-amide knobs and calcium ions. The inferred location of the γ–γ crosslink is shown by dashed line at top of model (from Ref. 3, with permission).

A. The junction is not tight, no salt links and hardly any hydrogen bonding being in evidence. There looks to be a layer of solvent between the complementary surfaces. The overall situation seems quite appropriate, since it would not be advantageous to have a strong end-to-end association of fibrinogen molecules in advance of fibrinopeptide removal.

On another front, the region of the α chain implicated in tPA activation remains inaccessible in the double-D structure. In this regard, an early report[18] of tPA activation by double-D has been superseded by more recent studies which show that, although D2E is an activator of tPA, double-D is not.[19]

STRUCTURE OF NATIVE FIBRINOGEN

The low-resolution structure of a native fibrinogen from chicken,[8] as well as that of a modified bovine fibrinogen,[7] provides a framework model that puts the fragment D structures in proper perspective. Both the chicken and bovine structures were solved by the method of molecular replacement, the human fragment D structure being used as the search model. The core fibrinogen molecule is obviously sigmoidal and flattened in one dimension (see FIGURE 3); it bears a remarkable resemblance to negative-stained preparations observed by electron microscopy twenty years ago.[20] As noted above, there are several highly mobile features in the native molecule, including the amino-terminal segments of the α and β chains and the always elusive carboxyl domain of the α chain.

In the case of chicken fibrinogen, it was possible to trace the chains in the region of the centrally located disulfide rings and establish the nature of the interchain connections (see FIGURE 4). The arrangement bears out early formulations based on evolutionary considerations;[21] the antiparallel nature of the interchain γ-chain connection, previously determined by chemical synthesis,[22] was also verified.

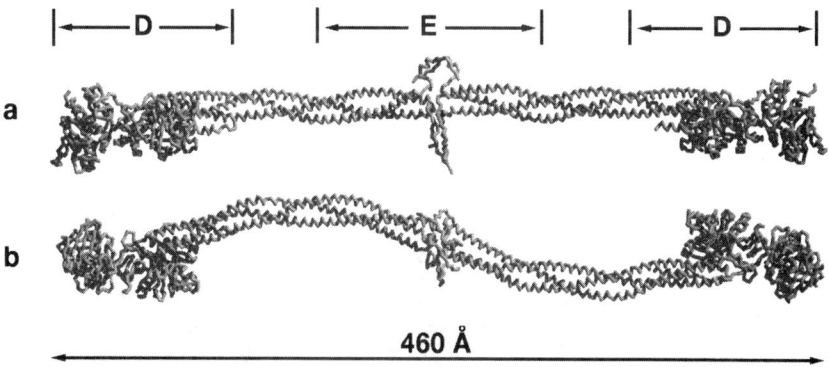

FIGURE 3. Model of native chicken fibrinogen (from Ref. 8, with permission).

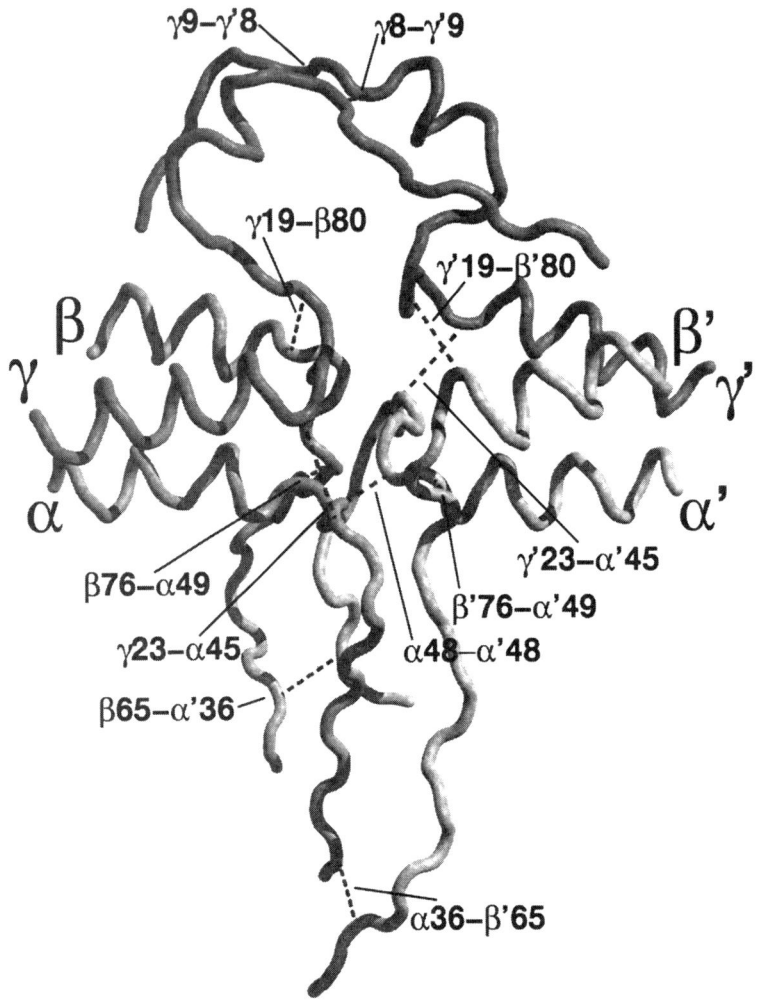

FIGURE 4. Backbone structure of central domain of fibrinogen showing disulfide connections (adapted from Ref. 8).

Even though it has not yet been possible to visualize the fibrinopeptides or the knobs they protect, the structure defines a localized neighborhood for where these entities must reside, mostly because of constraints imposed by disulfide bonds. In particular, the existence of an interdimeric disulfide between the cysteines at α-28 greatly restricts the whereabouts of the A-knobs. As such, it has been possible to make a very reasonable model of the staggered face-to-face connections that lead to the protofibril (see FIGURE 5).

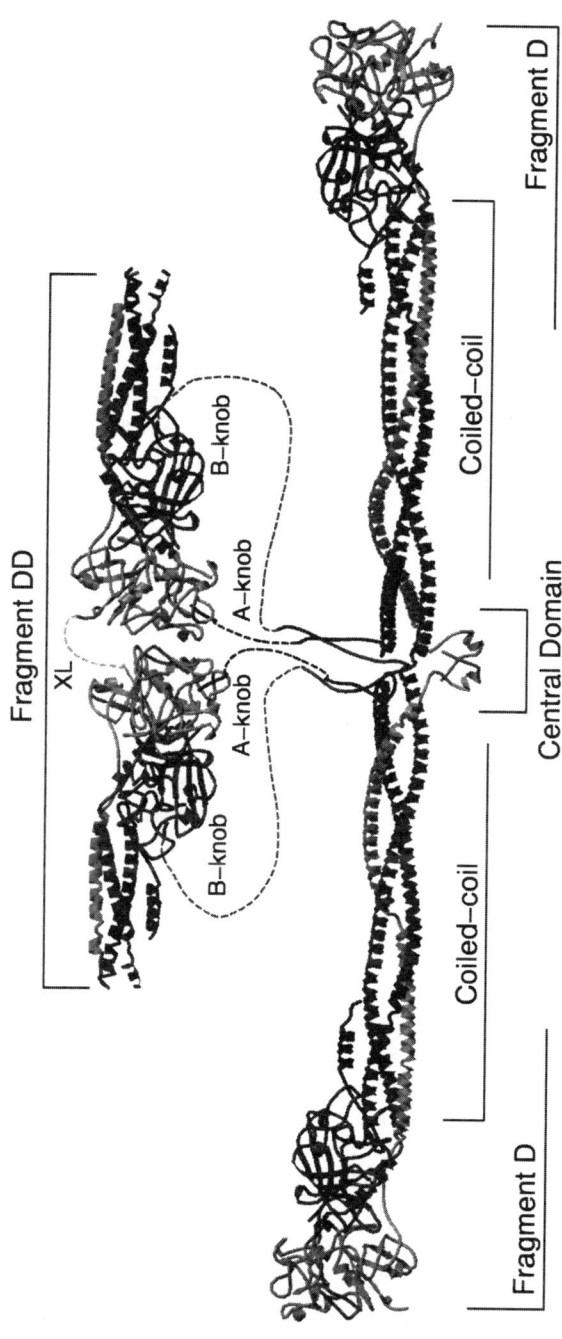

FIGURE 5. Combined models of double-D[4] and native fibrinogen[8] showing proposed knob–hole connections.

A KNOBS AND γ-CHAIN HOLES

The crystal structure of the crosslinked double-D fragments showed that the γ-chain holes on abutting molecules are only about 20 Å apart, just as would be expected if the two A knobs from another fibrinogen molecule were holding them together.[3] More precisely, the Cα carbons of the respective proline residues at positions four of the bound peptides are 22 Å apart. In native fibrinogen, the two α chains are held together by a disulfide bond between the two cysteine residues at position 28.[23] As such, it is a simple matter to model in the missing residues between Arg-19 (the Arg in Gly-Pro-Arg) and Cys-28 to form a bridge between knobs.

We also synthesized the 12-residue peptide GPRVVERHQSAC corresponding to the segment from the end of the A knob to Cysα28 and oxidized it to form the *bis*-dodecamer. The crystal structure of double-D complexed with this 24-residue peptide has been determined (unpublished work from this laboratory). The electron density of the peptide is only clear for the first five to seven residues on either side, including the bound Gly-Pro-Arg termini. The remainder of the peptide loop is not closely bound to the double-D fragment and must be mobile. This is an equivalent situation to that occurring on the opposite side of the D-D interface, where the crosslinked carboxyl-terminal segments of the γ chains could not be visualized either (FIG. 2). It must be concluded from these observations that not only are the end-to-end connections between fibrin units loose and mobile, but so are the interactions between D and E, the only tight attachments being the knob-hole interactions themselves.

Summing to this point, the first step of fibrin formation involves a natural association at the D–D interface between two abutting fibrinogen molecules pinned together by a pair of A knobs of another fibrinogen molecule. It is conceivable that the end-to-end interaction occurs in advance of the pinioning, and that the flexible knobs find their way into a predetermined orientation. Under physiological conditions, the union is further cemented by the action of factor XIII-induced crosslinks between the carboxyl-terminal segments of γ chains located diametrically opposite to the knob–hole sites. Both of these interactions are loose enough to insure an intrinsic flexibility in the ribbon-like protofibril chain (see FIGURE 6).

LIGANDS AND CONFORMATIONAL CHANGE

From the start it was clear that ligands influence the crystal structures, at least to the extent where different unit cells were observed when fragments D or double-D were cocrystallized with one or the other or both synthetic peptide knobs.[6,24] Indeed, if fragment D crystals prepared in the absence of a peptide are transferred to a solution containing the synthetic A knob, they shatter.[25] Nonetheless, the structural changes can be subtle, and initially no major changes were observed in or around the γ-chain hole.[4]

On the other hand, comparisons of several of these structures revealed that there is a major conformational change in the neighborhood of the β-chain hole.[6] In this regard, there is a big swing of side chains βGlu397 and βAsp398 from a position where they are pinned back against the coiled coil by a calcium bridge.[6] The

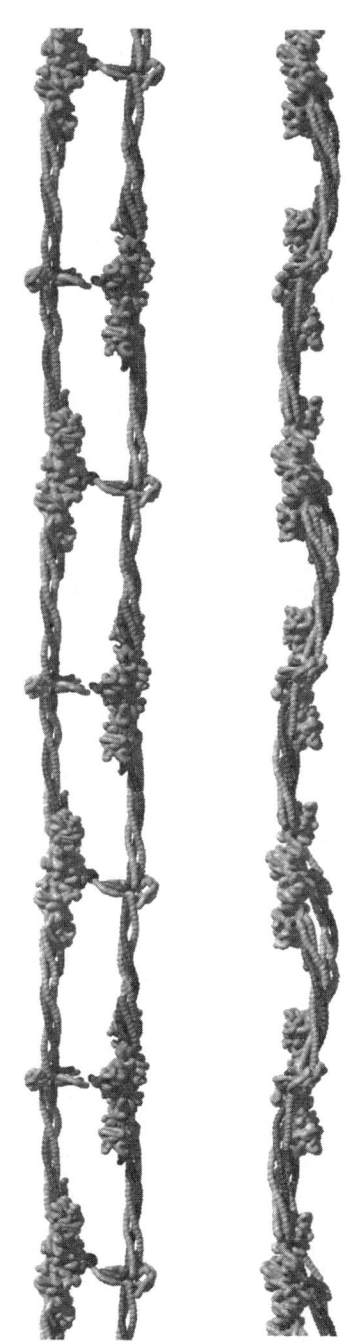

FIGURE 6. Top and side views of part of a protofibril composed of fibrin units arranged on basis of connections shown in Figure 5. A similar model appears in Reference 7.

conformational change may be the basis for the acceleration of fibrin formation in the presence of the GHRP-ligands.[11] The incomplete site that exists before removal of the fibrinopeptide B may insure that A knobs do not inadvertently bind to this site. The conformational change may also contribute to the exposure of the β-chain face, which is homologous to the region of γ chains involved in the end-to-end interaction, and which may be involved in the lateral association of protofibrils. The shift in side chain positions may also have a bearing on the exposure of the tPA activation site.

Some other aspects became apparent during the comparison of all these D and double-D structures. Among them was the discovery that the βC domain binds a calcium ion in the same position as does the homologous γC domain.[5] It should be stressed that this is a different calcium from the one that pins the βC domain back against the coiled coil.[6]

Returning to the involvements of the B knob, the observed location of the β chain hole aimed in quite the opposite direction to the γ-chain hole (FIG. 1) offers a possible explanation of why GHRP-affinity columns do not bind fibrinogen or fragment D, in that the arms in such columns may not have been long enough to allow the attached knobs to reach the somewhat hindered holes. Beyond that, its location apparently precludes a situation whereby a pair of B knobs from a single fibrinogen molecule can pin together two β domains in the same way that a pair of A knobs holds together two γ domains. Instead, it seems more likely that a pair of B knobs actually cement the same three-molecule construct held together by a pair of A knobs.

A model of how this interaction can occur is shown in FIG. 5. It is well known that fibrin exhibits a well defined periodicity reflecting the half-molecule overlap (230 Å), and as such the protofibrils must be in exact register. However, if the B knobs exhibit the same (or greater) flexibility than that we have attributed to the A knobs, how can this perfect registration of protofibrils occur? If a D–D interfacial association sets the stage for the involvement of A knobs, a similar precondition ought to exist for the participation of B knobs. The possibility of a β–β interaction equivalent to the γ–γ end-to-end interaction presents itself as a possibility for lateral association.

The B knobs, in contrast to the A knobs that are held close to each other by the disulfide connection between the cysteines at position α-28 mentioned above, must be extended from the parent molecule at a more obtuse angle. The B knobs (residues 15–17 of the β chain) are approximately 50 residues away from a constraining disulfide connection between the β and α chains, whereas the A knobs (residues 17–19 of the α chain) are only 20 residues away from that same disulfide and already more restrained by an even nearer disulfide at α-28. The possibility exists, of course, that different protofibrils are held together by some kind of B-knob involvement, but it seems unlikely in the light of the β-fibrin argument presented above.

The clasper-like involvement of B-knobs (FIG. 5) is also consistent with the observation that fibrin can be formed by the exclusive release of the fibrinopeptides B. β Fibrin is the product that ensues when the fibrinopeptides B are removed but not the fibrinopeptides A. It can be formed by the use of heterologous thrombins[26,27] or by the judicious employment of certain snake venom enzymes.[28,29] A key observation in these studies is that γ–γ crosslinking occurs in a mode indistinguishable from that when the fibrinopeptides A are removed. As such, the end-to-end packing in the fibrin must be the same, even though the A-knobs are not in the γ-chain holes. Modeling studies have convinced us that it is unreasonable to suppose that the B knobs can fit into the γ-chain holes ordinarily occupied by the A knobs and still pin together two molecules associated at the γ–γ interface. Our interpretation is that the abutting molecules are held together by the clasper-like action of the B knobs in the β-chain holes (FIG. 3).

ASSOCIATION OF PROTOFIBRILS

Models of fibrin formation have been proposed that invoke the association of αC domains.[30,31] It is reasonable to suppose a role for these domains in fibrin formation, if only because it is well known that under native conditions α chains are slowly crosslinked through the action of factor XIII.[32] It is also reasonable to suppose, as has been implied by this model, that there may be some kind of interaction between the fibrinopeptides B and the αC domains, and that removal of the fibrinopeptides B allows a relocation of the αC domains from some kind of intramolecular association to an intermolecular association.[30,31] Nonetheless, the disorder inherent in current electron density maps runs counter to a simple well defined initial structure. Further studies and higher resolution crystal structures of native fibrinogens may resolve the issue.

REFERENCES

1. YEE, V.C., K.P. PRATT, H.C. COTE, et al. 1997. Crystal structure of a 30 kDa C-terminal fragment from the γ chain of human fibrinogen. Structure **5:** 125–138.
2. PRATT, K.P., H.C.F. COTE, D.W. CHUNG, et al. 1997. The fibrin polymerization pocket: three-dimensional structure of a 30-kDA C-terminal γ chain fragment complexed with the peptide Gly-Pro-Arg-Pro. Proc. Natl. Acad. Sci. U.S.A. **94:** 7176–7181.
3. SPRAGGON, G., S.J. EVERSE & R.F. DOOLITTLE. 1997. Crystal structures of fragment D from human fibrinogen and its crosslinked counterpart from fibrin. Nature **389:** 455–462.
4. EVERSE, S.J., G. SPRAGGON, L. VEERAPANDIAN, et al. 1998. Crystal structure of fragment double-D from human fibrin with two different bound ligands. Biochemistry **37:** 8637–8642.
5. SPRAGGON, G., D. APPLEGATE, S.J. EVERSE, et al. 1998. Crystal structure of a recombinant extended $α_E$C domain from human fibrinogen-420. Proc. Natl. Acad. Sci. U.S.A. **95:** 1–9.
6. EVERSE, S.J., G. SPRAGGON, L. VEERAPANDIAN & R.F. DOOLITTLE. 1999. Conformational changes in fragments D and double-D from human fibrin(ogen) upon binding the peptide ligand Gly-His-Arg-Pro-amide. Biochemistry **38:** 2941–2946.
7. BROWN, J.H., N. VOLKMANN, G. JUN, et al. 2000. The crystal structure of modified bovine fibrinogen. Proc. Natl. Acad. Sci. U.S.A. **97:** 85–90.
8. YANG, Z., I. MOCHALKIN & R.F. DOOLITTLE. 2000. Crystal structure of a native chicken fibrinogen at 5.5Å resolution. Proc. Natl. Acad. Sci. U.S.A. **95:** 3907–3912.
9. LAURENT, T.C. & B. BLOMBACK. 1958. On the significance of the release of two different peptides from fibrinogen during clotting. Acta Chem. Scand. **12:** 1875–1977.
10. MATSUDA, M. 1996. The structure–function relationship of hereditary dysfibrinogens. Intern. J. Hematol. **64:** 167–179.
11. LAUDANO, A.P. & R.F. DOOLITTLE. 1978. Synthetic peptide derivatives that bind to fibrinogen and prevent the polymerization of fibrin monomers. Proc. Natl. Acad. Sci. U.S.A. **75:** 3085–3089.

12. LAUDANO, A.P. & R.F. DOOLITTLE. 1980. Studies on synthetic peptides that bind to fibrinogen and prevent fibrin polymerization. Structural requirements, number of binding sites, and species differences. Biochemistry **19:** 1013–1019.
13. SHIMIZU, A., G. NAGEL & R.F. DOOLITTLE. 1992. Photoaffinity labeling of the primary fibrin polymerization site: isolation and characterization of a labeled cyanogen bromide fragment corresponding to γ-chain residues 337–379. Proc. Natl. Acad. Sci. U.S.A. **89:** 2287–2892.
14. YAMAZUMI, K. & R.F. DOOLITTLE. 1992. Photoaffinity labeling of the primary fibrin polymerization site: Localization of the label to γ-chain Tyr-363. Proc. Natl. Acad. Sci. U.S.A. **89:** 2893–2896.
15. BLOMBACK, B., M. BLOMBACK & I.M. NILSSON. 1957. Coagulation studies on "Reptilase," an extract of the venom from Bothrops jararaca. Thromb. Diath. Hæm. **1:** 1–13.
16. DOOLITTLE, R.F. 1992. A detailed consideration of a principal domain of vertebrate fibrinogen and its relatives. Prot. Sci. **1:** 1563–1577.
17. SCHIELEN, W.J.G., H.P.H.M. ADAMS, M. VOSKUILEN, et al. 1991. Structural requirements of position Aa-157 in fibrinogen for the fibrin-induced rate enhancement of the activation of plasminogen by tissue-plasminogen activator. Biochem. J. **276:** 655–659.
18. NIEUWENHUIZEN, W., A. VERMOND, M. VOSKUILEN, et al. 1983. Identification of a site in fibrin(ogen) which is involved in the acceleration of plasminogen activation by tissue-type plasminogen activator. Biochim. Biophys. Acta **748:** 86–92.
19. YAKOVLEV, S., W. NIEUWENHUIZEN & L. MEDVED. 1999. The mechanism of the reversible exposure of the t-PA-binding sites in the fibrin-derived D-D fragment. Abstract 1225. 17th Intern. Cong. Thromb. Hæm., Washington, D.C.
20. WILLIAMS, R.C. 1981. Morphology of bovine fibrinogen monomers and fibrin oligomers. J. Mol. Biol. **150:** 399–408.
21. DOOLITTLE, R.F., D.M. GOLDBAUM & L.R. DOOLITTLE. 1978. Designation of sequences involved in the "coiled coil" interdomainal connections in fibrinogen: construction of an atomic scale model. J. Mol. Biol. **120:** 311–325.
22. HOEPRICH, P.D. & R.F. DOOLITTLE. 1983. Dimeric half-molecules of human fibrinogen are joined through disulfide bonds in an antiparallel orientation. Biochemistry **22:** 2049–2055.
23. BLOMBACK, B. 1971. Selectional trends in the structure of fibrinogen of different species. *In* Biochemical Evolution and the Origin of Life. E. Schoffeniels, Ed.: 112–119. North Holland Publishing Company, Amsterdam.
24. EVERSE, S.J., H. PELLETIER & R.F. DOOLITTLE. 1995. Crystallization of fragment D from human fibrinogen. Prot. Sci. **4:** 1013–1016.
25. DOOLITTLE, R.F., S.J. EVERSE & G. SPRAGGON. 1996. Human fibrinogen: anticipating a 3-dimensional structure. FASEB J. **10:** 1464–1470.
26. DOOLITTLE, R.F. 1965. Differences in the clotting of lamprey fibrinogen by lamprey and bovine thrombins. Biochem. J. **94:** 735–741.
27. COTTRELL, B.A. & R.F. DOOLITTLE. 1976. Amino acid sequences of lamprey fibrinopeptides A and B and characterization of the junctions split by lamprey and mammalian thrombins. Biochim. Biophys. Acta **453:** 426–438.
28. SHAINOFF, J.R. & B.N. DARDIK. 1979. Fibrinopeptide B and aggregation of fibrinogen to fibrin. Science **204:** 200–204.
29. DYR, J.E., B. BLOMBACK, B. HESSEL & F. KORNALIK. 1989. Conversion of fibrinogen to fibrin induced by preferential release of fibrinopeptide B. Biochim. Biophys. Acta **990:** 18–24.
30. VEKLICH, Y.I., O.V. GORKUN, L.V. MEDVED, et al. 1993. Carboxyl-terminal portions of the α chains of fibrinogen and fibrin. Localization by electron microscopy and the effects of isolated αC fragments on polymerization. J. Biol. Chem. **268:** 13577–13585.
31. GORKUN, O.V., Y.I. VEKLICH, L.V. MEDVED, et al. 1994. Role of the αC domains of fibrin in clot formation. Biochemistry **33:** 6986–6997.

32. MCKEE, P.A., P. MATTOCK & R.L. HILL. 1970. Subunit structure of human fibrinogen, soluble fibrin and cross-linked insoluble fibrin. Proc. Natl. Acad. Sci. U.S.A. **66:** 738–744.

Contribution of the α_EC Domain to the Structure and Function of Fibrinogen-420

GERD GRIENINGER

Lindsley F. Kimball Research Institute of the New York Blood Center, 310 East 67th Street, New York, New York 10021, USA

ABSTRACT: In addition to the conventional fibrinogen with its α, β, and γ subunit chains, there is a subclass of fibrinogen molecules, accounting for one percent of the total in human adults, in which both α chains have been replaced by extended α chains (α_E) that sport a globular C-terminal domain (α_EC) comparable to βC and γC. Using nomenclature based on molecular weight, the subclass of α_E-containing molecules has been named fibrinogen-420 to differentiate it from the better known fibrinogen, now referred to as fibrinogen-340. Review of the events leading to the discovery of fibrinogen-420 in the early 1990s and its subsequent characterization, culminating in the crystal structure of its unique α_EC domains, highlights special aspects of its evolutionary history, outstanding features of its structure, and the perplexities of its biology. Various working hypotheses that have driven prior investigation are evaluated and practical insights are offered to spur further research into the role of fibrinogen-420.

KEYWORDS: X-ray structure; Protein evolution; Blood clotting; Calcium binding site.

INTRODUCTION

Fibrinogen-420 is an intriguing, alternative form or subclass of human fibrinogen that had escaped detection throughout many decades of intensive research. During that time it was well established that all three fibrinogen subunits (α, β, and γ) share structural homologies at their N-termini and significant differences at their carboxyl ends. The common form of the α chain ends in a random coil (αC), quite distinct from the highly similar globular C-termini of the β and γ chains (βC and γC).[1,2] In the early 1990s our laboratory began to accumulate evidence for the existence of an α chain sporting a β- and γ-like C-terminus. We have named this isoform the extended α chain (α_E) and denote its globular C-terminus as α_EC. Fibrinogen-420, a nomenclature related to size, refers to the α_E-containing subclass of human fibrinogen molecules. These fibrinogens have the structure $(\alpha_E\beta\gamma)_2$.

Our laboratory was privileged to make the initial discoveries by virtue of the simple fact that we had been using cultures of chicken hepatocytes as a model system for studying the regulation of fibrinogen synthesis.[3] Here we review the scientific journey and its major implications, from the revelation of an additional exon in the

Address for correspondence: Gerd Grieninger, Ph.D., Lindsley F. Kimball Research Institute of the New York Blood Center, 310 East 67th Street, New York, NY 10021-6295, USA. Voice: 212-570-3124; fax: 212-288-5087.
ggrien@nybc.org

fibrinogen α gene, to the identification of the spliced transcript encoding α_E, expression of the α_E subunit, evaluation of its assembly into hexameric fibrinogen, analyses of its special C-terminus, and studies of its distribution in blood during health and disease. Clues to the still elusive function of fibrinogen-420 will undoubtedly be found in this story that encompasses elements of the molecule's evolution, structure, and biology.

FORTUITOUS EXPERIMENTS ENLARGE THE FIBRINOGEN α GENE AND ITS HOST OF TRANSCRIPTS

It was the unexpected detection of an irregular, large messenger RNA for the *α subunit* of chicken fibrinogen in a hybridization experiment using a probe for the

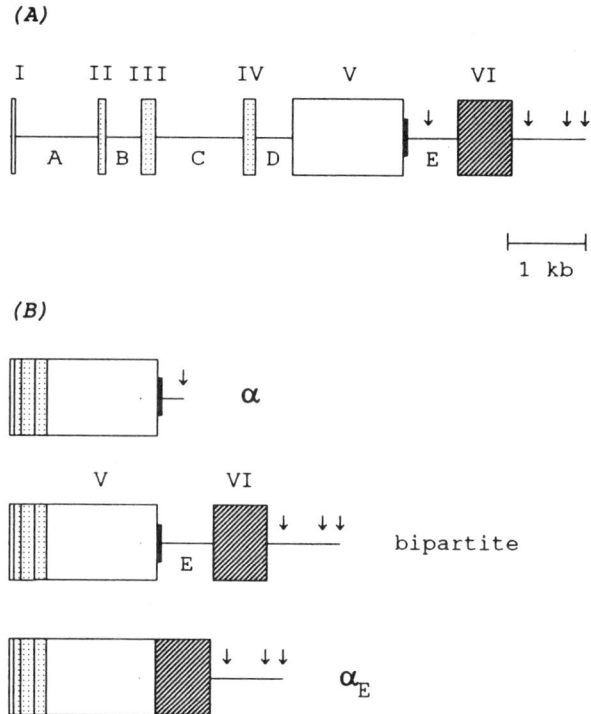

FIGURE 1. The human fibrinogen α gene and its transcripts. (**A**) Organization of the gene. The term *exon* is restricted to protein coding sequences only. Exons are depicted as *boxes*: exons I to IV, *light gray*; exon V, *white*; and exon VI, *dark gray*. Introns A to E and the 3' untranslated region are drawn as *lines*. The region at the end of exon V that is alternately translated (to generate the C-terminus of the predominant α chain) or spliced out (as part of intron E for formation of α_E) is depicted half-height in *black*. *Arrows* show polyadenylation sites. (**B**) Schematic presentation of transcripts. The symbols used are the same as in (**A**). Reprinted from Reference 5.

β subunit that gave us the first tantalizing glimpse.[4] A "bipartite" cDNA was isolated that contained, not only the coding region for the major form of the α chain but also a second open reading frame downstream with high homology to the β (and γ) chain C-terminal domain. The importance of this finding was not recognized until its parallel in humans was characterized.[5]

The cDNA sequence revealed the second open reading frame to be an additional α gene exon (exon VI) encoding a new C-terminus that, when appropriately spliced, extends the α chain. At least three transcription products are derived from the gene, as summarized in FIGURE 1, and proof of their existence necessitated revision of the earlier five-exon structure of the human fibrinogen α gene.[5,6] The relative prevalence of the transcripts is: predominant α, ca. 90%; bipartite, ca. 5%; and $α_E$, ca. 1–2%. Thus, the human $α_E$ isoform and the conventional α chain are identical in sequence through residue Val610, but in $α_E$ this valine is followed by Arg611 and the 236 residues encoded by exon VI (Asp612-Gln847) that form the $α_E$C domain. The apparent molecular mass of $α_E$ is more than 50% greater than that of α.[7] The additional mass is due to both the extra residues and glycosylation, the latter accounting in large measure for $α_E$ migration on SDS-PAGE at about 110 kDa, rather than at its calculated mass of 92,843 Da.

Oddly, the region at the end of exon V can be either translated (to generate the C-terminus of the predominant α chain) or spliced out as part of intron E (to yield the transcript for $α_E$). The predominant α transcript makes use of a poly(A) signal in the intron E sequence and thus bears only the first open reading frame. The transcript designated bipartite is incompletely spliced, with intron E still present. Its existence raises a number of questions, among them, whether the alternative splicing is regulated (it appears not to be; see sections on physiological context below). The common α chain can be translated from both the α and the bipartite transcripts and, in both cases, the coding sequence for its C-terminal residues is generated by continuing exon V translation past the 5′ splice site of intron E for another several codons.

LUCK AND THE CHICKEN: EVOLUTION OF THE FIBRINOGEN GENES IS AT THE HEART OF THE MATTER

The basic six-exonic structure of the α gene is conserved among mammals and birds, as are the intron positions.[8] Furthermore, there are highly conserved nucleotide sequences present at the 3′-end of intron E, which apparently render the splicing process inefficient. Bipartite transcripts are found, with levels low, but significant, relative to that of the most abundant α mRNA in all higher vertebrates examined.

Thus far, *only in chicken* has the bipartite transcript been found to serve as the major species of α mRNA (exceeding 90%).[4] The absence of a poly(A) signal in chicken intron E precludes a shorter mRNA, and transcription through exon VI to reach a downstream AATAAA site is apparently required to secure ample production of the common α chain of the chicken. Of note, there is no bipartite transcript in lamprey, which generates its α and $α_E$ chains by two separate genes, α-I and α-II.[9] The designation $α_E$, as for the extended α chains of birds and mammals, is not technically applicable to lamprey, in which the alternative chain is actually shorter than the major type. For simplicity, however, we here refer to the lamprey domain also as an $\alpha_E C$ domain.

Earlier fibrinogen evolutionary schemes were based on the strong similarities between the carboxy ends of the β and γ chains.[4,10,11] With the discovery of the α_E chain and the revision of the α gene, the common ancestry shared by all three chains became manifest. It was further underscored by the precision with which intron E, demarcating the 5' end of exon VI, aligns with intron positions (all type I splice junctions) at the very beginning of the corresponding homologous regions of the β and γ genes.[12] The sequence of the C-terminal extension of α_E is as similar to the carboxy ends of the human fibrinogen β and γ chains as the latter two are to each other (about 40% identity).[5] In light of this symmetry of homology we proposed a revision of earlier evolutionary schemes (see FIGURE 2) to indicate that the gene duplications leading to development of the three separate subunits occurred at a similar time, more than 700 million years ago.[5]

The exon VI-derived domain is the most conserved region of α_E. From dot matrix comparison of the α_E sequences of humans and chicken (see FIGURE 3) it can be seen that very few stretches outside the exon VI-encoded C-terminus share 60% or more amino acid identity, demonstrating a rather considerable divergence of the predominant α chains of these two species since separation of their lineages.[4]

The $\alpha_E C$ domains in mammals and chicken contain exactly 236 amino acids and the sequences align without a single gap.[8] In TABLE 1, which compares changes in the fibrinogen α_E-, β-, and γ-chain C-termini during evolution, it can be seen that the

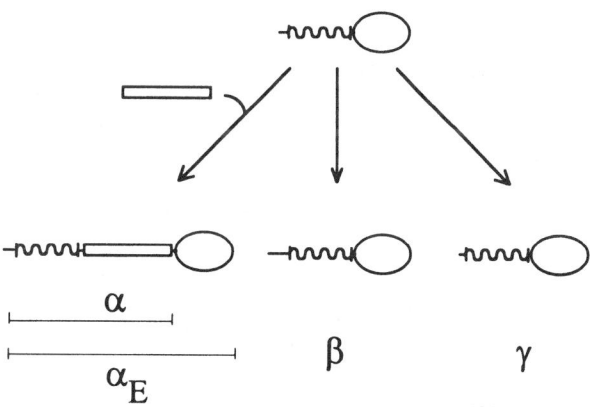

FIGURE 2. Proposed scheme for evolution of the fibrinogen genes. The three subunits are derived by gene duplication from a common ancestor. Shapes are used to symbolize the coding regions for the various domains: *wavy line,* the α-helical segment bordered by cysteine brackets (*short vertical lines*); *ellipse,* the globular C-terminal domain; *rectangle,* the long segment of the α chain (αC) that contains oligopeptide repeats and bears no sequence homology to the β and γ chains. The rectangular section, which also encodes the carboxy end of the common α chain, may have been a subsequent insertion (as depicted) or part of the ancestral gene that was lost later from the β and γ genes. Reprinted from Reference 5.

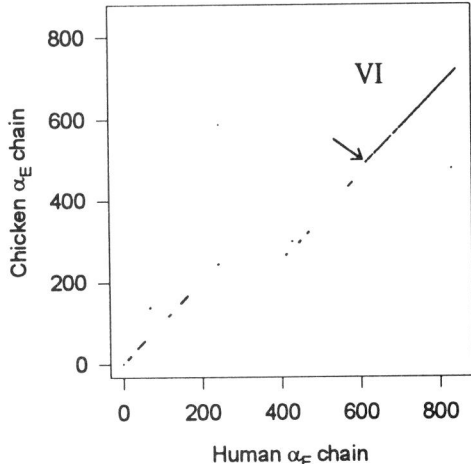

FIGURE 3. Exon VI-encoded domain is the most conserved region of α_E: dot-matrix comparison of chicken and human α_E chains. Analysis of the mature proteins (human Ala1 to Gln847 and chicken Gln1 to Leu723) was performed with check size 10, matching amino acids 6 (i.e. a *dot* appears when six out of ten amino acids are identical). An *arrow* indicates the beginning of the α_EC domain encoded by exon VI. The diagonal line formed by most of the dots is displaced by a large gap of about 150 residues because the chicken protein contains no oligopeptide repeats in the central region.[4] Reprinted from Reference 8.

α_EC domain is as well conserved as the homologous βC and γC domains. This degree of evolutionary preservation is difficult to ignore, high conservation being generally accepted as a key indicator of functional importance. Since evolutionary pressure to preserve the protein sequence (and fold; see section on structure below) is similar for all three globular domains of fibrinogen-420, it follows that their functions are equally important, a conclusion that has motivated our laboratory to seek out the specific contribution of the α_EC domain to the repertoire of fibrinogen activities.

TABLE 1. α_EC, βC, and γC domains: conservation among vertebrate species[a]

Species	Amino acid identity (%)		
	α_EC	βC	γC
Rat	93.2	91.7	88.1
Chicken	75.9	80.6	78.5
Xenopus	70.3	80.2	72.0
Lamprey	52.2	62.3	63.0

[a]Carboxy termini of the fibrinogen subunits of rat, chicken, Xenopus and lamprey were aligned with their human counterparts and their overall amino acid identity to the human sequence determined. The C-terminal regions of the human α_E, β, and γ chains begin at the conserved intron position with Asp612,[5] Glu210, and Asp152,[12] respectively. Reprinted from Reference 8.

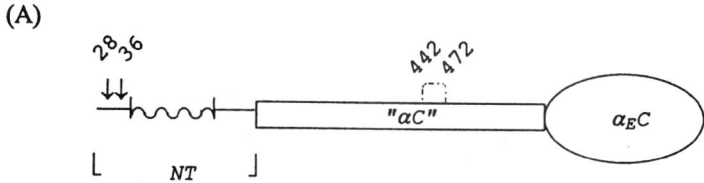

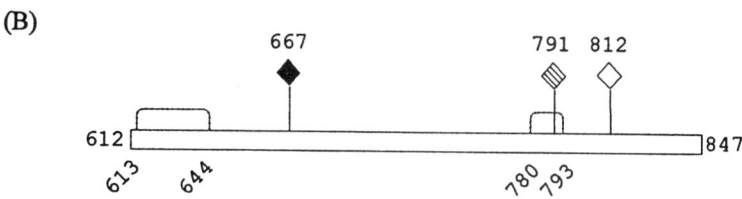

FIGURE 4. Schematic depiction of targeted amino acid positions in the α_E subunit for site-directed mutagenesis. *Arrows* or simple *vertical lines* indicate positions of those cysteines converted, either singly or in combination, to serine or alanine by site-directed mutagenesis. **(A)** the three regions of the α_E chain: the N-terminus (NT) containing a large α-helical segment, the so-called αC region, and the C-terminal globular α_EC domain; **(B)** the α_EC domain: potential glycosylation sites in the wildtype sequence at Asn667 and Asn812 are marked, respectively, with *closed* and *open diamonds*. The glycosylation site at Asn791 introduced by the Cys793→Ser change is marked with a *striped diamond*. Bonds between the cysteine pairs at α_E613/α_E644 and at α_E780/α_E793 are illustrated as *loops*. Reprinted from Reference 15.

As an aside, it may be noted that the α_E chain differs significantly in topology from the β and γ chains by virtue of the large αC region that tethers the globular C-terminus of the subunit (α_EC) to its α-helical N-terminus. Overall, the αC tether, encoded by exon V of the α gene, is poorly conserved among the fibrinogens of higher vertebrates.[13] An exception is its disulfide loop, analogous to the human cysteines at positions 442/472 (see FIGURE 4), that is well conserved although it does not play a role in intracellular assembly of either fibrinogen subclass. By contrast, the dependence of intracellular assembly on the integrity of the two homologous intradomain disulfide bridges in the globular C-termini is distinctly different for each subunit. As revealed by mutational analysis, in βC only the first loop is required; in α_EC it is the second loop; and in γC both loops play a critical part.[14,15]

α_E AND α_E-FIBRINOGEN: STOICHIOMETRIC SURPRISES

The development of α_EC-specific antibody probes enabled us to characterize α_E and examine its assembly into fibrinogen. In light of the low prevalence of α_E transcripts (about 1–2% of all α gene transcripts), we expected α_E incorporation to follow the pattern of γ', that is, yielding primarily heterodimeric molecules.[16] Much to our astonishment, that did not turn out to be the case.

In early studies, the α_E chain was found to be incorporated into hexameric fibrinogen[5] and α_E-containing fibrinogen was found to be present at substantial levels in human blood.[7] Remarkably, α_E was present primarily as part of $(\alpha_E\beta\gamma)_2$, a homodimeric (i.e., symmetrical) molecule in immunoprecipitation analyses of α_E-fibrinogen secreted by hepatocarcinoma (HepG2) cells.[7]

Based on predicted mass, the term fibrinogen-420 was coined for this subclass of molecules to distinguish it from the predominant 340-kDa $(\alpha\beta\gamma)_2$ subclass, similarly referred to as fibrinogen-340. Although the notion that both α subunits are replaced by α_E in fibrinogen-420 was received with some skepticism, it has been borne out by additional studies: the α_E-containing molecules in fibrinogen synthesized by primary adult liver cells and transfected COS cells, and particularly in fibrinogen purified from human plasma, are mostly homodimeric. One of these studies, using COS cells transfected with different sets of fibrinogen subunit cDNAs, is reviewed in FIGURE 5. $\alpha_E/\beta/\gamma$-transfectants secreted a high molecular weight species (fibrinogen-420) significantly larger than that of the $\alpha/\beta/\gamma$-transfectants (fibrinogen-340). Strikingly, only one α_E-containing species was secreted when all four chains were cotransfected ($\alpha_E/\alpha/\beta/\gamma$-transfectants); this product comigrates with fibrinogen-420. Mixed α/α_E-containing molecules, expected to be migrating at a position intermediate between fibrinogen-420 and fibrinogen-340, were not detected.

The mechanism that gives rise to symmetric α_E-containing fibrinogen against all stoichiometric odds remains an enigma. Our first hypothesis concerning fibrinogen-

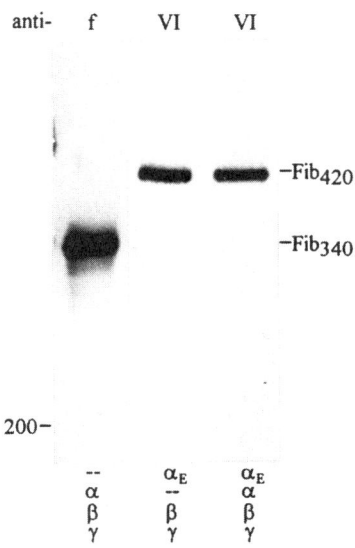

FIGURE 5. Secretion of fibrinogen-420 by transfected COS cells in the presence and absence of the common α chain. Cells were transfected with fibrinogen subunit cDNAs in combination as indicated below each lane. Fibrinogen was immunoprecipitated from culture medium with either antifibrinogen (*f*) or anti-VI (also called anti-α_EC) as indicated and run under non-reducing conditions. The positions of fibrinogen-420 (Fib420) and fibrinogen-340 (Fib340) are indicated. Reprinted from Reference 15.

420 structure portrayed the molecule as a fibrinogen-340 with two $\alpha_E C$ domains tethered by their random coil αC regions at each distal node, yet proximally situated with respect to one another. We speculated that the four cysteines in each $\alpha_E C$ domain might become involved in $\alpha_E C$-to-$\alpha_E C$ bonding that could energetically favor assembly of the symmetrical molecular structure.[7,15] In such a configuration, the $\alpha_E C$ domains might also form alternative disulfide bond connections to the central core of the fibrinogen molecule, thereby affording steric protection to the notoriously protease-prone αC regions of the molecule. Although the homodimeric composition of α_E-containing fibrinogen has held up under closer scrutiny, neither the reinforcement by interchain disulfide bond formation, nor greater resistance to protease proved to be true (see below). Thus, a different mechanism, at present unidentified but based on non-covalent interactions, must be advanced to explain the counterintuitive, symmetric incorporation of α_E chains into fibrinogen-420 during intracellular assembly.

PURIFICATION OF PLASMA FIBRINOGEN-420: ONE MYSTERY SOLVED

Could the double headed structure of homodimeric fibrinogen-420 be a clue to the importance of coordinated action of the two α_E domains in fibrinogen-420 function? Our first thoughts turned to their possible influence on the primary role of fibrinogen in clot formation, but a definitive answer awaited breakthroughs in efforts to obtain purified preparations.

Isolation of plasma fibrinogen-420 poses a formidable challenge, given that it comprises only one out of every 100 fibrinogen molecules in normal adult blood,[17] the remaining 99 sharing the same composition but for their lack of the two $\alpha_E C$ globular domains. It is no small wonder that such a minor "contaminant" of fibrinogen preparations was overlooked for so long. The representation of fibrinogen-420 in plasma is in close agreement with the estimate of α_E mRNA in adult human liver[5] and the relative secretion of fibrinogen-420 by adult hepatocytes in primary culture (unpublished). However, for reasons that remain obscure, in umbilical cord blood (fetal circulation) the molar proportion of fibrinogen-420 is three times higher than in adults, making it a more preferred source. Most of the α_E-containing molecules detectable in plasma, whether from neonate or adult, were found to be homodimeric fibrinogen-420[17] like those produced in culture, but the nearly universal presence *in vivo* of small amounts of lower molecular weight species also suggested the enigmatic production of heterodimers.

By 1999, building on the paucity of positively charged amino acids in the $\alpha_E C$ domains[5,18] indicating that fibrinogen-420 is more negatively charged than the predominant fibrinogen-340, we developed a protocol for the complete separation of the two species from fraction I-2 fibrinogen.[19] Using MonoQ anion exchange chromatography, we obtained peak A, containing most of the fibrinogen but with no trace of fibrinogen-420, and peak B, containing all of the α_E-fibrinogen (see FIGURE 6).

This accomplishment enabled us, for the first time, to definitively ascertain the subunit composition of fibrinogen-420 as well as the slightly smaller α_E-containing forms. Whereas fibrinogen-420 proved to be truly homodimeric, the smaller forms

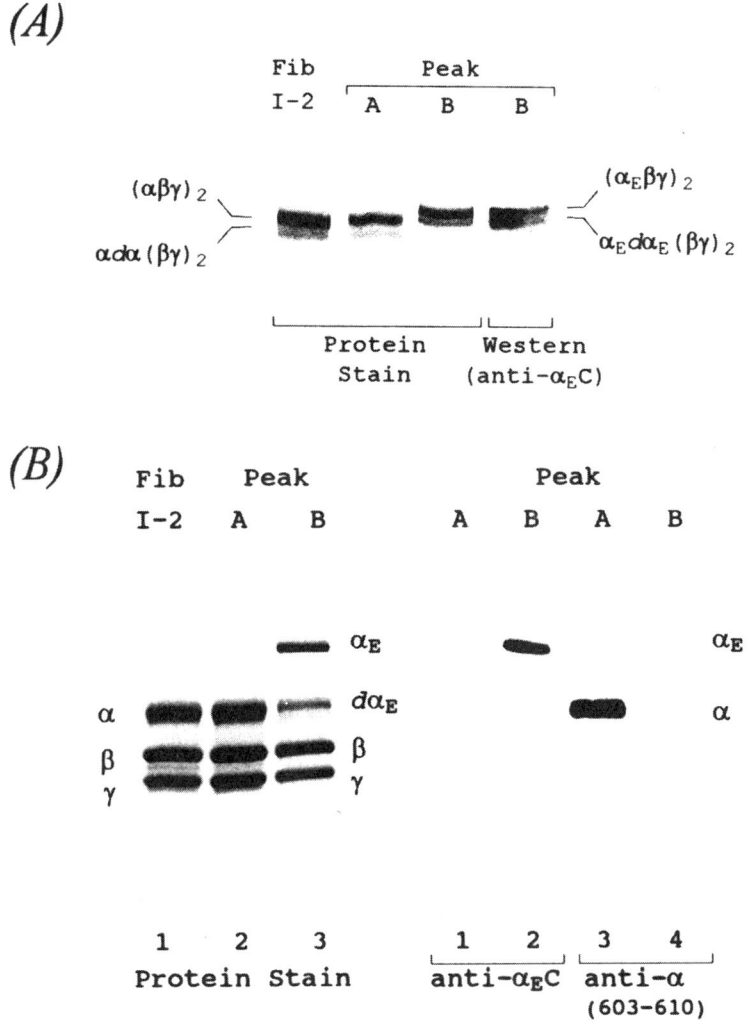

FIGURE 6. Characterization of fibrinogen species in peaks A and B. **Panel A** (unreduced samples) and **panel B** (reduced samples). Fibrinogen (Fib I-2) represents the material added and peak A and peak B the material eluted from the MonoQ anion exchange column. Western analysis was performed with either polyclonal anti-α_EC #9395 or monoclonal anti-α(603–610). Positions of various hexamers (**Panel A**) and individual chains (**Panel B**) are indicated; $d\alpha$ and $d\alpha_E$ refer to degraded α and α_E, respectively. Reprinted from Reference 19.

were found to be heterodimeric degradation products of fibrinogen-420. Proteolytic cleavage of nearly the entire $\alpha_E C$ domain from one of the α_E chains had yielded a degraded α_E ($d\alpha_E$) that was only slightly larger than—and easily misidentified as— intact conventional α chains. The degradation, whether due to plasmin or other proteases, occurred before the column chromatography step, either *in vivo* or during preparation of fibrinogen fraction I-2. A comparable degradation of the α chains was also evident, as might be expected given the well known heterogeneity of α chains[20] in adult plasma fibrinogen. On a practical note, we have found that those fibrinogen preparations from adult plasma displaying considerable α chain heterogeneity, as is common among commercial preparations, serve particularly poorly as starting material for the purification of intact fibrinogen-420.

With the purified fibrinogen subclasses in hand, we undertook to compare them in terms of clotting parameters. Alas, when incubated separately with human thrombin, purified preparations of fibrinogen-420 and fibrinogen-340 exhibited similar kinetics of clot formation,[19] and the clot networks were indistinguishable under the electron microscope (Mosesson and Grieninger, unpublished). There was also no significant difference in the kinetics of factor XIIIa-catalyzed crosslinking.[19] This observation contrasts with findings on lamprey fibrinogen where crosslinking of the α_E was considerably more efficient than for the α chain.[21] The disparity may be due to differences in the αC regions of the lamprey fibrinogen α and α_E chain, which are atypically derived from separate genes.[8,9]

PROTEOLYTIC DEGRADATION OF FIBRINOGEN-420: THE PARENT MOLECULE AS VEHICLE

Given the lack of a clear fibrinogen-420 influence on the clotting properties of fibrinogen, our working hypotheses regarding its particular mission have turned to the potential of the molecule for specialized binding properties. Among the possibilities we have entertained has been an, as yet unproven, capacity of the intact molecule to interconnect cells or macromolecules, much as fibrinogen links platelets by simultaneously engaging its distal γ chain C-termini.[22] This scenario would likely involve the fibrinogen-420 molecule in an extended topology with the two negatively charged $\alpha_E C$ domains occupying its distal ends, a position that is both remote from the core where the βC and γC reside and highly exposed because of the long, hydrophilic αC tether. However, since that tether is a region with a reputation for susceptibility to proteolytic attack, we first investigated resistance of fibrinogen-420 to proteases.

SDS-PAGE analysis of the pattern and kinetics of the digestion of fibrinogen-420 by plasmin showed it to be comparable in all respects to that of conventional fibrinogen molecules but for the presence of at least two additional small products that contained the $\alpha_E C$ domain (see FIGURE 7). One was a stable fragment ($\alpha_E CX$) comigrating with the 34-kDa yeast recombinant $\alpha_E C$ domain ($r\alpha_E C$), the other a slightly larger apparent precursor (pre-$\alpha_E CX$). These fragments were released early, before generation of fragment D, and the smaller remained intact even under extreme conditions. Two bands of the same mobility and antibody reactivity were found in Western blots of plasma collected from myocardial infarct patients shortly after initiation

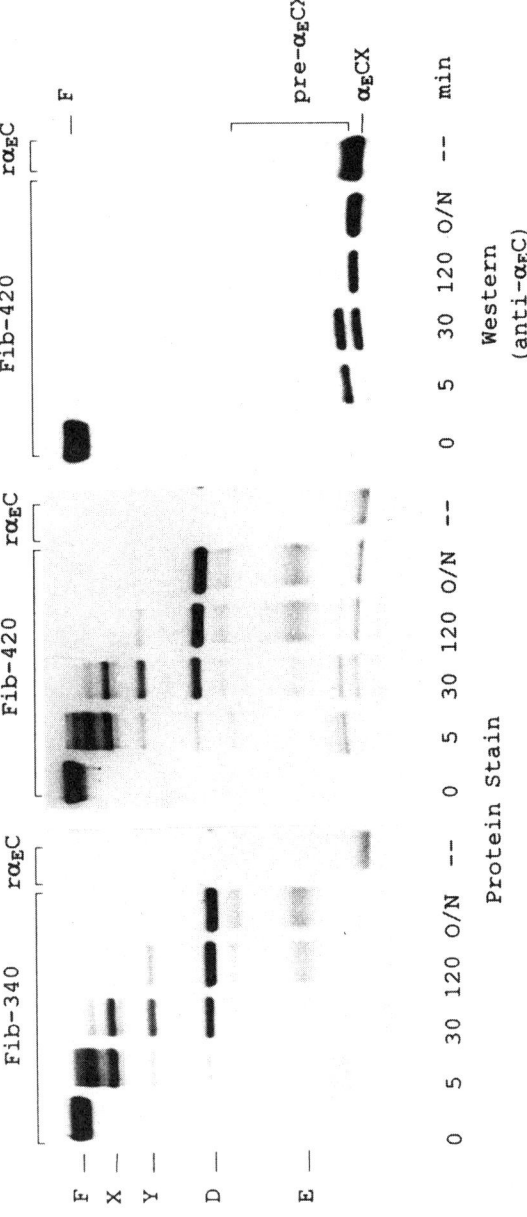

FIGURE 7. Plasmin degradation of fibrinogen-420 and fibrinogen-340. The first five lanes in each panel contain purified fibrinogen, either fibrinogen-420 or fibrinogen-340; the sixth contains recombinant human $\alpha_E C$ (r$\alpha_E C$) which migrates at 34 kDa. Proteins were separated on SDS-PAGE gels under non-reducing conditions. **Left** and **middle panels**, protein stain. **Right panel**, Western blot analysis of fibrinogen-420 using monoclonal anti-$\alpha_E C$ #29-1. Positions of fibrinogen (F) and fragments X, Y, D, and E are indicated, as are those of the α_E-containing cleavage products $\alpha_E CX$ and its precursors (pre-$\alpha_E CX$). Reprinted from Reference 19.

of thrombolytic therapy (see FIGURE 8). In these blood plasma samples, drawn by catheter from the right atrium of treated patients, we have observed accumulation of $\alpha_E CX$ and its survival with half life of up to three hours.

Matrix metalloproteinases (e.g., MMP3) were also found to cleave fibrinogen-420 yielding a final stable degradation product comparable to that generated by plasmin (unpublished). Within the context of the potential role of $\alpha_E C$ in cell interactions (see below), this observation may be significant. MMPs are a family of related zinc endopeptidases that function in the turnover of components of the extracellular matrix and are secreted by a wide variety of cell types including fibroblasts, epithelial cells, phagocytes, and lymphocytes. Expression of these MMPs is highly regulated, in particular, by inflammatory cytokines.[23]

Incidentally, proteolytic release of *monomeric* $\alpha_E CX$ (FIG. 7) provides definitive evidence that the $\alpha_E C$ domains of fibrinogen-420 have no disulfide attachments, either to each other or to the core of the molecule, a finding consistent with the results of mutational analysis of recombinant fibrinogen-420[15] and trypsin digests of α_E-fibrinogen in lamprey.[21]

More importantly, the results have compelled us to consider whether important function(s) are discharged by the $\alpha_E C$ domain independent of its parent fibrinogen-420 molecule. In the circulation, plasma fibrinogen-420 levels range from 20–150 µg/ml, the upper end reflecting the higher percentage characteristic of the neonate.[17] These values correspond to $\alpha_E C$ molar concentrations in the range of 0.1–1 µM. Under pathological conditions such as myocardial infarction, particularly after thrombolytic therapy, the $\alpha_E C$ domain containing plasmic degradation products ($\alpha_E CXs$) can accumulate, and, where blood flow is blocked, local levels may reach even higher concentrations. That fibrinogen-420 participates fully in clot formation

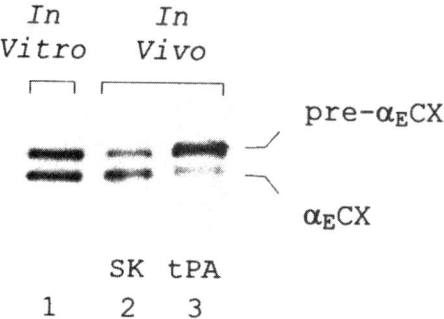

FIGURE 8. Presence of α_E-containing plasmin cleavage products *in vivo*. *In vitro*, a 30-min time point from plasmin digestion of purified fibrinogen-420 (see FIG. 7). *In vivo*, plasma samples were collected from myocardial infarction patients 30 min into treatment with either streptokinase (*SK*) or tissue plasminogen activator (*tPA*). Proteins were separated on SDS-PAGE gels under non-reducing conditions, Western blotted, and detected using monoclonal anti-$\alpha_E C$ #29-1. Positions of the split products, $\alpha_E CX$ and its precursor (pre-$\alpha_E CX$), are indicated. Reprinted from Reference 19.

and it releases $\alpha_E C$ relatively early in response to proteases (see above), together suggest that it may have evolved for the purpose of delivering the untethered $\alpha_E C$ in the vicinity of clot activity.

PROTEOLYSIS: SEEDS FOR THE FUTURE

Such considerations, among others, prompted us to generate a soluble recombinant $\alpha_E C$ domain (r$\alpha_E C$), expressed in the yeast *Pichia pastoris,* as a model for biochemical characterization of the domain and for initial studies of its cell binding capacity as an independent entity.[24] In yet another fortuitous event, crystals that were suitable for X-ray diffraction were obtained courtesy of sacrificing the first 37 residues of the domain to incipient proteolysis during long incubations of this material at room temperature before streak seeding.[18] As reviewed below, those regions of the molecule most likely to play an interactive role were spared.

STRUCTURE OF THE $\alpha_E C$ DOMAIN OF FIBRINOGEN-420

A rational investigation of $\alpha_E C$-specific binding properties is greatly facilitated by an intimate knowledge of the domain structure. It has been most fortunate that, by the end of the decade, our work and that of several other laboratories culminated in determining, at high resolution, the crystal structures of all three globular C-terminal domains of fibrinogen.[18,25,26] With the three carbon backbones giving rise to a common globular fold, the features unique to each domain became apparent.

The basic domain structure contains three subdomains: A, B, and P, terminology originally introduced for the γC fold.[25] The N-terminal subdomain A is the smallest (40–45 residues), followed by the central B subdomain with a six-stranded antiparallel β-sheet (about 100 residues) and the final P subdomain (about 90 residues) containing a cleft. Subdomain P, topologically distal to the upstream connection at subdomain A, appears to be the "business end" of the fold. It contains the calcium binding site, which is contiguous with the binding pocket, as well as loops (P1 and P3), that differ in sequence considerably among the fibrinogen subunits. For $\alpha_E C$, the atomic structure confirmed[18] and extended a host of earlier mutational studies in heterologous host cells and biochemical analyses with soluble r$\alpha_E C$. The salient features of this domain (see FIGURE 9) are reviewed below.

Carbohydrate

Although mamalian fibrinogen α chains have no carbohydrate moiety, human α_E chains are N-glycosylated, contributing significantly to their apparent molecular weight on SDS-PAGE.[7] Of the two sites on human $\alpha_E C$ (at Asn667 and Asn812; see FIG. 4), it has been shown—by mutagenesis[15] and by analysis of cyanogen bromide fragments[24]—that only the tripeptide consensus sequence at Asn667 is used. A putative carbohydrate binding site is conserved at this position, not only in all vertebrate $\alpha_E C$ domains,[4,8,9] but in several other members of the family of fibrinogen-related C-terminal domains (FReDs),[27] including those of the *Drosophila* scabrous

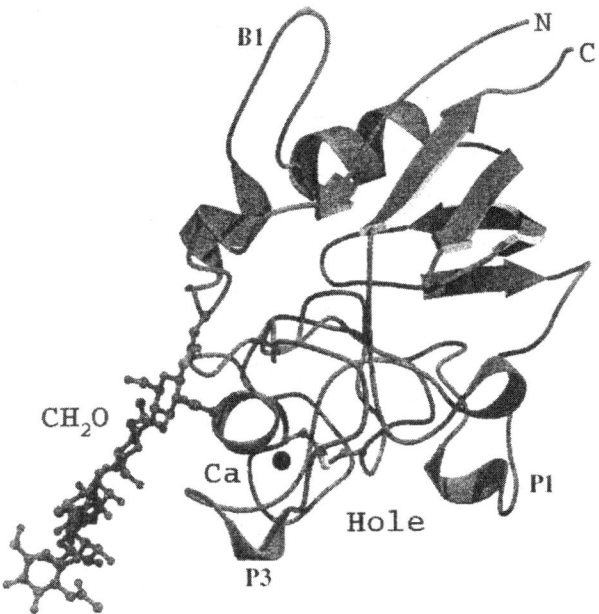

FIGURE 9. Ribbon diagram of the $\alpha_E C$ domain. The sugar (CH_2O) is attached to the asparagine at $\alpha_E 667$. The B1 loop is indicated as well as the calcium, shown as a *ball*. The binding cleft (*hole*) is surrounded by loops P1 and P3. The first 37 residues of the domain were removed in the process of obtaining crystals and are not shown. Diagram courtesy of R.F. Doolittle.

protein,[28] the mammalian protein known as pT49,[29] and the lamprey γC.[30] In the 2.1-Angstrom crystal structure of r$\alpha_E C$, the carbohydrate cluster at α_EAsn667 can be seen to be sufficiently close to the binding cleft so as to possibly block it under some circumstances. By comparison, the βC sugar, at βAsn364, is positioned right at the mouth of the binding cleft,[26] whereas in γC there is none.

Disulfide Bridges

The formation of correct disulfide bonds is essential for the assembly of fibrinogen from its constituent subunits and is ultimately responsible for its three-dimensional structure. A series of elegant studies, from groups led by Redman and Davies, has pinpointed the essential bonds for intracellular assembly of fibrinogen-340.[31–34] Whereas fibrinogen-340 has 29 disulfide bonds per hexamer, the fibrinogen-420 molecule has four more, contributed by the two cysteine pairs in each $\alpha_E C$ domain, for a total of 33. The positions of these cysteines are invariant among α_E homologs throughout vertebrate evolution and align precisely with cysteines in the βC and γC domains,[5,8] each of which has two intrachain disulfide bridges. Predictably, several lines of experimentation[15,24] as well as the crystal structure indicate that similar bonds exist between the cysteine pairs at $\alpha_E 613/\alpha_E 644$ and at $\alpha_E 780/\alpha_E 793$ (FIG. 4).

TABLE 2. Calcium binding sites: alignment of residues for the three fibrinogen chains[a]

α_E772–778	**D**	A	**D**	Q	W	**E**	**E**		
β381–385	**D**	N	**D**	G	W	*L*	*T*	*S*	**D**	*P*
γ318–324	**D**	N	**D**	K	F	**E**	**G**		

[a]Coordinating residues are shown in bold type. The sequence insertion at position 385 in the βC site is shown in italics.

Calcium Binding Site

Recombinant α_EC is highly susceptible to plasmin degradation in the presence of EDTA, but is resistant in the presence of calcium ions,[24] so it came as no surprise that the X-ray structure of rα_EC confirmed calcium bound in a position equivalent to sites in the other globular fibrinogen domains. The roughly octahedral calcium binding structural motif of these domains is unique among known calcium binding site structures.[18,25,35] Equivalent residues for the three chains are shown in TABLE 2. Deviations in the βC coordinating groups due to a five-residue insert in the domain sequence (italics; see also the alignment in FIGURE 10) provide a direct explanation for its weaker binding of calcium relative to γC[35] and, by extension, predict tight binding of calcium to α_EC.

```
                    340                                                          405
        FBEH   ..ISVNKYR-GTAGNALMDGASQLMGE---NRTMTIHNGMFFSTYDRDNDGWLTSDPRKQCSKEDGGGWWYN..
        FBER   ..VSVNKYK-GTAGNALMEGASQLVGE---NRTMTIHNGMFFSTYDRDNDGWVTTDPRKQCSKEDGGGWWYN..
  β     FBEC   ..LSVSNYK-GNAGNALMEGASQLYGE---NRTMTIHNGMYFSTYDRDNDGWLTTDPRKQCSKEDGGGWWYN..
        FBEL   ..LWVEDYS-GNAGNALLEGATQLMGD---NRTMTIHNGMQFSTFDRDNDNWNPGDPTKHCSREDAGGWWYN..
                    276                                                          337
        FGAH   ..LTYAYFAGGDAGDA-FDGFD--FGDDPSDKFFTSHNGMQFSTWDNDNDKF-----EGNCAEQDGSGWWMN..
        FGAR   ..LTYAYFIGGDAGDA-FDGYD--FGDDPSDKFFTSHNGMHFSTWDNDNDKF-----EGNCAEQDGSGWWMN..
  γ     FGAC   ..LTYAYFIGGERGDA-FDGFN--FGDDPSDKSYTYHNGMRFSTFDNDNDNF-----EGNCAEQDGSGWWMN..
        FGAL   ..LFYSMYLDGDAGNA-FDGFD--FGDDPQDKFYTTHLGMLFSTPERDNDKY-----EGSCAEQDGSGWWMN..
                    734                                                          791
        FAH2   ..LQVSSYE-GTAGDALIEGSVEEGAE------YTSHNNMQFSTFDRDADQW-----EENCAEVYGGGWWYN..
        FAR2   ..LQVSSYQ-GTAGDALMEGSVEEGTE------YTSHSNMQFSTFDRDADQW-----EENCAEVYGGGWWYN..
  αE    FAC2   ..LTVSSYE-GTAGDALVAGWLEEGSE------YTSHAQMQFSTFDRDQDHW-----EESCAEVYGGGWWYN..
        FAL2   ..LQVSSDYR-GTAGNALVSG VADDPE------LTSHGGMTFSTYDRDTDKW----SDGSCAEWYGGGWWIN..
        
        SLUG   ..LQVGGYS-GNAGDA---------------LTFHNGMAFSTNDRDNDAD-----SIDXAKVYHGAWWYK..
                    406                                                          461
        FBEH   ..RCHAANPNGRYYWGGQY-TWDMAKHGTDDGVVWMNWKGSWYSMRKMSMKIRPFFPQQ
        FBER   ..RCHAANPNGRYYWGGLY-SWDMSKHGTDDGVVWMNWKGSWYSMRRMSMKIRPVFPQQ
  β     FBEC   ..RCHAANPNGRYYWGGTY-SWDMAKHGTDDGIVWMNWKGSWYSMRKMSMKIKPYFPD
        FBEL   ..RCHAANPNGRYYWGGIY-TKEQADYGTDDGVVWMNWKGSWYSMRQMAMKLRPKWP
                    338                                                          411
        FGAH   ..KCHAGHLNGVYYQGGTY-SKASTPNGYDNGIIWATWKTRWYSMKKTTMKIIPFNRLTIGEGQQHHLGGAKQAGDV
        FGAR   ..KCHAGHLNGVYYQGGTY-SKSSTPNGYDNGIIWATWKTRWYSMKETTMKIIPFNRLSIGDGQQHHMGGSKQVGDM
  γ     FGAC   ..KCHAGHLNGVYYQGGVY-SRDTGTNSYDNGIIWATWRDRWYSMKKTTMKIIPFNRLSID-GQQH-SGGLKQVGDS
        FGAL   ..RCHAGHLNGKYYFGGNY-RKTDVEFPYDDGIIWATWHDRWYSLKMTTMKLLPMGRDLSGHGGQQQSKG-NSRGDN
                    792                                                          847
        FAH2   ..NCQAANLNGIYYPGGSYDPRNNSPYEIENGVVWVSFRGADYSLRAVRMKIRPLVTQ
        FAR2   ..SCQAANLNGIYYPGGSYDPRNNSPYEIENGVLWVPFRGADYSLWAVRMKIRPLVGQ
  αE    FAC2   ..SCQAANLNGIYYPGGHYDPRYNVPYEIENGVVWIPFRASDYSLKVVRMKIRPLETL
        FAL2   ..ACQAANLNGVYYQGGPYDPREKPPYEVENGVVWATYRGSDYSLKRTAVRFRRVQIPIVE
                                 **********
```

FIGURE 10. Sequence alignments of regions from various fibrinogen β, γ, and α_E chains and the partial sequence of a lectin from the slug *Limax flavus*. *H*-human, *R*-rat, *C*-chicken, *L*-lamprey. Only the human sequences are numbered. Those residues in which all 12 nonslug sequences are identical are emboldened, as are the two key residues (**) from the binding pocket in which the slug lectin is the same as three of the α_EC domains. The skein at α_E809–818 (the P3-loop) is denoted by a string of *asterisks*. Reprinted from Reference 18.

Neutral Binding Cleft

Among the reported members of the FReD family, only the βC and γC domains are known to bind Gly-Pro-Arg- and Gly-His-Arg-containing peptides that are surrogates for the thrombin-generated α- and β-chain knobs that permit polymerization of fibrin monomers.[36] The key residues in this peptide binding are the negatively charged cleft residues Gln-Asp and Glu-Asp, at γ329–330[37] and β397–398,[35] respectively (FIG. 10; aligned with the pair of asterisks). The aligned residues in the α_EC domain, as well as in other reported FReDs, are neutral (an exception being HHFREP[38]), with a tyrosine being most prevalent. The resultant lack of negative charge within the binding cleft of the α_EC domain (see FIGURE 11) is clearly visualized by GRASP depictions comparing equivalent projections of the domains.[18] The α_EC cleft neutrality explains why it does not bind the βC- and γC-binding peptides[24] and why an excess of rα_EC domains has no effect on thrombin-induced fibrin polymerization (unpublished). Based in part on the fact that the key residues of the α_EC cleft, Val and Tyr at α_E783–784, are identical to those at the aligned position in a sialic acid-binding lectin from the slug *Limax flavus,* it has been suggested that the α_EC domains may participate in binding some kind of sugar.[18]

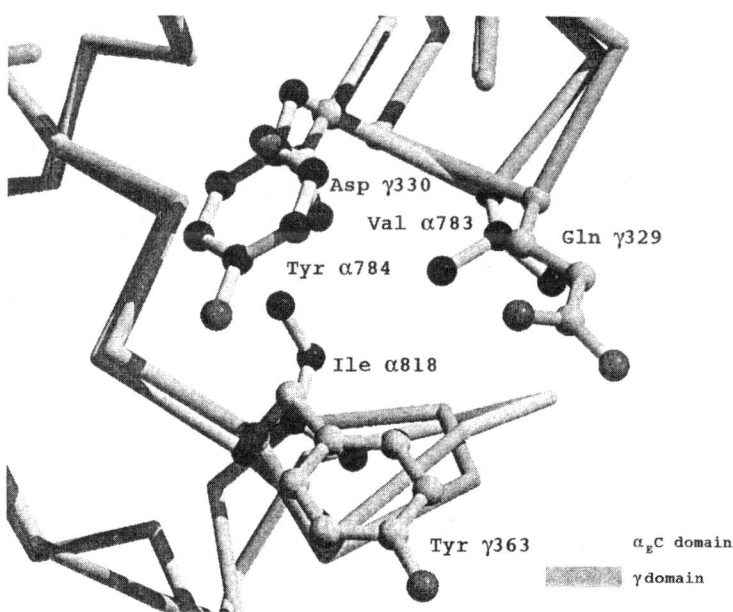

FIGURE 11. Ball and stick model of the binding cleft (*hole*). Note the position of the equivalent residues in α_EC and γC: α_EVal783/γGln329, α_ETyr784/γAsp330 and α_EIle818/γTyr383. The side chain of α_EIle818, which is in the center of the pocket, is the C-terminal residue of the α_EC P3 loop. Model courtesy of R.F. Doolittle.

$\alpha_E C$-Specific Loop

The skein at $\alpha_E 809$–818 (forming the P3 loop) is identical in the $\alpha_E C$ domains of all mammals examined.[8] It differs significantly from the P3 loop of the βC and γC domains (FIG. 10) and appears to be unique among the other related FReD domains. The segment forms a well defined, flexible loop that could easily be involved in specific binding to other macromolecules.

Thus, the picture that emerges is of an $\alpha_E C$ domain with an overall protein fold and calcium binding site in common with the βC and γC domains but with a sufficient number of differentiating features, some shared by other members of the FReD family (e.g., the neutral binding cleft and carbohydrate attachment), and others not (e.g., P3 loop), to predict a capacity for unique function(s).

IS MORE FIBRINOGEN-420 GOOD OR BAD FOR US?

A complete lack of fibrinogen-420 may be problematic but not lethal, as deduced from a homozygous case of dysfibrinogenemia known as fibrinogen Marburg.[39] In the α gene coding for this abnormal fibrinogen, a single base substitution (A→T) changes codon $\alpha 461$ AAA (lysine) to TAA (stop). As a result, the carboxy-terminal segment 461 to 625 of the common α chain is lacking and formation of α_E is not possible. Symptoms displayed by the homozygous propositus consisted of severe hemorrhage after delivery followed by repeated thrombotic events that occur, paradoxically, despite unusually low fibrinogen levels. The extent to which these specific symptoms derive from abnormal behavior of the truncated fibrinogen-340 and/or lack of fibrinogen-420 is not clear.

In an epidemiologic study, we determined the plasma level of fibrinogen-420 in healthy adults to be relatively consistent at 34 ± 7 µg/ml, or less than 0.1 µM, comprising 1% of the total circulating fibrinogen.[17] To set these numbers in context, factor XIII circulates at a similar concentration, whereas α_2-antiplasmin and fibronectin levels are about ten times higher.[40]

Pilot surveys of the preponderance of fibrinogen-420 in a variety of human diseases uncovered only minimal changes from healthy baseline proportions. On a technical note, for these studies we made extensive use of $\alpha_E C$-specific antibodies, monoclonal versions of which are also available now through the ATCC. Without such tools, it would be all too easy to confuse the γ-γ dimer band on reduced SDS-PAGE with that of the closely migrating but slower α_E band (e.g., see FIG. 4 in Ref. 19).

As has already been noted, proportional representation of fibrinogen-420 in the fetal circulation is three times higher than in adults.[17] As it turns out, a comparable difference exists between the relative rates of fibrinogen-420 synthesis in commercially available primary cultures of adult hepatocytes and in the hepatocarcinoma cell line HepG2: 1% and 5% of total fibrinogen, respectively. Fibrinogen-420 also comprised 5% of total fibrinogen synthesis in several carcinoma cell lines of non-hepatic origin that evidence fibrinogen synthesis, albeit at overall rates far lower than that of HepG2 (unpublished). It is not known whether similar mechanisms drive the higher percentage of fibrinogen-420 in these cell lines versus primary cells and in fetal versus adult plasma. The correlation in culture may not extend to the clinical setting, however, since the ratio of fibrinogen-420 to total fibrinogen in patients with carcinoma of the liver was closer to that of healthy adults than to that of neonates (unpublished).

Fibrinogen is an acute phase protein, for which interleukin-6 is recognized as the major regulator.[41] In HepG2 cell culture, synthesis of fibrinogen-420 and of fibrinogen-340 was stimulated similarly by Il-6,[5] which is not surprising since production of both α_E- and α-chain isoforms is under control of the same promoter. To examine whether additional factors *in vivo*, (e.g., circulating half life), govern the peripheral level of fibrinogen-420, we measured pre- and postoperative samples (provided by C.W. Francis) from a mixed cohort of otherwise healthy patients undergoing elective knee surgery. Consistent with our findings in cell culture, the two sets of samples in this clinical model of the acute phase had similar relative proportions of the fibrinogen subspecies: post surgery, fibrinogen-420 levels rose in parallel with total fibrinogen (unpublished).

Circulating fibrinogen is taken up by platelets via the integrin $\alpha_{IIb}\beta_3$ and stored in α granules for release upon platelet activation. In collaboration with B. Coller we have found that platelets store fibrinogen, which is more than twice as enriched in fibrinogen-420 compared to plasma from the same individual (Hertzberg *et al.*, unpublished). Neither the mechanism behind, nor the significance of, this finding is apparent, but the phenomenon might be consistent with the goal of releasing the independent α_EC domain in the vicinity of clots.

BINDING INTERACTIONS: SEARCHING IN THE DARK

Under normal conditions, we expect fibrinogen-420 to circulate freely along with fibrinogen-340. What then are the signals that trigger its α_EC domains, either as part of the parent molecule or dissociated from it, to exercise their capacity to interact, presumably with other elements in the vasculature? Do subtle changes in the domain configuration have major ramifications? Are the interactions sufficiently strong to withstand the repeated washings and long incubations commonly used to detect them experimentally? By what criteria can the interactions be deemed unique given the fundamental similarities of fibrinogen-420 to fibrinogen-340 and of α_EC to βC and γC? Thus far, only a handful of studies have attempted to illuminate these questions.

In our initial efforts, the soluble yeast-expressed recombinant protein, rα_EC, has been used as a model. Gel-filtered human platelets, free of plasma proteins, failed to bind this α_EC in solution or to adhere to the immobilized domain, even after stimulation with ADP or thrombin (Peerschke and Grieninger, unpublished). Using recombinant forms expressed in *E. coli*, others have found that α_EC also failed to bind to $\alpha_{IIb}\beta_3$-transfected cells; yet α_EC and γC were equivalent in supporting adhesion of $\alpha_v\beta_3$ transfectants.[42] In our hands, however, binding of the soluble α_EC to cultured endothelial cells well endowed with $\alpha_v\beta_3$ was low, and adhesion of the cells to the plastic-immobilized domain was weak compared to adhesion to γC. By contrast, α_EC consistently supported stronger adhesion than γC with transfected cells expressing only $\alpha_M\beta_2$ (Ugarova and Grieninger, unpublished).

CONCLUSION

Evolutionary considerations based on homology to βC and γC provide a compelling argument that the $\alpha_E C$ domains of fibrinogen-420 add an important dimension to the repertoire of the fibrinogen molecule. Through elucidation of fibrinogen-420 structure, it has become apparent that the $\alpha_E C$ domain must be specialized for an as yet poorly understood function rather different from that of βC and γC: (1) While still attached to the fibrinogen core via its αC tether, $\alpha_E C$ undoubtedly enjoys more degrees of spatial freedom than βC or γC and, consequently, greater availability of its potential binding sites to other macromolecules. (2) This location also appears designed to insure more rapid release of the $\alpha_E C$ domain, given the extreme susceptibility of the αC region to proteolysis. (3) During fibrin(ogen)olysis, the $\alpha_E C$ domains are released as monomers, unlike the βC and γC domains, which remain anchored together in the proteolytic fragment D. (4) Finally, the binding clefts of the βC and γC domains contain charged/polar amino acid pairs that engage the polymerization "knobs" during fibrin assembly, whereas the corresponding cleft in the $\alpha_E C$ domain has neutral residues at its center, suggesting a different purpose. With the tools that have now been developed, it is not unreasonable to expect that the nature of this purpose will be deciphered in the first decade of the new millennium.

ACKNOWLEDGMENTS

It gives me great pleasure to thank the many colleagues and collaborators who contributed to this research, and K.M. Hertzberg in particular for the discussions that were so essential to our progress. The work in my laboratory has been supported by the National Institutes of Health, the American Heart Association, The Hugoton Foundation, and the Abby R. Mauze Charitable Trust.

REFERENCES

1. DOOLITTLE, R.F. 1984. Fibrinogen and fibrin. Annu. Rev. Biochem. **53:** 195–229.
2. BLOMBACK, B. 1996. Fibrinogen and fibrin—proteins with complex roles in hemostasis and thrombosis. Thromb. Res. **83:** 1–75.
3. GRIENINGER, G., P.W. PLANT, T.J. LIANG, *et al.* 1983. Hormonal regulation of fibrinogen synthesis in cultured hepatocytes. Ann. N.Y. Acad. Sci. **408:** 469–489.
4. WEISSBACH, L. & G. GRIENINGER. 1990. Bipartite mRNA for chicken α-fibrinogen potentially encodes an amino acid sequence homologous to β- and γ-fibrinogens. Proc. Natl. Acad. Sci. U.S.A. **87:** 5198–5202.
5. FU, Y., L. WEISSBACH, P.W. PLANT, *et al.* 1992. Carboxy-terminal-extended variant of the human fibrinogen α subunit: a novel exon conferring marked homology to β and γ subunits. Biochemistry **31:** 11968–11972.
6. CHUNG, D.W. & G. GRIENINGER. 1994. Fibrinogen DNA and protein sequences. *In* Index of Variant Human Fibrinogens, 3rd edit. R.F. Ebert, Ed.: 13–24. CRC Press, Boca Raton.
7. FU, Y. & G. GRIENINGER. 1994. Fib_{420}: a normal human variant of fibrinogen with two extended α chains. Proc. Natl. Acad. Sci. U.S.A. **91:** 2625–2628.
8. FU, Y., Y. CAO, K.M. HERTZBERG & G. GRIENINGER. 1995. Fibrinogen α genes: conservation of bipartite transcripts and carboxy-terminal-extended α subunits in vertebrates. Genomics **30:** 71–76.

9. PAN, Y. & R.F. DOOLITTLE. 1992. cDNA sequence of a second fibrinogen α chain in lamprey: an archetypal version alignable with full-length β and γ chains. Proc. Natl. Acad. Sci. U.S.A. **89:** 2066–2070.
10. DOOLITTLE, R.F. 1983. The structure and evolution of vertebrate fibrinogen. Ann. N.Y. Acad. Sci. **408:** 13–27.
11. HENSCHEN, A., F. LOTTSPEICH, M. KEHL & C. SOUTHAN. 1983. Covalent structure of fibrinogen. Ann. N.Y. Acad. Sci. **408:** 28–43.
12. CHUNG, D.W., J.E. HARRIS & E.W. DAVIE. 1990. Nucleotide sequences of the three genes coding for human fibrinogen. Adv. Exp. Med. Biol. **281:** 39–48.
13. MURAKAWA, M., T. OKAMURA, T. KAMURA, et al. 1993. Diversity of primary structures of the carboxy-terminal regions of mammalian fibrinogen Aα-chains. Characterization of the partial nucleotide and deduced amino acid sequences in five mammalian species; rhesus monkey, pig, dog, mouse and Syrian hamster. Thromb. Hæmost. **69:** 351–360.
14. ZHANG, J.Z. & C. REDMAN. 1996. Fibrinogen assembly and secretion. Role of intrachain disulfide loops. J. Biol. Chem. **271:** 30083–30088.
15. FU, Y., J.-Z. ZHANG, C.M. REDMAN & G. GRIENINGER. 1998. Formation of the human fibrinogen subclass Fib_{420}: disulfide bonds and glycosylation in its unique (α_E chain) domains. Blood **92:** 3302–3308.
16. MOSESSON, M.W., J.S. FINLAYSON & R.A. UMFLEET. 1972. Human fibrinogen heterogeneities. 3. Identification of chain variants. J. Biol. Chem. **247:** 5223–5227.
17. GRIENINGER, G., X. LU, Y. CAO, et al. 1997. Fib_{420}, the novel fibrinogen subclass: newborn levels are higher than adult. Blood **90:** 2609–2614.
18. SPRAGGON, G., D. APPLEGATE, S.J. EVERSE, et al. 1998. Crystal structure of a recombinant $\alpha_E C$ domain from human fibrinogen-420. Proc. Natl. Acad. Sci. U.S.A. **95:** 9099–9104.
19. APPLEGATE, D., L.S. STEBEN, K.M. HERTZBERG & G. GRIENINGER. 2000. The $\alpha_E C$ domain of human fibrinogen-420 is a stable and early plasmin cleavage product. Blood **95:** 2297–2303.
20. MOSESSON, M.W. 1983. Fibrinogen heterogeneity. Ann. N.Y. Acad. Sci. **408:** 97–113.
21. SHIPWASH, E., Y. PAN & R.F. DOOLITTLE. 1995. The minor form α′ chain from lamprey fibrinogen is rapidly crosslinked during clotting. Proc. Natl. Acad. Sci. U.S.A. **92:** 968–972.
22. HAWIGER, J., S. TIMMONS, M. KLOCZEWIAK, et al. 1982. γ and α chains of human fibrinogen possess sites reactive with human platelet receptors. Proc. Natl. Acad. Sci. U.S.A. **79:** 2068–2071.
23. NAGASE, H. 1996. Matrix metalloproteinases. In Zinc Metalloproteases in Health and Disease. N.M. Hooper, Ed.: 153–204. Taylor & Francis, London.
24. APPLEGATE, D., L. HARAGA, K.M. HERTZBERG, et al. 1998. The $\alpha_E C$ domains of human $Fibrinogen_{420}$ contain calcium binding sites but lack polymerization pockets. Blood **92:** 3669–3674.
25. YEE, V.C., K.P. PRATT, H.C. COTE, et al. 1997. Crystal structure of a 30 kDa C-terminal fragment from the γ chain of human fibrinogen. Structure **5:** 125–138.
26. SPRAGGON, G., S.J. EVERSE & R.F. DOOLITTLE. 1997. Crystal structures of fragment D from human fibrinogen and its crosslinked counterpart from fibrin. Nature **389:** 455–462.
27. DOOLITTLE, R.F. 1992. A detailed consideration of a principal domain of vertebrate fibrinogen and its relatives. Protein Sci. **1:** 1563–1577.
28. BAKER, N.E., M. MLODZIK & G.M. RUBIN. 1990. Spacing differentiation in the developing Drosophila eye: a fibrinogen-related lateral inhibitor encoded by scabrous. Science **250:** 1370–1377.
29. KOYAMA, T., L.R. HALL, W.G. HASER, et al. 1987. Structure of a cytotoxic T-lymphocyte-specific gene shows a strong homology to fibrinogen β and γ chains. Proc. Natl. Acad. Sci. U.S.A. **84:** 1609–1613.
30. STRONG, D.D., M. MOORE, B.A. COTTRELL, et al. 1985. Lamprey fibrinogen γ chain: cloning, cDNA sequencing, and general characterization. Biochemistry **24:** 92–101.

31. ZHANG, J.Z., B. KUDRYK & C.M. REDMAN. 1993. Symmetrical disulfide bonds are not necessary for assembly and secretion of human fibrinogen. J. Biol. Chem. **268:** 11278–11282.
32. HUANG, S., Z. CAO & E.W. DAVIE. 1993. The role of amino-terminal disulfide bonds in the structure and assembly of human fibrinogen. Biochem. Biophys. Res. Commun. **190:** 488–495.
33. ZHANG, J.Z. & C.M. REDMAN. 1994. Role of interchain disulfide bonds on the assembly and secretion of human fibrinogen. J. Biol. Chem. **269:** 652–658.
34. ZHANG, J.Z. & C.M. REDMAN. 1996. Assembly and secretion of fibrinogen. Involvement of amino-terminal domains in dimer formation. J. Biol. Chem. **271:** 12674–12680.
35. EVERSE, S.J., G. SPRAGGON, L. VEERAPANDIAN, et al. 1998. Crystal structure of fragment double-D from human fibrin with two different bound ligands. Biochemistry **37:** 8637–8642.
36. LAUDANO, A. & R. DOOLITTLE. 1978. Synthetic peptide derivatives that bind to fibrinogen and prevent the polymerization of fibrin monomers. Proc. Natl. Acad. Sci. U.S.A. **75:** 3085–3089.
37. PRATT, K.P., H.C.F. COTE, D.W. CHUNG, et al. 1997. The primary fibrin polymerization pocket: three-dimensional structure of a 30-kDa C-terminal γ chain fragment complexed with the peptide Gly-Pro-Arg-Pro. Proc. Natl. Acad. Sci. U.S.A. **94:** 7176–7181.
38. YAMAMOTO, T., M. GOTOH, H. SASAKI, et al. 1993. Molecular cloning and initial characterization of a novel fibrinogen-related gene, HFREP-1. Biochem. Biophys. Res. Commun. **193:** 681–687.
39. KOOPMAN, J., F. HAVERKATE, J. GRIMBERGEN, et al. 1992. Fibrinogen Marburg: a homozygous case of dysfibrinogenemia, lacking amino acids Aα 461–610 (Lys 461 AAA→stop TAA). Blood **80:** 1972–1979.
40. DOOLITTLE, R.F. 1994. The molecular biology of fibrin. In The Molecular Basis of Blood Diseases, 2nd edit. G. Stamatoyannopoulos, A.W. Nienhuis, P.W. Majerus & H. Varmus, Eds.: 701–723. W.B. Saunders Company, Philadelphia.
41. Sehgal, P.B., G. Grieninger & G. Tosato, Eds. 1989. Regulation of the acute phase and immune responses: interleukin-6. Ann. N.Y. Acad. Sci. **557:** 1–583.
42. YOKOYAMA, K., X.P. ZHANG, L. MEDVED & Y. TAKADA. 1999. Specific binding of integrin $\alpha_v\beta_3$ to the fibrinogen γ and α_E chain C-terminal domains. Biochemistry **38:** 5872–5877.

Hereditary Disorders of Fibrinogen

MICHIO MATSUDA AND TERUKO SUGO

Division of Cell and Molecular Medicine, Center for Molecular Medicine, Jichi Medical School, Tochigi-Ken 329-0498, Japan

ABSTRACT: Fibrinogen, a 340-kDa plasma protein, is composed of two identical molecular halves each consisting of three non-identical Aα-, Bβ- and γ-chain subunits held together by multiple disulfide bonds. Fibrinogen is shown to have a trinodular structure; that is, one central nodule, the E domain, and two identical outer nodules, the D-domains, linked by two coiled-coil regions. After activation with thrombin, a pair of binding sites comprising Gly-Pro-Arg is exposed in the central nodule and combines with its complementary binding site *a* in the outer nodule of another molecules. By using crystallographic analysis, the α-amino group of αGly-1 is shown to be juxtaposed between γAsp-364 and γAsp-330, and the guanidino group of αArg-3 between the carboxyl group of γAsp-364 and γGln-329 in the *a* site. Half molecule-staggered, double-stranded protofibrils are thus formed. Upon abutment of two adjacent D domains on the same strand, D–D self association takes place involving Arg-275, Tyr-280, and Ser-300 of the γ-chain on the surface of the abutting two D domains. Thereafter, carboxyl-terminal regions of the α-chains are untethered and interact with those of other protofibrils leading to the formation of thick fibrin bundles and networks. Although many enigmas still remain concerning the exact mechanisms of these molecular interactions, fibrin assembly proceeds in a highly ordered fashion. In this review, these molecular interactions of fibrinogen and fibrin are discussed on the basis of the data provided by hereditary dysfibrinogens on introducing representative molecules at each step of fibrin clot formation.

KEYWORDS: Fibrinogen; Hereditary dysfibrinogen; Fibrin polymerization; Functional abnormality; Ultrastructure of fibrin clot.

INTRODUCTION

This review focuses on the structure–function relationships of human fibrinogen inferred from those of hereditary dysfibrinogens elucidated at the molecular or gene level, or both, as summarized in TABLE 1. Their relevance to clinical symptoms and insights into clinical implications are also discussed. The structure–function relationships are overviewed first in accordance with the successive steps of fibrin gel formation and then with interactions with other substances, such as thrombin and plasmin based on biochemical, electron microscopic, and crystallographic analysis data provided from normal and representative abnormal fibrinogen molecules and fibrin clots.

Address for correspondence: Michio Matsuda, M.D., Division of Cell and Molecular Medicine, Center for Molecular Medicine, Jichi Medical School, 3311-1 Minamikawachi-Machi, Kawachi-Gun, Tochigi-Ken 329-0498, Japan. Voice: 81-285-58-7397; fax: 81-285-44-7817.
thmichi@jichi.ac.jp

TABLE 1. Structurally elucidated dysfibrinogens

α-chain			β-chain			γ-chain		
Position	Substitution	Name	Position	Substitution	Name	Position	Substitution	Name
7	D→N	Lille	G9-L72	Deletion	New York I	139	C→Y	Pretoria[a]
11	E→G	Mitaka II	(Exon 2)			165	G→R	Milano XII[a]
12	G→V	Rouen	14	R→C	Christchurch	268	G→E	Kurashiki
16	R→C	Metz and others	15	G→C	Ise and other	275	R→C	Tokyo II and others
16	R→H	Manchester and others	44	R→C	Nijmegen		R→H	Bergamo and others
17	G→V	Bremen	68	A→T	Naples		R→S	Kamogawa
18	P→L	Kyoto II and other	160	N→S	Niigata	280	Y→C	Banks Peninsula
19	R→S	Detroit	335	A→T	Pontoise	284	G→R	Brescia
19	R→N	Munich	462	stop→K	Osaka VI	292	G→V	Baltimore I
	R→G	Aarhus and others		12 AA elongation		308	N→K	Kyoto I and others
20	V→D	Canterbury and others					N→I	Baltimore III
141	R→S	Lima				310	M→T	Asahi
434	S→N	Caracas II				318	D→G	Giessen IV
526	E→V	found in US					D→Y	Bastia
532	S→C	Caracas V				319–320	deletion	Vlissingen I
554	R→C	Dusart and others				327	A→T	Tokyo V[a]
554	R→L	Mexican family				329	Q→R	Ngoya

TABLE 1/continued.

α-chain			β-chain			γ-chain		
Position	Substitution	Name	Position	Substitution	Name	Position	Substitution	Name
						330	D→Y	Kyoto III
							D→V	Milano I
						337	N→K	Bern I
						15AA addition after 350 Q		Paris I
						358	S→C	Milano VII
						361	N→K	Poissy II[a]
						364	D→V	Melun I
							D→H	Matsumoto I
						375	R→G	Osaka V
						380	K→N	Kaiserslautern

α-chain truncation due to frame shift.
268 RNPS→QEP stop Otago I
452 GPD→WS-Stop Milano III
461–610 Deletion Marburg I
476 MDLG→HCLA Stop Lincoln
524 ERE→ELS—found in USA

[a]unpublished data.

DEFECTS IN FIBRIN GEL FORMATION

Fibrin clot formation is a series of highly ordered molecular interactions,[1,2] although many enigmas remain relating to the detailed mechanisms of these interactions. Recent electron microscopic analyses of individual molecules of fibrinogen and fibrin monomer, fibrin fibers and networks,[3-8] and crystal structure studies on particular domains and segments of human fibrinogen and fibrin,[9-13] have shed new light on the mechanisms of fibrin clot formation, and have highlighted the structure–function relationships of hitherto characterized hereditary dysfibrinogens.

Fibrin clot formation consists of three major steps, (1) transition of fibrinogen to fibrin monomer by thrombin, (2) construction of half-molecule overlapping double-stranded fibrin protofibrils, and (3) lateral association of protofibrils to form thick fibrin bundles and networks.

Defects in Transition of Fibrinogen to Fibrin

The transition of fibrinogen to fibrin monomer may be separated into two steps, (1) binding with thrombin and (2) cleavage of fibrinopeptides A and B (FPA and FPB) by fibrinogen-bound thrombin.

(1) Impaired Binding of Fibrinogen with Thrombin

Impaired binding with thrombin has been reported in three dysfibrinogens with a point mutation in the FPA segment, that is, Aα Asp-7 to Asn (Lille),[14] Aα Glu-11 to Gly (Mitaka II),[15] and Aα Gly-12 to Val (Rouen).[14]

The coagulation enzyme thrombin binds to fibrinogen, and cleaves the amino-terminal 16-residue FPA from each Aα-chain by catalyzing hydrolysis of the Arg-16·Gly-17 peptide bond. In the interaction with thrombin, Aα Glu-11 is shown to play a pivotal role by providing its carboxyl group side chain to form a salt bridge with the guanidino group of Arg-173 of thrombin on the basis of crystallography of the complex of thrombin with a synthetic peptide corresponding to the Aα (7–16) residue segment.[16,17] Indeed, a mutation of Aα Glu-11 to Gly is identified in fibrinogen Mitaka II associated with the defective binding with thrombin.[15] In this molecule, the release of FPA by thrombin is delayed, whereas that by a thrombin-like snake venom enzyme, ancrod, is normal. Binding of ^{125}I-thrombin to the immobilized aberrant FPA species is distinctly reduced in comparison with that of the normal FPA species, both derived from fibrinogen of the proband—a heterozygote for this abnormality (see FIGURE 1).

The crystallographic data provided lines of evidence that the residue at Aα-12 should be Gly in order for the thrombin-cleavage site to fit into the enzyme pocket of thrombin. In fact, a Gly to Val substitution is reported in fibrinogen Rouen that is characterized by the delayed release of FPA.[14] Despite an Aα Asp-7 to Asn substitution in fibrinogen Lille manifesting delayed release of FPA, replacement of the Asp residue to an Asn is suggested to be indifferent to the interaction with thrombin by using synthetic Aα (7–16) and Aα (7–20) residue peptides with an Asp-7 to Asn substitution.[18]

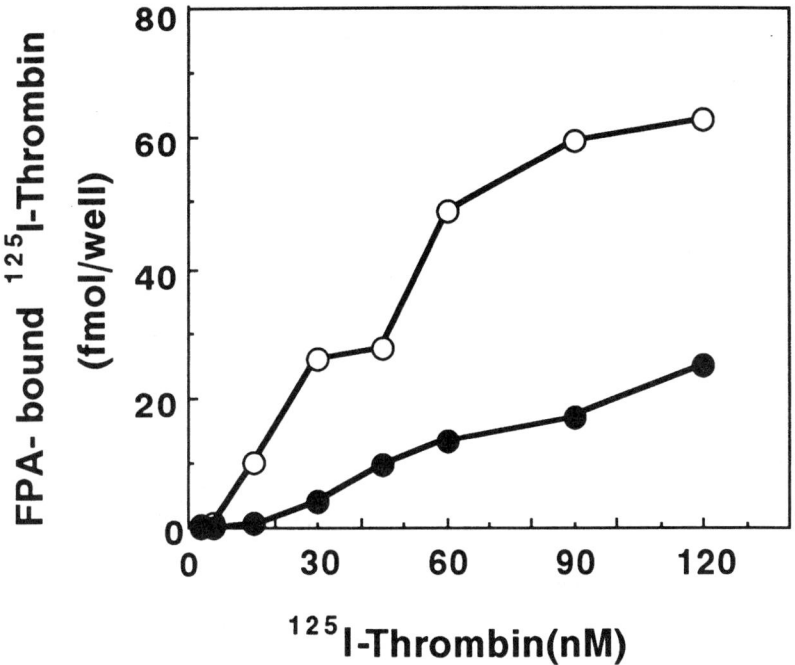

FIGURE 1. Binding of ^{125}I-thrombin to the immobilized normal (○) and aberrant (●) FPA species derived from fibrinogen of the patient. (Reproduced from Ref. 15 with permission.)

(2) Impaired Cleavage of FPA and FPB by Thrombin

Cleavage by thrombin of the carboxyl-side peptide bond of Aα Arg-16 is highly specific, and the replacement of Aα Arg-16 by other amino acids may lead to the delayed (His)[19] or defective (Cys)[20,21] release of FPA. These two types are the most common among the mutations thus far identified in hereditary dysfibrinogens.[22-24]

Hydrolysis of the Bβ Arg-14·Gly-15 bond is also catalyzed by thrombin, but much more slowly than that of the Aα Arg-16·Gly-17 bond.[1,2] For Bβ Arg-14, only a Cys substitution has so far been reported in association with the defective release of FPB[25] (see TABLE 1).

Defect in the Construction of Double Stranded Fibrin Protofibrils

Immediately after the release of paired FPAs by thrombin, fibrin monomers lacking solely the paired FPAs (des AA-fibrin monomers) aggregate spontaneously, and half-molecule overlapping oligomers are formed.[1,2] These oligomers then develop into elongated double strand protofibrils. Two independent molecular interactions are involved in this protofibril formation, (1) binding of the thrombin-activated polymerization site in the central E domain (*A* site, or the α-chain knob) with its complementary site residing in the D domain of adjacent fibrin monomers (*a* site, or the γ-chain bole),[26,27] and (2) self-association of abutting two D domains of different molecules on the same strand of double strand protofibrils.[28] Indeed, structural alterations have been identified in all these functional sites.

Alteration in the A Site (the α-Chain Knob)

After removal of FPA by thrombin, a new amino-terminal segment consisting of Gly-Pro-Arg (GPR) is exposed in the fibrin α-chain. This tripeptide corresponding to Aα (17–19) residues constitutes the *A* site, and functions much like a two-pronged plug with two positively charged side chains, that is, the α-amino group of α Gly-1 (Aα Gly-17) and the guanidino group of α Arg-3 (Aα Arg-19). It is likely that the positively charged knob plugs into a segment enriched with negatively charged carboxyl groups supplied by suitably juxtaposed Asp and/or Glu residues on the basis of recent biochemical analysis data.[1,2] The unoccupied *a* site enriched with negative electrostatic potentials residing on the γ-chain in the D domain interacts with the thrombin-exposed new amino-terminal sequence of GPR in the fibrin α-chain carrying two positive charges. The interaction of these clusters of positive and negative charges should play an important role in the alignment of the rod-like fibrin monomers into half-molecule overlapping protofibrils. In fact, this hypothesis has recently been confirmed by crystallographic analyses reported independently from two laboratories.[9–13] When bound to its complementary *a* site located on the D domain of another molecule, the α-amino group of α Gly-1 (Aα Gly-17) is situated between the side chain carboxyl groups of γ Asp-330 and γ Asp-364. The guanidino group of α Arg-3 (Aα Arg-19) lies nearby, between the side chain carboxyl group of γ Asp-330 and the carboxamide of γ Gln-329.[10,11]

Polymerization defective variants with an amino acid substitution at the three residues in the *A* site have all been reported. Interestingly, these mutant molecules are associated with one or more clinical symptoms (see TABLE 2).

TABLE 2. **Clinical manifestations associated with mutations in the *A* site**

Mutation	Name	Symptom[a]		
		B	T	W
αGly1→Val	Bremen	Y		Y
αPro2→Leu	Kyoto II	Y		
αArg3→Ser	Detroit	Y		
→Asn	Munich	Y		
→Gly	Aarhus		Y	
	Kumamoto		Y	
	Mannheim I	Y		

[a]B, bleeding; T, transient ischemic attack; W, would healing disturbance; Y, yes.

Alteration in the a *Site (the γ-Chain Hole)*

By crystal structure studies, the critical residues involved in the *a* site are shown to be γ Gln-329, γ Asp-330, and γ Asp-364.[10,11] In fact, replacements at any of these three positions are known in association with defective fibrin polymerization, that is, γ Gln-329 to Val (Nagoya[38]), γ Asp-330 to Val (Milano I[39]) or Tyr (Kyoto III[40]), and γ Asp-364 to His (Matsumoto I[41]) or Val (Melun I[42]). Besides these three positions in the γ-chain, γ Arg-375 may also be involved in the stabilization of the γ-chain hole by providing the guanidino group that forms a hydrogen bond with γ Asp-364 and a salt link with the carboxyl group of γ Asp-297.[43] To support this hypothesis, replacement of γ Arg-375 by Gly is identified in fibrinogen Osaka V associated with impaired fibrin monomer polymerization in the absence of Ca^{2+}.[44] Loss of the side chain of γ Arg-375 may lose a Ca^{2+}-dependent stabilizing effect on the entire region of the *a* site inferred from crystallographic analysis data.[43]

Although not directly involved in the *a* site hole, several other variant molecules with an amino acid substitution in the vicinity of the *a* site hole may also be classified into this group: γ Ala-327 to Thr (Tokyo V, unpublished data), γ Asp-337 to Lys (Bern I[45]) and γ Ser-358 to Cys (Milano VII[46]) (see TABLE 1).

Alteration in the D:D Association

There is one distinct mechanism of fibrin assembly designated as D:D association that promotes the association of two D domains of adjacent fibrin monomers in the same strand. The defect in the D:D association is demonstrated by an electron microscopic study on fibrinogen Tokyo II[28] with impaired fibrin polymerization despite normal E-D binding and factor XIIIa-crosslinking of the fibrin γ-chain.[47] This dysfibrinogen has a γ Arg-275 to Cys substitution, and this type of Cys substitution is shown to be linked with a single Cys residue via a disulfide bridge, as shown by fast atom bombardment mass spectrometry in fibrinogen Osaka II.[48] Crystal structures of fragment D and factor XIIIa-crosslinked fragment DD complexed with a GPR-containing peptide make it clear that γ Arg-275 is not involved in binding with the GPR-containing peptide, but instead occurs at the DD interface.[11] Dysfibrinogens with a replacement of this residue by His (Bergamo II, Essen, and Perugia,[49] Saga,[50] and others[24]) or Ser (Kamogawa[51]) are also classified into this group. Interestingly, each γ Arg-275 residue in the DD fragment has a different set of contacts, that is, with γ Tyr-280 in one direction and with γ Ser-300 in the other.[11] To support this finding, a γ Tyr-280 to Cys substitution has been reported in association with defective fibrin polymerization in fibrinogen Banks Peninsula.[52] Impaired D:D association may also take place in variant fibrinogens with a mutation located close to these residues, including γ Gly-268 to Glu (Kurashiki I[53]), γ Asn-308 to Lys (Kyoto I[54] and others[43]) or Ile (Baltimore III[55]) and γ Met-310 to Thr (Asahi[56]). Although not a γ-chain mutant, there is a unique dysfibrinogen that has a 12-residue extension at the carboxyl terminus of the Bβ-chain (Osaka VI). This abnormal molecule should also be classified into this group. Among them, I would like to discuss two unique abnormal molecules, fibrinogens Asahi and Osaka VI.

Fibrinogen Asahi with a γ Met-310 to Thr substitution is the first molecule, in which extra N-glycosylation has been identified at an Asn residue (γ Asn-308) due to a newly created glycosylation sequence of Asn-X-Thr.[56] The extra oligosaccharide has a biantennary structure as those in normal fibrinogen that are N-linked to

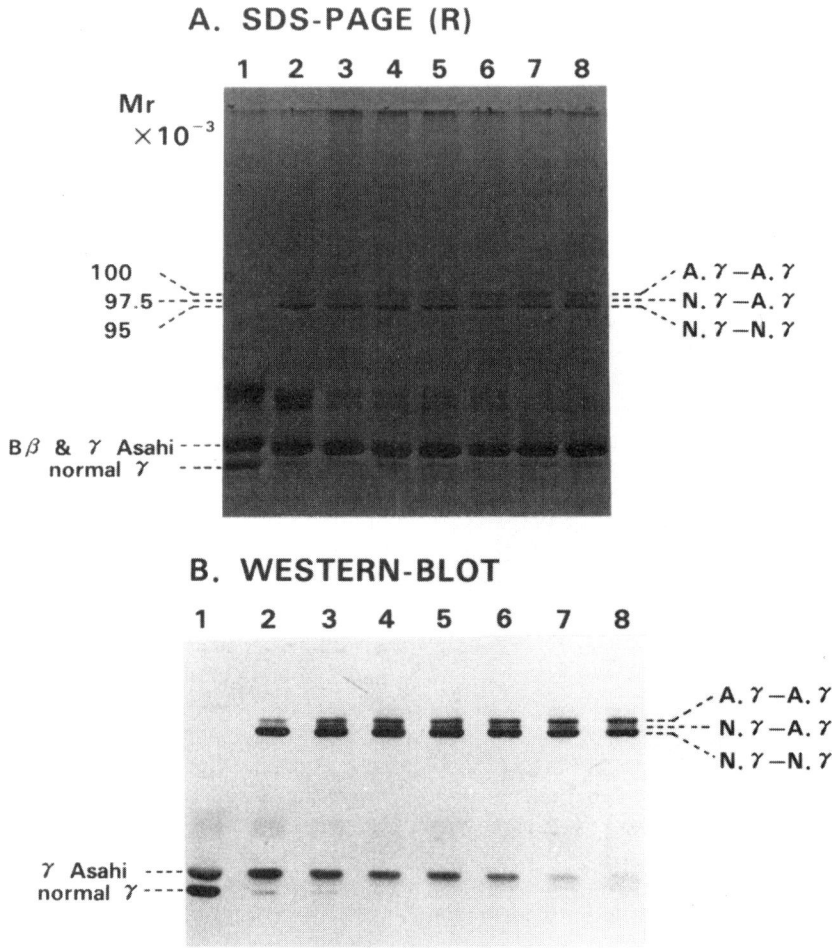

FIGURE 2. Factor XIIIa-catalyzed crosslinking of fibrin γ-chains analyzed by SDS-PAGE and Western blotting. From left to right (lane 1 to lane 8): 0 min, 2 min, 5 min, 10 min, 15 min, 30 min, 2 h, and 24 h. (Reproduced from Ref. 56 with permission.)

BβAsn-335 and γAsn-52.[57] This dysfibrinogen shows markedly impaired fibrin polymerization and distinctly delayed factor XIIIa-catalyzed crosslinking of the γ-chains (see FIGURE 2). Furthermore, binding of the isolated abnormal fragment D to immobilized fibrin monomer is reduced (profile not shown). Based on crystal structure studies, the γ Met-310 to Thr substitution, per se, may be benign structurally,[43] but both *a* and factor XIIIa-cross-linking sites may reside within the reach of the flexible extra oligosaccharide backbone linked to γ Asn-308. Thus, the *A–a* binding and reciprocal alignment of the γ-chains may be affected, resulting in severe defects in fibrin polymerization and factor XIIIa-crosslinking of the γ-chains. Although enzymatic removal of the extra oligosaccharides failed to restore the thrombin clotting time completely, the fibrin clot architecture became almost normal as compared with irregular and very porous fibrin clots with many fiber ends formed from the wild Asahi fibrinogen (see Figure 3).

Fibrinogen Osaka VI has a 12 residue carboxyl-terminal extension of SPMR-RFKLLFCM in a dysfibrinogen derived from a woman heterozygotic for this abnormality and associated with severe bleeding.[58] This extension is due to a T to A mutation that creates (AAG) encoding Lys at the stop (TAG) codon, thus translating 36-base pairs in the non-coding region of the Bβ-gene (see FIGURE 4, *upper panel*). The extra Cys residues appear to be involved in one or two disulfide bonds between two adjacent abnormal fibrinogen molecules, forming a fibrinogen homodimer as indicated by SDS-PAGE. Indeed, about half of the fibrinogen molecules exist as end-linked dimers oriented in parallel or with an angle, as observed by transmission electron microscopy (FIG. 4, *middle* and *lower panels*). These end-linked dimers may well alter the conformations of D and DD regions on fibrin assembly, leading to increased fiber branching at their sites in the growing protofibrils (see FIGURE 5). By scanning electron microscopy, the Osaka VI fibrin network appears to have a lace-like structure, composed of highly branched, thinner fibers than the normal fibrin architecture. Such fibrin networks may be easily damaged to form large pores, when fluids are allowed to pass through the gels. The fragility of Osaka VI fibrin clots, further confirmed by permeation and compaction studies, may account for the massive bleeding observed in this patient.

Impaired Lateral Association of Protofibrils

When double-stranded fibrin protofibrils propagate longitudinally and reach certain lengths, they associate laterally with one another to form thick twisted fibrin fibers and bundles,[5,6] which then branch at several points developing interwoven fibrin networks.[6-8] Lateral association of protofibrils proceeds in a two-step reaction.[5,6] First, the interconnected carboxyl-terminal segments of the fibrin α-chains (the αC-αC domain) are untethered upon release of paired fibrinopeptides B (FPBs) that proceeds much faster from desAA fibrin in the protofibril than from a single fibrinogen molecule.[6] Second, the untethered αC domains of one protofibril interact with the αC domains of another and form thick fibrin fibers and bundles.[5-8] For example, lack of interaction sites in the αC domain is thought to cause impaired lateral association of fibrinogen Marburg with 150-residue truncated Aα-chain, part of which is disulfide linked with serum albumin.[59,60] On activation with thrombin, double stranded fibrin protofibrils are normally formed in this dysfibrinogen, as evidenced by nearly normal tPA-catalyzed activation of plasminogen in the presence of polymerizing fibrin monomers (profile not shown), although virtually no solid fibrin gels are formed during the reaction.[60]

Although the mechanisms may not be identical with those for fibrinogen Marburg, two other types of dysfibrinogens with abnormal Aα-chains are also characterized by impaired lateral association of fibrin protofibrils. They are fibrinogens Dusart[61] and Chapel Hill III[62] linked with serum albumin at the Cys substitution in the Aα-drain via a disulfide bond, and fibrinogen Caracas II with the Aα-chains linked with a highly negatively charged extra oligosaccharide.[63] In fibrinogens

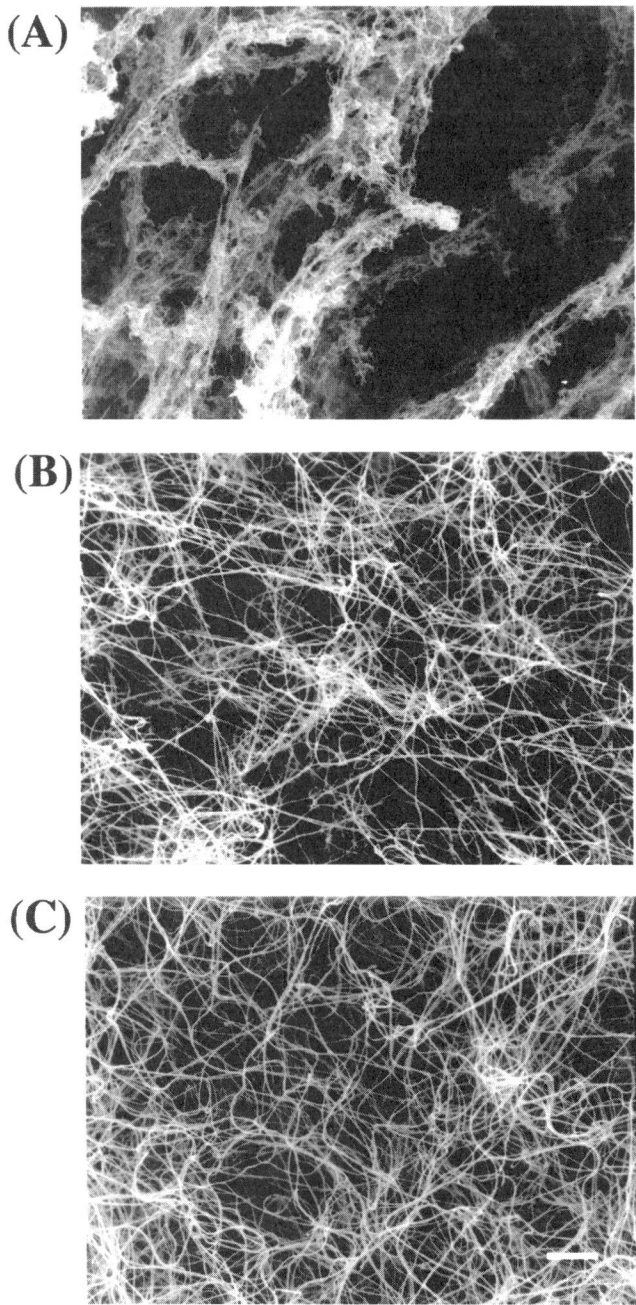

FIGURE 3. Scanning electron micrographs of the wild (**A**) and deglycosylated (**B**) Asahi fibrin clots in comparison with normal fibrin clot (**C**). *Bar*: 5 μm.

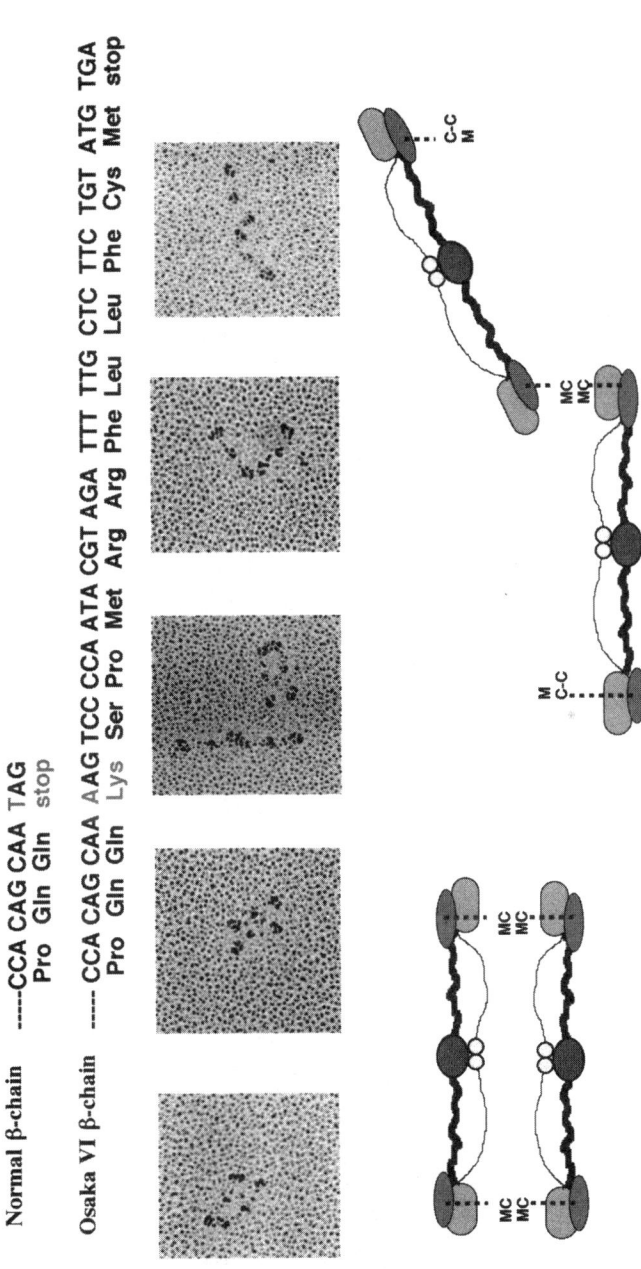

FIGURE 4. Alterations in the Bβ-chain gene and polypeptide showing a 12-amino acid extension at the carboxy-terminus of the aberrant Bβ-chain of fibrinogen Osaka VI. **Upper panel**: schematic representation of a single base exchange of A to T in the stop codon that creates a codon for Lys in the Bβ-chain gene. Thus, 36 base pairs in the non-coding region of the Bβ-chain gene were translated. **Middle panel**: two types of end-linked fibrinogen dimers observed by transmission electron microscopy. **Lower panel**: models of end-linked Osaka VI fibrinogen dimers, a bilayer dimer linked at both ends by two disulfide bonds (*left*) and a longitudinally aligned dimer linked at either end of the molecule via a single disulfide bond (*right*). (See Ref. 58.)

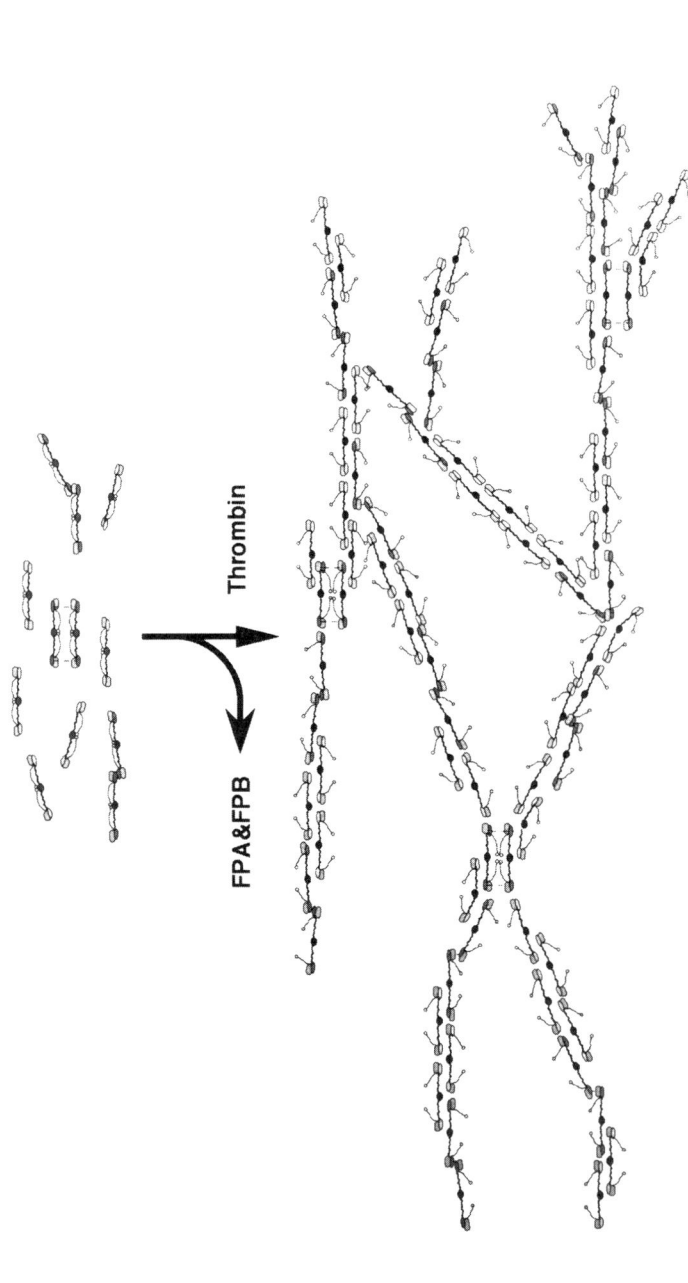

FIGURE 5. Schematic representation of increased branching of the Osaka VI fibrin networks that incorporate end-linked dimers at the branching points. This diagram seems to provide a molecular basis for construction of thin and highly branched fibrin fibers leading to formation of fragile fibrin clots. (Reproduced from Ref. 58 with permission.)

Dusart and Chapel Hill III, both defined to have an Aα Arg-554 to Cys substitution, the fibrin fibers are thinner and curvilinear, and far more highly branched than normal fibrin fibers as observed by scanning electron microscopy.[64] However, the compactness of fibrin clots is much higher, and thus, the permeability of the clots is greatly reduced. Furthermore, increased αC domain dissociation due to the attachment of serum albumin leads to increased intermolecular association and, concomitantly, or consequently, to enhanced interaction between reciprocally aligned two γ-chains. These findings may partly account for the thrombotic complications in the propositi of these two dysfibrinogens.[65] On the other hand, fibrinogen Caracas II has an Aα Ser-434 to Asn substitution, and the Asn residue itself is N-glycosylated due to a newly created Asn-X-Thr type glycosylation sequence.[63] The extra oligosaccharide moieties are represented mostly by a disialylated oligosaccharide accounting for 81.9% of the total extra oligosaccharides. By contrast, disialylated oligosaccharide linked to normal fibrinogen accounts for only 22.4% of the total oligosaccharides. Because of the strongly negative-charged extra oligosaccharide

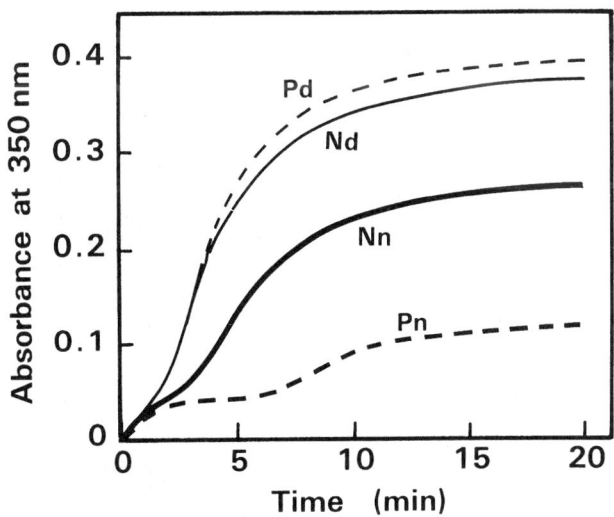

A. Thrombin Time (sec)

	Normal	Patient
Before	16.2	97.4
After	8.4	8.5

B. Fibrin Monomer Polymerization

FIGURE 6. Marked shortening of the prolonged thrombin time and acceleration of fibrin polymerization after desialylation of fibrinogen Caracas II. (See Ref. 60.)

linked to the αC-region, the double-stranded protofibrils of Caracas II may display extraordinary repulsive forces, when the untethered Aα-chains are aligned in such a way that the negatively charged structural alterations face one another on lateral association. Indeed, desialylation of fibrinogen Caracas II results in complete normalization of the thrombin time and fibrin monomer polymerization (see FIGURE 6). Electron microscopic analysis of rotary shadowed fibrinogen Caracas II reveals that nearly half of the heterozygous fibrinogen molecules manifest one or two small globular domains that project from the outer globular D domains.[66] This indicates that the αC domains of the mutant fibrinogen molecule linked with a strongly negatively charged extra oligosaccharide neither associate with each other nor with the negatively charged central part of the E domain. In contrast, the other half has no additional globular domains, as observed in normal fibrinogen.[6] In contrast to very compact fibrin meshes made from fibrinogens, Dusart and Chapel Hill III, that are associated with thromboembolic diseases, whole fibrin clots formed from fibrinogen Caracas II are very different in appearance when examined by scanning electron microscopy. Although considerable variation exists, many fibers are thinner than normal in diameter, and there is less aggregation to form bundles. However, the most striking feature is the presence of large pores or open areas bounded by local fiber networks, just like caves or tunnels bounded by long, curved bundles of fibers. Furthermore, free fiber ends are commonly observed (see FIGURE 7). Consistent with

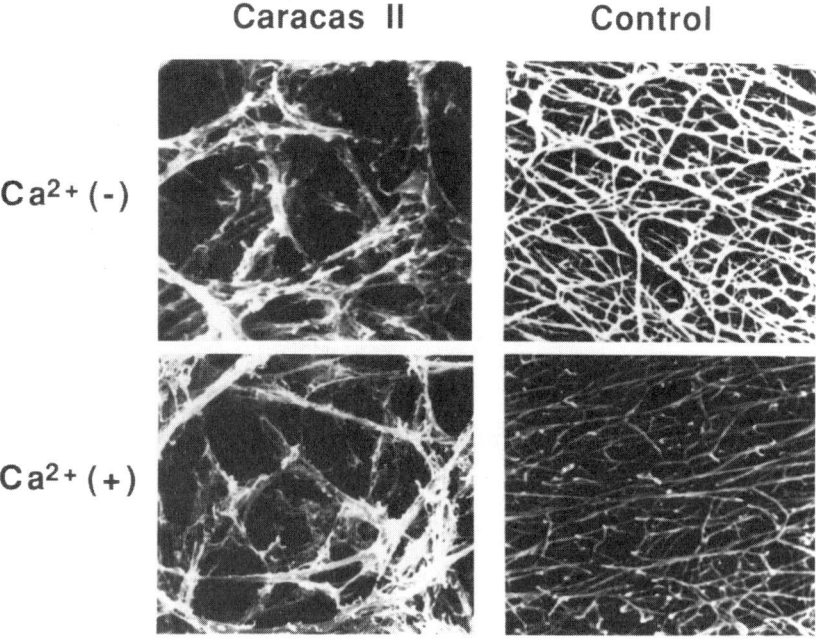

FIGURE 7. Scanning electron micrographs of normal and the Caracas II fibrin clots. The Caracas II fibrin clots show distinct large pores bounded by small secondary networks. (See Ref. 66.)

these findings, the flow rate in the Caracas II fibrin clots determined by the permeation study is much higher than in the control clots, which may account for the absence of thrombotic diseases in the propositus.[66]

Although the mechanisms may not be exactly identical, repulsive forces generated by the strong negative electric charges of extra oligosaccharides are thought to disturb the lateral association of protofibrils in two other dysfibrinogens linked with highly disialylated extra oligosaccharides. One is fibrinogen Lima derived from a homozygote that has a highly disialylated extra oligosaccharides (68.6%) at Aα Asn-139 due to a mutation of Aα Arg-141 to Ser substitution due to creation of an Asn-X-Ser type glycosylation sequence.[67] Although the extra oligosaccharide is located distant from the a site in the D domain, fibrin clot formation is substantially delayed. However, double stranded fibrin protofibrils are normally created, as evidenced by normal tPA-catalyzed plasminogen activation in the presence of fibrin monomers polymerizing far more slowly than normal fibrin monomers. As anticipated, desialylation normalized the prolonged thrombin time (from 32.4 sec to 10.2 sec; control, from 16.2 sec to 8.4 sec) and altered fibrin monomer polymerization. The other is fibrinogen Kaiserslautern, also associated with a highly disialylated extra oligosaccharide (92.1%) at the γ Asn-380 mutation site itself due to creation of an Asn-X-Thr type glycosylation sequence.[68] The location of this oligosaccharide is also remote from the sites involved in fibrin assembly, and the strong negative electric charges are thought to be responsible for delayed fibrin formation.

Miscellaneous Types

There are several structurally well-characterized dysfibrinogens associated with delayed fibrin gel formation (see TABLE 1), but the exact mechanisms of how fibrin assembly is disturbed in these molecules may not necessarily be attributed to the impairment of the established binding sites or molecular interactions. Close inspection of these molecules may provide us with new insights into other aspects of structure–function relationships of fibrinogen, which have not as yet been described.

DEFECTS IN BINDING WITH OTHER SUBSTANCES IN BLOOD AND CELL SURFACE PROTEINS

After conversion of fibrinogen to fibrin, a variety of binding sites are exposed in polymerizing fibrin monomers. They include binding sites with Ca^{2+},[69–71] thrombin,[72,73] tPA,[74,75] plasminogen,[76,77] and integrins on the cell surfaces.[78–81]

Impaired High Affinity Calcium Binding Site

The high affinity calcium binding site has been shown to appear on the γ (311–336) residues by terbium fluorescence studies.[71] Indeed, recent crystal structure studies of a 30-kDa carboxyl terminal recombinant fragment of the fibrinogen γ-chain show that the calcium ion is liganded by two aspartate side chains of Asp-318 and Asp-320, two carboxyl oxygen atoms of Phe-322 and Gly-324, and two water molecules.[9,10,12] This calcium binding site is distinct from the a polymerization site.[12] There is a unique dysfibrinogen, fibrinogen Vlissingen, lacking the calcium binding capacity due to a deletion of γ Asn-319·Asp-320 due to a six base (AAT-GAT) deletion in the γ-chain gene.[82] Loss of the calcium binding capacity seems to affect the stabilized structure of the *a* site, resulting in altered fibrin polymerization.[12,13,43]

Impaired Binding of Fibrin with Thrombin

Two dysfibrinogens with a structural alteration in the amino-terminal region of the fibrin β-chain are known to have reduced affinity for thrombin. One is fibrinogen New York I with a deletion of the Bβ (9–72) residue segment that exactly corresponds to exon 2 of the fibrin Bβ-chain gene,[83] and the other is fibrinogen Naples I with a Bβ Ala-68 to Thr substitution.[84] Consistent with the impaired localization of thrombin onto newly formed fibrin clots, the propositi for these two dysfibrinogens manifested recurrent thrombosis associated with a pulmonary embolism and an arterial thrombosis, respectively.

Impaired Binding with Plasminogen

When fibrinogen is converted to fibrin and binds with tPA, fibrinogen-bound plasminogen is effectively activated to plasmin on the solid fibrin fibers. In an early study of factor XIIIa-crosslinked fibrin derived from the fibrinogen Dusart patient associated with severe thrombotic disease, reduced affinity for Lys-plasminogen was thought to account for the thrombophilia at least partly via impaired fibrinolysis.[87] This dysfibrinogen has been shown to have an Aα Arg-554 to Cys substitution due to a C to T exchange in the codon (CGT) for Aα Arg-554, and part of the Cys-substitutes is linked with albumin via a disulfide bridge.[61] Further studies on this type of dysfibrinogen revealed additional abnormalities related to impaired fibrinolysis, which are discussed in the following section.

Impaired tPA-Catalyzed Plasminogen Activation in the Presence of Polymerizing Fibrin Monomers, and Digestion of Fibrin Clots by Plasmin

During fibrin formation, tPA binds to its specific binding sites on the fibrin molecule, and catalyzes plasminogen to plasmin conversion efficiently on the fibrin fibers. The minimally required structure for fibrin to promote tPA-catalyzed plasminogen activation is shown to be a double stranded protofibril.[88] In fibrinogen Dusart, the fibrin-mediated enhancement of tPA-catalyzed plasminogen activation is shown to be strongly decreased, although the binding of tPA to the Dusart fibrin is normal.[89] In the same type of dysfibrinogen, fibrinogen Chapel Hill III,[62] the patient-derived thrombin- and reptilase-clotted fibrin gels are highly resistant against plasmin; essentially no plasmic cleavage having occurred.[90] Thus, the structure of the patient's fibrin clot may also be responsible for the defective degradation by plasmin. Indeed, the fibrin fibers are significantly thinner but stiffer than normal in these dysfibrinogens,[64,65] being relevant to extreme resistance of fibrin clots against plasmin.[87,89,90]

Similarly increased stiffness, decreased permeation rate, and nearly complete resistance against plasmin are observed also in fibrinogen Marburg with the 150 residue-truncated Aα (1–460) chains partly linked with serum albumin at Aα Cys 442 that has lost its disulfide partner Aα Cys-472.[60] In this molecule, approximately one molecule of albumin is linked to every three molecules of fibrinogen. Actually, there are three species of Marburg fibrinogen molecules—those linked with one or two molecules of albumin, or free of albumin. SDS-PAGE and amino acid sequence analyses of the resolved subunit multimers of factor XIIIa-crosslinked fibrin shows the presence of heterozygous multimers ($\alpha m \cdot \gamma n$, $m \neq n$) and part of the disulfide-linked serum albumin cross-linked to the γ-chain. Thus, the Marburg fibrin seems to undergo critical structural alterations by factor XIIIa, and thereby forms fine but compact clots and acquires resistance against plasmin[60] (see FIGURE 8). The acquisition of plasmin resistance may have contributed to pelvic vein thrombosis and recurrent

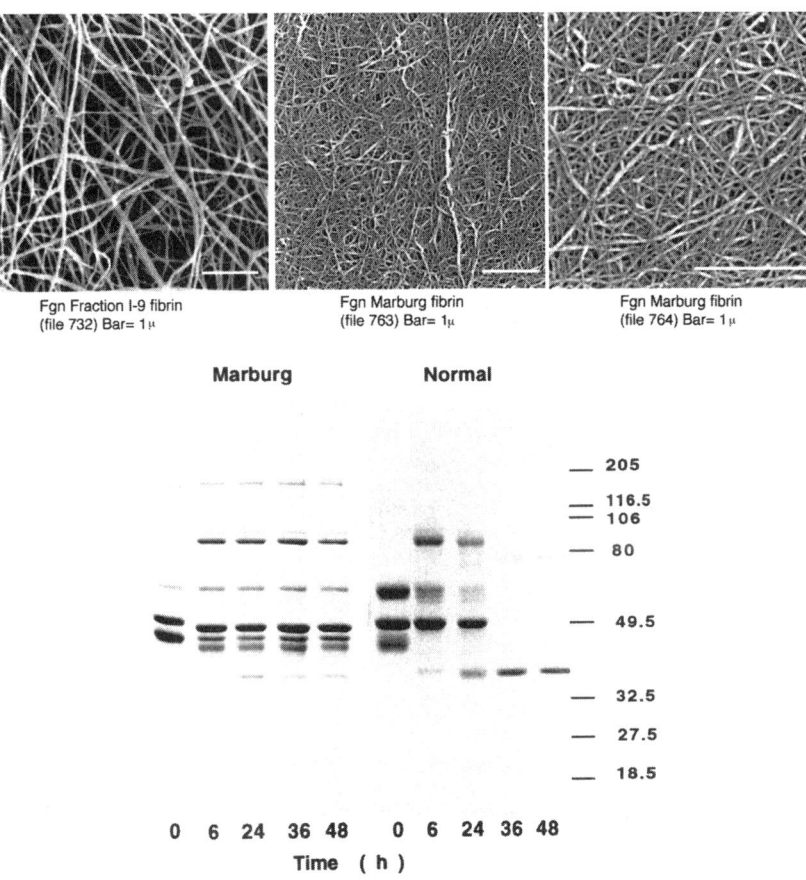

FIGURE 8. Characteristic features of the Marburg fibrin clots relevant to thrombotic disease. **Upper panel**: scanning electron micrographs showing the compact Marburg fibrin clots composed of extremely thin fibers as compared with the normal counterpart, fraction I-9 that lacks the carboxy-terminal region of the Aα-chain. **Lower panel**: SDS-PAGE showing highly resistant subunit polypeptides of factor XIIIa-crosslinked Marburg fibrin against tPA-catalyzed plasmin digestion. (Reproduced from Ref. 60 with permission.)

pulmonary embolisms after Cæsarian section for the first delivery of the propositus at the age of 20.[59,60]

Reduced Affinity for Integrins on the Cell Surfaces

Each Aα-chain has two Arg-Gly-Asp (RGD) sequences, Aα (95–97) and Aα (572–574) that may be essential for binding with integrins.[80,91] Thus, untethering of the αC domains on conversion to fibrin appears to expose one or both of these RGD segments. In fibrinogen Marburg reduced affinity to endothelial cells has been noted,[59] most probably due to the missing Aα (572–574) RGD segment.

Relation to Hereditary Renal Amyloidosis

Hereditary dysfibringens have been described in relation to hereditary renal amyloidosis (HRA), in which the aberrant peptide segments derived from the fibrinogen Aα-chains have been identified as amyloid deposits in the kidney. HRA is a subtype of hereditary systemic amyloidosis that is an autosomal trait characterized by the extracellular accumulation of protein fibrils having β-sheet structure such as transerythretin, gelsolin, apoprotein A1, and lysozyme in various organs.

So far, four types of the Aα-chain mutants of fibrinogen have been characterized in relation to renal amyloidosis; single amino acid substitutions of AαArg 554 to Leu[92] and AαGlu 526 to Val,[93,94] and two types of frame shift at positions Aα524[95] and Aα522,[96] both resulting in premature termination of the polypeptide at position Aα547. In the AαArg 554 to Leu substitution, the abnormal fibrinogen species is not present in plasma, but a peptide segment corresponding to Aα(500–580) residues containing the Leu substitution was isolated from the kidney.[92] On the other hand, the aberrant fibrinogen species with an AαGlu 526 to Val substitution is present in plasma at an equal amount with normal fibrinogen species.[93] In the kindred with a frame shift at position Aα 524 of fibrinogen, no abnormal fibrinogen species was detectable in plasma, but immunohistological studies indicated the presence of a predicted aberrant peptide in amyloid deposits.[95] In the kindred having a frame shift at position Aα522, a unique 49 amino acid residue hybrid segment was identified in the kidney, consisting of 23 amino acids corresponding to normal Aα(499–521) residues followed by 26 amino acids with a unique sequence to the termination at position Aα 522.[96]

CLINICAL IMPLICATIONS DERIVED FROM THE STUDY OF DYSFIBRINOGENS

On conversion to fibrin, fibrinogen serves as the major constituent of hemostatic thrombi, physiologically at the damaged tissue and under some pathological conditions, intravascular clots. However, these normal and pathologic fibrin clots would subsequently undergo digestion mediated by fibrinolysis. The wound healing thus proceeds and circulation of blood is guaranteed. Failure to form hemostatic thrombi and to digest intravascular fibrin clots necessitates to manifest bleeding and thrombosis, respectively. Indeed, these altered functions are observed in certain types of dysfibrinogens that have been discussed in individual molecules in this review.

Unless underlying mechanisms are clearly elucidated in direct relation to bleeding, thrombosis, disturbed wound healing, or any other clinical manifestations, the relevance of dysfibrinogens to these clinical symptoms should be carefully concluded. Some of these clinical signs and symptoms include delayed fibrin polymerization, impaired factor XIIIa-cross-linking of the α- and γ-chains, excessive fibrin clot formation, extraordinarily compact and/or stiff fibrin clot composed of fine fibrin fibers, and reduced fibrin degradation by fibrinolysis due to decreased affinity for tPA or plasminogen, acquisition of resistance against plasmin of fibrin clots, as observed in a variety of hereditary dysfibrinogens discussed in the previous sections.

Although not directly related to the process of fibrinogen to fibrin transition, rapid degradation of a certain types of abnormal molecule with a structural alteration in the carboxy terminal region of the Aα-chain may cause the release of the amyloid peptides and lead to amyloid deposition in the kidney.

Readers are encouraged to refer to other recently published review articles for a better understanding of fibrinogen, a true heavy weight among human plasma proteins.[23,24,43,97,98]

ACKNOWLEDGMENTS

We thank all the colleagues in our laboratory who conducted analyses of the hereditary dysfibrinogens sent to us for structure and function studies, and Drs. Carmen L. Arocha-Piñango, Norma B. de Bosch, Günter Auerswald, Manfred Popp, Rudolf Egbring, R.C. Franz, and many clinicians throughout Japan, though not individually mentioned, who kindly provided the abnormal fibrinogens and their genes. We are also indebted to Dr. Noriko Takahashi for the structure studies of extra oligosaccharides found in several mutant molecules, Drs. John W. Weisel and Michael W. Mosesson for their collaboration in electron microscopic studies, and Drs. Russell F. Doolittle and Kathleen P. Pratt for discussion on the crystal structures of dysfibrinogens reported from our laboratory.

This work was supported in part by Grants-in-Aid for Scientific Research 06404043, 08407034, 09671132, and 11470250, and for International Scientific Research Program, Joint Research 06044196, 09044329, 10044316, and 11694308 from the Ministry of Education, Science and Culture of the Government of Japan and from the Research Foundation for Community Medicine.

REFERENCES

1. DOOLITTLE, R.F., H. BOUMA, III, B.A. COTTRELL, et al. 1979. The covalent structure of human fibrinogen. In The Chemistry and Physiology of the Human Plasma Proteins. D.H. Bing, Ed.: 77–95. Pergamon Press, New York.
2. DOOLITTLE, R.F. 1981. Fibrinogen and fibrin. In Hæmostasis and Thrombosis, 2nd edit. A.L. Bloom & D.P. Thomas, Eds.: 163–191. Churchill Livingstone, Edinburgh.
3. WEISEL, J.W. 1986. Lateral aggregation and the role of the two pairs of fibrinopeptides. Biophys. J. **50:** 1079–1093.
4. WEISEL, J.W., Y. VEKLICH & O. GORKUN. 1993. The sequence of cleavage of fibrinopeptides from fibrinogen is important for protofibril formation and enhancement of lateral aggregation in fibrin clots. J. Mol. Biol. **232:** 285–297.

5. GORKUN, O.V., Y.I. VEKLICH, L.V. MEDVED, et al. 1994. Role of the αC domains of fibrin in clot formation. Biochemistry **33:** 6986–6997.
6. VEKLICH, Y.I., O.V. GORKUN, L.V. MEDVED, et al. 1993. Carboxyl-terminal portions of the α chains of fibrinogen and fibrin. Localization by electron microscopy and the effects of isolated αC fragments on polymerization. J. Biol. Chem. **268:** 13577–13585.
7. MOSESSON, M.W., J.P. DIORIO, K.R. SIEBENLIST, et al. 1993. Evidence for a second type of fibrin branch point in fibrin polymer networks, the trimolecular branch junction. Blood **82:** 1517–1521.
8. BARADET, T.C., J.C. HASELGROVE & J.W. WEISEL. 1995. Three-dimensional reconstruction of fibrin clot networks from stereoscopic intermediate voltage electron microscopic images and analysis of branching. Biophys. J. **68:** 1551–1560.
9. YEE, V.C., K.P. PRATT, H.C.F. CÔTÉ, et al. 1997. Crystal structure of a 30 kDa C-terminal fragment from the γ chain of human fibrinogen. Structure **5:** 125–138.
10. PRATT, K.P., H.C.F. CÔTÉ, D.W. CHUNG, et al. 1997. The fibrin polymerization pocket: three-dimensional structure of a 30 kDA C-terminal γ chain fragment complexed with the peptide Gly-Pro-Arg-Pro. Proc. Natl. Acad. Sci. U.S.A. **94:** 7176–7181.
11. SPRAGGON, G., S. EVERSE & R.F. DOOLITTLE. 1997. Crystal structures of fragment D from human fibrinogen and its crosslinked counterpart from fibrin. Nature **389:** 455–462.
12. CÔTÉ, H.C.F., K.P. PRATT, E.W. DAVIE & D.W. CHUNG. 1997. The polymerization pocket 'a' within the carboxyl-terminal region of the γ chain of human fibrinogen is adjacent to but independent from the calcium-binding site. J. Biol. Chem. **272:** 23792–23798.
13. EVERSE, S.J., G. SPRAGGON & R.F. DOOLITTLE. 1998. A three dimensional consideration of varian human fibrinogens. Thromb. Hæmost. **80:** 1–9.
14. SOUTHAN, C. 1998. The elucidation of molecular defects in congenital dysfibrinogenemia. *In* Fibrinogen, Fibrin Stabilisation, and Fibrinolysis. J.L. Francis, Ed.: 100–127. Ellis Horwood, Chichester.
15. NIWA, K., A. YAGINUMA, M. NAKANISHI, et al. 1993. Fibrinogen Mitaka II: a hereditary dysfibrinogen with defective thrombin binding caused by an Aα Glu-11 to Gly substitution. Blood **82:** 3658–2663.
16. MARTIN, P.D., W. ROBERTSON, D. TURK, et al. 1992. The structure of residues 7–16 of the Aα-chain of human fibrinogen bound to bovine thrombin at 2.3-Å resolution. J. Biol. Chem. **267:** 7911–7920.
17. STUBB, M.T., H. ASCHKINAT, I. MAYR, et al. 1992. The interaction of thrombin with fibrinogen. A structural basis for its specificity. Eur. J. Biochem. **206:** 187–195.
18. ZHENG, Z., R.W. ASHTON, F. NI & H.A. SCHERAGA. 1992. Thrombin hydrolysis of an N-terminal peptide from fibrinogen Lille: kinetic and NMR studies. Biochemistry **31:** 4426–4431.
19. HIGGINS, D.L. & J.A. SHAFER. 1981. Fibrinogen Petoskey, a dysfibrinogenemia characterized by replacement of Arg-Aα16 by a histidyl residue. Evidence for thrombin-catalyzed hydrolysis at a histidyl residue. J. Biol. Chem. **256:** 12013–12017.
20. HENSCHEN, A., M. KEHL & S. SOUTHAN. 1984. Genetically abnormal fibrinogens—strategies for structure elucidation, including fibrinopeptide analysis. *In* Variants of Human Fibrinogen. E.A. Beck & M. Furlan, Eds.: 273–320. Hans Huber Verlag, Bern.
21. MATSUDA, M. 1990. Molecular abnormalities of fibrinogen—the present status of structure elucidation. *In* Fibrinogen 4. Current Basis and Clinical Aspects. M. Matsuda, S. Iwanaga, A. Takada & A. Henschen, Eds.: 139–152. Excerpta Medica, Amsterdam.
22. GALANAKIS, D. 1993. Inherited dysfibrinogenemia: emerging abnormal structure associations with pathologic and nonpathologic dysfunctions. Semin. Thromb. Hemostas. **19:** 386–395.
23. MATSUDA, M. 1996. The structure–function relationship of hereditary dysfibrinogens. Intern. J. Hematol. **64:** 167–179.
24. EBERT, R.F., Ed. 1994. Index of Variant Human Fibrinogens. CRC Press, Boca Raton.

25. KAUDEWITZ, H., A. HENSCHEN, C. SORIA, et al. 1986. The molecular defect of the genetically abnormal fibrinogen Christchurch II. In Fibrinogen and Its Derivatives. G. Müller-Berghaus, V. Scheefers-Borchel, E. Selmayr & A. Henschen, Eds.: 31–36. Elsevier, Amsterdam.
26. KUDRYK, B.J., D. COLLEN, K.R. WOODS & B. BLOMBÄCK. 1974. Evidence for localization of polymerization sites in fibrinogen. J. Biol. Chem. **249:** 3322–3325.
27. OLEXA, S.A. & A.Z. BUDZYNSKI. 1980. Evidence for four different polymerization sites involved in human fibrin formation. Proc. Natl. Acad. Sci. U.S.A. **77:** 1374–1378.
28. MOSESSON, M.W., K.R. SIEBENLIST, J.P. DIOLIO, et al. 1995. The role of fibrinogen D domain intermolecular association sites in the polymerization of fibrin and fibrinogen Tokyo II (γ 275 Arg→Cys). J. Clin. Invest. **96:** 1053–1058.
29. LAUDANO, A.P. & R.F. DOOLITTLE. 1978. Synthetic peptide derivatives that bind to fibrinogen and prevent the polymerization of fibrin monomers. Proc. Natl. Acad. Sci. U.S.A. **75:** 3085–3089.
30. LAUDANO, A.P. & R.F. DOOLITTLE. 1980. Studies on synthetic peptides that bind to fibrinogen and prevent fibrin polymerization. Structural requirements, number of binding sites, and species differences. Biochemistry **19:** 1013–1019.
31. WADA, Y., K. NIWA, H. MAEKAWA, et al. 1993. A new type of congenital dysfibrinogen, fibrinogen Bremen, with an Aα Gly-17 to Val substitution associated with hemorrhagic diathesis and delayed wound healing. Thromb. Hæmost. **70:** 397–403.
32. YOSHIDA, N., M. OKUMA, H. HIRATA, et al. 1991. Fibrinogen Kyoto II, a new congenitally abnormal molecule, characterized by the replacement of Aα proline-18 by leucine. Blood **78:** 149–153.
33. UOTANI, C., T. MIYATA, I. KUMABASHIRI, et al. 1991. Fibrinogen Kanazawa: a congenital dysfibrinogenemia with delayed polymerization having a replacement of proline-18 by leucine in the Aα-chain. Blood Coag. Fibrinol. **2:** 413–417.
34. BLOMBÄCK, M., B. BLOMBÄCK, E.F. MAMMEN & A.S. PRASAD. 1968. Fibrinogen Detroit-A molecular defect in the N-terminal disulphide knot of human fibrinogen? Nature **218:** 134–137.
35. HESSEL, B., S. STENBJERG, J. DYR, et al. 1986. Fibrinogen Aarhus—a new case of dysfibrinogenemia. Thromb. Res. **42:** 21–37.
36. DEMPFLE, C.E.H. & A. HENSCHEN. 1990. Fibrinogen Mannheim I-identification of an Aα 19 Arg→Gly substitution in dysfibrinogenemia associated with bleeding tendency. In Fibrinogen 4. Current Basic and Clinical Aspects. M. Matsuda, S. Iwanaga, A. Takada & A. Henschen, Eds.: 159–166. Elsevier Science Publ., Amsterdam.
37. YAMAGUCHI, S., T. SUGO, Y. HASHIMOTO, et al. 1997. Fibrinogen Kumamoto with an Aα Arg-19 to Gly substitution has reduced affinity for thrombin: possible relevance to thrombosis. Jpn. J. Thromb. Hemost. **8:** 382–392.
38. MIYATA, T., K. FURUKAWA, S. IWANAGA, et al. 1989. Fibrinogen Nagoya, a replacement of glutamine-329 by arginine in the γ-chain that impairs the polymerization of fibrin monomer. J. Biochem. **105:** 10–14.
39. REBER, P., M. FURLAN, C. RUPP, et al. 1986. Characterization of fibrinogen Milano I: amino acid exchange γ 330 Asp→Val impairs fibrin polymerization. Blood **67:** 1751–1756.
40. TERUKINA, S., K. YAMAZUMI, K. OKAMOTO, et al. 1989. Fibrinogen Kyoto III: a congenital dysfibrinogen with a γ aspartic acid-330 to tyrosine substitution manifesting impaired fibrin monomer polymerization. Blood **74:** 2681–2687.
41. OKUMURA, N., K. FURIHATA, F. TERASAWA, et al. 1996. Fibrinogen Matsumoto I: a γ 364 Asp→His (GAT→CAT) substitution associated with defective fibrin polymerization. Thromb. Hæmost. **75:** 887–891.
42. BENTOLILA, S., M.M. SAMAMA, J. CONARD, et al. 1995. Association of dysfibrinogenemia and thrombosis. Apropos of a family (fibrinogen Melun) and review of the literature (in French). Annalen. Med. Interne. **146:** 575–580.
43. CÔTÉ, H.C.F., S.T. LORD & K.P. PRATT. 1998. γ-Chain dysfibrinogenemias: molecular structure-function relationships of naturally occurring mutations in the γ chain of human fibrinogen. Blood **92:** 2195–2212.

44. YOSHIDA, N., H. HIRATA, Y. MORIGAMI, et al. 1992. Characterization of an abnormal fibrinogen Osaka V with the replacement of γ-arginine 375 by glycine. J. Biol. Chem. **267:** 2753–2759.
45. STEINMANN, C., P. REBER, M. JUNGO, et al. 1993. Fibrinogen Bern I: substitution γ 337 Asn→Lys is responsible for defective fibrin monomer polymerization. Blood **82:** 2104–2108.
46. STEINMANN, C., C. BÖGLI, M. JUNGO, et al. 1994. A new substitution, γ 358 Ser→Cys, in fibrinogen Milano VII causes defective fibrin polymerization. Blood **84:** 1874–1880.
47. MATSUDA, M., C. NAKAMIKAWA, M. BABA & K. MORIMOTO. 1983. "Fibrinogen Tokyo II": an abnormal fibrinogen with an impaired polymerization site on the aligned DD domain of fibrin molecules. J. Clin. Invest. **72:** 1034–1041.
48. TERUKINA, S., M. MATSUDA, H. HIRATA, et al. 1988. Substitution of γ Arg-275 by Cys in an abnormal fibrinogen, "fibrinogen Osaka II". Evidence for a unique solitary cystine structure at the mutation site. J. Biol. Chem. **263:** 13579–13587.
49. REBER, P., M. FURLAN, A. HENSCHEN, et al. 1986. Three abnormal fibrinogen variants with the same amino acid substitution (γ 275 Arg→His): fibrinogens Bergamo II, Essen and Perugia. Thromb. Hæmost. **56:** 401–406.
50. YAMAZUMI, K., S. TERUKINA, S. ONOHARA & M. MATSUDA. 1988. Normal plasmic cleavage of the γ chain variant of "fibrinogen Saga" with an Arg-275 to His substitution. Thromb. Hæmost. **60:** 476–480.
51. MIMURO, J., Y. KAWATA, K. NIWA, et al. 1999. A new type of Ser substitution for γ Arg-275 in fibrinogen Kamogawa I characterized by impaired fibrin assembly. Thromb. Hæmost. **81:** 940–944.
52. FELLOWES, A.P., S.O. BRENNAN, H.J. RIDGWAY, et al. 1998. Electrospray ionization mass spectrometry identification of fibrinogen Banks Peninsula (γ 280 Tyr→Cys): a new variant with defective polymerization. Brit. J. Hæmat. **101:** 24–31.
53. NIWA, K., M. TAKEBE, T. SUGO, et al. 1996. A γ Gly-268 to Glu substitution is responsible for impaired fibrin assembly in a homozygous dysfibrinogen Kurashiki I. Blood **87:** 4686–4694.
54. YOSHIDA, N., S. TERUKINA, M. OKUMA, et al. 1988. Characterization of an apparently lower molecular weight γ-chain variant in fibrinogen Kyoto I. The replacement of γ Asn-308 by Lys which caused an accelerated cleavage of fragment D_1 by plasmin and the generation of a new plasmin cleavage site. J. Biol. Chem. **263:** 13848–13856.
55. BANTIA, S., W.R. BELL & C.V. DANG. 1990. Polymerization defect of fibrinogen Baltimore III due to a gamma Asn-308→Ile mutation. Blood **75:** 1659–1663.
56. YAMAZUMI, K., K. SHIMURA, S. TERUKINA, et al. 1989. A γ methionine-310 to threonine substitution and consequent N- glycosylation at γ asparagine-308 identified in a congenital dysfibrinogenemia associated with posttraumatic bleeding, fibrinogen Asahi. J. Clin. Invest. **83:** 1590–1597.
57. TOWNSEND, R.R., E. HILLIKER, Y.-T. LI, et al. 1982. Carbohydrate structure of human fibrinogen. J. Biol. Chem. **257:** 9704–9710.
58. SUGO, T., C. NAKAMIKAWA, N. YOSHIDA, et al. 2000. End-linked homodimers in fibrinogen Osaka VI with a Bβ-chain extension lead to fragile clot structure. Blood **96:** 3779–3785.
59. KOOPMAN, J., F. HAVERKATE, J. GRIMBERGEN, et al. 1992. Fibrinogen Marburg: a homozygous case of dysfibrinogenemia, lacking amino acids Aα 461–610 (Lys 461 AAA→Stop TAA). Blood **80:** 1972–1979.
60. SUGO, T., C. NAKAMIKAWA, M. TAKEBE, et al. 1998. Factor XIIIa-cross-linking of the Marburg Fibrin: Formation of αm·γn-heteromultimers and the α-chain-linked albumin·γ complex, and disturbed protofibril assembly resulting in acquisition of plasmin-resistance relevant to thrombophilia. Blood **91:** 3282–3288.
61. KOOPMAN, J., F. HAVERKATE, J. GRIMBERGEN, et al. 1993. Molecular basis for fibrinogen Dusart (Aα 554 Arg→Cys) and its association with abnormal fibrin polymerization and thrombophilia. J. Clin. Invest. **91:** 1637–1643.

62. WADA, Y. & S.T. LORD. 1994. A correlation between thrombotic disease and a specific fibrinogen abnormality (Aα 554 Arg→Cys) in two unrelated kindred, Dusart and Chapel Hill III. Blood **84:** 3709–3714.
63. MAEKAWA, H., K. YAMAZUMI, S. MURAMATSU, et al. 1991. An Aα Ser-434 to N-glycosylated Asn substitution in a dysfibrinogen, fibrinogen Caracas II, characterized by impaired fibrin gel formation. J. Biol. Chem. **266:** 11575–11581.
64. COLLET, J.P., J.L. WOODHEAD, J. SORIA, et al. 1996. Fibrinogen Dusart: electron microscopy of molecules, fibers and clots, and viscoelastic properties of clots. Biophys. J. **70:** 500–510.
65. MOSESSON, M.W., K.R. SIEBENLIST, J.F. HAINFELD, et al. 1996. The relationship between the fibrinogen D domain self-association/cross-linking site (γXL) and the fibrinogen Dusart abnormality (Aα R554C-albumin). Clues to thrombophilia in the "Dusart syndrome". J. Clin. Invest. **97:** 2342–2350.
66. WOODHEAD, J.L., C. NAGASWAMI, M. MATSUDA, et al. 1996. The ultrastructure of fibrinogen Caracas II molecules, fibers and clots. J. Biol. Chem. **271:** 4946–4953.
67. MAEKAWA, H., K. YAMAZUMI, S. MURAMATSU, et al. 1992. Fibrinogen Lima: a homozygous dysfibrinogen with an Aα-arginine-141 to serine substitution associated with extra N-glycosylation at Aα-asparagine-139. J. Clin. Invest. **90:** 67–76.
68. RIDGWAY, H.J., S.O. BRENNAN, R.M. LORETH & P.M. GEORGE. 1997. Fibrinogen Kaiserslautern (γ 380 Lys to Asn): A new glycosylated fibrinogen variant with delayed polymerization. Br. J. Hæmatol. **99:** 562–569.
69. MARGUERIE, G., G. CHAGNIEL & M. SUSCILLION. 1977. The binding of calcium to bovine fibrinogen. Biochim. Biophys. Acta **490:** 94–103.
70. HAVERKATE, F. & G. TIMAN. 1977. Protective effect of calcium in the plasmin degradation of fibrinogen and fibrin fragments. Thromb. Res. **10:** 803–812.
71. DANG, C.V., R.F. EBERT & W.R. BELL. 1985. Localization of a fibrinogen calcium binding site between γ-subunit positions 311 and 336 by terbium fluorescence. J. Biol. Chem. **260:** 9713–9719.
72. LIU, C.Y., H.L. NOSSEL & K.L. KAPLAN. 1979. The binding of thrombin to fibrin. J. Biol. Chem. **254:** 10421–10425.
73. HAVARKATE, F., J. KOOPMAN, C. KLUFT, et al. 1986. Fibrinogen Milano II: a congenital dysfibrinogenemia associated with juvenile arterial and venous thrombosis. Thromb. Hæmost. **55:** 131–135.
74. RIJKEN, D.C., M. HOYLAERTS & D. COLLEN. 1982. Fibrinolytic properties of one-chain and two-chain human extrinsic (tissue-type) plasminogen activator. J. Biol. Chem. **257:** 2920–2925.
75. NIEUWENHUIZEN, W., A. VERMOND, M. VOSKUILEN, et al. 1983. Identification of a site in fibrin(ogen) which is involved in the acceleration of plasminogen activation by tissue-type plasminogen activator. Biochim. Biphys. Acta **748:** 86–92.
76. LUCAS, M.A., L.J. FRETTO & P.A. MCKEE. 1983. The binding of human plasminogen to fibrin and fibrinogen. J. Biol. Chem. **258:** 4249–4256.
77. VÁRADI, A. & L. PATTHY. 1983. Location of plasminogen binding sites in fibrin(ogen). Biochemistry **22:** 2440–2446.
78. HAWIGER, J., S. TIMMONS, M. KLOCZEWIAK, et al. 1982. Gamma and alpha chains of human fibrinogen possess sites reactive with human platelet receptors. Proc. Natl. Acad. Sci. U.S.A. **79:** 2068–2071.
79. STRONG, D.D., A.P. LAUDANO, J. HAWIGER & R.F. DOOLITTLE. 1982. Isolation, characterization, and synthesis of peptides from human fibrinogen that block the staphylococcal clumping reaction and construction of a synthetic clumping principle. Biochemistry **21:** 1414–1420.
80. DEJANA, E., L.R. LANGUINO, N. POLENTARUTTI, et al. 1985. Interaction between fibrinogen and cultured endothelial cells. Induction of migration and specific binding. J. Clin. Invest. **75:** 11–18.
81. GONDA, S.R. & J.R. SHAINOFF. 1982. Adsorptive endocytosis of fibrin monomer by macrophages: evidence of a receptor for the amino terminus of the fibrin α chain. Proc. Natl. Acad. Sci. U.S.A. **79:** 4565–4569.

82. KOOPMAN, J., F. HAVERKATE, E. BRIËT & S.T. LORD. 1991. A congenitally abnormal fibrinogen (Vlissingen) with a 6-base deletion in the γ chain gene, causing defective calcium binding and impaired fibrin polymerization. J. Biol. Chem. **266:** 13456–13461.
83. LIU, C.Y., J.A. KOEHN & F.J. MORGAN. 1985. Characterization of fibrinogen New York I. A dysfunctional fibrinogen with a deletion of Bβ (9–72) corresponding exactly to exon 2 of the gene. J. Biol. Chem. **260:** 4390–4396.
84. KOOPMAN, J., F. HAVERKATE, S.T. LORD, et al. 1992. Molecular basis of fibrinogen Naples associated with defective thrombin binding and thrombophilia. Homozygous substitution of Bβ 68 Ala→Thr. J. Clin. Invest. **90:** 238–244.
85. VOSKUILEN, M., A. VERMOND, G.H. VEENEMAN, et al. 1987. Fibrinogen lysine residue Aα 157 plays a crucial role in the fibrin-induced acceleration of plasminogen activation, catalyzed by tissue-type plasminogen activator. J. Biol. Chem. **262:** 5944–5946.
86. YONEKAWA, O., M. VOSKUILEN & W. NIEUWENHUIZEN. 1992. Localization in the fibrinogen γ-chain of a new site that is involved in the acceleration of the tissue-type plasminogen activator-catalyzed activation of plasminogen. Biochem. J. **283:** 187–191.
87. SORIA, J., C. SORIA & J.P. CAEN. 1983. A new type of congenital dysfibrinogenemia with defective fibrin lysis—Dusard syndrome: possible relation to thrombosis. Brit. J. Hæmatol. **53:** 575–586.
88. SUENSON, E., P. BJERRUM, A. HOLM, et al. 1990. The role of fragment X polymers in the fibrin enhancement of tissue plasminogen activator-catalyzed plasmin formation. J. Biol. Chem. **265:** 22228–22237.
89. LIJNEN, H.R., J. SORIA, C. SORIA, et al. 1984. Dysfibrinogenemia (Fibrinogen Dusart) associated with impaired fibrin-enhanced plasminogen activation. Thromb. Hæmost. **51:** 108–109.
90. CARRELL, N., D.A. GABRIEL, P.M. BLATT, et al. 1983. Hetedetary dysfibrinogenemia in a patient with thrombotic disease. Blood. **62:** 439–447.
91. ASAKURA, S., K. NIWA, T. TOMOZAWA, et al. 1997. Fibroblasts spread on immobilized fibrin monomer by mobilizing a $β_1$-class integrin, together with a vitronectin receptor $α_vβ_3$ on their surface. J. Biol. Chem. **272:** 8824–8829.
92. BENSON, M.D., J. LIEPNIEKS, T. UEMICHI, et al. 1993. Hereditary renal amyloidosis associated with a mutant fibrinogen α-chain. Nature Genet. **3:** 252–255.
93. UEMICHI, T., J.J. LIEPNIEKS & M.D. BENSON. 1994. Hereditary renal amyloidosis with a novel variant fibrinogen. J. Clin. Invest. **93:** 731–736.
94. UEMICHI, T., J.J. LIEPNIEKS, F. ALEXANDER & M.D. BENSON. 1996. The molecular basis of renal amyloidosis in Irish-American and Polish-Canadian kindreds. Q. J. Med. **89:** 745–750.
95. UEMICHI, T., J.J. LIEPNIEKS, T. YAMADA, et al. 1996. A frame shift mutation in the fibrinogen Aα chain gene in a kindred with renal amyloidosis. Blood **87:** 4197–4203.
96. ASL, L.H., J.J. LIEPNIEKS, T. UEMICHI, et al. 1997. Renal amyloidosis with a frame shift mutation in fibrinogen Aα-chain gene producing a novel amyloid protein. Blood **90:** 4799–4805.
97. MATSUDA, M., T. SUGO, N. YOSHIDA, et al. 1999. Structure and function of fibrinogen: insights from dysfibrinogens. Thromb. Hæmost. **82:** 291–297.
98. MATSUDA, M. 2000. Structure and function of fibrinogen inferred from hereditary dysfibrinogens. Fibrinolysis & Proteolysis **14:** 187–197.

A Database for Human Fibrinogen Variants

M. HANSS AND F. BIOT

Laboratoire d'hématologie, Hôpital Cardiologique, Lyon, France

ABSTRACT: Identifying and studying abnormal human fibrinogens is a source of much information, and helps in taking care of the affected patients. To permit exhaustive numbering and easy updates, an extensive register has been compiled and made available on the internet. Known molecular abnormalities are mentioned with the essential clinical features.

KEYWORDS: Database; Fibrinogen variant.

Human fibrinogen may be impaired in a constitutional manner. This leads to dysfibrinogenaemia, hypofibrinogenaemia, or afibrinogenaemia according to the residual plasma molecule quality and to its amount. Many different molecular abnormalities have been shown at protein and/or gene levels. Furthermore, the associated clinical symptoms are highly variable, ranging from severe hemorrhages to pulmonary embolism, eventually following fibrinogen supplementation. The existence of any possible relationship between the abnormality and clinical manifestations is important to know, for instance when choosing optimal treatment in case of surgery that enhances both bleeding and thrombotic risks. Such a connection may be suggested or established by means of familial studies and molecular models.[1–4] Comparison of identical mutations can also be meaningful. For this purpose, we have grouped all the known molecular abnormalities and made the compilation available on the internet.

Data has been collected from the reports available in the literature, including previous reviews, referenced articles and abstracts, searches through Entrez PubMed© and Current Contents©. Cited works have been classified according to the position in the protein. The name of the abnormality is that quoted either in the referred articles, or in a subsequent reference. If not available, the town from which the work originates is mentioned. Some cases have been reported under two names. For optimal clarity, only the reference describing gene or protein abnormality is reported. Clinical symptoms correspond to the propositus presentation, when mentioned in the report, and discriminate between bleeding features, thrombotic features, familial amyloidosis, and no particular hemostatic sign. In 1994, Ebert found 95 mutation reports.[5] In June 2000, we found 174 reports (including 7 a- and hypofibrinogenemias), among which 59 are original mutations. They occur, first in the Aα chain ($n = 107$), then in the γ chain ($n = 55$), finally in the Bβ chain ($n = 18$), and are mainly punctual missense mutations ($n = 163$). As displayed in FIGURE 1, the two hot spots known to occur at positions Aα 16 Arg and γ 275 Arg still remain high, corresponding, respectively, to 59 and 21 cases. In order to present an extensive register that can be easily updated, and to permit exhaustive numbering, this database has been made

Address for correspondence: M. Hanss, Laboratoire d'hématologie, Hôpital L. Pradel, BP Lyon, Montchal 69394, Lyon, Cedex 03, France. Voice: 33(0)4 7235 7436; fax: 33(0)4 7235 7991.
michel.hanss@chu-lyon.fr

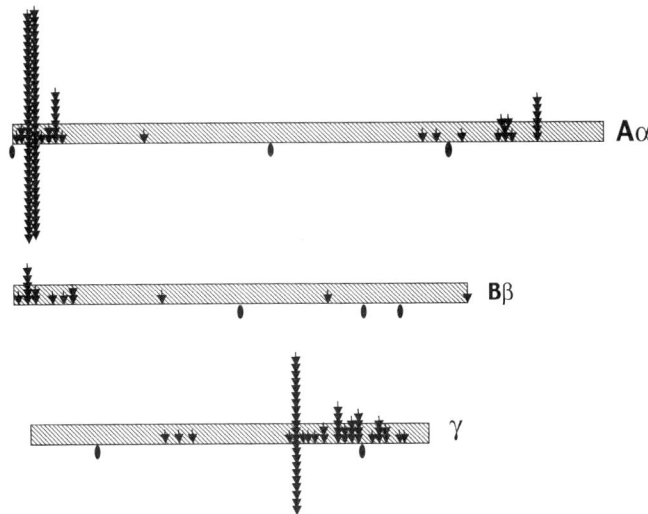

FIGURE 1. Schematic exon gene map of the three expressed human fibrinogen chains. The reported mutations (or mutation initiation) are represented by *triangles* for dysfibrinogenaemias and by *circles* for a/hypofibrinogenaemias. Scale for total exon region and mutation localization is respected.

available on an internet server, like several others for human hemostatic defects (factors VII, VIII, IX, Von Willebrand, antithrombin, protein C, and platelet glycoproteins IIb and IIIa). It is available at the address <http://www.geht.org>, or from the ISTH home page <http://www.med.unc.edu/isth>. Only the collection and patient description of new cases will provide further evidence to allow a better medical appraisal. This should profit both from original mutations, but also from already known mutations, for which a regular publication may not be justified. Therefore, an on line registration of new cases is provided, so that a representative distribution of abnormalities in the population can be expected.

REFERENCES

1. HAVERKATE, F. & M. SAMAMA. 1995. Familial dysfibrinogenemia and thrombophilia: report on a study of the SSC Subcommittee on Fibrinogen. Thromb. Hæmost. **73:** 151–161.
2. MOSESSON, M.W. 1999. Dysfibrinogenemia and thrombosis. Sem. Thromb. Hæmost. **3:** 311–319.
3. MATSUDA, M. 2000. Structure and function of fibrinogen inferred from hereditary dysfibrinogens. Fibrinol. Proteol. **14:** 187–197.
4. CÔTÉ, H.C.F., S.T. LORD & K.P. PRATT. 1998. γ-chain dysfibrinogenemias: molecular structure–function relationships of naturally occurring mutations in the γ chain of human fibrinogen. Blood **92:** 2195–2212.
5. EBERT, R.F. 1994. Index of Variant Human Fibrinogen. CRC Press, Boca Raton.

Molecular Mechanisms of Hypo- and Afibrinogenemia

STEPHEN O. BRENNAN, ANDREW P. FELLOWES, AND PETER M. GEORGE

Molecular Pathology Laboratory, Canterbury Health Laboratories, Christchurch, New Zealand

ABSTRACT: Point mutations responsible for hypo- and afibrinogenemia are yielding new insights into amino acid side chains involved in the molecular processing, assembly, secretion, and domain stability of fibrinogen. Reverse phase chromatography, isoelectric focussing, electrospray mass spectrometry, and tryptic peptide mass mapping have shown that chains with heterozygous mutations of γ284 Gly→Arg, Bβ316 Asp→Tyr and γ371 Thr→Ile are absent from plasma fibrinogen. The nonexpression of these mutations appears to result from perturbation of the five-stranded β sheet of the D domain. We propose that this is due to retention of the variant in the endoplasmic reticulum and that in turn this leads to hypofibrinogenemia. Other mutations effect intracellular proteolysis and chain assembly. For example the mutation, Aα20 Val→Asp, makes the protein a substrate for furin, which removes the first 19 residues of the Aα chain as the mature molecule transits the trans golgi complex. Transient expression of γ153 Cys→Arg chains together with Aα and Bβ chains suggests this mutation might perturb chain assembly, and the incorporation of mutations of Bβ353 Leu→Arg or Bβ400 Gly→Asp into intracellular fibrinogen precludes its subsequent export from host cells expressing fibrinogen genes. The graded severity of the hypo- and afibrinogenemias associated with homozygous Aα chain truncations suggest the absolute minimal requirement for molecular assembly is the formation of the C terminal disulfide ring of the coiled coil.

KEYWORDS: Hypofibrinogenemia; Afibrinogenemia; Mutation; Protein structure.

INTRODUCTION

Classical dysfibrinogenemias result from an impairment of polymerization. They are associated with prolonged thrombin times and low functional, but normal antigenic, fibrinogen concentrations and are usually caused by point mutations in the coding region of one of the three fibrinogen genes. Hypofibrinogenemias, on the other hand, are associated with low functional and antigenic concentrations (less than 1.5 mg/ml) and can result from a variety of different mutations. These can affect either, transcription, mRNA processing, translation, polypeptide chain processing and assembly, export from the hepatocyte, or the stability of the mature protein. Although heterozygosity for any such mutations might reduce fibrinogen

---

Address for correspondence: S.O. Brennan, Molecular Pathology Laboratory, Canterbury Health Laboratories, Christchurch Hospital, P.O. Box 151, Christchurch, New Zealand. Voice: 64 3-3640-549; fax: 64 3-3640-545.

steve.brennan@chmeds.ac.nz

levels below the normal range, they would not usually be expected to produce a significant clinical condition unless inherited in a homozygous, or compound heterozygous state. In this case afibrinogenemia may result, and lead to a serious bleeding condition.

Surprisingly however, a total lack of plasma fibrinogen is typically no more severe than hæmophilia. Afibrinogenemia has been described in over 150 families and is usually detected at birth, when it causes uncontrolled bleeding from the umbilical cord. Spontaneous intracerebral hæmorrhage and splenic rupture can occur throughout life and there is often significant bleeding after minor trauma, but patients respond well to replacement therapy. As in cases of severe hypofibrinogenemia, fibrinogen replacement may be necessary to prevent spontaneous abortion during pregnancy.

The investigation of dysfibrinogenemias and the association of defined mutations with specific functional defects has provided unique insights into the location of functionally important sites in the fibrinogen molecule. Many of these sites have now been portrayed in exquisite detail in the emerging crystal structures. Point mutations responsible for hypofibrinogenemia are also yielding new insights into amino acid residues that are involved in molecular assembly, secretion, and domain stability. In the last year there has been a remarkable expansion in the number of hypo- and afibrinogenemias that have been defined at the molecular level. These cases, which are listed in TABLE 1, are reviewed here in relation to the insights they provide into protein processing and molecular stability.

Mutations Affecting Intracellular Processing

Fibrinogen is synthesized in the liver and transits the export pathway of the hepatocyte before entering the circulation. During this passage the nascent chain, or protein, encounters at least two endoproteases.[1,2] In the endoplasmic reticulum the signal peptidase cotranslationally removes the prepeptide from the N-terminal of all three chains and later, in the golgi/secretory vesicles, a furin like protease cleaves the propeptide from the C-terminal of the Aα chain.[3,4] The specificity of furin is for paired basic residues in either an X Y Arg Arg, or better still an Arg X Y Arg configuration and cleavage normally occurs after the -Arg-Pro-Val-Arg- in the C-terminal region of the Aα chain (see FIGURE 1). Subsequent carboxypeptidase H pruning of the P_1 Arg gives rise to the mature protein with its Val C-terminal. Furin however cannot cleave after dibasic sites that are followed by a large alkyl side chain[5] and although the exposed thrombin susceptible sequence -Arg[16]-Gly-Pro-Arg-Val- conforms to the dibasic motif, the P_1' Val protects the site from furin cleavage. In fibrinogen Canterbury the Aα20 Val→Asp substitution revokes this protection and the first 19 residues of the chain are cleaved within the golgi/secretory vesicles.[4]

Mutations Causing Endoplasmic Retention

The hypofibrinogenemia caused by the Brescia mutation involves defective intracellular transport. This γ284 Gly→Arg substitution causes retention of the protein within the rough endoplasmic reticulum and leads to chronic liver cirrhosis.[6] Ironically, the Brescia patient had a prolonged thrombin time despite the fact that no molecules with mutant chains entered the circulation. This functional defect was

TABLE 1. Mutations that cause hypo- or afibrinogenemia

Mutation[a]	Name	Dose[b]	Level	Ratio	Structure affected	Ref.
γIVS-2 gt→at	Waikato	he	1.1		Terminates before 1st SS ring	21
γ82 Ala→Gly	Dunedin	co	0.80	0.5:1	Affects center of coiled coil	20
γ153 Cys→Arg	Matsumoto	he	0.87	n ex	S-S bridge in γD (γ153–γ182)	17
γ284 Gly→Arg	Brescia	he	0.70	0:1	D-domain 5-stranded β sheet	6
γ371 Thr→Ile	Muncie	he	0.80	0:1	H-bond to 5-stranded β sheet	
Bβ316 Asp→Tyr	Hamilton	he	1.10	0:1	D-domain 5-stranded β sheet	15
Bβ353 Leu→Arg		he	1.3	n ex	Distal P region of D-domain	18
		ho	<0.00			
Bβ400 Gly→Asp		he	1.5		Distal P region of D-domain	18
		ho	0.02	n ex		
Aα codon-18 ins c		co	0		Terminates in prepeptide	23
Aα IVS-1 gta→gtg					Prepeptide	23
Aα20 Val→Asp	Canterbury	he	1.3	0.7:1	Golgi cleavage at Arg19	4
Aα Δ exons 2–5		ho	0		Δ 11 kb (residues 1–625)	22
Aα125 Δaa		co	0		Terminates in coiled coil	23
Aα149 Arg→.		ho	0		Terminates in coiled coil	28
Aα IVS-4 gt→tt		ho	0		Terminates in coiled coil	23
Aα268 Arg→QEP.	Otago[c]	ho	0.1		Truncates in αC after 270	27
		he	1.6	0:1		
Aα297 Gly→.		co	0		Terminates in αC after 296	23
Aα315 Trp→.		co	0		Terminates in αC after 314	23
Aα333 Pro→L...H.		co	0		Terminates in αC after 394	23
Aα452 Gly→WS.	Milano III[c]	ho	2.6		Terminates in αC after 453	26
		he	2.8	0.1:1[d]		
Aα461 Lys→.	Marburg[c]	ho	0.6		Truncates in αC after 460	25
		he	1.9	0.1:1[d]		
Aα476 Met→HCLA.	Lincoln[c]	he	3.0	0.2:1	Truncates in αC after 479	24

[a]Residue numbering is from the mature N-terminal. Nucleotide substitutions are represented in lower case whereas amino acids are in upper case. The character (·) is used to denote a termination codon.
[b]*he* (heterozygous), *ho* (homozygous), *co* (compound heterozygous).
[c]These truncations also cause dysfibrinogenemia.
[d]Estimated from published SDS acrylamide gels. *Ratio*, ratio of variant to normal chain in plasma fibrinogen; *n ex*, not expressed in transfected cell supernatant; *level*, antigenic fibrinogen concentration (mg/ml).

	P_6	P_5	P_4	P_3	P_2	P_1	P_1'	P_2'	P_3'
profibrinogen[606-614]	Lys	Ser	Arg	Pro	Val	Arg	Gly	Ile	His
proprothrombin	Leu	Gln	Arg	Val	Arg	Arg	Ala	Asn	Thr
profactor X	Leu	Gln	Arg	Val	Thr	Arg	Ala	Asn	Ser
proalbumin	Arg	Gly	Val	Phe	Arg	Arg	Asp	Ala	His
albumin Blenheim*	Arg	Gly	Val	Phe	Arg	Arg	**Val**	Ala	His
fibrinogen Aα[14-22]*	Gly	Val	Arg	Gly	Pro	Arg	Val	Val	Glu
fib Canterbury	Gly	Val	Arg	Gly	Pro	Arg	**Asp**	Val	Glu
fibrinogen α_E[606-614]*	Lys	Ser	Arg	Pro	Val	Arg	Asp	Cys	Asp

FIGURE 1. Amino acid sequence of representative human proproteins processed at dibasic motifs in liver golgi/secretory vesicles. Cleavage occurs after either X Y R R or R X Y R sites unless followed by a large alkyl residue. The 1Asp→Val mutation in proalbumin Blenheim abolishes a cleavage site and the converse Aα20 Val→Asp mutation creates a new site in fibrinogen Canterbury. * sequences refractile to furin like proteases, the non cleavage of α_E chains is possibly due to a disulfide bond to the P_2' Cys.

subsequently found to be due to hypersialylation of structurally normal Bβ and γ chains;[6] with the increased sialylation presumably resulting from the chronic liver disease.[7] Heterozygosity for the mutation is sufficient to produce hepatic inclusion bodies and, although two family members have now died from cirrhosis, six other carriers have hypofibrinogenemia, but no obvious signs of liver diseases. This may however be like Z antitrypsin deficiency, where there are similar inclusion bodies, but only about 15% of homozygous ZZ individuals develop significant liver injury.[8] Gly 284 is absolutely conserved in all fibrinogen chains and all homologous molecules in all species.[9] It lies at the end of the five-stranded β sheet in the γD domain (see FIGURE 2). The fifth strand starts at γ280 Tyr and extends to Ala 286.[10,11] It appears that this strand might be quite mobile, since in the double D structure residues 280–284 appear as a random coil.[12]

The β sheet structure is reminiscent of the A sheet in antitrypsin that goes from a metastable-five to a stable-six stranded structure on cleavage within its reactive center loop.[13] The Z mutation is at the hinge point of this loop sheet insertion and the 342Glu→Lys substitution provokes intermolecular loop to sheet insertion leading to antitrypsin Z fibril formation and the blockage of secretion.[14] Like antitrypsin, the fibrinogen molecule is in a metastable state since its D:D polymerization sites preexist, and do not require, thrombin activation. The γ284 Gly→Arg mutation may perturb the β sheet array and facilitate intramolecular strand insertions, leading to fibrinogen fibril formation and the blockage of its secretion. Certainly the EM images of Fibrinogen Brescia inclusions[6] show a degree of order similar to Z antitrypsin fibrils.

Although the fibrinogen Hamilton mutation[15] involves the Bβ chain, the mechanism of hypofibrinogenemia may be similar to that of fibrinogen Brescia since the Bβ316 Asp→Tyr mutation occurs at the end of a strand in the five-stranded sheet of the Bβ chain. The structural importance of the residue is again evidenced by its high interchain and interspecies conservation. The corresponding γ chain position, Asp252, and its location within strand 3 is shown in FIGURE 2. Heterozygotes for the mutation have no variant Bβ chains in their plasma fibrinogen, but no biopsy material was available to probe for inclusion bodies that would confirm a blockage of secretion. It is however entirely probable that a number of different mutations might cause hepatic retention since Callea has described four different morphological types of fibrinogen inclusion bodies in liver biopsies.[16]

Mutations Affecting Intracellular Assembly

The Matsumoto IV (γ153 Cys→Arg) mutation affects the proximal A region of the γD domain.[17] Transient expression of 153 Arg γ chains with Aα and Bβ chains in Chinese hamster ovary cells demonstrated that, although the variant chain was synthesized, no variant fibrinogen molecules were secreted into the culture medium. Pulse chase experiments further showed that neither assembly intermediates (αγ or βγ dimers) nor mature fibrinogen were produced in cells expressing the variant γ chain suggesting an impairment of chain assembly. Since γ153 Cys forms an inter chain S-S bond to γ182 Cys it appears that either, γ153 Cys has a direct role in

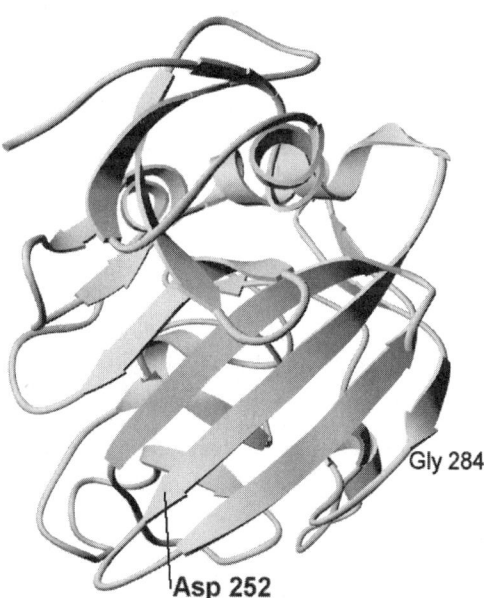

FIGURE 2. Central region of the γD domain showing the five-stranded β sheet structure. The location of the Brescia γ284 Gly→Arg mutation is indicated, as is γ252 Asp. The latter position is equivalent to the Bβ316 site of the Hamilton Bβ316 Asp→Tyr mutation.

complex formation or, that the freed thiol at γ182 impairs the formation of S-S bonds in the αγ or βγ complexes.[17]

The Bβ353 Leu→Arg and Bβ400 Gly→Asp mutations[18] are located in the distal P region of the βD domain.[11] Heterozygosity for either lowers fibrinogen levels by approximately 50%, whereas homozygosity produces afibrinogenemia. The Bβ400 homozygotes do, however, have some residual plasma fibrinogen. The Bβ353 site is conserved as a large hydrophobic residue (Leu, Phe, or Met) between species, chains, and fibrinogen-like proteins, whereas the Bβ400 residue is similarly conserved as a small side chain (Gly or Ser). Both residues are located in the hydrophobic core of the subdomain and both substitutions introduce bulky charged side chains into this hostile environment. The charge and steric effects of these changes would be expected to distort the packing of the region to result in protein instability, as occurs in the unstable hemoglobin Hb Volga (β27Ala→Asp).[19] Cotransfection in COS-1 cells expressing normal α and γ chains showed that, although the variant β chains were synthesized and assembled into fibrinogen molecules, no fibrinogen reached the culture media.

Mutations Expressed in Circulatory Fibrinogen

The Dunedin mutation (γ82 Ala→Gly) is actually expressed in the circulatory fibrinogen of heterozygotes, albeit at a reduced ratio (1:2) to γ^A chains.[20,21] The lesion occurs near the plasmin sensitive site in the coiled coil connecting the E and D domains. Glycine is grossly under-represented in the triple helix, accounting for only three of 337 residues and destabilized coil packing together with increased proteolytic sensitivity may account for the decreased fibrinogen levels.

Truncations of the Aα Chain

Partial Aα gene deletions[22] and Aα splice site mutations[23] are the most frequent cause of afibrinogenemia. The Aα intron-4 3′-splice sight mutation (gt→tt) is the single most common cause, having been found homozygously in five unrelated individuals with afibrinogenemia and as a compound heterozygous allele in a further four cases.[23] The 11-kb deletion of exons 2–5 (which removes all the coding sequence of the mature chain) also causes afibrinogenemia when inherited homozygously[22] or together with the Aα intron-4 mutation.[23]

All 14 Aα mutations in TABLE 1 result in a shortening of the predicted mature protein. Apart from fibrinogen Canterbury with its N-terminal cleavage, the remaining 11 involve C terminal truncations ranging from molecules with only the prepeptide to the substantially complete 479-residue chains of fibrinogen Lincoln. These mutations help define a minimal Aα chain length necessary for assembly and secretion (FIG. 3).

Although heterozygotes for the Lincoln truncation after residue 479 do not actually have hypofibrinogenemia, the low ratio (0.2:1) of variant to $A\alpha^A$ chains suggests that Lincoln chains compete less well than $A\alpha^A$ chains when forming complexes with other chains, or that they are degraded intracellularly.[23] Heterozygotes for the Marburg[25] truncation after residue 460 do not have hypofibrinogenemia either, but again they do have a low expression ratio (0.1:1) and homozygotes have marked hypofibrinogenemia with plasma levels of 0.6 mg/ml (TABLE 1). Surprisingly

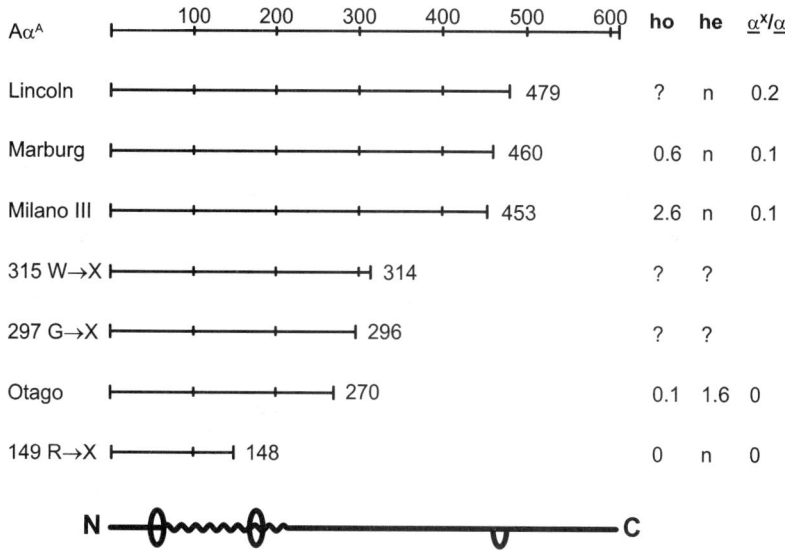

FIGURE 3. Aα chain truncations showing relative length of chain in relation to the disulfide rings. Homozygote (*ho*) and heterozygote (*he*) fibrinogen concentrations are shown, together with the estimated ratio of normal to variant chain (α^X/α^A) in the heterozygotes.

however, the Milano III homozygote with an earlier termination (after residue 453) had a normal fibrinogen concentration. The Milano III heterozygotes, however, appeared very like the Marburg and Lincoln heterozygotes in that they also had very few variant chains incorporated into their plasma fibrinogen.

When coinherited with another defective Aα allele, truncations after 394, 341, and 296 induce afibrinogenemia,[23] but the shorter truncation at 270 does not. The homozygote for the latter (fibrinogen Otago) had a plasma fibrinogen level of 0.1 mg/ml, whereas her heterozygous son was only marginally hypofibrinogenemic with concentrations measured at 1.5 and 1.6 mg/ml on two occasions. Notably, however, the son had no detectable variant chains in his plasma fibrinogen,[27] again suggesting that the variant chain is either degraded or out-competed by Aα^A chains during molecular assembly. The five homozygous mutations that cause truncation before the C-terminal S-S ring totally prevent the expression of fibrinogen in plasma,[22,28] indicating that formation of the coiled coil is an absolute minimal requirement for successful assembly and secretion.

DISCUSSION

Seventeen of the 22 mutations in TABLE 1 were reported within the last year and although advances in DNA analysis have made mutation detection simpler, it is not always clear whether or not the new mutation actually causes the presenting phenotype. Family studies showing segregation of the mutation with the disease, together with allele frequency analysis showing exclusion from the general population are essential prerequisites to establishing cause and effect. For example, whereas four new DNA mutations were identified in a fibrinogen Dunedin (γ82 Ala→Gly) patient, two non-coding mutations were excluded by their occurrence in an unaffected daughter, and the Bβ235 Pro→Leu substitution was excluded when it was found at a polymorphic frequency (5%) in the community.[19]

Where possible, additional protein and/or coexpression studies should be undertaken to determine if (and at what level) the predicted new chain is actually incorporated into plasma or culture media fibrinogen. Again in the Dunedin example, protein analysis showed equal amounts of variant (Bβ235 Pro→Leu) and normal Bβ chains in plasma fibrinogen, but reverse phase analysis indicated a decreased ratio of variant to normal γ chains suggesting the γ82 Ala→Gly mutation as the cause of the hypofibrinogenemia. When there is a well-established mechanism for diminished expression, such the introduction of a stop codon, or a change in a 3′ (gt) or 5′ (ag) mRNA splice site,[21,23] these additional investigations may not be necessary. However, when a single amino acid substitution is identified as a putative cause of hypofibrinogenemia, further protein or expression studies are indicated. In these cases fibrinogen chain composition can be examined directly by SDS PAGE, reverse phase HPLC, isoelectric focussing, or by electrospray or matrix assisted laser desorption mass spectrometry.[6,15,20,29] If these prove inconclusive, then various peptide mapping procedures can be employed. In these cases mass spectrometry offers a major advantage over traditional mapping methods because it is possible to predict the exact m/z location of both the variant and normal peptides, permitting a definitive conclusion concerning expression level.[15,20]

Caution is needed in assessing how a DNA mutation might induce hypofibrinogenemia. For example: (1) Point mutations in exons, as well as introns, might activate cryptic splice sites. (2) Nonsense mediated mRNA decay suggests some nonsense mutations may primarily affect mRNA stability rather than protein stability.[30] (3) Standard PCR methods might not detect large deletions. (4) Point mutations might introduce novel cleavage sequences for intracellular as well as extracellular proteolysis. Mutation nomenclature is another concern; although it is quite legitimate to express base changes in terms of the Genbank sequence, confusion arises when this is directly translated into a protein location. For example, the G316X and W334X mutations reported by Neerman-Arbez et al.[23] are numbered from the start of the signal peptide and have been redesignated G297X and W315X in TABLE 1 to indicate their locations in the mature circulatory protein.

Although heterozygosity for a particular mutation might reduce fibrinogen levels below the normal range, it would not usually produce a significant clinical condition unless inherited in a homozygous or compound heterozygous state. There are, however, exceptions; heterozygosity for the Brescia mutation can lead to liver cirrhosis and heterozygous Aα amyloidotic mutations provoke fatal renal disease.[31] The latter, however, are not usually associated with hypofibrinogenemia. Gene deletions, early stop codons, and splice site mutations would be expected to have a lesser impact on fibrinogen levels than amino acid substitutions. Total non-expression of a gene should only decrease chain availability by 50%, however, the random incorporation of a variant chain into mature fibrinogen would potentially affect 75% of molecules. For example, in fibrinogen Brescia the mean fibrinogen levels in seven carriers of the γ284 mutation was 0.70 mg/ml, but nine unaffected family members had an average level of 2.4 mg/ml.

REFERENCES

1. BRENNAN, S.O., et al. 1990. Albumin Redhill (-1 Arg, 320 Ala→Thr). A glycoprotein variant of human albumin whose precursor has an aberrant signal peptide cleavage site. Proc. Natl. Acad. Sci. U.S.A. **87:** 26–30.
2. BRENNAN, S.O. & K. NAKAYAMA. 1994. Furin has the proalbumin substrate specificity and serpin inhibitory properties of an *in situ* hepatic convertase. FEBS Lett. **338:** 147–151.
3. FARRELL, D., et al. 1993. Processing of the carboxyl 15 amino acid extension in the Aα of fibrinogen. J. Biol. Chem. **268:** 10351–10355.
4. BRENNAN, S.O., et al. 1995. Aberrant hepatic processing causes removal of activation peptide and primary polymerisation site from fibrinogen Canterbury (Aα20 Val→Asp). J. Clin. Invest. **96:** 2854–2858.
5. BRENNAN, S.O., et al. 1989. Novel human proalbumin variant with intact dibasic sequence facilitates identification of its converting enzyme. Biochim. Biophys. Acta **993:** 48–50.
6. BRENNAN, S.O., et al. 2000. Fibrinogen Brescia; hepatic endoplasmic reticulum storage and hypofibrinogenemia because of a γ284 Gly→Arg mutation. Am. J. Pathol. **157:** 189–196.
7. MARTINEZ, J., et al. 1983. The role of sialic acid in the dysfibrinogenima associated with liver disease, distribution of sialic acid on the constituent chains. Blood **61:** 1196–1202.
8. PERLMUTTER, D.H. 1998. Alpha-1-antitrypsin deficiency. Seminars in Liver Disease **18**(3): 217–225.
9. http://brie.bmsc.washington.edu/people/teller/
10. SPRAGGON, G., et al. 1997. Crystal structures of fragment D from human fibrinogen and its crosslinked counterpart from fibrin. Nature **389:** 455–462.
11. YEE, V.C., et al. 1997. Crystal structure of a 30 kDa C-terminal fragment from the γ chain of human fibrinogen. Structure **5:** 125–138.
12. SPRAGGON, G., et al. 1998. Crystal structure of a recombinant αEC domain from human fibrinogen-420. Proc. Natl. Acad. Sci. U.S.A. **95:** 9099–9104.
13. CARRELL, R.W. & P.E. STEIN. 1996. The biostructural pathology of the serpins: critical function of sheet opening mechanism. Biol. Chem. Hoppe-Seyler **377:** 1–17.
14. LOMAS, D.A., et al. 1992. The mechanism of Z antitrypsin accumulation in the liver. Nature **357:** 605–607.
15. BRENNAN, S.O., et al. 2001. Hypofibrinogenemia due to novel 316 Asp→Tyr substitution in the fibrinogen Bβ chain. Thromb. Hæmost. In press.
16. CALLEA, F., et al. 1992. Hepatic endoplasmic reticulum storage diseases. Liver **12:** 357–362.
17. TERASAWA, F., et al. 1999. Hypofibrinogenemia associated with a heterozygous missense mutation γ153Cys to Arg (Matsumoto IV): *in vitro* expression demonstrates defective secretion of the variant fibrinogen. Blood **94:** 4122–4131.
18. DUGA, S., et al. 2000. Missense mutations in the human β fibrinogen gene cause congenital afibrinogenemia by impairing fibrinogen secretion. Blood **95:** 1336–1341.
19. OCKELFORD, P.A., et al. 1980. Hemoglobin Volga β27 (B9) Ala→Asp. Functional and clinical correlations of an unstable hemoglobin. Hemoglobin **4:** 295–306.
20. BRENNAN, S.O., et al. 2000. Hypofibrinogenemia in an individual with two coding (γ82 A→G and Bβ235 P→L) and two non-coding mutations. Blood **95:** 1709–1713.
21. WYATT, J., et al. 2001. Hypofibrinogenæmia with compound heterozygosity for two γ chain mutations; γ82 Ala→Gly and an intron two GT→AT splice site mutation. Thromb. Hæmost. **95:** 1709–1713.

22. NEERMAN-ARBEZ, M., *et al.* 1999. Deletion of the fibrinogen α chain gene (FGA) causes congenital afibrinogenæmia. J. Clin. Invest. **103:** 215–218.
23. NEERMAN-ARBEZ, M., *et al.* 2000. Mutations in the fibrinogen Aα gene account for the majority of cases of congenital afibrinogenemia. Blood **96:** 149–152.
24. RIDGWAY, H., *et al.* 1996. Fibrinogen Lincoln: a new truncated α chain variant with delayed clotting. Brit. J. Hæmatol. **93:** 117–184.
25. KOOPMAN, J., *et al.* 1992. Fibrinogen Marburg: a homozygous case of dysfibrinogenemia, lacking amino acids Aα 461–610 (Lys461 AAA→Stop TAA). Blood **80:** 1972–1979.
26. FURLAN, M., *et al.* 1994. A frameshift mutation in exon V of the Aα gene leading to truncated Aα chains in the homozygous dysfibrinogenemia Milano III. J. Biol. Chem. **269:** 33129–33134.
27. RIDGWAY, H.J., *et al.* 1997. Fibrinogen Otago: a major α chain truncation associated with severe hypofibrinogenæmia and recurrent miscarriage. Br. J. Hæmatol. **98:** 632–639.
28. FELLOWES, A.P., *et al.* 2001. Homozygous truncation of the fibrinogen Aα chain within the coiled coil causes congenital afibrinogenæmia. Blood. **96:** 773–775.
29. BRENNAN, S.O., *et al.* 2000. Defective fibrinogen polymerisation associated with novel γ279 Ala→Asp mutation. Brit. J. Hæmatol. **108:** 236–240.
30. ANDERSON, B.D., *et al.* 1996. Two mutations in exon XII of the protein S alpha gene in four thrombophilic families resulting in premature stop codons and depressed levels of mutated mRNA. Thromb. Hæmost. **76:** 143–50.
31. UEMICHI, T., *et al.* 1996. A frameshift mutation in the fibrinogen Aα chain gene in a kindred with renal amyloidosis. Blood **87:** 4197–4203.

Insight from Studies with Recombinant Fibrinogens

SUSAN T. LORD[a] AND OLEG V. GORKUN[b]

[a]*Departments of Pathology and Laboratory Medicine, and Chemistry, Curriculum in Genetics and Molecular Biology, University of North Carolina at Chapel Hill, Chapel Hill, North Carolina, USA*

[b]*Department of Pathology and Laboratory Medicine, University of North Carolina at Chapel Hill, Chapel Hill, North Carolina, USA*

ABSTRACT: Using a two-step cloning strategy, we have synthesized more than 20 variant human fibrinogens for biochemical studies. In preliminary experiments we showed that normal fibrinogen produced in CHO cells serves as an accurate model for plasma fibrinogen. We focus here on those variants whose characterization has provided insight into the mechanism of thrombin-catalyzed polymerization. Analysis of N-terminal variants showed that thrombin specificity dictates the ordered release of fibrinopeptides. Nevertheless, analysis of C-terminal variants indicated that fibrinopeptide B (FpB) release is dependent on polymerization. Changes in the *a* polymerization site and the high-affinity calcium-binding site were associated with a complete loss of polymerization. These experiments showed that alterations in the calcium-binding site influenced function of the *a* site; in contrast, alterations in the *a* site did not alter calcium binding. Analysis of variants in the N-terminus of the Bβ chain provided the first direct evidence that this region impacts predominantly on lateral aggregation, as has long been presumed. These experiments also suggested that lateral aggregation facilitated by this region proceeds without the release of FpB. From these studies we learned that individual sites within fibrinogen do not function in isolation. We conclude that thrombin-catalyzed polymerization is mediated by a continuum of concerted interactions.

KEYWORDS: Recombinant; Fibrinogen; Fibrin; Polymerization; Lateral aggregation; Fibrinopeptide; Clot.

INTRODUCTION

We described the successful synthesis of fully assembled, functional recombinant human fibrinogen in cultured Chinese Hamster Ovary cells in 1993.[1] Using this system, we have been able to produce milligrams of recombinant fibrinogen, sufficient for efficient purification and subsequent biochemical analysis. By incorporating a two-step cloning strategy, this system permits the expedient synthesis of new variant fibrinogens.

Address for correspondence: Susan T. Lord, Ph.D., Department of Pathology and Laboratory Medicine, University of North Carolina, CB#7525, Brinkhous-Bullitt 603, Chapel Hill, NC 27599-7525, USA. Voice: 919-966-3548; fax: 919-966-6718.
stl@med.unc.edu

An important preliminary experiment for analysis of recombinant fibrinogen variants was the comparison of normal recombinant fibrinogen to normal plasma fibrinogen.[2] Plasma fibrinogen is a heterogeneous mixture of molecules due to variations in mRNA processing, posttranslational modification and proteolysis in the circulation.[3] Because the significance of this heterogeneity is not known, it was necessary to compare the recombinant and plasma proteins for each biochemical trait under study. In multiple experiments, we have found that in essence these molecules are the same. Transmission electron microscopy studies showed trinodular structures for both fibrinogens, and scanning electron microscopy studies showed that the structures of the two fibrin clots were indistinguishable. We found that thrombin-catalyzed fibrinopeptide release, FXIIIa-catalyzed cross-link formation, and fibrin monomer polymerization were remarkably similar for both proteins. We saw small differences in thrombin-catalyzed polymerization, suggesting that the ordered assembly of protofibrils and fibers was not identical. We speculate that these subtle differences reflect the heterogeneity of plasma fibrinogen. We conclude that the normal recombinant fibrinogen produced in CHO cells serves as an accurate model for plasma fibrinogen, and thus is an appropriate control for studies with recombinant variant fibrinogens.

We have subsequently synthesized and characterized more than 20 variant fibrinogens. We designed our variants to test implications based on published data. Our results have demonstrated that the availability of planned variant fibrinogens provides a novel approach to examine current ideas relevant to the many functions of fibrinogen. A list of the variants that have been described, including those first described at this conference, is presented in TABLE 1. In this communication we focus on variants made at sites that have provided insight into the details of thrombin-catalyzed polymerization: sites that contribute to thrombin specificity and sequential fibrinopeptide release; sites that contribute to protofibril formation and high-affinity calcium binding; a site that contributes to D:D interactions; and a site that contributes to lateral aggregation.

TABLE 1. Described variant fibrinogens

Aα chain	Bβ chain	γ chain
F8Y[8]	R14H[a]	D318A[a]
G14V[9]	A′β[9]	D320A[a]
D97E[b27]	A68T[28]	D318A,D320A[a]
D574E[b27]	P70S[29]	Δ319,320[18]
Aα251[30]	L72S[29]	Y363A[14]
	BβH16A,R17A,P18A[a]	D364A[14]
		D364H[14]
		Δ408–411[31]

NOTE: In addition to the changes in single chains, we have also examined one variant with changes in two chains, AαD97E,D597E,$\gamma\Delta$408-411.[27]

[a]Described in this paper.
[b]Synthesized in BHK cells as described by Farrell et al.[32]

In most instances we chose to synthesize variants with the intention to interrupt one specific interaction, and thereby one specific function. Instead we found that these variants influence multiple activities. From these studies we learned something that might be considered intuitively obvious from the beginning: because they are joined in a single molecule, individual sites do not function in isolation. Thus, the steps in thrombin-catalyzed polymerization are a continuum and only a very few, if any, individual steps can be examined alone.

RESULTS

Thrombin Specificity and Sequential Fibrinopeptide Release

Thrombin catalyses the hydrolysis of four peptide bonds in fibrinogen releasing two fibrinopeptides A (FpA) and two fibrinopeptides B (FpB) from the N-termini of the Aα and Bβ chains, respectively. The release of FpA follows simple first order kinetics.[4] The release of FpB is more complex, but can be described as a first order process under the assumption that FpB release occurs subsequent to FpA release.[4] The efficient release of FpB has been linked to fibrin polymerization, in part because the release of FpB is slower in the presence of the peptide GPRP, which inhibits polymerization.[5] To examine the basis for the sequential release of fibrinopeptides, and the correlation of sequential release with polymerization,[6] we synthesized four variants. To reverse the ordered release, such the FpB release precedes FpA release, we synthesized AαF8Y fibrinogen where Phe 8 in the Aα chain is replaced with Tyr. To eliminate the ordered release, such that all four peptides are released concomitantly, we synthesized A′β fibrinogen, where we replaced FpB with an FpA-like substrate, FpA′ where Gly 14 in FpA is replaced by Val; we also synthesized the required control A′α fibrinogen, with FpA′ replacing FpA. To exaggerate the ordered release, such that FpB release is substantially delayed, we synthesized BβR14H fibrinogen where Arg14 in the Bβ chain is replaced with His.

Based on prior results,[7] we anticipated that the AαF8Y substitution would impair thrombin-catalyzed hydrolysis of the Aα Arg16-Gly17 bond. Indeed, kinetic analysis demonstrated that AαF8Y1-16 was a very poor thrombin substrate, with a specificity constant 280-fold lower than the normal fibrinopeptide.[8] These experiments confirmed that Phe8 plays a critical role in thrombin-catalyzed FpA release. The release of FpB was also impaired, but less so than FpA, such that FpB release preceded FpA release. Thrombin-catalyzed polymerization of AαF8Y fibrinogen, monitored by the change in turbidity at 350 nm, was markedly delayed relative to normal. These data suggest that exposure of the N-terminus of the α chain is required for normal polymerization. Thus, FpA release occurs first, and then polymerization begins.

Also based on prior results,[7] we anticipated that FpA′ release from the N-termini of the β chains of A′β fibrinogen would occur concomitantly with FpA release from the α chains. In control experiments with Gly14 replaced with Val in the Aα chain, we found that this variant fibrinopeptide was a good thrombin substrate, with release of FpA′ essentially identical to FpA.[9] Kinetic analysis of thrombin-catalyzed fibrinopeptide release from A′β fibrinogen showed that the release of both peptides followed simple first order kinetics, and that the specificity constants for both substrates were similar. We concluded that the substrate specificity directs the ordered release of FpA prior to FpB. Thrombin-catalyzed polymerization of A'β fibrinogen was the same as normal fibrinogen, and the two clot structures, examined by scanning electron microscopy, were indistinguishable. We concluded that early exposure of the N-terminus of the β chain does not affect polymerization. We also examined the release of fibrinopeptides in the presence of 1 mM GPRP. Consistent with previous reports,[5] we found that the rate of FpB release from normal fibrinogen was threefold slower in the presence of GPRP. In contrast, the rate of FpA' release from A'β fibrinogen was unaffected. This remarkable result indicates that it is the structure of the fibrinopeptide *per se* that links the influence of GPRP to both arrested polymerization and delayed FpB release. Further experiments are clearly needed to appreciate the link between FpB release and polymerization.

Based on the analysis of dysfibrinogen Petrosky, AαR16H fibrinogen,[10] we anticipated that the thrombin-catalyzed release of the variant fibrinopeptide from BβR14H fibrinogen would be dramatically slower than the release of FpB from normal fibrinogen. Indeed, the rate was reduced 300-fold. Thrombin-catalyzed polymerization of BβR14H fibrinogen was also significantly impaired, with a lower final turbidity than normal, as expected when lateral aggregation is impaired. This result suggests that the timely release of FpB is critical for normal lateral aggregation. Further experiments, however, did not support this suggestion. We examined Batroxobin-catalyzed fibrinopeptide release and fibrin polymerization. Neither the

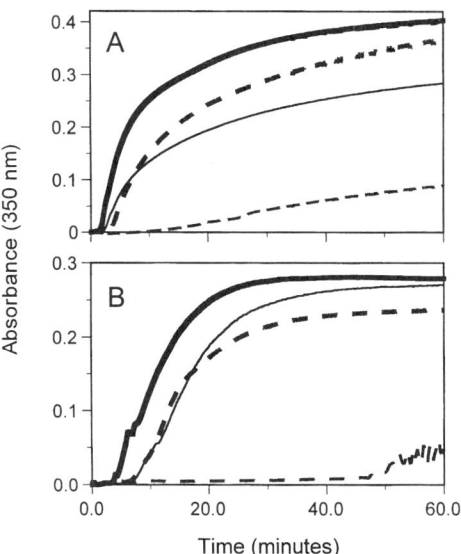

FIGURE 1. Representative polymerization curves. The change in turbidity with time of normal (*solid lines*) and BβR14H (*dashed lines*) fibrinogen with 0.2 mg/mL fibrinogen, 0.1 U/mL enzyme in 20 mM HEPES, pH 7.4, and ionic strength held constant at 0.15 with NaCl. *Thin lines* represent polymerization in the absence of added calcium and *thick lines* represent polymerization with 10 mM $CaCl_2$. **Panel A**, thrombin-catalyzed polymerization. **Panel B**, batroxobin-catalyzed polymerization. From Reference 11.

normal nor the variant FpB was released by Batroxobin. However, the polymerization curves for these fibrinogens were different. Moreover, as is shown in FIGURE 1,[11] the magnitude of this difference varied with the calcium concentration, such that at mM calcium the variant and normal curves were quite similar. We concluded that the N-terminus of the variant Bβ chain participates in polymerization in a calcium-dependent association, probably during lateral aggregation. This result suggests that the N-terminus of the Bβ chain normally participates in lateral aggregation, and that the normal association is calcium-dependent.

Protofibril Formation and the Influence of Calcium

The formation of half-staggered, two-stranded protofibrils is mediated through *A:a* interactions, between the *A* site in the central domain exposed by FpA release and *a* site in the C-terminal domain of the γ chain. Recent high-resolution structural data of the C-terminal domain with bound GPRP, an *A* site analog, have identified specific *A:a* interactions, including the side chain of Asp364.[12,13] The structural data also identified the residues that compose the high-affinity calcium binding site, including the side chains of Asp318 and Asp320. These two binding sites are juxtaposed in the C-terminal domain. We have synthesized three variants with substitutions in the *a* site, Y363A fibrinogen where Tyr363 is replaced with Ala, D364A and D364H fibrinogens where Asp364 is replaced with either Ala or His; and four variants with substitutions in the calcium binding site, Δ319,320 fibrinogen where Asn319 and Asp 320 have been deleted; D318A fibrinogen where Asp 318 is replaced with Ala; D320A fibrinogen where Asp 320 is replaced with Ala; and D318A,D320A fibrinogen where both Asp318 and Asp320 are replaced with Ala.

We focused our *a* site studies,[14] which were initiated prior to the availability of the high resolution structure, on Tyr γ363, identified by photoaffinity labeling of fibrinogen with a GPRP analog,[15] and Asp γ364, identified in the dysfibrinogen Matsumoto I.[16] As is shown in FIGURE 2, thrombin-catalyzed polymerization of variants with substitutions at either site was significantly impaired, but polymerization was evident for both Y363A and D364A. In contrast, with D364H there was no change in turbidity under these conditions. These results indicated that the *A:a* interactions were compromised by the two Ala substitutions and eliminated by the His substitution. Recently, we have used plasmin protection assays to independently assess function of the *a* site in the γ364 variants. As shown by Lounes *et al.* (these proceedings), plasmin digestion of either D364A or D364H was not changed by the presence of 1 mM GPRP; in contrast, as previously reported,[17] plasmin cleavage of normal fibrinogen was limited by the presence of GPRP. This result confirmed that the *a* site was defective in both variants. The polymerization and plasmin protection experiments together suggest that both Ala and His substitutions cripple the *A:a* interaction, but the D364H substitution has more profound consequences than the D364A substitution. We conclude that multiple interactions, including protofibril formation, were impacted by the histidine substitution. In contrast, whereas normal *A:a* interactions may be lost with the alanine substitution, remaining interactions were able to support polymer formation. Further experiments are needed to characterize the differences between these variants.

We initiated studies of the calcium binding domain by synthesizing a variant analogous to the dysfibrinogen Vlissingen/Frankfurt IV.[18] Using turbidity at 350 nm,

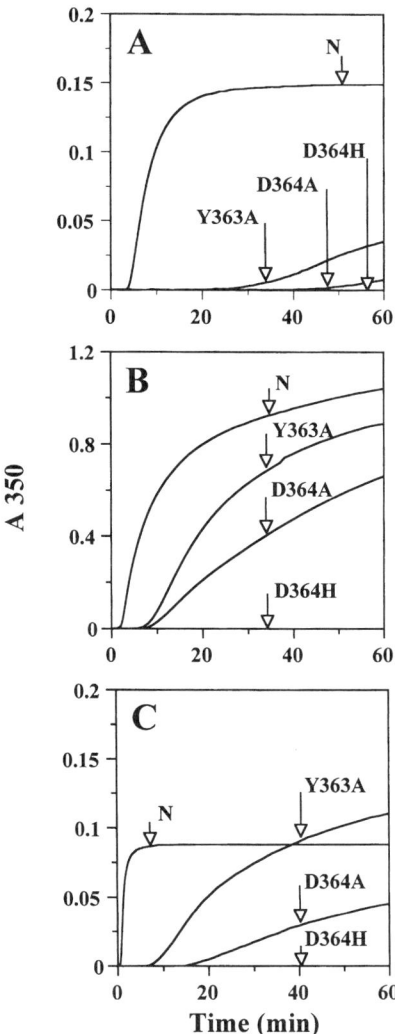

FIGURE 2. Thrombin-catalyzed fibrinogen polymerization. The change in turbidity over time with normal recombinant fibrinogen (N) and three γ chain variant fibrinogens, Y363A, D364A, and D364H. Thrombin (**Panels A** and **B**, 0.05 unit/ml; **Panel C**, 0.5 unit/ml) was added at 0 min to fibrinogen at 0.09 mg/ml (**Panels A** and **C**) or to fibrinogen at 0.45 mg/ml (**Panel B**). From Reference 14.

we have never detected polymerization with this variant, even when the thrombin concentration was increased to 10 U/ml, the fibrinogen concentration increased to 1.3 mg/ml, the calcium concentration increased to 10 mM, the temperature lowered to 14°C, or the experiment monitored for 14 hours. Using dynamic light scattering, we confirmed that this variant fibrin does not polymerize. The experiments shown in FIGURE 3 were performed under conditions where within a few seconds normal fibrinogen would show a six- to 10-fold increase in mass and a 10-fold drop in normalized diffusion coefficient, consistent with the formation of protofibrils. With γΔ319,320 fibrinogen the mass increased about twofold and the diffusion coefficient dropped about 1.5-fold after four hours consistent with the formation of small aggregates. The formation of small aggregates was also seen by transmission electron microscopy. We concluded that this variant could not form any ordered polymer structure. In additional experiments, we were unable to detect either ADP-induced platelet aggregation or FXIIIa-catalyzed fibrinogen cross-linking with the γΔ319, 320 variant. These results suggest that this two-residue deletion affected the overall structure of the C-terminal domain of the γ chain. Therefore, it is not appropriate to link these multiple changes in function specifically with the alteration at the high-affinity calcium binding site.

We synthesized two single residue substitutions in the calcium binding site, D318A and D320A, because we thought the single changes would not impact the

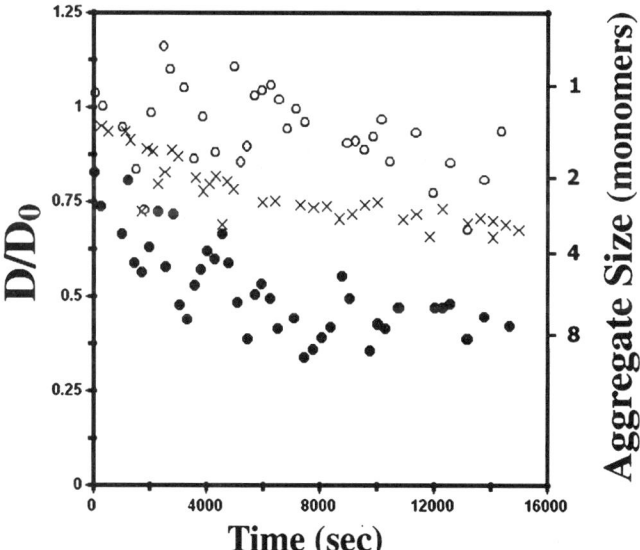

FIGURE 3. Dynamic light scattering. Fractional change in translational diffusion coefficient over time. Thrombin (0.3 U/ml) was added to purified γΔ319,320 fibrinogen (0.2 mg/ml) and monitored for 4 h. D/D_0, the normalized diffusion coefficient, is shown on one ordinate-axis and size distribution (units are monomers of fibrin) are shown on the opposite ordinate-axis. Three reactions are shown. From Reference 18.

overall structure of the C-terminal domain of the γ chain. We also synthesized the variant where both residues are changed, D318A,D320A. In preliminary studies with these variants (Lounes *et al.*, this volume), we found that both D318A and D320A did polymerize, as detected by changes in turbidity, although in both cases polymerization was impaired. No change in turbidity was seen with the D318A,D320A variant. We also examined plasmin digests with these variants in the presence of 1 mM GPRP, 5 mM calcium, and 1 mM EDTA. The results from these plasmin protection studies indicate that the function of the *a* site depends upon the structure and function of the high-affinity calcium binding site. This result contrasts with the inverse interdependence, that is, the structure and function of the high-affinity calcium binding site is independent from the *a* site. We (Lounes *et al.*, this volume) and others[19] have shown in plasmin protection assays with variants at the *a* site, that GPRP protection was lost and calcium protection was normal.

Experiments are in progress to determine whether the D318A and D320A substitutions alter other functions associated with C-terminal domain of the γ chain.

The Connection between FpB Release and Polymerization

We synthesized three variants at residue γ308—N308A, N308I, and N308K—because several dysfibrinogens have been identified at this site. The crystallographic structures show that this residue likely participates in D:D interactions and thereby influences polymerization.[13] In preliminary experiments, we have found that polymerization was impaired with all three variants, but the severity of the defect varied with the specific substitution. Moreover, the differences were markedly dependent on the concentration of calcium. As is shown in FIGURE 4A and B in the absence of added calcium the lag time was longer and the rise in turbidity was slower for all three variants relative to normal. In contrast, in the presence of 1 mM calcium, only the N308K variant remained abnormal; both the lag time and the rise in turbidity were normal for N308A and N308I. These data indicate that the charged substitution disrupts the alignment of monomers in protofibrils, irrespective of calcium concentration. In contrast, the uncharged substitutions apparently influence calcium binding and thereby influence protofibril formation. It is notable that our findings are consistent with data from analysis of the dysfibrinogen N308I, or Baltimore III. This dysfibrinogen was identified as a heterozygous mutation in an asymptomatic woman who had a prolonged thrombin time that was corrected by excess calcium.[20] Thus, the connection between the substitution and the calcium dependent polymerization was detected in both plasma fibrinogen, which likely contained some normal γ chains, and purified recombinant fibrinogen.

Furthermore, we found a similar association between calcium concentration and the kinetics of thrombin-catalyzed FpB release. As is shown in FIGURE 4C, in the absence of added calcium the rate of FpB release was reduced for all three variants. In the presence of 1 mM calcium the rate of FpB release from N308K remained abnormal, whereas FpB release from N308A and N308I was normal. The coincident calcium dependence of the turbidity data and the kinetics of FpB release demonstrate the interdependence of these two events. We conclude that defective alignment of the D:D interface is associated with impaired release of FpB, because the addition of calcium not only normalized polymerization, but also normalized FpB release. Alternatively, it is reasonable to conclude that the Ala and Ile substitutions impaired

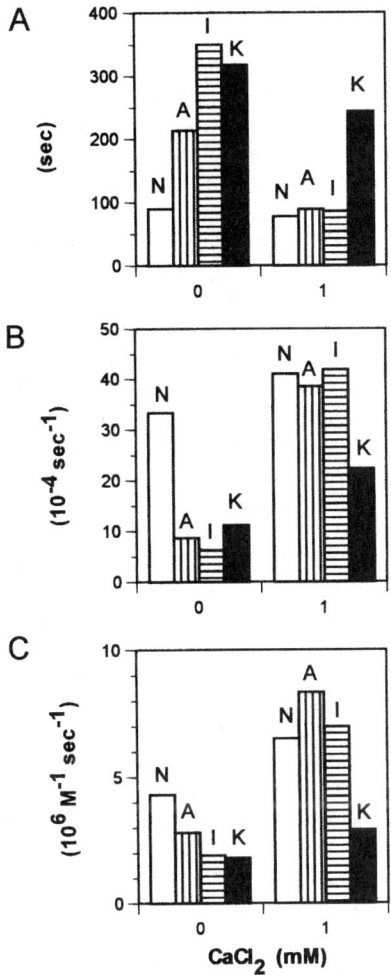

FIGURE 4. Calcium dependence of polymerization and FpB release. The lag time (**Panel A**) and V_{max} (**Panel B**) determined from turbidity curves[2] with normal (N), γN308A (A), γN308I (I), and γN308K (K) fibrinogens at 0.4 mg/ml fibrinogen, 0.05 U/ml thrombin in 20 mM HEPES, pH 7.4, in the absence of added calcium (0) or with 1 mM $CaCl_2$ (1) and 0.12 M NaCl. The specificity constants (**Panel C**) determined[2] from progress curves of fibrinopeptide release with normal (N), γN308A (A), γN308I (I), and γN308K (K) fibrinogens at 0.1 mg/ml fibrinogen and 0.005 U/ml thrombin, with and without added calcium as for **Panel A**.

calcium binding, such that in the absence of added calcium the D:D interactions and the release of FpB were independently impaired.

Bβ Chain N-Terminal Residues and Lateral Aggregation

The release of FpB and exposure of the *B* site is thought to mediate lateral aggregation of protofibrils. High-resolution structures of fragment D and crosslinked fragment D crystallized in the presence of the peptide GHRP, which represents the *B* site, showed that GHRP binds to the C-terminal domain of the β-chain, in a pocket whose geometry is analogous to the γ-chain GPRP binding pocket.[21] Thus, the structural data show that the *B:b* interaction can occur. Nevertheless, biochemical data to support a model where this *B:b* interaction mediates lateral aggregation are limited. Studies with fibrinogen lacking Bβ chain residues 1–42 and synthetic peptides derived from the N-terminus of the Bβ chain support the conclusion that a polymerization center is located within this extended region.[22,23] Both of these studies suggested, however, that residues in N-terminus of the β chain are involved with both lateral aggregation and protofibril formation. In addition, photooxidation of BβH16 specifically impaired protofibril formation.[24] Finally, although GHRP binds to fibrinogen, it does not prevent clot formation.[25] Together these studies demonstrate that residues in the *B* site influence polymerization, but the specific association of *B:b* interactions with lateral aggregation remains ambiguous.

We synthesized BβH16A,R17A,P18A fibrinogen to study the role of the *B* site. Thrombin catalyzed polymerization of BβH16A,R17A,P18A showed no change in turbidity under conditions where normal fibrinogen reached a maximum turbidity within 10 min. For example, with 0.1 mg/ml fibrinogen, 0.1 U/ml thrombin in HEPES/NaCl with 0.1 mM $CaCl_2$, no change in turbidity was seen with BβHRP16-18AAA fibrinogen, even when monitored overnight. Under these conditions we found that thrombin catalyzed release of FpA was similar to normal, while release of FpB was delayed about 20-fold relative to normal. We also monitored polymerization at 20 mM $CaCl_2$ and 110 mM NaCl. As is shown in FIGURE 5, polymerization of BβH16A,R17A,P18A was apparent, although still remarkably impaired relative to normal (FIG. 5A). We also examined Ancrod-catalyzed polymerization (FIG. 5B) and fibrin monomer polymerization (data not shown) with this variant. In both experiments polymerization was markedly impaired relative to normal, even at 20 mM $CaCl_2$. We noted that Ancrod catalyzed polymerization of BβH16A,R17A,P18A was similar to thrombin-catalyzed polymerization of BβH16A,R17A,P18A (compare FIG. 5A and 5B), suggesting that polymerization proceeded by the same path irrespective of which enzyme was used. We conclude that the N-terminal region of the Bβ-chain is important for polymerization, with or without FpB release. Furthermore, because high calcium concentrations were essential for any polymerization, bound calcium must affect a structure or structures that facilitate polymerization. That is, bound calcium either changed a facilitating structure in the distal domain, so it could interact with the altered central domain, or changed the structure of the variant site itself, or both.

To determine whether protofibril formation or lateral aggregation or both were impaired, we monitored polymerization by static light scattering. In contrast to the turbidity experiments, we saw changes in light scattering even in the absence of calcium. We performed several experiments at varying the calcium concentrations, plotted the normalized intensity of scattered light versus time, and determined the lag

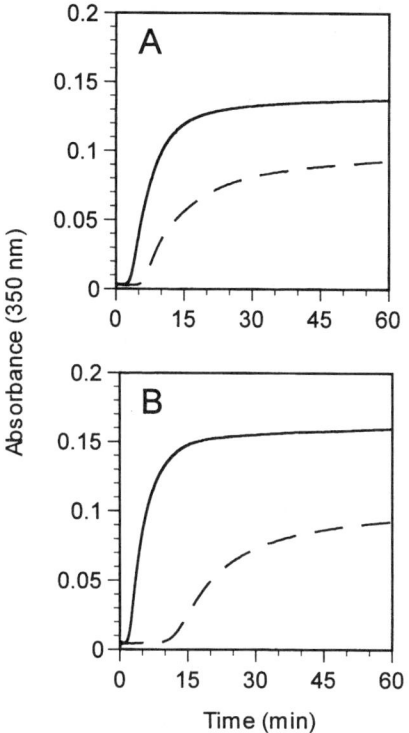

FIGURE 5. Average polymerization curves. The change in turbidity with time of normal (*solid lines, n* = 3) and BβH16A,R17A,P18A (*dashed lines, n* = 3) fibrinogen with 0.1 mg/mL fibrinogen and 0.1 U/mL thrombin (**Panel A**) or 0.1 U/ml Ancrod (**Panel B**) in 20 mM HEPES, pH 7.4, 0.11 M NaCl, 20 mM $CaCl_2$.

time and average rate constants from exponential fits of the data. We found that the rate of polymerization was increased and the lag time decreased in the presence of calcium (1–5 mM) relative to no added calcium. We also performed dynamic light scattering experiments and these likewise showed a calcium dependent increase in the rate of protofibril formation. Aside from the rate effects seen on adding calcium, protofibril formation was normal such that normal sized aggregates were formed. Because the turbidity experiments with no added calcium showed no polymerization, we concluded that lateral aggregation of the BβH16A,R17A,P18A was defective. Thus, these experiments provide direct evidence that the N-terminal region of the Bβ chain participates in lateral aggregation. In the presence of calcium we saw a slow rise in turbidity, demonstrating that elevated calcium facilitated polymerization of this variant, although lateral aggregation likely remained compromised. Additional experiments, such as transmission electron microscopy, are needed to confirm that the light scattering data is consistent with normal protofibril formation. Nevertheless, these preliminary data clearly demonstrate that changes in the hypothesized *B* site markedly altered lateral aggregation, in a calcium dependent manner, with little or no change in protofibril formation. These experiments are the first to link the B site specifically to lateral aggregation.

DISCUSSION

The variants described here were designed to examine the three steps in thrombin catalyzed polymerization: fibrinopeptide release, protofibril formation, and lateral aggregation. Analysis of each of these variants has proven to be informative. Moreover, when considered together, these analyses provide novel insight into the mechanism of thrombin catalyzed polymerization. Furthermore, by considering these biochemical data alongside the high-resolution structural data, it has been possible to achieve a more detailed description of the interactions that support the conversion of fibrinogen to a fibrin clot.

With AαF8Y and A'β fibrinogens, we have seen that the substrate specificity of thrombin dictates the order of fibrinopeptide release, suggesting that enzyme alone controls the timing of FpB release. Yet, analyses of the γ308 variants suggest that polymerization, specifically proper alignment of the D:D interface in protofibrils, controls the timing of FpB release. Perhaps the enzyme specificity ensures that FpA release precedes FpB release, whereas protofibril formation controls the length of the interval between FpA and FpB release. Thus, when we examined fibrinopeptide release from A'β fibrinogen, where all four fibrinopeptides are equivalent, the substrate specificity preempted the influence of polymerization and the addition of GPRP was inconsequential. Alternatively, perhaps the link between polymerization of the γ308 variants and FpB release is coincidental. Perhaps calcium controls both the order and timing of fibrinopeptide release. Thus, if we altered a calcium binding site when we replaced FpB with FpA', and this is the same calcium binding site that affected polymerization of N308A and N308I, then the timing of FpB release would be changed in both variants. Clearly more experiments are needed to resolve the significance of the timing of FpB release.

We have seen that lateral aggregation occurs without the release of FpB. Of course, this has been known for some time, since Batroxobin- or Ancrod-catalyzed polymerization, which removes only FpA, yields essentially normal fibers, albeit with somewhat slower kinetics. Our studies with N-terminal Bβ chain variants (BβR14H and BβH16A,R17A,P18A) suggest that irrespective of the release of FpB, the N-terminus of Bβ chain participates in lateral aggregation. With both variants, the turbidity studies with of desA fibrin were abnormal. With both variants the magnitude of the differences in polymerization, catalyzed by either thrombin or the thrombin-like protease, varied with calcium concentration. We conclude that the N-terminus of the variant Bβ chain participates in polymerization in a calcium dependent association. Because the BβH16A,R17A,P18A light scattering data demonstrated apparently normal protofibril formation, we surmise that protofibril formation is normal for both N-terminal Bβ chain variants. Thus, with both variants, lateral aggregation is defective. Therefore, we conclude that the N-terminus of the Bβ chain normally participates in lateral aggregation, and that the normal association is calcium dependent. Previous studies have also linked this N-terminal domain with calcium dependent polymerization. Polymerization of the dysfibrinogen New York I, lacking Bβ chain residues 9–72, occurred only in the presence of calcium.[26] Reptilase-catalyzed polymerization of desBβ1–42 fibrinogen increased significantly in the presence of 10 mM calcium.[23] Taken all together, these experiments suggest a specific role for calcium in the lateral aggregation of the N-terminal Bβ chain variants. These analyses are not sufficient to understand the link between calcium and lateral aggregation, but they do connect calcium with the N-terminal domain of the Bβ chain. We suggest that in these variants, the high-affinity calcium binding site in the E domain has been changed to a low-affinity site. Thus, higher calcium concentrations were required for lateral aggregation of these altered fibrinogens.

Our experiments have also demonstrated that measurements of turbidity at 350 nm do correlate with polymerization, such that a change in the lag time or the rate of turbidity rise does demonstrate a change in polymerization. As expected, however, abnormal turbidity analyses do not distinguish between abnormal protofibril formation and abnormal lateral aggregation. In particular, we demonstrated that a complete loss of turbidity can be associated with either a loss of protofibril formation, as shown with γΔ319,320, or normal protofibril formation and a loss of lateral aggregation, as shown with BβH16A,R17A,P18A in the absence of added calcium.

FUTURE DIRECTIONS

The results described here suggest two ambiguous aspects of thrombin catalyzed polymerization that can be addressed by further analysis of variant fibrinogens. First, what is the mechanism by which bound calcium so profoundly influences thrombin catalyzed polymerization? Do both high-affinity and low-affinity sites regulate polymerization? Second, where is the *b* site and does it have a role in lateral aggregation? Does the role of the N-terminus of the Bβ chain depend on the release of FpB, such that *B:b* interactions mediate lateral aggregation? Based on the recently published crystallographic structure determined in the presence of GPRP and GHRP, we plan to synthesize new variants to facilitate our understanding of these two aspects.

Our findings suggest that both strong and weak calcium binding sites have an instrumental role in polymerization. We plan further analysis of existing variants with substitutions in the strong, γ chain calcium binding sites; these experiments will likely demonstrate whether this site participates directly in polymerization, or participates only by influencing the overall conformation of the D domain. We also plan to measure calcium binding to A′β, BβR14H, and BβH16A,R17A,P18A fibrinogens. In addition, we plan to synthesize variants with substitutions in the three weak calcium binding sites that were recently identified in the structural data.[21] One of these sites, which is altered in the presence of GHRP, lies in the C-terminal module of the Bβ chain. The second is found in the γ chain when GHRP is bound to the *a* site, and the third is found in the C-terminus of the β chain at a position homologous to the high-affinity site in the γ chain. Biochemical analyses of these variants will address whether calcium bound at specific sites is critical to thrombin catalyzed polymerization, in particular, to lateral aggregation.

Our findings suggest that the N-terminus of the Bβ chain has a role in lateral aggregation, though it is not clear whether the release of FpB is critical. Thus, the role of the $B:b$ interactions remains ambiguous. The structural data obtained in the presence of GHRP and GPRP identified residues within the C-terminal module of the Bβ chain that participate in GHRP binding. Thus, to locate the b site and determine whether $B:b$ interactions are important for polymerization, specifically for lateral aggregation, we plan to synthesize variants that will impair GHRP binding. We plan to incorporate single residue substitutions in the b site that are analogous to a site substitutions that are known to impair GPRP binding. Based on our previous work, we anticipate that studies with these new variants will not only address the two targeted areas (calcium binding and $B:b$ interactions), but will also contribute to our understanding of the concerted interactions in the continuum of events that participate in normal thrombin-catalyzed polymerization.

ACKNOWLEDGMENTS

We thank Li Fang Ping, who assisted in essentially every aspect of this work, and in particular for her help in generating all the cell lines and maintaining the cultures required for synthesis of the variant fibrinogens. Culture medium with normal recombinant fibrinogen was purchased from the Cell Culture Center (Minneapolis, MN). We thank Kasim McLain for her assistance with the purification of multiple recombinant fibrinogens. We thank John Weisel and Roy Hantgan, who collaborated on the electron microscopy and light scattering studies, respectively, and moreover, who contributed to our appreciation of the complex processes of polymerization. Finally, we thank our laboratory colleagues—Cameron Binnie, Andrew Coates, Kelly Hogan, Karim Lounes, Jennifer Mullin, Nobuo Okumura, Michael Rooney, and Elizabeth Steele—whose experiments and ideas have provided the foundation and inspiration for this paper. This work was supported in part by NIH grants HL31048 and HL52706.

REFERENCES

1. BINNIE, C.G., J.M. HETTASCH, E. STRICKLAND, *et al.* 1993. Characterization of purified recombinant fibrinogen: partial phosphorylation of fibrinopeptide A. Biochemistry **32:** 107–113.
2. GORKUN, O.V., Y.I. VEKLICH, J.W. WEISEL, *et al.* 1997. The conversion of fibrinogen to fibrin: recombinant fibrinogen typifies plasma fibrinogen. Blood **89:** 4407–4414.
3. HENSCHEN, A.H. 1993. Human fibrinogen—structural variants and functional sites. Thromb. Hæmostasis **70:** 42–47.
4. NG, A.S., S.D. LEWIS & J.A. SHAFER. 1993. Quantifying thrombin-catalyzed release of fibrinopeptides from fibrinogen using high-performance liquid chromatography. Meth. Enzymol. **222:** 341–358.
5. HIGGINS, D.L., S.D. LEWIS & J.A. SHAFER. 1983. Steady state kinetic parameters for the thrombin-catalyzed conversion of human fibrinogen to fibrin. J. Biol. Chem. **258:** 9276–9282.
6. WEISEL, J.W., Y. VEKLICH & O. GORKUN. 1993. The sequence of cleavage of fibrinopeptides from fibrinogen is important for protofibril formation and enhancement of lateral aggregation in fibrin clots. J. Mol. Biol. **232:** 285–297.
7. LORD, S.T., P.A. BYRD, K.L. HEDE, *et al.* 1990. Analysis of fibrinogen Aα-fusion proteins. Mutants which inhibit thrombin equivalently are not equally good substrates. J. Biol. Chem. **265:** 838–843.

8. ROONEY, M.M., J.L. MULLIN & S.T. LORD. 1998. Substitution of tyrosine for phenylalanine in fibrinopeptide A results in preferential thrombin cleavage of fibrinopeptide B from fibrinogen. Biochemistry 37: 13704–13709.
9. MULLIN, J.L., O.V. GORKUN, C.G. BINNIE, et al. 2000. Recombinant fibrinogen studies reveal that thrombin specificity dictates order of fibrinopeptide release. J. Biol. Chem. 275: 25239–25246.
10. HIGGINS, D.L. & J.A. SHAFER. 1981. Fibrinogen Petoskey, a dysfibrinogenemia characterized by replacement of Arg-Aα 16 by a histidyl residue. Evidence for thrombin-catalyzed hydrolysis at a histidyl residue. J. Biol. Chem. 256: 12013–12017.
11. MULLIN, J.L. 2000. An Expanded and Refined Model of Fibrin Polymerization Based on Studies with Variant Recombinant Fibrinogens. Ph.D. Thesis, Department of Chemistry, University of North Carolina, Chapel Hill.
12. PRATT, K.P., H.C.F. COTE, D.W. CHUNG, et al. 1997. The primary fibrin polymerization pocket 3-dimensional structure of a 30-Kda C-terminal γ chain fragment complexed with the peptide Gly-Pro-Arg-Pro. Proc. Natl. Acad. Sci. U.S.A. 94: 7176–7181.
13. SPRAGGON, G., S.J. EVERSE & R.F. DOOLITTLE. 1997. Crystal structures of fragment D from human fibrinogen and its crosslinked counterpart from fibrin. Nature 389: 455–462.
14. OKUMURA, N., O.V. GORKUN & S.T. LORD. 1997. Severely impaired polymerization of recombinant fibrinogen γ364 Asp—His, the substitution discovered in a heterozygous individual. J. Biol. Chem. 272: 29596–29601.
15. YAMAZUMI, K. & R.F. DOOLITTLE. 1992. Photoaffinity labeling of the primary fibrin polymerization site: localization of the label to γ-chain Tyr-363. Proc. Natl. Acad. Sci. U.S.A. 89: 2893–2896.
16. OKUMURA, N., K. FURIHATA, F. TERASAWA, et al. 1996. Fibrinogen Matsumoto I: a γ364 Asp→His (GAT→CAT) substitution associated with defective fibrin polymerization. Thromb. Hæmostasis 75: 887–891.
17. YAMAZUMI, K. & R.F. DOOLITTLE. 1992. The synthetic peptide Gly-Pro-Arg-Pro-amide limits the plasmic digestion of fibrinogen in the same fashion as calcium ion. Prot. Sci. 1: 1719–1720.
18. HOGAN, K.A., O.V. GORKUN, K.C. LOUNES, et al. 2000. Recombinant fibrinogen Vlissingen/Frankfurt IV. The deletion of residues 319 and 320 from the γ chain of fibrinogen alters calcium binding, fibrin polymerization, cross-linking, and platelet aggregation. J. Biol. Chem. 275: 17778–17785.
19. COTE, H.C.F., K.P. PRATT, E.W. DAVIE, et al. 1997. The polymerization pocket "a" within the carboxyl-terminal region of the γ chain of human fibrinogen is adjacent to but independent from the calcium-binding site. J. Biol. Chemistry 272: 23792–23798.
20. EBERT, R.F. & W.R. BELL. 1988. Fibrinogen Baltimore III: congenital dysfibrinogenemia with a shortened gamma-subunit. Thromb. Res. 51: 251–258.
21. EVERSE, S.J., G. SPRAGGON, L. VEERAPANDIAN, et al. 1999. Conformational changes in fragments D and double-D from human fibrin(ogen) upon binding the peptide ligand Gly-His-Arg-Pro-amide. Biochemistry 38: 2941–2946.
22. PANDYA, B.V., J.L. GABRIEL, et al. 1991. Polymerization site in the β chain of fibrin: mapping of the Bβ 1–55 sequence. Biochemistry 30: 162–168.
23. SIEBENLIST, K.R., J.P. DIORIO, A.Z. BUDZYNSKI, et al. 1990. The polymerization and thrombin-binding properties of des-(Bβ 1–42)-fibrin. J. Biol. Chem. 265: 18650–18655.
24. HENSCHEN-EDMAN, A.H. 1997. Photo-oxidation of histidine as a probe for aminoterminal conformational changes during fibrinogen-fibrin conversion. Cell Mol. Life Sci. 53: 29–33.
25. LAUDANO, A.P., B.A. COTTRELL, R.F. DOOLITTLE. 1983. Synthetic peptides modeled on fibrin polymerization sites. Ann. N.Y. Acad. Sci. 408: 315–329.
26. LIU, C.Y., J.A. KOEHN & F.J. MORGAN. 1985. Characterization of fibrinogen New York I. A dysfunctional fibrinogen with a deletion of Bβ (9–72) corresponding exactly to exon 2 of the gene. J. Biol. Chem. 260: 4390–4296.

27. ROONEY, M.M., D.H. FARRELL, B.M. VAN HEMEL, *et al.* 1998. The contribution of the three hypothesized integrin-binding sites in fibrinogen to platelet-mediated clot retraction. Blood **92:** 2374–2381.
28. MULLIN, J., O. GORKUN & S. LORD. 2000. Decreased lateral aggregation of a variant recombinant fibrinogen provides insight into the polymerization mechanism. Biochemistry **39:** 9843–9849.
29. LORD, S.T., E. STRICKLAND & E. JAYJOCK. 1996. Strategy for recombinant multichain protein synthesis: fibrinogen Bβ-chain variants as thrombin substrates. Biochemistry **35:** 2342–2348.
30. GORKUN, O.V., A.H. HENSCHEN-EDMAN, L.F. PING, *et al.* 1998. Analysis of Aα251 fibrinogen: the αC domain has a role in polymerization, albeit more subtle than anticipated from the analogous proteolytic fragment X. Biochemistry **37:** 15434–15441.
31. ROONEY, M.M., L.V. PARISE & S.T. LORD. 1996. Dissecting clot retraction and platelet aggregation. Clot retraction does not require an intact fibrinogen γ chain C terminus. J. Biol. Chem. **271:** 8553–8555.
32. FARRELL, D.H., P. THIAGARAJAN, D.W. CHUNG, *et al.* 1992. Role of fibrinogen α and γ chain sites in platelet aggregation. Proc. Natl. Acad. Sci. U.S.A. **89:** 10729–10732.

Synthesis of a Mouse Model of the Dysfibrinogen Vlissingen/Frankfurt IV

KELLY A. HOGAN, NOBUYO MAEDA, KIMBERLY D. KLUCKMAN, AND SUSAN T. LORD

Department of Pathology and Laboratory Medicine, University of North Carolina, Chapel Hill, North Carolina, USA

ABSTRACT: The dysfibrinogen Vlissingen/Frankfurt IV is characterized as a deletion of Asn319 and Asp320 from the C-terminus of the γ-chain of fibrinogen. This dysfibrinogen, which was identified in several family members that are all heterozygous for the in-frame 6-bp deletion, is associated with both venous and arterial thrombosis. Here, we describe the generation of a murine model of the V/F IV dysfibrinogen using gene targeting of mouse γ-chain DNA. Preliminary analysis shows that the human and mouse variant fibrinogens are similar: analogous to the human V/F IV protein, the D1 fragment of the variant mouse fibrinogen is partially protected from digestion in the presence of calcium or Gly-Pro-Arg-Pro. These heterozygous mice provide the first opportunity to examine the association of thrombophilia and dysfibrinogenemia in a controlled genetic background.

KEYWORDS: Dysfibrinogen; Mouse; Thrombosis.

INTRODUCTION

V/F IV is a heterozygous dysfibrinogen identified in patients with thrombosis.[1] We have synthesized the homogeneous recombinant protein with the analogous mutation, a 6-bp deletion in exon 8 coding for γAsn319 and γAsp320. Unexpectedly, the recombinant protein did not polymerize, did not support FXIII crosslink formation and did not support platelet aggregation *in vitro*.[2] These three abnormalities are typically associated with bleeding, not thrombosis. By comparing polymerization profiles for the recombinant fibrinogen with the plasma fibrinogen, we obtained data indicating that the plasma fibrinogen molecules contained one normal and one abnormal chain; that is, heterodimers of V/F IV.[3] We suggest that the incorporation of the heterodimers into the plasma clot can result in an abnormal structure that does not lyse normally, thus resulting in thrombosis. In the study reported here, we have generated a mouse with the analogous mutation, so we will be able to examine plasma clot lysis *in vivo*. Based on our previous data, we expect homozygous mice will have a bleeding disorder whereas heterozygous mice will have a thrombotic disorder.

Address for correspondence: Susan T. Lord, Department of Pathology and Laboratory Medicine, University of North Carolina, Chapel Hill, NC 27599-7525, USA. Voice: 919-966-3548; fax: 919-966-6718.
  stl@med.unc.edu

METHODS

Construction of the Mouse Fibrinogen γ Chain Gene Expression Vector

A 129/SVEV strain mouse library was screened with PCR products from conserved regions of human Aα, Bβ, and γ fibrinogen genes. The P1 phage, P1-35, which contains all three murine fibrinogen genes, was digested with Sal I and Sma I and the multiple fragments were cloned into plasmid Bluescript (pBS). The mouse γ gene spanning from intron 4 to beyond exon 10 was contained in one plasmid, p746.8 and was entirely sequenced. Digestion of p746.8 with Hind III produced two fragments; the smaller fragment, approximately 2 kb, contained exons 7 and 8. This small fragment was ligated into the Hind III site of pBS, resulting in p758.6, which was altered by oligonucleotide-directed mutagenesis using the Transformer TM site-directed mutagenesis kit from CLONTECH Laboratories, (Palo Alto, CA). One oligonucleotide used for the selection site, disrupted a unique Xho I site in p758.6 ($^{5'}$ CGA TAC CGT CGA CCT TGA GGG $^{3'}$). A second oligonucleotide deleted the 6-bp and added a unique Sca I site in exon 8 ($^{5'}$ GCA GTT CAG TAC TTG GGA CAA CGA CAA GTT TGA AGG CAA CTG TGC G $^{3'}$). A third oligonucleotide added a new Bgl II site 68 bp upstream from the start of exon 8 ($^{5'}$ GCA TTG CCT AGA TCT GCA GAG $^{3'}$). The resulting clone p760.8 was identified by restriction digest, was sequenced in its entirety, digested with Hind III, and ligated back to the large fragment from a p746.8 digest with Hind III (resulting in p761.3). Plasmid 761.3 was digested with Bsr G I and ligated to the large fragment of a Bsr G I digest of the γ Δ 5 construct,[4] kindly provided by Dr. Jay L. Degen at the Children's Hospital Research Foundation, Cincinnati, OH, to construct the targeting vector (see FIGURE 1).

Targeted Replacement of the Mouse Fibrinogen γ Chain Gene

E14TG2a embryonic stem (ES) cells, 129/Ola ES cell lines that are HPRT deficient, were electroporated (250 mFd, 300 volts) with Sal I linearized targeting vector (10 μg). The cells were selected for 14 days with gancyclovir and hypoxanthine-aminopterin-thymidine (HAT). After 12 days of selection, colonies were isolated and transferred to 24-well culture dishes. On confluency, some cells were removed by trypsinization. Clones that had undergone homologous recombination were identified by the presence of an 800-bp PCR product with primers A and B (FIG. 1). Primer A is complementary to the HPRT cassette ($^{5'}$ CCT GAA GAA CGA GAT CAG CAG CCT CTG TTC $^{3'}$) and primer B is complementary to a downstream region of the γ chain not included in the targeting construct ($^{5'}$ ATA CAT GGA TAT TAG CCA GGC AGT AGT GAC $^{3'}$). The 274-bp product with primers C and D was sequenced with primer C to determine if the deletion was present. A Southern blot of a Bgl II genomic digest was performed with all colonies, identified by PCR as being targeted, to confirm that the mutation was present (FIG. 1). The number of chromosomes were also analyzed. Cells were injected into C57BL6 blastocysts and implanted into pseudopregnant mice. Chimeric mice were back crossed to C57BL6 mice.

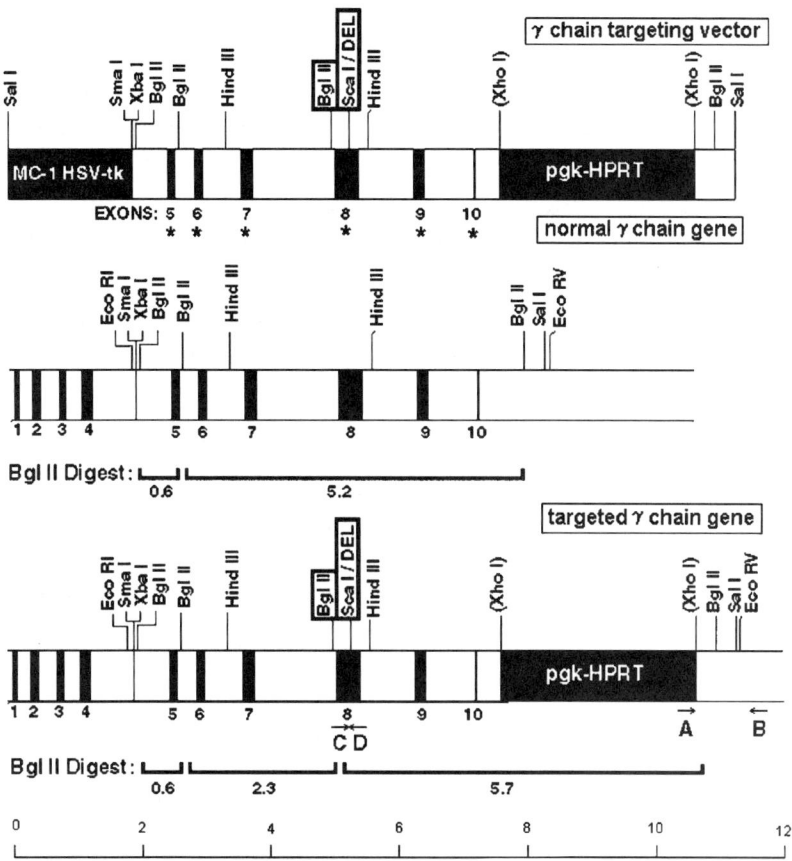

FIGURE 1. Targeted disruption and homologous recombination of the mouse γ-chain fibrinogen gene. **Top**: the targeting vector contains a negative selection gene, tk, and a positive selection gene, HPRT. The introduced changes, the deletion and two silent and new restriction sites are shown *boxed*. **Middle**: normal γ-chain gene. **Bottom**: targeted γ-chain gene. *Arrows*, primers used for PCR; *asterisks*, indicate the exons hybridized to the probe used in a Southern blot analysis of BgII digested DNA.

RESULTS AND DISCUSSION

Amino acid sequence analysis of the mouse γ-chain gene showed 90.1% identity between human and mouse exons 5–10. Exon 8, the exon in which the V/F IV deletion is located, was 94.6% identical. Due to the high degree of homology, we suspected that the six-bp deletion mimicking the human V/F IV deletion, would result in a similar phenotype in mice to humans.

Approximately 800 ES colonies were screened by PCR and eight targeted colonies were detected. From these, two contained the mutation, confirmed by Southern blot, PCR, and direct sequencing. One of these colonies was missing a chromosome and, therefore, was not pursued. Blastocyst injection of the remaining colony resulted in four chimeric mice, two females and two males. Germline transmission occurred twice, once from each female chimera. Both resulting heterozygotes were male and were shown to contain the mutation by PCR and direct sequencing. The first male was sterile, presumably due to a sex chromosome abnormality, that males from female chimeras often have.[5] The second male recently reached sexual maturity and is breeding.

Analysis of plasma from the first heterozygote by plasmin digest showed that plasmin degradation of mouse fibrinogen is similar to human fibrinogen for both the normal and the variant fibrinogens (see FIGURE 2). The D1 fragment is degraded to D3 when not protected (lanes 1 and 4). In contrast to normal mouse fibrinogen, which is protected by calcium or Gly-Pro-Arg-Pro (GPRP, lanes 2 and 3), fibrinogen from the heterozygote mouse was not fully protected from plasmin cleavage in the presence of calcium or GPRP (lanes 5 and 6). These results are consistent with a mouse heterozygous for the mutation. Moreover, they indicate that the deletion disrupts both the *a* site and the calcium-binding site in mice, as it does in humans. Currently, the first male is six months old and has no obvious abnormalities or problems.

In summary, we have synthesized a mouse model of the dysfibrinogen Vlissingen/Frankfurt IV. When more mice are obtained through breeding, phenotypic characterization can be continued. We expect that mice heterozygous for the deletion will have a prothrombotic phenotype, similar to that in humans with the deletion.[1] Based on the studies with the recombinant protein, we expect homozygous mice to have bleeding problems.

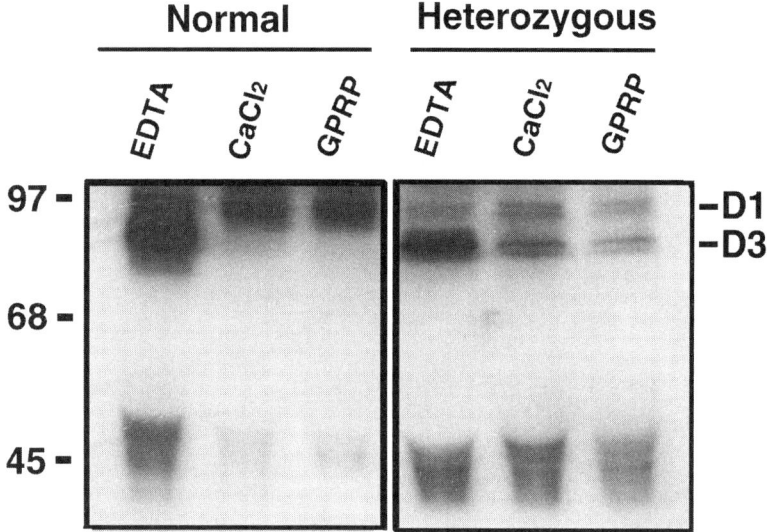

FIGURE 2. Plasmin digestion of mouse plasma. Plasma was incubated with 5 mM EDTA (*lanes* 1 and 4), 2 mM $CaCl_2$ (*lanes* 2 and 5), or 2 mM GPRP (*lanes* 3 and 6), run on a 7.5% SDS-PAGE gel (non-reduced) and blotted with rabbit-antihuman fibrinogen polyclonal antibody (DAKO).

ACKNOWLEDGMENT

This work was supported by NIH Grant HL 52706.

REFERENCES

1. KOOPMAN, J., et al. 1991. A congenitally abnormal fibrinogen (Vlissingen) with a 6-base deletion in the γ chain gene, causing defective calcium binding and impaired fibrin polymerization. J. Biol. Chem. **266:** 13456–13461.
2. HOGAN, K.A., et al. 2000. Recombinant fibrinogen Vlissingen/Frankfurt IV. The deletion of residues 319 and 320 from the γ chain of fibrinogen alters calcium binding, fibrin polymerization, cross-linking, and platelet aggregation. J. Biol. Chem. **275:** 17778–17785.
3. HOGAN, K.A., et al. 2000. A functional assay suggests that heterodimers exist in two C-terminal γ chain dysfibrinogens: Matsumoto I and Vlissingen/Frankfurt IV. Thromb. Hæmost. **83:** 592–597.
4. HOLMBACK, K., et al. 1996. Impaired platelet aggregation and sustained bleeding in mice lacking the fibrinogen motif bound by integrin $\alpha_{IIb}\beta3$. EMBO J. **15:** 5760–5771.
5. BRONSON, S.K., et al. 1995. High incidence of XXY and XYY males among the offspring of female chimeras from embryonic stem cells. Proc. Natl. Acad. Sci. U.S.A. **92:** 3120–3123.

Structural and Functional Role of the β-Strand Insert (γ381–390) in the Fibrinogen γ-Module

A "Pull Out" Hypothesis

SERGEI YAKOVLEV, DMITRY LOUKINOV, AND LEONID MEDVED

Biochemistry Department, The Holland Laboratory, American Red Cross, Rockville, Maryland, USA

ABSTRACT: Study of the folding status of the fibrinogen γ-module (residues γ148–411) revealed that its COOH-terminal β-strand (residues γ381–390), that is normally inserted into its central domain, can be removed without destroying its compact structure. Based on this and other observations we propose a "pull out" hypothesis that suggests a mechanism for the formation of transverse γ-γ crosslinks in fibrin.

KEYWORDS: Fibrinogen; γ-Module; β-Strand insert; Pull out hypothesis; γ-γ Crosslinking.

The crystal structure of the fibrinogen γ-module,[1,2] consisting of three domains, NH$_2$-terminal (residues γ148–191), central (γ192–286), and COOH-terminal (γ287–379), revealed an unusual structural feature. Part of its functionally important COOH-terminal portion (residues γ381–390) forms a β-strand that is inserted into an antiparallel β-sheet of the central domain, whereas the rest (γ393–411) seems to be flexible (see FIGURE 1A). The major goal of this study was to clarify the structure/functional importance of this β-strand insert.

It is commonly accepted that conversion of human fibrinogen fragment D$_1$ into fragment D$_3$ by plasmin is accompanied by the removal of the COOH-terminal region of its γ chain (γ303–411) that includes most of the COOH-terminal domain. However, we found that the native D$_3$ fragment retains part of this region, the γ374–405 peptide, associated non-covalently with the bulk of the molecule, suggesting that its γ381–390 β-strand is an integral part of the central domain.[3] To test the role of this β-strand insert in the structural integrity of D$_3$, we prepared peptide-free D$_3$ in denaturing conditions, refolded it in the presence and absence of this peptide, and compared the thermal stability of the peptide-loaded and peptide-free D$_3$ fragment. In fluorescence experiments both native and peptide-loaded D$_3$ exhibited a similar heat induced denaturation transition with a midpoint (T_m) at 53°C, whereas peptide-free D$_3$ was destabilized ($T_m = 40$°C). This result suggested that without this peptide the central and/or COOH-terminal domains of the γ-module, that melt in this transition (the NH$_2$-terminal domain melts at much higher temperature[4]), were destabilized or unfolded; addition of the peptide on refolding restored their original

---

Address for correspondence: Sergei Yakovlev, Ph.D., The Holland Laboratory, American Red Cross, 15601 Crabbs Branch Way, Rockville, MD 20855, USA. Voice: 301-738-0724; fax: 301-738-0794.

yakovles@usa.redcross.org

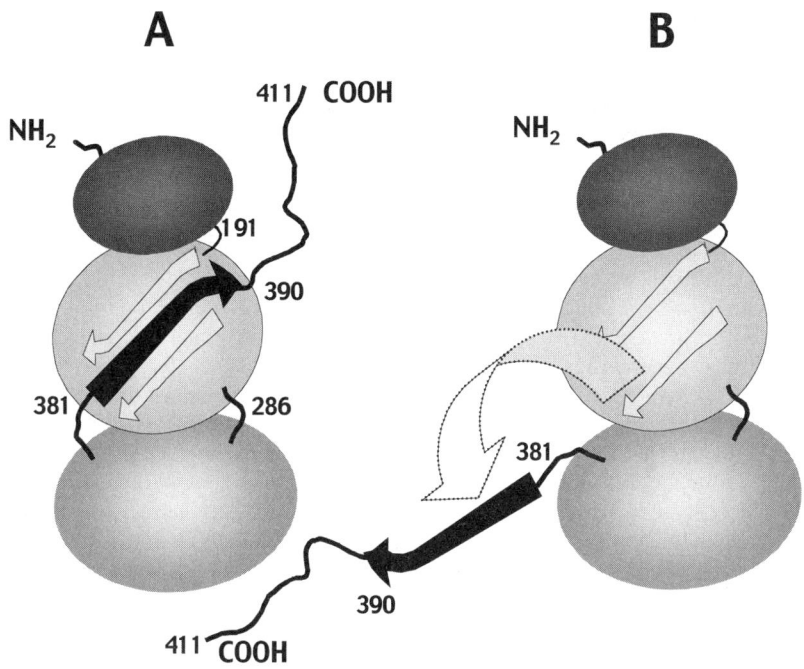

FIGURE 1. Schematic representation of the domain structure of the γ-module (**A**) and the "pull out" hypothesis (**B**). The γ381–390 β-strand insert in the central domain in (**A**) is represented by a *black arrow*. The *curved dotted arrow* shows how the β-strand insert could be pulled out from the central domain to form transverse crosslink.

stability. To test directly which domain is affected and how, we compared the folding status and stability of the wildtype γ-module (γ-wt, residues γ148–411) and its truncated variants, γ-t373 (γ148–373) and γ-t286 (γ148–286), lacking, respectively, either the β-strand insert or both the insert and the COOH-terminal domain. The variant γ-t392 (γ148–392) lacking only the extreme COOH-terminus was also tested.

All species were expressed in *E. coli*, purified and then refolded by slow dialysis from 8 M urea. The majority of both γ-wt and γ-t392 were found in oligomeric form, whereas γ-t373 and γ-286 were mainly monomeric, suggesting that the β-strand insert is responsible for the formation of oligomers on refolding. All fragments exhibited fluorescence spectra with maxima at 344 nm in comparison with 352 nm in denaturing conditions, suggesting that they all were folded into a compact structure. The presence of a compact structure in all variants was confirmed in chemical- and heat-induced denaturation experiments. Monomeric γ-t392 exhibited a sigmoidal transition between 2 and 5 M of GdmCl that was the same as in the monomeric wildtype γ-module, indicating that it is folded into a compact structure and that the extreme COOH-terminal residues γ393–411 do not contribute to the structural integrity of the γ-module. In contrast, γ-t373 and γ-t286 variants exhibited a sigmoidal transition at substantially lower concentration of GdmCl. The same pattern of relative stability for all fragments was also observed in the heat-induced denaturation experiments. These results indicate directly that deletion (removal) of the β-strand insert from the γ-module results in destabilization but not unfolding of its central domain.

The results obtained demonstrate that the β-strand insert could be removed (pulled out) from the γ-module without the loss of the compact fold of the latter (FIG. 1B). Based on this and other observations we hypothesize that such a "pull out" mechanism would allow alterations in the conformation of the γ-module in fibrin(ogen) and for the extreme COOH-terminal portion of the γ chain (γ393–411) that is involved in γ-γ crosslinking and interaction with some proteins and cells to be stretched from the molecule.

There is ongoing controversy as to whether the factor XIIIa-catalyzed crosslinks between fibrin γ-chains are longitudinal; that is, between γ-chains aligned end-to-end in the same protofibril strand,[5] or transverse, between γ-chains of two adjacent strands.[6] Electron microscopic data suggest that both may occur. At the same time a rough calculation of the distance between the γ-modules in fibrin[1] and in the dimeric D-D fragment[2] led to the suggestion that transverse crosslinking is unlikely. Our hypothesis reconciles the basis for formation of transverse γ-γ crosslinks. Namely, if the β-strand insert is pulled out of the γ-module, each COOH-terminus containing reactive Gln398/399 and Lys406 could easily extend an additional 35 Å, a distance sufficient to form a transverse crosslink.

ACKNOWLEDGMENT

This work was supported by National Institutes of Health Grant HL-56051 (to L.M.).

REFERENCES

1. YEE, V.C., K.P. PRATT, H.C. COTE, et al. 1997. Crystal structure of a 30 kDa C-terminal fragment from the gamma chain of human fibrinogen. Structure **5**: 125–138.
2. SPRAGGON, G., S.J. EVERSE & R.F. DOOLITTLE. 1997. Crystal structures of fragment D from human fibrinogen and its crosslinked counterpart from fibrin. Nature **389**: 455–462.
3. MEDVED, L., S. YAKOVLEV & S. LITVINOVICH. 1998. Fibrinogen D_3 fragment contains non-covalently associated γ374–396 peptide (Abst.). Blood Coag. Fibrinolysis **9**: 672.
4. MEDVED, L., S. LITVINOVICH, T. UGAROVA, et al. 1997. Domain structure and functional activity of the recombinant human fibrinogen gamma-module (gamma 148–411). Biochemistry **36**: 4685–4693.
5. WEISEL, J.W., C.W. FRANCIS, C. NAGASWAMI & V.J. MARDER. 1993. Determination of the topology of factor XIIIa-induced fibrin gamma-chain cross-links by electron microscopy of ligated fragments. J. Biol. Chem. **268**: 26618–26624.
6. MOSESSON, M.W., K.R. SIEBENLIST, J.F. HAINFELD & J.S. WALL. 1995. The covalent structure of factor XIIIa crosslinked fibrinogen fibrils. J. Struct. Biol. **115**: 88–101.

Structure and Properties of Clots from Fibrinogen Bicêtre II (γ308 Asn→Lys)

Increased Permeability due to Larger Pores, Thicker Fibers, and Decreased Rigidity

RITA MARCHI,[a,b] STÉPHANE LOYAU,[c] EDUARDO ANGLÉS-CANO,[c] AND JOHN W. WEISEL[b]

[a]*Laboratorio de Fisiopatología, Centro de Medicina Experimental, IVIC, Caracas, Venezuela*

[b]*Department of Cell and Developmental Biology, University of Pennsylvania School of Medicine, Philadelphia, Pennsylvania, USA*

[c]*Inserm U.143, Hémostase-Biologie Vasculaire, Bicêtre, France*

ABSTRACT: Fibrinogen Bicêtre II is a dysfibrinogenemia in which there is a substitution of Lys for Asn at γ308. We have studied the polymerization of this abnormal fibrinogen by measurement of turbidity and have characterized clot structure by scanning electron microscopy, permeation, and viscoelastic measurements. The results of these studies demonstrate that this amino acid substitution has substantial effects on the structure and properties of the clot, resulting in clots made up of thick fibers and large pores with greatly reduced stiffness and increased slippage of protofibrils.

KEYWORDS: Structure and properties of clots; Fibrinogen Bicêtre II; Permeability; Pores; Fibers; Rigidity; Dysfibrinogenemia.

INTRODUCTION

Fibrinogen Bicêtre II is a dysfibrinogenemia with a γ 308 Asn→Lys substitution in which the propositus had thrombotic episodes not clearly associated with the molecular anomaly.[1] The molecular anomaly found in fibrinogen Bicêtre II is shared with three other abnormal fibrinogens reported in the literature: Kyoto I and Matsumoto II with the same amino acid substitution, and Baltimore III where asparagine is replaced by isoleucine.[2] The clinical outcome does not seem to be related to the γ 308 Asn mutation, since Baltimore III and Kyoto I are asymptomatic, and Matsumoto II is a bleeder.

It was found by X-ray crystallography of rFbgγC30, which contains residues 143 Val to 411 Val (30-kDa fragment from the C-terminal region of the human γ chain), that the residue γ 308 Asn is located on the surface of rFbgγC30.[3] The side chain is hydrogen bonded both to γ Asn 278 and to the backbone carbonyl of γ 309 Gly. In

---

Address for correspondence: John W. Weisel, Ph.D., Department of Cell and Developmental Biology, University of Pennsylvania School of Medicine, 1054 BRB II/III, 421 Curie Blvd, Philadelphia, PA 19104-6058, USA. Voice: 215-898-3573; fax: 215-898-9871.

weisel@mail.cellbio.med.upenn.edu

double D (isolated from fibrin), this residue occurs at the staggered interface involved in DD association. The X-ray structure indicates that the native hydrogen bonded network would be disrupted by the change of Asn to Lys, and a charge repulsion with γ 321 Lys is also apparent.[4]

Due to the position of the residue on the surface at which the DD interactions occur, several functional impairments have been hypothesized; such as, disruption of the proper alignment of fibrin monomers during protofibril formation, the introduction of a new Lys residue on an accessible location that could alter the pattern of cleavage by plasmin, and finally, disruption of Factor XIII crosslinking, since γ 308 Asn is only 15–20 Å from γ 398 Gln.[2,3]

In the work described here, we have determined the functional effects of the structural alterations introduced by this type of mutation.

MATERIALS AND METHODS

Fibrinogen Purification. Fibrinogen was purified by the method described by Grailhe.[1]

Fibrin Polymerization. 0.2 mg/ml of purified fibrinogen solution was clotted with 0.2 NIH units/ml of thrombin in the presence of 10 mM $CaCl_2$. The change in the optical density was monitored at 350 nm for 30 min. Three clots were polymerized both for the control and the patient and the averages were calculated.

Permeability and Scanning Electron Microscopy. Clotting conditions were 1 mg/ml of purified fibrinogen in 0.05 M Tris buffer, 0.15 M NaCl, pH 7.4, clotted with 1 NIH units/ml of thrombin (American Diagnostica, Greenwich, CT) incubated previously with 10 mM $CaCl_2$ for 1 min. The mixture (100 µl) was transferred to a pre-etched plastic tube and incubated for 2 h at room temperature in a moist environment to complete the polymerization process. Then, the tubes were filled with buffer and the flux (g/min) at a given pressure was measured. The clots were then processed for scanning electron microscopy.

Viscoelastic Properties. The clotting conditions were the same as described for permeability and the storage modulus (G′) and loss modulus (G″) were recorded using a torsion pendulum.

Fibrin Crosslinking. Clots were prepared in the same conditions as those used for viscoelastic measurements and analyzed by SDS/PAGE with a 4–10% gradient in a Mini Protean II Dual Slab Cell (BioRad, Hercules, CA).

RESULTS AND DISCUSSION

The polymerization process of fibrin Bicêtre II has a longer lag period; there is a delay in the protofibril formation, revealed by the prolonged lag time that speaks in favor of an impairment of the fibrin monomer associations. Since the anomaly in this case is not located in the thrombin cleavage site, after this stage is overcome, the rate of fiber formation (lateral association of protofibrils) is the same as that of the control and the final turbidity is increased, as is expected when the lag time is prolonged.[5]

It has been proposed, based on X-ray crystallographic studies of rFbgγC30 and fragments DD[3,4] that γ 308 Asn is located on the surface of the γ nodule in the vicinity

TABLE 1. Permeation of fibrinogen Bicêtre II clots

	Control	Bicêtre II
K_s (10^{-9} cm^2)	8.7 ± 0.7	28.5 ± 5.3
d (μ)	0.197	0.357
μ (10^{12} dalton/cm^2)	28	91

NOTE: K_s, Darcy constant; d, fibrin fiber diameter; μ, mass/length ratio.

of the region interacting with a contiguous γ nodule of another fibrin monomer. Therefore, an altered fibrin monomer association during protofibril formation has been proposed and this is corroborated by the behavior of Bicêtre II fibrin polymerization.

When the Darcy constant (K_s) was calculated from the flux (g/sec) at a constant pressure, it was found that Bicêtre II fibrin has approximately four times more area available for perfusion than control fibrin (see TABLE 1). After finishing the permeation measures, these same clots were processed for scanning electron microscopy, and the pictures of Bicêtre II clots reveal thicker fibers than those of control clots and larger pores, consistent with the values of the Darcy constant obtained by permeation studies.

Unexpectedly, when rheological or viscoelastic properties of this abnormal fibrinogen were determined, the G' values for clots made from fibrinogen Bicêtre II were approximately half (Bicêtre II, 38 dyne/cm^2 vs. control, 83 dyne/cm^2) and the loss tangent was increased 36% over that of the control. In the light of this finding, crosslinked fibrin was analyzed by SDS-PAGE and it was found that Bicêtre II clot γ chain crosslinking was lower than that of the control. The change in clot structure and the extent of crosslinking could explain the decrease in G' values and the increase in the loss tangent.

It might be expected that this type of mutation, a single amino acid substitution, would not affect the γ-γ crosslinking, since the residues crosslinked are quite far away (10–20 Å) from Gln 398. For example, previous results with recombinant fibrinogen containing γ 308 Asn→Ile and γ 308 Asn→Lys showed normal Factor XIIIa crosslinking.[6] On the other hand, in fibrinogen Asahi, where there is extra glycosylation due to the creation of a new consensus sequence, this bulky group could span such a distance and it was found that the γ chain crosslinking is seriously affected, as it is for fibrinogen Vlissingen/Frankfurt IV (studies performed with recombinant fibrinogen),[7] in which γ Asn 319 and Asp 320 are deleted. However, the increase in the value of the loss tangent for fibrinogen Bicêtre II clots could indicate that the γ-γ crosslinking is diminished, since unligated clots have a loss tangent greater than that of ligated clots.[8]

In summation, the mutation of fibrinogen Bicêtre II does not severely disturb the polymerization process, but it seems to influence the rheological properties of the fibrin network, due to the decrease in the formation of γ-γ crosslinking, and the architecture of the clots, which have larger pores and thicker fibers when compared to normal fibrin. Studies on Factor XIIIa crosslinking are in progress to clarify the contrasting results obtained with recombinant fibrinogens with the same structural anomaly.

ACKNOWLEDGMENTS

We thank Mark Allan for fibrin fiber diameter measurements using the NIH IMAGE program and Dr. Catherine Boyer-Neumann (Hôpital A. Béclère, Clamart, France) for the collection of plasma from the propositus. We acknowledge the support of NIH grant HL30954 and INSERM grant 4M101C.

REFERENCES

1. GRAILHE, P., et al. 1993. Blood Coag. Fibrinol. **4:** 679–687.
2. CÔTE, H.C.F., S.T. LORD & K.P. PRATT. 1998. Blood **92:** 2195–2212.
3. YEE, V.C., et al. 1997. Structure **5:** 125–138.
4. DOOLITTLE, R.F., G. SPRAGGON & S.J. EVERSE. 1999. Thromb. Hæmostas. **82:** 271–276.
5. WEISEL, J.W. & C. NAGASWAMI. 1992. Biophys. J. **63:** 111–128.
6. OKUMURA, N., F. TERASAWA & S.T. LORD. 1999. Comparison of functions between recombinant variant fibrinogens, r308K, r308I and r308A (Abst.). Thromb. Hæmostas. Suppl. 321.
7. HOGAN, K.A., et al. 2000. J. Biol. Chem. **275:** 17778–17785.
8. ROBERTS, W.W., L. LORAND & L.F. MOCKROS. 1973. Biorheology **10:** 29–42.

Fibrinogen Longmont

A Heterozygous Abnormal Fibrinogen with Bβ Arg-166 to Cys Substitution Associated with Defective Fibrin Polymerization

KARIM C. LOUNES,[a] JERRY B. LEFKOWITZ,[b] ANDREW I. COATES,[c] ROY R. HANTGAN,[d] AGNES HENSCHEN-EDMAN,[e] AND SUSAN T. LORD[a,c]

[a]*Departments of Pathology and Laboratory Medicine and* [c]*Chemistry, University of North Carolina, Chapel Hill, North Carolina*

[b]*Department of Pathology, University of Colorado School of Medicine, Denver, Colorado*

[d]*Department of Biochemistry, Wake Forest University School of Medicine, Winston-Salem, North Carolina*

[e]*Department of Molecular Biology and Biochemistry, University of California, Irvine, California*

> ABSTRACT: BβArg166 to Cys substitution was identified in an abnormal fibrinogen named fibrinogen Longmont. The proband, a young woman, and her mother were heterozygous; both experienced episodes of severe hemorrhage at childbirth. The *neo*-Cys residues were found to be disulfide-bridged to either an isolated Cys amino acid or to the corresponding Cys residue of another abnormal fibrinogen molecule, forming dimers. Thrombin and batroxobin induced fibrin polymerization were impaired, despite normal release of fibrinopeptides A and B. Moreover, the polymerization defect was not corrected by removing the dimeric species or adding calcium. Fibrinogen Longmont had normal polymerization site *a*, as evidenced by normal GPRP-peptide binding. Thus, the sites *A* and *a* can interact to form protofibrils, as evidenced by dynamic light scattering measurements. These protofibrils, however, do not associate laterally in a normal manner, leading to an abnormal clot formation.
>
> KEYWORDS: Fibrin polymerization; Lateral aggregation; Human abnormal fibrinogen; Bleeding disorders.

Human fibrinogen variants provide a valuable tool to study the function of different domains of the molecule, and to correlate the dysfibrinogenemias with hæmostatic disorders. Fibrinogen Longmont was discovered in a young woman after routine laboratory screening. She and her mother had a mild history of bleeding and each had severe hemorrhagic episodes with child birth. The coagulation data of the propositus were uncommon for a dysfibrinogenemia diagnosis. Unlike the majority of abnor-

---

Address for correspondence: Susan T. Lord, Department of Pathology and Laboratory Medicine, CB 7525, University of North Carolina, Chapel Hill, NC 27599, USA. Voice: 919-966-3548; fax: 919-966-6718.
stl@med.unc.edu

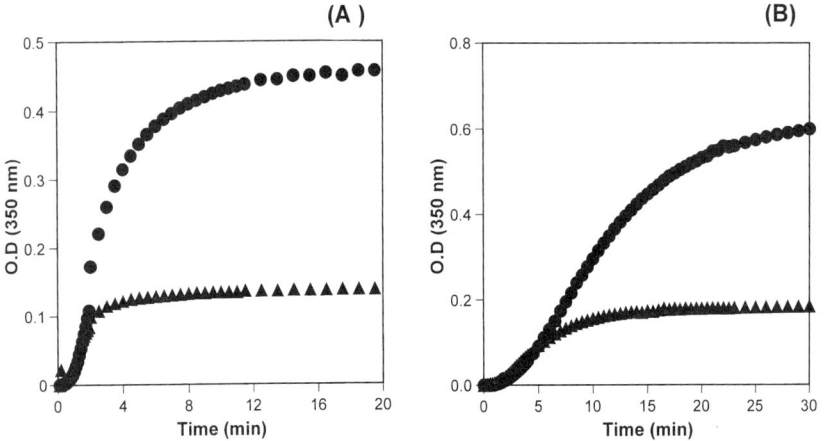

FIGURE 1. Fibrin polymerization profiles of normal fibrinogen and fibrinogen Longmont. Fibrin polymerization reaction was initiated by adding 0.1 U/ml thrombin (**A**) or 0.1 U/ml batroxobin (**B**) to a purified fibrinogen solution at 0.5 mg/ml in HEPES buffer (20 mM HEPES, pH 7.4, 150 mM NaCl). The increase of turbidity was recorded continuously at 350 nm. Normal fibrinogen, ● and fibrinogen Longmont, ▲.

mal fibrinogens, the functional and immunologic fibrinogen levels were both within the normal range. Nevertheless, a discrepancy was found, depending on the analysis method. Both reptilase and thrombin clotting times were prolonged with a spectrophotometric method, but both were normal with an electromechanical method.[1]

The defect in fibrin polymerization was confirmed with the purified protein. Both thrombin- and reptilase-catalyzed polymerization were impaired as measured by turbidity at 350 nm (see FIGURE 1), despite normal release of fibrinopeptides A and B. Since the amino-terminus of the fibrinogen molecule appeared to be normal, we analyzed the *a* polymerization site located in the C-terminus part of the molecule. Analysis of plasmin digests of fibrinogen Longmont by SDS-PAGE demonstrated that both calcium ions and the peptide GPRP provided full protection against further degradation of the C-terminal part of the γ-chain. Thus, the polymerization site *a* and its nearby high affinity calcium binding site were normal. Surprisingly, a fragment of 200 kDa was found in the plasmin digests of fibrinogen Longmont in the presence of calcium or GPRP. In the presence of EDTA the 200-kDa fragment was absent, and a smaller fragment with molecular weight (MW) greater than fragment D1 was seen. These unusual plasmin fragments were likely derived from the digestion of a novel fibrinogen species with MW greater than 340 kDa. The larger fibrinogen species, which was identified by electrophoresis in a 2% SDS-agarose gel, represented about 10% of the total fibrinogen.

These functional and structural studies narrowed the location of the fibrinogen anomaly to the fragment D1. DNA sequence analysis of the Bβ-gene of the propositus and her mother revealed a single substitution changing the codon for arginine 166 (CGT) to a cysteine (TGT). Mass spectroscopic analysis confirmed the mutation at the protein level, and demonstrated that the higher MW species seen on SDS-agarose gels resulted from disulfide bridged molecules, forming dimers. We purified the normal molecular weight fraction by size exclusion chromatography, and found that fibrin polymerization of this fraction was also impaired. This implies that only a fraction of the mutated molecules were disulfide bridged, and that the presence of dimeric molecules was not the sole cause of the defective polymerization.

Since thrombin released fibrinopeptide A normally, and the polymerization site *a* was functional, we examined protofibril formation and lateral aggregation. By SDS-PAGE, we found normal factor XIII-catalyzed crosslinking of the γ chains. By dynamic light scattering analysis, we found a normal time course of protofibril formation. Therefore, the impaired polymerization indicates that lateral aggregation of the protofibrils was abnormal. Thus, defective lateral aggregation resulted in a translucent fibrin clot, characterized by reduced turbidity.

On the basis of recent crystal structures of the D domain and the crosslinked D-dimer, the altered residue Bβ-166 is located in the coiled-coil region. This region is interposed between the E and D domains, and distant from the polymerization site *a*.[2] Two other abnormal fibrinogens, Lima (Aα Arg141→Ser), and Niigata (Bβ Asn160→Ser) have been identified in this region of the coiled-coil.[3,4] These abnormal fibrinogens were both characterized by abnormal polymerization related to an extra N-glycosylation. In combination, these data suggest that the distal part of the coiled-coil region contains sites that are necessary for lateral aggregation. This interpretation supports the fibrin polymerization model proposed by Fowler *et al.*[5] The authors suggested the existence of polymerization sites in the D domain, different than the polymerization sites *a* and *b*, that are necessary for lateral interaction between protofibrils (DD-lateral). The strength of each of the single interactions is weak; however, in the association of protofibrils, these interactions form simultaneously in a cooperative fashion.

In summary, we identified a new region of the fibrinogen molecule that might play an important role in the lateral association step of fibrin polymerization. The data reported showed that a single mutation of the residue Bβ Arg 166 to a Cys induced a dramatic effect on fibrin polymerization by affecting the lateral aggregation of the protofibrils.

ACKNOWLEDGMENTS

This work was supported by NIH Grant HL-31048. We thank Dr. D. Lavrinets for providing us blood samples.

REFERENCES

1. LEFKOWITZ, J. B., T. DEBOOM, A. WELLER, *et al.* 2000. Fibrinogen Longmont: a dysfibrinogenemia that causes prolonged clot-based test results only when using an optical detection method. Amer. J. Hematol. **63:** 149–155.
2. SPRAGGON, G., S.J. EVERSE & R.F. DOOLITTLE. 1997. Crystal structures of fragment D from human fibrinogen and its crosslinked counterpart from fibrin. Nature **389:** 455–462.

3. MAEKAWA, H., K. YAMAZUMI, S. MURAMATSU, et al. 1992. Fibrinogen Lima: a homozygous dysfibrinogen with an Aα-arginine-141 to serine substitution associated with extra N-glycosylation at Aα-asparagine-139. Impaired fibrin gel formation but normal fibrin-facilitated plasminogen activation catalyzed by tissue-type plasminogen activator. J. Clin. Invest. **90:** 67–76.
4. SUGO, T., C. NAKAMIKAWA, H. TAKANO, et al. 1999. Fibrinogen Niigata with impaired fibrin assembly: an inherited dysfibrinogen with a Bβ Asn-160 to Ser substitution associated with extra glycosylation at Bβ Asn-158. Blood. **94:** 3806–3813.
5. FOWLER, W.E., R.R. HANTGAN, J. HERMANS & H.P. ERICKSON. 1981. Structure of the fibrin protofibril. Proc. Natl. Acad. Sci. U.S.A. **78:** 4872–4876.

Determinants of Thrombin Specificity

ENRICO DI CERA AND ANGELENE M. CANTWELL

*Department of Biochemistry and Molecular Biophysics,
Washington University School of Medicine, St. Louis, Missouri, USA*

ABSTRACT: Thrombin recognizes a number of natural substrates that are responsible for important physiologic functions. Its high specificity is controlled by residues within the active site, and by separate recognition sites located on the surface of the enzyme. A number of studies have addressed the question of how thrombin changes its specificity from fibrinogen to protein C, switching from a procoagulant to an anticoagulant enzyme. Site directed mutagenesis studies have revealed important aspects of how this switch takes place. Specifically, residues W215 and E217 have emerged as key residues in controlling the interaction with fibrinogen in that mutation of these residues compromises the procoagulant function of the enzyme up to 500-fold. The loss of fibrinogen clotting reaches 20,000-fold in the double mutant W215A/E217A, whereas protein C activation is compromised less than sevenfold. These findings demonstrate that thrombin specificity can be dissected at the molecular level using Ala-scanning mutagenesis and the procoagulant function of the enzyme can be abrogated rationally and selectively. It is now possible to extend this strategy to the study of other interactions of thrombin, as well as to related serine proteases.

KEYWORDS: Thrombin specificity; Fibrinogen clotting; Serine proteases.

INTRODUCTION

Thrombin, a serine protease of the chymotrypsin family, plays two important and opposing functions in the blood. It acts as a procoagulant when it converts fibrinogen into an insoluble fibrin clot that anchors platelets to the site of lesion and initiates processes of wound repair. This action is reinforced and amplified by activation of the transglutaminase factor XIII that covalently stabilizes the fibrin clot, the inhibition of fibrinolysis, the proteolytic activation of factors V, VIII, and XI, and the cleavage of the protease-activated receptor-1 (PAR-1) on the platelet surface that leads to platelet activation and aggregation.[1] In contrast, thrombin acts as an anticoagulant when it activates protein C. This function unfolds upon binding to thrombomodulin, a receptor on the membrane of endothelial cells. Binding of thrombomodulin competitively suppresses the ability of thrombin to cleave fibrinogen or PAR-1, but enhances by more than 1,000-fold the specificity of the enzyme toward the zymogen protein C.[2] Activated protein C cleaves and inactivates factors Va and VIIIa, two essential cofactors of coagulation factors Xa and IXa that are required for thrombin generation, thereby downregulating both the amplification and progression

---

Address for correspondence: Enrico Di Cera, M.D., Department of Biochemistry and Molecular Biophysics, Washington University School of Medicine, 660 S. Euclid Avenue, Box 8231, St. Louis, MO 63110, USA. Voice: 314-362-4185; fax: 314-362-7183.
enrico@caesar.wustl.edu

of the coagulation cascade.[3] Scavenging of thrombin by thrombomodulin and activation of protein C in the microcirculation constitute the natural anticoagulant pathway that prevents massive intravascular conversion of fibrinogen into an insoluble clot upon thrombin generation. In addition, thrombin is irreversibly inhibited at the active site by the serine protease inhibitor antithrombin III with the assistance of heparin. Besides its roles in the extracellular milieu, thrombin elicits a variety of important effects in a number of cell types via cleavage and signaling by PAR-1, PAR-3, and PAR-4 receptors.[4,5]

The remarkable ability of thrombin to interact with a variety of substrates, inhibitors, and effectors has long stimulated interest in understanding the molecular basis of its specificity. This interest is borne out by the possibility of dissociating thrombin functions to better understand structure–function relations and the role of each specific interaction *in vivo*. Here, we review the current knowledge on thrombin specificity and the epitopes for substrate recognition. We also describe new data about the role of W215 and E217 in specifically recognizing fibrinogen over protein C.

THROMBIN STRUCTURE

Thrombin is composed of two polypeptide chains of 36 (*A* chain) and 259 (*B* chain) residues that are covalently linked through a disulfide bond.[6] The *A* chain runs in the back of the molecule, opposite to the front hemisphere of the *B* chain that hosts the entrance to the active site and all functional epitopes of the enzyme. The *B* chain has the typical fold of serine proteases,[7] with two six-stranded β-barrels of similar structure that pack together asymmetrically to accommodate at their interface the residues of the catalytic triad H57, D102, and S195. The P1-P3 specificity[8] of thrombin has been known for a long time[9] and has been the subject of recent systematic studies.[10,11] The trypsin-like specificity for Arg residues at P1 is conferred to thrombin by the presence of D189 in the S1 site occupying the bottom of the catalytic pocket. The Arg→Lys replacement at P1 in chromogenic substrates causes the value of k_{cat}/K_m to drop approximately 10-fold.[10] Thrombin has a preference for small and hydrophobic side chains at P2 (especially Pro) that pack tightly against the hydrophobic wall of the S2 site. This site is defined by L99 and residues P60b-P60c-W60d of the W60d loop that restrict access to the active site. Aromatic and hydrophobic residues at P3-P4 make favorable interactions with a site defined by W215, Y60a, and L99 immediately to the left of the W60d loop.

A structural domain of thrombin that controls specificity toward fibrinogen, PAR-1 and protein C is the Na$^+$ binding site located between the two surface loops 220 and 180.[12] The site is located 15–20 Å away from the catalytic triad and lies within 5 Å from D189 in the S1 site within a cylindrical cavity occupied by up to sixteen water molecules. The bound Na$^+$ is coordinated octahedrally by two carbonyl O atoms from R221a and K224, and four buried water molecules. One of these water molecules hydrogen bonds to the side chain of D189 establishing a direct communication between the cation and the S1 site. The Na$^+$ site is stabilized by three ion pairs, R221a-E146, K224-E217, and D222-R187. The first ion pair connects the Na$^+$ site to the autolysis loop. The second ion pair bridges two of the three β-strands that define the Na$^+$ site and the S1 site. The third ion pair connects the loops 180 and 220.

The Na^+ loop (220-loop) is also present with similar architecture in all serine proteases of the chymotrypsin family and plays a role in defining the primary specificity of the enzyme.[13,14]

Na^+ is the major procoagulant cofactor of thrombin. The fast (Na^+-bound) form of thrombin has higher catalytic activity toward fibrinogen and the PARs compared to the slow (Na^+-free) form, whereas the two forms cleave the anticoagulant protein C with similar values of k_{cat}/K_m.[1] Mutations that reduce Na^+ binding to thrombin are therefore expected to elicit an anticoagulant effect, leading to possible bleeding phenotypes. This expectation has been confirmed by several mutations introduced in the Na^+ binding environment to selectively compromise the procoagulant activity of thrombin. The effects produced by these mutations are exquisitely allosteric because none of the mutated residues contacts fibrinogen or PAR1 in the available crystal structures.[15,16] These mutants also explain the effects observed with other mutations reported in the literature. The R221aA substitution breaks the R221a-E146 ion pair and reduces Na^+ binding. The clotting activity of this mutant is affected more than protein C activation thereby enhancing the anticoagulant potency of the enzyme.[17] The R221aA mutant behaves like the naturally occurring mutant thrombin Salakta (E146A) where the ion pair is broken from the side of the autolysis loop.[18] The properties of thrombin Salakta were puzzling before the identification of the Na^+ binding site, because E146 was known not to interact with fibrinogen.[15] The K224A substitution breaks the K224-E217 ion pair and reduces Na^+ binding. This results in a 30-fold loss of clotting activity with a modest threefold loss of protein C activation.[17] The properties of this mutant are similar to those of the anticoagulant mutant E217A reported by Gibbs et al.,[19] where the ion pair is broken from its proximal side. The D222A substitution impairs Na^+ binding and the clotting activity of the enzyme, which results in properties similar to those of the naturally occurring thrombin, Greenville (R187Q), that is associated with a bleeding phenotype of moderate gravity.[20]

In addition to residues within the active site and the Na^+ binding site, thrombin specificity depends on other regions of the molecule (see FIGURE 1). The loop centered on K70 defines exosite I and is homologous to the Ca^{2+} binding loop of trypsin and chymotrypsin. In these pancreatic proteases, Ca^{2+} stabilizes the fold and confers increased resistance to proteolytic digestion. In thrombin, the need for Ca^{2+} is eliminated by insertion of the K70 moiety into the cavity available for binding the divalent cation. Exosite I contains several positively charged residues that give rise to an intense electrostatic field and flank a hydrophobic patch that spans the entire length of the exosite, from I24 to M84. The field provides electrostatic steering and optimal preorientation for fibrinogen, PAR-1, thrombomodulin and the natural inhibitor hirudin to facilitate formation of a productive complex upon binding. Structural and functional data support exosite I as a binding epitope for fibrinogen,[21] thrombomodulin,[22] and the thrombin receptors PAR-1[23] and PAR-3.[24] Proteolytic cleavage of exosite I with excision of the fragment I68-R77a produces γ-thrombin and abrogates fibrinogen, PAR-1, PAR-3, and thrombomodulin binding, but has no effect on the catalytic activity of the enzyme toward small chromogenic substrates that bind only to the active site.[21,23,24] Residues R67, R73, R75, and R77a make ionic interactions with hirudin.[25] Of these, R75 and R77a contact thrombomodulin, together with K110 and

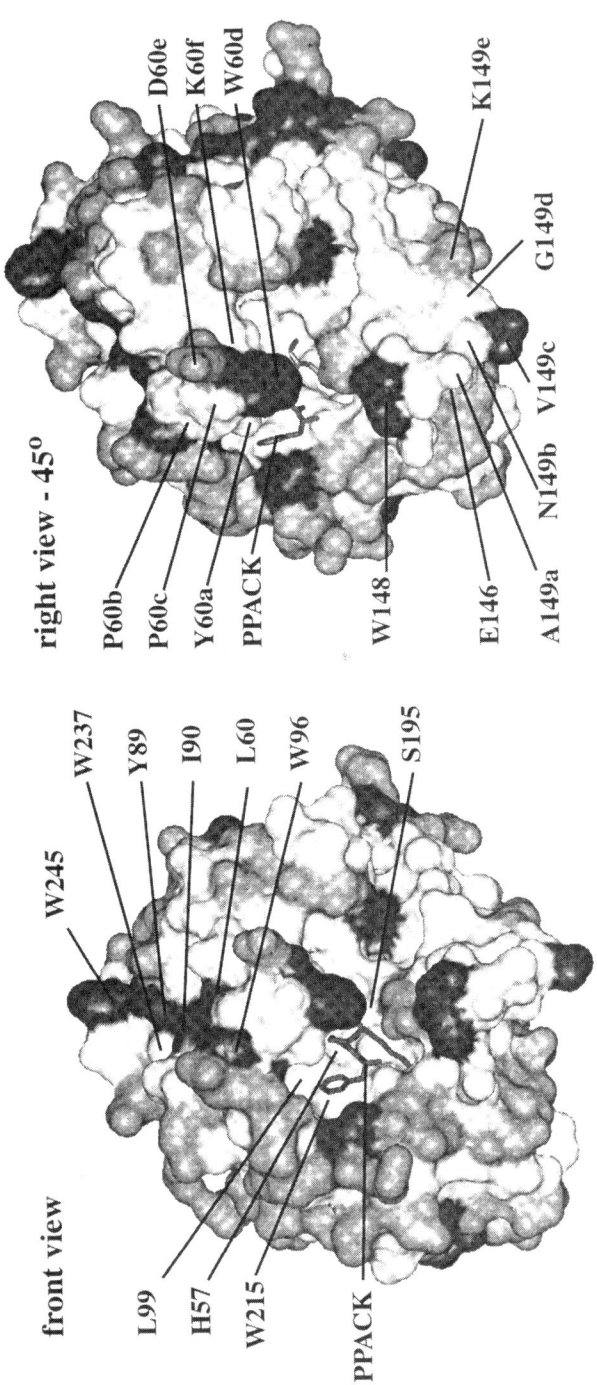

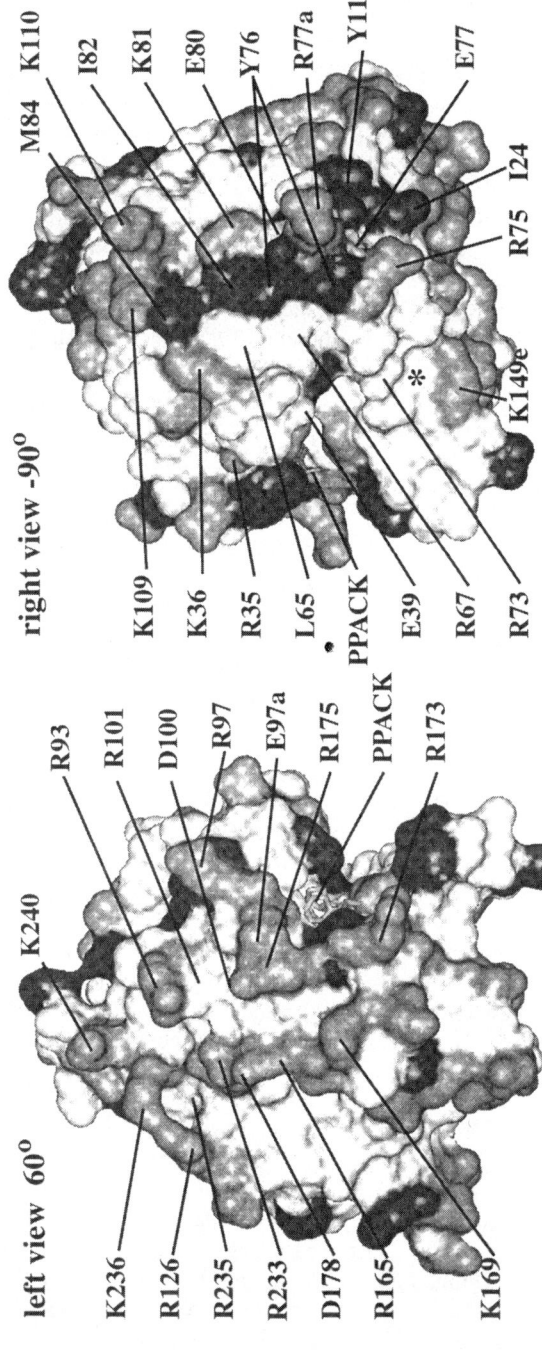

FIGURE 1. Solvent accessible surface of thrombin in various orientations (rotated as indicated, relative to the standard front view). Residues on the surface are shaded according to their nature. Charged residues greater than 25% exposed to the solvent are shown in *gray*. Aromatic and hydrophobic residues greater than 25% exposed to the solvent are shown in *black*. All other residues and those less than 25% exposed to the solvent (charged, aromatic, hydrophobic, polar, and neutral) are shown in *white*. Important residues are labeled. The putative Cl⁻ binding site (*) is located between K149e and R73 in exosite I. The active site inhibitor PPACK is shown as a *stick model*.

a hydrophobic patch including Y76 and I82.[22] Y76 is also important for fibrinogen binding.[26]

The autolysis loop spans nine residues (E146–K149e), and contains a five residue insertion (A149a–K194e) relative to chymotrypsin. The loop is strategically located between the 220-loop where Na^+ binds and exosite I and likely provides a means of communicating changes between these domains. K149e is close to exosite I and there is strong evidence that this residue may be part of a binding site for Cl^- together with R73. Binding of Cl^- promotes Na^+ binding (and vice versa), but opposes fibrinogen and hirudin binding in a competitive manner.[27] The effect of Cl^- is much reduced in the proteolytic derivative γ-thrombin, where a cleavage occurs after K149e and the entire segment I68–R77a in exosite I is excised, as well as in bovine thrombin, where residue 149e is Glu. Furthermore, R73 makes an ion pair with D55 of hirudin[25] and binding of the hirudin mutant D55N is not inhibited by Cl^-. Cl^- bound between K149e and R73 may function like a switch to communicate changes between the autolysis loop and exosite I. This switch would act in concert with the R221a–E146 ion pair to propagate changes from the Na^+ site to exosite I. The role of other residues in the loop remains to be defined. The W148S mutation is inconsequential on thrombin function,[26] as is the deletion of the five residue insertion.[28] When the ETWT proximal sequence is swapped with the SSGT sequence of trypsin,[29] the mutant behaves like R221aA[17] suggesting that only E146 is important due to its ion pair with R221a. Deletion of E146-T147-W148 compromises clotting and protein C activation 20-fold.[30] However, deletion of the entire loop compromises selectively fibrinogen and PAR-1 binding and reduces Na^+ binding.[28,31] These results suggest that the autolysis loop contains residues important for recognition of fibrinogen and PAR-1, but not protein C and thrombomodulin.

On the side of the enzyme opposite to exosite I, a C-terminal helix and its neighbor domains host a number of positively charged residues and define exosite II. Exosite II carries a strong electrostatic field due to the presence of numerous basic residues and is thought to be the locale for binding of heparin and the chondroitin sulfate moiety of thrombomodulin. Heparin enhances inhibition of thrombin by antithrombin III via a template mechanism in which a high affinity heparin–antithrombin III complex is first formed, then is docked into exosite II and is finally bound to the thrombin active site by electrostatic coupling.[32] Notably, exosite II contains no hydrophobic patches on its surface, unlike exosite I. On the other hand, it contains two ion clusters that stabilize the tertiary structure of the site. These clusters contain partially buried residues organized in the quartet E97a-D100-R101-R175 and the triplet R165-D178-R233 that divide the site in two domains. The upper domain contains K236, K240, R93, R235, R101, R233, and R126. The lower domain contains R97, R165, R175, R173, and K169. Residues R93, R233, K236, and K240 in the upper domain have been shown to be involved in heparin binding from chemical modifications of Lys residues[33] or charge reversal mutagenesis.[34,35] R175 has been included in this epitope in one study,[35] but ruled out in another study involving Ala mutants.[36] The latter study defined a larger epitope for heparin binding that also includes R97 and R101. Furthermore, it ruled out exosite II as a binding epitope for antithrombin III. Ala mutants of R93, R97, and R101 show that R93 is more important for heparin binding, whereas R97 and R101 play a dominant role in thrombomodulin binding.[37]

More recently, exosite II has been shown to provide the locale for thrombin interaction with the platelet receptor GpIb.[38]

FUNCTIONAL MAPPING OF EPITOPES

Even when structural information is available about the binding epitopes for a given substrate, the question remains about the importance of specific residues in the recognition process. A contact documented in the crystal structure may or may not be associated with a significant binding energy. On the other hand, a residue may contribute substantially to ligand recognition even though it makes no apparent contact with the ligand in the crystal structure. For example, Ala substitutions of R221a, K224, and Y225 in thrombin affect fibrinogen recognition up to 100-fold,[17,39] but none of these residues makes contact with fibrinogen in the crystal structure. On the other hand, Ala replacement of R173 has only a modest effect on fibrinogen binding,[26] although this residue makes an ion pair interaction with E11 of the fibrinogen Aα chain. Epitopes should be defined from a combination of structural and functional information that is derived from site directed mutagenesis of specific residues. The strategy enables the quantification of the energetic contribution of contacts identified in the crystal structure.

Numerous mutagenesis studies of thrombin have been reported in the literature providing a great deal of information on how the enzyme recognizes fibrinogen and protein C. Of particular importance has been the observation that some residues appear to be involved in the binding of different substrates, whereas others recognize preferentially only a single substrate. The first report that thrombin functions could be dissociated came from the observation that R73 in exosite I is important for recognition of fibrinogen and thrombomodulin, whereas the neighboring R75 is more important for thrombomodulin binding and K60f in the W60d loop is more important for fibrinogen binding.[40] Another pioneering study identified E192 near the entrance to the active site as a residue responsible for the poor activation of protein C by thrombin in the absence of thrombomodulin.[41] The isosteric replacement of E192 with Gln drastically improves specificity toward protein C in the absence of the cofactor, but has no effect on the interaction with small substrates or fibrinogen.

A systematic Ala scan of thrombin residues has been carried out by Tsiang et al.[36,42] In this study, more than 70 Ala mutants of solvent exposed residues were characterized for their interaction with fibrinogen, protein C, and antithrombin III. A striking discovery emerging from these studies was that the balance between procoagulant and anticoagulant activities of thrombin, measured respectively as the ability to cleave fibrinogen or protein C (in the presence of thrombomodulin), could be altered substantially by mutation of a single residue, E217.[19] The E217A mutant showed 40-fold reduced activity toward fibrinogen and compromised protein C activation only twofold. More recently, Hall et al.[43] have shown that the binding epitopes for protein C and the thrombin activated fibrinolytic inhibitor (TAFI) do not overlap completely. The region around D60e in the W60d loop is more important for TAFI binding, whereas regions around the Na$^+$ binding site are more important for the binding of protein C. W60d appears to be important for both protein C and TAFI binding. W60d, together with E217 and R221a, have recently been assigned to the binding epitope for antithrombin III,[36] showing that residues around the 220-loop of thrombin may be directly involved in recognition of protein C and antithrombin III.

RATIONAL DESIGN OF THROMBIN SPECIFICITY

The structural and functional information gathered on thrombin enables the rational engineering of specificity of this enzyme. Of particular importance is the recent observation that the anticoagulant properties of thrombin can be enhanced by selectively compromising the cleavage of fibrinogen. This can be accomplished either by targeting E217 near the entrance to the active site,[19] or residues that control Na^+ binding and whose mutation stabilize the anticoagulant slow form.[17] Practically all mutants of thrombin defective for Na^+ binding reported to date show a gain in anticoagulant activity because of the larger loss of fibrinogen clotting relative to protein C activation.[31] A mutation of thrombin causes a pure allosteric effect only if it shifts the slow↔fast equilibrium without compromising other properties of the enzyme. In this case, the properties of the mutant can be predicted from those of the slow and fast forms of the wildtype and the Na^+ binding affinity of the mutant. Mutation of a residue that not only affects Na^+ binding, but also protein stability or direct ligand recognition, may generate effects in addition to the perturbation of the slow↔fast equilibrium. This is the case for residue W215, that is involved in direct binding of small chromogenic substrates and fibrinogen and has been shown recently to play a key role in the procoagulant function of thrombin.[44] W215 is absolutely conserved in thrombin from hagfish to human[45] and highly conserved in the chymotrypsin family. The aromatic nature of residue 215 is considered essential for high affinity substrate binding and, indeed, ablation of this property in the W215A mutant produces a significant increase in K_m, with a smaller effect on k_{cat} in the chromogenic substrate H-D-Phe-Pro-Arg-p-nitroanilide (FPR).[44] The crystal structure of thrombin inhibited with H-D-Phe-Pro-Arg-CH_2Cl (PPACK), that structurally resembles FPR, shows an edge-to-face interaction between H-D-Phe at P3 and W215.[6] Similarly, residue F8 of the fibrinogen Aα chain makes an edge-to-face interaction with W215.[15] On the other hand, the structure of thrombin inhibited at the active site with a fragment of PAR-1, carrying the sequence LDPR into the active site, shows Asp at P3 pointing away from W215 and making a water mediated contact with R221a in the Na^+ binding loop.[16] Similarly, protein C carries an Asp residue at P3 and may not require a contact with W215.[22]

The foregoing results presage improved anticoagulant properties of mutants of W215. Indeed, mutation of W215 to Ala affects the ability of thrombin to release fibrinopeptide A from fibrinogen 500-fold (see TABLE 1). Although W215A shows weak Na^+ binding, its loss of fibrinogen clotting largely exceeds that of the slow form[44] and therefore the W215A substitution generates more than a simple shift of the slow↔fast equilibrium. Na^+ binding does not restore activity because the release of fibrinopeptide A by the fast form of W215A occurs with a specificity constant 300-fold lower than that of wildtype.

An interesting feature of the role of W215 in the interaction of thrombin with fibrinogen is that the W215A mutant releases fibrinopeptide A from fibrinogen and fibrinopeptide B from fibrin with similar kinetics and rate constants. The Shafer

TABLE 1. Specificity of wildtype and mutant thrombins[a]

	wt	W215A[b]	E217A[c]	W215A/E217A
Fibrinopeptide A k_{cat}/K_m ($\mu M^{-1}s^{-1}$)	17 ± 1	0.034 ± 0.002	0.33 ± 0.01	0.00089 ± 0.00007
Fibrinopeptide B k_{cat}/K_m ($\mu M^{-1}s^{-1}$)	8.1 ± 0.5	0.053 ± 0.003	n.d.	0.0021 ± 0.0001
Protein C k_{cat}/K_m ($\mu M^{-1}s^{-1}$)[d]	0.22 ± 0.01	0.075 ± 0.006	0.095 ± 0.009	0.033 ± 0.002
RAP[e]	1	170 ± 20	22 ± 3	2800 ± 300

[a]Experimental conditions: 5 mM Tris, 0.1% PEG, 145 mM NaCl, pH 7.4, 37°C.
[b]From Reference 44.
[c]From Reference 19.
[d]In the presence of 100 nM rabbit thrombomodulin and 5 mM CaCl$_2$.
[e]Relative anticoagulant potency calculated as the ratio of the rate for protein C activation over the rate for fibrinopeptide A release, relative to the same ratio of wildtype thrombin.

mechanism of release of fibrinopeptides[46] states that fibrinopeptide A is released first from fibrinogen leading to formation of fibrin I monomers. These monomers aggregate to form fibrin I protofibrils, from which fibrinopeptide B is released to give rise to fibrin II protofibrils that form the scaffold of the fibrin clot. Under conditions where thrombin concentration is rate limiting, a lag phase occurs after the release of fibrinopeptide A before appreciable amounts of fibrinopeptide B can be detected. This mechanism does not hold for the W215A mutant that releases fibrinopeptide B without delay and with a rate constant comparable to that of fibrinopeptide A. This demonstrates that steric constraints in the S3 and S4 sites of thrombin oppose the release of fibrinopeptide B directly from fibrinogen and that these constraints are removed with the W215A substitution. As a result, fibrinopeptide B is released directly from fibrinogen, like fibrinopeptide A, and starts to accumulate before fibrin I protofibrils are formed. The drastic perturbation of the S3 site therefore unravels the ability of thrombin to cleave fibrinogen at the Bβ chain, although the rate of cleavage is too small to measure under physiologic conditions. The drastic perturbation of substrate binding seen in the case of fibrinogen is not matched by protein C in the presence of thrombomodulin. This substrate experiences only a modest loss of specificity, as the value of k_{cat}/K_m drops threefold in the W215A mutant.

The effects observed in the W215A mutant and those previously reported for the E217A mutant[19] provided an opportunity to rationally design a thrombin derivative where fibrinogen binding is practically abolished, but protein C activation is retained. The mutation of W215 was combined with that of E217 in the double mutant W215A/E217A. Mutations that cause distinct structural and functional perturbations are expected to show additive effects when combined. E217 makes a contact with G12 of the fibrinogen Aα chain, and ion pairs with K224 to stabilize the Na$^+$ binding environment. The E217A mutant has compromised fibrinogen clotting 40-fold, whereas protein C activation is decreased only twofold.[19] The structural perturbation in the case of the W215A mutant is distinct because W215 contacts F8 of the fibrinogen Aα chain. Again, fibrinogen clotting is compromised more than

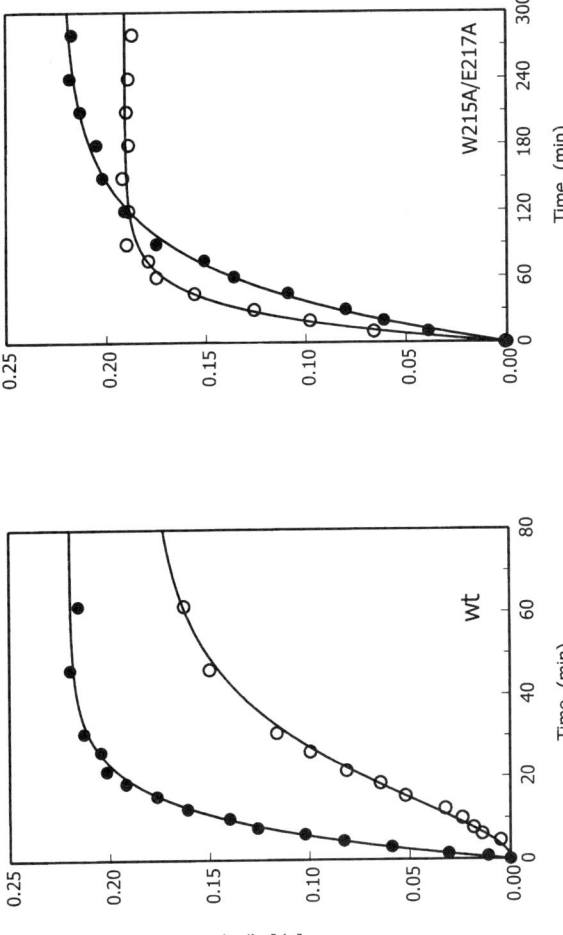

FIGURE 2. Progress curves of the release of fibrinopeptides A (●) and B (○) by wildtype and the W215A/E217A mutant of thrombin. *Continuous lines* were drawn using Equations 3a and 3b of Vindigni and Di Cera,[46] with κ_1 and κ_2 expressed as the value of $s = k_{cat}/K_m$ for the release of fibrinopeptide (s_1 for fibrinopeptide A, s_2 for fibrinopeptide B) times the concentration of thrombin e_T. The best-fit parameter values are: wt, $[FpA]_\infty = 0.22 \pm 0.01$ μM, $s_1 = 17 \pm 1$ μM^{-1}s^{-1}, $f[FpB]_\infty = 0.18 \pm 0.01$ μM, $s_2 = 8.1 \pm 0.5$ μM^{-1}s^{-1}, $e_T = 0.1$ nM; W215A/E217A, $[FpA]_\infty = 0.22 \pm 0.01$ μM, $s_1 = 0.00089 \pm 0.00007$ μM^{-1}s^{-1}, $f[FpB]_\infty = 0.19 \pm 0.01$ μM, $s_2 = 0.0021 \pm 0.0001$ mM^{-1}s^{-1}, $e_T = 300$ nM (see also TABLE 1). In the case of W215A/E217A, the parameters for the release of fibrinopeptide B refer to an equation of the same form as Equation 3a of Vindigni and Di Cera[46] because no lag phase was observed. Experimental conditions were: 5 mM Tris, 0.1% PEG, 145 mM NaCl, pH 7.4, 37°C. Note that the release of fibrinopeptides by the W215A/E217A is approximately sevenfold slower compared to wildtype, although the concentration used in the assay is 3,000-fold higher.

protein C activation, in this case 500-fold versus threefold.[44] The double mutant W215A/E217A has properties that are additive relative to the individual mutations. The value of k_{cat}/K_m for the release of fibrinopeptide A from fibrinogen is decreased 20,000-fold, whereas protein C activation in the presence of thrombomodulin is compromised less than sevenfold (TABLE 1). This gives the W215A/E217A mutant a relative anticoagulant potency (RAP) over 2,800-fold that of wildtype, which is by far the largest RAP value reported to date. As for the W215A mutant, the release of fibrinopeptide B occurs without delay and is twice as fast as that of fibrinopeptide A. The data documenting these striking properties are shown in FIGURES 2 and 3. Finally, the order of fibrinopeptide release is inconsequential on the shape of the clotting curve, and most likely on the structure of the clot, because the curve obtained with wildtype can be reproduced exactly with the mutant W215A/E217A used at concentrations approximately 20,000-fold higher (data not shown).

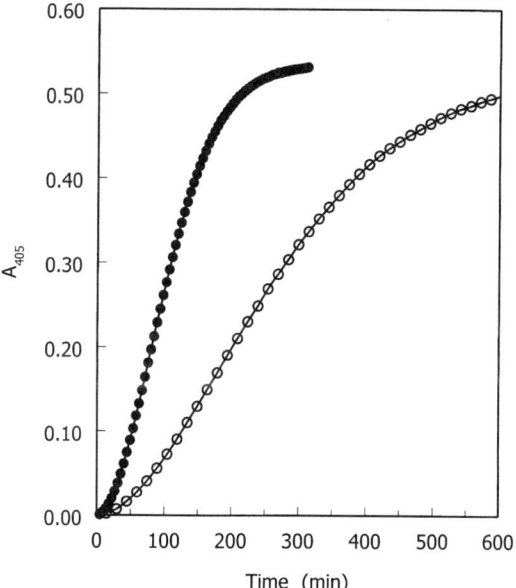

FIGURE 3. Progress curves of the activation of protein C by wildtype (●) and the W215A/E217A mutant (○) of thrombin, in the presence of thrombomodulin. The data depict the absorbance change at 405 nm due to the release of *p*-nitroaniline from the chromogenic substrate H-D-Asp-Arg-Arg-*p*-nitroanilide (DRR) by activated protein C, after activation by thrombin.[47] *Continuous lines* were drawn from integrated rate equations using the values of k_{cat}/K_m for the hydrolysis of protein C reported in TABLE 1. Experimental conditions were: 5 mM Tris, 0.1% PEG, 145 mM NaCl, 5 mM $CaCl_2$, 100 mM rabbit thrombomodulin, 400 nM protein C, 50 μM DRR, pH 7.4, 37°C. The concentration of thrombin wildtype or W215A/E217A is 0.2 nM. Note that the curve relative to the activation of protein C by the mutant W215A/E217A is only sevenfold slower compared to wildtype, as opposed to the effect shown in FIGURE 2 for the cleavage of fibrinogen.

CONCLUSIONS

We currently have a fairly complete understanding of the structural basis of thrombin specificity. Functional data emerged from recent site directed mutagenesis studies have complemented the structural information and revealed important determinants that control thrombin interaction with physiologic ligands. The knowledge gained from these studies can now be used to alter thrombin specificity in a predictable manner and to generate derivatives that lack specific functions of the enzyme. Anticoagulant thrombins that are defective for fibrinogen cleavage and retain protein C activation can be engineered and optimized, as it has been shown for mutations of W215 and E217. Preliminary data from our laboratory indicate that other functions of thrombin, such as the cleavage of PAR-1, can be dissociated from clotting, protein C activation, and the cleavage of the other thrombin receptors PAR-3 and PAR-4. These exciting developments validate the approach based on Ala-scanning mutagenesis of critical residues of the enzyme and the combination of individual mutations to achieve additive functional effects. Extension of this strategy to the study of other enzymes involved in blood coagulation, or other multifunctional proteins, will determine whether the rules for molecular recognition being unraveled for thrombin are of general validity.

ACKNOWLEDGMENTS

This work was supported in part by NIH research grants HL49413 and HL58141.

REFERENCES

1. DI CERA, E., Q.D. DANG & Y.M. AYALA. 1997. Molecular mechanisms of thrombin function. Cell Mol. Life Sci. **53:** 701–730.
2. ESMON, C.T. 1989. The roles of protein C and thrombomodulin in the regulation of blood coagulation. J. Biol. Chem. **264:** 4743–4746.
3. DAVIE, E.W., K. FUJIKAWA & W. KISIEL. 1991. The coagulation cascade: initiation, maintenance and regulation. Biochemistry **30:** 10363–10370.
4. XU, W.F., et al. 1998. Cloning and characterization of human protease-activated receptor 4. Proc. Natl. Acad. Sci. U.S.A. **95:** 6642–6646.
5. COUGHLIN, S.R. 1999. How the protease thrombin talks to cells. Proc. Natl. Acad. Sci. U.S.A. **96:** 11023–11027.
6. BODE, W., D. TURK & A. KARSHIKOV. 1992. The refined 1.9-Å X-ray crystal structure of D-Phe-Pro-Arg-chloromethylketone-inhibited human α-thrombin: structure analysis, overall structure, electrostatic properties, detailed active-site geometry, and structure-function relationships. Protein Sci. **1:** 426–471.
7. LESK, A.M. & W.D. FORDHAM. 1996. Conservation and variability in the structures of serine proteinases of the chymotrypsin family. J. Mol. Biol. **258:** 501–537.
8. SCHECHTER, I. & A. BERGER. 1967. On the size of the active site in proteases. I. Papain. Biochem. Biophys. Res. Commun. **27:** 157–162.
9. LOTTENBERG, R., et al. 1983. The action of thrombin on peptide p-nitroanilide substrates. Substrate selectivity and examination of hydrolysis under different reaction conditions. Biochim. Biophys. Acta **742:** 539–557.
10. VINDIGNI, A., Q.D. DANG & E. DI CERA. 1997. Site-specific dissection of substrate recognition by thrombin. Nature Biotechnol. **15:** 891–895.

11. BACKES, B.J., et al. 2000. Synthesis of positional-scanning libraries of fluorogenic peptide substrates to define the extended substrate specificity of plasmin and thrombin. Nat. Biotechnol. **18:** 187–193.
12. DI CERA, E., et al. 1995. The Na^+ binding site of thrombin. J. Biol. Chem. **270:** 22089–22092.
13. CRAIK, C.S., et al. 1985. Redesigning trypsin: alteration of substrate specificity. Science **228:** 291–297.
14. HEDSTROM, L., L. SZILAGYI & W.J. RUTTER. 1992. Converting trypsin to chymotrypsin: the role of surface loops. Science **255:** 1249–1253.
15. STUBBS, M., et al. 1992. The interaction of thrombin with fibrinogen: a structural basis for its specificity. Eur. J. Biochem. **206:** 187–195.
16. MATHEWS, I.I., et al. 1994. Crystallographic structures of thrombin complexed with thrombin receptor peptides: existence of expected and novel binding modes. Biochemistry **33:** 3266–3279.
17. DANG, Q.D., E.R. GUINTO & E. DI CERA. 1997. Rational engineering of activity and specificity in a serine protease. Nature Biotechnol. **15:** 146–149.
18. MIYATA, T., et al. 1992. Prothrombin Salakta: substitution of Glu466 by Ala reduced the fibrinogen clotting activity and the esterase activity. Biochemistry **31:** 7457–7462.
19. GIBBS, C.S., et al. 1995. Conversion of thrombin into an anticoagulant by protein engineering. Nature **378:** 413–416.
20. HENRIKSEN, R.A., et al. 1998. Prothrombin Greenville, $Arg^{517}{\rightarrow}Gln$, identified in an individual heterozygous for dysprothrombinemia. Blood **91:** 2026–2031.
21. HOFSTEENGE, J., P.J. BRAUN & S.R. STONE. 1988. Enzymatic properties of proteolytic derivatives of human α-thrombin. Biochemistry **27:** 2144–2151.
22. FUENTES-PRIOR, P., et al. 2000. Structural basis for the anticoagulant activity of the thrombin-thrombomodulin complex. Nature **404:** 518–525.
23. VU, T.K.H., et al. 1991. Domains defining thrombin receptor interactions. Nature **353:** 674–677.
24. ISHIHARA, H., et al. 1997. Protease-activated receptor 3 is a second thrombin receptor in humans. Nature **386:** 502–505.
25. RYDEL, T.J., et al. 1991. Refined structure of the hirudin-thrombin complex. J. Mol. Biol. **221:** 583–601.
26. GUINTO, E.R., et al. 1995. Identification of residues linked to the slow→fast transition of thrombin. Proc. Natl. Acad. Sci. U.S.A. **92:** 11185–11189.
27. AYALA, Y.M. & E. DI CERA. 1994. Molecular recognition by thrombin. Role of the slow→fast transition, site-specific ion binding energetics and thermodynamic mapping of structural components. J. Mol. Biol. **235:** 733–746.
28. DANG, Q.D., M. SABETTA & E. DI CERA. 1997. Selective loss of fibrinogen clotting in a loop-less thrombin. J. Biol. Chem. **272:** 19649–19651.
29. DI BELLA, E.E. & H.A. SCHERAGA. 1996. The role of the insertion loop around tryptophan 148 in the activity of thrombin. Biochemistry **35:** 4427–4433.
30. LE BONNIEC, B.F., E.R. GUINTO & C.T. ESMON. 1992. Interaction of des-ETW with antithrombin III, the Kunitz inhibitors, thrombomodulin and protein C. J. Biol. Chem. **267:** 19341–19348.
31. DI CERA, E. 1998. Anticoagulant thrombins. Trends Cardiovasc. Med. **8:** 340–350.
32. OLSON, S.T., H.R. HALVORSON & K.I. BJÖRK. 1991. Quantitative characterization of the thrombin-heparin interaction. Discrimination between specific and nonspecific binding models. J. Biol. Chem. **266:** 6342–6352.
33. CHURCH, F.C., et al. 1989. Structural and functional properties of human α-thrombin, phosphopyridoxylated α-thrombin, and γ_T-thrombin. Identification of lysyl residues in α-thrombin that are critical for heparin and fibrin(ogen) interactions. J. Biol. Chem. **264:** 18419–18425.
34. SHEEHAN, J.P. & J.E. SADLER. 1994. Molecular mapping of the heparin-binding exosite of thrombin. Proc. Natl. Acad. Sci. U.S.A. **91:** 5518–5522.
35. GAN, Z.R., et al. 1994. Identification of basic amino acid residues in thrombin essential for heparin-catalyzed inactivation by antithrombin III. J. Biol. Chem. **269:** 1301–1305.

36. TSIANG, M., A.K. JAIN & C.S. GIBBS. 1997. Functional requirements for inhibition of thrombin by antithrombin III in the presence and absence of heparin. J. Biol. Chem. **272:** 12024–12029.
37. HE, X., et al. 1997. Influence of Arginines 93, 97, and 101 of thrombin to its functional specificity. Biochemistry **36:** 8969–8976.
38. DE CRISTOFARO, R., et al. 2000. The Asp272-Glu282 region of platelet glycoprotein Ibα interacts with the heparin-binding site of α-thrombin and protects the enzyme from the heparin-catalyzed inhibition by antithrombin III. J. Biol. Chem. **275:** 3887–3895.
39. GUINTO, E.R., et al. 1999. Unexpected crucial role of residue 225 in serine proteases. Proc. Natl. Acad. Sci. U.S.A. **96:** 1852–1857.
40. WU, Q., et al. 1991. Single amino acid substitutions dissociate fibrinogen clotting and thrombomodulin binding activities of human thrombin. Proc. Natl. Acad. Sci. U.S.A. **88:** 6775–6779.
41. LE BONNIEC, B.F. & C.T. ESMON. 1991. Glu192→Gln substitution in thrombin mimics the catalytic switch induced by thrombomodulin. Proc. Natl. Acad. Sci. U.S.A. **88:** 7371–7375.
42. TSIANG, M., et al. 1995. Functional mapping of the surface residues of human thrombin. J. Biol. Chem. **270:** 16854–16863.
43. HALL, S.W., et al. 1999. Thrombin interacts with thrombomodulin, protein C, and thrombin-activatable fibrinolysis inhibitor via specific and distinct domains. J. Biol. Chem. **274:** 25510–25516.
44. AROSIO, D., Y.M. AYALA & E. DI CERA. 2000. Mutation of W215 compromises thrombin cleavage of fibrinogen, but not of PAR-1 or protein C. Biochemistry **39:** 8095–8101.
45. BANFIELD, D.K. & R.T.A. MACGILLIVRAY. 1992. Partial characterization of vertebrate prothrombin cDNAs: amplification and sequence analysis of the B chain of thrombin from nine different species. Proc. Natl. Acad. Sci. U.S.A. **89:** 2779–2783.
46. VINDIGNI, A. & E. DI CERA. 1996. Release of fibrinopeptides by the slow and fast forms of thrombin. Biochemistry **35:** 4417–4426.
47. DANG, Q.D., A. VINDIGNI & E. DI CERA. 1995. An allosteric switch controls the procoagulant and anticoagulant activities of thrombin. Proc. Natl. Acad. Sci. U.S.A. **92:** 5977–5981.

The Fibrin Intermediate, Its Place in the Fibrinogen–Fibrin Transformation

JOHN R. SHAINOFF,[a] GARY B. SMEJKAL,[a] PATRICIA M. DIBELLO,[a] BARBARA CHASE,[a] OLGA V. MITKEVICH,[b] AND HELMUT LILL[c]

[a]*Department of Chemistry, Cleveland State University, Cleveland, Ohio 44115, USA*

[b]*Institute of Experimental Cardiology, Cardiology Research Center of Russia, Moscow 121552, Russia*

[c]*Roche Diagnostics, R&D Coagulation, D-82377 Penzberg, Germany*

ABSTRACT: Our preceding study indicated that, in course of coagulation of human fibrinogen by thrombin, substantial production of the fibrin intermediate (α-profibrin) lacking only one fibrinopeptide A (FPA) precedes the formation of α-fibrin monomer lacking both FPAs. The plateau concentration of α-profibrin (20% of initial fibrinogen) appearing in reactions indicated, however, that the second FPA is released four times faster than the first. The study reported here confirms those findings, and provides new insight into the significance of differing rate constants for the production of α-profibrin and its conversion to α-fibrin. The intermediate could be isolated in a distinct electrophoretic band by electrophoresing partial thrombin digests at high concentrations. Its identity was verified by digesting it with CNBr and by demonstrating that its N-terminal domain, the NDSK fragment, both lacks an FPA and contains an FPA, unlike the NDSKs of the bands from fibrin which contained no FPA or the fibrinogen band that lacked no FPA. The single step isolation also enabled us to confirm the 15–20% plateau level of α-profibrin in course of thrombin reactions, well below the 37% maximum that would be expected if release of the first and second FPA proceeded independently with no difference in rate. The 37% maximum is observed in reactions with atroxin, and it is suggested that the abundant production of α-profibrin underlies the therapeutic utility of atroxin as a defibrinating agent. Gel chromatography procedures were optimized for isolation of α-profibrin/fibrin mixtures free of fibrinogen, the final step of which involves literal use of agarose gel as a filter to remove fibrin aggregates from the fibrinogen free fractions (aggregates are left behind in gel filtration, rather than their moving ahead in gel chromatography). Unlike human fibrinogen, rabbit fibrinogen does not yield much α-profibrin in course of its conversion to fibrin, less than 10% as determined by electrophoresis and comparison with abundant production with atroxin. This low production of α-profibrin conformed with conclusions from our early studies on the generalized Shwartzman reaction in rabbits, and we now infer that the low production of α-profibrin and rapid conversion to fibrin by rabbit fibrinogen underlies the unparalleled susceptibility of these animals toward fibrinoid formation in the generalized Shwartzman reaction.

KEYWORDS: Fibrin intermediate; Fibrinogen–fibrin transformation; Profibrin; Shwartzman reaction.

Address for correspondence: John R. Shainoff, Ph.D., Cleveland State University, Department of Chemistry, 2351 Euclid Ave., Cleveland, OH 44115-2406, USA. Voice: 216-687-2463; fax: 216-687-9298.

j.shainoff@csuohio.edu

INTRODUCTION

The existence of fibrin intermediates was first predicted by Blombäck and Laurent, who observed formation of soluble aggregates accompanying partial release of fibrinopeptide A (FPA).[1] While pursuing fibrinopeptides as potential indicators of intravascular coagulation, we found a cryoprecipitable, FPA-deficient derivative of fibrinogen that was deemed to be an intermediate in the blood of endotoxin treated rabbits.[2,3] Follow-up characterizations showed that the derivative was largely separable into fibrinogen with FPA intact and fibrin monomer nearly devoid of FPA. It was an aggregation intermediate, not a chemical intermediate partially stripped of FPA. Using gel chromatography, Benabid et al.[4] and Smith[5] described separations of remarkably stable and soluble thrombin products that were deemed to be dimers formed of molecules lacking one FPA and intermediary polymers with FPA largely missing. However, their reactions were conducted without removing factor XIII or adding EDTA, and the products appeared crosslinked. Subsequent studies using a variety of biophysical methods raised considerable doubt whether a profibrin lacking one FPA even existed.[6–12] Prior to the last proclamation of its nonformation, Meh et al.[13] showed that thrombin digests of plasmin-derived fragment E transiently yielded a product with isoelectric point between the fragments from fibrinogen and α-fibrin, and they were also able to partially resolve a CNBr-derived fragment from thrombin/fibrinogen reactions with molecular size and N-termini between the NDSKs[14] of fibrinogen and α-fibrin.[15] That study convincingly indicated to us that the intermediate was indeed being formed, but was not being found because it was not aggregating as anticipated, as was subsequently established.[16]

To distinguish aggregating from non-aggregating derivatives of FPA release from fibrinogen, we performed electrophoresis in a zone of buffer containing GPR-peptide analogs of the fibrin aggregation site to keep fibrin from aggregating until it crossed into buffer lacking the peptide inhibitors. On migrating into the GPR-deficient buffer, the fibrin molecules begin to aggregate and form a sharp band of immobile strands. The intermediate, which we call α-profibrin, was then found comigrating with fibrinogen largely unaffected by the absence of the aggregation inhibitors. It was distinguished from the fibrinogen by its positive immunostaining with anti-fibrin (α 17–23) monoclonal antibody, which reacts with the α-chain N-terminus that becomes exposed upon release of FPA.[17] The weak aggregation characteristics of the intermediate, α-profibrin, explained why it had not been found in previous searches using biophysical methods that relied upon protein aggregation for separation. We now find, however, that α-profibrin can be electrophoretically separated from fibrinogen by running the thrombin/fibrinogen reaction mixtures at high concentration (greater than 12 mg/ml). This one-step separation has enabled us to verify its identity more acutely without questions about intervening steps. The fidelity of this separation and larger scale purification of α-profibrin are subjects of this communication, subjects of importance because of the long standing controversy over its existence and significance in the fibrinogen/fibrin conversion. By not depending on monoclonal antibodies for detection and measurement of α-profibrin, these methods also enabled us to readdress the question of whether rabbit fibrinogen yields substantial α-profibrin, as does human fibrinogen. The answer to that question is a hallmark of these studies.

MATERIALS AND METHODS

The sources of materials were as follows:
Enzyme Research Laboratories <http://www.enzymeresearch.com> for thrombin and fibrinogen.
XC Corporation <http://www.xcbeads.com> for agarose beads and glyoxyl agarose.
Peptide Synthesiz, Ltd. (3rd Cherepkoskaya Str., Moscow 121552, Russia) for $GPRP_{NH_2}$.
Research Organics <http://www.resorg.com> for buffer components.
Pharmacia <http://www.apbiotech.com> for Protein G Sepharose.
Bachem <http://www.bachem.com> Phe-Pro-Arg-chloromethylketone (PPACK).

The fibrinogen was purified by gel chromatography on 6% agarose in Tris-buffered saline-EDTA-azide (TBSEA, 0.144 M NaCl, 0.06 M Tris, 0.2 mM EDTA, 0.1% NaAzide at pH 8.4 to remove aggregates and a trailing shoulder). The antifibrin α17–23 monoclonal antibody was a generous gift from Roche Diagnostics, R&D Coagulation, D-82377 Penzberg, Germany.

All solutions contained 0.2 mM EDTA, and chromatographic buffers also contained 0.1% sodium azide. Partial reactions between thrombin and fibrinogen in Tris-buffered saline/EDTA (TBSE) were halted with PPACK (50 μM added in two 2-min interval aliquots), and also treated with 0.5 mM iodoacetic acid. Chromatography was performed with flow rates of 3 cm/h at ambient temperature or 1 cm/h at 2°C. Sample loads were generally 1/20 of the column volumes.

GPRphoresis and direct immunoprobing of electropherograms were as described elsewhere.[16] Just prior to applying the samples, they were either diluted with 1/3 volume of 9 M urea, or precipitated on ice with 1/7 volume of cold ethanol and redissolved at about 12 mg/ml in electrophoresis buffer containing 3 M urea. Samples were applied and electrophoresis commenced within an hour of adding the urea. Gels were either stained with Coomassie Crystal Violet, or fixed by immersion in hot (60°C) PBS for immunoprobing. Stained gels were occasionally destained by washing in water, and subsequently immunoprobed. Direct immunoprobing was used instead of Western blotting, because fibrinogen and fibrin do not blot transfer well.

Immunostaining of dot blots on cellulose nitrate were compared with standards ranging from 10–0.3 μg of fibrinogen and fibrin. Albumin (1%) was used for blocking, and the blots were heated 20 minutes at 65°C to denature the fibrin(ogen) prior to immunoprobing. Denaturation was required for immunostaining with the antifibrin antibody.

CNBr digests were made on denatured protein extracted from electropherograms. Coomassie-revealed bands were excised, transferred to a glass centrifuge tube to melt the agarose and denature the protein in 5 ml water by heating in a boiling water bath. The precipitated protein was pelleted while hot, and resuspended twice in boiling water. The pellet was then dissolved in 0.3 ml 70% formic acid, digested overnight with 15 mg CNBr, dialyzed exhaustively against 10% acetic acid, then for two hours against 0.2% acetic acid, and stored frozen at −70°C. Just prior to immunoaffinity chromatography, it was admixed with an equal volume of 50 mM NaBorate at pH 8.

Immunoaffinity chromatography of NDSKs was performed with 3.5 mg antifibrin α17–23 immobilized and crosslinked with dimethylpimelimidate on 1 ml of protein

G Sepharose as prescribed by the Protein G agarose manufacturer, Pharmacia. The protein in 50 mM NaBorate at pH 8 was percolated through the adsorbent (4 mm dia X 3 cm h column) with two column volumes of borate buffer at a flow rate of 1 cm/h, and the retained protein was eluted with 0.3 M citric acid at pH 3.5.

All gel chromatography of partial thrombin digests of fibrinogen were performed on 6% non-crosslinked agarose beads in 0.3 M NaCl buffered with 0.015 M Tris HCl supplemented with 0.1 mM EDTA and 0.1% NaN_3 at pH 8.4. After initial chromatography, selected fractions were precipitated and rechromatographed. To avoid losses that were frequently encountered from clot formation during precipitation of protein in the chromatographic fractions, precipitations were optimally performed by first acidifying the fractions, by admixing one-third volume of 0.3 M Tris acetate at pH 5.3, and quickly readjusting to that pH with a few drops of 0.3 M citric acid at pH 3.5. The acidification to pH 5.3 also served to dissolve clot like precipitates that would form in fractions that were stored frozen prior to precipitation. After acidifying and dissolving any precipitate by stirring in a 39°C bath, one-third volume of saturated ammonium sulfate was added. A flocculent precipitate formed over a 30 minute period. It was pelleted, and redissolved by stirring in about 1/15 volume of 0.45 M NaCl buffered with 0.02 M acetate at pH 5.3. The resultant solutions were quite clear, and remained clear for rechromatography when adjusted to pH 8.4 by admixing 1/5 volume of 0.3 M Tris-HCl containing 0.1 mM EDTA at pH 8.6.

Removal of fibrin from chromatographic fractions was effected by gel filtration in the literal sense, where a 1.5 mm thick 3% agarose gel served as filter for trapping the fibrin complexes. The α-profibrin was washed out of the filter gel into a 4.5 mm thick 3% agarose capture gel. (Whereas fibrin moves ahead of the fibrinogen/profibrin and is constantly redissolved by them during agarose gel chromatography, it is left behind them to form clots on the agarose filter.) The separation was performed on a gel dryer, on which a sheet of polyethylene food wrap was laid over the porous polyethylene platform of the gel dryer, and held in place by vacuum. A window was cut in the food wrap with a razor, so that the window overlapped the capture gel by a few milimeters on each side. After opening the window the capture gel was laid over it, and the capture gel, in turn, overlaid by filtering gel. After suctioning a 0.5 mm high volume (by calculation) of profibrin/fibrin solution into the filtering gel, a 6 mm thick slab of 0.5% agarose in TBSEA was overlaid as a reservoir to feed TBSE through the filter- and capture-gel under moderate vacuum of 20 cm Hg. Buffer (TBSE-azide) was intermittently added at the top of the reservoir as soon as its upper surface became somewhat indented, and the buffer additions were continued for about three hours, until the volume added reached seven times the volume of the filter gel. That volume of wash flushed the profibrin to the center of the capture gel, leaving all fibrin behind in the filter gel. The α-profibrin was recovered by dicing the capture gel with a series of 1-mm interspaced razors, freezing the diced gel to fragment it further, and then transferring the gel to a column for elution with TBSE-azide.

RESULTS

Electrophoretic Separation of α-Profibrin

Our preceding studies on thrombin/fibrinogen reactions conducted with fibrinogen at physiologic concentration showed that α-profibrin aggregated so weakly that it was electrophoretically inseparable from fibrinogen, not forming dimers or aggregates required for separation from fibrinogen. Such comigration of α-profibrin and fibrinogen is illustrated here with reactions conducted with atroxin instead of thrombin (see FIGURE 1). The atroxin reaction is illustrated because, unlike the thrombin reactions, a greater plateau concentration of α-profibrin (30–35% instead of 20% of the initial fibrinogen) was produced in course of the reactions. That no fibrin comigrates with fibrinogen had been demonstrated previously.[16]

While studying thrombin reactions with concentrated fibrinogen we had observed that α-profibrin became electrophoretically separable from fibrinogen. It formed a distinct band just behind fibrinogen when reaction mixtures were run at 12 mg/ml (see FIGURE 2). Immunostaining with antifibrin α17–23 showed it to be clearly separated from fibrinogen, revealed by Coomassie-staining and ^{125}I-fibrinogen added to the samples as a tracer. The separation of α-profibrin from fibrinogen and fibrin in a single step on a preparative gel (see FIGURE 3) enabled us to acutely test its identity

^{125}I-Normal Fibrinogen / Atroxin Reactions
Phosphorimager

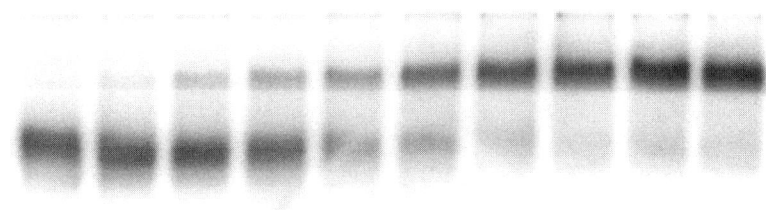

Anti-fibrin #17-23 immunostaining

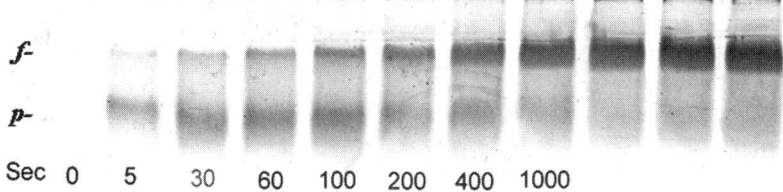

f-
p-
Sec 0   5   30   60   100   200   400   1000

FIGURE 1. GPRphoresis, showing the distribution of fibrinogen and its derivatives in the coarse of fibrin production by atroxin reacting with ^{125}I-labeled human fibrinogen at 3 mg/ml. The **upper panel** shows total iodinated protein visualized with the PhosphorImager. The **lower panel** shows levels of fibrin monomer (*f*) and profibrin (*p*) revealed by immunostaining with peroxidase-labeled antifibrin (α17–23) antibody prior to imaging the gel on the PhosphorImager. The α-profibrin comigrated with fibrinogen (φ), as observed previously in thrombin reactions. That no fibrin comigrates with fibrinogen/α-profibrin band had been demonstrated previously.

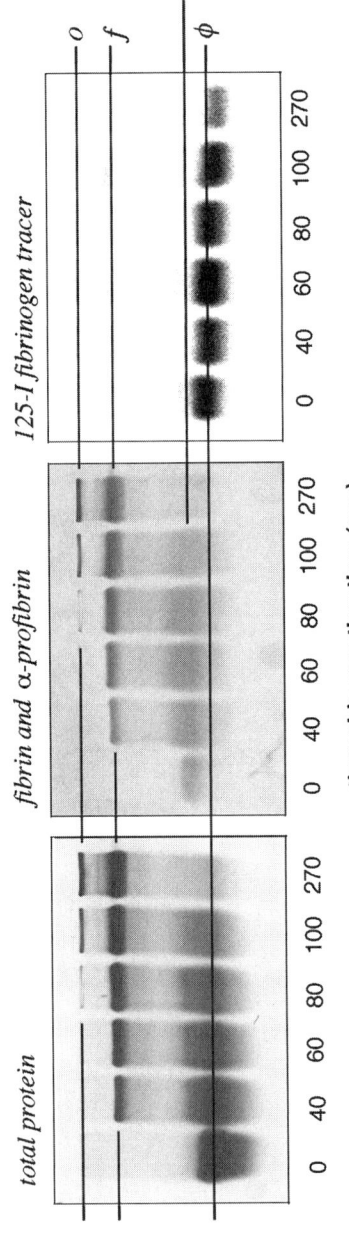

FIGURE 2. Electropherograms showing separation of α-profibrin (p) from fibrinogen (φ) and fibrin monomer (f) in concentrated thrombin/fibrinogen reaction mixtures. The **left panel** shows the distribution of total protein revealed by Coomassie-staining. The **central panel** shows the distribution of fibrin and profibrin revealed by immunostaining with antifibrin (α17–23) antibody. The **right panel** shows the distribution of fibrinogen revealed by PhosphorImager scan of ^{125}I-fibrinogen added to the reaction mixtures after inactivating thrombin with PPACK. A composite made by overlaying the gray scale PhosphorImager and the amber immunostained scans showed little overlap between the α-profibrin and fibrinogen bands, but the clear separation was not as self-evident in gray scale alone.

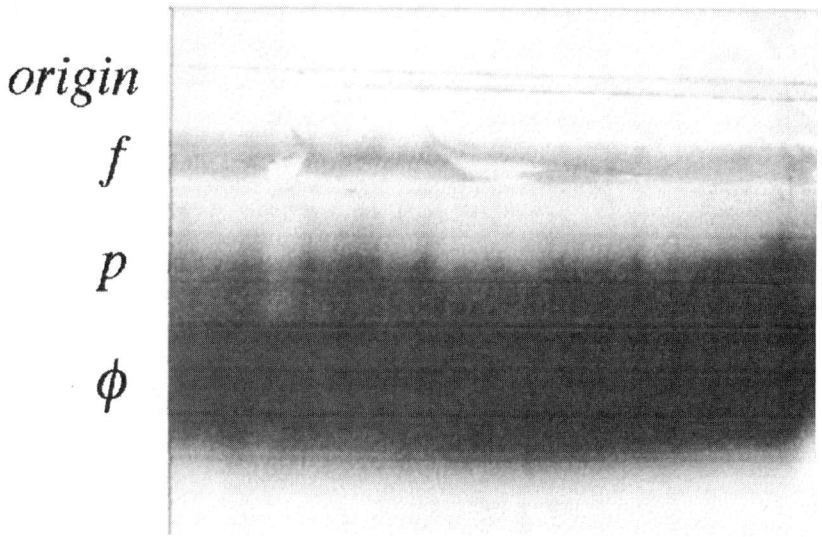

FIGURE 3. Coomassie stained cellulose nitrate surface blot of a preparative gel used to isolate the α-profibrin band for preparation of its NDSK by CNBr digestion, and subsequent examination of its composition (FIGS. 4 and 5).

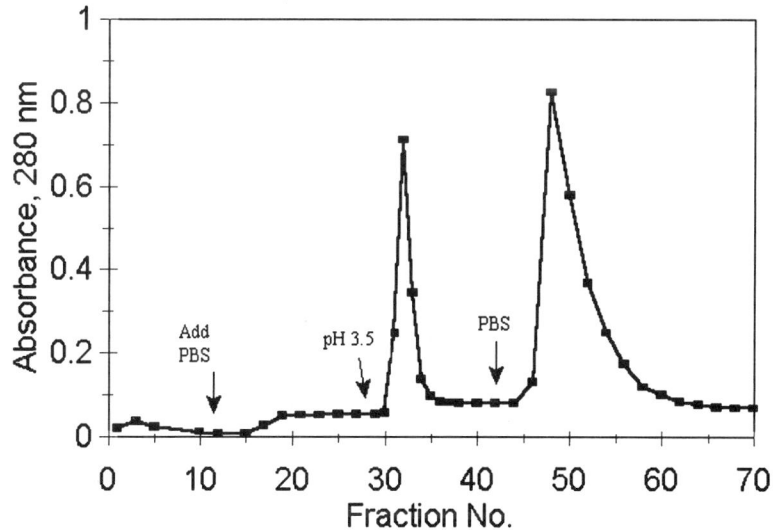

FIGURE 4. Retention and elution of the α-profibrin NDSK on an antifibrin (α17–23) antibody immunoaffinity column.

by excising the band, digesting it with CNBr, assessing retention of its NDSK fragment by immobilized antifibrin α17–23 and measuring the FPA content of eluates from the immunosorbent.

Verification of Identity of α-Profibrin

As shown in FIGURE 4, the NDSK fragment from the CNBr digest of the α-profibrin band was eluted from the immobilized antifibrin α17–23 immunosorbent with pH 3.5 citrate, and migrated as a pure NDSK band on SDS-PAGE electrophoresis (see FIGURE 5). In parallel experiments, fibrin NDSK was also retained by the immunosorbent, whereas fibrinogen NDSK was not. Dot blots of the eluted NDSK showed that it contained the expected level of reactivity towards both antifibrin α17–23 and anti-FPA (see FIGURE 6). Thus, its retention by the antifibrin α17–23 immunosorbent and its content of FPA indicated that all molecules contained both an N-terminal segment lacking FPA and one containing FPA.

Purification by Gel Chromatography

Our initial rationale for large scale purification of α-profibrin was to separate α-profibrin and α-fibrin free of fibrinogen by gel chromatography at 4°C on 6% agarose beads using large columns, and then remove the fibrin from the profibrin as described.[16] However, numerous trials failed to yield a profibrin/fibrin fraction

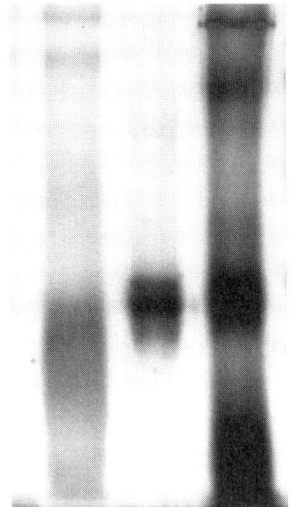

20-25 31-33 Precol

FIGURE 5. SDS-PAGE electropherogram showing composition of the pre-elution (*lane 1*) and acid-eluted (*lane 2*) fractions from the affinity chromatography (FIG. 4). The large peak eluted with PBS after the acid was a blur of high molecular weight material along with some antibody due to instability of the immunosorbent (not shown).

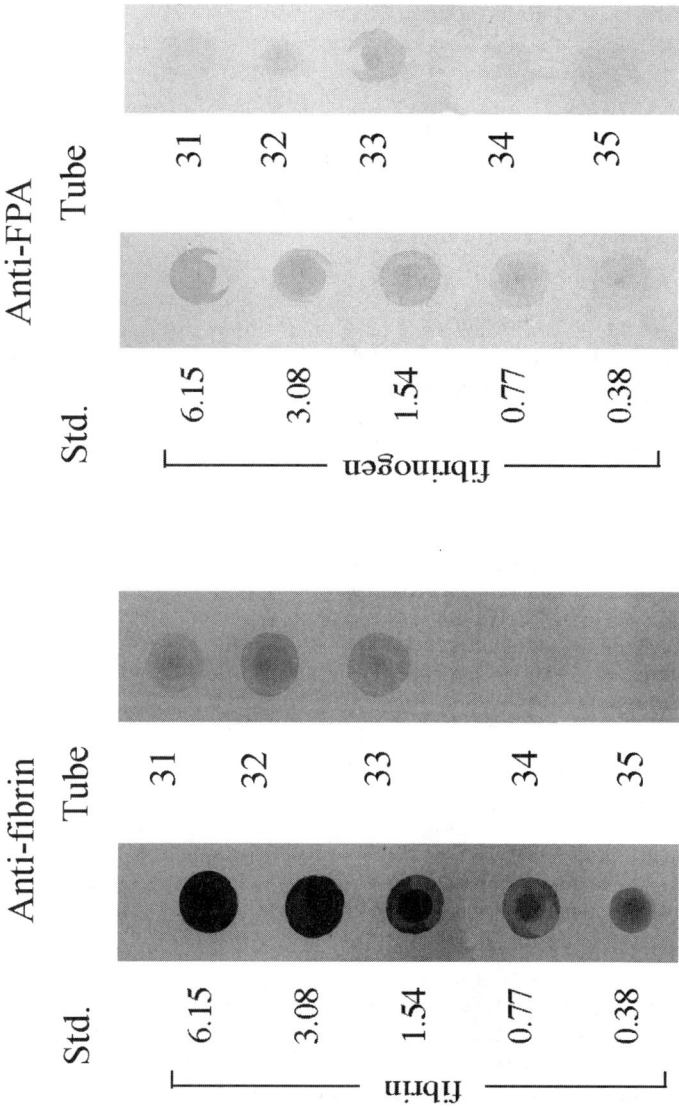

FIGURE 6. Immunostained dot blots of standards and fractions from the immunoaffinity chromatography (FIG. 4) of α-profibrin NDSK stained with indicated antibodies showing that the NDSK both contains and lacks FPA.

greatly enriched over fibrinogen on the large columns. Subsequently, we remembered that Becker had long ago showed that fibrinogen undergoes considerable self-aggregation on prolonged storage at low temperatures.[19] Our initial studies had been carried out with short 25-cm columns requiring only a few hours for separation, whereas the meter-long columns were taking a full day, plus a variable length of cold storage. We verified that fibrinogen free of immunodetectable α-profibrin formed aggregates in course of prolonged low temperature chromatography on the large columns, and consequently changed the conditions of chromatography.

Chromatographic Reduction of Fibrinogen in Reaction Mixtures

Gel chromatography of fibrinogen/thrombin partial reaction mixtures at pH 7.4 at ambient temperature yielded poor separations when fibrin and α-profibrin comprised less than 25% of reaction mixtures. The fibrin and α-profibrin would elute only slightly ahead as a leading shoulder on the fibrinogen peak and, as judged from immunostaining of dot blots using antifibrin (α17–23), they trailed well into the fibrinogen. The migration of α-fibrin and profibrin ahead of fibrinogen is due to the presence of α-profibrin, for we had shown that fibrin (in the absence of profibrin) comigrates with fibrinogen, aggregating into immobile strands as it pulls ahead of fibrinogen, and redissolving as fibrinogen catches up with it.[20] We then tried chromatography at pH 8.4, near the pH 8.9 used long ago by Benabid et al.[4] With moderately high (12 mg/ml) concentrations of fibrinogen/thrombin reaction mixtures, we found three peaks that were similar to theirs (see FIGURE 7). Profiles for elution of fibrinogen were calculated from the elution of ^{125}I-fibrinogen added as a tracer prior to chromatography, and normalizing the peak cpm with the peak absorbance of the fibrinogen where no immunoreactive α-profibrin was present. The difference between total absorbance and the calculated contribution of fibrinogen was equated to the fibrin and α-profibrin in the fractions, and the amount of fibrin was determined by GPRphoresis and immunostaining (FIG. 7, inset). The lead peak consisted α-fibrin monomer carried through the column as complexes with α-profibrin. The second peak consisted mainly of α-profibrin, along with some trailing fibrin complexes and a small amount of tracer fibrinogen. Numerous chromatographic profiles regularly showed that α-profibrin consistently comprised 15–20% of the total protein, essentially the same as determined in our preceding study based on the difference between fibrin monomer production and fibrinopeptide release.[16] The third peak consisted of fibrinogen itself except for slight trailing of α-profibrin at the front. The fibrinogen and small amount of profibrin in the second peak was concentrated by ethanol precipitation for production of additional product with thrombin. The front half of the leading peak was discarded, and all fractions back to the descending slope of the second peak were combined, concentrated by precipitation, redissolved to about 10 mg/ml at pH 5.3, and rechromatographed after adjusting back to pH 8.4 as described in the METHODS section.

We believe that the improved separation of soluble complexes at pH 8.4 is due to the poor aggregation of fibrin at that pH, because it is outside the pH range for optimal polymerization.[21] The separations depend in part on fibrin monomer pulling α-profibrin well ahead of the fibrinogen. At pH 7.4, the fibrin does not get far ahead because it forms immobile strands as it moves ahead of fibrinogen and α-profibrin, and redissolves as the fibrinogen and α-profibrin catches up to it. At pH 8.4, the

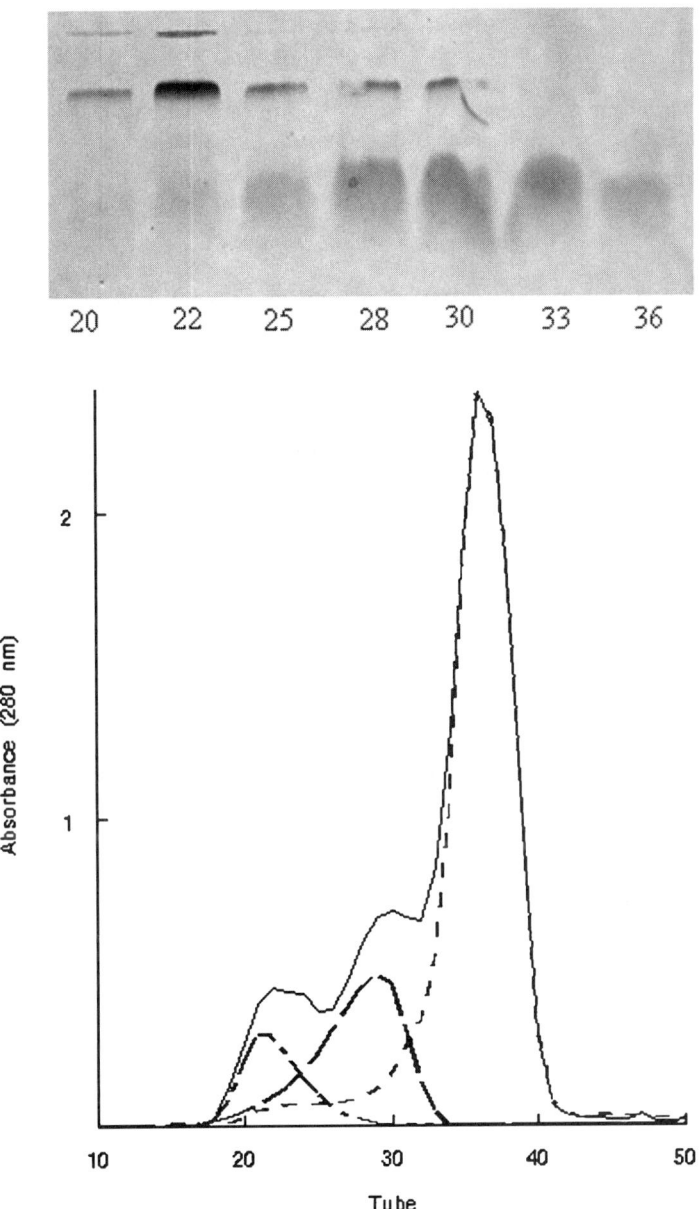

FIGURE 7. Gel chromatogram showing distributions of fibrin (— · —), profibrin (— —), and fibrinogen (- - - -) in fractions based on (1) Coomassie staining to reveal the content of fibrin in the fractions, as shown in the **upper panel**, and (2) the content of fibrinogen from radioactivity from added tracer.

fibrin complexes precipitate so slowly that they can move well ahead of the fibrinogen, and elute in the void volume. That the improved separations are not due to greater stability of the complexes at the alkaline pH, as was suggested by Benabid *et al.*, and by Smith,[4,5] is indicated by observation that the complexes break down to monomeric fibrinogen/α-profibrin and insoluble fibrin on electrophoresis (not shown). The stability of the complexes that they isolated was probably due to crosslinking by factor XIIIa.

Rechromatography

Rechromatography of concentrates (12 mg/ml) of pooled fractions from the α-profibrin rich peaks failed to separate the ^{125}I-labeled tracer that had been added to track fibrinogen in the original chromatography (see FIGURE 8). None of the tracer eluted in the normal position for fibrinogen, but migrated well ahead of the normal elution position in the latter half of the α-profibrin/fibrin peak. We believe that the isotopic tracer is actually α-profibrin converted from the fibrinogen tracer by traces of thrombin that escaped inactivation by PPACK. That traces of thrombin did escape inactivation, was indicated by increases in α-profibrin in the fibrinogen peak of the original chromatography after storing those fractions in the cold for several days. Whatever, we selected from the rechromatographies only those fractions that were deemed to consist only of α-profibrin and fibrin with no more 2–5% fibrinogen, determined from the tracer, and we removed the fibrin from the fractions. Since our original method for removing fibrin by reverse affinity chromatography, as described,[16] required a large excess of immobilized anti-FPA relative to the profibrin, which would be costly for large scale purifications, we devised a less costly filtration method. The filtrational method was based on observation that fibrin monomer is filtered out at the surface of 3% agarose gels when samples are applied to them for electrophoresis with no $GPRP_{NH_2}$ added.

Purification of Profibrin by Filtering out the Fibrin

The removal by filtration of fibrin from the α-profibrin/fibrin mixtures was effected using a 3% agarose gel as a filter, as described (in METHODS), and is illustrated in FIGURE 9, where an underlaid 3% gel is used to capture the washed out α-profibrin, and an overlaid weak agarose gel (0.6%) is used as a buffer reservoir. Prior to filtration, the fractions selected from rechromatographies were concentrated (to about 8 mg/ml) by precipitation and redissolution at pH 5.3, adjusted to pH 7.4, and a volume corresponding to one-third that of the filter gel was subjected to filtration. As is shown in FIGURE 10, the fibrin was trapped right at the surface of the filter gel, and peeled away in course of staining with Coomassie crystal violet to reveal protein distributions after flushing a calibrated volume (seven times the filter gel volume) through it. The wash sufficed to transfer the α-profibrin completely out of the filter gel, and halfway through the underlying capture gel. Due in part to the high content of fibrin (30%) in the applied solution, some incorporation of α-profibrin into the fibrin film, and incomplete extraction of the α-profibrin from the capture gel, we recovered only 10 mg of α-profibrin from the 18 mg of applied protein. Despite losses, the yield provided sufficient material to pursue our long term goals.

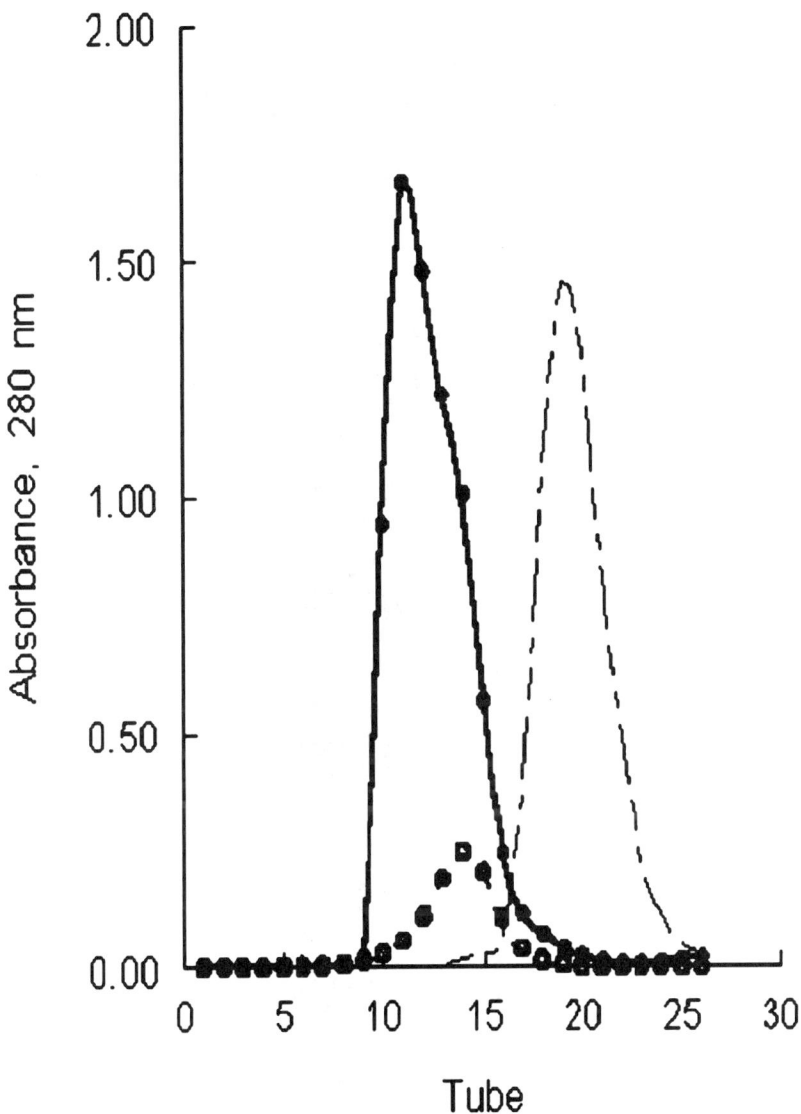

FIGURE 8. Rechromatography of pooled and reconcentrated fractions eluting ahead of fibrinogen in the initial chromatography (FIG. 7). None of the contaminating fibrinogen (○) deemed present based on radioactivity of isotopically labeled fibrinogen in the initial chromatography eluted in the expected position, illustrated by the *dashed tracing* from a calibrational run.

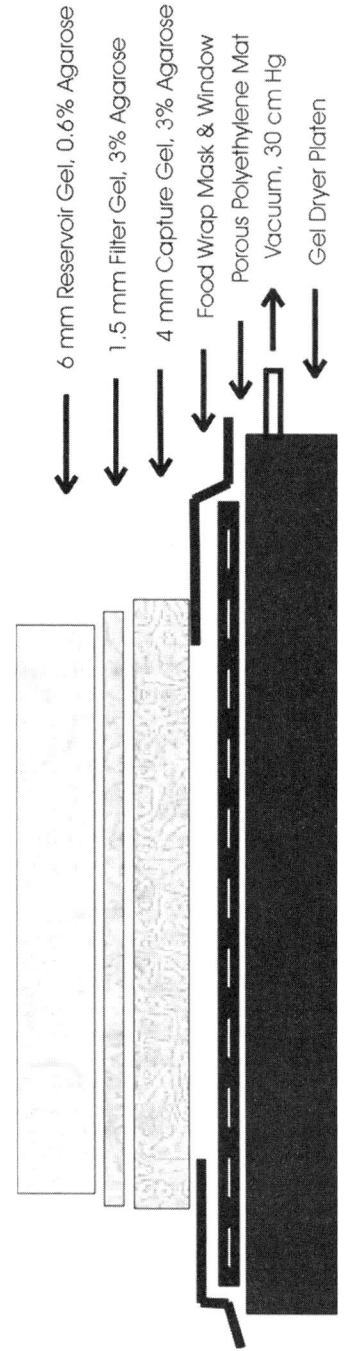

FIGURE 9. Arrangement for gels for filtering fibrin from fibrin/profibrin mixtures.

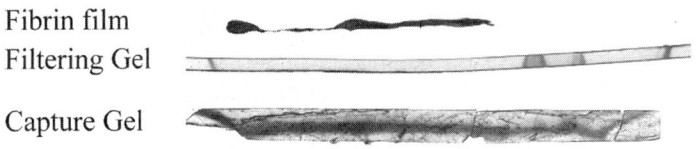

Fibrin film
Filtering Gel
Capture Gel

FIGURE 10. Coomassie-stained cross section of the gels after the filtration. The fibrin film became detached from the filter gel during processing. The filter gel is virtually free of protein, and the capture gel contains eluted α-profibrin near the center of the gel.

Differences in α-Profibrin Production Producing Sharp Contrasts between Thrombin Reactions and Human and Rabbit Fibrinogen

The high solubility of fibrin monomer (20–25% of total protein) in partial thrombin reactions with human fibrinogen differed from the low (8%) solubility observed in our old studies with rabbit fibrinogen. The 8% threshold for fibrin strand formation in course of thrombin reactions corresponded closely to the solubility of purified fibrin monomer in solutions of fibrinogen. Furthermore, cryoprecipitates, formed after partial thrombin reactions with human fibrinogen, remained soluble when redissolved and could even be put back in the cold without reforming a cryoprecipitate because of the better fibrin dissolving power of α-profibrin in the absence of a large excess of fibrinogen. On the other hand, cryoprecipitates from rabbit fibrinogen reactions separated into fibrinogen and fibrin almost immediately after redissolving. All of these considerations led us to suspect that rabbit fibrinogen, unlike human, was producing little or no α-profibrin. We could not use the antifibrin (α17–23) antibody to test for α-profibrin in electropherograms, because it does not crossreact with rabbit fibrin, but could examine the question using the electrophoretic and chromatographic methods that resolved it in the studies with human fibrinogen.

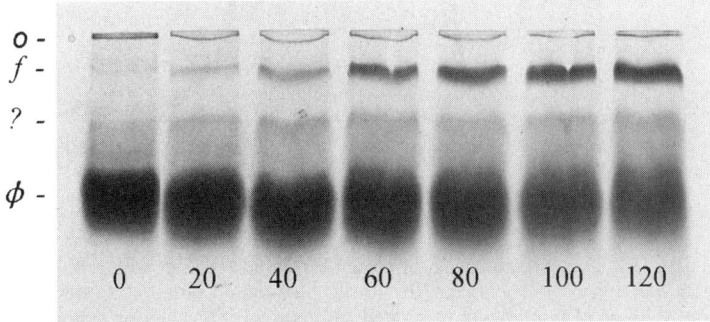

FIGURE 11. GPRphoresis of reactions between human thrombin and rabbit fibrinogen. Unlike reactions with human fibrinogen, very little derivative was produced with mobility between fibrin (f) and fibrinogen (ϕ). The intermediate (labeled ? because it was not clearly identified in this gel) showed up more clearly when enriched by chromatography (FIG. 13).

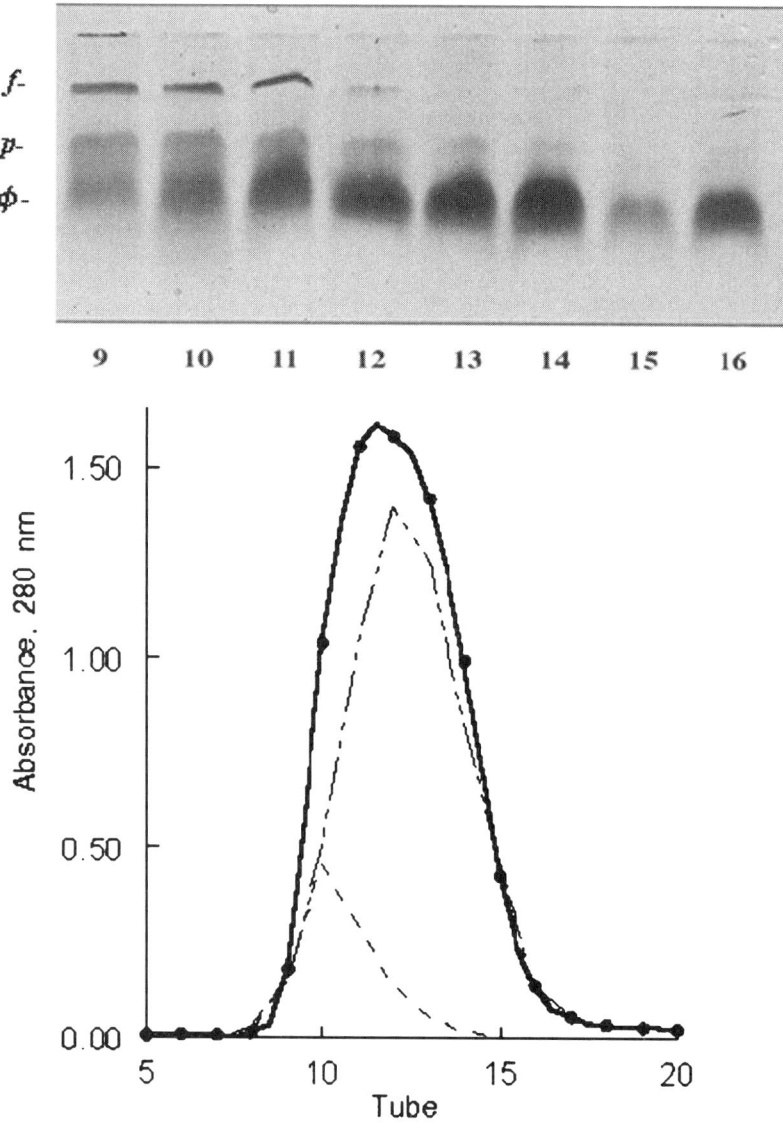

FIGURE 12. Gel chromatogram of partial thrombin reaction with rabbit fibrinogen. The reaction mixture corresponded to the 40-sec reaction in FIGURE 11. Coomassie-stained electropherograms of fractions (precipitated and redissolved to approximately 12 mg/ml) are shown in the **upper panel.** The component deemed to be the intermediate (p) was better resolved here because of its enrichment in the lead fractions; yet, it was small compared to fibrin (f) in all fractions. The chromatogram (**lower panel**) shows the absorbance of the fractions (●), and the contribution of fibrin (dashed line), profibrin (dotted line), and fibrinogen (dash-dotted line) to the absorbance as determined from the relative amounts in the electropherogram.

GPRphoresis of partial thrombin digests of rabbit fibrinogen with both bovine and human thrombin showed only a slight increase in derivative migrating between the fibrinogen and fibrin band, and no mobility shift or broadening of the fibrinogen band (see FIGURE 11). All of the protein migrating between fibrinogen and fibrin comprised, maximally, 8% of the total. Chromatography of a partial thrombin digest showed only a small leading shoulder for the fibrinogen band, a profile that one would expect for elution of fibrin together with fibrinogen in absence of substantial levels of α-profibrin (see FIGURE 12). Electrophoresis of concentrated (12 mg/ml) chromatographic fractions showed the "intermediate" more clearly due to enrichment in the lead shoulder (FIG. 12, upper panel), but its concentration was small compared to fibrin monomer. We calculate from the low levels found in comparison to fibrin over the course of reaction that release of the 2nd FPA from rabbit fibrinogen is on the order of 10-times faster than release of the first ($k_2/k_1 \approx 10$).

Rabbit α-Profibrin Produced with Atroxin

Unlike thrombin, atroxin produced greater quantities of α-profibrin than fibrin in the early stages of reaction (see FIGURE 13). The level of profibrin plateaued at approximately 30% of the initial fibrinogen, and migrated at the same rate as the small amounts produced by thrombin. The ability to separate the profibrin confirmed that only small amounts on the order of 8% of the total protein was being produced by thrombin acting on rabbit fibrinogen.

DISCUSSION

Isolation and Characterization of α-Profibrin

Isolation of α-profibrin had been difficult, because it is amphoteric. GPRphoresis of thrombin reaction mixtures at high concentrations enabled us to isolate α-profibrin in a single step, and confirm its identity by showing that its NDSK fragment both

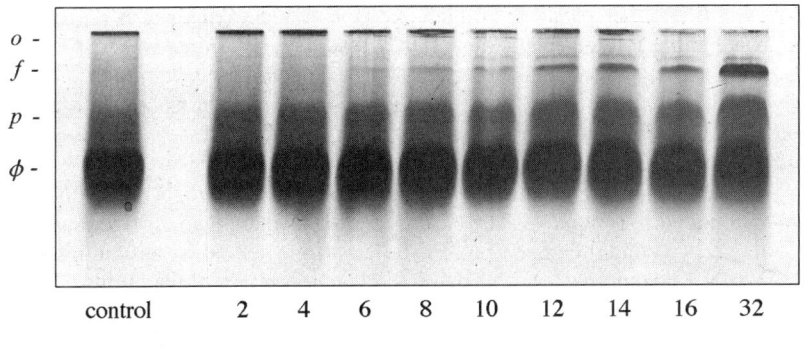

FIGURE 13. GPRphoresis showing substantial production of α-profibrin from rabbit fibrinogen using atroxin instead of thrombin. Note that, unlike the thrombin reactions (FIG. 11), profibrin (*p*) levels are high compared to the fibrin (*f*) over much of the reaction.

lacks and contains an FPA. Its NDSK was retained by an antifibrin (α17–23) affinity column, and FPA was found in expected level in the specific eluates. The isolation methods described here appear general, and not dependant on the monoclonal antibodies used previously.

Significance of Differing Rates of Release of the First and Second FPA

High levels of α-profibrin were produced in the course of conversion of fibrinogen to fibrin by atroxin, where it rose to about 35% of the initial fibrinogen concentration. This high level conforms with the level that would be expected if release of the first and second FPA proceeded independently with equal rate constants. Independent release of the two FPAs agreed with prior observations by Siebenlist *et al.* and by Meh *et al.*[13,15,22] However, in contrast to Meh's studies, we concluded that the release of the second FPA was not independent of release of the first, and that the second was released with rate constant four times that of the first, because the level of α-profibrin rises maximally to only 15–20% of the initial fibrinogen concentration. All of the measurements in the current study, direct measurement of α-profibrin after GPRphoresis of concentrated reaction mixtures and measurements by gel chromatography, conform with the 20% maximum. From ongoing studies with Di Cera,[23] we suspect that the faster release of the second FPA is due to a potentiation of thrombin by fibrinogen in a manner analogous to the allosteric thrombin potentiation induced by sodium ions. The difference between 20% and the theoretic 37% maximal production of α-profibrin is, in our opinion, very important physiologically. If thrombin produced 37% levels, we might undergo intravascular coagulation, or lack of it, in a manner analogous to the "defibrination" that accompanies atroxin infusions. The high production of α-profibrin would not only suppress fibrin aggregation, but might even prevent fibrin aggregation altogether if the α-profibrin is cleared rapidly.

The low production of α-profibrin from rabbit fibrinogen paints an entirely different picture. We suspect that the low production of α-profibrin is one of the reasons why rabbits are so eminently susceptible to intravascular fibrin deposition in the generalized Shwartzman reaction, in which injection of endotoxin induces massive disseminated intravascular coagulation (DIC) resulting in fibrinoid necrosis of the lungs and kidneys. Such DIC is very difficult to induce in rats and mice, and also humans. Endotoxin injections were used in the 1950s to treat malignant hypertension without such untoward effects, and that is one of the reasons Irvine H. Page asked us to look into the generalized Shwartzman reaction.[2,3]

With respect to its metabolism, we anticipate that α-profibrin is cleared from the circulation almost as rapidly as the clearance of α-fibrin monomer.[24] Otherwise, we would expect to find it comprising 35% of the fibrinogen during atroxin infusions. Furthermore, it comprises only a small fractional percentage of the fibrinogen without concurrent presence of measurable fibrin monomer in normal subjects and in subjects with inactive peripheral vascular disease, an indication that it is removed before fibrin can build to detectable levels.

Finally, we will address the question whether purified α-profibrin releases FPA faster than fibrinogen. That question will answer whether α-profibrin is really a better substrate, or whether the more rapid release of the second FPA results from reseating of thrombin after release of the first FPA. We are confident that the faster release of FPA is not due to occasional concurrent release of the two FPAs, because we have observed no difference in levels of α-profibrin, whether we stop reactions with PPACK (which inactivates the active site whether or not thrombin is bound to fibrinogen) or with hirudin (which requires access to both the active- and exosite-binding domains for efficient inhibition). We are engaged in ongoing studies with Di Cera that indicate that high concentrations of fibrinogen potentiate the activity of thrombin in a manner analogous to the potentiation induced by sodium ion binding described by his group.[25,26] Once potentiated, thrombin can either reseat or revert to its slow form before encountering other fibrinogen molecules in dilute but not concentrated fibrinogen. It is doubtful that rabbit fibrinogen potentiates thrombin more than human fibrinogen; thus, we suspect that the low production of α-profibrin in thrombin reactions with rabbit fibrinogen is due to a low k_1 (release of the first FPA) rather than a high value of k_2 (release of the second).

In summary, drawing correct converse conclusions with two different species of fibrinogen in asserting that (1) little intermediate is produced in rabbits,[2,3] whereas (2) substantial but less than uncoupled production occurs with human fibrinogen,[16] has provided new insight into the significance of α-profibrin in hemostasis.

ACKNOWLEDGMENT

This work was supported in part by grant HL-60896 from the National Heart, Lung and Blood Institute, National Institutes of Health.

REFERENCES

1. BLOMBÄCK, B. & T.C. LAURENT. 1958. N-terminal and light-scattering studies on fibrinogen and its transformation to fibrin. Arkiv. Kemi. **12:** 137–146.
2. SHAINOFF, J.R. & I.H. PAGE. 1960. Cofibrins and fibrin intermediates as indicators of thrombin activity *in vivo*. Circ. Res. **8:** 1013–1022.
3. SHAINOFF, J.R. & I.H. PAGE. 1962. Significance of cryoprofibrin in fibrinogen-fibrin conversion. J. Exp. Med. **116:** 687–707.
4. BENABID, Y., E. CONCORD & M. SUSCILLON. 1977. Soluble fibrin complexes: separation as a function of pH and characterization. Thromb. Hæmostas. **37:** 144–153.
5. SMITH, G.F. 1980. Fibrinogean–Fibrin conversion. Biochem. J. **185:** 1–11.
6. JANMEY, P.A. 1982. Kinetics of formation of fibrin oligomers. Biopolymers **21:** 2253–2264.
7. JANMEY, P.A., M.D. BALE & J.D. FERRY. 1983. Polymerization of fibrin: analysis of light-scattering data and relation to A peptide release of ligated oligomers. Biopolymers **22:** 2017–2019.
8. JANMEY, P.A., L. ERDILE, M.D. BALE & J.D. FERRY. 1983. Kinetics of fibrin oligomer formation observed by electron microscopy. Biochemistry **22:** 4336–4340.
9. BALE, M.D., P.A. JANMEY & J.D. FERRY. 1982. Kinetics of formation of fibrin oligomers. II. Size distribution of ligated oligomers. Biopolymers **21:** 2265–2277.
10. WILF, J. & A.P. MINTON. 1986. Soluble fibrin-fibrinogen complexes as intermediates in fibrin gel formation. Biochemistry **25:** 3124–3133.
11. HENSCHEN, A. 1990. On the identity of fibrin(ogen) oligomers appearing during fibrin polymerization. Adv. Exper. Med. Biol. **281:** 49–53.
12. LI, X., D. GALANAKIS & D.A. GABRIEL. 1996. Transient intermediates in the thrombin activation of fibrinogen. Evidence for only the DesAA species. J. Biol. Chem. **271:** 11767–11771.

13. MEH, D.A., K.R. SIEBENLIST, G. BERGTROM & M.W. MOSESSON. 1993. Comparison of the sequence of fibrinopeptide A cleavage from fibrinogen fragment E by thrombin, atroxin, or batroxobin. Thrombosis Res. **70:** 437–449.
14. BLOMBÄCK, B., M. BLOMBÄCK, A. HENSCHEN, *et al.* 1968. N-terminal disulphide knot of fibrinogen. Nature **218:** 130–134.
15. MEH, D.A., K.R. SIEBENLIST, G. BERGTROM & M.W. MOSESSON. 1995. Sequence of release of fibrinopeptide A by thrombin or atroxin. J. Lab. Clin. Med. **125:** 384–391.
16. SHAINOFF, J.R., G.B. SMEJKAL, P.M. DIBELLO, *et al.* 1996. Isolation and characterization of the fibrin-intermediate arising from cleavage of one fibrinopeptide A from fibrinogen. J. Biol. Chem. **271:** 24129–24137.
17. LILL, H., M. SPANNAGL, A. TRAUNER, *et al.* 1993. A new immunoassay for soluble fibrin enables a more sensitive detection coagulation *in vivo*. Blood Coag. Fibrinol. **4:** 97–102.
18. SHAINOFF, J.R. 1993. Electrophoresis and direct immunoprobing on glyoxyl agarose and polyacrylamide composites. Adv. Electrophor. **6:** 161–177.
19. BECKER, C.M. 1986. Separation of human plasma fibrin stabilizing factor (factor XIII) from fibrinogen by fibrinogen poymer formation. Biochem. Biophys. Res. Commun. **134:** 678–684.
20. SHAINOFF, J.R. & B.N. DARDIK. 1980. Role of fibrinogen in fibrin transport: chromatographic studies. Thrombosis Res. **17:** 491–500.
21. STURTEVANT, J.M., M. LASKOWSKI, JR., T.J. DONNELLY & H.A. SCHERAGA. 1955. Equilibria in the fibrinogen-fibrin conversion III. Heats of polymerization and clotting of fibrin monomer. J. Am. Chem. Soc. **77:** 6168–6172.
22. SIEBENLIST, K.R., J.P. DIORIO, A.Z. BUDZYNSKI & M.W. MOSESSON. 1990. The polymerization and thrombin-binding properties of Des-(Bβ1–42)-fibrin. J. Biol. Chem. **265:** 18650–18655.
23. SHAINOFF, J.R., G.B. SMEJKAL, P.M. DIBELLO, *et al.* 1999. Fibrinogen potentiates thrombin in a manner analogous to the allosteric potentiation induced by sodium ions; implications for the rapidity of fibrinopeptide release (Abst.). Thromb. Hæmostas. **82**(Suppl.): 40.
24. DARDIK, B.N. & J.R. SHAINOFF. 1985. Kinetic characterization of a saturable pathway for rapid clearance of circulating fibrin monomer. Blood **65:** 680–688.
25. DANG, O.D., A. VINDIGNI & E. DI CERA. 1995. An allosteric switch controls the procoagulant and anticoagulant activities of thrombin Proc. Natl. Acad. Sci. U.S.A. **92**(13): 5977–5981.
26. AYALA, Y. & E. DI CERA. 1994. Molecular recognition by thrombin. Role of the slow–fast transition, site-specific ion binding energetics and thermodynamic mapping of structural components. J. Mol. Biol. **235**(2): 733–746.

Early Events in the Polymerization of Fibrin

MATTIA ROCCO,[a] SIMONETTA BERNOCCO,[a,b] MARCO TURCI,[a,b] ALDO PROFUMO,[a] CARLA CUNIBERTI,[b] AND FABIO FERRI[c]

[a]*U.O. Biologia Strutturale, Istituto Nazionale per la Ricerca sul Cancro (IST), c/o Centro Biotecnologie Avanzate (CBA), Genova, Italy*

[b]*Dipartimento di Chimica e Chimica Industriale (DCCI), Università di Genova, Italy*

[c]*Dipartimento di Scienze Chimiche, Fisiche e Matematiche & INFM, Università dell'Insubria, Como, Italy*

ABSTRACT: The early events in the thrombin-induced formation of fibrin have been studied by the use of stopped-flow multiangle laser light scattering (SF-MALLS). This technological advancement has allowed the recovering, as a function of time with a resolution of about 0.5 sec, of the mean square radius of gyration $\langle R_g^2 \rangle_z$ and of the molecular weight M_w, and to place an upper bound to the values of the mass/unit length M_L. The ionic strength, pH and salt type conditions investigated were all close to physiological, starting with a 50 mM Tris, 104 mM NaCl, pH 7.4 buffer (TBS), to which either 1 mM EDTA-Na_2 or 2.5 mM $CaCl_2$ were also added. Fibrinogen was 0.2–0.3 mg/ml and rate-limiting concentrations of thrombin were used (0.05–0.25 NIH units/mg fibrinogen). By plotting $\langle R_g^2 \rangle_z$ and M_L versus M_w on log–log scales, runs proceeding at different velocities and under different solvent conditions could be compared and confronted with model curves. It was found that: (1) within this thrombin range, the mechanism of association does not depend on its concentration, nor on the buffers employed; (2) the $\langle R_g^2 \rangle_z$ versus M_w curves could all be reasonably fitted with a bifunctional polycondensation scheme involving semiflexible worm-like, double-stranded, half-staggered polymers with persistence length between 200–600 nm, provided that a ratio $Q = 16$ between the rate of release of the two fibrinopeptides A was employed; (3) the M_L versus M_w data seemed more compatible with lower Q values ($4 < Q < 8$), but their uncertainty prevented a better assessment of this issue; the formation of fibrinogen–fibrin monomer complexes may also play a role in the polymer distributions; (4) in the very early stages (e.g., when $M_w < 7 \times 10^5$), the $\langle R_g^2 \rangle_z$ versus M_w data were fitted well only in TBS and at the lowest thrombin concentration, suggesting that a transient, either sequential or concurrent fast second mechanism, involving longer and thinner polymers, may be at work.

KEYWORDS: Fibrin polymerization; Thrombin induced; Mean square radius of gyration; Polymer distribution; Light scattering

INTRODUCTION

Much is now known about fibrinogen (FG) and its conversion to fibrin[1–4] (see also the other articles in this volume). The formation of half-staggered, double-stranded filaments (the "protofibrils") has been shown to be an important early

Address for correspondence: Dr. Mattia Rocco, U.O. Biologia Strutturale, IST-CBA, Largo R. Benzi 10, I-16132 Genova, Italy. Voice: +39-0105737-310; fax: +39-0105737-325.
rocco@ermes.cba.unige.it

intermediate in the assembly process, their lateral aggregation subsequently giving rise to the thicker, branched fibers that constitute the final, gelled network. Yet, the fine details of this process are still unclear, leaving important questions, such as the onset of branching, open to debate. A great deal of information on this subject has been provided by the complementary methods of electron microscopy (EM) and light scattering (LS).[1–3] Although EM has the distinctive advantage of providing visual snapshots of "frozen" moments during a polymerization, it suffers, however, from poor time resolution, and it might be somewhat biased, especially if dealing with a mixture of stable and labile structures. On the other hand, LS techniques can be fast and are noninvasive, but they unavoidably yield average values so that detailed modeling is necessary for their interpretation. However, most of the LS literature data on fibrin formation does not exploit the full potential of the technique, which in the so called "static" mode can yield the z-average mean square radius of gyration $\langle R_g^2 \rangle_z$, the weight-average molecular weight M_w, and the (weight/z)-average mass/length ratio M_L, provided that the angular dependence of the scattered light is recorded. In the past, this was done with LS goniometers measuring one angle at the time, and thus only very slow reactions could be followed.[5–9]

By using a multiangle laser light scattering (MALLS) photometer connected to a manually operated stopped-flow (SF) device, we have been able to measure $\langle R_g^2 \rangle_z$ and M_w, and to obtain an estimate of M_L, as a function of time, with a resolution of about 0.5 sec. A full description of the technique and related data analysis procedures was published very recently.[10] The measured parameters were compared with model curves, generated according to various hypotheses concerning both the nature and distribution of the polymers involved.[10]

We report here the first results of an extension of our recent work[10] on the thrombin induced polymerization of fibrin, in which the influence of different, rate-limiting, thrombin:substrate ratios, on one hand, and of the presence/absence of EDTA-Na$_2$ and CaCl$_2$, on the other hand, were investigated.

MATERIALS AND METHODS

All chemical were reagent grade from Merck (Bracco, Milano, Italy), unless otherwise indicated. Fibrinogen was plasminogen- and von Willebrandt factor-depleted from Enzyme Research Laboratories (South Bend, IL; cat. FIB2L, lot 1720PL). The details of FG stock solution preparation and conservation, of the final purification before polymerization by size exclusion chromatography, and of sample characterization by SDS-PAGE followed by Western blot analysis, can be found in Reference 10, with one noticeable difference in that 1 mM EDTA-Na$_2$ was not added to the buffers. As an additional step, to inhibit the activity of potentially contaminating Factor XIIIa, the reconstituted FG solution was treated, prior to the dialysis step, for 30' in the dark with 0.2 mM of p-hydroxymercuibenzoic acid (PCMB, Sigma-Aldrich, Milano, Italy). FG concentrations were determined from the absorbance at 280 nm using a specific absorption coefficient[11] E of 1.51 ml mg^{-1} cm^{-1}, after correcting for scattering contributions by subtracting the absorbance at 320 nm. Human thrombin was from Sigma-Aldrich (cat T-6884, lot 88H7611, 1260 nominal NIH units, 2610 NIH units/mg protein). It was reconstituted with exactly 1 ml of water, and its activity immediately assayed in semimicrocuvettes (about 0.8 ml) in a Beckman DU-640 spectrophotometer at $\lambda = 405$ nm, employing the chromogenic substrate N-benzoyl-L-Phe-L-Val-L-Arg p-nitroanilide (Sigma-Aldrich Cat. B-7632), essentially as described by Lottenberg et al.[12] In short, the reaction was performed at 25°C in Tris 50 mM, NaCl 100 mM, pH 8, using about 0.4 nominal NIH units/ml of thrombin and about 32 nmol ml^{-1} of the synthetic substrate. The thrombin solution was then subdivided in small aliquots, quick frozen in liquid nitrogen, and stored at −20°C. Each time an aliquot was thawed (at 37°C) for polymerization studies, its activity was rechecked as described above, and then it was immediately used or kept at 4°C until used (no significant variations in thrombin activity were found after up to 4–5 h at 4°C).

The release of the fibrinopeptides A and B (FPA and FPB) from about 0.5 mg ml^{-1} FG solutions following thrombin addition was measured by reverse-phase HPLC, with a modification of the procedure of Kehl et al.,[13] as will be described in more detail elsewhere (Profumo et al., in preparation). The data were fitted to the Equations 6 and 8 of Mihaly,[14] using the nonlinear curve fit option of the Origin 5.0 software (Microcal Software Inc., Northampton, MA) (Profumo et al., in preparation). The two constants, k_1 and k_2, were then normalized to the actual thrombin concentrations in the SF-MALLS experiments, and reentered in the above mentioned equations to calculate the release curves of FPA and FPB as a function of reaction time.

SF-MALLS was performed at 25 ± 0.2°C by linking two commercially available instruments, a RX-1000 rapid mixing device from Applied Photophysics (Leatherhead, UK), operated manually (dead time 2–3 sec), and a DAWN-DSP-F multiangle laser light scattering photometer (Wyatt Technology Corp., Santa Barbara, CA, USA) equipped with a glass flow cell and Peltier control of the temperature, in which the light source is a 5-mW He–Ne laser operating at $\lambda = 633$ nm. A full description of this set up can be found in,[10] together with its calibration and normalization procedures, the relevant light scattering theory, and data analysis routines. Here, it suffices to say that in this photometer, the true scattering angles seen by each detector depend on the refractive index n of the medium, and under our conditions ($n = 1.3338$–1.3342 at $\lambda = 633$ nm for our buffers) 15 scattering angles, ranging from about 22° to about 158°, were simultaneously employed. The K_c/R_θ versus $\sin^2(\theta/2)$ plots were fitted with third- and fourth-degree polynomials, from which M_w and $\langle R_g^2 \rangle_z$ where extracted as the reciprocal of the intercept, and from the ratio between the initial slope (first coefficient) and the intercept, respectively, as described in more detail elsewhere.[10,15] Here K is the optical constant (cm^2 mol g^{-2}), equal to $4\pi^2 n^2 (dn/dc)^2/(N_A \lambda^4)$, n being the refractive index of the solvent, dn/dc the refractive index increment of the solute in that particular solvent (0.192 ml mg^{-1} for fibrinogen[10,16]), λ the wavelength *in vacuo*, N_A the Avogadro number, c the sample concentration (g cm^{-3}), and where the Rayleigh ratio R_θ (cm^{-1}) is proportional to the excess intensity of scattered light recorded as function of the angle θ between the primary and the scattered beams. The M_L were estimated by unweighted linear regression, from the slope of the linear portions of K_c/R_θ versus $\sin(\theta/2)$ plots (Casassa plots[17]), as fully described elsewhere,[10,15] and their associated error bars were derived from the standard deviation in slope. The 15 detectors were normalized, to correct for their different responses, using the unreacted fibrinogen baseline as described in,[10] with the following modification: the $(\langle R_g^2 \rangle_z)^{1/2}$ employed for the normalization was measured on the pooled peak fractions of the chromatographed FG solution in macrobatch mode with the DAWN-DSP-F (software DAWN 3.3, Wyatt Technology Corp.), using a 20-ml glass scintillation vial (Wheaton) carefully washed to a dust free condition with 0.22 μm-filtered water and then dried under vacuum. In this configuration, the scattering angles seen by the detectors are independent of the solvent refractive index, and could be normalized using a vial filled with dust free toluene. Finally, in an attempt to at least partially overcome the problem of thrombin gradually losing its activity on standing between runs in the glass syringe of the stopped-flow apparatus (see Ref. 10), the enzyme-carrying syringe was precoated with a thin film of silicone (Sigmacote, Sigma-Aldrich).

Computer simulations of the $\langle R_g^2 \rangle_z$ versus M_w plots were also performed as described elsewhere.[10] In short, both the fibrinogen and the fibrin monomeric units were supposed to be rigid cylinders with length $L_0 = 50$ nm, diameter $d_0 \sim 3$ nm, and molecular weight $M_0 = 300{,}000$ g mol^{-1} (to account for the partial degradation of the C-terminal portions of the α-chains[10]). The polymers were supposed to be semiflexible half-staggered, double-stranded worm-like chains (WLDS), characterized by a molecular weight $M_i = iM_0$ (i being the number of monomers inside the ith polymer), by a contour length $L_c(i) = L_0 + (iL_0/2)$, and by a persistence length l_p equal for all the polymers. The square radius of gyration $\langle R_g^2(i) \rangle$ of each polymer was then calculated according to Benoit and Doty.[18] Size distributions were calculated in the framework of the Janmey theory,[19,20] according to which from each FG molecule the second FPA is released faster than the first by a factor Q. A full description of the procedure used to calculate the size-distributions can be found in our previous work.[10]

RESULTS AND DISCUSSION

In FIGURE 1 are reported, as a function of time, the Rayleigh ratios R_θ at seven selected detectors (whose corresponding scattering angles are indicated only in panel B), for FG (about 0.2 mg ml^{-1}) polymerizations induced by 0.25 NIH units/mg FG of thrombin in Tris 50 mM, NaCl 104 mM, pH 7.4 (TBS, panel A), to which was added either 1 mM EDTA-Na$_2$ (TBE, panel B), or 2.5 mM CaCl$_2$ (TBC, panel C). The portion of the curves at negative time points represent the *baseline* signal of unreacted FG, before the addition of thrombin (the fact that in panel B this baseline does not join smoothly to the beginning of the polymerization, could be due either to a slight difference in the FG concentration between these two zones, or to a very fast event not caught with the present set up). In each panel, the middle curve corresponds to the 90° scattering angle, which has been often used to follow FG polymerizations. The first two panels (A and B, run in TBS and TBE, respectively) present similar curves, the curves in panel B, however, developing over nearly twice the time required for those in panel A. The marked inflection point occurring for the 90° detector at about 400 and about 750 sec, respectively, represent the end of the so called *lag time* observed with turbidity measurements, and shortly precede the gel point. In this early phase, there is a sigmoidal growth of the scattering signal, appearing in the four lower curves, being more pronounced in the presence of EDTA-Na$_2$

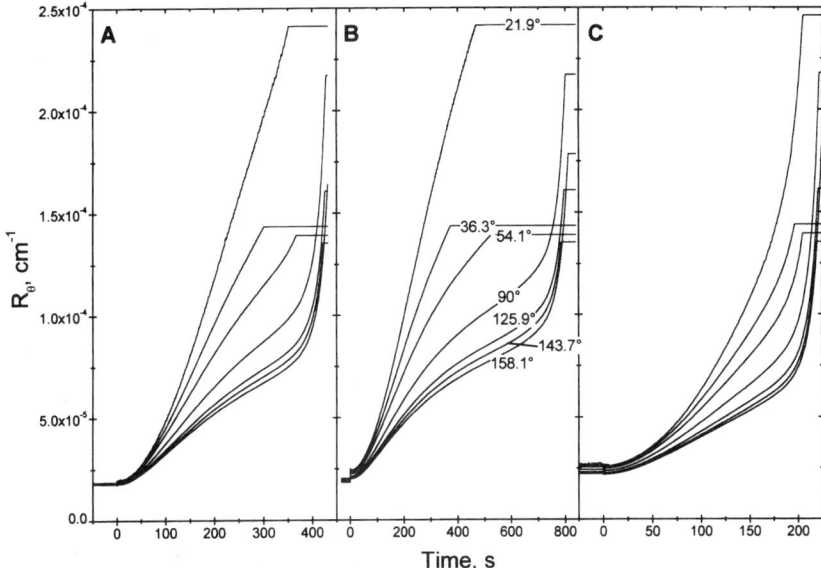

FIGURE 1. Rayleigh ratios R_θ for seven selected detectors (out of 15) of the MALLS photometer as a function of time after thrombin (0.25 NIH units/mg FG) addition to FG solutions in TBS (0.216 mg ml^{-1}, **Panel A**), in TBE (0.216 mg ml^{-1}, **Panel B**), and in TBC (0.248 mg ml^{-1}, **Panel C**).

(panel B). However, a plateau is not reached, since the second fast transition commences before this phase can end. A similar behavior has been previously reported for 90° scattering by Hantgan and Hermans[6] at similar ionic strength conditions, albeit on a much faster time scale due to the non-rate limiting thrombin amounts (50–200 NIH units/mg FG) employed by them. The sigmoidal phase is instead almost not existent when physiological amounts of Ca^{2+} are present during the polymerization (panel C), whose main effect seems to be the anticipation of the onset of the second step. All these observations are fully consistent with the mechanism proposed by Hantgan and Hermans,[6] according to which, in the lag phase, there is mainly an elongation of the protofibrils, and the subsequent rapid rise of the scattering intensity is due to their lateral aggregation, which is known to be enhanced by the presence of Ca^{2+} ions.

Since, contrary to previous studies,[5–9] we have recorded the intensity signal at many scattering angles *simultaneously,* we are now in a position to investigate the structural details of protofibrils formation in the lag phase. However, due to their different responses, the time at which the signal saturates (flat portions at the end of each curve) is different for each detector. Therefore, the determination of M_w and $\langle R_g^2 \rangle_z$ is possible only up to the time at which once of the detectors saturates, and in our system this correspond to the detector collecting the light scattered at 36.3°. Conversely, to estimate M_L only 4–5 detectors, corresponding to angles from 102.9° to 143.7°, were employed (see below), and since they saturate much later, this parameter can be determined up to the marked inflection point that precedes the gel point. Unfortunately, in polydisperse systems of semiflexible polymers, M_L can be reliably determined, for reasons explained in detail elsewhere,[10,15] only when a significant portion of the scatterers obey the relations $qL_c > 3.8$ and $ql_p > 1.9$ (where $q = (4\pi n/\lambda)\sin(\theta/2)$), conditions that in our set up are met for scattering angles above 90° only for polymers with $L_c > 200$ nm and $l_p > 100$ nm. Although it is difficult to ascertain when this truly happens, our simulations[10,15] indicate that the determination of M_L for WLDS polymers with $l_p > 100$ nm, and polydisperse according to a Janmey[19] distribution with $Q = 16$, starts to be reliable at conversion degrees between 0.2–0.3. This roughly corresponds to times greater than about 150 sec for the curves in FIGURE 1, panels A and B, and greater than about 110 sec for the curves in panel C.

As described in MATERIALS AND METHODS, the determination of M_w and $\langle R_g^2 \rangle_z$ was made by polynomial fitting of the K_c/R_θ versus $\sin^2(\theta/2)$ plots. As an improvement

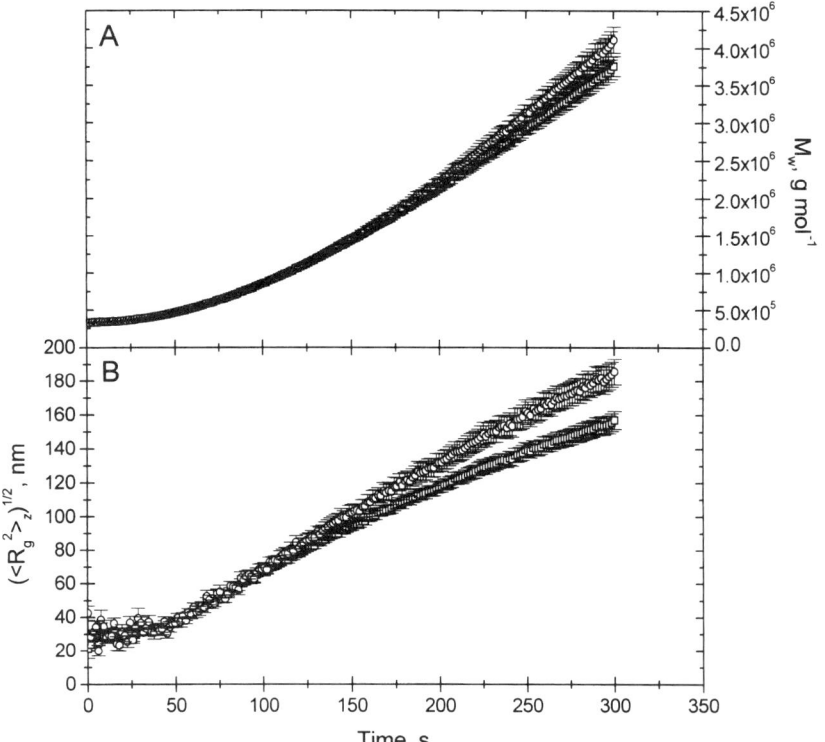

FIGURE 2. Effect of the polynomial degree in the extrapolation of M_w (**Panel A**) and $\langle R_g^2 \rangle_z$ (**Panel B**) values from the K_c/R_θ versus $\sin^2(\theta/2)$ plots, as a function of time (*squares,* third-degree polynomials; *circles,* fourth-degree polynomials). FG (0.216 mg ml^{-1}) polymerization induced by 0.25 NIH units/mg FG of thrombin in TBS. For reasons of clarity, only one third of the data points are shown.

in respect to our recently published work on FG polymerization,[10] we have used both third- and fourth-degree polynomials, since more simulations have shown the theoretical advantage of the use of higher degree polynomials as the polymerization proceeds.[15] In FIGURE 2, the results of the fits with a third- (*squares*) and a fourth- (*circles*) degree polynomial on the same set of K_c/R_θ versus $\sin^2(\theta/2)$ time-dependent data are reported, for a FG polymerization in TBS (panel A, M_w; panel B, $[\langle R_g^2 \rangle_z]^{1/2}$). It is clear that the third- and fourth-degree polynomials yield similar values in both cases for times less than about 100–120 sec, after which the fourth-degree polynomial yields increasingly higher values (more evident for the $(\langle R_g^2 \rangle_z)^{1/2}$ versus time plot). Since the third-degree polynomials are more stable, and yield less "noisy" data points at early times, they were used up to the point at which the curves diverge, after which the fourth-degree polynomial data were used. It must be pointed out that not all the runs yielded such smooth transitions, especially for the $(\langle R_g^2 \rangle_z)^{1/2}$ versus time plots, in which case a small discontinuity resulted on switching between

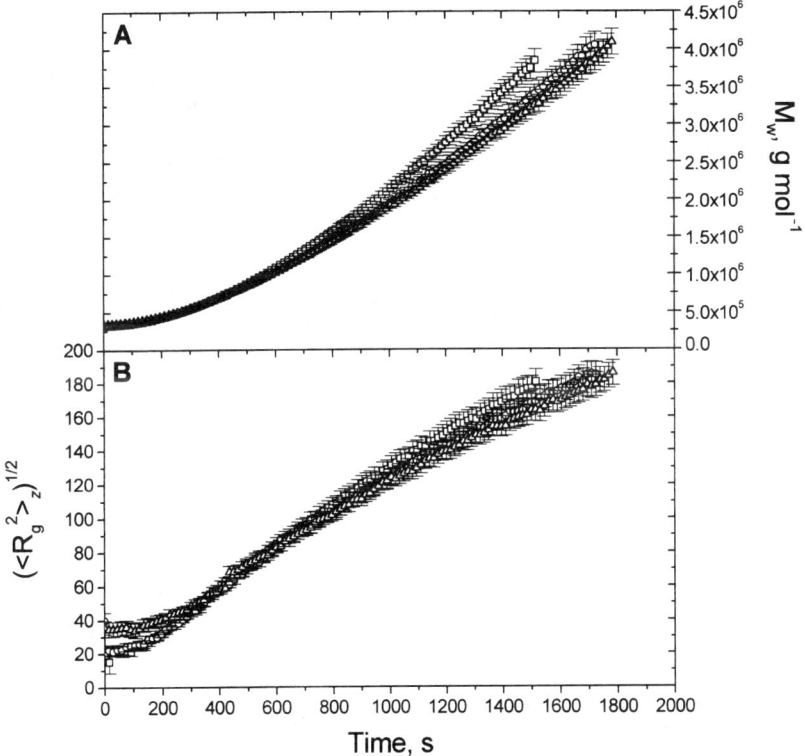

FIGURE 3. Reproducibility of the extrapolated M_w (**Panel A**) and $\langle R_g^2 \rangle_z$ (**Panel B**) values, as a function of time, between successive polymerization runs (*squares,* first run; *circles,* second run; *triangles,* third run). FG (0.216 mg ml^{-1}) polymerization induced by 0.05 NIH units/mg FG of thrombin in TBS. For each run, only 1/30 of the data points are shown.

the two polynomials (see below), but the advantages of this a procedure clearly outweighed the disadvantages.

Another issue that was not fully tackled in our previous work,[10] was the reproducibility of our data. In FIGURE 3 are shown the results of three consecutive polymerizations in which FG at 0.21 mg ml^{-1} in TBS was activated by 0.05 NIH units/mg FG of thrombin. As can be best seen in panel A, between the first (*squares*) and the second (*circles*) runs, there is a small but noticeable difference in the time evolution of M_w, which is much less between the second and the third (*triangles*) runs. This behavior was consistently observed, albeit with varying magnitudes, at other thrombin concentrations and in the other buffers employed, and most likely results from a residual interaction of thrombin with the glass syringe of the stopped-flow mixer, our precautions, coating the syringe with silicone, notwithstanding. The fact that the difference between the first and second runs was always greater than that between successive runs, further support this hypothesis, since thrombin adsorption probably saturates with time (note that in the experiments reported in FIG. 3, more than 30 min elapsed between runs). The same effect is also evident in panel B of FIGURE 3, where it is also noticeable that one of the runs (*triangles,* corresponding to the third in the series) yielded, in the very first phases ($t < 300$ sec), sensibly higher $(\langle R_g^2 \rangle_z)^{1/2}$ values than the other two runs. This happened occasionally, and can either be an artifact (residues from previous polymerizations, trapped microbubbles), or result from a difference in the time of mixing, since this is done manually. In the latter case, it will also be interesting to study this effect in the light of what follows (see below). In any case, both problems should be resolved in the future by the use of a stepping motor-driven mixing system with PEEK syringes, but for the time being, it was not possible to average successive runs to improve the accuracy of our data.

We examined the differences in the time evolution of the three parameters measured in this study as a function of solution conditions and thrombin concentration. In FIGURE 4 are reported the time-dependent values of M_w (panels A and D), of $(\langle R_g^2 \rangle_z)^{1/2}$ (panels B and E), and of M_L (panels C and F), for the FG polymerization induced by 0.25 NIH units of thrombin/mg FG in TBS (□, all panels), in TBE (△, panels A–C), and in TBC (○, panels A–C), whereas in panels D–F are also shown the effects of reducing the thrombin:FG ratio (in TBS) to 0.1 NIH units/mg FG (▽) and to 0.05 NIH units/mg FG (◇). The data of panels A–C are derived from the scattering curves partially reported in FIGURE 1. For the polymerizations in TBS and in TBE, there is a noticeable difference between the M_w versus time and $(\langle R_g^2 \rangle_z)^{1/2}$ versus time curves. The former have an initial exponential portion followed by a linear portion, but the latter appear to be mildly sigmoidal. In addition, the difference between the $(\langle R_g^2 \rangle_z)^{1/2}$ values in the two buffers at the later times are smaller than the corresponding difference between the M_w values, although at the beginning they markedly differ, those in TBE (△) starting at nearly twice the initial value measured in TBS (□). On the other hand, in TBC both curves appear to have a pure exponential shape, reflecting again the fact that in this buffer the second phase starts much earlier than in the other two curves (see FIG. 1).

Another relevant difference between the effects of TBS and TBE can be observed in FIGURE 4, panel C, where the estimated M_L values are reported as a function of time. In TBE, between 150–200 sec (although the M_w and $(\langle R_g^2 \rangle_z)^{1/2}$ values remain comparable), the average width of the protofibrils seems to be smaller than that in

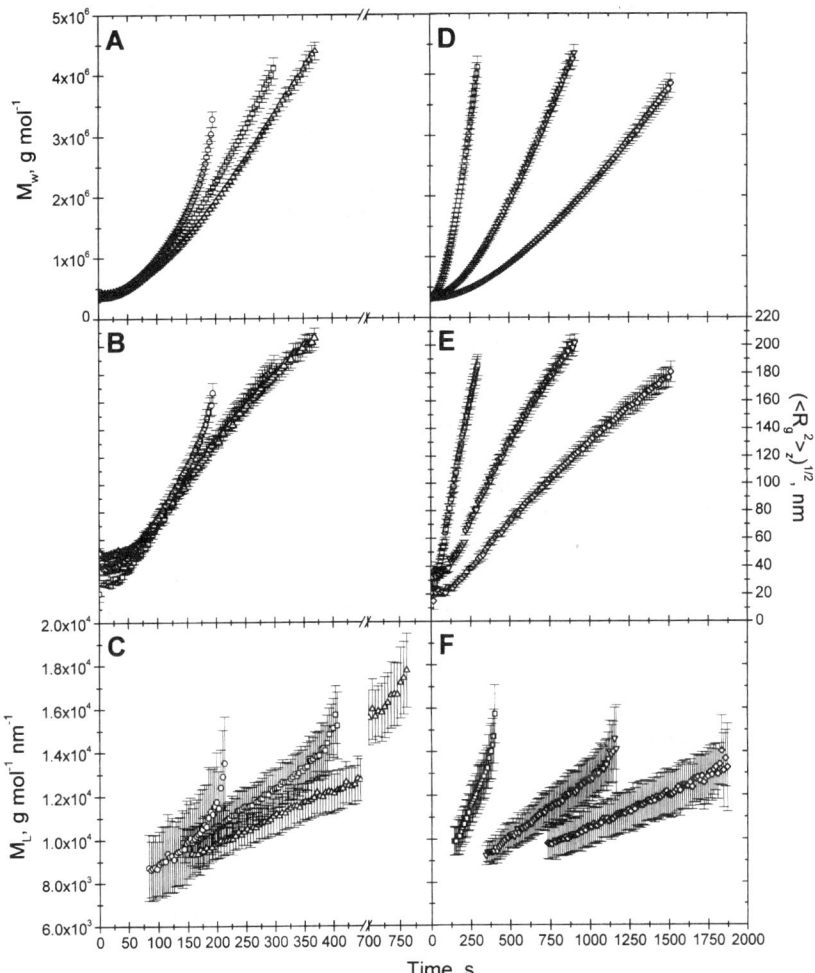

FIGURE 4. Effects of the buffers employed (**Panels A–C**), and of the thrombin:FG ratio (**Panels D–F**) on the time-dependent values of M_w (**Panels A** and **D**), $(\langle R_g^2 \rangle_z)^{1/2}$ (**Panels B** and **E**), and M_L (**Panels C–F**), for FG polymerizations in TBS ($\square$, $\triangledown$, and $\diamond$), in TBE ($\triangle$), and TBC ($\bigcirc$). In **Panels A–C**, 0.25 NIH units/mg FG of thrombin for all the polymerizations. In **Panels D–F**, 0.25 ($\square$), 0.1 ($\triangledown$), and 0.05 ($\diamond$) NIH units/mg FG of thrombin. The data points shown are: in **Panels A–B**, only 1/5 ($\bigcirc$), 1/8 ($\square$), and 1/10 ($\triangle$); in **Panel C**, only 1/7 ($\square$ and $\bigcirc$) and 1/10 ($\triangle$); in **Panels D** and **E**, only 1/10 ($\square$), 1/20 ($\triangledown$), and 1/30 ($\diamond$); in **Panel F**, only 1/15 ($\square$), 1/20 ($\triangledown$), and 1/25 ($\diamond$).

TBS. This could result either from the presence of single-stranded species, or from a complete inhibition of the lateral aggregation between the double-stranded protofibrils, assuming that such an event happens that early during the polymerization. Later, the difference between the M_L values increases, but also the M_w show a similar trend, reflecting the different polymerization velocities between the two solution conditions. We also stress that, unfortunately, it is very difficult to determine with sufficient precision the M_L in the early phases (see above; note also the span of the error bars), and the values reported here should be taken only as an upper bound on this parameter. This is due to the fact that only 4–5 angles (about 103–144°, but excluding the last two angles, about 151° and about 158°, which were judged to be unreliable) were used in the fittings, instead of the last 8–9 angles used in our previous study.[10] This procedure was chosen after a closer look at the Casassa plots has revealed a mild upward curvature (data not shown), and results in lower M_L values (and a greater span in the error bars) as higher (and fewer) angles are used. Although the average M_L values recovered for the polymerization in TBS at 0.25 NIH units thrombin/mg FG are similar to those that we have previously reported,[10] the present data sets represent more conservative estimates overall. These caveats notwithstanding, it interesting to note that the average M_L values at the end of the sigmoidal lag phase are different for the three buffers investigated at equal thrombin/FG ratio, ranging from about 10,000 g mol^{-1} nm^{-1} in TBC to about 14,000 and about 15,000 g mol^{-1} nm^{-1} in TBS and in TBE, respectively (see FIG. 4, panel C). This does not seem to be a simple direct consequence of the different polymerization velocities, since the transition M_L values for the three runs in TBS, but at different thrombin/FG ratios (FIG. 4, panel F) are instead quite similar (see below). In any case, the behavior of the M_L versus time plots clearly indicates that at the end of the sigmoidal, lag phase, side-by-side aggregation indeed starts to be prominent.

As well as affecting the kinetics of polymerization, the sudden addition of EDTA-Na$_2$ or Ca^{2+} can also have an effect on the shape and state of aggregation of FG. In fact, the experiments described here were all started from the same pooled peak fractions of a FG solution chromatographed in TBS, and the additional salts were added in the mixing phase. When this was done in the absence of thrombin, to record the FG baseline values, the sudden addition of 2 mM EDTA-Na$_2$ (to bring its final value to 1 mM) clearly perturbed the FG state, as indicated by mildly curved Zimm plots, reminiscent of those observed in the initial phases of the polymerization (data not shown). In respect to the values determined prior to mixing, the resulting, stable changes were from 320,000 to 351,000 g mol^{-1} for M_w, and from 15.1 to 34 nm for $(\langle R_g^2 \rangle_z)^{1/2}$, indicating that the sudden stripping of the tightly bound Ca^{2+} ions from FG may induce a conformational change leading to the formation of some aggregates. This effect could well contribute to the quite high $(\langle R_g^2 \rangle_z)^{1/2}$ values observed in the very early phases of the thrombin induced polymerization in the presence of EDTA-Na$_2$ (FIG. 4, panel B). An effect of similar magnitude was observed when 5 mM CaCl$_2$ (to bring its final value to 2.5 mM) was suddenly added in the mixing phase, with M_w and $(\langle R_g^2 \rangle_z)^{1/2}$ increasing to 356,000 g mol^{-1} and 29 nm, respectively. In this case, however, the solutions were not stable, with the M_w and $(\langle R_g^2 \rangle_z)^{1/2}$ values slowly increasing with time, a phenomenon for which we have no immediate explanation, since contamination by thrombin is unlikely and Factor XIIIa, if present, should have been completely inhibited by the treatment with

PCMB. In any case, this effect can again partially explain the high initial values of $(\langle R_g^2 \rangle_z)^{1/2}$ observed in the thrombin induced polymerization (FIG. 4, panel B). Future experiments should therefore be carried out after chromatography of the FG solutions already equilibrated with the buffers used for the polymerizations.

We can now briefly analyze the effects of changing the enzyme:substrate ratio by looking at FIGURE 4, panels D–F, where a series of experiments in TBS are shown. The general aspect of the M_w, $(\langle R_g^2 \rangle_z)^{1/2}$, and M_L versus time plots remains nicely unaltered on varying the thrombin:FG ratio from 0.25 (□) to 0.1 (▽) and to 0.05 (◇) NIH units/mg FG. It is interesting to note that the second, fast transition, seems to take place when the M_L has reached a certain value (about 14,000 g mol^{-1} nm^{-1}) irrespective of the polymerization rate, further supporting the hypothesis that the lateral aggregation of the protofibrils is a cooperative process highly dependent of their degree of polymerization.

As in our previous work,[10] we have further analyzed our data as $\langle R_g^2 \rangle_z$ versus M_w double logarithmic plots, with the aid of model curves. For instance, this is shown in FIGURE 5, panel A (left axis scale, open symbols), for the data of FIGURE 4, panels D–E, that we just examined. It is immediately evident that all the curves are almost superimposable, notwithstanding the great differences between the velocities of the runs, about 6 times between the slowest (◇) and the fastest (□). Only in the very early phases are there differences (best seen in FIG. 5, inset B), but the uncertainties connected to the injection procedure with the present set up prevent further speculation on these data. The fact that the polymerization activated by 0.1 NIH units of thrombin/mg FG (▽) presents higher $\langle R_g^2 \rangle_z$ values for any given M_w value above about 800,000 g mol^{-1} than those at the other two thrombin:substrate ratios is likely due to noisier primary data, as evidenced by the discontinuity in the $(\langle R_g^2 \rangle_z)^{1/2}$ versus time plot (FIG. 4, panel E) on switching between the third and fourth-degree polynomials.

The corresponding M_L versus M_w estimates are also reported in FIGURE 5, panel A (right axis scale, center dotted symbols), and are also quite superimposable, given the uncertainties associated with these data. Overall, all these plots clearly demonstrate that in the lag phase, the mechanism of fibrin polymerization is the same irrespective of thrombin:FG ratio, at least within this range where the activation of the monomers is the rate limiting step.

Superimposed on the data in FIGURE 5 are also shown model curves for both $\langle R_g^2 \rangle_z$ versus M_w and M_L versus M_w plots, generated as described in MATERIALS AND METHODS. Initially, a $Q = 16$ ratio between the rate of release of the two FPA from the same FG molecule was assumed, because of previous supporting evidence (see Ref. 10 and references therein). The solid lines were generated according to a worm-like, double-stranded, half-staggered model with $l_p = 200$ nm (WLDSQ16-200), and clearly fit well the $\langle R_g^2 \rangle_z$ versus M_w data, as we had already shown.[10] However, we had also realized[10] that the M_L versus time behavior of such a simulation was not in agreement with the real data, as clearly confirmed in FIGURE 5 (lower solid curve), and it was thus supposed that fibrin monomer-FG interactions could play a relevant role, as had already been suggested.[21–22] As we have done previously,[10] we can simulate the effect of the immediate, complete formation of fibrin monomer:FG complexes upon FPA cleavage. Such a M_L versus M_w curve, also shown in FIGURE 5 (WLDSQ16-200M, lower dash-dot curve), exhibits a trend similar to, but is clearly

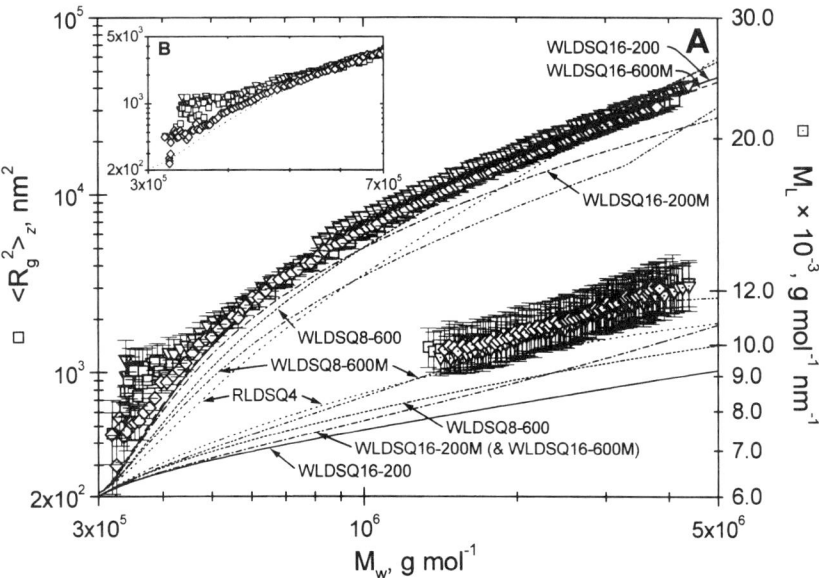

FIGURE 5. Double logarithmic plots of $\langle R_g^2 \rangle_z$ versus M_w (○, □, ▽, left-hand ordinate) and of M_L versus M_w (⊙, ⊡, ▽, right-hand ordinate) for the data of FIGURE 4, **Panels D–F**. FG polymerizations in TBS induced by 0.25 (□), 0.1 (▽), and 0.05 (◇) NIH units/mg FG of thrombin. Only 1/6 (□), 1/15 (▽), and 1/25 (◇) of the data points are shown. In **Inset B**, the $\langle R_g^2 \rangle_z$ versus M_w data for $M_w < 700{,}000$ g mol^{-1} are replotted on an expanded scale, where only 1/3 (□), 1/5 (▽), and 1/10 (◇) of the data points are shown. Overlaid in **Panel A** and labeled are the theoretical curves calculated for different bifunctional polycondensations of rod-like, 50-nm long, 3-nm thick, 300,000 g mol^{-1} monomers, generating either rod-like (RLDS, *dotted lines*) or worm-like (WLDS), double-stranded, half-staggered polymers according to Janmey distributions with $Q = 4$, 8, or 16. For the WLDS polymers, the persistence lengths employed were 200 nm (WLDSQ16–200, *solid lines*) and 600 nm (WLDSQ8–600, *short dashed lines*), and the effect of supposing the total formation of FG-fibrin monomer complexes is indicated by the letter *M* (WLDSQ8–600M, *dash-dot-dot lines*; WLDSQ16–200M, *dash-dot lines*; WLDSQ16–600M, *dashed line*, which is shown also in **Panel B**).

still below the experimental data. Moreover, this simulation no longer fits well the $\langle R_g^2 \rangle_z$ versus M_w data (upper dash-dot curve), and an increase of the persistence length to 600 nm is required to restore a proper fit (WLDSQ16–600M, upper dash curve, best visible for $M_w > 4 \times 10^6$ g mol^{-1}). It must be pointed out that the WLDSQ16–600M curve has a behavior identical to that of the WLDSQ16–200M curve on the M_L versus M_w plots, as indicated by the double label on the lower dash-dot curve in FIGURE 5. We explored the possibility of lower Q values, as suggested by other studies.[23–24] In particular, in the work of Shainoff *et al.*,[24] Q values between about 2 and 5 were determined (see also the article by Shainoff in this volume). However, such low Q values were found to be rather incompatible with the $\langle R_g^2 \rangle_z$ versus M_w data, as exemplified by the behavior of the dotted curve labeled RLDSQ4 in FIGURE 5, representing a simulation with rigid rod-like polymers and $Q = 4$ (semiflexible, worm-like polymers fared even worse, data not shown). The same simulation compared better with the M_L versus M_w data, as shown in FIGURE 5, but inclusion of the fibrinogen:fibrin monomer complexes produced a curve that leveled too early at the 12,000 g mol^{-1} nm^{-1} value (not shown). Moreover, the corresponding $\langle R_g^2 \rangle_z$ versus M_w simulation was totally incompatible with the experimental data (not shown). We then generated a series of curves at an intermediate $Q = 8$ value, again reported in FIGURE 5 (short dash, WLDSQ8–600; dash-dot-dot, WLDSQ8–600M). Although a reasonably matching curve could be found for the M_L versus M_w data (WLDSQ8-600M), its counterpart was still far removed in respect to the $\langle R_g^2 \rangle_z$ versus M_w data. Given the high uncertainty associated with the M_L versus M_w data, and recalling that their values likely represent only an upper bound, we still favor the high Q simulations as more compatible with our present experimental data. A better set of M_L data and at least an estimation of the above mentioned equilibrium constants for the formation of fibrinogen:fibrin monomer complexes are needed to better clarify this issue.

Similarly, we can compare the effects of the different buffers employed, as is done in FIGURE 6. In panel A and inset B, the three $\langle R_g^2 \rangle_z$ versus M_w plots are shown for the polymerizations of FIGURE 4, panels A–B. Time points are also shown for the three data sets (*arrows*), as well as the "baseline" values (slightly enlarged filled symbols) before thrombin addition. A subset of the model curves of FIGURE 5 (WLDSQ16–200, solid lines; WLDSQ16–200M, dash-dot lines; WLDSQ16–600M, dash lines) is also superimposed in both panels. It is again immediately evident that all the experimental data exhibit a very similar behavior (the one in TBE being always slightly above the other twos), and are fitted well by the WLDSQ16–200 and WLDSQ16–600M simulations, especially for M_w greater than about 700,000 g mol^{-1}. However, at the earlier times (e.g., for $t < 70$–80 sec, or M_w less than about 700,000 g mol^{-1}), there are marked differences, better seen in inset B. In particular, both the polymerizations in TBC and TBE exhibit a marked jump in the $\langle R_g^2 \rangle_z$ values, followed by a phase in which this parameter grows more slowly in respect to the M_w values. This effect could be at least partially due to the sudden addition of CaCl$_2$ or EDTA-Na$_2$ together with the thrombin solution, as indicated by the high "baseline" values in these two buffers (see above).

In an attempt to better understand this issue, we have replotted in FIGURE 6, panel C, these three data sets (open symbols, left-hand ordinate scale), together with the corresponding M_L versus M_w estimates (center-dotted symbols, inner right-hand ordinate scale), and the related entire set of WLDS simulated polymerization curves. In addition, we have also computed the percentage of fibrinopeptides released as a function of M_w, starting from the determination of the kinetic constants for the thrombin-catalyzed removal of the FPAs and FPBs under conditions very similar to those employed for the SF-MALLS experiments, as will be reported in detail elsewhere (Profumo *et al.*, in preparation; see also MATERIALS AND METHODS). These plots are also overlaid (and labeled) in FIGURE 6, panel C (upper curves, outer right-hand ordinate scale), as solid and short-dotted curves (FPA and FPB in TBS, respectively), dashed and short dashed curves (FPA and FPB in TBE, respectively), and dash-dotted and short-dash-dotted curves (FPA and FPB in TBC, respectively). As can be seen, there is a mild correlation between the rate of release of the fibrinopeptides and both

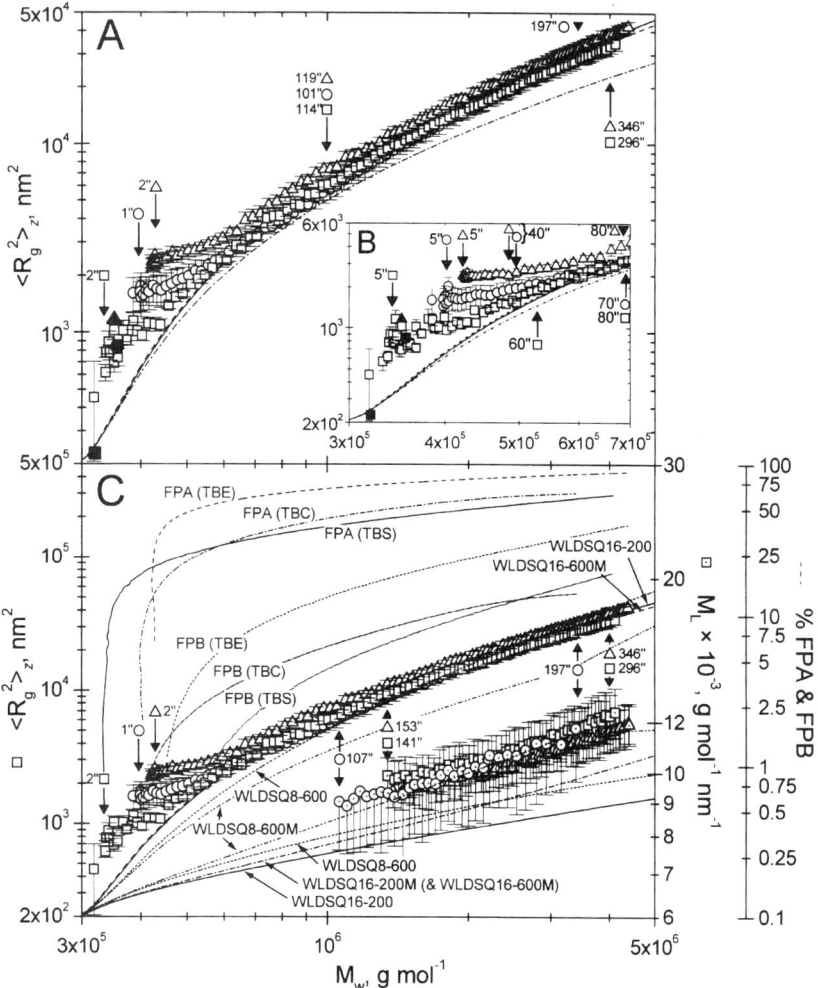

FIGURE 6. Double logarithmic plots of $\langle R_g^2 \rangle_z$ versus M_w (○, □, △, left-hand ordinate, **Panels A–C**) and of M_L versus M_w (⊙, ⊡, △, inner right-hand ordinate, **Panel C**) for the data of FIGURE 4, **Panels A–C**. FG polymerizations induced by 0.25 NIH units/mg FG of thrombin in TBS (□), in TBE (△) and in TBC (○). In **Panels A** and **C**, only 1/6 (□ and ○) and 1/8 (△) of the data points are shown. In **Panel B**, only 1/3 (□ and ○) and 1/6 (△) of the data points are shown. Indicative time points are overlaid in all panels (*arrows* with symbols). In **Panel C**, the theoretical curves for the percentage FPA and FPB release are also shown in double logarithmic plots as a function of the experimental M_w (*labeled curves,* outer right-hand ordinate scale). Some of the same theoretical curves calculated for different bifunctional polycondensations models already shown in FIGURE 5 are also overlaid here in all panels (WLDSQ8-600, *short dash lines;* WLDSQ8–600M, *dash-dot-dot lines;* WLDSQ16–200, *solid lines;* WLDSQ16–200M, *dash-dot line;* WLDSQ16–600M, *dash lines).*

the $\langle R_g^2 \rangle_z$ versus M_w and M_L versus M_w behavior. For instance, in TBE the FPA and FPB releases are the fastest, reaching more than 75% and about 20%, respectively, already for M_w about 1.5×10^6, the $\langle R_g^2 \rangle_z$ initial jump is the highest, and the M_L estimates are the lowest. Conversely, in TBS the FPA and FPB releases are the slowest, reaching about 45% and about 7%, respectively, for the same M_w as above (about 1.5×10^6), the $\langle R_g^2 \rangle_z$ initial jump is the smallest, and the M_L estimates are the highest. As for TBC, all the parameters appear to be close to those in TBS, but they still confirm the trend.

From these data, it is tempting to speculate about the reasons for these differences. For instance, the lower M_L estimates in TBE, coupled with the high $\langle R_g^2 \rangle_z$ values observed immediately after mixing, could result from the presence of long, single-stranded structures, which then coalesce into the classical double-stranded fibrils. In this case, the formation of fibrinogen:fibrin monomers complexes should also be less relevant, perhaps due to the presence of EDTA-Na$_2$, but direct evidence for this is unfortunately unavailable at the moment. It is also difficult to separate the effect of protofibril formation from their lateral aggregation, due also to the non-irrelevant release of FPB, especially for $M_w > 1.5 \times 10^6$, and in fact the low M_L estimates for the TBE data could be due to a stronger inhibition of lateral aggregation, again fostered by the absence of free and bound Ca^{2+} ions. However, in this case it should be possible to describe the data with a Flory or Janmey type model, but thus far we have been unable to find a suitable simulation. In fact, changing the Q value in Janmey type simulations resulted in curves that either required very low, unlikely l_p values (higher Q) to fit the $\langle R_g^2 \rangle_z$ versus M_w plots (not shown), or that could not match them (lower Q), as exemplified by the curves reported in the FIGURE 6, lower part of panel C. The apparently better fit of the WLDS8–600M curve to the M_L versus M_w data should be interpreted with care, due to the higher uncertainty associated with these data (see also above). In any case, the initial jumps in the $\langle R_g^2 \rangle_z$ values remain difficult to interpret with a pure double-stranded, half-staggered model.

CONCLUSIONS

In this paper, we have presented an extension of our previous work on the early stages of the thrombin-induced fibrin polymerization using stopped-flow, multi-angle laser light scattering technology (SF-MALLS). In particular, we have investigated the effect of different thrombin:substrate ratios in TBS, and the effect of adding EDTA-Na$_2$ (TBE) or CaCl$_2$ (TBC) at a fixed thrombin:substrate ratio. In the process, we have made improvements on the determination of the three parameters, M_w, $\langle R_g^2 \rangle_z$, and M_L, from the SF-MALLS primary data, and we have assessed their reproducibility with the current set up. Our findings can be summarized as follows.

1. At these rate-limiting thrombin:substrate ratios, the mechanism of polymerization is basically the same, irrespective of thrombin concentration, and in the so called lag time, the data are compatible with a model in which double-stranded, half-staggered semiflexible protofibrils, with a persistence length between 200–600 nm, are the main species. A better fitting to the $\langle R_g^2 \rangle_z$ versus M_w data is obtained when the ratio Q between the release of the two FPAs is high ($Q = 16$), whereas the opposite seems to apply for the M_L versus M_w data ($4 < Q < 8$). However, the latter

data sets are to be taken as at best an upper bound to the real M_L distributions. The formation of fibrin monomer:fibrinogen complexes may also play a role.

2. Small amounts of EDTA-Na$_2$ (1 mM) or physiological amounts of CaCl$_2$ (2.5 mM) do not alter this general scheme in respect to polymerizations carried out in TBS. The rapid addition of these salts to a fibrinogen solution in TBS is, however, able to perturb its conformation, and induce the formation of aggregates.

3. The very early stages of the polymerization are not fitted well by the above mentioned model, except in TBS at the lowest thrombin:substrate ratio investigated (0.05 NIH units/mg FG). Unfortunately, the reproducibility of our hand driven mixer is not good enough to better assess these very fast events, but a clear trend seem to emerge, in which the reaction in TBE seems to produce the longest, thinner species for a given M_w, whereas those in TBS show the opposite behavior, with those in TBC being closer to the latter than to the former. A correlation with the rate of fibrinopeptides release also emerges from our data.

In order to further and better investigate these early events in fibrin polymerization, it will be necessary to: (1) assess the role of fibrinogen:fibrin monomer complexes, possibly estimating their association constants in the various buffers employed; (2) improve on the accuracy of the M_L determination, and verify its reliability on a system in which the lateral aggregation is totally inhibited; (3) improve the speed and reproducibility of the mixing phase, while at the same time preserving, as much as possible, the enzymatic activity in between the runs; (4) expand the current modeling schemes, to include possible different, either sequential or concurrent mechanisms (single-stranded, double-stranded), as well as the effects of branching on the model curves (for instance, using the computer modeling schemes pioneered by Weisel and coworkers[25–26]). More accurate FG models, based on the recently published crystal structures of its main body,[27–28] could also be employed in the simulations.

Further work along these lines is currently being carried out in our laboratories.

ACKNOWLEDGMENTS

This work was partially supported by grants from the Agenzia Spaziale Italiana (ASI) to M.R. and F.F., from the Consiglio Nazionale delle Ricerche (CNR) to F.F., and from the Ministero Italiano della Ricerca Scientifica e Tecnologica (MURST) to C.C. We thank Michael Mosesson for stimulating discussions, and for allowing us to revise the original manuscript in order to include the updated M_L data sets and the new simulations with the lower Q values.

REFERENCES

1. MOSESSON, M.W. & R.F. DOOLITTLE, Eds. 1983. Molecular Biology of Fibrinogen and Fibrin. Ann. N.Y. Acad. Sci. **406**.
2. MOSESSON, M.W. 1990. Fibrin polymerization and its regulatory role in hemostasis. J. Lab. Clin. Med. **116:** 8–17.
3. BLOMBÄCK, B. 1996. Fibrinogen and fibrin—proteins with complex roles in hemostasis and thrombosis. Thromb. Res. **83:** 1–75.

4. DOOLITTLE, R.F., G. SPRAGGON & S.J. EVERSE. 1999. Three-dimensional structural studies on fragments of fibrinogen and fibrin. Curr. Opin. Struct. Biol. **80:** 792–798.
5. MÜLLER, M. & W. BURCHARD. 1978. Fibrinogen-fibrin transformations characterized during the course of reaction by their intermediate structures. A light scattering study in dilute solution under physiological conditions. Biochim. Biophys. Acta **537:** 208–225.
6. HANTGAN, R.R. & J. HERMANS. 1979. Assembly of fibrin. A light scattering study. J. Biol. Chem. **254:** 11272–11281.
7. MÜLLER, M., H. LASARCYK & W. BURCHARD. 1981. Fibrinogen-fibrin transformation: 2. Influence of temperature, pH and of various enzymes on the intermediates. Int. J. Biol. Macromol. **3:** 19–24.
8. WILTZIUS, P., G. DIETLER, W. KÄNZIG, et al. 1982. Fibrin aggregation before sol-gel transition. Biophys. J. **38:** 123–132.
9. WILTZIUS, P., G. DIETLER, W. KÄNZIG, et al. 1982. Fibrin polymerization studied by elastic and dynamic light-scattering as a function of fibrinopeptide A release. Biopolymers **21:** 2205–2223.
10. BERNOCCO, S., F. FERRI, A. PROFUMO, et al. 2000. Polymerization of rod-like macromolecular monomers studied by stopped-flow, multiangle light scattering: set-up, data processing, and application to fibrin formation. Biophys. J. **79:** 561–583.
11. MIHALYI, E. 1968. Physicochemical studies of bovine fibrinogen. IV. Ultraviolet absorption and its relation to the structure of the molecule. Biochemistry **7:** 208–223.
12. LOTTEMBERG, R., U. CHRISTENSEN, C.M. JACKSON & P.L. COLEMAN. 1981. Assay of coagulation proteases using peptide chromogenic and fluorogenic substrates. Meth. Enzymol. **80:** 341–361.
13. KEHL, M., F. LOTTSPEICH & A. HENSCHEN. 1981. Analysis of human fibrinopeptides by high-performance liquid chromatography. Hoppe-Seyler's Z. Physiol. Chem. **362:** 1661–1664.
14. MIHALY, E. 1988. Clotting of bovine fibrinogen. Kinetic analysis of the release of fibrinopeptides by thrombin and of the calcium uptake upon clotting at high fibrinogen concentrations. Biochemistry **27:** 976–982.
15. FERRI, F., M. GRECO & M. ROCCO. 2000. On the determination of the average molecular weight, radius of gyration, and mass/length ratio of polydisperse solutions of polymerizing rod-like macromolecular monomers by multi-angle static light scattering. Macromol. Symp. **162:** 23–43.
16. CARR, JR., M.E., L.L. SHEN & J. HERMANS. 1977. Mass-length ratio of fibrin fibers from gel permeation and light scattering. Biopolymers **16:** 1–15.
17. CASASSA, E.F. 1955. Light scattering from very long rod-like particles and an application to polymerized fibrinogen. J. Chem. Phys. **23:** 596–597.
18. BENOIT, H. & P. DOTY. 1953. Light scattering from non-Gaussian chains. J. Phys. Chem. **57:** 958–963.
19. JANMEY, P.A. 1982. Kinetics of formation of fibrin oligomers. I. Theory. Biopolymers **21:** 2253–2264.
20. BALE, M.D., P.A. JANMEY & J.D. FERRY. 1982. Kinetics of formation of fibrin oligomers. II. Size distributions of ligated oligomers. Biopolymers **21:** 2265–2277.
21. BRASS, E.P., W.B. FORMAN, R.V. EDWARDS & O. LINDAN. 1976. Fibrin formation: the role of the fibrinogen-fibrin monomer complex. Thromb. Heamost. **36:** 37–48.
22. BROSSTAD, F., P. KIERULF & H.C. GODAL. 1979. The fibrin-solubilizing effect of fibrinogen as studied by light scattering. Thromb. Res. **14:** 705–712.
23. MEH, D.A., K.R. SIEBENLIST, G. BERGTROM & M.W. MOSESSON. 1995. Sequence of release of fibrinopeptide A from fibrinogen molecules by thrombin or Atroxin. J. Lab. Clin. Med. **125:** 384–391.
24. SHAINOFF, J.R., G.B. SMEJKAL, P.M. DIBELLO, et al. 1996. Isolation and characterization of the fibrin intermediate arising from cleavage of one fibrinopeptide A from fibrinogen. J. Biol. Chem. **271:** 24129–24137.
25. WEISEL, J.W. & C. NAGASWAMI. 1992. Computer modeling of fibrin polymerization kinetics correlated with electron microscope and turbidity observations: clot structure and assembly are kinetically controlled. Biophys. J. **50:** 1079–1093.

26. WEISEL, J.W., Y. VEKLICH & O. GORKUN. 1993. The sequence of cleavage of fibrinopeptides from fibrinogen is important for protofibril formation and enhancement of lateral aggregation in fibrin clots. J. Mol. Biol. **232:** 285–297.
27. BROWN, J.H., N. VOLKMANN, G. JUN, *et al.* 2000. The crystal structure of modified bovine fibrinogen. Proc. Natl. Acad. Sci. U.S.A. **97:** 85–90.
28. YANG, Z., I. MOCHALKIN, L. VEERAPANDIAN, *et al.* 2000. Crystal structure of native chicken fibrinogen at 5.5-Å resolution. Proc. Natl. Acad. Sci U.S.A. **97:** 3907–3912.

Conformational Changes upon Conversion of Fibrinogen into Fibrin

The Mechanisms of Exposure of Cryptic Sites

LEONID MEDVED, GALINA TSURUPA, AND SERGEI YAKOVLEV

*Biochemistry Department, The Holland Laboratory,
American Red Cross, Rockville, Maryland, USA*

ABSTRACT: Conformational changes upon conversion of fibrinogen to fibrin result in the exposure of multiple binding sites that provide its interaction with various proteins and cells and, thus, its participation in a number of physiological and pathological processes. Here we focus on conformational changes in the fibrinogen D regions (domains) and αC-domains that are directly involved in intermolecular interactions upon fibrin assembly. According to the current view, two αC-domains that interact intramolecularly in fibrinogen undergo an intra- to intermolecular switch to form αC-polymers in fibrin. The availability of recombinant fragments that correspond to the αC-domain made it possible to further clarify this mechanism and to reveal novel cryptic sites in this domain for plasminogen and its activator tPA, whose exposure may play an important role in the regulation of fibrinolysis. To elucidate the mechanism of exposure of cryptic sites in the D regions, we tested the accessibility of their fibrin specific epitopes (Aα148–160 and γ312–324) that are also involved in binding of plasminogen and tPA, in several fragments derived from fibrinogen (fragment D), and crosslinked fibrin (fragment D-D and its non-covalent complex with the E_1 fragment, D-D:E_1). Neither D nor D-D bound tPA, plasminogen, or anti-Aα148–160 and anti-γ312–324 monoclonal antibodies. At the same time both epitopes became accessible in the D-D:E1 complex. Melting of D and D-D revealed that their domains have the same stability while in the D-D:E_1 complex they became more stable. These results indicate that upon fibrin assembly, driven primarily by the interaction between complementary binding sites of the E and two D regions, the latter undergo conformational changes that cause the exposure of their cryptic sites. They also suggest that the fibrin specific conformation of the D regions is preserved in the D-D:E_1 complex.

KEYWORDS: Fibrinogen; Fibrin; Conversion; Cryptic sites; Plasminogen; tPA.

INTRODUCTION

Fibrinogen is rather inert in circulation but it becomes reactive toward itself and toward a number of proteins and cell types upon conversion into fibrin. This reactivity provides its spontaneous polymerization, formation of insoluble fibrin clot, and its subsequent participation in a number of important physiological and pathological

---

Address for correspondence: Leonid Medved, Ph.D., The Holland Laboratory, American Red Cross, 15601 Crabbs Branch Way, Rockville, MD 20855, USA. Voice: 301-738-0719; fax: 301-738-0794.

medvedl@usa.redcross.org

processes, including fibrinolysis, inflammation, angiogenesis, atherogenesis, and tumorigenesis. The reactivity also suggests the exposure of multiple cryptic binding sites in fibrin. The fibrin assembly process has been well studied and many of its aspects are well established.[1,2] It is known that cleavage by thrombin of two pairs of fibrinopeptides from the central part of fibrinogen generates new NH_2-terminal sequences (knobs A and B) starting with Gly-Pro-Arg and Gly-His-Arg, whose interaction with complementary sites (polymerization pockets or holes a and b) of the adjacent molecules is critical for fibrin polymer formation. Thus, the exposure of cryptic polymerization sites (knobs) may be described by a limited proteolysis mechanism in which liberation of active sequences occurs after removal of the peptide regions that keep them inactive. Less is known about the exposure of those cryptic sites that are located away from the points of thrombin cleavage. Obviously, their activation cannot be connected directly with the limited proteolysis, but is rather a result of conformational changes that occur upon fibrin assembly. However, there is still no clear understanding of the mechanism(s) of these changes despite numerous structural studies of fibrinogen and fibrin. In this presentation we review briefly published data and present new data that shed light on this aspect of the fibrin assembly process, with an emphasis on the mechanisms of conformation dependent exposure of cryptic sites in those fibrin(ogen) regions that are directly involved in intermolecular interactions upon formation of fibrin clot.

MATERIALS AND METHODS

Fibrinogen degradation products, fragment D_1, D_H, D_Y, and TSD were prepared from plasmin digest of human and bovine fibrinogen as described elsewhere.[3,4] Fibrin degradation products, the D-D:E_1 complex and its components, dimeric D-D fragment and fragment E_1, were prepared from plasmin digest of factor XIIIa-crosslinked human fibrin.[5] Fibrin specific murine monoclonal antibodies (Mabs) against Aα148–160 and γ312–324 regions of fibrin(ogen)[6,7] were obtained from Dr. Nieuwenhuizen.

Binding studies were performed by enzyme linked immunoassay (ELISA) and by surface plasmon resonance method (SPR) using the IAsys biosensor (Fisons, Cambridge, UK) as described in References 5, 8, and 9.

Differential scanning calorimetry (DSC) measurements were made with DASM-4M calorimeter at a scan rate of 1°C/min as described elsewhere.[10] Protein concentrations varied from 1.0 to 2.0 mg/ml. Fibrin des-AB devoid of fibrinopeptides A and B and fibrin des-A devoid of fibrinopeptide A only were prepared as follows. Plasminogen depleted human fibrinogen (Calbiochem) at 1 mg/ml was mixed with thrombin at 0.1 NIH unit/ml or thrombin-like enzyme Batroxobin at 4°C and a portion of the mixture was immediately placed into calorimetric capillary cell that was at the same temperature. The mixture then was heated in calorimeter at 1°C/min up to 37°C to allow the formation of fibrin gel and then cooled to 4°C to equilibrate the instrument. The endotherms were recorded in a temperature range 10–90°C. The other portion of the mixture was placed simultaneously into a tube and subjected to a similar heating–cooling procedure to control fibrin formation. Amino acid sequence analysis of the control samples withdrawn after the heating–cooling procedure revealed that the removal of fibrinopeptides A and B, or A alone, by thrombin and Batroxobin, respectively, was complete; no removal of the fibrinopeptide B by the latter was observed. Deconvolution analysis was performed as described elsewhere,[10,11] using a dependent scheme that assumes an ordered process in which constituent domains unfold sequentially, implying the occurrence of interactions between domains such that the unfolding of any given domain depends on the status of its neighbors.

Protein structure analysis including determination of the solvent accessible surface area and interatomic contacts, and computer graphics were performed using the commercially available MOLE™ program (Applied Thermodynamics, Hunt Valley, MD).

CONFORMATIONAL CHANGES IN THE αC-DOMAINS UPON CONVERSION OF FIBRINOGEN INTO FIBRIN

Fibrinogen consists of two identical subunits, each of which is formed by three non-identical polypeptide chains, Aα, Bβ, and γ. They are linked together by disulfide bonds and assembled to form multiple domains that are grouped into three major structural regions, the central E and two identical terminal regions D.[1,2,12,13] Fibrinogen also contains two surface-oriented αC-domains formed by the COOH-terminal two-thirds (residues Aα220–610) of the two Aα chains. According to the current view the COOH-terminal half of each αC-domain forms a compact structure that is attached to the bulk of the molecule by the remaining polypeptide part that is presumably unordered; both compact parts interact with each other and with the central region E (see FIGURE 1 A).[13–15]

Based on structural/functional studies it was hypothesized that upon conversion of fibrinogen into fibrin, the αC-domains, which interact with each other intramolecularly in the former, dissociate and switch to intermolecular interaction in the latter to promote fibrin self-association.[16,17] This hypothesis was subsequently confirmed by electron microscopy studies.[18,19] It was also shown that the dissociation of the αC-domains from the central domain requires the removal of both fibrinopeptides A and B.[19] Since fibrinopeptide B is cleaved preferentially when protofibrils are formed, the intermolecular interaction between the αC-domain seems to be more significant at the stage of lateral aggregation of protofibrils.[19] Numerous biochemical data revealed that in fibrin the α chains are crosslinked by factor XIIIa resulting in α-polymers and that the crosslinking occurs between the regions comprising the αC-domains.[20–22] This implies that in fibrin the αC-domains form not the intermolecular dimers, like the intramolecular dimers in fibrinogen, but rather polymeric structures. In agreement, the formation of long, linear branching arrays by an isolated bovine fibrinogen αC-fragment derived proteolytically from the αC-domain was observed by electron microscopy.[18] This also suggests that the formation of αC-polymers may occur via interaction sites that differ from those that form the αC–αC dimer in fibrinogen. Thus, upon conversion of fibrinogen into fibrin, the αC-domains switch from intra- to intermolecular interaction, resulting in the formation of the αC-polymers that become crosslinked by factor XIIIa.

The αC-domains contain various interaction sites. First, as just described they contain polymerization sites that promote this process, especially in dilute solutions.[16,18,19] Second, they interact with some integrins via their Aα572–574

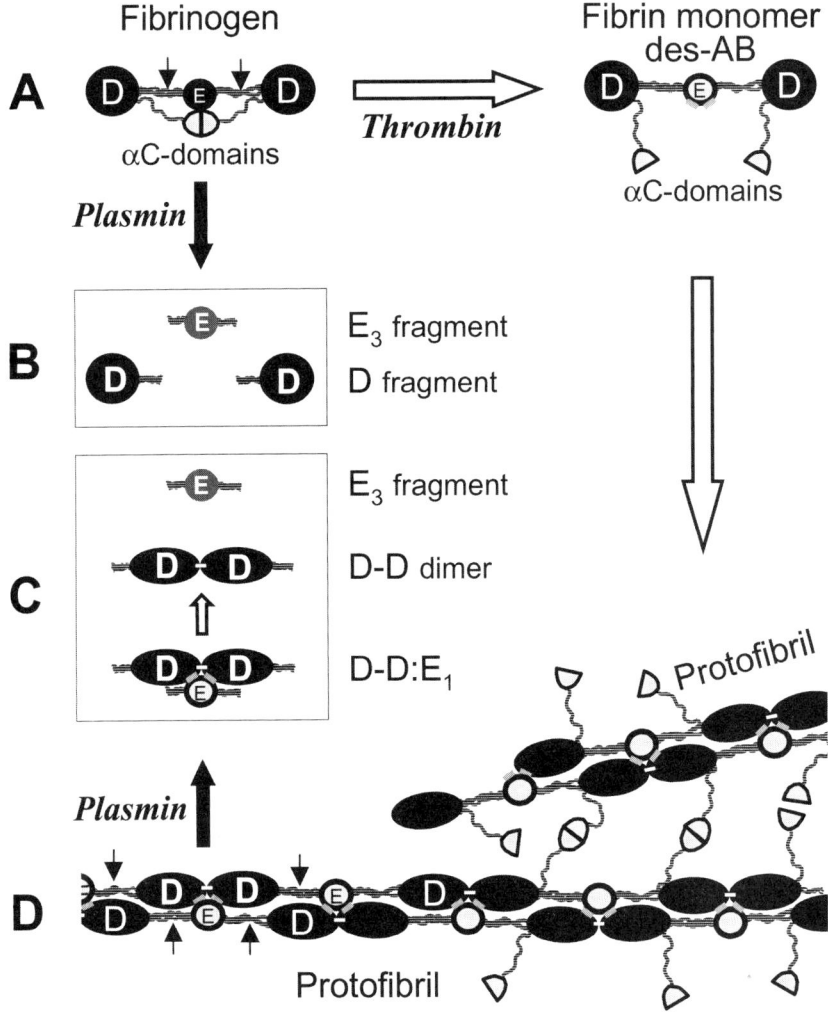

FIGURE 1. Schematic diagram representing location of the αC-domains in fibrinogen and their switch from intra- to intermolecular interaction in fibrin. **Panel A** represents intramolecular interactions of the αC-domains and their dissociation on removal of fibrinopeptides A and B (FPA and FPB) by thrombin and **Panel D** represents intermolecular interaction of the αC-domains that promote lateral aggregation of protofibrils into thicker fibers. Note that normally thrombin removes FPAs first resulting in the formation of protofibril whereas the removal of FPBs coincides with their lateral aggregation (see text). **Panels B** and **C** represent plasmin generated fibrinogen and fibrin degradation products, respectively. *Small arrows* in **Panels A** and **D** indicate the points of the plasmin attack between fibrin(ogen) D and E regions.

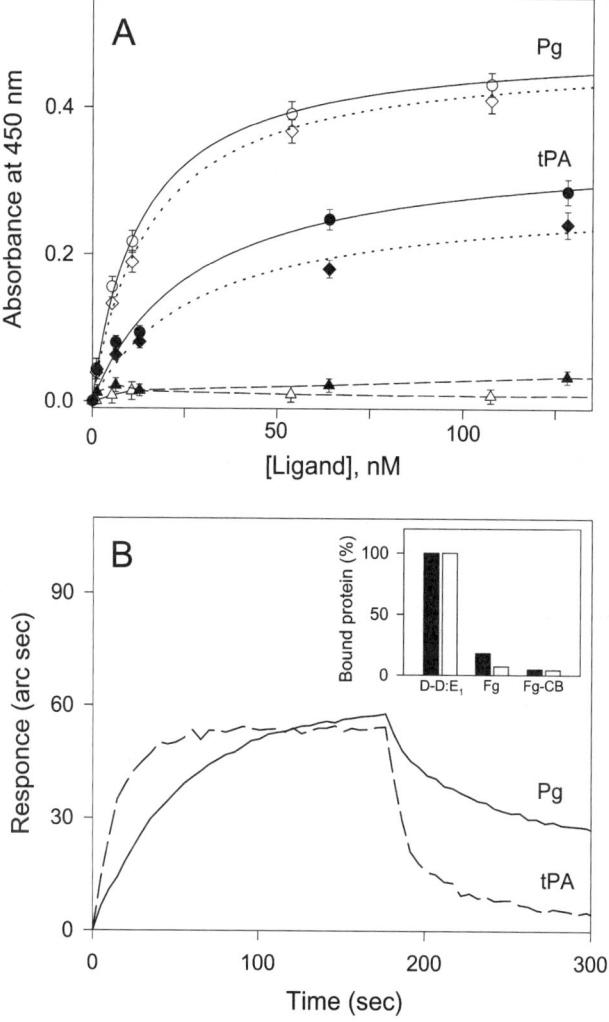

FIGURE 2. Analysis of binding of plasminogen and tPA to immobilized αC-fragment. **Panel A** represents ELISA results for which increasing concentrations of ligand, plasminogen (*open symbols*) or tPA (*closed symbols*), were incubated with microtiter wells coated with the αC-fragment (*circles*) and its truncated variants, Aα221–391 (*triangles*), or Aα392–610 (*diamonds*), and bound species were detected with polyclonal antibodies against plasminogen or tPA. **Panel B** represents surface plasmon resonance-detected binding of plasminogen and tPA with the αC-fragment. Plasminogen at 20 nM (*solid line*) and tPA at 40 nM (*dashed line*) were added to immobilized αC-fragment and their association/dissociation was monitored in real time while registering the resonance signal (response). The **Inset** shows binding of immobilized plasminogen (*filled bars*) and tPA (*open bars*) with fibrinogen (Fg) or fibrinogen treated with carboxypeptidase B (Fg-CB); the binding of the D-D:E_1 complex (100%), which mimics fibrin, is given for comparison.

Arg-Gly-Asp(RGD) sequence.[23,24] They also interact with factor XIII, PAI-1, fibronectin, and thrombospondin, and become crosslinked by factor XIIIa to the latter two and to α_2-antiplasmin, PAI-2, and von Willebrand factor.[25–34] This list of interactions is not complete. For example, the impaired fibrin mediated activation of plasminogen by tissue plasminogen activator (tPA) caused by a mutation in the αC-domains[35,36] suggests the involvement of these domains in fibrinolysis. It is also not clear whether these interaction sites are active in both fibrinogen and fibrin or conversion of fibrinogen into fibrin is required for these interactions to occur.

Most of the known interactions of the αC-domains were identified by comparison of intact fibrin(ogen) with the proteolytically modified derivatives lacking these domains, namely fragment X and fibrinogen fraction I-9 or their analogues. Proteolytic isolation of the αC-domain in sufficient quantity is not a simple task since this portion of fibrin(ogen) is highly susceptible to proteolysis and is rapidly degraded into several smaller polypeptides.[18,22,37] This is probably the major reason why practically no binding studies have been performed with the isolated αC-fragment. To overcome this problem, we expressed in *E. coli* recombinant αC-fragments corresponding to the entire αC-domain (residues Aα221–610) and its truncated variants, Aα221–391 and Aα392–610, corresponding respectively to its NH_2- and COOH-terminal halves,[8,38] and tested directly their interaction with plasminogen and tPA. The experiments revealed that the αC-fragment binds both proteins with dissociation constants in the range of 16–33 nM.[8] An example demonstrating binding of plasminogen and tPA to immobilized αC-fragment detected by two independent methods, ELISA and surface plasmon resonance (SPR), is presented in FIGURE 2. Both sites were localized in the COOH-terminal half of the αC-domain since of the two truncated variants only Aα392–610 was active. In agreement, both αC-fragment and Aα392–610 but not Aα221–391, stimulated activation of plasminogen by tPA in a chromogenic substrate assay.[8] These data indicate that fibrinogen αC-domains contain novel high affinity tPA and plasminogen binding sites that may play an important role in fibrinolysis. It is known that, in contrast to fibrin, intact fibrinogen does not bind tPA and plasminogen and does not stimulate activation of the latter. In agreement, we did not observe any substantial binding of fibrinogen to these proteins, especially when it was treated with carboxipeptidase-B to remove possible COOH-terminal lysine residues that may be generated upon partial degradation of its αC-domains by plasmin (FIG. 2 B, Inset). This indicates that tPA and plasminogen binding sites are cryptic in fibrinogen αC-domains, as are those identified earlier in the D regions;[39] they are exposed only in the isolated αC-domains and presumably in fibrin. Their exposure is, most probably, connected with the intra- to intermolecular switch mechanism described above.

CONFORMATIONAL CHANGES IN THE D REGIONS

The other candidates for conformational changes upon conversion of fibrinogen into fibrin are the two terminal D regions that contain polymerization pockets and are involved directly in the interaction with the central regions of adjacent fibrin molecules via complementary knobs (FIG. 1 D). It is well established that each D region contains tPA- and plasminogen-binding sites that include sequences γ312–324 and Aα148–160, respectively.[39] It was also demonstrated that these sites are cryptic in

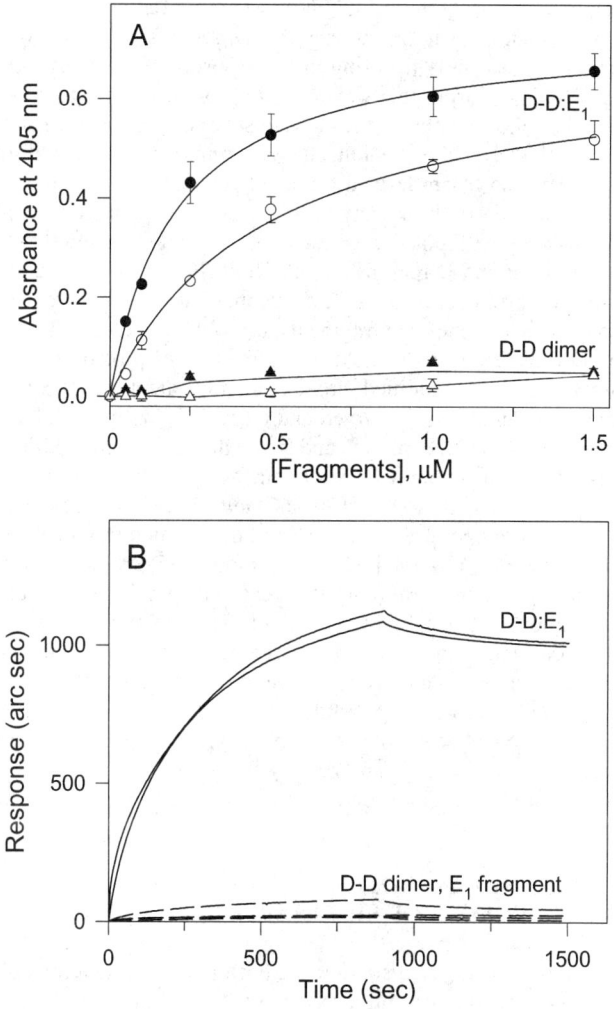

FIGURE 3. Analysis of binding of fibrin derived D-D:E_1 complex and its individual components to immobilized anti-γ312–324 and anti-Aα148–160 monoclonal antibodies. **Panel A** represents results of ELISA, in which increasing concentrations of D-D:E_1 (*circles*) or D-D (*triangles*) were incubated with microtiter wells coated with anti-γ312–324 Mab (*filled symbols*) and anti-Aα148–160 Mabs (*open symbols*), and bound species were detected with polyclonal antibodies against γ-module. **Panel B** represents surface plasmon resonance-detected binding of the D-D:E_1 complex and its components, E_1 and D-D, to the immobilized Mabs. The fragments and the complex were added at 1 μM and their association/dissociation was monitored in real time. The curves for binding of D-D:E_1 (*solid lines*) to both antibodies essentially coincide. The amount of binding observed with E_1 and D-D (*broken lines*) in both cases was negligible.

fibrinogen since specific anti-γ312–324 and anti-Aα148–160 monoclonal antibodies (Mabs) recognize them only in fibrin;[6,7] here we will refer to them as fibrin specific epitopes. These findings suggested that the D regions undergo conformational changes upon fibrin assembly resulting in the exposure of their tPA- and plasminogen-binding sites. The fibrin specific Mabs also provided useful tools that could be used as molecular probes to clarify the mechanism of this exposure. Toward this end, in collaboration with Dr. Nieuwenhuizen, we compared the accessibility of these sites in fibrinogen- and fibrin-derived fragments that presumably preserve the conformation and properties of the corresponding regions of fibrinogen and fibrin.[5] Degradation of fibrinogen with plasmin results in the E_3 fragment and two D fragments that derive from its E and D regions, respectively (FIG. 1B). Degradation of factor XIIIa-crosslinked fibrin occurs in the same manner with the appearance of E_3; however, instead of two monomeric D fragments, one obtains a crosslinked dimeric analogue, the D-D fragment (FIG. 1C). At the early stage of proteolysis, before the E_1 fragment is proteolytically modified into E_3, a non-covalent complex of D-D and E_1 (D-D:E_1 complex) can also be observed and purified.[40] We prepared all these fragments, as well as the D-D:E_1 complex, and tested the accessibility of their fibrin specific epitopes.[5] The results obtained are summarized in FIGURE 3. Neither D nor D-D bound to immobilized anti-Aα148–160 and anti-γ312–324 Mabs indicating that their fibrin specific epitopes are inaccessible in the isolated fragments as in fibrinogen. No binding was also observed with the immobilized tPA (not shown). This is in agreement with the recent X-ray data that revealed very little difference between the conformation of the D fragment and that of the same region in the dimeric D-D fragment.[41] At the same time both epitopes became accessible in the non-covalent D-D:E_1 complex, which bound well to immobilized Mabs (FIG. 3) and tPA. The exposure of these epitopes was reversible since dissociation of D-D:E_1 made them unavailable while reconstitution of the complex exposed them again.[5] These results clearly indicate that conformational changes resulting in the exposure of the fibrin specific epitopes in the D regions are triggered by the DD:E interaction.

FURTHER INSIGHT INTO THE MECHANISM OF EXPOSURE OF FIBRIN SPECIFIC EPITOPES

To further clarify the mechanism of the exposure of fibrin specific epitopes in the D regions we analyzed the crystal structure of the D fragment.[5] Crystallographic data[41] indicate that this fragment consists of a triple helical coiled coil domain, formed by all three chains, followed by the β- and γ-modules, each formed by the COOH-terminal portions of the Bβ and γ chains, respectively (see FIGURE 4A). The γ312–324 epitope is located in the γ-module, namely, in a surface-oriented loop that is adjacent to the polymerization pocket *a* complementary to the Gly-Pro-Arg-knob A. This epitope is largely exposed as revealed by calculation of its solvent accessible surface area[5] and may require not further exposure, but rather a rearrangement, to adopt an active fibrin specific conformation upon DD:E interaction, which is known to occur mainly via *A–a* sites.[42] By contrast, the other epitope, Aα148–160, that is situated on the coiled coil domain, is almost totally buried by the β-module and the fourth helix of the Aα166–195 region, suggesting that these structural elements should be moved to expose this epitope. Our calculations[5] suggest that removal of

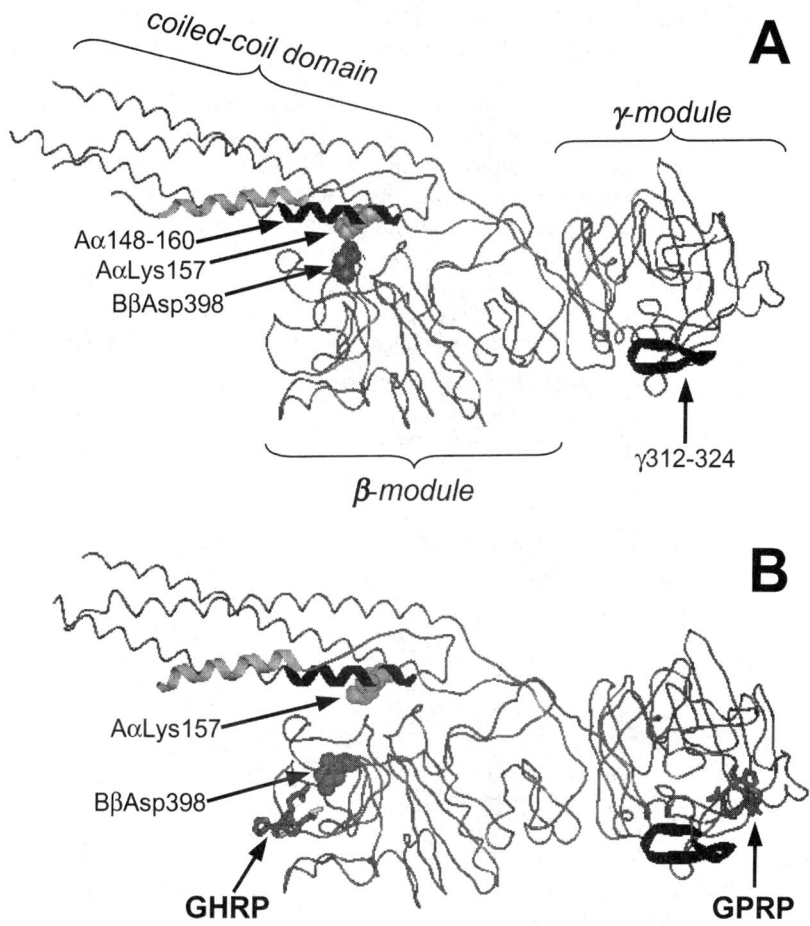

FIGURE 4. Three-dimensional structure of human fibrinogen D region. **Panel A** represents a wire diagram of the D region of the D-D fragment, whereas **Panel B** represents that of D-D loaded with Gly-Pro-Arg-Pro (GPRP) and Gly-His-Arg-Pro (GHRP) peptides based upon their crystal structure.[41,47] The D region includes the coiled coil domain formed by all three chains and two homologous modules, β- and γ-, formed by the COOH-terminal portions of the β and γ chains, respectively. The Aα148–160 and γ312–324 regions are represented by *black ribbons*, the fourth α-helix of the Aα chain 166–195 region is represented by *gray ribbon*. **Panel A** also shows the side chain contacts between Lys157 of the Aα chain (*gray balls*) and Asp398 of the β-module (*dark gray balls*). **Panel B** represents the position of these residues in D-D loaded with GPRP and GHRP peptides[50] (see text).

the fourth helix, proposed to be involved in the exposure,[41] is not sufficient and that the β-module must also move away (dissociate) from the coiled coil domain to allow access to this region.

To test the above suggestion we studied the exposure of the Aα148–160 epitope in the bovine fibrinogen D_Y and TSD fragments from which the γ- and β-modules were removed proteolytically.[5] In SPR experiments both D_Y and TSD bound to immobilized anti-Aα148–160 Mab (see FIGURE 5) in contrast to the D_H fragment and its human analogue D_1 fragment that exhibited no binding to this antibody; both also bound to immobilized plasminogen and tPA.[5] The exposure of this epitope in TSD, in which both modules are absent, was expected whereas its exposure in the D_Y fragment, consisting of the coiled coil domain and the β-module, was surprising. However, taking into account our previous data indicating interaction between domains in the D region[4,43,44] as well as crystallographic evidence for the interaction between the β- and γ-modules,[41] one can provide following explanation for this phenomenon. If there is an interaction dependent communication between domains in

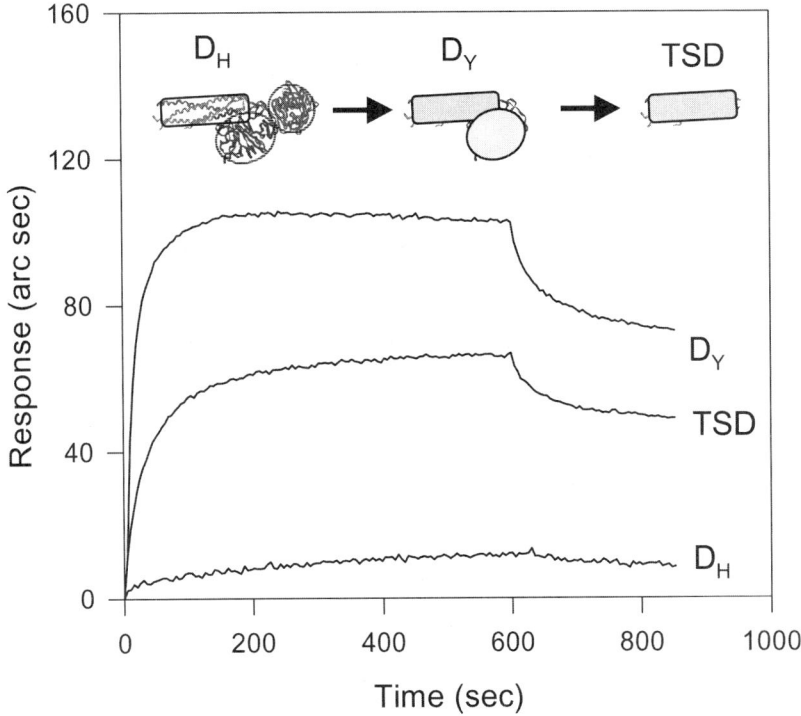

FIGURE 5. Analysis of binding of various fibrinogen fragments to immobilized anti-Aα148–160 Mab by surface plasmon resonance. The D_H, D_Y, and TSD fragments at 2 μM were added to the immobilized Mab and their association/dissociation was monitored in real time while registering the resonance signal (response). A diagram that defines the fragments used in these experiments is presented at the top.

the D region, proteolytic removal of the γ-module may modulate the interaction between the β-module and the coiled coil domain resulting in their dissociation and exposure of the Aα148–160 epitope in D_Y. One can also speculate that the exposure of this epitope in the D-D:E_1 complex and in fibrin may occur in a similar manner; the disruption of interaction between the β-module and coiled coil in this case may be triggered by conformational changes in the γ-module upon its interaction with the E region. This suggests that upon fibrin formation the intramolecular interaction between domains in the D region is altered.

DIFFERENTIAL SCANNING CALORIMETRY ANALYSIS OF THE INTERACTION BETWEEN DOMAINS IN THE D REGION OF FIBRINOGEN AND FIBRIN

To clarify relationships between domains in the D regions upon fibrin assembly we performed a differential scanning calorimetry (DSC) study of fibrin and fibrin-derived fragments. This method was previously applied to analyze the domain structure and domain–domain interactions in fibrinogen and its fragments.[12,43,45,46] It was found that fibrinogen D regions are independent and entirely preserve their structure and thermal stability in the D fragments.[12,45] It was also shown that the first and second heat absorption peaks of the complex endotherm of the D fragment reflect unfolding (melting) of three domains in each of the β- and γ-modules.[46] The first high amplitude peak was assigned to the melting of their COOH-terminal and central domains and the second peak to the melting of the remaining NH_2-terminal domains in both modules[46] (see FIGURE 6). Since interaction between domains often alters their stability, it is possible to learn about interaction between those domains in the D region that are directly involved in fibrin assembly, that is, the COOH-terminal domains in each module containing polymerization pockets *a* and *b* and the neighboring central domains that interact with the COOH-terminals. This analysis should be straightforward since in the D fragment these four domains melt in the first peak, which is easily fitted by four two-state transitions corresponding to their unfolding (FIG. 6).

We first compared the stability of the central and COOH-terminal domains of the β- and γ-modules of the D_1 fragment and the fibrin derived D-D dimer (see FIGURE 7 A and B). Note that since D-D is a symmetric structure, we present the molar excess heat capacity for only its symmetric half to simplify its comparison with the D_1 fragment. Both species exhibited a prominent heat absorption peak in the range 45–65°C reflecting the melting of four domains. No noticeable difference between the peaks was observed indicating that the lateral DD contacts revealed by crystallography[41] do not influence the stability of these domains, in agreement with the above mentioned data suggesting that conformation of the D region is similar in both species. In the presence of Ca^{2+}, which binds to both modules,[41,47] the peak in D-D (FIG. 7C) and in D (not shown) became much sharper indicating cooperation between domains. As a result, it was possible to fit this peak with only two two-state transitions. Note that this result does not necessarily mean that two domains in each module merge to form one cooperative unit. Rather it reflects increased interaction between all domains that make it impossible to fit the endotherm with more than two

two-state transitions. The formation of the D-D:E_1 complex (in which E_1 melts at much higher temperature[12,45]) in the presence of Ca^{2+} results in the stabilization and broadening of the peak that can then be fitted by three two-state transitions (see FIGURE 8A) suggesting decreased interaction between domains. A similar picture was observed when fibrin des-A was made in the calorimetric cell (see MATERIALS AND METHODS) and then melted under the same conditions as D-D:E_1 (FIG. 8B). Finally, fibrin des-AB exhibited an even broader peak with a center of mass at higher temperature, that was easily fitted by four two-state transitions (FIG. 8C) clearly indicating that these four domains are becoming much more independent. In addition, at least two of them are becoming more stable.

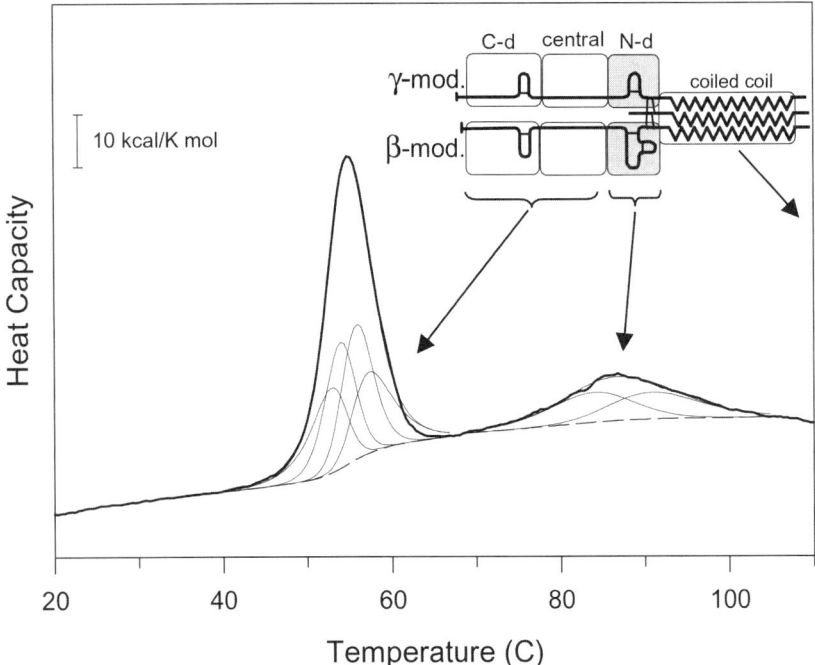

FIGURE 6. Differential scanning calorimetry-detected endotherm reflecting melting of the β- and γ-modules in human fibrinogen D_1 fragment (adapted from Ref. 46). The experimental curve (*solid thick line*) was obtained by heating the protein in 20 mM Gly, pH 8.6, while registering its heat capacity. The *broken lines* indicate the manner in which the excess heat capacities for the first and second peaks were determined. The *solid thin lines* represent the component two state transitions obtained by the deconvolution analysis (see MATERIALS AND METHODS) and the best fit that essentially coincides with the experimental curve. A diagram that shows domain structure of D_1 is presented in the *right corner*. The first peak that is fitted by four two-state transitions reflects melting of the COOH-terminal (C-d) and central domains of the β- and γ-modules. The remaining NH_2-terminal domains (N-d) melt in the second peak that is fitted by two transitions; the coiled coil domain at this pH melts above 110°C.[46]

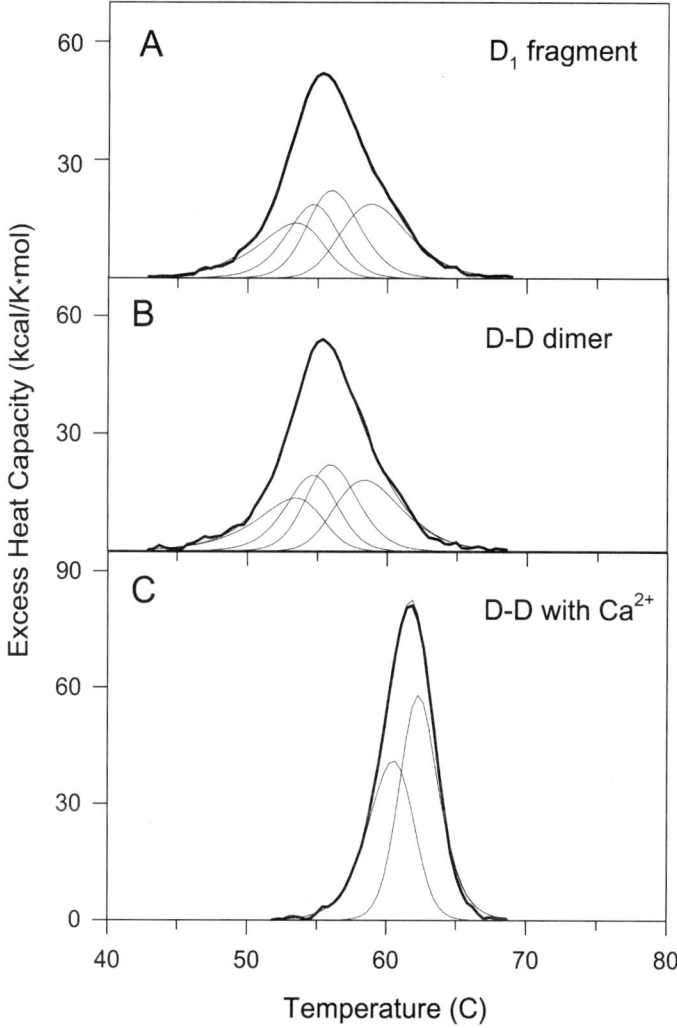

FIGURE 7. Deconvolution analysis of the excess heat capacity functions of the D_1 fragment (**A**) and the D-D dimer without (**B**) and with Ca^{2+} (**C**). The peaks represent heat capacity functions determined in a similar manner to that presented in FIGURE 7. Note that the molar excess heat capacity of the D-D dimer in **B** and **C** is presented for only one D region to simplify its comparison with D_1. The *solid thick curves* represent experimental endotherms and the *thin curves* represent the component two-state transitions obtained by the deconvolution analysis (see MATERIALS AND METHODS) and the best fit that essentially coincides with the experimental curve. The experiments were performed in 20 mM Gly buffer, pH 8.6, without (**A** and **B**) and with 0.5 mM Ca^{2+} (**C**).

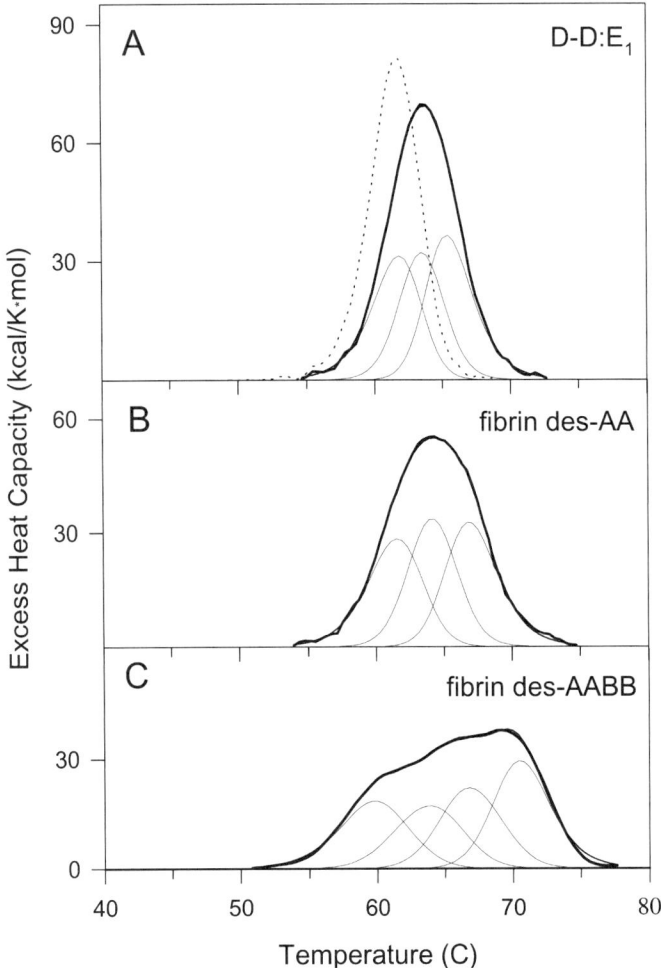

FIGURE 8. Deconvolution analysis of the excess heat capacity functions of the D-D:E_1 complex (**A**) and fibrin des-A (**B**), and des-AB (**C**). The peaks represent heat capacity functions determined in a similar manner to that presented in FIGURE 7. The molar excess heat capacities in all panels are presented for only one D region. The *solid thick curves* represent experimental endotherms, whereas the *thin curves* represent the component two-state transitions obtained by the deconvolution analysis (see MATERIALS AND METHODS) and the best fit that essentially coincides with the experimental curve. The endotherm of the D-D fragment with Ca^{2+} is represented by *dotted line* in **A** for comparison. The fibrins were formed as described in MATERIALS AND METHODS. All experiments were performed in 20 mM Gly buffer, pH 8.6, with 0.5 mM Ca^{2+}.

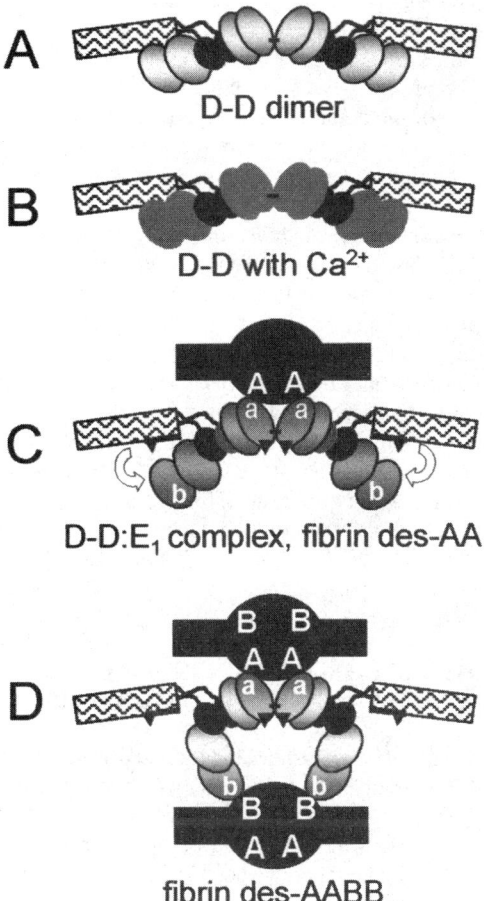

FIGURE 9. Schematic representation of the interaction induced conformational changes in the D regions of the D-D:E_1 complex and fibrin upon their interaction with the E region. **Panel A** represents D dimer in which two symmetrical D are crosslinked. Each D consists of the coiled coil domain (*rectangle*) and three domains in each γ- and β-module, NH_2-terminal (*black filling*), central and COOH-terminal (both are shown in *gray*); the β-module interacts with the coiled coil domain. In the presence of Ca^{2+} (**B**) interaction between domains of the β- and γ-modules dramatically increases. The interaction of the D regions with the E region via *a–A* polymerization sites in the D-D:E_1 complex and in fibrin des-A (**C**) results in the reduction of interaction between domains, dissociation of the γ-module from the coiled coil (shown by *bent arrows*), exposure of tPA- and plasminogen-binding sites (shown by *triangles*), and formation of polymerization pockets *b* (see text). Subsequent intermolecular *b–B* interaction in fibrin des-AB (**D**) further reduces intramolecular interaction between domains. The E region containing polymerization knobs *A* and *B* is represented in *black*.

The results of DSC can be interpreted as follows. (1) Binding of Ca^{2+} to both modules dramatically increases the stability of their domains and interaction between them suggesting that these ions play an important role in the regulation of domain–domain interactions in the D region. (2) The interaction between the γ-modules and the E region via a–A sites in the D-D:E_1 complex and in fibrin des-A results in stabilization of the domains in both modules and in the reduction of interaction between them. This indicates that both modules should communicate with each other in an interaction dependent manner; otherwise only the domain(s) in the γ-module that are directly involved in a–A interaction would be affected. (3) Subsequent interaction between the β-modules and the E region via b–B sites in fibrin des-AB further reduces interaction between all four domains and further stabilizes some of them. These events are presented schematically in FIGURE 9.

The above results indicate that there is a strong Ca^{2+}-dependent interaction between domains in the D region of fibrinogen, which is substantially reduced in fibrin where they switch from intramolecular to intermolecular interaction and are stabilized. Such relationships between domains are in a good agreement with the interaction dependent communication between them, hypothesized above, and may be responsible for the dissociation of the β-module from the coiled coil domain described in the previous section. Interestingly, the stabilization of individual domains and the reduction of interaction between them seem to occur in a two step manner, first on a–A interaction and then on b–B interaction. Is there any connection between these events and a two stage process of fibrin fiber formation?

ROLE OF DOMAIN–DOMAIN INTERACTION IN THE D REGION IN LATERAL AGGREGATION OF PROTOFIBRILS

According to the current view, formation of fibers occurs in two stages. Thrombin first cleaves two fibrinopeptides A to open primary polymerization sites (knobs A). Their interaction with complementary pockets a located in the γ-modules of two adjacent molecules results in the formation of two-stranded protofibrils in which individual molecules are half staggered (FIG. 1 D). Subsequent removal of two fibrinopeptides B opens two secondary polymerization sites (knobs B) that interact with complementary pockets b located in the β-modules to promote the second stage, lateral aggregation of protofibrils into thicker fibers. The removal of fibrinopeptide B also liberates αC-domains that contribute to the lateral aggregation[19] as discussed above. However, such aggregation, although less efficient, can occur without the removal of fibrinopeptides B.[48,49] In addition, polymerization of monomeric fibrin des-AB and αC-domainless fibrin des-AB (X_2 monomer), in which both fibrinopeptides are removed prior to initiation of polymerization, also occurs in two stages.[19] These facts suggest that some other mechanism(s) besides the delayed cleavage of fibrinopeptides B may be involved in governing this two stage process. Could it be connected with a possible cryptic nature of pockets b?

Crystal structures of D-D variants revealed that pocket b consists of several residues including BβGlu397 and BβAsp398.[47] Although there is no direct evidence that this pocket is hidden in fibrinogen, comparison of crystal structures of D-D variants revealed that in the D-D dimer and in the D fragment this site is not fully developed, since the above mentioned residues are pointed away from the pocket b

whereas in D-D loaded with the Gly-His-Arg-Pro peptide that mimics complementary knob B, they undergo a flap movement to become part of the pocket[50] (FIG. 4A and B). We analyzed side chain contacts between the β-module and coiled coil domain in the D fragment and found that the AαLys157 residue of the Aα140–160 fibrin specific epitope interacts with BβAsp398[5] preventing the latter from moving to the pocket. Obviously this interaction must be disrupted upon dissociation of the β-module from the coiled coil domain described above. This suggests that the a–A interaction at the first stage that triggers dissociation of the β-module, not only exposes the Aα148–160 region, but also liberates Bβsp398 to complete the formation of the polymerization pocket β, which becomes more accessible for the b–B interaction at the second stage. Thus, in addition to the removal of the fibrinopeptides B that exposes knobs B and liberates αC-domains, conformational changes in the D regions that occur due to a–A interaction also govern lateral aggregation of protofibrils.

CONCLUDING REMARKS

The results summarized above allow one to suggest the following scenario of structural changes upon conversion of fibrinogen to fibrin. After removal of fibrinopeptides A the intermolecular DD:E interaction via a–A sites, that provides protofibril formation, also causes conformational changes in the γ-module resulting in the exposure of its cryptic tPA-binding site including γ312–324 region. These changes also disrupt the intramolecular interaction between the coiled coil domain and the β-module, which dissociate from each other to open up access to the Aα148–160 plasminogen binding site (FIG. 9) and to liberate BβAsp398. The latter undergoes a flap movement together with BβGlu397 to complete the formation of pocket b (FIG. 4), whose interaction with complementary knob B that is exposed only after removal of fibrinopeptide B, promotes lateral aggregation of protofibrils. The removal of fibrinopeptides B also liberates the αC-domains that in fibrinogen interact with each other and with the central region E. These domains undergo an intra- to intermolecular switch, that is, they dissociate and reassociate in a new manner, intermolecularly, also contributing to the lateral aggregation of protofibrils (FIG. 1A and D). This reassociation also results in the formation of αC-polymers that become crosslinked by factor XIIIa and in the exposure of some cryptic sites including those that bind plasminogen and tPA.

The above scenario provides a reasonable explanation for the mechanism of exposure of cryptic tPA and plasminogen binding sites in fibrin αC-domains and in the D regions. It also suggests a new additional mechanism, the DD:E-induced liberation of BβAsp398 that, together with the delayed removal of the fibrinopeptides B, governs lateral aggregation of protofibrils. This mechanism is carried out by changes in domain–domain interactions upon fibrin assembly, the evidence for which is presented in this paper (FIG. 8). In addition, it suggests further experiments to seek a more comprehensive understanding of these mechanisms. First, the structural analysis of the αC-polymers may provide further insight into the mechanism of the intra- to intermolecular switch and the exposure of cryptic binding sites in the αC-domains. The availability of the recombinant αC-domain and its truncated variants make such a study possible. Second, the fact that the D regions adopt new conformation in the D-D:E_1 complex, which seems to mimic that in fibrin, suggests that crystallization of this complex and solution of its structure would allow us to elucidate the mechanisms of the conformational changes upon conversion of fibrinogen to fibrin in greater detail than presented here.

ACKNOWLEDGMENT

This work was supported by National Institute of Health Grant HL-56051 (to L.M.).

REFERENCES

1. DOOLITTLE, R.F. 1984. Fibrinogen and fibrin. Annu. Rev. Biochem. **53:** 195–229.
2. MOSESSON, M.W. 1998. Fibrinogen structure and fibrin clot assembly. Semin. Thromb. Hæmostas. **24:** 169–174.
3. CIERNIEWSKI, C.S., M. KLOCZEWIAK & A.Z. BUDZYNSKI. 1986. Expression of primary polymerization sites in the D domain of human fibrinogen depends on intact conformation. J. Biol. Chem. **261:** 9116–9121.
4. LITVINOVICH, S.V, A.H. HENSCHEN, K.G. KRIEGLSTEIN, et al. 1995. Structural and functional characterization of proteolytic fragments derived from the C-terminal regions of bovine fibrinogen. Eur. J. Biochem. **229:** 605–614.
5. YAKOVLEV, S., E. MAKOGONENKO, N. KUROCHKINA, et al. 2000. Conversion of fibrinogen to fibrin: mechanism of exposure of tPA- and plasminogen-binding sites. Biochemistry **39:** 15730–15741.
6. SCHIELEN, W.J.G., M. VOSKUILEN, G.I. TESSER & W. NIEUWENHUIZEN. 1989. The sequence Aα-(148–160) in fibrin, but not in fibrinogen, is accessible to monoclonal antibodies. Proc. Natl. Acad. Sci. U.S.A. **86:** 8951–8954.
7. SCHIELEN, W.J.G., H.P.H.M. ADAMS, K. VAN LEUVEN, et al. 1991. The sequence γ-(312–324) is a fibrin-specific epitope. Blood **77:** 2169–2173.
8. TSURUPA, G. & L. MEDVED. 2001. Identification and characterization of novel tPA- and plasminogen-binding sites within fibrin(ogen) αC-domains. Biochemistry **40:** 801–808.
9. GORGANI, N.N., C.R. PARISH & J.G. ALTIN. 1999. Differential binding of histidine-rich glycoprotein (HRG) to human IgG subclasses and IgG molecules containing kappa and lambda light chains. J. Biol. Chem. **274:** 29633–29640.
10. PRIVALOV, P.L. & S.A. POTEKHIN. 1986. Scanning microcalorimetry in studying temperature-induced changes in proteins. Methods Enzymol. **131:** 4–51.
11. FILIMONOV, V.V., S.A. POTEKHIN, S.V. MATVEEV & P.L. PRIVALOV. 1982. Thermodynamic analysis of scanning microcalorimetry data. 1. Algorithms for deconvolution of heat absorption curves. Molec. Biol. (USSR) **16:** 551–562.
12. PRIVALOV, P.L. & L.V. MEDVED. 1982. Domains in the fibrinogen molecule. J. Mol. Biol. **159:** 665–683.
13. WEISEL, J.W., C.V. STAUFFACHER, E. BULLIT & C. COHEN. 1985. A model for fibrinogen: domains and sequence. Science **230:** 1388–1391.
14. MEDVED, L.V., O.V. GORKUN & P.L. PRIVALOV. 1983. Structural organization of C-terminal parts of fibrinogen Aα-chains. FEBS Lett. **160:** 291–295.
15. ERICKSON, H.P. & W.E. FOWLER. 1983. Electron microscopy of fibrinogen, its plasmic fragments and small polymers. Ann. N.Y. Acad. Sci. **408:** 146–163.
16. MEDVED, L.V., O.V. GORKUN, V.F. MANYAKOV & V.A. BELITSER. 1985. The role of fibrinogen αC-domains in the fibrin assembly process. FEBS Lett. **181:** 109–112.
17. MEDVED, L.V., O.V. GORKUN, V.F. MANYAKOV & V.A. BELITSER. 1986. Fibrinogen αC-domains as the structure accelerating the fibrin assembly process. Molec. Biol. (USSR) **20:** 461–470.

18. VEKLICH, Y.I., O.V. GORKUN, L.V. MEDVED, et al. 1993. Carboxyl-terminal portions of the α chains of fibrinogen and fibrin. Localization by electron microscopy and the effect of isolated αC fragments on polymerization. J. Biol. Chem. **268:** 13577–13585.
19. GORKUN, O.V., Y.I. VEKLICH, L.V. MEDVED, et al. 1994. Role of the αC domains of fibrin in clot formation. Biochemistry **33:** 6986–6997.
20. COTTRELL, B.A., D.D. STRONG, K.W. WATT & R.F. DOOLITTLE. 1979. Amino acid sequence studies on the α chain of human fibrinogen. Exact location of cross-linking acceptor sites. Biochemistry **18:** 5405–5410.
21. FRETTO, L.J., E.W. FERGUSON, H.M. STEINMAN & P.A. MCKEE. 1978. Localization of the α-chain cross-link acceptor sites of human fibrin. J. Biol. Chem. **253:** 2184–2195.
22. HENSCHEN, A. & J. MCDONAGH. 1986. Fibrinogen, fibrin and factor XIII. In Blood Coagulation. R.F.A. Zwaal & H.C. Hemker, Eds.: 171–241. Elsevier Science Publishers, Amsterdam.
23. RUOSLAHTI, E. & M.D. PIERSCHBACHER. 1987. New perspectives in cell adhesion: RGD and integrins. Science **238:** 491–497.
24. CHERESH, D.A., S.A. BERLINER, V. VINCENTE & Z.M. RUGGERI. 1989. Recognition of distinct adhesive sites on fibrinogen by related integrins on platelets and endothelial cells. Cell **58:** 945–953.
25. PROCYK, R., P.D. BISHOP & B. KUDRYK. 1993. Fibrin—recombinant human factor XIII a-subunit association. Thromb. Res. **71:** 127–138.
26. REILLY, C.F. & J.E. HUTZELMANN. 1992. Plasminogen activator inhibitor-1 binds to fibrin and inhibits tissue-type plasminogen activator-mediated fibrin dissolution. J. Biol. Chem. **267:** 17128–17135.
27. STATHAKIS, N.E., M.W. MOSESSON, A.B. CHEN & D.K. GALANAKIS. 1978. Cryoprecipitation of fibrin-fibrinogen complexes induced by the cold-insoluble globulin of plasma. Blood **51:** 1211–1222.
28. KIRSCHBAUM, N.E. & A.Z. BUDZYNSKI. 1990. A unique proteolytic fragment of human fibrinogen containing the Aα COOH-terminal domain of the native molecule. J. Biol. Chem. **265:** 13669–13676.
29. TUSZYNSKI, G.P., S. SRIVASTAVA, H.I. SWITALSKA, et al. 1985. The interaction of human platelet thrombospondin with fibrinogen. Thrombospondin purification and specificity of interaction. J. Biol. Chem. **260:** 12240–12245.
30. IWANAGA, S., K. SUZUKI & S. HASHIMOTO. 1978. Bovine plasma cold-insoluble globulin: gross structure and function. Ann. N.Y. Acad. Sci. **312:** 56–73.
31. BALE, M. D., L.G. WESTRICK & D.F. MOSHER. 1985. Incorporation of thrombospondin into fibrin clots. J. Biol. Chem. **260:** 7502–7508.
32. SAKATA, Y. & N. AOKI. 1980. Cross-linking of α_2-plasmin inhibitor to fibrin by fibrin-stabilizing factor. J. Clin. Invest. **65:** 290–297.
33. RITCHIE, H., L.C. LAWRIE, P.W. CROMBIE, et al. 2000. Cross-linking of plasminogen activator inhibitor 2 and α_2-antiplasmin to fibrin(ogen). J. Biol. Chem. **275:** 24915–24920.
34. HADA, M., M. KATO, S. IKEMATSU, et al. 1982. Possible cross-linking of factor VIII related antigen to fibrin by factor XIII in delayed coagulation process. Thromb. Res. **25:** 163–168.
35. LIJNEN, H.R., J. SORIA, C. SORIA, et al. 1984. Dysfibrinogenemia (fibrinogen Dusard) associated with impaired fibrin-enhanced plasminogen activation. Thromb. Hæmost. **51:** 108–109.
36. KOOPMAN, J., F. HAVERKATE, J. GRIMBERGEN, et al. 1993. Molecular basis for fibrinogen Dusard (Aα 554 Arg→Cys) and its association with abnormal fibrin polymerization and thrombophilia. J. Clin. Invest. **91:** 1637–1643.
37. RUDCHENKO, S., I. TRAKHT & J.H. SOBEL. 1996. Comparative structural and functional features of the human fibrinogen αC domain and the isolated αC fragment. Characterization using monoclonal antibodies to define COOH-terminal Aα chain regions. J. Biol. Chem. **271:** 2523–2530.

38. MATSUKA, Y.V., L.V. MEDVED, M.M. MIGLIORINI & K.C. INGHAM. 1996. Factor XIIIa-catalyzed cross-linking of recombinant αC fragments of human fibrinogen. Biochemistry **35:** 5810–5816.
39. NIEUWENHUIZEN, W. 1994. Sites in fibrin involved in the acceleration of plasminogen activation by t-PA. Possible role of fibrin polymerisation. Thromb. Res. **75:** 343–347.
40. OLEXA, S.A. & A.Z. BUDZYNSKI. 1979. Primary soluble plasmic degradation product of human cross-linked fibrin. Isolation and stoichiometry of the (DD)E complex. Biochemistry **18:** 991–995.
41. SPRAGGON, G., S.J. EVERSE & R.F. DOOLITTLE. 1997. Crystal structures of fragment D from human fibrinogen and its crosslinked counterpart from fibrin. Nature **389:** 455–462.
42. BUDZYNSKI, A.Z. 1986. Fibrinogen and fibrin: biochemistry and pathophysiology. Crit. Rev. Oncol. Hematol. **6:** 97–146.
43. MEDVED, L.V., S.V. LITVINOVICH & P.L. PRIVALOV. 1986. Domain organization of the terminal parts in the fibrinogen molecule. FEBS Lett. **202:** 298–302.
44. MEDVED, L.V., T.N. PLATONOVA, S.V. LITVINOVICH & N.I. LUKINOVA. 1988. The cleavage of β-chain in bovine fibrinogen D_H fragment (95 kDa) leads to a significant increase in its anticlotting activity. FEBS Lett. **232:** 56–60.
45. DONOVAN, J.W. & E. MIHALYI. 1974. Conformation of fibrinogen: calorimetric evidence for a three-nodular structure. Proc. Natl. Acad. Sci. U.S.A. **71:** 4125–4128.
46. MEDVED, L., S. LITVINOVICH, T. UGAROVA, et al. 1997. Domain structure and functional activity of the recombinant human fibrinogen γ-module (γ148–411). Biochemistry **36:** 4685–4693.
47. EVERSE, S.J., G. SPRAGGON, L. VEERAPANDIAN, et al. 1998. Crystal structure of fragment double-D from human fibrin with two different bound ligands. Biochemistry **37:** 8637–8642.
48. WEISEL, J.W. 1986. Fibrin assembly. Lateral aggregation and the role of the two pairs of fibrinopeptides. Biophys. J. **50:** 1079–1093.
49. WEISEL, J.W., Y.I. VEKLICH & O.V. GORKUN. 1993. The sequence of cleavage of fibrinopeptides from fibrinogen is important for protofibril formation and enhancement of lateral aggregation in fibrin clots. J. Mol. Biol. **232:** 285–297.
50. EVERSE, S.J., G. SPRAGGON, L. VEERAPANDIAN & R.F. DOOLITTLE. 1999. Conformational changes in fragments D and double-D from human fibrin(ogen) upon binding the peptide ligand Gly-His-Arg-Pro-amide. Biochemistry **38:** 2941–2946.

Polymerization Site *a* Function Dependence on Structural Integrity of Its Nearby Calcium Binding Site

KARIM C. LOUNES,[a] NOBUO OKUMURA,[b] KELLY A. HOGAN,[a] LIFANG PING,[a] AND SUSAN T. LORD[a]

[a]*Department of Pathology and Laboratory Medicine, University of North Carolina, Chapel Hill, North Carolina, USA*

[b]*Department of Medical Technology, School of Allied Medical Sciences, Shinshu University, Matsumoto, Japan*

ABSTRACT: To explore the functional relationship between the polymerization site *a* and the nearby high affinity calcium binding site, we analyzed four variant fibrinogens with substitutions at these sites: γD364A in the *a* site and γD318A, γD320A, and γD318+γD320A in the Ca^{2+} site. In all cases fibrinopeptide A release was normal and thrombin catalyzed polymerization was markedly impaired (unpublished observations). We examined the functional connection between the Ca^{2+} site and the *a* site by testing for plasmin protection in the presence of Ca^{2+} or the *a* site peptide ligand GPRP. SDS-PAGE analysis of the products showed that γD364A fibrinogen was protected from plasmin cleavage by Ca^{2+} but not by the GPRP peptide. In contrast, neither Ca^{2+} nor the GPRP peptide protected γD318A, γD320A, or γD318+γD320A fibrinogens from complete plasmin cleavage. These results suggest that the structural integrity of the calcium binding site is required for expression of the *a* site. In contrast, the structural integrity of the *a* site has no functional consequence on Ca^{2+} binding to this high affinity site.

KEYWORDS: Fibrin polymerization; Polymerization sites; High affinity calcium binding site.

INTRODUCTION

The fibrinogen molecule is a trinodular molecule consisting of a pair of three different polypeptide chains, Aα, Bβ, and γ that are held together by disulfide bonds. The N-termini of the six chains form a central globular domain (E domain), whereas the C-termini of the Bβ and γ chains and a short fragment of Aα chain form the D domains. The two D domains are linked to the central E domain by a coiled-coil connector. The C-termini of the Aα chains extend outward from the D domain and fold back onto the E-domain, forming the αC-domains.

On injury, the activation of coagulation ends in the conversion of soluble fibrinogen to insoluble fibrin. This reaction is catalyzed by cleavage of fibrinopeptides A

Address for correspondence: Susan T. Lord, Ph.D., Department of Pathology and Laboratory Medicine, University of North Carolina, Chapel Hill, NC 27599-7525, USA. Voice: 919-966-3548; fax: 919-966-6718.

stl@med.unc.edu

and B (FpA and FpB) by thrombin from the N-termini ends of the Aα and Bβ chains, unmasking the polymerization sites A and B at the *neo* N-termini extremities of the α and β chains, respectively.[1] The widely accepted model of fibrin formation can be described in two steps: (1) the polymerization site A interacts spontaneously with its complementary site a, located in the C-terminus of the γ-chain forming protofibrils, and (2) after reaching the appropriate size, the protofibrils laterally aggregate into thicker fibrin fibers that then branch to form a fibrin network.[2] The fibrinogen molecule has three high affinity calcium binding sites (μM) and between 11 and 20 low affinity sites (mM). High resolution structural data have shown that the two high affinity sites located in the C-terminus of the γ-chain are composed of the side chains of the residues Asp318 and Asp320 and the main chain carbonyl oxygens of Phe322 and Gly324.[3] The third site has been tentatively localized to the central E-domain.[4]

The identification of dysfibrinogens with alterations in the γ-chain high affinity sites has highlighted the importance of these sites in the fibrin polymerization process.[5,6] The structural data show that the one site known to be important for polymerization, the a site, is located near the high affinity calcium binding site. Thus, based on the structural data and the analyses of these dysfibrinogens, one can hypothesize that the two sites are dependent on one another. Other data, however, do not support this hypothesis. Binding studies with the peptides GPRP and GHRP, which mimic, respectively, the polymerization sites A and B in the central domain of fibrinogen, have shown that Ca^{2+} ions influence the binding affinity of GHRP to fibrinogen but do not influence the binding of GPRP to the a site.[7] Furthermore, using a recombinant C-terminal domain of the γ-chain, Côté *et al.* have shown by analysis of normal and variant fibrinogen fragments that the a polymerization site and the nearby high affinity Ca^{2+} site are independent of each another.[8] They clearly demonstrated that a substitution that impaired the a site did not impair calcium binding to the high affinity site. To further examine the functional relationship between these two sites, we synthesized recombinant fibrinogen variants with substitutions in either the high affinity calcium binding site or the a site.

RESULTS AND DISCUSSION

Using site directed mutagenesis, we have generated three variant recombinant fibrinogens with residues γAsp318, γAsp320, or both γAsp318 and γAsp320 changed to Ala; these residues directly participate in calcium binding. To study a variation of the a site, we have used the recombinant fibrinogen, γD364A that was previously shown to exhibit impaired fibrin polymerization and normal FpA release.[9] Normal recombinant fibrinogen was used as a control. The method of introducing these substitutions in the expression vector γ-pMLP has been described elsewhere.[10]

Before studying the function of the proteins, they were analyzed by SDS-PAGE under reduced and non-reduced conditions. All the variants were found to be properly assembled and consisted of the three polypeptide chains Aα, Bβ, and γ. The γ-chains of the variants displayed different electrophoretic migration profile from normal γ-chain (see FIGURE 1). This difference in mobility is consistent with data

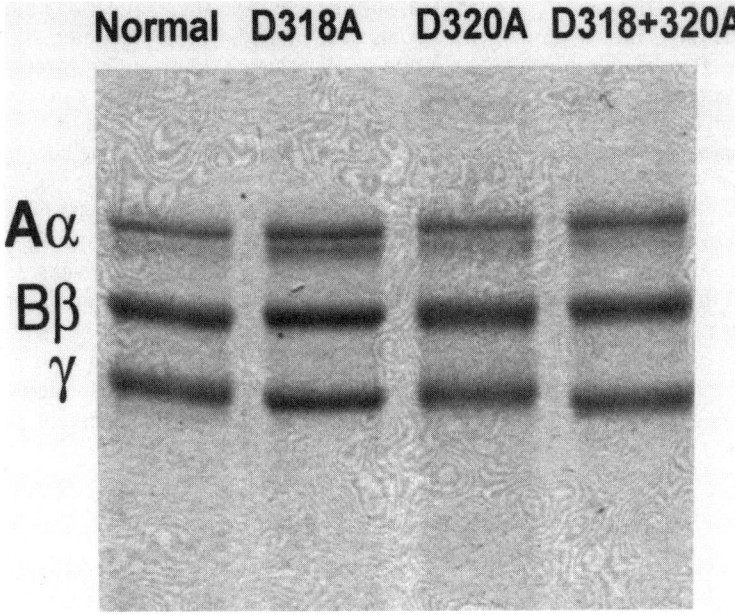

FIGURE 1. SDS-PAGE analysis of the polypeptide chains of normal and fibrinogen variants. The purified fibrinogens were reduced in the Laemmli sample buffer containing 10% beta-mercaptoethanol then run on 10% SDS-PAGE gel. The gel was stained with Coomassie Brilliant Blue. The fibrinogens variants are indicated on the *top* and the polypeptides chains on the *left*.

reported for a number of human fibrinogen variants, all having a mutation in the C-terminal part of the γ-chain.

Thrombin catalyzed fibrin polymerization was impaired for γAsp318Ala, γAsp320Ala, and γAsp318+320Ala fibrinogens, and the extent of these defects varied. The two single substitution variants showed a significant delay in the lag period, whereas the double mutant showed no increase in turbidity even after four hours. These data are in agreement with reports from fibrinogen Bastia and Vlissingen, two human variants characterized by impaired fibrin polymerization due to mutations in the high affinity binding site on the γ-chain.[5,6] The kinetics of FpA release from all three variants were found to be normal, implying that the defect in fibrin polymerization resulted from an anomaly in the polymerization site *a*.

To evaluate the effect of the substitutions on the functional properties of site *a* and *vice versa*, the variant fibrinogens were tested for the described protective effect of Ca^{2+} ions or the peptide Gly-Pro-Arg-Pro (GPRP) on plasmin degradation.[11,12] Plasmin digestion of normal fibrinogen in the presence of Ca^{2+} ions or GPRP results in the fragments E and D1. In the presence of EDTA, the γ-chain fragment remnant in D1 is further degraded to fragments D2 and D3. SDS-PAGE analysis of plasmin digests showed that γD364A was protected by Ca^{2+} ions, but was not protected by GPRP. In contrast, both Ca^{2+} ions and GPRP peptide failed to completely protect γD318A, γD320A, and γD318+γD320A fibrinogens. The altered function of the site *a* in the three variant fibrinogens was also examined by Factor XIIIa crosslinking studies. Because fibrin is a better substrate for Factor XIIIa than fibrinogen, it has been suggested that the arrangement of monomers in protofibrils enhances the crosslinking of the γ-chains.[13] All three variants showed a delay in the formation of γ–γ dimers implying that protofibril formation was impaired. We conclude that the substitutions in the calcium binding site altered the polymerization site *a* and interfered with protofibril formation by hindering critical *A–a* interactions.

These data indicate that the relationship between the high affinity calcium binding site and the polymerization site *a* is not reciprocal. In fact, the results demonstrated that the structural integrity of the calcium binding site is required for the preservation of the *a* site. In contrast, as with the data from Côté *et al.*, changes altering the *a* site had no consequence on the binding properties of the high affinity calcium binding site.[8] We conclude that the calcium binding site located in the C-terminus of the γ-chain participates in maintaining the functional structure of the polymerization site *a*.

ACKNOWLEDGMENTS

This work was supported by the American Heart Association grant AH-40047 and the National Institutes of Health grant HL-31048. Culture medium with normal recombinant fibrinogen was purchased from the Cell Culture Center (Minneapolis, MN).

REFERENCES

1. OLEXA, S.A. & A.Z. BUDZYNSKI. 1981. Localization of a fibrin polymerization site. J. Biol. Chem. **256:** 3544–3549.
2. WEISEL, J.W. 1986. Fibrin assembly. Lateral aggregation and the role of the two pairs of fibrinopeptides. Biophys. J. **50:** 1079–1093.
3. PRATT, K.P., H.C.F. CÔTÉ, D.W. CHUNG, *et al.* 1997. The primary fibrin polymerization pocket—three-dimensional structure of a 30-Kda C-terminal gamma chain fragment complexed with the peptide Gly-Pro-Arg-Pro. Proc. Natl. Acad. Sci. U.S.A. **94:** 7176–7181.
4. NIEUWENHUIZEN, W., A. VERMOND & J. HERMANS. 1983. Evidence for the localization of a calcium-binding site in the amino-terminal disulphide knot of fibrin(ogen). Thromb. Res. **31:** 81–86.
5. LOUNES, K.C., C. SORIA, A. VALOGNES, *et al.* 1999. Fibrinogen Bastia (gamma 318 Asp→Tyr) a novel abnormal fibrinogen characterized by defective fibrin polymerization. Thromb. Hæmostas. **82:** 1639–1643.
6. KOOPMAN, J., F. HAVERKATE, E. BRIET & S.T. LORD. 1991. A congenitally abnormal fibrinogen (Vlissingen) with a 6-base deletion in the gamma-chain gene, causing defective calcium binding and impaired fibrin polymerization. J. Biol. Chem. **266:** 13456–13461.
7. LAUDANO, A.P. & R.F. DOOLITTLE. 1981. Influence of calcium ion on the binding of fibrin amino terminal peptides to fibrinogen. Science **212:** 457–459.
8. COTE, H.C.F., K.P. PRATT, E.W. DAVIE & D.W. CHUNG. 1997. The polymerization pocket a within the carboxyl-terminal region of the gamma chain of human fibrinogen is adjacent to but independent from the calcium-binding site. J. Biol. Chem. **272:** 23792–23798.

9. OKUMURA, N., O.V. GORKUN & S.T. LORD. 1997. Severely impaired polymerization of recombinant fibrinogen gamma-364 Asp-]His, the substitution discovered in a heterozygous individual. J. Biol. Chem. **272:** 29596–29601.
10. BINNIE, C.G., J.M. HETTASCH, E. STRICKLAND & S.T. LORD. 1993. Characterization of purified recombinant fibrinogen: partial phosphorylation of fibrinopeptide A. Biochemistry **32:** 107–113.
11. HAVERKATE, F. & G. TIMAN. 1977. Protective effect of calcium in the plasmin degradation of fibrinogen and fibrin fragments D. Thromb. Res. **10:** 803–812.
12. YAMAZUMI, K. & R.F. DOOLITTLE. 1992. The synthetic peptide Gly-Pro-Arg-Pro-amide limits the plasmic digestion of fibrinogen in the same fashion as calcium ion. Protein Science **1:** 1719-1720.
13. LORAND, L., K.N. PARAMESWARAN & S.N. MURTHY. 1998. A double-headed Gly-Pro-Arg-Pro ligand mimics the functions of the E domain of fibrin for promoting the end-to-end crosslinking of gamma chains by factor XIIIa. Proc. Natl. Acad. Sci. U.S.A. **95:** 537–541.

Fibrin Formation and Proteolysis during Ancrod Treatment

Evidence for Des-A-Profibrin Formation and Thrombin Independent Factor XIII Activity

CARL-ERIK DEMPFLE,[a] SOTIRIA ARGIRIOU,[a] SONJA ALESCI,[a] KLAUS KUCHER,[b] H. MÜLLER-PELTZER,[b] KLAUS RÜBSAMEN,[b] AND DIETER L. HEENE[a]

[a]*First Department of Medicine, University of Heidelberg, Mannheim University Hospital, Theodor Kutzer Ufer, D-68167 Mannheim, Germany*

[b]*Knoll AG, D-67008 Ludwigshafen, Germany*

ABSTRACT: Ancrod is a purified fraction of venom from the Malayan pit viper *Calloselasma rhodostoma,* containing a serine protease that cleaves fibrinopeptides A from fibrinogen. We report on a study that involved intravenous and subcutaneous application of ancrod in healthy subjects in which it was shown that ancrod induces the formation of desAA–fibrin complexes that are partially crosslinked by factor XIII proenzyme, and act as cofactor in tPA induced plasminogen activation. The plasmin generated degrades fibrin, as well as fibrinogen, leading to the appearance of large amounts of fibrinogen and fibrin degradation products in the circulation, including fragment D-dimer. At low concentrations of ancrod, formation of desAA–fibrin is preceded by production of desA–profibrin, lacking only one fibrinopeptide A.

KEYWORDS: Fibrin formation; Ancrod; Proteolysis; DesA-profibrin; Factor XIII.

Ancrod is a purified fraction of venom from the Malayan pit viper, *Calloselasma rhodostoma,* containing a serine protease that cleaves fibrinopeptides A from fibrinogen.[1] In contrast to thrombin, the enzyme does not cleave fibrinopeptides B,[2] and does not activate factor XIII.[3] Ancrod has been shown to possess therapeutic effects in thrombotic diseases, including cerebral ischemic stroke,[4] and peripheral limb ischemia.[5,6] In addition, ancrod may serve as a model for studying the process of fibrin formation and dissolution in a human *in vivo* system.

The investigations reported here consisted of two phases:

Phase 1. Included twelve healthy volunteers who received a standardized dose of 1.0 unit of ancrod per kg body weight as intravenous infusion over a period of six hours,[7] and

Phase 2. Included 32 healthy volunteers who received ancrod by single subcutaneous injections.

Address for correspondence: Carl-Erik Dempfle, M.D., First Department of Medicine, University of Heidelberg, Mannheim University Hospital, Theodor Kutzer Ufer, D-68167 Mannheim, Germany.

The Phase 2 population was separated into four groups: Group 1, receiving placebo; Group 2, receiving ancrod at a dose of 1.0 unit/kg body weight; Group 3, receiving 1.5 units/kg body weight; and Group 4, receiving 2.0 units/kg body weight. Blood samples were drawn before ancrod infusion, and after 1, 2, 4, 6, 8, 12, 15, 24, 48, and 72 hours in Phase 1, and before subcutaneous injection, and after 3, 6, 12, 24, 48, 72, and 96 hours in Phase 2.

Monoclonal antibody-based ELISA systems, and immunoblotting procedures were used to measure specific fibrinogen and fibrin derivatives.

After ancrod therapy initiation, fibrinopeptides A were cleaved from fibrinogen. At low concentrations of ancrod present in circulation, as observed in the initial phase of ancrod treatment, as well as at the decline of ancrod plasma levels after the end of ancrod infusion, predominantly desA-profibrin[8] was formed. At higher concentrations of ancrod, desA-profibrin was rapidly transformed to desAA-fibrin monomer, by release of the second fibrinopeptide A. Both desA-profibrin, as well as desAA-fibrin monomer, formed complexes that prevent the binding of a monoclonal antibody against the α-chain N-terminus without prior denaturation of the fibrin compound.

In the presence of physiological concentrations of calcium, factor XIII proenzyme catalyzes the formation of covalent bonds between the γ-chains of adjacent fibrin monomer units within the fibrin polymer.[9] Factor XIIIa is not formed, since ancrod does not display proteolytic activity toward factor XIII, and does not lead to the formation of thrombin, which would be able to cleave the activation peptide of factor XIII. The ordered structure of fibrin monomer units within the fibrin polymer allows non-proteolytically activated factor XIII to induce the formation of covalent links between adjacent γ-chain C-termini.

Fibrin complexes act as cofactors in the activation of plasminogen by tPA.[10] Since ancrod therapy does not lead to any significant changes in the levels of tPA and PAI-1, leaving the circadian rhythms of activity and antigen unchanged, the amount of tPA present in circulation appears to suffice for efficient plasminogen activation in the presence of ancrod-induced fibrin. This confirms earlier speculations that increased activity of tPA is of no consequence in the absence of a critical level of soluble fibrin.[11] Plasmin proteolysis of fibrin generates fibrin degradation products (FbDP), whereas proteolysis of fibrinogen results in the formation of fibrinogen degradation products (FgDP). Plasmin is inhibited by α-2-antiplasmin, which forms a covalent complex, the PAP complex, with the enzyme. Fibrin and fibrinogen degradation, as well as generation of PAP complexes occur after a lag phase exceeding one hour in the case of intravenous ancrod therapy. This lag phase indicates that desA-profibrin does not effectively support tPA-induced plasminogen activation, whereas desAA-fibrin does. The effect may be related to the size of complexes formed. The predominant structure formed from desA-profibrin is presumably a dimer, consisting of two desA-profibrin molecules, or one desA-profibrin molecule and one fibrinogen molecule. In contrast, desAA-fibrin monomers associate to extended fibrin protofibrils. Partially crosslinked fibrin oligomers have also been shown to support tPA-induced plasminogen activation.[12]

The fibrinolytic response induced by ancrod therefore is determined by the following factors:
- plasma concentration of desAA-fibrin complexes, and
- plasminogen concentration.

The plasma concentration of fibrin complexes in turn depends upon the activity of ancrod introduced into the blood, and the amount of the substrate, fibrinogen. In addition, fibrinogen serves as solubilizing agent for fibrin monomers,[13] by limiting the size of fibrin polymers formed.[14] Plasminogen levels fall during ancrod therapy, due to plasmin formation, mirroring the production of PAP complexes.

Plasmin degrades ancrod induced fibrin, as well as fibrinogen. Degradation of fibrin leads to the formation of fibrin degradation products, including crosslinked fibrin derivatives such as fragment D-dimer. Compared with thrombin induced fibrin, the degree of fibrin crosslinking is lower, since factor XIII proenzyme less effectively crosslinks the γ-chains of adjacent fibrin monomer units within the polymer, compared with proteolytically activated factor XIIIa. Therefore, the proportion of fragment D-dimer over fragment D is lower in ancrod induced fibrin formation. Due to the fibrin crosslinking activity of factor XIII proenzyme, appearance of fibrin fragment D-dimer during ancrod therapy cannot be considered an indicator for the proteolysis of thrombin induced fibrin clots. Since fibrin crosslinking has an impact on the resistance of fibrin towards plasmin degradation, ancrod induced fibrin is more rapidly proteolysed, compared with thrombin induced fibrin. The process of fibrin formation and proteolysis induced by ancrod treatment is summarized in FIGURE 1.

From a therapeutic viewpoint, ancrod treatment is an effective means for the induction of a systemic fibrinolytic response, although the mechanism leading to this effect is basically procoagulant, involving fibrin formation. The therapeutic benefit may be ascribed to the combination of

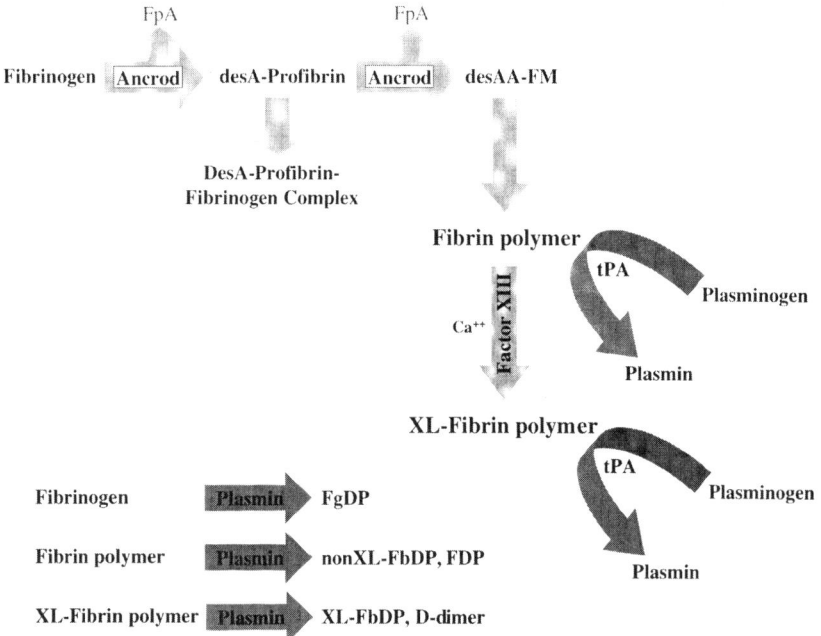

FIGURE 1.

- anticoagulant effect by reduction of substrate fibrinogen,
- platelet inhibition by reduction of fibrinogen needed for platelet aggregation via GPIIb/IIIa, and interference of monovalent fibrinogen/fibrin degradation products with platelet aggregation, and
- proteolytic effect toward preexisting fibrin clots.

On the other hand, the initial phase of ancrod treatment may be associated with hypercoagulability, as exemplified by a reduction in batroxobin clotting time in the initial phase of therapy. Finally, increased levels of PAI-1, as observed in clinical conditions associated with an acute phase response, or other inhibitors of fibrinolytic enzyme systems, may reduce the therapeutic effect of ancrod.

In conclusion, treatment with ancrod induces the production of large amounts of plasmin by providing tPA-induced plasminogen activation with its necessary cofactor, polymeric fibrin. In contrast to thrombin-related disseminated intravascular coagulation, which also induces the production of fibrin polymers, ancrod therapy is not associated with the additional sequels of coagulation activation, such as platelet activation, generation of factor XIIIa, thrombin induced feedback activation of coagulation processes, and endothelial perturbation.

REFERENCES

1. REID, H.A. 1971. Therapeutic defibrination by ancrod (Arvin). Folia Hæmatol. Int. Mag. Klin. Morphol. Blutforsch. **95:** 209–215.
2. EDGAR, W. & C.R.M. PRENTICE. 1973. The proteolytic actions of ancrod on human fibrinogen and polypeptide chains. Throm. Res. **2:** 85–96.
3. BARLOW, G.H., W.H. HOLLEMAN & L. LORAND. 1970. The action of Arvin on fibrin stabilizing factor (factor XIII). Res. Comm. Chem. Path. Pharm. **83:** 39–42.
4. ATKINSON, R.P. 1997. Ancrod in the treatment of acute ischaemic stroke. Drugs **54**(Suppl 3): 100–108.
5. LOWE, G.D., J.J. MORRICE, C.D. FORBES, et al. 1979. Subcutaneous ancrod therapy in peripheral arterial disease: improvement in blood viscosity and nutritional blood flow. Angiology **30:** 594–599.
6. SOUTAR, R.L. & J.S. GINSBERG. 1993. Anticoagulant therapy with ancrod. Crit. Rev. Oncol. Hematol. **15:** 23–33.
7. DEMPFLE, C.E., S. ARGIRIOU, K. KUCHER, et al. 2000. Analysis of fibrin formation and proteolysis during intravenous administration of ancrod. Blood **96:** 2793–2802.
8. SHAINOFF, J.R., G.B. SMEJKAL, P.M. DIBELLO, et al. 1996. Isolation and characterization of the fibrin intermediate arising from cleavage of one fibrinopeptide A from fibrinogen. J. Biol. Chem. **271:** 24129–24137.
9. SIEBENLIST, K.R. & M.W. MOSESSON. 1997. The plasma factor XIII zymogen crosslinks fibrinogen. Thromb. Hæmost. **77**(Suppl. 1): 757.
10. SUENSON, E. & L.C. PETERSEN. 1986. Fibrin and plasminogen structures essential to stimulation of plasmin formation by tissue-type plasminogen activator. Biochim. Biophys. Acta **870:** 510–519.
11. REICHL, A.P. & F.R. MANDELBAUM. 1993. Proteolytic specificity of moojeni protease A isolated from the venom of Bothrops moojeni. Toxicon **31:** 187–194.
12. SUENSON, E., P. BJERRUM, A. HOLM, et al. 1990. The role of fragment X polymers in the fibrin enhancement of tissue plasminogen activator-catalyzed plasmin formation. J. Biol. Chem. **265:** 22228–22237.
13. BROSSTAD, F., P. KIERULF & H.C. GODAL. The fibin-solubilizing effect of fibrinogen as studied by light scattering. Thromb. Res. **14:** 795–712.

14. PREISSNER, K.T., J. ROTKER, E. SELMAYR, *et al.* 1985. Influence of fibrinogen on fibrin polymerization. Ultracentrifugation studies. Biochim. Biophys. Acta **829:** 358–364.

Characterization of Crosslinking Sites in Fibrinogen for Plasminogen Activator Inhibitor 2 (PAI-2)

HELEN RITCHIE,[a] LAURA C. LAWRIE,[a] MICHAEL W. MOSESSON,[b] AND NUALA A. BOOTH[a]

[a]*Department of Molecular and Cell Biology, University of Aberdeen, Aberdeen, United Kingdom*

[b]*Blood Research Institute, Blood Center of Southeastern Wisconsin, Milwaukee, Wisconsin, USA*

> ABSTRACT: PAI-2 is a serpin that can be crosslinked to fibrin(ogen) via the Gln-Gln-Ile-Gln sequence (residues 83–86). We have characterized the lysine residues in fibrinogen to which PAI-2 is crosslinked by tissue transglutaminase and factor XIIIa. There was no competition with the crosslinking of α_2-antiplasmin, another inhibitor of fibrinolysis, which was specific for Lys 303 in the Aα chain. PAI-2 was crosslinked to several lysine residues, all in the Aα chain, 148, 176, 183, 230, 413, and 457, but not to Lys 303. The contrast with α_2-antiplasmin was clear from studies with truncated fibrinogens and competition by peptides. This was confirmed and extended by mass spectrometry of peptides after protease digestion of crosslinked products, which identified the lysine residues to which the inhibitors were crosslinked. PAI-2 remained active after cross-linking and inhibited fibrin breakdown, even by two-chain t-PA. Thus, a second inhibitor of fibrinolysis, in addition to α_2-antiplasmin, is crosslinked to fibrin and protects it from lysis.
>
> KEYWORDS: Transglutaminase; Crosslinking; Aα chain; Fibrinogen; Plasminogen activator inhibitor 2; PAI-2; Fibrinolysis.

Plasminogen activator inhibitor 2 (PAI-2) is an inhibitor of the serpin family; it inhibits u-PA and, to a lesser extent, t-PA.[1] It is produced by monocytes after stimulation with several agents, of which thrombin may be one of the most important in terms of local function in disease.[2] It is primarily an intracellular protein, lacking a classical signal sequence, but monocytes secrete it via a pathway independent of the endoplasmic reticulum-Golgi network.[3] PAI-2 is unique among serpins in the length of an additional loop between helix C and D; this contains a glutamine-rich sequence that has been shown to make it a substrate for transglutaminases.[4] We have shown that Gln 83 and 86 are involved in the crosslinking of PAI-2 to fibrin and fibrinogen.[5] It is well-known that α_2-antiplasmin (α_2-AP) is crosslinked by Gln 2 to Lys 303 of the Aα chain of fibrinogen.[6] In this report, we identified the lysine

Address for correspondence: Nuala A. Booth, Ph.D., Department of Molecular & Cell Biology, University of Aberdeen, Institute of Medical Sciences, Aberdeen AB25 2ZD, UK. Voice: 44-1224-273118; fax: 44-1224-273144.

n.a.booth@aberdeen.ac.uk

residues to which PAI-2 is crosslinked by tissue transglutaminase (tTG) and factor XIIIa.[7]

All crosslinking of PAI-2 was to the fibrinogen Aα chain. Both PAI-2 and α_2-AP could be crosslinked simultaneously; they did not compete with each other and neither did PAI-1, which binds to fibrin but is not crosslinked.[5] A 30-residue peptide containing Lys 303 specifically competed with fibrinogen in crosslinking α_2-AP, but not in crosslinking PAI-2.[7] Further evidence that PAI-2 was not crosslinked to Lys 303 was the crosslinking of PAI-2 to I-9 and desαC fibrinogens that lack 100 and 390 amino acids from the C-terminus of the Aα chain, respectively. Definition of the Lys residues to which PAI-2 was crosslinked was accomplished by digesting crosslinked fibrinogen with trypsin or endopeptidase Glu-C from *Staphylococcus aureus* (V8) and analysis of the peptides by mass spectrometry. The crosslinking reaction was carried out in the presence of PAI-2, α_2-AP or no inhibitor.[7] Only the high molecular mass species specific for each reaction were analyzed.

The mass spectrometry data were searched for peptides of the Aα chain that had disappeared after crosslinking. We also searched for peptides that correspond to the combined mass of the inhibitor and Aα digestion products. These approaches confirmed that α_2-AP was only crosslinked to Lys 303. PAI-2, in contrast, was crosslinked to a variety of Lys residues in the Aα chain, 148, 176, 183, 230, 413, and 457 all being identified (see TABLE 1). Digestion with V8 was useful in distinguishing between Lys 176 and 224, both of which might have been involved on the basis of the trypsin digestion.[7] PAI-2 did not interfere with fibrinogen crosslinking, which is consistent with the fact that different Lys residues are involved.

Several of the Lys residues to which PAI-2 is crosslinked (148, 176, and 183) are in the four-helix coiled coil of fibrinogen, between residue 160, where the Aα chain reverses due to a disulfide ring, and residue 220.[8,9] This is close to the plasmin sensitive hinge at residue 220. Lys 230 and 303, involved in crosslinking of PAI-2 and α_2-AP, respectively, are located on the flexible C-terminal domain, as are Lys 413 and 457. This domain of the Aα chain is a primary target for proteolysis of fibrinogen.

TABLE 1. Lysine residues identified in crosslinking PAI-2 to the Aα chain of fibrinogen by mass spectrometry of peptides after crosslinking by tTG or factor XIIIa

tTG	Factor XIIIa
148	*148*
176 or *224*	*230*
183 or *457*	*413*
230	
148	148
176	176
183	
457	

NOTE: Residues derived from tryptic digestion are shown in italics. Residues after V8 digestion are shown in plain text.

Crosslinking to Aα 148 may be functionally important, since it is close to the sequence that is cryptic in fibrinogen but exposed in fibrin; this site is key in binding t-PA and stimulating the activation of plasminogen.[10] Consistent with this, we found that fibrin prepared from fibrinogen to which PAI-2 was crosslinked resisted lysis, by both u-PA (not shown) and t-PA (see FIGURE 1), despite the fact that t-PA is normally only poorly inhibited by PAI-2.[1] It may be that interference by crosslinked PAI-2 with binding of t-PA and/or plasminogen, and subsequent effects on plasmin generation is the explanation for this finding, rather than a classical protease–inhibitor interaction.

Both tissue transglutaminase and factor XIIIa crosslinked PAI-2 and α_2-AP to fibrinogen. That α_2-AP is a substrate for tissue transglutaminase has been independently reported recently.[11,12] Our study shows that the same Gln residue in α_2-AP and Lys residue in fibrinogen Aα were involved, whichever transglutaminase was used. Some differences were observed in the Lys residues to which PAI-2 was crosslinked by the two enzymes. Notably residues 148, 176, and 230 were identified irrespective of the transglutaminase; 183 and 457 were additionally identified with tissue transglutaminase, whereas 413 was also identified with factor XIIIa (TABLE 1).[7] This

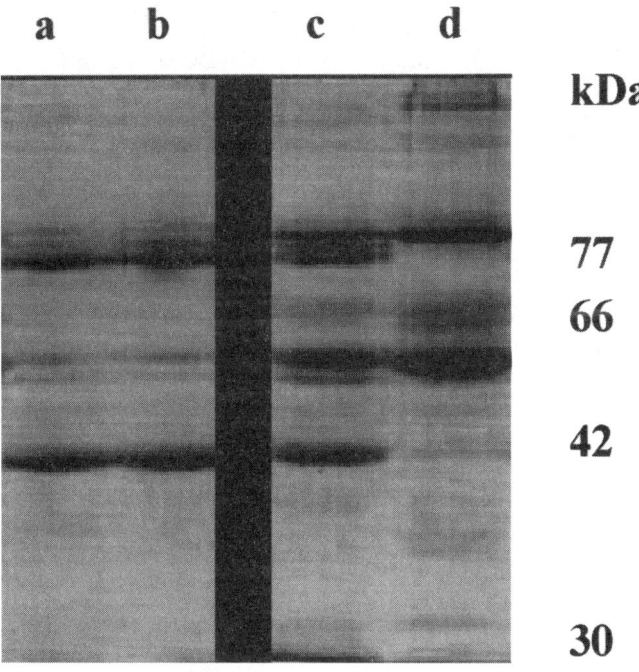

FIGURE 1. Fibrinogen (10 μg/track) was incubated for 30 min at 37°C in the presence of calcium, without (*tracks a* and *b*) or with (*tracks c* and *d*) 200 ng PAI-2, and without (*tracks a* and *c*) and with (*tracks b* and *d*) 50 ng tissue transglutaminase. Each tube was clotted after addition of 200 ng plasminogen and 50 pg of t-PA, incubated for a further two hours and analyzed by SDS-PAGE under reducing conditions and stained with Coomassie Blue.

method of analysis does not allow us to reach quantitative conclusions on preferences for Lys residues, as would be possible by sequencing approaches.

Crosslinking of inhibitors to fibrin is relevant not just to thrombus stability but also to vessel wall disease, where fibrin is a key component.[13] Crosslinking of PAI-2 and α_2-AP to the Aα chain of fibrinogen, by tissue transglutaminase or factor XIIIa, allows these two serpins to be localized on the surface of fibrin strands, where efficient inhibition of plasmin generation and activity can occur.

REFERENCES

1. KRUITHOF, E.K.O., M.S. BAKER & C.L. BUNN. 1995. Biological and clinical aspects of plasminogen activator inhibitor type 2. Blood **86:** 4007–4024.
2. RITCHIE, H., A. JAMIESON & N.A. BOOTH. 1995. Thrombin modulates synthesis of plasminogen activator inhibitor 2 (PAI-2) by human peripheral blood monocytes. Blood **86:** 3428–3435.
3. RITCHIE, H. & N.A. BOOTH. 1998. Secretion of plasminogen activator inhibitor 2 by human peripheral blood monocytes occurs via an endoplasmic reticulum-Golgi-independent pathway. Exp. Cell. Res. **242:** 439–450.
4. JENSEN, P.H., et al. 1993. Type-2 plasminogen activator inhibitor is a substrate for trophoblast transglutaminase and factor XIIIa-transglutaminase-catalyzed cross linking to cellular and extracellular structures. Eur. J. Biochem. **214:** 141–146.
5. RITCHIE, H., et al. 1995. Monocyte plasminogen activator inhibitor 2 (PAI-2) inhibits u-PA-mediated fibrinolysis and is cross-linked to fibrin. Thromb. Hæmost. **81:** 96–103.
6. KIMURA, S. & N. AOKI. 1986. Cross-linking site in fibrinogen for α_2-plasmin inhibitor. J. Biol. Chem. **261:** 15591–15595.
7. RITCHIE, H., et al. 2000. Cross-linking of plasminogen activator inhibitor 2 and α_2-antiplasmin to fibrin(ogen). J. Biol. Chem. **275:** 24915–24920.
8. BROWN, J.H., et al. 2000. The crystal structure of modified bovine fibrinogen. Proc. Natl. Acad. Sci. U.S.A. **97:** 85–90.
9. SPRAGGON, G., et al. 1997. Crystal structures of fragment D from human fibrinogen and its crosslinked counterpart from fibrin. Nature **389:** 455–462.
10. NIEUWENHUIZEN, W. 1994. Sites in fibrin involved in the acceleration of plasminogen activation by tPA. Possible role of fibrin polymerization. Thromb. Res. **75:** 343–347.
11. LEE, K.N., et al. 2001. Cross-linking of wild-type and mutant α_2-antiplasmins to fibrin by activated factor XIII and by a tissue transglutaminase. J. Biol. Chem. **275:** 37382–37389.
12. HEVESSY, Z., et al. 2001. α_2-Plasmin inhibitor is a substrate for tissue transglutaminase: an in vitro study. Thromb. Res. **99:** 399–406.
13. SMITH, E.B. & W.D. THOMPSON. 1994. Fibrin as a factor in atherogenesis. Thromb. Res. **73:** 1–19.

The Formation of β Fibrin Requires a Functional *a* Site

KELLY A. HOGAN,[a] BETTINA BOLLIGER,[a] NOBUO OKUMURA,[b] AND SUSAN T. LORD[a]

[a]*Department of Pathology and Laboratory Medicine, University of North Carolina, Chapel Hill, North Carolina, USA*

[b]*Laboratory of Clinical Chemistry, Department of Medical Technology, School of Allied Medical Sciences, Shinshu University, Matsumoto, Japan*

ABSTRACT: We used recombinant fibrinogens in which the *a* site is disrupted to examine β-fibrin formation in the absence of a functional *a* site. Our variants have only *b* sites available, and they showed no evidence of fibrin polymer formation after cleavage of FpB with venzyme. We conclude that *B–b* interactions are not strong enough to induce clot formation. Our studies do not rule out the involvement of *b* in the formation of β-fibrin, yet they do provide evidence that *a* is likely to be essential in the formation of β-fibrin.

KEYWORDS: β-fibrin; Polymerization sites.

INTRODUCTION

Fibrin in which only fibrinopeptide B (FpB) is cleaved, is termed β-fibrin. Venzyme, purified from the Southern copperhead snake, has been shown to release FpB predominantly at 14°C,[1,2] thus making it possible to form β-fibrin. Fibrinogen Metz (Aα16 Arg→Cys), a homozygous variant with no FpA release, can polymerize at 14°C with the addition of either thrombin or venzyme.[3] Although it is somewhat widely accepted that β-fibrin occurs through *B–b* interactions, there are some data that suggest that the *a* site may also interact with *B* in the formation of β-fibrin (see FIGURE 1). First, Everse *et al.* crystallized a portion of fibrin with only the peptide that mimics the *B* site, Gly-His-Arg-Pro-Amide (GHRP-am), and both *a* and *b* bound the tetrapeptide.[4] These data suggest that when only *B* is exposed, as in the case for β-fibrin, *B–a* interactions occur. In another study prior to the crystallography data, Weisel examined β-fibrin fibers by transmission electron microscopy and reported that the striations were identical to those in thrombin formed clots.[5] He determined that the *b* site, not yet located to the β chain by crystallography, would need to be in the C-terminus of the γ chain for the striation pattern to occur identically to that in the thrombin induced fibers. Although Weisel proposed that the *b* site was near the *a* site in the γ-chain, we suggest that what he described was actually the *a* site.

Address for correspondence: Susan T. Lord, Department of Pathology and Laboratory Medicine, University of North Carolina, Brinkhaus-Bullitt 603, Chapel Hill, NC 27599-7525, USA. Voice: 919-966-3548; fax: 919-966-6718.
stl@med.unc.edu

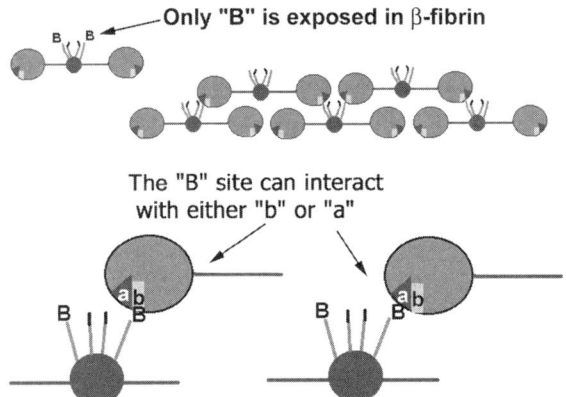

FIGURE 1. Schematic of *B–b* and *B–a* interactions.

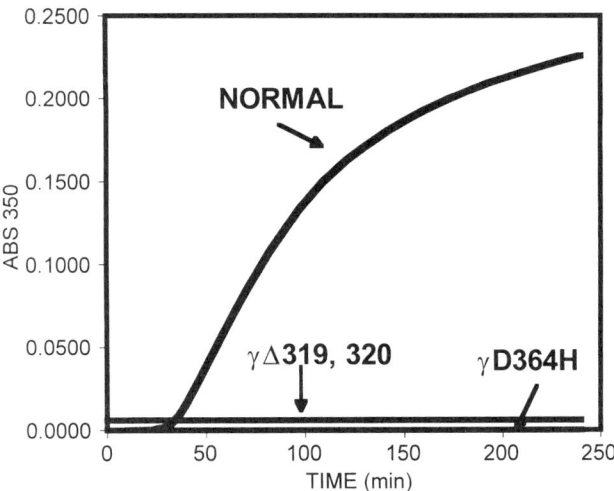

FIGURE 2. Polymerization of *a* site variants at 14°C with venzyme. All polymerization reactions were carried out at 14°C and 350 nm using a Biospec-1601 spectrophotometer (Shimadzu Corp., Tokyo, Japan) equipped with a jacketed specimen holder. The final concentration of fibrinogen was 0.09 mg/ml in a 100 µl reaction with 1mM $CaCl_2$ and 10 µl venzyme in 20 mM HEPES, pH 7.4, 150 mM NaCl. Turbidity was followed for four hours and three experiments were performed for each fibrinogen. One additional experiment with γΔ 319, 320 fibrinogen was performed at 0.5 mg/ml fibrinogen and 5 mM $CaCl_2$; the final concentration of venzyme was unchanged (data not shown).

In the study we report here, we use recombinant fibrinogens in which the a site is disrupted to examine β-fibrin formation in the absence of a functional a site. The fibrinogens used for this study were γΔ319, 320, which has a deletion of Asn-319 and Asp-320 from the γ-chain[6] and γD364H, which has a single amino acid substitution in the γ-chain.[7]

RESULTS

When normal recombinant fibrinogen was mixed with venzyme at 14°C; the turbidity rose steadily over a 240 min period (see FIGURE 2). In contrast, no rise in turbidity was evident with either γΔ 319, 320 or γD364H. The fibrinogen and calcium concentration were increased with the γΔ 319, 320 variant to determine if altering these parameters would enhance the reaction. Again, no rise in turbidity was observed. To be certain that the loss of polymerization was not due to impaired release of FpB, we measured fibrinopeptide release, as described elsewhere,[6] from the fibrinogens after four hours of incubation with venzyme at 14°C under the same conditions, as were used for polymerization. The venzyme cleaved FpB, but also a large amount of FpA. When the area under each peak was normalized, averaged, and compared, the variant fibrinogens were shown to release a similar or greater amount of peptide than normal (data not shown). Thus, impaired release of FpB was not the cause of the polymerization defect. A kinetic experiment with normal recombinant fibrinogen showed that FpB was released faster than FpA (data not shown). However, the enzyme was not as specific as previously reported[1–3] because FpA was also released at significant amounts from the beginning.

CONCLUSIONS

The polymerization observed with fibrinogen Metz at 14°C where only B is exposed may occur either through interactions of B–b, B–a, or both. Our variants have only b sites available, and they showed no increase in turbidity. We conclude that B–b interactions are not strong enough to induce clot formation, and therefore, the rise in turbidity in fibrinogen Metz was supported by B–a interactions. Our studies do not rule out the involvement of b in the formation of β-fibrin, yet they do provide some evidence that a is likely to be essential in the formation of β-fibrin.

ACKNOWLEDGMENTS

We would like to thank Dr. John Shainoff from the Cleveland Clinic Foundation, Cleveland, OH for his kind gift of venzyme. We also thank Li Fang Ping and Kasim McLain for their help with protein expression and purification. This study was supported by NIH Grant HL31048 and the Swiss National Science Foundation.

REFERENCES

1. HERZIG, R.H., *et al.* 1970. Studies on a procoagulant fraction of southern copperhead snake venom: the preferential release of fibrinopeptide B. J. Lab. Clin. Med. **76:** 451–465.
2. SHAINOFF, J.R., *et al.* 1979. Fibrinopeptide B and aggregation of fibrinogen. Science **204:** 200–202.
3. MOSESSON, M.W., *et al.* 1987. Studies on the ultrastructure of fibrin lacking fibrinopeptide B (β-fibrin). Blood **69:** 1073–1081.
4. EVERSE, S.J., *et al.* 1999. Conformational changes in fragments D and double D from human fibrin(ogen) upon binding the peptide ligand Gly-His-Arg-Pro-Amide. Biochemistry **38:** 2941–2946.
5. WEISEL, J.W. 1986. Fibrin assembly. Lateral aggregation and the role of the two pairs of fibrinopeptides. Biophys. J. **50:** 1079–1093.
6. HOGAN, K.A., *et al.* 2000. Recombinant fibrinogen Vlissingen/Frankfurt IV. The deletion of residues 319 and 320 from the γ-chain of fibrinogen alters calcium binding, fibrin polymerization, cross-linking, and platelet aggregation. J. Biol. Chem. **275:** 17778–17785.
7. OKUMURA, N., *et al.* 1997. Severely impaired polymerization of recombinant fibrinogen γ-364 Asp→His, the substitution discovered in a heterozygous individual. J. Biol. Chem. **272:** 29596–29601.

Mode of Perturbation of Asahi Fibrin Assembly by the Extra Oligosaccharides

TERUKO SUGO,[a] OSAMU SEKINE,[a] CHIZUKO NAKAMIKAWA,[a]
HITOSHI ENDO,[b] CARMEN L. AROCHA-PIÑANGO,[c] AND MICHIO MATSUDA[a]

[a]*Center for Molecular Medicine, Jichi Medical School, Tochigi, Japan*

[b]*Deparment of Biochemistry, Jichi Medical School, Tochigi, Japan*

[c]*Instituto Venezolano de Investigaciones Cientificas, Caracas, Venezuela*

ABSTRACT: Steric hindrance by the backbone of extra oligosaccharides at γ-Asn 308 may cause the repulsive force to widen the junction at the D:D interface, and thus, interfere with the longitudinal elongation and lateral association of protofibrils.

KEYWORDS: Steric hindrance; Extra oligosaccharides; D:D interface; Perturbation of fibrin assembly; Fibrinogen Asahi.

The extra carbohydrate in six dysfibrinogens so far characterized, are known to have a higher sialic acid content than normal fibrinogen (see Matsuda and Sugo elsewhere in this volume). It provides strong negative charges that could affect fibrin assembly, leading to the formation of thinner fibrin fiber networks than the normal control. Indeed, three fibrinogens, Caracas II,[1] Asahi, and Lima (see FIGURES 1 and 2) are shown to form highly branched and thinner fibrin fibers, although the minute ultrastructures are different from one another. Another dysfibrinogen, fibrinogen Niigata,[2] with a low sialic acid content is found to form fibrin networks with a rather normal appearance. In the study reported here we focus on the mode of perturbation by the extra oligosaccharide of Asahi fibrin assembly.

Fibrinogen Asahi is a dysfibrinogen possessing an extra carbohydrate linked to γ-Asn-308 due to an Asn-X-Thr type glycosylation sequence created by a γ-Met-310 to Thr substitution.[3] Although fibrinogen Asahi is derived from a heterozygote for this abnormality, the patient fibrinogen, (Asahi fibrinogen) is hard to clot with thrombin in the range of thrombin time measurement. Addition of fibrinogen Asahi to normal fibrinogen resulted in the two-phase prolongation of thrombin time. By confocal laser scanning microscopy (CLSM) of fibrin clots formed from the mixtures of Asahi and normal fibrinogen, fibers were found to become shorter, and less ordered areas were enlarged in accordance with increasing the ratio of Asahi fibrinogen in the mixture. In CLSM images of the Asahi fibrin, normal species in fibrinogen Asahi were found to align with each other, forming short fibers initially, and thereby, appearing to exclude the abnormal molecules from fibrin assembly. To determine whether the function of abnormal fibrinogen species could be restored by

---

Address for correspondence: Teruko Sugo, Ph.D., Division of Cell and Molecular Medicine, Center for Molecular Medicine, Jichi Medical School, Yakushiji 3311-1, Tochigi, 329-0498, Japan. Voice: 81-285-58-7397; fax: 81-285-44-7817.

sugosan@jichi.ac.jp

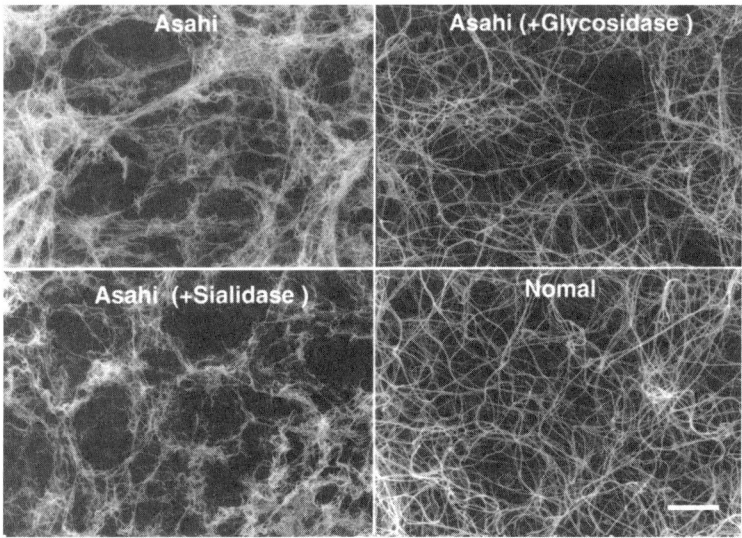

FIGURE 1. SEM images of Asahi fibrin clots. The specimens were prepared from the clots formed by fibrinogen Asahi (Asahi), sialidase treated (+Sialidase) or glycosidase treated (+Glycosidas) Asahi fibrinogens together with normal fibrinogen (Normal) as a control. *Bar*, 3 μm.

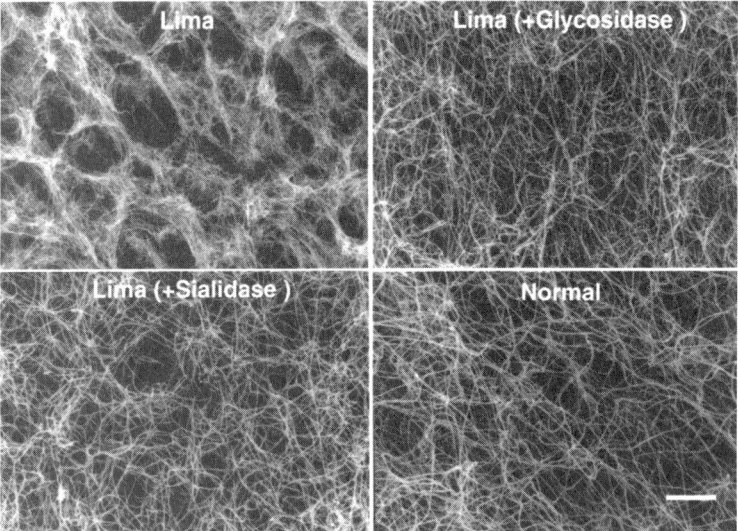

FIGURE 2. SEM images of Lima fibrin clots. The fibrin clots for SEM were: Lima fibrinogen (Lima), sialidase treated (+Sialidase) or glycosidase treated (+Glycosidase) Lima fibrinogens, and normal fibrinogen (Normal). *Bar,* 3 μm.

modifying the extra carbohydrate moieties, we treated the Asahi fibrinogen separately with glycosidase or sialidase. Although sialidase treated fibrinogen Asahi did not clot at all, the glycosidase treatment brought about distinct shortening of the thrombin time, though still incompletely. By using scanning electron microscopy, the network of Asahi fibrin was found to have sponge like appearance, composed of sticky thinner fibers with large pores, as compared with a normal fibrin architecture. Many fibers seemed to be fused to form the boundary between the networks and the pores, and the free fiber ends were often observed at the boundary (FIG. 1). Removal of all the carbohydrate moieties, seemed to be able to restore fibrin architecture with normal appearance, but removal of sialic acids failed to do so. On the other hand, removal of sialic acids alone was sufficient for normalization of both thrombin time and the abnormal networks that have a close resemblance to Caracas II fibrin[1] in the case of fibrinogen Lima with a highly disialilated extra carbohydrate at Aα Asn-139[4] (FIG. 2). These results indicate that the extra carbohydrate backbone in Asahi fibrinogen, *per se,* is predominantly responsible for the perturbation of fibrin assembly.

From the established crystal structure of crosslinked D-dimer of fibrin, the position of γ-Asn-308 (Asahi) is not directly involved in the D:D interface but is located close to the interface and not far from the *a* site.[5] The steric hindrance by the backbone of extra carbohydrate at γ Asn-308 may cause the repulsive force to widen the junction of D:D interface and probably to weaken the E:D binding as well. Thus, the longitudinal elongation of the protofibril may also be disturbed. The presence of mobile negatively charged carbohydrate backbones around the D:D interface and the crosslinking site could also interfere with the lateral association of protofibrils and factor XIIIa catalyzed crosslinking, as previously reported.[3] The steric hindrance by the extra carbohydrate backbone itself in the Niigata fibrin formation,[2] and strong negative charges of sialic acids in the extra carbohydrate in the Lima fibrin formation may have functioned as major disturbing factors.

REFERENCES

1. WOODHEAD, J.L., *et al.* 1996. The ultrastructure of fibrinogen Caracas II molecules, fibers, and clots. J. Biol. Chem. **271:** 4946–4953.
2. SUGO, T., *et al.* 1999 Fibrinogen Niigata with impaired fibrin assembly: An inherited dysfibrinogen with a Bβ Asn-160 to Ser substitution associated with extra glycosylation at Bβ Asn-158. Blood **94:** 3806–3813.
3. YAMAZUMI, K., *et al.* 1989. A γ methionine-310 to threonine substitution and consequent N-glycosylation at γ asparagine-308 identified in a congenital dysfibrinogenemia associated with posttraumatic bleeding, fibrinogen Asahi. J. Clin. Invest. **83:** 1590–1597.
4. MAEKAWA, H., *et al.* 1992 Fibrinogen Lima: a homozygous dysfibrinogen with an Aα-arginine-141 to serine substitution associated with extra N-glycosylation at Aα-asparagine-139. J. Clin. Invest. **90:** 67–76.
5. SPRAGGON, G., *et al.* 1997. Crystal structures of fragment D from human fibrinogen and its crosslinked counterpart from fibrin. Nature **389:** 455–462.

Elements of the Fibrinolytic System

H.R. LIJNEN

Center for Molecular and Vascular Biology, University of Leuven, Belgium

ABSTRACT: The blood fibrinolytic system comprises an inactive proenzyme, plasminogen, that can be converted to the active enzyme, plasmin. Plasmin degrades fibrin into soluble fibrin degradation products, by two physiological plasminogen activators (PA), the tissue type PA (t-PA) and the urokinase type PA (u-PA). t-PA mediated plasminogen activation is mainly involved in the dissolution of fibrin in the circulation. u-PA binds to a specific cellular receptor (u-PAR), resulting in enhanced activation of cell bound plasminogen. Inhibition of the fibrinolytic system may occur either at the level of the PA, by specific plasminogen activator inhibitors (PAI), or at the level of plasmin, mainly by α_2-antiplasmin. Several molecular interactions have been observed between the fibrinolytic and the matrix metalloproteinase (MMP) system; both systems may cooperate in generating proteolytic activity. Thus, stromelysin-1 (MMP-3) cleaves a 55-kDa kringle 1-4 fragment, containing the lysine binding site(s) involved in cellular binding, from plasminogen and removes a 17-kDa NH_2-terminal fragment, containing the cellular receptor binding site, from urokinase (u-PA). Thereby, MMP-3 may downregulate cell associated plasmin activity by decreasing the amount of activatable plasminogen, without affecting cell bound u-PA activity.

KEYWORDS: Blood fibrinolytic system; Plasminogen activators; Fibrinolysis.

INTRODUCTION

The blood fibrinolytic (plasminogen/plasmin) system comprises an inactive proenzyme, plasminogen, that can be converted to the active enzyme, plasmin, which in turn degrades fibrin into soluble fibrin degradation products. Two immunologically distinct physiologic plasminogen activators (PA) have been identified: the tissue-type PA (t-PA) and the urokinase type PA (u-PA). t-PA–mediated plasminogen activation is mainly involved in the dissolution of fibrin in the circulation.[1] u-PA binds to a specific cellular receptor (u-PAR) resulting in enhanced activation of cell bound plasminogen. The main role of u-PA appears to be in the induction of pericellular proteolysis via the degradation of matrix components or via activation of latent proteases or growth factors.[2] Inhibition of the fibrinolytic system may occur either at the level of the PA, by specific plasminogen activator inhibitors (PAI), or at the level of plasmin, mainly by α_2-antiplasmin. Physiological fibrinolysis is regulated by specific molecular interactions between its main components as well as by controlled synthesis and release, primarily from endothelial cells, of PAs and PAIs.[1]

Address for correspondence: H.R. Lijnen, Ph.D., Center for Molecular and Vascular Biology, K.U. Leuven, Campus Gasthuisberg, O & N Herestraat 49, B-3000 Leuven, Belgium. Voice: 32-16-34.57.71; fax: 32-16-34.59.90.
   roger.lijnen@med.kuleuven.ac.be

The physiological importance of the fibrinolytic system is demonstrated by the association between abnormal fibrinolysis and a tendency towards bleeding or thrombosis. The generation of transgenic mice over- or underexpressing components of the plasminogen/plasmin system has allowed to establish its causal role in several biological processes.[3] Recently, molecular interactions between the fibrinolytic and matrix metalloproteinase (MMP) systems have been recognized at different levels, suggesting that both systems may cooperate in processes requiring extracellular matrix degradation or remodelling.[4]

STRUCTURE–FUNCTION RELATIONS OF THE MAIN COMPONENTS OF THE FIBRINOLYTIC SYSTEM

Plasminogen

Human plasminogen is a single-chain glycoprotein with M_r 92,000, present in plasma at a concentration of 1.5 to 2 µM. It consists of 791 amino acids and contains five homologous triple-loop structures or "kringles".[5] These kringles contain structures, called lysine binding sites and aminohexyl binding sites, that mediate the specific binding of plasminogen to fibrin and the interaction of plasmin with α_2-antiplasmin, and thereby play a crucial role in the regulation of fibrinolysis.[6]

Native plasminogen (Glu-plasminogen) is easily converted by limited plasmic digestion to modified forms commonly designated "Lys-plasminogen". This conversion occurs by hydrolysis of the Arg^{68}-Met^{69}, Lys^{77}-Lys^{78}, or Lys^{78}-Val^{79} peptide bonds. Plasminogen is converted to plasmin by cleavage of the Arg^{561}-Val^{562} peptide bond.[7] The plasmin molecule is a two-chain trypsin-like serine proteinase with an active site composed of His^{603}, Asp^{646}, and Ser^{741}.[5]

Tissue-Type Plasminogen Activator (t-PA)

Human t-PA was first isolated as a single-chain serine proteinase with M_r 70,000, consisting of 527 amino acids with Ser as the NH_2-terminal amino acid;[8] it was subsequently shown that native t-PA contains an NH_2-terminal extension of three amino acids, but in general the initial numbering system has been maintained. Limited plasmic hydrolysis of the Arg^{275}-Ile^{276} peptide bond converts t-PA to a two-chain molecule held together by one interchain disulfide bond. The t-PA molecule contains four domains: (1) an NH_2-terminal region of 47 residues (residues 4 to 50) that is homologous with the finger domains mediating the fibrin affinity of fibronectin; (2) residues 50 to 87 that are homologous with epidermal growth factor; (3) two kringle regions, comprising residues 87 to 176 and 176 to 262, that share a high degree of homology with the five kringles of plasminogen, and (4) a serine proteinase domain (residues 276 to 527) with the active site residues His^{322}, Asp^{371}, and Ser^{478}.[8] The t-PA molecule comprises three potential N-glycosylation sites, at Asn^{117}, Asn^{184}, and Asn^{448}. In contrast to the single-chain precursor form of most serine proteinases, single-chain t-PA is enzymatically active. The X-ray crystal structure of the catalytic domain of recombinant human t-PA is available.[9]

t-PA is a poor enzyme in the absence of fibrin, but the presence of fibrin strikingly enhances the activation rate of plasminogen.[10] During fibrinolysis, fibrinogen and fibrin are continuously modified by cleavage with thrombin or plasmin, yielding a diversity of reaction products.[11] Optimal stimulation is only obtained after early plasmin cleavage in the COOH-terminal Aα-chain and the NH_2-terminal Bβ-chain of fibrin, yielding fragment X-polymer.[11] Kinetic data support a mechanism in which fibrin provides a surface to which t-PA and plasminogen adsorb in a sequential and ordered way yielding a cyclic ternary complex.[10] Formation of this complex results in an enhanced affinity of t-PA for plasminogen, yielding up to three orders of magnitude higher efficiencies for plasminogen activation. This is mediated, at least in part, by COOH-terminal lysine residues generated by plasmin cleavage. Plasmin formed at the fibrin surface has both its lysine binding sites and active site occupied and is thus only slowly inactivated by $α_2$-antiplasmin (halflife of about 10–100 sec); in contrast, free plasmin, when formed, is rapidly inhibited by $α_2$-antiplasmin (half-life of about 0.1 s).[6]

Thus, proteins that remove lysine residues from the fibrin surface, such as the thrombin activatable fibrinolysis inhibitor (TAFI) may have an antifibrinolytic action. TAFI is a M_r 60,000 single-chain protein, identical to plasma procarboxypeptidase B, occurring at a concentration of 75 nM.[12,13] Thrombin, trypsin, or plasmin convert the protein to an active carboxypeptidase B. Activated TAFI suppresses fibrinolysis, most likely by removing COOH-terminal lysine residues from (partially degraded) fibrin, thereby preventing additional binding of plasminogen and/or t-PA to fibrin.[12]

Urokinase-Type Plasminogen Activator (u-PA)

u-PA is secreted as a single-chain molecule (scu-PA, pro-urokinase) that can be converted to a two-chain form (tcu-PA). u-PA is a serine proteinase of 411 amino acids, with active site triad His^{204}, Asp^{255}, and Ser^{356}. The molecule contains an NH_2-terminal growth factor domain and one kringle structure homologous to the five kringles found in plasminogen and the two kringles in t-PA.[14] u-PA contains only one N-glycosylation site (at Asn^{302}), and contains a fucosylated threonine residue at position 18. Conversion of scu-PA to tcu-PA occurs after proteolytic cleavage at position Lys^{158}-Ile^{159} by plasmin, but also by kallikrein, trypsin, cathepsin B, human T cell associated serine proteinase-1, and thermolysin. A fully active tcu-PA derivative is obtained after additional proteolysis by plasmin at position Lys^{135}-Lys^{136}. A low molecular weight form of scu-PA (32 kDa) can be obtained by selective cleavage at position Glu^{143}-Leu^{144}.[15] In contrast, scu-PA is converted to an inactive two-chain molecule by thrombin after proteolytic cleavage at position Arg^{156}-Phe^{157}. This inactivation is strongly enhanced in the presence of thrombomodulin, and is dependent on the 0-linked glycosaminoglycan of thrombomodulin (see citations in Ref. 16).

In contrast to tcu-PA, scu-PA displays very low activity toward low molecular weight chromogenic substrates, but it appears to have some intrinsic plasminogen activating potential, which represents no more than 0.5% of the catalytic efficiency of tcu-PA.[17,18] In plasma, in the absence of fibrin, scu-PA is stable and does not activate plasminogen; in the presence of a fibrin clot, scu-PA, but not tcu-PA, induces fibrin specific clot lysis.[17] Fibrin fragment E-2 selectively promotes the activation of plasminogen by scu-PA, mainly by enhancing the catalytic rate constant of the activation.[19] Scu-PA is an inefficient activator of plasminogen bound to internal lysine residues on intact fibrin, but has a higher activity toward plasminogen bound to newly generated COOH-terminal lysine residues on partially degraded fibrin.[19,20]

The binding of u-PA to its receptor (u-PAR) at the cell surface is believed to be crucial for its activity under physiological conditions. Binding results in an enhanced plasmin generation, due to effects on both the activation of plasminogen[21] and on the feedback activation of scu-PA to tcu-PA by generated plasmin.[22] Both of these effects are also critically dependent on the cellular binding of plasminogen. Cell associated plasmin is protected from rapid inhibition by α_2-antiplasmin, which further favours the activation of receptor-bound scu-PA. Concentration of proteolytic activity at the cell surface may also occur via vitronectin, trapping soluble u-PAR-u-PA complexes.[23] Both u-PA and u-PAR have been found in a complex with β_1-, β_2-, and β_3-integrins, thereby allowing mutual interactions and regulatory processes between cell adhesion and proteolysis (reviewed in Ref. 24).

u-PA Receptor (u-PAR)

The specific cell surface receptor for u-PA is a heterogeneously glycosylated protein of M_r 50,000 to 60,000, synthesized as a 313-amino acid polypeptide, anchored to the plasma membrane by a glycosyl-phosphatidylinositol (GPI) moiety. The u-PAR molecule is composed of three distantly related structural domains, of which the NH_2-terminal domain is involved in binding u-PA; it binds all forms of u-PA containing an intact growth factor domain (see references cited in Ref. 25).

Alpha2-Antiplasmin

α_2-Antiplasmin is the main physiological plasmin inhibitor in human plasma, whereas plasmin formed in excess of α_2-antiplasmin may be neutralized by α_2-macroglobulin. α_2-Antiplasmin, a M_r 67,000 single-chain glycoprotein containing 464 amino acids and 13% carbohydrate, is present in plasma at a concentration of about 1 µM. The reactive site of the inhibitor is the Arg^{376}-Met^{377} peptide bond.[26] Two forms of α_2-antiplasmin were detected in about equal amounts in purified preparations of the inhibitor[27]: a native 464-residue long inhibitor with NH_2-terminal methionine (Met^1-α_2-antiplasmin) and a 12-amino acid shorter form with NH_2-terminal asparagine (Asn^{13}-α_2-antiplasmin). It is not known whether Asn^{13}-α_2-antiplasmin is present in the circulating blood or whether it is generated *in vitro*. The NH_2-terminal Gln^{14}-residue of α_2-antiplasmin can crosslink to Aα-chains of fibrin, in a process that requires Ca^{2+} and is catalyzed by activated coagulation factor XIII.[28]

α_2-Antiplasmin forms an inactive 1:1 stoichiometric complex with plasmin. The inhibition of plasmin by α_2-antiplasmin involves two consecutive reactions: a fast, second order reaction producing a reversible inactive complex, which is followed by a slower first-order transition resulting in an irreversible complex. The second-order rate constant of the inhibition is very high (2 to 4×10^7 $M^{-1}sec^{-1}$), but this high inhibition rate depends both on the presence of a free lysine binding site and active site in the plasmin molecule, and on availability of a plasminogen binding site and reactive site peptide bond in the COOH-terminal end of the inhibitor.[29]

Plasminogen Activator Inhibitor-1

Rapid inhibition of both t-PA and u-PA in normal human plasma occurs primarily by plasminogen activator inhibitor-1 (PAI-1).[30] PAI-1 is a single-chain glycoprotein with M_r about 52,000 consisting of 379 amino acids. The cDNA sequence revealed that PAI-1 is a member of the serpin family with reactive site peptide bond Arg^{346}-Met^{347}.[31,32]

PAI-1 reacts very rapidly with single-chain and two-chain t-PA and with two-chain u-PA, with second-order inhibition rate constants of the order of 10^7 $M^{-1}.sec^{-1}$, and it does not react with scu-PA.[33] Like other serpins, PAI-1 inhibits its target proteinases by formation of a 1:1 stoichiometric reversible complex, followed by covalent binding between the hydroxyl group of the active site serine residue of the proteinase and the carboxyl group of the P1 residue at the reactive center ("bait region") of the serpin. Modeling approaches suggest that sequence 350–355 of PAI-1, which contains three negatively charged amino acids, interacts with highly positively charged regions in t-PA (residues 296–304)[34] or in u-PA (residues 179–184).[35]

PAI-1 occurs as an active inhibitory form that spontaneously converts to a latent form, which can be partially reactivated by denaturing agents.[36] Another molecular form of intact PAI-1 has been isolated that does not form stable complexes with t-PA, but is cleaved at the P1–P1′ peptide bond ("substrate PAI-1").[37] The X-ray structure of the cleaved substrate variant shows that it has a new β-strand (s4A) formed by insertion of the NH_2-terminal portion of the reactive site loop into β-sheet A subsequent to cleavage.[38]

PAI-1 is stabilized by binding to a plasminogen activator inhibitor binding protein identified as S-protein or vitronectin.[39] The PAI-1 binding motif is localized to residues 12–30 of the somatomedin B domain of vitronectin; this motif is anchored in the active conformation by disulfide bonds.[40] PAI-1 may play a role in cell adhesion and migration, independently from its antiproteolytic activity, as a result of its high affinity binding to vitronectin, thereby competing with u-PAR-dependent or integrin-dependent binding of cells to the extracellular matrix.[41–45] The blocking effect of PAI-1 on both u-PAR- and integrin mediated cell adhesion was shown to be fully reversible with u-PA.[41,45,46] The exact role in cell migration of PAI-1 inhibitory activity versus its adhesion properties remains to be established.

MOLECULAR INTERACTIONS BETWEEN THE FIBRINOLYTIC AND MMP SYSTEMS

Overview

Matrix metalloproteinases (MMPs) constitute a tightly regulated family of enzymes that degrade most components of the extracellular matrix. They are classified on the basis of substrate specificity. More than 23 members of the MMP family have been identified, among which there are at least six membrane-type (MT) MMPs. MMPs are generally secreted as zymogens, which are extracellularly activated by organomercurial compounds, by several proteinases, oxygen radicals, or association with the cell surface. Active MMPs are inhibited by specific tissue inhibitors (TIMPs) via formation of a non-covalent stoichiometric complex; four members of the TIMP family have been identified (see citations in Ref. 4). *In vitro*, plasmin directly activates proMMP-1, proMMP-3, proMMP-9, proMMP-10, and proMMP-13.[47–50] Membrane type-1 MMP converts the 72-kDa proMMP-2 to an intermediate 64-kDa species, which is activated by plasmin following conversion to a 62-kDa molecule.[51]

Role of the Plasminogen/Plasmin System in MMP Activation

Plasmin may play an important role in the *in vivo* activation of proMMPs, but the physiological relevance has not been clearly established, neither has the role of the physiological plasminogen activators t-PA and u-PA. The availability of mice with targeted inactivation of the main components of the plasminogen/plasmin system has permitted study of its physiological role in MMP activation in more detail.

The distribution of latent (pro-) and active MMP-2 or MMP-9 was monitored in aorta extracts and in serum free conditioned cell culture medium obtained from wildtype (WT) mice and from mice with deficiency of t-PA (t-PA$^{-/-}$), u-PA (u-PA$^{-/-}$), PAI-1 (PAI-1$^{-/-}$), or plasminogen (Plg$^{-/-}$). This was done by zymography on gelatin containing gels that detect molecular forms of latent MMP-2 (70-kDa and 65-kDa) and of active MMP-2 (61-kDa and mainly 58-kDa), in addition to 94-kDa proMMP-9 and 83-kDa MMP-9. *In vivo* activation of proMMP-2 was observed independently of plasmin(ogen).[52] This was substantiated by the findings that: (1) total levels of MMP-2 (active plus latent) were comparable in fibroblasts, smooth muscle cells (SMC) or aorta extracts of the different gene-deficient mice; (2) the contribution of active 58-kDa MMP-2 to the total MMP-2 level was comparable in WT, t-PA$^{-/-}$, u-PA$^{-/-}$, PAI-1$^{-/-}$, and also in Plg$^{-/-}$ mice; and (3) addition of plasmin(ogen) to the cell culture medium did not significantly affect activation of proMMP-2. These data also suggest that direct activation of proMMP-2 by u-PA or t-PA does not play a major physiological role. Active MMP-9 was detected in samples from WT, t-PA$^{-/-}$, u-PA$^{-/-}$, and PAI-1$^{-/-}$ mice, indicating that its activation does not require both physiological plasminogen activators. In contrast, no active MMP-9 was detected in macrophages or smooth muscle cells derived from Plg$^{-/-}$ mice, indicating that efficient proMMP-9 activation in these cells is plasmin(ogen)-dependent. However, active MMP-9 was detected in Plg$^{-/-}$ fibroblasts indicating that proMMP-9 activation may also occur independently of plasminogen. These data obtained with gene deficient mice thus indicate that *in vivo* activation of proMMP-2 occurs independently of plasmin(ogen), whereas activation of proMMP-9 may occur via plasmin-dependent or plasmin-independent (MMP-mediated) mechanisms.[52]

Because MMP-3 can activate proMMP-9, it was hypothesized that plasmin may be involved in proMMP-9 activation indirectly through activation of proMMP-3. Therefore, a potential physiological role of MMP-3 in the expression or activation of MMP-2 or MMP-9 in the wall of injured arteries was studied with the use of homozygous MMP-3-deficient (MMP-3$^{-/-}$) mice.[53] One week after perivascular electric injury of the carotid or femoral artery in wildtype (MMP-3$^{+/+}$) or MMP-3$^{-/-}$ mice, 70-kDa and 65-kDa proMMP-2 levels were enhanced by two- to fourfold, with corresponding increases of 20- to 40-fold for active 61-kDa and 58-kDa MMP-2. Active MMP-2 species represented approximately one third of the total MMP-2 concentration for both MMP-3$^{+/+}$ and MMP-3$^{-/-}$ mice. ProMMP-9 levels were enhanced 10- to 80-fold; MMP-9 was not detected in non-injured carotid or femoral arteries, whereas one week after injury its contribution to the total MMP-9 level was 11 to 18% for MMP-3$^{+/+}$ and MMP-3$^{-/-}$ mice. Cell culture experiments confirmed comparable ratios of active versus latent MMP-2 in skin fibroblasts and smooth muscle cells derived from MMP-3$^{+/+}$ and MMP-3$^{-/-}$ mice. Addition of plasmin(ogen) did not significantly affect activation of proMMP-2. In MMP-3$^{+/+}$ and MMP-3$^{-/-}$ macrophages, comparable levels of 94-kDa proMMP-9 were detected, and plasmin(ogen)-mediated conversion to 83-kDa MMP-9 was obtained in both genotypes. ProMMP-9 activation thus appears to be possible via plasmin–dependent and MMP-3–independent mechanisms. MMP-3 or other MMPs, which critically depend on MMP-3 for their activation, thus do not play a critical role in physiological activation of proMMP-9.[53]

Effect of the MMP System on Plasminogen/Plasmin System Function

Plasminogen/plasmin system function was studied in mice deficient in MMP-3, MMP-11, or TIMP-1.[54,55] We have found that: (1) urinary u-PA antigen and activity levels were similar in wildtype and deficient mice; (2) vascular plasminogen activator activity, both t-PA- and u-PA-mediated, was comparable; (3) u-PA activity in *in vivo* thioglycollate stimulated macrophages of wildtype and deficient mice was comparable, resulting in comparable invasion in the peritoneal cavity and comparable u-PA-mediated activation of plasminogen and degradation of fibrin or extracellular matrices; (4) the spontaneous thrombolytic potential was similar in wildtype and MMP-3$^{-/-}$ or TIMP-1$^{-/-}$ mice; and (5) no fibrin deposition was detected in the livers of MMP-3$^{-/-}$, MMP-11$^{-/-}$, or TIMP-1$^{-/-}$ mice, indicating that the endogenous plasminogen/plasmin system functions adequately in the absence of these MMP system components. A fragment of MMP-11 containing the catalytic domain was able to degrade the Aα-chain of fibrinogen, but did not significantly lyse fibrin.[55] Thus, MMP-11 may participate in matrix degradation in an environment where its catalytic domain is released from the mature protein. Taken together, these data suggest, however, that MMP-3, MMP-11, TIMP-1, or other MMPs that critically depend on these components for their activation or inhibition, do not play a major role in physiological fibrinolysis.

Effect of MMPs on Cellular Proteolytic Activity

MMP-3 (stromelysin-1) specifically hydrolyzes the Glu^{143}-Leu^{144} peptide bond in u-PA, yielding a 17-kDa NH$_2$-terminal fragment containing the receptor (u-PAR)-binding sequence and a 32-kDa COOH-terminal fragment containing the intact serine proteinase domain.[56] Thus, on the one hand, u-PA may play a role in the activation of proMMP-3 via generation of plasmin, and on the other hand, active MMP-3 may prevent cellular receptor-binding of u-PA. Subsequent studies revealed that an angiostatin-like fragment (containing kringles 1–4) can be generated from plasminogen with MMP-3. MMP-3 indeed hydrolyzes the Glu^{59}-Asn^{60}, Pro^{447}-Val^{448}, and Pro^{544}-Ser^{545} peptide bonds in plasminogen, yielding a 55-kDa NH$_2$-terminal fragment comprising kringles 1–4, a 14-kDa domain corresponding to kringle 5, and a 30-kDa domain comprising the serine proteinase domain.[57] Antiangiogenic activity of angiostatin was first described by O'Reilly *et al.*,[58] in a plasminogen fragment consisting of the first four kringle domains, and confirmed an elastase-derived fragment containing kringles 1–3.[58,59]

Because of these specific molecular interactions between MMP-3 and plasminogen or u-PA, we investigated a potential role of MMP-3 in the regulation of cellular fibrinolytic activity by affecting binding and/or activation of plasminogen and/or single-chain u-PA.[60] Human plasminogen bound specifically to human monocytoid THP-1 cells, to murine MMP-3 deficient smooth muscle cells (SMC) and fibroblasts. Treatment with MMP-3 of cells saturated with bound plasminogen, resulted in a dose dependent reduction of the amount of u-PA activatible plasminogen (reduction to 25–40% of the value in the absence of MMP-3). Immunoblotting with specific monoclonal antibodies and autoradiography of eluates of the cells treated with MMP-3 revealed cleavage of plasminogen into the 55-kDa fragment and miniplasminogen (kringle 5 plus the proteinase domain). Treatment with MMP-3 of cell bound u-PA (THP-1 or HT1080 cells) did not alter cell associated u-PA activity, measured in a direct chromogenic substrate assay or in a plasminogen coupled chromogenic substrate assay (residual u-PA activity always at least 85% of that without MMP-3 treatment). Autoradiography of ^{125}I-labeled u-PA moieties, removed from the cells by treatment with acid or with phosphatidylinositol phospholipase C, confirmed that u-PA remained essentially intact after MMP-3 treatment. These data indicate that MMP-3 may downregulate cell associated plasmin activity by decreasing the amount of activatible plasminogen, without affecting cell bound u-PA activity.[60]

REFERENCES

1. COLLEN, D. & H.R. LIJNEN. 1991. Basic and clinical aspects of fibrinolysis and thrombolysis. Blood **78**: 3114–3124.
2. BLASI, F. 1993. Urokinase and urokinase receptor: A paracrine/autocrine system regulating cell migration and invasiveness. BioEssays **15**: 105–111.
3. CARMELIET, P. & D. COLLEN. 1998. Development and disease in proteinase-deficient mice: role of the plasminogen, matrix metalloproteinase and coagulation system. Thrombos. Res. **91**: 255–285.
4. LIJNEN, H.R. & D. COLLEN. 1999. Matrix metalloproteinase system deficiencies and matrix degradation. Thromb. Hæmost. **82**: 837–845.
5. FORSGREN, M., B. RADEN, M. ISRAELSSON, et al. 1987. Molecular cloning and characterization of a full-length cDNA clone for human plasminogen. FEBS Lett. **213**: 254–260.
6. COLLEN, D. 1980. On the regulation and control of fibrinolysis. Thromb. Hæmost. **43**: 77–89.
7. ROBBINS, K.C., L. SUMMARIA, B. HSIEH, et al. 1967. The peptide chains of human plasmin. Mechanism of activation of human plasminogen to plasmin. J. Biol. Chem. **242**: 2333–2342.
8. PENNICA, D., W.E. HOLMES, W.J. KOHR, et al. 1983. Cloning and expression of human tissue-type plasminogen activator cDNA in E. coli. Nature **301**: 214–221.
9. RENATUS, M., R.A. ENGH, M.T. STUBBS, et al. 1997. Lysine 156 promotes the anomalous proenzyme activity of tPA: X-ray crystal structure of single-chain human tPA. EMBO J. **16**: 4797–4805.
10. HOYLAERTS, M., D.C. RIJKEN, H.R. LIJNEN, et al. 1982. Kinetics of the activation of plasminogen by human tissue plasminogen activator. Role of fibrin. J. Biol. Chem. **257**: 2912–2919.
11. THORSEN, S. 1992. The mechanism of plasminogen activation and the variability of the fibrin effector during tissue-type plasminogen activator-mediated fibrinolysis. Ann. N.Y. Acad. Sci. **667**: 52–63.

12. NESHEIM, M., W. WANG, M. BOFFA, et al. 1997. Thrombin, thrombomodulin and TAFI in the molecular link between coagulation and fibrinolysis. Thromb. Hæmost. **78:** 386–391.
13. EATON, D.L., B.E. MALLOY, S.P. TSAI, et al. 1991. Isolation, molecular cloning, and partial characterization of a novel carboxypeptidase B from human plasma. J. Biol. Chem. **266:** 21833–21838.
14. HOLMES, W.E., D. PENNICA, M. BLABER, et al. 1985. Cloning and expression of the gene for pro-urokinase in Escherichia coli. Biotechnology **3:** 923–929.
15. STUMP, D.C., H.R. LIJNEN & D.COLLEN. 1986. Purification and characterization of a novel low molecular weight form of single-chain urokinase-type plasminogen activator. J. Biol. Chem. **261:** 17120–17126.
16. DE MUNK, G.A., J.F. PARKINSON, E. GROENEVELD, et al. 1993. Role of the glycosaminoglycan component of thrombomodulin in its acceleration of the inactivation of single-chain urokinase-type plasminogen activator by thrombin. Biochem. J. **290:** 655–659.
17. GUREWICH, V., R. PANNELL, S. LOUIE, et al. 1984. Effective and fibrin-specific clot lysis by a zymogen precursor form of urokinase (pro-urokinase). A study *in vitro* and in two animal species. J. Clin. Invest. **73:** 1731–1739.
18. LIJNEN, H.R., B. VAN HOEF, L. NELLES, et al. 1990. Plasminogen activation with single-chain urokinase-type plasminogen activator (scu-PA). Studies with active site mutagenized plasminogen (Ser740→Ala) and plasmin-resistant scu-PA (Lys158→Glu). J. Biol. Chem. **265:** 5232–5236.
19. LIU, J.N. & V. GUREWICH. 1992. Fragment E-2 from fibrin substantially enhances pro-urokinase–induced Glu-plasminogen activation. A kinetic study using the plasmin-resistant mutant pro-urokinase Ala-158-rpro-UK. Biochemistry **31:** 6311–6317.
20. FLEURY, V., H.R. LIJNEN & E. ANGLÈS-CANO. 1993. Mechanism of the enhanced intrinsic activity of single-chain urokinase-type plasminogen activator during ongoing fibrinolysis. J. Biol. Chem. **268:** 18554–18559.
21. ELLIS, V., N. BEHRENDT & K.DANØ. 1991. Plasminogen activation by receptor-bound urokinase. A kinetic study with both cell-associated and isolated receptor. J. Biol. Chem. **266:** 12752–12758.
22. ELLIS, V., M.F. SCULLY & V.V. KAKKAR. 1989. Plasminogen activation initiated by single-chain urokinase-type plasminogen activator. Potentiation by U937 monocytes. J. Biol. Chem. **264:** 2185–2188.
23. CHAVAKIS, T., S.M. KANSE, B. YUTZY, et al. 1998. Vitronectin concentrates proteolytic activity on the cell surface and extracellular matrix by trapping soluble urokinase receptor-urokinase complexes. Blood **91:** 2305–2312.
24. MAY, A.E., S.M. KANSE, T. CHAVAKIS, et al. 1998. Molecular interactions between the urokinase receptor and integrins in the vasculature. Fibrinol. Proteol. **12:** 205–210.
25. LIJNEN, H.R., F. BACHMANN, D. COLLEN, et al. 1994. Mechanisms of plasminogen activation. J. Intern. Med. **236:** 415–424.
26. HOLMES, W.E., L. NELLES, H.R. LIJNEN, et al. 1987. Primary structure of human α_2-antiplasmin, a serine protease inhibitor (serpin). J. Biol. Chem. **262:** 1659–1664.
27. BANGERT, K., A.H. JOHNSEN, U. CHRISTENSEN, et al. 1993. Different N-terminal forms of α_2-plasmin inhibitor in human plasma. Biochem. J. **291:** 623–625.
28. KIMURA, S. & N. AOKI. 1986. Cross-linking site in fibrinogen for α_2-plasmin inhibitor. J. Biol. Chem. **261:** 15591–15595.
29. WIMAN, B. & D. COLLEN. 1979. On the mechanism of the reaction between human α_2-antiplasmin and plasmin. J. Biol. Chem. **254:** 9291–9297.
30. KRUITHOF, E.K.O., C. TRAN-THANG, A. RANSIJN, et al. 1984. Demonstration of a fast-acting inhibitor of plasminogen activators in human plasma. Blood **64:** 907–913.
31. NY, T., M. SAWDEY, D. LAWRENCE, et al. 1986. Cloning and sequence of a cDNA coding for the human β-migrating endothelial-cell-type plasminogen activator inhibitor. Proc. Natl. Acad. Sci. U.S.A. **83:** 6776–6780.
32. PANNEKOEK, H., H. VEERMAN, H. LAMBERS, et al. 1986. Endothelial plasminogen activator inhibitor (PAI): a new member of the serpin gene family. EMBO J. **5:** 2539–2544.

33. KRUITHOF, E.K.O. 1988. Plasminogen activator inhibitors—a review. Enzyme **40:** 113–121.
34. MADISON, E.L., E.J. GOLDSMITH, R.D. GERARD, et al. 1989. Serpin-resistant mutants of human tissue-type plasminogen activator. Nature **339:** 721–724.
35. ADAMS, D.S., L.A. GRIFFIN, W.R. NACHAJKO, et al. 1991. A synthetic DNA encoding a modified human urokinase resistant to inhibition by serum plasminogen activator inhibitor. J. Biol. Chem. **266:** 8476–8482.
36. HEKMAN, C.M. & D.J. LOSKUTOFF. 1985. Endothelial cells produce a latent inhibitor of plasminogen activators that can be activated by denaturants. J. Biol. Chem. **260:** 11581–11587.
37. DECLERCK, P.J., M. DE MOL, D.E. VAUGHAN, et al. 1992. Identification of a conformationally distinct form of plasminogen activator inhibitor-1, acting as a non-inhibitory substrate for tissue-type plasminogen activator. J. Biol. Chem. **267:** 11693–11696.
38. AERTGEERTS, K., H.L. DE BONDT, C.J. DE RANTER, et al. 1995. Mechanisms contributing to the conformational and functional flexibility of plasminogen activator inhibitor-1. Nature Struct. Biol. **2:** 891–897.
39. DECLERCK, P.J., M. DE MOL, M.C. ALESSI, et al. 1988. Purification and characterization of a plasminogen activator inhibitor-1 binding protein from human plasma. Identification as a multimeric form of S protein (Vitronectin). J. Biol. Chem. **263:** 15454–15461.
40. DENG, G., G. ROYLE, S. WANG, et al. 1996. Structural and functional analysis of the plasminogen activator inhibitor-1 binding motif in the somatomedin B domain of vitronectin. J. Biol. Chem. **271:** 12716–12723.
41. DENG, G., S.A. CURRIDEN, S.J. WANG, et al. 1996. Is plasminogen activator inhibitor-1 the molecular switch that governs urokinase receptor-mediated cell adhesion and release? J. Cell Biol. **134:** 1563–1571.
42. KANSE, S.M., C. KOST, O.G. WILHELM, et al. 1996. The urokinase receptor is a major vitronectin-binding protein on endothelial cells. Exp. Cell Res. **224:** 344–353.
43. STEFANSSON, S. & D.A. LAWRENCE. 1996. The serpin PAI-inhibits cell migration by blocking integrin $\alpha_v\beta_3$ binding to vitronectin. Nature **383:** 441–443.
44. KJOLLER, L., S.M. KANSE, T. KIRKEGAARD, et al. 1997. Plasminogen activator inhibitor-1 represses integrin- and vitronectin-mediated cell migration independently of its function as an inhibitor of plasminogen activation. Exp. Cell Res. **232:** 420–429.
45. WALTZ, D.A., L.R. NATKIN, R.M. FUJITA, et al. 1997. Plasmin and plasminogen activator inhibitor type 1 promote cellular motility by regulating the interaction between the urokinase receptor and vitronectin. J. Clin. Invest. **100:** 58–67.
46. LOSKUTOFF, D.J., S.A. CURRIDEN, G. HU, et al. 1999. Regulation of cell adhesion by PAI-1. APMIS **107:** 54–61.
47. OKADA, Y., Y. GONOJI, K. NAKA, et al. 1992. Matrix metalloproteinase 9 (92-kDa gelatinase/type IV collagenase) from HT1080 human fibrosarcoma cells. Purification and activation of the precursor and enzymic properties. J. Biol. Chem. **267:** 21712–21719.
48. SUZUKI, K., J.J. ENGHILD, T. MORODOMI, et al. 1990. Mechanisms of activation of tissue procollagenase by matrix metalloproteinase 3 (stromelysin). Biochemistry **29:** 10261–10270.
49. EECKHOUT, Y. & G. VAES. 1977. Further studies on the activation of procollagenase, the latent precursor of bone collagenase. Effects of lysosomal cathepsin B, plasmin and kallikrein, and spontaneous activation. Biochem. J. **166:** 21–31.
50. HE, C.S., S.M. WILHELM, A.P. PENTLAND, et al. 1989. Tissue cooperation in a proteolytic cascade activating human interstitial collagenase. Proc. Natl. Acad. Sci. U.S.A. **86:** 2632–2636.
51. BARAMOVA, E.N., K. BAJOU, A. REMACLE, et al. 1997. Involvement of PA/plasmin system in the processing of proMMP-9 and in the second step of proMMP-2 activation. FEBS Lett. **405:** 157–162.
52. LIJNEN, H.R., J. SILENCE, G. LEMMENS, et al. 1998. Regulation of gelatinase activity in mice with targeted inactivation of components of the plasminogen/plasmin system. Thromb. Hæmost. **79:** 1171–1176.

53. LIJNEN, H.R., J. SILENCE, B.VAN HOEF, *et al.* 1998. Stromelysin-1 (MMP-3)-independent gelatinase expression and activation in mice. Blood **91:** 2045–2053.
54. LIJNEN, H.R., B. VAN HOEF, P. SOLOWAY, *et al.* 1998. Plasminogen/plasmin system function in mice deficient in stromelysin-1 (MMP-3) or in tissue inhibitor of metalloproteinases type 1 (TIMP-1). Fibrinol. Proteol. **12:** 1–8.
55. LIJNEN, H.R., F. UGWU, M.C. RIO, *et al.* 1998. Plasminogen/plasmin and matrix metalloproteinase system function in mice with targeted inactivation of stromelysin-3 (MMP-11). Fibrinol. Proteol. **12:** 155–164.
56. UGWU, F., B. VAN HOEF, A. BINI, *et al.* 1998. Proteolytic cleavage of urokinase-type plasminogen activator by stromelysin-1 (MMP-3). Biochemistry **37:** 7231–7236.
57. LIJNEN, H.R., F. UGWU, A. BINI, *et al.* 1998. Generation of an angiostatin-like fragment from plasminogen by stromelysin-1 (MMP-3). Biochemistry **37:** 4699–4702.
58. O'REILLY, M.S., L. HOLMGREN, Y. SHING, *et al.* 1994. Angiostatin: a novel angiogenesis inhibitor that mediates the suppression of metastases by a Lewis lung carcinoma. Cell **79:** 315–328.
59. CAO, Y., R.W. JI, D. DAVIDSON, *et al.* 1996. Kringle domains of human angiostatin. Characterization of the anti-proliferative activity on endothelial cells. J. Biol. Chem. **271:** 29461–29467.
60. UGWU, F., G. LEMMENS, D. COLLEN, *et al.* 1999. Modulation of cell-associated plasminogen activation by stromelysin-1 (MMP-3). Thromb. Hæmost. **82:** 1127–1131.

Fibrin-Mediated Plasminogen Activation

WILLEM NIEUWENHUIZEN

Gaubius Laboratory, TNO Prevention and Health, Leiden, the Netherlands

ABSTRACT: Fibrin, but not fibrinogen, enhances the rate of activation of plasminogen by tissue type plasminogen activator (t-PA). Studies with enzymatic and chemical fragments of fibrinogen showed that several sites in fibrinogen are involved in this rate enhancement; these are, Aα148–160 (located in CNBr fragment FCB-2), and FCB-5 (a CNBr fragment comprising γ312–324), and recently discovered sites in the fibrinogen αC domains. All these sites are buried in fibrinogen, but exposed in fibrin and some fibrinogen fragments. For the first two of these, located in the D-domains, this was shown by the fact that monoclonal antibodies against Aα148–160 and γ312–324 bind to fibrin and rate enhancing fibrin(ogen) fragments, but not to fibrinogen. Direct binding studies indicate that at physiological concentrations plasminogen binds to FCB-2, and t-PA to FCB-5. More detailed studies have demonstrated the importance of residues Aα-157 and Aα-152, and that the minimum stretch with rate enhancing properties is Aα154–159. The sites in the αC domains await further identification. With the recently reported three-dimensional structure of fragments D and D-dimer it is now possible to explain these findings at the molecular level. Molecular calculations and experimental data show that the site Aα148–160 in fibrinogen is covered among others by a part of the Aα chain (Aα166–195) that forms an α-helix, and by a globular domain formed by the β-chain. On fibrin formation, the last two may move away, and give access to Aα148–160. It is conceivable that in the αC domain sites are involved in the early phases of fibrinolysis. The site Aα148–160 and that in FCB-5 may be more important at later stages. It is also clear that fibrin polymerization is important. This polymerization has probably several effects: exposure of the rate enhancing sites; mutual positioning of the t-PA and plasminogen binding sites; a concentrating effect of t-PA and plasminogen on the fibrin surface; effects on the kinetic properties of t-PA and plasminogen. These effects together explain the rate enhancement.

KEYWORDS: Fibrin; Fibrinogen; Plasminogen activation; Fragments; Fibrin polymerization; Rate enhancement; Rate enhancing sites; Monoclonal antibodies.

FIBRIN ASSEMBLY

Fibrinogen is a soluble disulfide-linked dimer of three non-identical polypeptide chains, the Aα chains (610 amino acids), the Bβ chains (461 amino acids), and the γ chains (411 amino acids). The amino-termini of the six chains are concentrated in a central domain, called the E-domain or amino-terminal disulfide knot. The carboxyl-terminal regions of the Bβ and γ chains are located at the distal ends of the molecule

---

Address for correspondence: Willem Nieuwenhuizen, Ph.D., Gaubius Laboratory, TNO Prevention and Health, P.O. Box 2215, 2301 CE Leiden, the Netherlands. Voice: +31.71.518.1464; fax: +31.71.518.1904.
w.nieuwenhuizen@pg.tno.nl

and have a globular structure. They are designated as D-domains. The carboxyl-terminal ends of the Aα chains also form globular domains (αC domains) that are located near the E domain and are non-covalently associated with each other.[1]

The E and the two D-domains are connected by irregular coiled coil regions in which α-helical parts of the Aα chain (Aα49–161), Bβ chain (Bβ80–193), and the γ-chain (γ23–135) are associated.[2] These coiled coil regions together span about 75% of the total length of the 450Å-length of the fibrinogen molecule, and have a short interruption in the middle,[2] which is a target for fibrin(geno)lysis by plasmin.

The formation of polymeric fibrin has been described in many studies (see Refs. 3 and 4, and the references cited therein). The assembly of fibrin is initiated by the release of fibrinopeptides A (Aα1–16) from the Aα chains of fibrinogen by thrombin. This release exposes the sequence Aα 17 and beyond (GPR…) (A sites or "knobs") that then bind to complementary sites (a sites or "holes") in the carboxyl-terminal region of the γ chain.[5–8]

These interactions lead to the formation of two-stranded protofibrils in which the subunits are arranged in a half-staggered overlap.[9] During this process the two αC domains dissociate from each other and move away from the centre of the fibrin monomers to distal positions.[1] Simultaneously, but somewhat more slowly, fibrinopeptides B (Bβ1–14) are released by thrombin, and this exposes the sequences Bβ 15 and further (GHR…) (b sites or knobs) that are involved in the lateral association of the protofibrils. This lateral association requires the interaction between the b knobs and B holes in the carboxyl-terminal regions of the Bβ chain.[8]

In the presence of activated factor XIII intermolecular isopeptide bonds are formed between the side chains of glutamine-γ 398 and lysine-γ 406 residues of two adjacent subunit fibrin monomers, of which the carboxyl-terminal ends of the γ chains are antiparallel oriented.

FIBRIN(OGEN) FRAGMENTS

In the course of the elucidation of the primary structure of fibrinogen, several fragments have been prepared and their chain remnant amino acid sequences have been determined.[2,10] These fragments have been prepared mainly by proteolysis

TABLE 1. Chain remnant composition of some fibrinogen fragments

Fragment	Amino acid stretches derived from			Ref.
	Aα chain	Bβ chain	γ chain	
D_{cate} (= D_1)	Val 111–Arg 197	Asp 134–Glu 461	Ser 80–Val410	48
D_{EGTA} (= D_3)	Val 111–Arg 197	Asp 134–Glu 461	Ser 86–Lys 303	48
E	(Ala 1–Lys 78)$_2$	(Lys 54–Lys 122)$_2$	(Tyr 1–Lys 58)$_2$	49
FCB-2	Lys 148–Met 207	(Glu 191–Met 224) + (Tyr 225–Met 242) + (Asn 243–Met 305)	Lys 95–Met 265	10, 50
FCB-5			(Glu 311–Met 336) + (Asn 337–Met 379)	10, 50

with plasmin and by chemical cleavage with CNBr. Their chain remnant compositions have been reviewed.[11] The most relevant fragments in the context of this overview are summarized in TABLE 1.

IDENTIFICATION OF SITES INVOLVED IN THE ENHANCEMENT OF THE ACTIVATION OF PLASMINOGEN BY t-PA

Fibrin, but not fibrinogen, enhances the rate of activation of plasminogen by t-PA.[12-15] However, the primary structures of fibrin and its precursor fibrinogen, differ only in that the two 16-amino acid fibrinopeptides A (and the two 14-amino acid fibrinopeptides B) are absent in fibrin. We put forward the hypothesis that during one or more of the steps in the formation of fibrin (release of fibrinopeptides, protofibril formation, lateral aggregation, and crosslinking) conformational changes occur, which expose sites that are involved in the rate enhancement of plasminogen activation. To prove this, we prepared plasmin fragments and CNBr digests of fibrinogen and fibrin and tested these fragments for their rate enhancing properties. We found that the smallest rate enhancing plasmin generated fragment is D_{EGTA} and that the E-fragment is not rate enhancing.[15,16] A whole CNBr digest of fibrinogen accelerated the glu-plasminogen activation[16] and we isolated two accelerating fragments from such digests, namely, fragments FCB-2[17,18] and FCB-5.[19]

Fragments D_{EGTA} and FCB-2 are related and consist of disulfide-linked remnants of all three chains of fibrinogen (see TABLE 1). To investigate whether the rate enhancing properties of these fragments could be assigned to one of the chain remnants, we separated the chain remnants and found that only the Aα chain remnants of both D_{EGTA} and FCB-2 are rate enhancing. Since the Aα chain remnants in D_{EGTA} and FCB-2 are Aα111-197 and Aα148-207, respectively, we concluded that a rate enhancing site must reside in the overlapping sequence Aα148-197.[18] This sequence comprises a remarkable charge pattern in Aα148-160:

$$\begin{array}{cccccccccccc} K & R & L & E & V & D & I & D & I & K & I & R & S \\ + & + & 0 & - & 0 & - & 0 & - & 0 & + & 0 & + & 0 \end{array}$$

We then investigated the importance of this charge motif and the minimum required length by synthesizing a number of analogues of this sequence. It was found that position Aα157 plays an important role, whereas lysine Aα148 is not important.[20] Synthetic analogues of Aα148-160, in which Lysine Aα157 was replaced by a valine or an aspartic acid residue, were inactive, but analogs with alanine or glutamic acid at this position were fully active.[21,22] The overall conclusion from that study was, that the side chain of the amino acid residue at Aα157 should be zero (glycine) or two carbon atoms long, with a polar group at position γ, the charge of which is less relevant.

We then studied the importance of the Val residue at position Aα152[22,23] and found that the rate enhancing properties of Aα148-160 are retained when Val Aα152 is replaced by other apolar amino acid residues such as Ala, Tyr, or the non-natural norleucine. Activity is lost with Gly, Glu, Arg, Lys, or Pro at this position. The minimum required stretch within Aα148-160 with rate enhancing properties was found to be Aα154-159.[22,24]

A second site involved in the acceleration of the t-PA mediated plasminogen activation was found in CNBr fragment FCB-5,[19] which consists of the γ chain stretches γ311–336 and γ337–379 linked by a single disulfide bond between Cys-γ-326 and Cys-γ-339. The individual stretches are not active. A charge pattern, similar to that seen in Aα148–160 can be observed in γ315–323; that is,

$$\begin{array}{cccccccc} \text{Leu} & \text{Glu} & \text{Val} & \text{Asp} & \text{Ile} & \text{Asp} & \text{Ile} & \text{Lys} & \text{Ile} \\ 0 & - & 0 & - & 0 & - & 0 & + & 0 \end{array}$$

In a recent study[25] (and see elsewhere in this volume) it was found that the αC domains (Aα221–610) in particular Aα392–610, also contain rate enhancing sites that bind t-PA and plasminogen with high affinity, and are cryptic in fibrinogen. This may be physiologically relevant, since the αC domains are the very first to be hydrolyzed during fibrinolysis, and the sites in the αC domains may thus be involved in the early phase of fibrinolysis.

ACCESSIBILITY OF THE RATE ENHANCING SITES

As stated above, our working hypothesis was, that sites exist that are involved in the enhancement of the t-PA mediated plasminogen activation and that are not accessible to t-PA and/or plasminogen in fibrinogen, but become accessible or operational on conversion of fibrinogen to fibrin.

The aforementioned results with purified fibrinogen fragments supported this hypothesis. To obtain more direct evidence we prepared monoclonal antibodies against Aα148–160[26] and γ312–324[27] and found that these antibodies indeed bind to fibrin and not to fibrinogen. Furthermore, a good correlation was found between the rate enhancing properties of fibrin(ogen) fragments and their reactions with the monoclonal antibodies (TABLE 2).

TABLE 2. Binding of fibrin specific monoclonal antibodies and rate enhancing properties of fibrin(ogen) fragments

	Binds		
Fragment	Anti Aα 148–160	Anti γ 312–324	Rate Enhancing
FCB-2	Yes	No[a]	Yes
D_{cate}	No	No	No
D_{EGTA}	Yes	No[a]	Yes
D-dimer	Yes/No[b]	No	Yes/No[b]
FCB-5	No[a]	Yes	Yes

[a]The fragment obviously does not bind this antibody, since the epitope of the latter is not present in the fragment.

[b]Some of our D-dimer preparations were rate enhancing,[16,51] whereas others were not. In previous studies it was found that anti-Aα148–160 binds to D-dimer.[52] In a later study,[35,36] this was not confirmed. As pointed out in References 35 and 36, this may be due to the D-dimer preparation method, which was different in the two studies.

These findings confirmed our hypothesis and showed that not only the fibrinogen to fibrin conversion exposes the sites, but also enzymatic or CNBr fragmentation. Other studies have shown that partial denaturation of fibrinogen by freeze drying, heating at 47°C, EDTA and exposure to pH 10 also induce rate enhancing properties[28,29] and reactivity with the anti-Aα148–160 and γ312–324 monoclonal antibodies.[30]

A remarkable observation was that fibrinogen that had been crosslinked *in vitro* with activated factor XIII behaved like fibrin in that it exhibited a strong stimulatory activity on the plasminogen activation by t-PA, and was reactive with anti-Aα148–162 and anti-γ312–324.[31] It is known that in normal fibrin a certain degree of polymerization is required for optimal rate enhancement of the t-PA/plasminogen reaction, since dissociation with GPRP abolishes the stimulating effect of fibrin,[32–34] and that fibrin monomers kept in a monomeric state by GPRP do not expose Aα148–160 and γ312–324.[30,31] This indicates, that A–a interactions are important for the exposure of the fibrin specific epitopes Aα148–160 and γ312–324, but may not be sufficient.[31]

In a recent study,[35,36] it was demonstrated that DD/E complexes, but not their constituent moieties enhance the activation rate, and expose the fibrin specific epitopes. Addition of GPRP, which dissociates the complexes, abolished these properties. Although the complex was dissociated, the binding sites *a* are still occupied by GPRP (*A*). This means that A–a interactions alone are not sufficient for fibrin specific epitope exposure and that more, possibly secondary, interactions are needed.

In the same study,[35,36] the average solvent accessible surface area (ASA percentage) of the individual amino acid residues within Aα148–160 and γ312–324 was calculated, based on the recently obtained crystallographic data of fragment D_{cate} and D-dimer.[7] None of the residues in Aα148–160 except Arg Aα149 (ASA 16%) was significantly solvent accessible. Of the residues in γ312–324, all residues except Phe 312, Ser 313, Thr 314, and Asp 320 are accessible. In accordance with the notion that A–a interactions are not sufficient for the induction of rate enhancement, molecular calculations show that the accessibilities of residues in Aα148–160 and γ312–324 are not significantly modified by the presence of GPRP and GHRP.

INTERACTIONS WITH t-PA AND PLASMINOGEN

It is conceivable that the observed enhancement of the activation of the t-PA catalyzed plasminogen activation in the presence of fibrin is based on ternary complex formation between t-PA, plasminogen, and fibrin,[14] and that the sites described above are involved in this complex formation.

The binding of t-PA and plasminogen to fibrin and fibrin(ogen) fragments has been reviewed extensively[11] and the reader is referred to that review for details. It has been found that fragment FCB-2 (comprising Aα148–160) binds both plasminogen and t-PA[37] with about equal affinity. This means, that under physiological conditions, where there is a huge excess of plasminogen over t-PA, the FCB-2 site will bind only plasminogen. FCB-5 (comprising γ312–324) binds only t-PA.[19] The binding of t-PA to FCB-5 is lysine independent,[38] like the t-PA binding to fibrin. Glu plasminogen, Lys-plasminogen, the domain consisting of plasminogen kringles 1–3, kringle 4, and miniplasminogen all bind to fibrin.[39–41] The activation rates of miniplasminogen but also of microplasminogen (no fibrin binding) are increased in the presence of FCB-5.[19]

KINETIC CONSEQUENCES OF THE t-PA AND PLASMINOGEN BINDINGS TO FIBRIN

As reviewed in references 11 and 42, the reported effects on the kinetic parameters of fibrin and fibrin(ogen) fragments vary widely. The reason for these variations lies, in part, in the fact that the activation rate of plasminogen does not obey Michaelis–Menten kinetics at higher plasminogen concentrations. There is general agreement that the interactions of t-PA and plasminogen with fibrin and rate enhancing fibrin(ogen) fragments decrease the K_m and increase the k_{cat} of the reaction.

MECHANISTIC INSIGHTS BASED ON THE AVAILABLE THREE-DIMENSIONAL STRUCTURES

In recent studies the three-dimensional structures of fragments of human fibrin(ogen), fragments D and D-dimer,[7,8] an expressed 30-kDa γ domain,[6,43] modified bovine fibrinogen,[44] and chicken fibrinogen[45] were elucidated. The availability of these structures enable us to understand our earlier observations, and to put them into a three-dimensional molecular perspective.

In 1979, Doolittle[46] predicted that virtually all the α-helix content of the Aα chain would occur in Aα1–194. The three-dimensional structures now available confirm this, and show that Aα148–160 has an α-helical conformation, and is part of the triple-helical coiled coil region. It was a surprising observation that the Aα chain reverses its direction beyond Aα148–160, and forms a fourth α-helical stretch for at least the entire coiled coil of fragment D.[7]

As reported in Reference 7 the site Aα148–160 is surrounded by the β module, the β- and γ-helices of the coiled coil, and the fourth helix formed by Aα166–195. Studies by Yakovlev et al.[35,36] based on fragments derived from bovine fragment D_H, and on molecular calculations show that removal of the fourth α-helix is not sufficient for exposure of Aα148–160, but that the molecular interactions between Aα148–160 and the β module must also be disrupted. The latter interactions are partly based on contacts between the side chains of Aα-157 (Lys) and Bβ-398 (Asp).

The separation of the fourth helix and the β module appears to occur when D-dimer forms a complex with fragment E, as also happens during fibrin formation. This could explain why fibrin and DD/E complexes are rate enhancing, in contrast to fibrinogen, D-dimer (see footnote to TABLE 2), and fragment E.

In contrast to the residues of Aα148–160 most residues in γ312–324 are solvent accessible, except γ312–314. Yet, t-PA does not bind to fibrinogen. This indicates that either these three γ-chain residues are required for t-PA binding and buried, and become exposed upon fibrin formation; or that they are not buried and rearrange themselves into a fibrin specific conformation.[35,36]

On the basis of the above, and previously reported observations, the following elements may be of importance in the observed rate enhancing properties of fibrin.[47]

- When polymeric fibrin is formed DD/E interactions occur that expose or operationalize the sites Aα148–160, γ312–324, and sites as yet not fully identified in the αC domains.

- The sites in the αC domains bind t-PA and plasminogen; under physiological conditions plasminogen binds to FCB-2 (comprising Aα148–160) and t-PA to FCB-5 (comprising γ312–324).

- Because of the binding, plasminogen and t-PA are concentrated on the ordered fibrin surface in an ordered fashion.

- They are oriented such that they are in close contact with each other.

- As a consequence, the kinetic parameters of the activation reaction change, that is, K_m decreases and k_{cat} increases.[11,42]

REFERENCES

1. VEKLICH, Y.I., O.V. GORKUN, L.V. MEDVED, *et al.* 1993. Carboxyl-terminal portions of the alpha chains of fibrinogen and fibrin. Localization by electron microscopy and the effects of isolated alpha C fragments on polymerization. J. Biol. Chem. **268:** 13577–13585.
2. DOOLITTLE, R.F. 1981. Fibrinogen and fibrin. Sci. Am. **245:** 126–135.
3. DOOLITTLE, R.F. 1984. Fibrinogen and fibrin. Ann. Rev. Biochem. **53:** 195–229.
4. MOSESSON, M.W. 1998. Fibrinogen structure and fibrin clot assembly. Semin. Thromb. Hæmostas. **24:** 169–174.
5. SHIMIZU, A., G.M. NAGEL & R.F. DOOLITTLE. 1992. Photoaffinity labeling of the primary fibrin polymerization site: isolation and characterization of a labeled cyanogen bromide fragment corresponding to gamma-chain residues 337–379. Proc. Natl. Acad. Sci. U.S.A. **89:** 2888–2892.
6. PRATT, K.P., H.C. CÔTE, D.W. CHUNG, *et al.* 1997. The primary fibrin polymerization pocket: three-dimensional structure of a 30-kDa C-terminal gamma chain fragment complexed with the peptide Gly-Pro-Arg-Pro. Proc. Natl. Acad. Sci. U.S.A. **94:** 7176–7181.
7. SPRAGGON, G., S.J. EVERSE & R.F. DOOLITTLE. 1997. Crystal structures of fragment D from human fibrinogen and its crosslinked counterpart from fibrin. Nature **389:** 455–462.
8. EVERSE S.J., G. SPRAGGON, L. VEERAPANDIAN, *et al.* 1998. Crystal structure of fragment double-D from human fibrin with two different bound ligands. Biochemistry **37:** 8637–8642.
9. WEISEL, J.W., G.N. PHILLIPS, JR. & C. COHEN. 1983. The structure of fibrinogen and fibrin: II. Architecture of the fibrin clot. *In* Molecular Biology of Fibrinogen and Fibrin. M.W. Mosesson & R.F. Doolittle, Eds.: 367–379. Ann. N.Y. Acad. Sci. **408**.
10. HENSCHEN, A. 1981. Fibrinogen-Blutgerinnungsfaktor I. Biochemische Aspekte. Hæmostaseologie **1:** 49–61.
11. NIEUWENHUIZEN, W. 1988. Fibrinogen and its specific sites for modulation of t-PA-induced fibrinolysis. *In* Tissue-Type Plasminogen Activator (t-PA): Physiological and Clinical Aspects, Vol. 1. C. Kluft, Ed.: 171–187. CRC Press, Boca Raton.
12. WALLÉN, P. 1977. Activation of plasminogen with urokinase and tissue activator. *In* Thrombosis and Urokinase, Vol. 9. R. Paoletti & S. Sherry, Eds.: 91–102. Academic Press, New York.
13. ALLEN, R.A. & D.S. PEPPER. 1981. Isolation and properties of human vascular plasminogen activator. Thromb. Hæmostas. **45:** 43–50.

14. HOYLAERTS, M., D.C. RIJKEN, H.R. LIJNEN, et al. 1982. Kinetics of the activation of plasminogen by human tissue plasminogen activator. J. Biol. Chem. **257:** 2912–2919.
15. VERHEIJEN, J.H., W. NIEUWENHUIZEN & G. WIJNGAARDS. 1982. Activation of plasminogen by tissue factor is increased specifically in the presence of certain soluble fibrin(ogen) fragments. Thromb. Res. **27:** 337–385.
16. VERHEIJEN, J.H., W. NIEUWENHUIZEN, D.W. TRAAS, et al. 1985. Fibrin and plasminogen structures involved in the tissue-type plasminogen activator catalyzed activation of plasminogen. *In* Fibrinogen—Structural Variants and Interactions, Vol. 3. A. Henschen, B. Hessel, J McDonagh & T. Saldeen, Eds.: 323–330. Walter de Gruyter & Co, Berlin.
17. NIEUWENHUIZEN, W., J.H. VERHEIJEN, A. VERMOND, et al. 1983. Plasminogen activation by tissue activator is accelerated in the presence of fibrin(ogen) cyanogen bromide fragment FCB-2. Biochim. Biophys. Acta **755:** 531–533.
18. NIEUWENHUIZEN, W., A. VERMOND, M. VOSKUILEN, et al. 1983. Identification of a site in fibrin(ogen) which is involved in the acceleration of plasminogen activation by tissue-type plasminogen activator. Biochim. Biophys. Acta **748:** 86–92.
19. YONEKAWA, O., M. VOSKUILEN & W. NIEUWENHUIZEN. 1992. Localization in the fibrinogen γ–chain of a new site that is involved in the acceleration of the tissue-type plasminogen activator-catalysed activation of plasminogen. Biochem. J. **283:** 187–191.
20. VOSKUILEN, M., A. VERMOND, G.H. VEENEMANS, et al. 1987. Fibrinogen lysine residue Aα 157 plays a crucial role in the fibrin-induced acceleration of plasminogen activation, catalyzed by tissue-type plasminogen activator. J. Biol. Chem. **262:** 5944–5946.
21. SCHIELEN, W.J.G., M. VOSKUILEN, H.P.H.M. ADAMS, et al. 1990. Structural requirements of fibrinogen Aα-(148–160) for the enhancement of the rate of plasminogen activation by t-PA. Blood Coag. Fibrinol. **1:** 521–524.
22. SCHIELEN, W.J.G., H.P.H.M. ADAMS, M. VOSKUILEN, et al. 1991. Structural requirements of position Aα-157 in fibrinogen for the fibrin-induced rate enhancement of the activation of plasminogen by tissue-type plasminogen activator. Biochem. J. **276:** 655–659.
23. SCHIELEN, W.J.G., H.P.H.M. ADAMS, M. VOSKUILEN, et al. 1993. The role of [152]Val of the fibrinogen Aα-chain in the fibrin-induced rate enhancement of the plasminogen activation by t-PA. Fibrinolysis **7:** 63–67.
24. SCHIELEN, W.J.G., H.P.H.M. ADAMS, M. VOSKUILEN, et al. 1991. The sequence Aα-(154–159) of fibrinogen is capable of accelerating the t-PA catalysed activation of plasminogen. Blood Coag. Fibrinol. **2:** 465–470.
25. TSURUPA, G. & L. MEDVED. 2000. Identification and characterization of novel t-PA and plasminogen-binding sites within fibrin(ogen) αC domains (Abst.). Fibrinol. Proteol. **14:** O-77.
26. SCHIELEN, W.J.G., M. VOSKUILEN, G.I. TESSER, et al. 1989. The sequence Aα-(148–160) in fibrin, but not in fibrinogen, is accessible to monoclonal antibodies. Proc. Natl. Acad. Sci. U.S.A. **86:** 8951–8954.
27. SCHIELEN, W.J.G., H.P.H.M. ADAMS, C. VAN LEUVEN, et al. 1991. The sequence γ-(312–324) is a fibrin-specific epitope. Blood **77:** 2169–2173.
28. HADDELAND, U., K. SLETTEN, A. BENNICK, et al. 1994. Freeze-dried fibrinogen or fibrinogen in EDTA stimulate the tissue-type plasminogen activator-catalysed conversion of plasminogen to plasmin. Blood Coagul. Fibrinolysis **5:** 575–581.
29. HADDELAND, U., A. BENNICK & F. BROSSTAD. 1995. Stimulating effect on tissue-type plasminogen activator—a new and sensitive indicator of denatured fibrinogen. Thromb. Res. **77:** 329–336.
30. HADDELAND, U., K. SLETTEN, A. BENNICK, et al. 1996. Aggregated, conformationally changed fibrinogen exposes the stimulating sites for t-PA-catalysed plasminogen activation. Thromb. Hæmostas. **75:** 326–331.
31. MOSESSON, M.W., K.R. SIEBENLIST, M. VOSKUILEN, et al. 1998. Evaluation of the factors contributing to fibrin-dependent plasminogen activation. Thromb. Hæmostas. **79:** 796–801.

32. SUENSON, E. & L.C. PETERSEN. 1986. Fibrin and plasminogen structures essential to stimulation of plasmin formation by tissue-type plasminogen activator. Biochim. Biopys. Acta **870:** 510–519.
33. KOOPMAN, J., L. ENGESSER, M. NIEVEEN, et al. 1986. Fibrin polymerization associated with tissue-type plasminogen activator (t-PA) induced glu-plasminogen activation. *In* Fibrinogen and Its Derivatives. G. Müller-Berghaus, U. Scheefers-Borchel, E. Selmayr & A. Henschen, Eds.: 315–318. Elsevier, Amsterdam.
34. KACZMAREK, E., M.H. LEE & J. MCDONAGH. 1993. Initial interaction between fibrin and tissue plasminogen activator (t-PA). The Gly-Pro-Arg-Pro binding site on fibrin(ogen) is important for t-PA activity. J. Biol. Chem. **268:** 2474–2479.
35. YAKOVLEV, S., E. MAKOGONENKO, N. KUROCHKINA, et al. 2000. Conversion of fibrinogen to fibrin: the mechanism of exposure of t-PA– and plasminogen-binding sites. Biochemistry **39:** 15730–15741.
36. YAKOVLEV, S., E. MAKOGONENKO, N. KUROCHKINA, et al. 2001. Ann. N.Y. Acad. Sci. **936:** this volume.
37. BOSMA, P., D.C., RIJKEN & W. NIEUWENHUIZEN. 1988. Binding of tissue-type plasminogen activator to fibrinogen fragments. Eur. J. Biochem. **172:** 399–404.
38. GRAILHE, P., W. NIEUWENHUIZEN & E. ANGLÈS-CANO. 1994. Study of tissue-type plasminogen activator binding sites on fibrin using distinct fragments of fibrinogen. Eur. J. Biochem. **219:** 961–967.
39. THORSEN, S., I. CLEMMENSEN, L. SOTTRUP-JENSEN, et al. 1981. Adsorption to fibrin of native fragments of known primary structure from human plasminogen. Biochim. Biophys. Acta **668:** 377–387.
40. LIJNEN, H.R., B. VAN HOEF & D. COLLEN. 1981. On the role of carbohydrate side chains of human plasminogen in its interaction with α_2-antiplasmin and fibrin. Eur. J. Biochem. **120:** 149–154.
41. FLEURY, V., S. LOYAU, H.R. LIJNEN, et al. 1993. Molecular assembly of plasminogen and tissue-type plasminogen activator on an evolving fibrin surface. Eur. J. Biochem. **216:** 549–556.
42. NIEUWENHUIZEN, W., M. VOSKUILEN, A. VERMOND, et al. 1988. The influence of fibrin(ogen) fragments on the kinetic parameters of the tissue-type plasminogen-activator-mediated activation of different forms of plasminogen. Eur. J. Biochem. **174:** 163–169.
43. YEE, V.C., K.P. PRATT, H.C. CÔTE, et al. 1997. Crystal structure of a 30 kDa C-terminal fragment from the gamma chain of human fibrinogen. Structure (Lond.) **5:** 125–138.
44. BROWN, J.H., N. VOLKMANN, G. JUN, et al. 2000. The crystal structure of modified bovine fibrinogen. Proc. Natl. Acad. Sci. U.S.A. **97:** 85–90.
45. YANG, Z, I. MOCHALKIN, L. VEERAPANDIAN, et al. 2000. Crystal structure of native chicken fibrinogen at 5.5-Å resolution. Proc. Natl. Acad. Sci. U.S.A. **97:** 3907–3912.
46. DOOLITTLE, R.F., K.W.K. WATT, B.A. COTTRELL, et al. 1979. The amino acid sequence of the alpha-chain of human fibrinogen. Nature **280:** 464–468.
47. NIEUWENHUIZEN, W. 1994. Sites in fibrin involved in the acceleration of plasminogen activation by t-PA. Possible role of fibrin polymerisation. Thromb. Res. **75:** 343–347.
48. VAN RUIJVEN-VERMEER, I.A.M., W. NIEUWENHUIZEN, F. HAVERKATE, et al. 1979. A novel method for the rapid purification of human and rat fibrin(ogen) degradation products in high yield. Hoppe-Seyler's Z. Physiol. Chem. **360:** 633–637.
49. OLEXA, S.A., A.Z. BUDZYNSKI, R.F. DOOLITTLE, et al. 1981. Structure of fragment E species from human cross-linked fibrin. Biochemistry **20:** 6139–6145.
50. GÅRDLUND, B., B. HESSEL, G. MARGUERIE, et al. 1977. Primary structure of human fibrinogen. Characterization of disulphide-containing cyanogen-bromide fragments. Eur. J. Biochem. **77:** 595–610.
51. VERHEIJEN, J.H., W. NIEUWENHUIZEN, E. MULLAART, et al. 1983. Evidence for specific sites in fibrin(ogen) that stimulate tissue activator catalyzed plasminogen activation. *In* Fibrinogen: Structure, Functional Aspects, Metabolism, Vol. 2. F. Haverkate, A. Henschen, W. Nieuwenhuizen & P.W. Straub, Eds.: 305–314. Walter de Gruyter & Co, Berlin.

52. NIEUWENHUIZEN, W., W.J.G. SCHIELEN, O.YONEKAWA, *et al.* 1990 Studies on the localization and accessibility of sites in fibrin which are involved in the acceleration of the activation of plasminogen by tissue-type plasminogen activator. *In* Fibrinogen, Thrombosis, Coagulation and Fibrinolysis. C.Y. Liu & S. Chien, Eds.: 83–91. Plenum Press, New York.

Modulation of Fibrin Cofactor Activity in Plasminogen Activation

MICHAEL NESHEIM, JOHN WALKER, WEI WANG, MICHAEL BOFFA, ANTON HORREVOETS, AND LASZLO BAJZAR

Department of Biochemistry, Queen's University, Kingston, Ontario, Canada

ABSTRACT: Fibrin is a cofactor for the formation of plasmin from plasminogen as catalyzed by tissue plasminogen activator. Initial cleavages of fibrin by plasmin upregulates the cofactor activity of fibrin by exposing carboxyl terminal lysine residues. This effect is eliminated by a carboxypeptidase B-like enzyme generated from the precursor, thrombin activatable fibrinolysis inhibitor (TAFI) that is generated by thrombin during the formation of fibrin. Thus, TAFI and its activation to TAFIa create a link between the coagulation and fibrinolytic cascade, such that activation of the former suppresses the latter. Complete solubilization of fibrin results in a family of very large fibrin degradation products. These also have very substantial tissue plasminogen activator cofactor activity that is very highly downregulated by TAFIa.

KEYWORDS: Fibrinolysis; Plasminogen; Plasmin; Tissue plasminogen activator; Fibrin; Thrombin activatable fibrinolysis inhibitor; Kinetics.

INTRODUCTION

The processes of fibrin formation from fibrinogen and subsequent solubilization of fibrin are vital to the integrity of the vascular system. Fibrin formation helps prevent the life threatening loss of vascular fluid in the event of an injury, and fibrin solubilization helps prevent life threatening lack of flow in the event of inappropriate fibrin deposition within the vasculature. Fibrin deposition occurs through the action of the coagulation cascade, and fibrin solubilization occurs through the action of the fibrinolytic cascade.[1–5] When these cascades are properly regulated the vasculature is protected from both excess fluid loss and impeded flow. When these cascades are not properly balanced, numerous pathological consequences occur marked by excessive bleeding or tendencies to clot excessively, such as in heart attacks or strokes.

FIGURE 1 depicts the balance between fibrin formation, which is catalyzed by the enzyme thrombin generated via the coagulation cascade, and fibrin degradation, which is catalyzed by the enzyme plasmin generated via fibrinolytic cascade. The activities of these cascades are regulated at numerous steps involving upregulation by positive feedback and downregulation through inhibitors. One of the processes deemed essential in the regulation of coagulation involves the binding of thrombin to thrombomodulin on the surface of endothelial cells.[6,7] This event changes

---

Address for correspondence: Michael Nesheim, Ph.D., Department of Biochemistry, Queen's University, Kingston, Ontario, Canada, K7L 3N6. Voice: 613-533-2957; fax: 613-533-2987.

nesheimm@post.queensu.ca

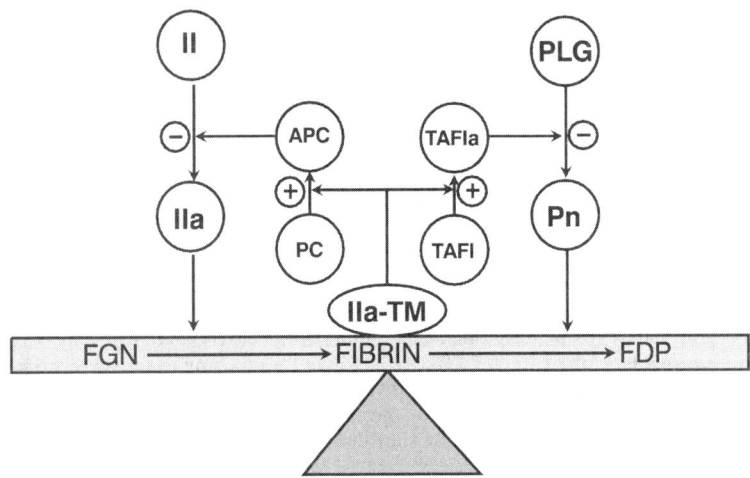

FIGURE 1. The balance between fibrin deposition and removal. The coagulation cascade generates thrombin (*IIa*), which forms fibrin from fibrinogen (*FGN*). The fibrinolytic cascade generates plasmin (*Pn*), which degrades fibrin to soluble fibrin degradation products (*FDP*). Thrombin binds to thrombomodulin (*TM*) and activates protein C (*PC*) to downregulate coagulation and TAFI to downregulate fibrinolysis.

the specificity of thrombin from fibrinogen to protein C, the activation of which generates activated protein C. This latter enzyme functions as an anticoagulant by proteolytically inactivating two essential cofactors of the coagulation cascades, factors Va and VIIIa. The net result is downregulation of thrombin formation. Recently, an analogous reaction has been identified in the fibrinolytic cascade.[4,8,9] In this reaction, the thrombin–thrombomodulin complex activates a zymogen, known by various names including thrombin activatable fibrinolysis inhibitor (TAFI), plasma procarboxypeptidase B, and procarboxypeptidase U.[8,10,11] The product of this reaction is a carboxypeptidase B-like enzyme (TAFIa) that suppresses the activity of the fibrinolytic cascade.

The coagulation and fibrinolytic cascades both depend on proteolytic zymogen to enzyme conversions. They also depend on the function of cofactor proteins that, although not enzymes, profoundly influence proteolytic conversions within the cascades. In coagulation, the cofactors are tissue factor, factor Va and factor VIIIa.[2] In the anticoagulant (and antifibrinolytic) pathway the main cofactor is thrombomodulin.[6] In the fibrinolytic pathway the essential cofactor is fibrin itself.[3] Fibrin not only is the substrate for plasmin, it also profoundly influences the rate of plasminogen activation by tissue type plasminogen activator (tPA). Fibrin functions as a template to which both tPA and plasminogen bind, events that enhance the catalytic efficiency of the reaction by about three orders of magnitude.[12] The binding interactions are mediated by both a fibronectin-like finger domain and a kringle domain on tPA, and one or more kringle domains on plasminogen.[13] The kringle dependent interactions are dependent on both internal and COOH-terminal lysine residues on fibrin. As fibrin is acted upon by plasmin, new COOH terminal lysines appear at sites cleaved in the presence of plasmin.[14] As a consequence, the cofactor activity of fibrin is modulated by plasmin during the process of fibrinolysis, and these new COOH terminal lysine residues are targeted for removal by TAFIa.[15] This article comprises a review of the modulation of fibrin cofactor activity by plasmin, the attenuation of cofactor activity by TAFIa, and cofactor activities of a family of large, soluble fibrin degradation products produced during plasmin catalyzed fibrin breakdown.

CHARACTERISTICS OF PLASMINOGEN ACTIVATION

An example of the time course of plasminogen consumption within a clot is shown in FIGURE 2.[15] The process, as shown here and by others, can be separated into three phases. In the first, plasminogen consumption exhibits a characteristic lag. This is followed by a second phase, in which plasminogen activation accelerates. Eventually plasminogen activation stops. At this point about 50% of the plasminogen has been consumed and the clot has solubilized. Although not shown here, the plasmin generated from plasminogen is captured as the plasmin α_2-antiplasmin complex. The transition from the slow first phase to the second rapid phase coincides with initial cleavages in fibrin to expose COOH terminal lysine residues. In the second phase, especially toward the end of this phase, considerable conversion of glu-plasminogen to lys-plasminogen occurs.[16] The switch from slow to rapid activation

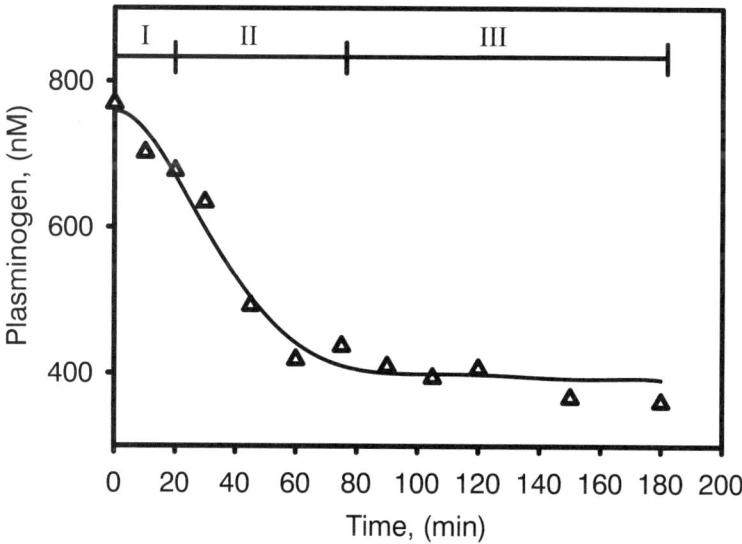

FIGURE 2. Plasminogen consumption during fibrinolysis. Three phases are recognized. *Phase I* is a lag phase. This is followed by *Phase II*, in which plasminogen activation is accelerated due to feedback modification of both fibrin and plasminogen by plasmin. Near the end of *Phase II* the clot dissolves and plasminogen activation ceases (*Phase III*).

kinetics is typical of the reaction in fibrin and reflects positive feedback due to the modification of the tPA cofactor activity of fibrin in plasminogen activation and the cofactor activity of partially degraded fibrin in the plasmin catalyzed conversion of glu-plasminogen to lys-plasminogen.

ATTENUATION OF PLASMINOGEN ACTIVATION AND FIBRINOLYSIS BY TAFIa

The ability of TAFIa to attenuate tPA induced fibrinolysis is depicted in FIGURE 3. In these experiments[9] the time for a clot formed from fibrinogen by adding it, plus plasminogen and α_2-antiplasmin, to a mixture of tPA and thrombin (plus or minus TAFIa), increases with the concentration of TAFIa. The effect exhibits saturation with a half-maximal TAFIa concentration of 1.0nM and a maximum threefold increase. As these experiments indicate, TAFIa attenuates but does not completely eliminate fibrinolysis. In a similar set of experiments, a series of identical clots were formed and at regular intervals clots were solubilized and deproteinated. Free arginine and lysine then were measured in the protein free supernatant.[15] The results, depicted in FIGURE 4, show that arginine and lysine are continuously liberated in the system when TAFIa is present. Arginine is initially liberated in a burst following clotting, presumably from cleavage of fibrinopeptides A and B generated as a consequence of clot formation catalyzed by thrombin. Additional experiments were carried out, in which cleavage of a recombinant plasminogen derivative (S741C), labeled at C741 with fluorescein[17] was monitored within a clot by fluorescence. The

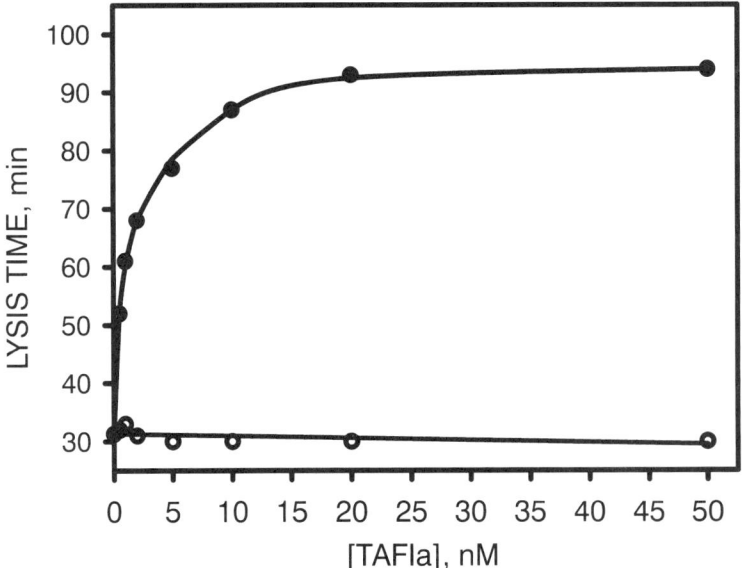

FIGURE 3. The effect of TAFIa on the time for lysis of clots. TAFIa at saturating levels triples the time needed for clot lysis to occur.

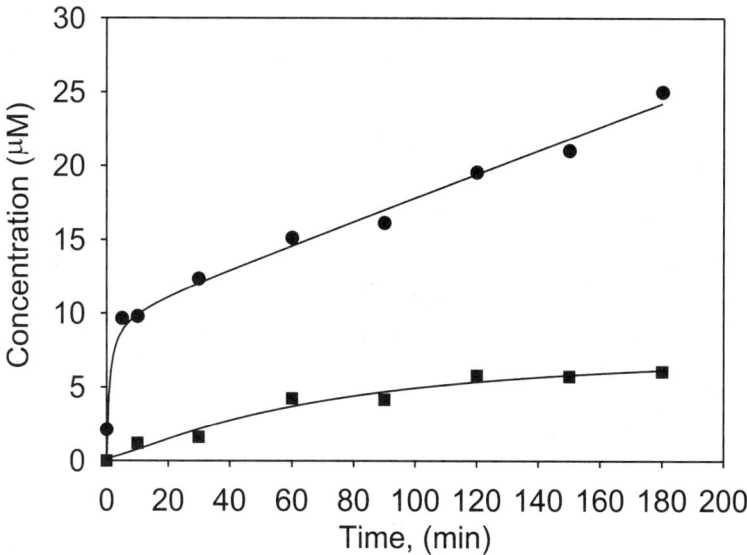

FIGURE 4. Release of free arginine (●) and lysine (■) from fibrin during the process of fibrinolysis in the presence of TAFIa.

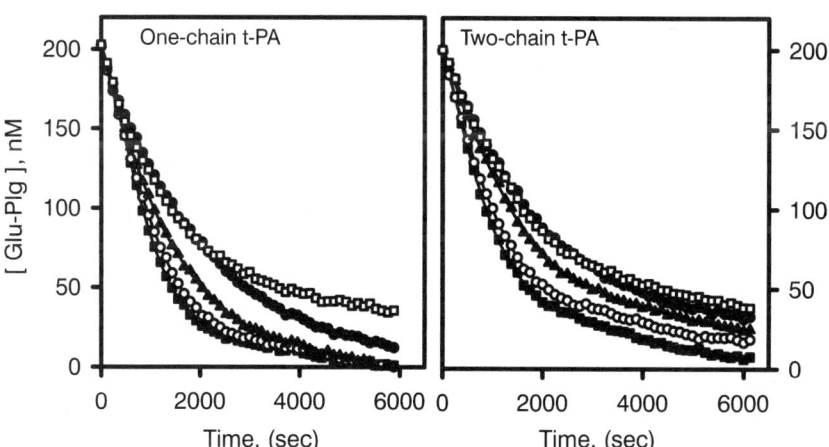

FIGURE 5. Time course of cleavage of a fluorescent plasminogen derivative. The data indicated by *open squares* were obtained with the derivative only. The data indicated by *filled squares* were obtained with the derivative plus a trace of native plasminogen to allow for active plasmin formation during plasminogen activation. The data indicated by *open circles, filled triangles,* and *filled circles* were obtained with TAFIa included at levels of 0.5, 1.0, and 5.0 nM, respectively. The data were obtained with both one-chain and two-chain tPA.

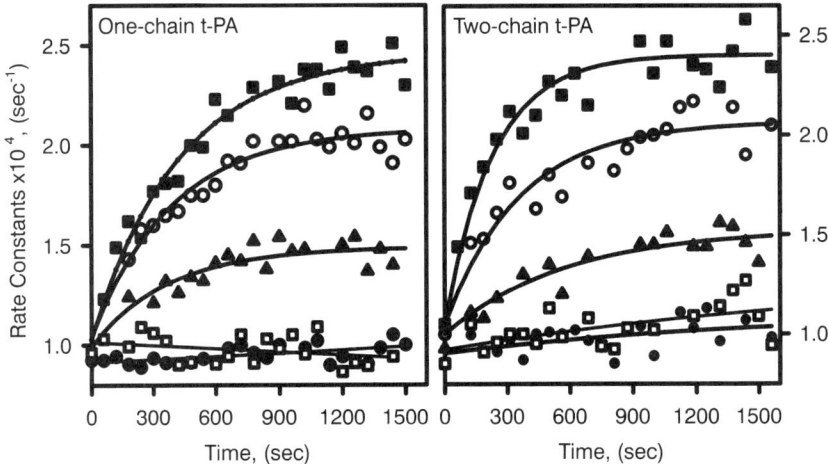

FIGURE 6. Rate constants for plasminogen activation during fibrinolysis. The rate constants were obtained by dividing point to point slopes from the data shown in FIGURE 5 by the plasminogen concentration on the intervals over which the slopes were measured. Inclusion of a trace of native plasminogen (*filled squares*) increases the rate constant 2.5-fold due to plasmin catalyzed modification of fibrin. This effect is eliminated by TAFIa. Experimental conditions were as indicated in FIGURE 5.

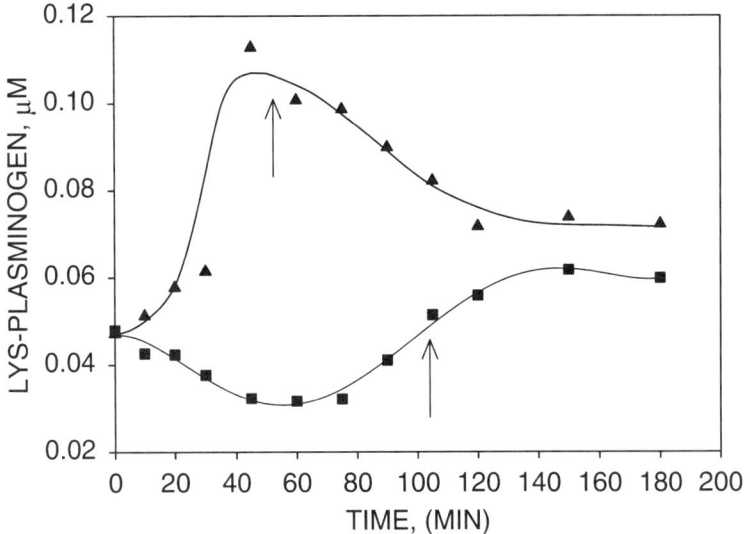

FIGURE 7. TAFIa eliminates the formation of lys-plasminogen during fibrinolysis. Lys-plasminogen formation was measured during the course of lysis of a clot in the absence (*filled diamonds*) and presence (*filled squares*) of TAFIa. The times at which the clots solubilized are marked by *arrows*.

experiments were performed with and without a trace of included plasminogen, and with and without TAFIa.[15] In the presence of native plasminogen, the consumption of the labeled plasminogen showed an accelerated cleavage time course, presumably due to the upregulation of the tPA cofactor activity by plasmin. This effect was progressively eliminated as the TAFIa concentration was increased (see FIGURE 5). The upregulation and its TAFIa-dependent elimination are exhibited quantitatively in FIGURE 6. These data are rate constants for plasminogen activation, calculated moment to moment from the point to point slopes and remaining substrate concentrations of FIGURE 5. This analysis indicates that the rate constant for plasminogen activation increases 2.5-fold as the reaction progresses. This effect is eliminated by TAFIa. Furthermore, as is shown in FIGURE 7, TAFIa prevents the formation of lys-plasminogen during the process of fibrinolysis.[15] Thus, TAFIa removes COOH terminal lysine and arginine residues formed in fibrin during fibrinolysis, and as a consequence, eliminates the positive feedback steps that normally occur due to plasmin catalyzed formation of COOH-terminal lysine and/or arginine residues during fibrinolysis.

CHARACTERIZATION OF THE PRODUCTS OF FIBRIN BREAKDOWN

In order to characterize the products solubilized during plasmin catalyzed fibrin breakdown, dilute plasmin was perfused through a clot formed in a small column in a stack of polyethylene disks.[18] The plasmin in the perfusate from the column was

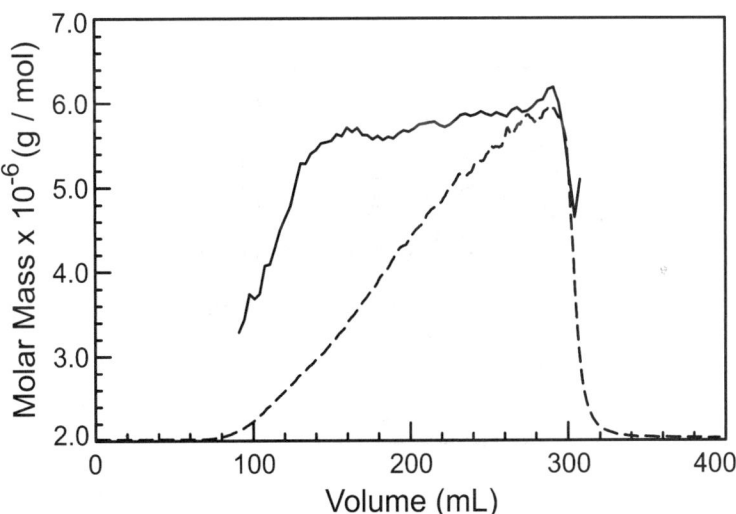

FIGURE 8. Analysis of soluble material eluted from a clot perfused with dilute plasmin. The absorbance profile is indicated by the *dashed line*. The weight-average molecular mass, as determined by on-line measurements of light scattering, are indicated by the *solid line*.

inhibited with a chloromethylketone. The effluent was monitored online with an absorbance detector and a multiangle laser light scattering detector arranged in tandem. Fractions were also collected for analysis by SDS-PAGE. The two detectors permitted measurement of the concentration and weight-average molecular mass of the soluble fibrin degradation products in the effluent. The concentration as a function of perfusate volume and the weight-average molecular weights are shown in FIGURE 8. These data indicate that, from the outset, the soluble materials have a molecular mass of about 5.5×10^6. Thus, the FDPs are very large, relative to the "building blocks" of fibrin (Fragment X, 260,000 kDa). SDS-PAGE under nonreducing conditions indicates identical compositions throughout lysis of the clot, and that a distinct pattern of covalent species is exhibited (see FIGURE 9). From the bottom of the gel these appear as triplets. The first triplet has the components D-D, D-Y, and Y-Y. The next triplet has components D-X-D, D-X-Y, and Y-X-Y. This pattern is repeated with triplets having the composition $D-X_n-D$, $D-X_n-Y$, and $Y-X_n-Y$, with n up to about 8 resolved in the gel. Under reducing conditions, the extents of cleavage of the α, β, and γ chains of the FDPs are revealed. These are shown in FIGURE 10. The cleavage products are discrete and relatively few in number. In addition, much less than quantitative cleavage is required to solubilize the clot. The solubilized materials were concentrated and subjected to gel filtration (see FIGURE 11). The results indicate that the sample was very heterogeneous with respect to molecular masses of components of the mixture. Gel filtration resolved the mixture into FDPs with molecular masses ranging from 7.0×10^6 to 2.5×10^5. Analysis by SDS-PAGE of fractions pooled according to molecular weight showed that larger covalent species were present in the higher molecular weight materials (see FIGURE 12, nonreduced) and that minor but systematic variations in the extents of cleavage of the

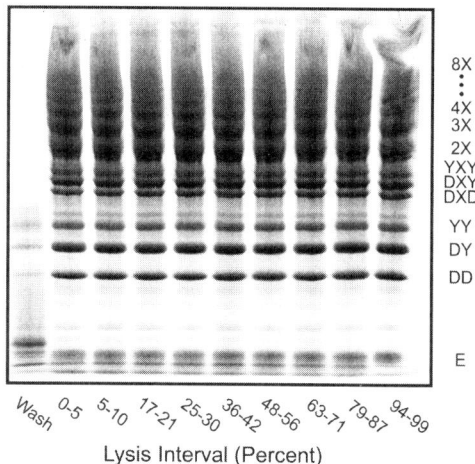

FIGURE 9. Non-reduced SDS-PAGE of materials eluted from a clot perfused with plasmin. The lysis interval indicates the extent to which the clot had been solubilized when the sample was taken. Characteristic triplets exist, the first of which consists of DD, DY, and YY.

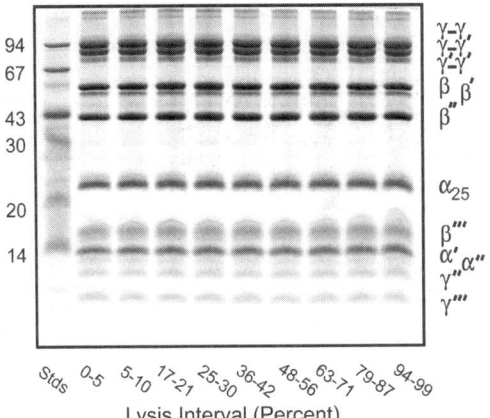

FIGURE 10. Reduced SDS-PAGE of materials eluted from a clot perfused with plasmin. The samples are as indicated in FIGURE 9. The species γ-γ has both γ-chains of the γ-dimer intact; γ-γ' and γ'-γ' have one or both chains cleaved, respectively. The designation β and $α_{25}$ represent intact β and $α_{25}$ chains, and the corresponding species designated with *primes* indicate plasmin cleavage products.

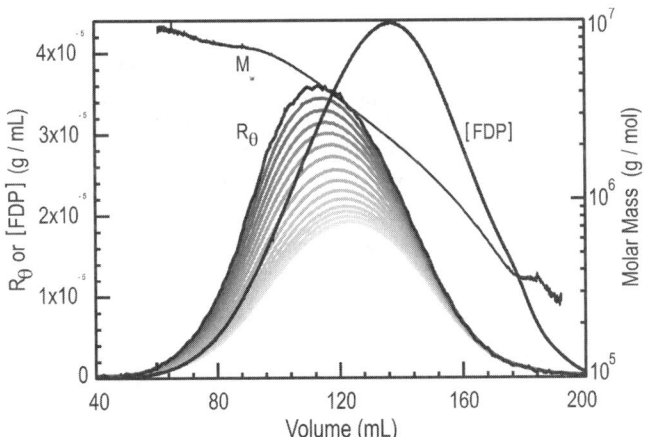

FIGURE 11. Gel filtration of material eluted from a clot perfused with plasmin. The materials indicated in FIGURE 8 were concentrated and gel-filtered. The *solid line* indicated by [FDP] is the absorbance profile. The series of curves indicated by $R_θ$ are light scattering intensities at numerous angles. The values of [FDP] and $R_θ$ are on the *left ordinate*. The *line* indicated by M_w gives the molecular weights of the material eluted from the column. Units are on the *right ordinate*. Materials were recovered with molecular weights ranging from 1×10^7 to 2×10^5 gm/mol.

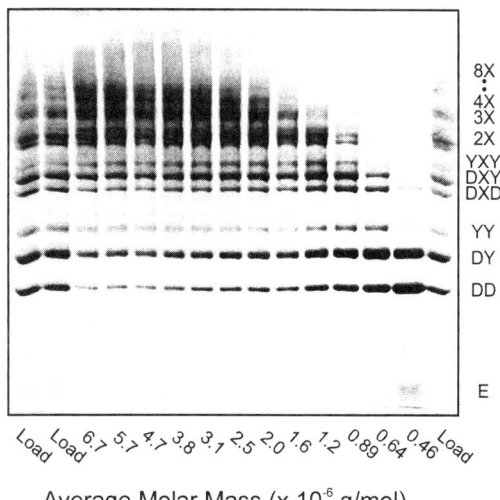

Average Molar Mass (x 10^{-6} g/mol)

FIGURE 12. Non-reduced SDS-PAGE of materials recovered after gel filtration. The materials depicted in FIGURE 11 were pooled according to molecular weight and analyzed by SDS-PAGE. The triplet patterns seen prior to size fractionation (FIG. 9) are retained, although the distribution changes somewhat from fraction to fraction.

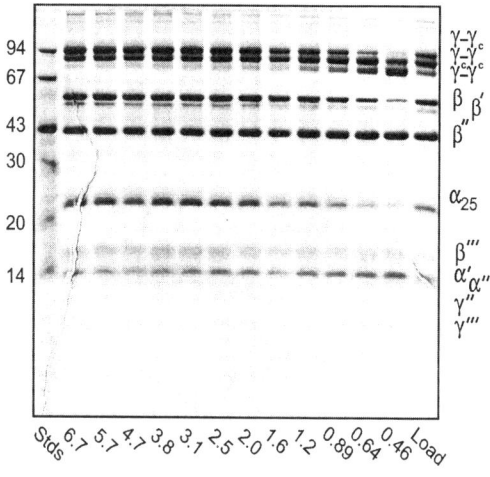

Average Molar Mass (x 10^{-6} g/mol)

FIGURE 13. Reduced SDS-PAGE of materials recovered after gel filtration. The samples depicted in FIGURE 12 were subjected to reduction in 2-mercaptoethanol. The species are as indicated in FIGURE 10.

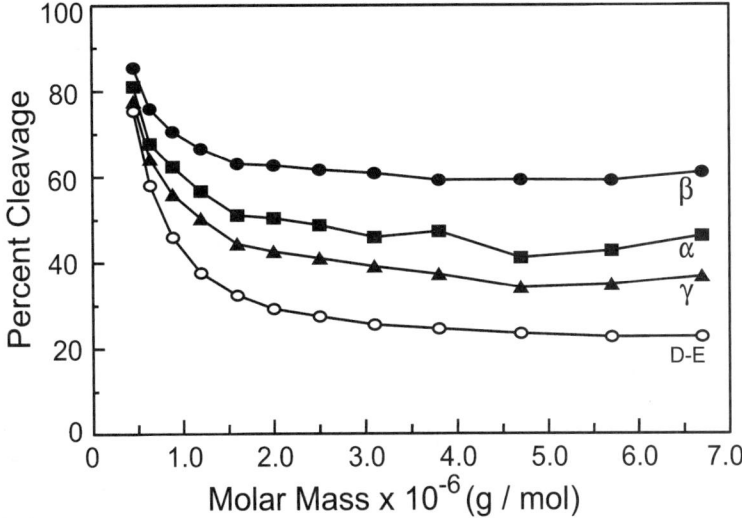

FIGURE 14. Extents of cleavage of fibrin chains. The data of FIGURES 12 and 13 were analyzed by densitometry to measure the extent to which the individual α, β, and γ chains in the solublilized materials had been cleaved. The cleavage of D to E domain connections was also measured. The extents of cleavage increased as the molecular weight of the soluble FDP decreased.

individual chains were evident (see FIGURE 13, reduced). The extent of cleavage of individual chains and relative amounts of intact D to E domain connections as a function of FDP molecular mass are shown in FIGURE 14. The smaller FDPs had more extensive cleavage of the α, β, and γ chains than the larger FDPs and more of the D to E domain connections had been severed.

COFACTOR ACTIVITIES OF FDPS

The kinetics of cleavage of the fluorescent plasminogen derivative were measured by fluorescence as the concentrations of it and six different FDPs were systematically varied.[19] The FDPs were obtained by gel filtration. Six pools were collected and concentrated. The respective molecular masses of the FDPs were 500, 1100, 1900, 2100, 3100, 4000, and 4900 kDa. The binding of plasminogen to these FDPs was measured by a decrement of fluorescence that occurred when the FDPs and the fluorescent plasminogen were mixed. All FDPs bound with equal affinity, with a dissociation constant of approximately 200 nM. Two different activators were used. One was tissue plasminogen activator (tPA) and the other was an activator from the saliva of the vampire bat (DSPAα1). tPA binds to fibrin by both its fibronectin finger-like domain and its kringle-2 domain.[20] DSPAα1, which does not have kringle-2 domains, binds only by its finger domain. At any fixed concentration of

TABLE 1. Catalytic efficiencies of FDPs before and after treatment with TAFIa ($k_{cat}/K_m \times 10^{-5}$)

FDP mass (mDa)	tPA		DSPAα1	
	−TAFIa	+TAFIa	−TAFIa	+TAFIa
0.5	5.6	0.19	0.14	0.011
1.1	16.9	0.18	1.00	0.012
1.9	13.0	0.17	2.41	0.021
3.1	16.3	0.18	3.64	0.019
4.0	17.1	—	4.09	—
4.9	6.7	—	2.33	—

FDP, regardless of its molecular mass, activation kinetics conformed to the Michaelis–Menten equation with respect to the free plasminogen concentration, yielding apparent k_{cat} and apparent K_m values. These values varied in characteristic ways with the fibrin concentration. The reaction kinetics were consistent with the template model described for intact fibrin by Horrevoets et al.,[12] and were interpreted accordingly. Data were fit globally by nonlinear regression to obtain the catalytic efficiencies (k_{cat}/K_m) of the reactions for both activators and each of the FDPs under optimal conditions (saturating cofactor). The FDPs were also treated with TAFIa and the measurements were repeated. TAFIa treatment eliminated high affinity plasminogen binding to the FDPs. The results of the analyses of kinetics are presented in TABLE 1. With tPA as the activator, the catalytic efficiency was only marginally dependent on the molecular mass of the FDP. On treatment of the FDPs with TAFIa, the catalytic efficiencies of all FDPs decreased by a factor of approximately 100. With DSPAα1 as the activator, the catalytic efficiency was strongly dependent on the molecular mass of the FDPs. This was especially pronounced for FDPs with mass less than or equal to 1,100 kDa. The catalytic efficiency of smallest FDP (500 kDa) was 29 times smaller than that of the 4,000-kDa FDP. All catalytic efficiencies were sensitive to treatment with TAFIa, decreasing by 10-fold for the smallest FDP and up to 200-fold for the 3,100-kDa FDP. Although the data are not shown here, the loss of activity for DSPAα1 with decreasing FDP molecular mass correlated with a loss of binding affinity or binding sites for DSPAα1. This is consistent with the data Stewart et al., who showed that the smallest FDP (DDE, 260 kDa) does not bind to, nor is active with, DSPAα1.[21] In order to associate finger dependent binding with structural features of the FDPs, attempts were made to correlate binding (and activity) with the relative proportions of intact D to E domain connections in each of the FDPs (see FIG.14). The activities and proportions of intact D-E connections did not correlate linearly. Activities did correlate linearly, however, with the square of the fraction of intact D to E connections. This suggests that finger dependent binding requires two adjacent intact D to E domain connections. This follows, since the probability of finding two adjacent regions intact is the square of finding just one intact. The fact that tPA binding and activity were largely retained with the smallest FDP suggests that kringle dependent binding suffices for activity with the small FDPs.

The catalytic efficiency values, in the absence of TAFIa, with large FDPs indicate that tPA is intrinsically about three-fold more efficient than DSPAα1. In addition, with tPA the catalytic efficiency with the FDPs is about 20-fold greater than that obtained with glu-plasminogen and intact fibrin.[12] In fact, the catalytic efficiency with glu-plasminogen and the FDPs is almost identical to that obtained with lys-plasminogen and intact fibrin.[12] This may reflect the fact that glu-plasminogen, although not binding to intact fibrin with high affinity, binds the FDPs as tightly as lys-plasminogen binds intact fibrin.[17]

These studies showed that soluble FDPs can be very effective cofactors in plasminogen activation. The cofactor activity is very sensitive to TAFIa. These data also suggest that two adjacent intact D to E domain connections are needed to present a finger binding site to the activator.

REFERENCE

1. DAVIE, E.W., K. FUJIKAWA & W. KISIEL. 1991. The coagulation cascade: initiation, maintenance, and regulation. Biochemistry **30:** 10363–10370.
2. MANN, K.G., E.G. BOVILL & S. KRISHNASWAMY. 1991. Surface-dependent reactions in the propagation phase of blood coagulation. Ann. N.Y. Acad. Sci. **614:** 63–75.
3. COLLEN, D. & H.R. LIJNEN. 1991. Molecular and cellular basis of fibrinolysis. *In* Hematology: Basic Principles and Practice. R. Hoffman, E. Benz, S. Shattil, B. Furie, & H. Cohen, Eds.: 1232–1242. Churchill Livingstone, New York.
4. NESHEIM, M., W. WANG, M. BOFFA, et al. 1997. Thrombin, thrombomodulin and TAFI in the molecular link between coagulation and fibrinolysis. Thromb. Hæmost. **78:** 386–391.
5. BACHMANN, F. 1987. Fibrinolysis. *In* Thrombosis and Hæmostasis. M. Verstraete, J. Vermylen, R. Lijnen & J. Arnout, Eds.: 227–265. Leuven University Press, Leuven, Belgium.
6. ESMON, C.T., A.E. JOHNSON & N.L. ESMON. 1991. Initiation of the Protein C Pathway. Ann. N.Y. Acad. Sci. **614:** 30–43.
7. SADLER, J.E., S.R. LENTZ, J.P. SHEEHAN, et al. 1993. Structure function relationships of the thrombin thrombomodulin interaction. Hæmostasis **23:** 183–193.
8. BAJZAR, L., R. MANUEL & M.E. NESHEIM. 1995. Purification and characterization of TAFI, a thrombin activatable fibrinolysis inhibitor. J. Biol. Chem. **270:** 14477–14484.
9. BAJZAR, L., J. MORSER & M. NESHEIM. 1996. TAFI, or plasma procarboxypeptidase B, couples the coagulation and fibrinolytic cascades through the thrombin-thrombomodulin complex. J. Biol. Chem. **271:** 16603–16608.
10. TAN, A.K. & D.L. EATON. 1995. Activation and characterization of procarboxypeptidase B from human plasma. Biochemistry **34:** 5811–5816.
11. WANG, W., D.F. HENDRIKS & S.S. SCHARPE. 1994. Carboxypeptidase U, a plasma carboxypeptidase with high affinity for plasminogen. J. Biol. Chem. **269:** 15937–15944.
12. HORREVOETS, A.J.G., H. PANNEKOEK & M.E. NESHEIM. 1997. A steady-state template model that describes the kinetics of fibrin-stimulated Glu1- and Lys78- plasminogen activation by native tissue type plasminogen activator and variants that lack either the finger or kringle 2 domain. J. Biol. Chem. **272:** 2183–2191.
13. HORREVOETS, A.J.G., A. SMILDE, C. DE VRIES & H. PANNEKOEK. 1994. The specific roles of finger and kringle 2 domains of tissue-type plasminogen activator during the *in vitro* fibrinolysis. J. Biol. Chem. **269:** 12639–12644.
14. SUENSON, E. & S. THORSEN. 1988. The course and prerequisites of Lys-plasminogen formation during fibrinolysis. Biochemistry **27:** 2435–2443.
15. WANG, W., M.B. BOFFA, L. BAJZAR, et al. 1998. A study of the mechanism of inhibition of fibrinolysis by activated thrombin-activable fibrinolysis inhibitor. J. Biol. Chem. **273:** 27176–27181.

16. FREDENBURGH, J.C. & M.E. NESHEIM. 1992. Lys-plasminogen is a significant intermediate in the activation of Glu-plasminogen during fibrinolysis *in vitro*. J. Biol. Chem. **267:** 26150–26156.
17. HORREVOETS, A.J.G., H. PANNEKOEK & M.E. NESHEIM. 1997. Production and characterization of recombinant human plasminogen (S741C-Fluorescein): a novel approach to study zymogen activation without generation of active protease. J. Biol. Chem. **272:** 2176–2182.
18. WALKER, J.B. & M.E. NESHEIM. 1999. The molecular weights, mass distribution, chain composition, and structure of soluble fibrin degradation products released from a fibrin clot perfused with plasmin. J. Biol. Chem. **274(8):** 5201–5212.
19. WALKER, J.B. & M.E. NESHEIM. 2001. A kinetic analysis of the tissue plasminogen activator and DSPAα1 cofactor activities of untreated and TAFIa-treated soluble fibrin degradation products of varying size. J. Biol. Chem. **276(5):** 3138–3148.
20. RENATUS, M., M.T. STUBBS, R. HUBER, *et al.* 1997. Catalytic domain structure of vampire bat plasminogen activator: a molecular paradigm for proteolysis without activation cleavage. Biochemistry **36:** 13483–13493.
21. STEWART, R.J., FREDENBURGH, J.C., LESLIE, *et al.* 2000. Identification of the mechanism responsible for the increased fibrin specificity of TNK-tissue plasminogen activator relative to tissue plasminogen activator. J. Biol. Chem. **275:** 10112–10120.

Inhibition of Fibrinolysis by Lipoprotein(a)

EDUARDO ANGLÉS-CANO,[a] AURORA DE LA PEÑA DÍAZ,[b] AND STÉPHANE LOYAU[a]

[a]*Institut National de la Santé et de la Recherche Médicale, U460, Faculté de Médecine Xavier-Bichat, France*

[b]*Instituto Nacional de Cardiología "Ignacio Chávez", Departamento de Hematología, México*

ABSTRACT: A high plasma concentration of lipoprotein Lp(a) is now considered to be a major and independent risk factor for cerebro- and cardiovascular atherothrombosis. The mechanism by which Lp(a) may favour this pathological state may be related to its particular structure, a plasminogen-like glycoprotein, apo(a), that is disulfide linked to the apo B100 of an atherogenic LDL-like particle. Apo(a) exists in several isoforms defined by a variable number of copies of plasminogen-like kringle 4 and single copies of kringle 5 and the catalytic region. At least one of the plasminogen-like kringle 4 copies present in apo(a) (kringle IV type 10) contains a lysine binding site (LBS) that is similar to that of plasminogen. This structure allows binding of these proteins to fibrin and cell membranes. Plasminogen thus bound is cleaved at Arg_{561}-Val_{562} by plasminogen activators and transformed into plasmin. This mechanism ensures fibrinolysis and pericellular proteolysis. In apo(a) a Ser-Ile substitution at the Arg-Val plasminogen activation cleavage site prevents its transformation into a plasmin-like enzyme. Because of this structural/functional homology and enzymatic difference, Lp(a) may compete with plasminogen for binding to lysine residues and impair, thereby, fibrinolysis and pericellular proteolysis. High concentrations of Lp(a) in plasma may, therefore, represent a potential source of antifibrinolytic activity. Indeed, we have recently shown that during the course of the nephrotic syndrome the amount of plasminogen bound and plasmin formed at the surface of fibrin are directly related to *in vivo* variations in the circulating concentration of Lp(a) (*Arterioscler. Thromb. Vasc. Biol.*, 2000, 20: 575–584; *Thromb. Hæmost.*, 1999, 82: 121–127). This antifibrinolytic effect is primarily defined by the size of the apo(a) polymorphs, which show heterogeneity in their fibrin-binding activity—only small size isoforms display high affinity binding to fibrin (*Biochemistry*, 1995, 34: 13353–13358). Thus, in heterozygous subjects the amount of Lp(a) or plasminogen bound to fibrin is a function of the affinity of each of the apo(a) isoforms and of their concentration relative to each other and to plasminogen. The real risk factor is, therefore, the Lp(a) subpopulation with high affinity for fibrin. According to this concept, some Lp(a) phenotypes may not be related to atherothrombosis and, therefore, high Lp(a) in some individuals might not represent a risk factor for cardiovascular disease. In agreement with these data, it has been recently reported that Lp(a) particles containing low molecular mass apo(a) emerged as one of the leading risk conditions in advanced stenotic atherosclerosis (*Circulation*, 1999, 100: 1154–1160). The predictive value of high Lp(a) as a risk factor, therefore,

Address for correspondence: E. Anglés-Cano, M.D., Sc.D., INSERM U460, Remodelage CardioVasculaire, 16 rue Henri Huchard - BP 416, F-75870-Cdx, Paris, France. Voice: 33-1 44 85 61 50; fax: 33-1 44 85 61 56.
angles@kb.infobiogen.fr

depends on the relative concentration of Lp(a) particles containing small apo(a) isoforms with the highest affinity for fibrin. Within this context, the development of agents able to selectively neutralise the antifibrinolytic activity of Lp(a), offers new perspectives in the prevention and treatment of the cardiovascular risk associated with high concentrations of thrombogenic Lp(a).

KEYWORDS: Fibrin surface; Fibrinogen; Plasminogen; Plasmin; Kringle domains; Lysine binding site; Tissue-plasminogen activator; Urokinase; Atherosclerosis; Thrombosis; Atherothrombosis; Monocytes; Macrophages; Fibrinolysis; Plasminogen activation; Plasminogen activator inhibitor.

INTRODUCTION

The presence of the lipoprotein Lp(a) in plasma was first identified by Berg in 1963 as an LDL-like particle.[1] Lp(a) composition is indeed similar to that of LDL in terms of cholesterol, triglycerides, phospholipids, and apo B100. It was not until 1987 when Eaton *et al.*[2] and Mc Lean *et al.*[3] identified apo(a), the unique and distinctive component of Lp(a), as a member of the plasminogen gene family,[4] both by amino acid sequence analysis and cDNA clonning. The apo(a) glycoprotein indeed showed a strong structural homology with plasminogen, the plasmin precursor. Apo(a) is disulfide linked to the apo B100 of the LDL-like particle. These findings stimulated the interest of different research groups that found in Lp(a) a common point between atherosclerosis and thrombosis.[5,6] Different epidemiological studies were performed from this perspective and identified high Lp(a) plasma concentration as a major and independent cardiovascular risk factor.[7] Although some studies did not find a relationship between Lp(a) and coronary heart disease,[8,9] most prospective studies (reviewed in Ref. 10) confirmed the initial results and extended its application to coronary restenosis, post-angioplastic occlusion, as well as cerebrovascular accidents, and premature development of atherosclerosis. The mechanism by which Lp(a) favors the atherogenic and/or thrombogenic processes is not clearly understood. However, Lp(a) has been found associated with fibrin deposits in atherosclerotic plaques.[11] Moreover, recent data indicate that the large structural heterogeneity of the apo(a) molecule is conducive to a functional heterogeneity as a plasminogen competitive inhibitor for fibrin binding that lowers the formation of plasmin and thereby decreases fibrinolysis.[12,13] In order to understand the molecular basis of this mechanism we analyze the structural aspects of plasminogen and apo(a) interactions with fibrin and consider recent evidence indicating that Lp(a) behaves as an inhibitor of fibrinolysis.

STRUCTURAL BASIS FOR THE ANTIFIBRINOLYTIC ACTIVITY OF LP(A)

Plasminogen and Apolipoprotein(a): General Structure

Apo(a) and plasminogen are constituted by structural domains called kringles and a serine-proteinase region (see FIGURE 1). The kringle structure was first described in prothrombin and is found in several copies in proteins that evolved from a common ancestral gene: plasminogen, apo(a) and hepatocyte growth factors.[4] Kringles

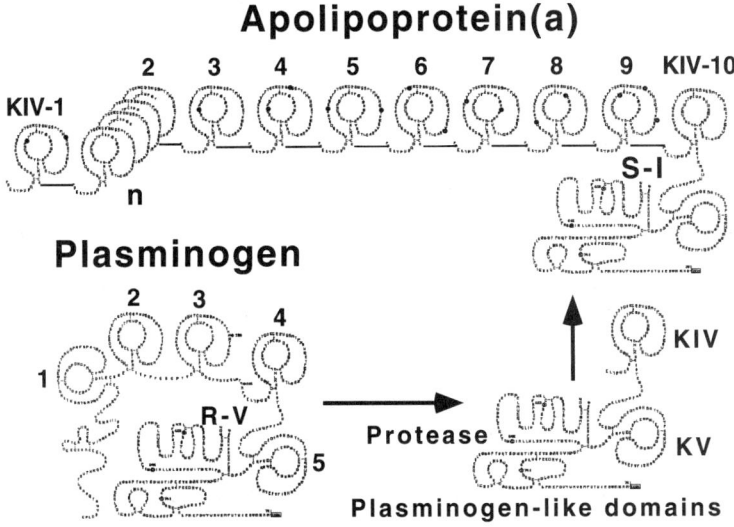

FIGURE 1. Structural homologies between apolipoprotein(a) and plasminogen (adapted from Petersen et al., J. Biol. Chem., 1990, **265**: 6104–6111). Plasminogen consist of five kringle domains and a serine-protease region. Apo(a) is constituted by single copies of plasminogen kringle 5 (KV) and of the protease region and by multiple copies of plasminogen kringle 4 (KIV). These KIV copies are not identical and have been classified[18] in ten different types (KIV-1 to KIV-10). A variable number of KIV-2 determines the existence of several apo(a) isoforms. All other KIV types are present in single copies. R-V indicates the Arg561-Val562 plasminogen cleavage site, which in apo(a) has been substituted by Ser-Ile, S-I (see FIG. 2).

are sequences of 80 to 90 amino acid residues organized in a triple-loop structure, stabilized by three disulfide bridge.[14] The kringle domains of plasminogen, designated 1 to 5, differ from each other. Kringles 1 and 4 contains lysine binding subsites that show high affinity for lysine residues in fibrin.[15,16] The specific interactions between lysine residues in fibrin or cell membrane proteins and the lysine binding subsites of plasminogen allow plasminogen binding and activation.[17]

Apo(a) contains a variable number of kringle domains that share 61–75% homology with kringle 4 of plasminogen.[2,3] The kringle 4-like repeats of apo(a) are followed by a single copy of plasminogen kringle 5 and a protease domain that shares 94% homology with the corresponding domain of plasminogen. Kringles are interconnected by regions of 26–36 serine-, proline-, and/or threonine-rich amino acids with six potential O-glycosilation sites. Not all plasminogen-like kringle 4 of apo(a) are alike; they are classified in 10 different types,[18] hereafter referred as KIV-1 to KIV-10 (see FIG. 1). Each of these kringles is present only once, except KIV-2, which appears in a variable number, originating structural heterogeneity and different size isoforms, that account for molecular weights between about 280 and 800 kDa including 30% weight due to glycosilation.[19] KIV-9 possesses an additional cysteine residue that ensures the covalent binding between apo(a) and Cys3734 of apo B-100 and, thereby, the formation of the Lp(a) particle.[20] By sequence comparison and molecular modeling[21] it has been shown that KIV-10 contains a lysine binding site (LBS) similar to that of plasminogen kringle 4, and that slightly modified LBS are present in KIV-5 to KIV-8. These kringle copies may confer to apo(a) binding capabilities similar to those of plasminogen.

Structure of the Lysine Binding Site

The structure of the lysine binding site, an elongated hydrophobic depression lined with aromatic residues that separate by 6.8 Å a cationic group from an anionic group positioned at opposite ends of the trough, has been well defined.[22] Its structural characteristics generate a relatively rigid geometry that allows selective access and binding of C-terminal lysine residues in fibrin, as well as aliphatic or aromatic ligands of ω-amino carboxylic acid type. Based on information derived from biochemical,[23] mutagenesis,[24] crystallographic,[25,26] and spectroscopic[27] studies, an appropriate set of amino acids that are key components for lysine binding has been identified (see TABLE 1). Thus, the LBS of plasminogen K1 consists of three aromatic residues, Tyr64/Trp62/Tyr72, lining the hydrophobic trough of the LBS and the double charged anionic, Asp55/Asp57, and cationic, Arg35/Arg71 centres located at opposite ends of the hydrophobic trough. In apo(a) only KIV-10 contains an LBS with similar characteristics that is identical to that of plasminogen kringle 4 except for a conservative Arg→Lys35 substitution also present in plasminogen kringle 1.[28]

TABLE 1. Key components of the lysine-binding site

Kringle	Anionic pole		Hydrophobic aromatic residues			Cationic pole	
Pg K1	Asp 55	Asp 57	Tyr 64	Trp 62	Tyr 72	Arg 35	Arg 71
Pg K4	Asp 55	Asp 57	Phe 64	Trp 62	Trp 72	Lys 35	Arg 71
Pg K5	Asp55	Asp 57	Tyr 64	Trp 62	Trp 72	Ile 35	Leu 71
KV	Asp 55	Asp 57	Tyr 64	Trp 62	Trp 72	Ser 35	Leu 71
KIV-10	Asp 55	Asp 57	Phe 64	Trp 62	Trp 72	Arg 35	Arg 71
KIV-9	Asp 55	Gly 57	Tyr 64	Trp 62	Trp 72	Arg 35	Arg 71
KIV-8	Asp 55	Glu 57	Tyr 64	Trp 62	Trp 72	Arg 35	Arg 71
KIV-7	Asp 55	Glu 57	Tyr 64	Trp 62	Trp 72	Arg 35	Arg 71
KIV-6	Asp 55	Glu 57	Tyr 64	Trp 62	Trp 72	Arg 35	Arg 71
KIV-5	Asp 55	Glu 57	Tyr 64	Trp 62	Trp 72	Arg 35	Arg 71
KIV-4	Asp 55	Val 57	Tyr 64	Trp 62	Trp 72	Arg 35	Arg 71
KIV-3	Asp 55	Val 57	Tyr 64	Tyr 62	Trp 72	Arg 35	Arg 71
KIV-2	Asp 55	Val 57	Tyr 64	Tyr 62	Trp 72	Arg 35	Arg 71
KIV-1	Asp 55	Val 57	Tyr 64	Tyr 62	Trp 72	Arg 35	Arg 71

NOTE: Pg. Plasminogen kringles are designated K1, K4, and K5. KV and KIV-1 to KIV-10 correspond to apo(a) kringles.

Both plasminogen K4 and apo(a) KIV-10 however, differ, from plasminogen kringle 1 by two important substitutions, Phe64 and Trp72, instead of Tyr64 and Tyr72, that are, therefore, specific structure requirements for the LBS of plasminogen K4 and apo(a) KIV-10. The critical role of this residue in lysine binding has been well documented; its hydrophobic indole is unusually exposed in the LBS of K4 and KIV-10 and serves as an obvious marker of the binding site.[25]

The Serine–Proteinase Domain in Plasminogen and Apo(a)

Serine-proteinases are synthesized and secreted as single-chain zymogens that are activated by proteolytic processing through specific interactions with active enzymes.[29] These include binding forces between subsites of the active site of the enzyme (S_1-S_n and $S_1'-S_n'$) and amino acid residues (P_1-P_n and $P_1'-P_n'$) of the cleavage site on the substrate[30] (see FIGURE 2). For instance, cleavage of the P_1-P_1', Arg_{561}-Val_{562}, peptide scissile bond in plasminogen by its activators yields two-chain active plasmin.[31] This cleavage site is located in a small disulfide-bounded loop that must restrict the conformation around this bond. It has been shown, indeed, that this disulfide bond is of importance for the specificity of plasminogen activation.[32] The sequence corresponding to this loop is comprised in the P_2-P_{17} / $P_2'-P_{23}'$ sequence of plasminogen that is strictly preserved in apo(a); in contrast, the P_1-P_1' peptide bond in apo(a) consist of Ser-Ile instead of Arg-Val,[2] a substitution that may impair recognition by activators (FIG. 2). Indeed, the introduction of these substitutions in plasminogen prevent its cleavage by activators.[33]

We have recently considered the possibility that introduction of both Arg and Val residues at, respectively, $P-P_1'$ by site directed mutagenesis together with the use of specific fibrinolytic tests may ensure potential recognition of this activation cleavage site by activators and generation of plasmin-like activity.[34] To test this hypothesis, r-apo(a) mutant plasmids encoding this sequence were expressed in 293 cells and HepG2 cells; culture supernatants containing wildtype r-apo(a), Arg-Val r-apo(a), and the corresponding r-Lp(a)-like particles were thus obtained. These products were treated with u-PA or t-PA in the presence of cofactors known to stimulate

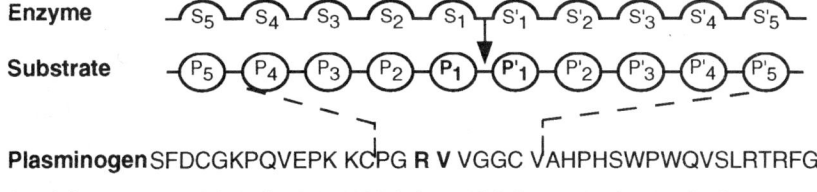

FIGURE 2. Scheme of the plasminogen activation cleavage site and the corresponding region in apo(a). Amino acid sequences of the cleavage site in plasminogen (residues 544–585, according to Petersen *et al. J. Biol. Chem.* 1990, **265**, 6104–6111) compared to the corresponding sequence in apo(a) (residues 4291–4332, according to Ref. 3). Plasminogen activators cleave plasminogen at the P_1-P_1' (Arg-Val) scissile peptide bond (indicated by the *arrow*). In apo(a), amino acid residues P_2-P_{17} and $P_2'-P_{23}'$ are identical to those of plasminogen, however the cleavage site is substituted by Ser-Ile.

plasminogen activation by t-PA (fibrin and CNBr-fragments of fibrinogen) or u-PA (6-Ahx, a lysine analogue). However, neither cleavage of apo(a) nor plasmin-like activity could be demonstrated on treatment of Arg-Val r-apo(a)/r-Lp(a) mutants with plasminogen activators, either in the presence or in the absence of cofactors known to induce changes in plasminogen conformation or to stimulate plasminogen activation.

These results strongly suggest that to promote efficient cleavage of plasminogen, t-PA recognises complex structural elements on the surface of native plasminogen that are distant from those at P4-P2'.[35] The exact nature of these secondary interactions remains unknown. Because r-apo(a) contains multiple copies of kringle 4 and lacks copies of kringles 1 to 3 of plasminogen, it may not possess the potential for such enzyme–substrate interactions, that are independent of the binding cleft around P1; r-apo(a) Arg-Val may fail, thereby, to induce a better fit of the cycle peptide loop containing the Arg-Val bond with the activators. The presence of multiple kringle repeats in apo(a) does not substitute the specific conformation present in plasminogen that is required for the selective and specific recognition by activators. It is, therefore, apparent that besides the absence of the Arg-Val site, apo(a) has undergone further evolutionary changes which prevent its activation towards a plasmin-like protease. Because of this important enzymatic difference, the ability of selected kringle-4 copies of apo(a) to compete with plasminogen for binding to fibrin and cell surfaces results in decreased fibrinolysis and impaired pericellular proteolysis.[36,37]

INHIBITION OF PLASMINOGEN BINDING TO FIBRIN, THE MAJOR MECHANISM OF ACTION OF LP(A)

Physiological fibrinolysis involves heterogeneously catalyzed reactions that proceed at the fibrin/plasma interface, where fibrin provides a surface to which tissue plasminogen activator (t-PA) and plasminogen bind specifically.[15,17] Molecular assembly of these proteins results in a ternary complex that efficiently generates plasmin on the surface of fibrin and thereby triggers the dissolution of a clot.[38]

Initial limited degradation of the surface of fibrin by plasmin unveils carboxy-terminal lysine residues and increases the local concentration of plasminogen, a process that amplifies and accelerates the degradation of fibrin.[39] This mechanism is mediated by specific interactions of lysine residues in fibrin with the kringle domains of plasminogen.[40] In a plasma milieu, the progression of this process is markedly influenced by α_2-antiplasmin, the specific plasmin inhibitor, and by active carboxy-peptidase B, an exopeptidase that cleaves Arg and Lys residues at C-terminal position.[41] The number of carboxy-terminal lysine residues is thus limited and, thereby, the amount of bound plasminogen.[17] On the other hand, the blockade of such residues by isolated plasminogen kringle 4 has been shown to interfere competitively with clot lysis by a mechanism involving binding to lysine–fibrin residues.[42] The presence in apo(a) of kringle modules structurally related to those of plasminogen endows Lp(a) with the ability to compete with plasminogen for binding to fibrin. The effect of Lp(a) on plasminogen binding to fibrin and cell surfaces has been studied by several groups.[37,43–45] A number of experimental *in vitro* studies resulted in convincing evidence that Lp(a) binds to the fibrin surface and cell membranes and, thereby, competes with plasminogen so as to inhibit its activation.[46,47] We have shown that, indeed, both free recombinant apo(a) (see FIGURE 3) and apo(a) in Lp(a) particles (see

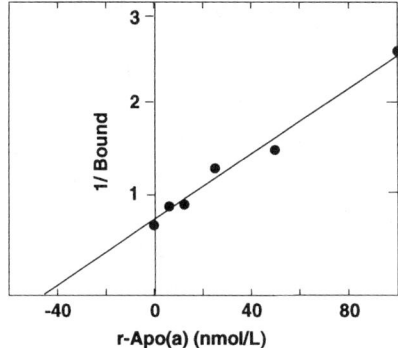

FIGURE 3. Determination of the inhibition constant of apo(a) for the binding of plasminogen to fibrin (as indicated in Ref. 48). Plasminogen binding experiments were performed in the absence and in the presence of various concentrations of recombinant apo(a) (18 kringles). For each concentration of apo(a) the binding parameters (B_{max}, the maximum bound, and K_d, the dissociation constant) were determined according to the Langmuir equation. The graph depicts the reciprocal of B_{max} as a function of the concentration of apo(a). The inhibition constant (K_i = 44 nmol/L) was calculated from the *intercept* of the straight line with the abcissa.

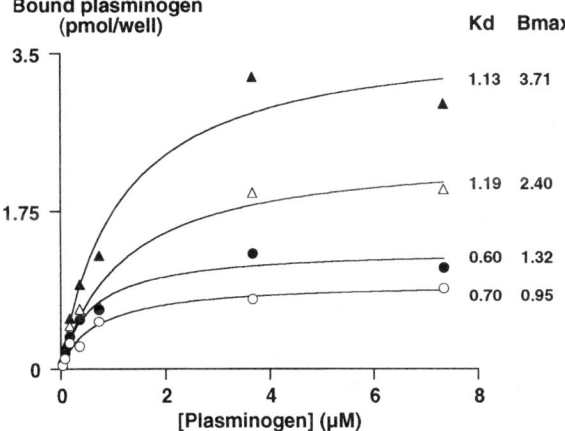

FIGURE 4. Inhibition of the binding of plasminogen to fibrin by purified 18 kringles Lp(a) (as indicated in Ref. 12). Increasing concentrations of plasminogen (0 to 8 μmol/L) containing a trace amount of radiolabeled plasminogen (5 nmol/L) were incubated with fibrin in the presence of 0 (▲), 50 (△), 100 (●) and 200 (○) nmol/L Lp(a). Data fitted to the Langmuir equation allowed calculation of B_{max} (maximum amount of plasminogen bound) and K_d (the dissociation constant of the interaction) for each Lp(a) concentration. By plotting the reciprocal of the B_{max} versus the concentration of added Lp(a) an inhibition constant K_i = 32 nmol/L was determined at the *intercept* of the linear regression curve with the abcissa (not shown).

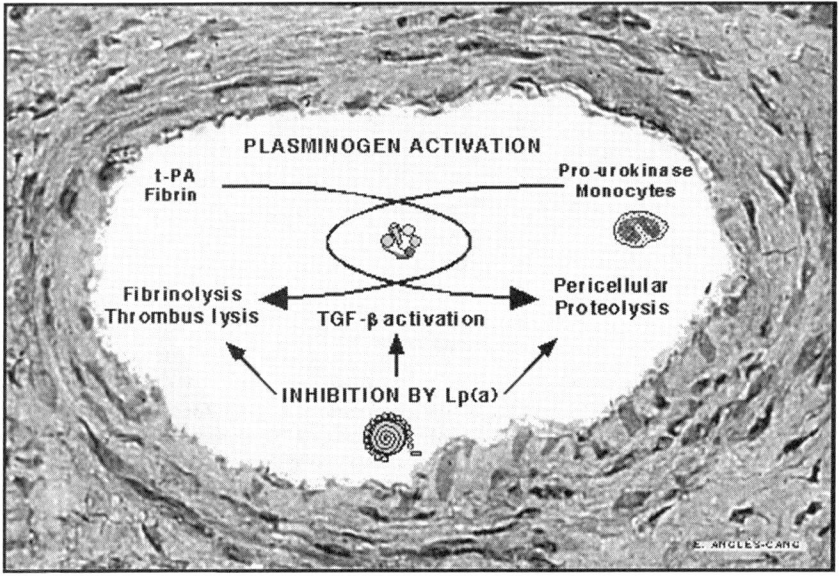

FIGURE 5. Competitive inhibition of the binding and activation of plasminogen by Lp(a) on fibrin and cell surfaces. This competitive mechanism results in inhibition of fibrinolysis, pericellular proteolysis and TGF-β activation. The different components of the plasminogen activation system and Lp(a) are represented in the intravascular space of a vessel section.

FIGURE 4) efficiently inhibit (K_i = 44 nmol/L and 30 nmol/L, respectively) the binding of plasminogen to fibrin[12,48] (see FIGURE 5). Such unique behaviour was attributed to the fibrin-binding properties conferred by the kringle 4 repeats of apo(a).[48] Plasminogen and the different Lp(a) isoforms also compete for lysine residues on the surface receptors of endothelial cells,[37] platelets,[49] and mononuclear cells.[45] Thus, Lp(a) interferes with the evolution of fibrinolysis at the surface of fibrin, endothelial cells, monocytes, and platelets through binding of apo(a), an eternal zymogen that decreases the local concentration of plasminogen and cannot be transformed into an active enzyme. Most of the effects of Lp(a)—persistence of fibrin deposits, accumulation of cholesterol and proliferation of smooth muscle cells in the intima, are related to a decrease in plasmin activity (FIG. 5). Hypofibrinolysis and cholesterol accumulation are a direct consequence of the presence of Lp(a) at the surface of fibrin and cell membranes: apo(a) inhibits plasmin formation and the LDL components favors cholesterol accumulation.

OTHER ANTIFIBRINOLYTIC AND ANTIPROTEOLYTIC EFFECTS OF LP(A)

Lp(a) may stimulate the expression of PAI-1 and inhibit the synthesis of t-PA by endothelial cells in culture.[50,51] Thus, inhibition of t-PA by PAI-1 and low t-PA antigen levels may enhance Lp(a)-dependent hypofibrinolysis by decreasing the amount of t-PA available for the activation of plasminogen. It has also been shown that Lp(a) inhibits the activation of TGF-β[52] and that, in transgenic mice expressing human apo(a), the generation of plasmin and thereby the activation of TFG-β are decreased.[53] Insufficient activation of TGF-β may result in migration and proliferation of smooth muscle cells into the intima, an important mechanism in atheroma plaque formation.

ANTIFIBRINOLYTIC ACTIVITY OF APOLIPOPROTEIN(A) *IN VIVO*

To date there is no definitive evidence that Lp(a) interferes with plasminogen binding and activation *in vivo*. Some evidence has been obtained from a few experimental studies. Occlusive arterial thrombosis with incorporation of Lp(a) into damage arterial segments was observed in cynomolgous monkeys with high Lp(a) plasma levels.[54] Biemond *et al.*[55] studied the effect of a recombinant form of apo(a) on endogenous and t-PA mediated lysis in an *in vivo* model of experimental thrombosis; endogenous thrombolysis but not t-PA induced lysis was significantly reduced in the presence of apo(a). In contrast, transgenic mice that express human apo(a) have been shown to exhibit reduced t-PA induced lysis of pulmonary emboli produced by injection of human platelet-rich plasma clots.[56] However, the most conclusive report to this respect concerns the elimination of the pathogenic activity of apo(a) by disruption of the KIV-10 lysine binding site using genetic engineering in transgenic mice.[57]

In humans Lp(a) plasma concentration, which depends on its hepatic synthesis,[58] varies from one individual to another within an approximate range of less than 10 mg/dl to more than 100 mg/dl; it is independent from other factors, such as diet, cholesterol, obesity, and smoking and is maintained within small variations during the life span. However, during the course of the nephrotic syndrome, large individual variations in Lp(a) plasma concentration have been observed.[59] Data have been obtained in this human model system,[60,61] indicating that Lp(a) interferes with plasminogen binding and activation under conditions fashioned *in vivo*. We have measured the effect of individual variations in the concentration of Lp(a) on plasminogen activation at the surface of fibrin and have determined the existence of an inverse relationship between plasmin formation and quantity of Lp(a) that is bound (see FIGURE 6). A direct relationship between binding of Lp(a) to fibrin and the decrease in plasmin formation due to modifications *in vivo* of the Lp(a) plasma concentration has thus been established.[60,61] These studies strongly suggest that apo(a) may decrease fibrinolysis *in vivo*.

LP(A) INHIBITS FIBRINOLYSIS DEPENDING ON APO(A) ISOFORMS

In general, an individual inherits, in a codominant autosomic fashion, two apo(a) isoforms; only a reduced number of individuals are homozygous for the apo(a) trait.[62] The size of the Apo(a) isoforms accounts for an inverse correlation with Lp(a) plasma concentration.[63] In general, the smaller the hypervariable region of the apo(a) allele and, therefore, the size of the apo(a) isoform, the higher is the plasma concentration of Lp(a). A difference in the distribution of apo(a) isoforms between patients with

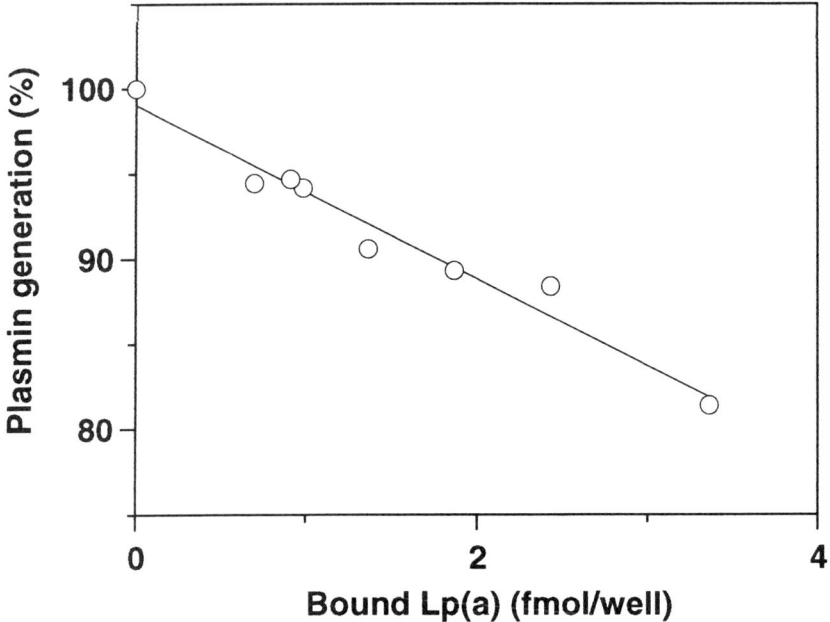

FIGURE 6. Inhibition of plasmin generation by Lp(a). Varying concentrations of Lp(a) (18 kringles) were incubated with 200 nmol/L of Glu-plasminogen (final concentration) on fibrin bound-t-PA. Plasminogen activation was followed at 37°C by measuring the amidolytic activity of plasmin with a synthetic substrate (CBS 00-65, Stago). At the end of the activation, the plate was washed and bound Lp(a) was detected with a radiolabeled antibody directed against apo(a); bound radioactivity was transformed in fmol of Lp(a). Plasminogen activation expressed as a percentage relative to the maximum of plasmin generated in the absence of Lp(a), is plotted against the amount of Lp(a) bound. Data represent the mean of three determinations. A linear correlation ($r = 0.985$) was found between the inhibition of plasmin generation and the increase in the amount of Lp(a) bound.

atherosclerosis and a control population has been recently reported.[64–66] Low molecular mass isoforms were found more frequently in subjects with high Lp(a) concentration and a history of myocardial infarction or intermitent claudication. Thus, short apo(a) alleles may favor atherogenesis by increasing the concentration of Lp(a). However, the difference in allele distribution between patients at risk and controls is not always observed, and the inverse relationship between apo(a) size and Lp(a) concentration is not linear, thus suggesting the existing of a functional diversity among apo(a) isoforms. Indeed, some plasmas with a high Lp(a) concentration may fail to induce a decrease in fibrinolysis.[44] The question remains as to whether short apo(a) isoforms are *per se* an atherothrombotic risk factor, especially when associated with high Lp(a) plasma concentration. Recent experimental and clinical evidence provide arguments that favor this hypothesis.

Since the atherogenic potential of Lp(a) is related to the lysine binding properties conferred by the kringle 4 repeats of apo(a), we have explored the possibility that Lp(a) phenotypes may have different functional properties with regard to their affinity for fibrin. We demonstrated that Lp(a) particles purified from homozygous subjects containing a single, distinct isoform of apo(a) display different antifibrinolytic effect and that this heterogeneity is related to apo(a) size polymorphism. Isoforms of low molecular mass showed the highest affinity for fibrin (see FIGURE 7). This finding suggests that the variable number of kringles in apo(a) influence its ability to bind to fibrin.[67] In heterozygous subjects, representing more than 94% of the population,[62] differences in affinity for fibrin of each of the Lp(a) particles have beeen demonstrated.[13] Under these conditions, the net antifibrinolytic effect of Lp(a) is a function of the affinity of each of the Lp(a) isoforms and of their concentration relative to each other and to plasminogen.[12] This mechanism is in agreement with the emergence of low molecular weight apo(a) phenotypes as a leading risk condition of advanced stenotic atherosclerosis.[68]

Our findings of an inverse relationship between isoform size and affinity for fibrin support the hypothesis that *the real risk factor is the Lp(a) population with high affinity for fibrin.* According to this concept some Lp(a) phenotypes might fail to be related to atherogenesis and, therefore, some individuals with high Lp(a) would not

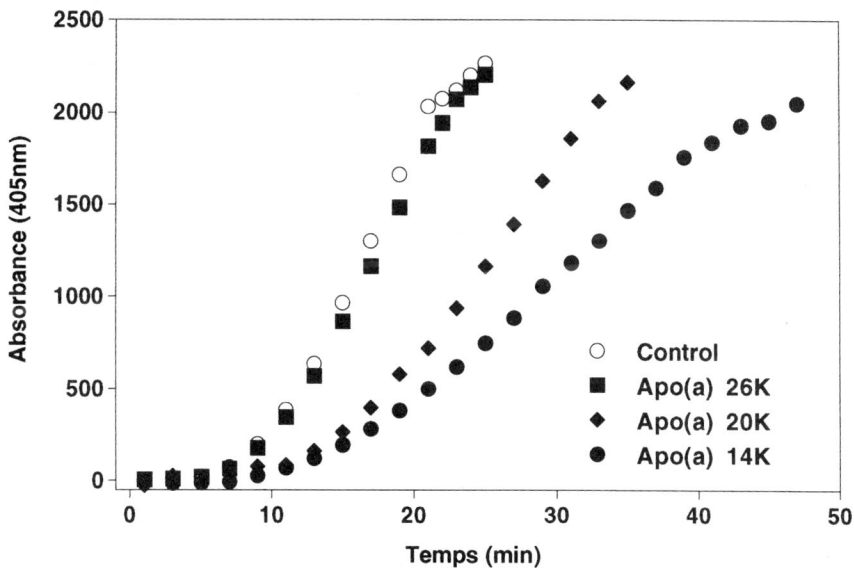

FIGURE 7. Inhibition of the formation of plasmin on fibrin by different Lp(a) isoforms. Plasminogen was activated by fibrin-bound t-PA in the presence of 0 or 300 nmol/L of disctinct Lp(a) particles containing either 26, 20, or 14 kringles. Chromogenic substrate detection of the amidolytic activity of plasmin is expressed by the change in absorbance at 405 nm as a function of time. The amount of plasmin formed was lower in the presence of apo(a) isoforms containing 14 and 20 kringles as indicated by lower initial velocities (ΔA_{405}/min).

be at risk of coronary heart disease. Taken altogether, these results suggest that, besides the quantitative factor, an important qualitative effect must be considered in the atherothrombogenic role of Lp(a). This new concept about the functional heterogeneity of Lp(a) adds a new dimension to the evaluation of the predictive value of Lp(a) as a risk factor for cardiovascular disease.

CONCLUSION

The similarity between plasminogen and apolipoprotein (a) allows different apo(a) isoforms to compete with plasminogen for fibrin affinity sites. The affinity of each isoform depends on its size and its plasma concentration.[19,47,67] Indeed, we have clearly shown that the influence of Lp(a) on fibrinolysis depends on the high affinity of small apo(a) isoforms for fibrin and on their concentration relative to plasminogen. Lp(a) particles containing low molecular weight apo(a) isoforms have the most profound influence on fibrinolysis by acting as a prominent competitive antagonist of plasminogen. As a result, the plasmin formed at the surface of fibrin may also vary with modifications of the concentration of Lp(a) *in vivo*.[60] An antifibrinolytic mechanism may therefore explain the pathophysiological effects of Lp(a) by relating both high concentrations of Lp(a) and small apo(a) isoforms. This mechanism is in agreement with the emergence of low molecular weight apo(a) phenotypes as a leading risk condition of advanced stenotic atherosclerosis.[68] These data indicate that, to evaluate the potential risk associated with Lp(a) levels in heterozygous subjects, it is the relative concentration of the Lp(a) subpopulation having an apo(a) isoform with the highest affinity for fibrin and not the absolute Lp(a) concentration that should be considered

ACKNOWLEDGMENTS

This research was supported by INSERM (Project 4N001B) and by Ministère de la Recherche (Grant ACC-SV9). We gratefully acknowledge the editorial assistance of Jérôme Lechevalier.

REFERENCES

1. BERG, K. 1963. A new serum type system in man—the Lp system. Acta Pathol. Microbiol. Scand. **59:** 369–382.
2. EATON, D.L., *et al.* 1987. Partial amino acid sequence of apolipoprotein(a) shows that is homologous to plasminogen. Proc. Natl. Acad. Sci. U.S.A. **84:** 3224–3228.
3. MCLEAN, J.W., *et al.* 1987. cDNA sequence of human apolipoprotein(a) is homologous to plasminogen. Nature **330:** 132–137.
4. ICHINOSE, A. 1992. Multiple members of the plasminogen-apolipoprotein(a) gene family associated with thrombosis. Biochemistry **31:** 3113–3118.
5. DAHLEN, G.H. 1994. Lp(a) Lipoprotein in cardiovascular disease. Atherosclerosis **108:** 111–126.
6. UTERMANN, G. 1989. The mysteries of lipoprotein(a). Science **246:** 904–910.

7. ROSENGREN, A., *et al.* 1990. Lipoprotein (a) and coronary heart disease: a prospective case-control study in a general population sample of middle aged men. Br. Med. J. **301:** 1248–1251.
8. JAUHIAINEN, M., *et al.* 1991. Lipoprotein(a) and coronary heart disease risk. A nested case-control study of the Helsinki Heart Study participants. Atherosclerosis **89:** 59–67.
9. RIDKER, P.M., C.H. HENNEKENS & M.J. STAMPFER. 1993. A prospective study of lipoprotein (a) and the risk of myocardial infarction. JAMA **270:** 2195–2199.
10. DJUROVIC, S., *et al.* 1997. Epidemiology of Lp(a) lipoprotein: its role in atherosclerotic/thrombotic disease. Clin. Genet. **52:** 281–292.
11. SMITH, E.B., *et al.* 1991. Does lipoprotein(a) (Lp(a)) compete with plasminogen in human atherosclerotic lesions and thrombi? Atherosclerosis **89:** 127–136.
12. HERVIO, L., *et al.* 1995. Multiple binding with identical linkage: a mechanism that explains the effect of lipoprotein(a) on fibrinolysis. Biochemistry **34:** 13353–13358.
13. HERVIO, L., *et al.* 1996. The antifibrinolytic effect of lipoprotein(a) in heterozygous subjects is modulated by the relative concentration of each of the apolipoprotein(a) isoforms and their affinity for fibrin. Eur. J. Clin. Invest. **26:** 411–417.
14. PATTHY, L., *et al.* 1984. Kringles: modules specialized for protein binding. FEBS Lett. **171:** 131–136.
15. VALI, Z., *et al.* 1984. The fibrin-binding site of human plasminogen. J. Biol. Chem. **25:** 13690–13694.
16. WU, T.P., *et al.* 1991. The refined structure of the e-aminocaproic acid complex of human plasminogen kringle 4. Biochemistry **30:** 10589–10594.
17. FLEURY, V. & E. ANGLES-CANO. 1991. Characterization of the binding of plasminogen to fibrin surfaces: the role of carboxy-terminal lysines. Biochemistry **30:** 7630–7638.
18. GUEVARA, J., *et al.* 1992. A structural assessment of the apo(a) protein of human lipoprotein(a). Proteins **12:** 188–199.
19. FLESS, G.M., M.E. ZUMMALLEN & S. SCANU. 1986. Physicochemical properties of apolipoprotein(a) and Lp(a) derived from the dissociation of human plasma Lp(a). J. Biol. Chem. **261:** 8712–8718.
20. BRUNNER, C., *et al.* 1993. Cys^{4057} of apolipoprotein(a) is essential for lipoprotein(a) assembly. Proc. Natl. Acad. Sci. U.S.A. **90:** 11643–11647.
21. GUEVARA, J., *et al.* 1993. Comparison of ligand-binding sites of modeled Apo(a) kringle-like sequences in Human lipoprotein(a). Arterioscler.Thromb. **13:** 758–770.
22. TULINSKY, A. 1991. The structures of domains of blood proteins. Thromb. Hæmost. **66:** 16–31.
23. TREXLER, M., Z. VALI & L. PATTHY. 1982. Structure of the ω-aminocarboxylic acid-binding sites of human plasminogen. J. Biol. Chem. **257:** 7401–7406.
24. HOOVER, G.J., *et al.* 1993. Amino acids of the recombinant kringle 1 domain of human plasminogen that stabilize its interaction with ω-amino acids. Biochemistry **32:** 10936–10943.
25. MATHEWS, I.I., *et al.* 1996. Crystal structures of the recombinant kringle 1 domain of human plasminogen in complexes with the ligands e-aminocaproic acid and trans-4-(aminomethyl)cyclohexane-1-carboxilic acid. Biochemistry **35:** 2567–2576.
26. WU, T.P., K.P. PADMANABHAN & A. TULINSKY. 1994. The structure of recombinant plasminogen kringle 1 and the fibrin binding site. Blood Coagul. Fibrinol. **5:** 157–166.
27. RAMESH, V., *et al.* 1987. Proton magnetic resonance study of lysine-binding to the kringle 4 domain of human plasminogen. The structure of the binding site. J. Mol. Biol. **198:** 481–98.
28. MOCHALKIN, I., *et al.* 1999. Recombinant kringle IV-10 modules of human apolipoprotein(a): structure, ligand binding modes, and biological relevance. Biochemistry **38:** 1990–1998.
29. NEURATH, H. 1999. Proteolytic processing and physiological regulation. TIBS **14:** 268–272.
30. BERGER, A. & I. SCHECHTER. 1970. Mapping the active site of papain with the aid of peptide substrates and inhibitors. Philos. Trans. R. Soc. Lond. B. Biol. Sci. **257:** 249–264.

31. ROBBINS, K.C., *et al.* 1967. The peptide chains of human plasmin. Mechanism of activation of human plasminogen to plasmin. J. Biol. Chem. **242:** 2333–2342.
32. LINDE, V., *et al.* 1998. Elimination of the Cys558-Cys566 bond in Lys78-plasminogen—effect on activation and fibrin interaction. Eur. J. Biochem. **251:** 472–479.
33. BUSBY, S.J., *et al.* 1991. Expression of recombinant human plasminogen in mammalian cells is augmented by suppression of plasmin activity. J. Biol. Chem. **266:** 15286–15292.
34. HERVIO, L., *et al.* 1999. Effect of plasminogen activators on human recombinant apolipoprotein(a) having the plasminogen activation cleavage site. Biochim. Biophys. Acta **1434:** 124–134.
35. COOMBS, G.S., *et al.* 1996. Distinct mechanisms contribute to stringent substrate specifity of tissue-type plasminogen activator. J. Biol. Chem. **271:** 4461–4467.
36. MILES, L.A., *et al.* 1989. A potential basis for the thrombotic risks associated with lipoprotein(a). Nature **339:** 301–303.
37. HAJJAR, K.A., *et al.* 1989. Lipoprotein(a) modulation of endothelial cell surface fibrinolysis and its potential role in atherosclerosis. Nature **339:** 303–305.
38. HOYLAERTS, M., *et al.* 1982. Kinetics of the activation of plasminogen by human tissue plasminogen activator. J. Biol. Chem. **257:** 2912–2919.
39. SUENSON, E., *et al.* 1984. Initial plasmin-degradation of fibrin as the basis of a positive feedback mechanism in fibrinolysis. Eur. J. Biochem. **140:** 513–522.
40. VARADI, A. & L. PATTHY. 1983. Location of plasminogen—binding sites in human fibrin (ogen). Biochemistry **22:** 2440–2446.
41. PLOW, E.F., K. ALLAMPALLAM & V. REDLITZ. 1997. The plasma carboxypeptidases and the regulation. Trends Cardiovasc. Med. **7:** 71–75.
42. SUGIYAMA, N., M. IWAMOTO & Y. ABIKO. 1987. Effects of kringles derived from human plasminogen on fibrinolysis *in vitro*. Thromb. Res. **47:** 459–468.
43. LOSCALZO, J., *et al.* 1990. Lipoprotein(a), fibrin binding, and plasminogen activation. Atherosclerosis **10:** 240–245.
44. ROUY, D., *et al.* 1991. Lipoprotein (a) impairs generation of plasmin by fibrin-bound tissue-type plasminogen activator. Arterioscler. Thromb. **11:** 629–638.
45. MILES, L.A., *et al.* 1995. Interaction of Lp(a) with plasminogen binding sites on cells. Thromb. Hæmost. **73:** 458–465.
46. HARPEL, P.C. & W. BORTH. 1992. Fibrin, lipoprotein(a), plasmin interactions: a model linking thrombosis and atherogenesis. Ann. N.Y. Acad. Sci. **667:** 233–238
47. ANGLÉS-CANO, E., *et al.* 1993. Effects of lipoprotein(a) on the binding of plasminogen to fibrin and its activation by fibrin-bound tissue-type plasminogen activator. Chem. Phys. Lipids **67/68:** 369–380.
48. ROUY, D., *et al.* 1992. Apolipoprotein(a) and plasminogen interactions with fibrin: a study with recombinant apolipoprotein(a) and isolated plasminogen fragments. Biochemistry **31:** 6333–6339.
49. EZRALTY, A., D.I. SIMON & J. LOSCALZO. 1993. Lipoprotein(a) binds to human platelets and attenuates plasminogen binding and activation. Biochemistry **32:** 4628–4633.
50. ETINGIN, O., *et al.* 1991. Lipoprotein(a) regulates plasminogen activator inhibitor-1 expression in endothelial cells. J. Biol. Chem. **266:** 2459–2465.
51. LEVIN, E.G., *et al.* 1994. Lipoproteins inhibit the secretion of tissue plasminogen activator from human endothelial cells. Arterioscler. Thromb. **14:** 438–442.
52. KOJIMA, S., P.C. HARPEL & D.B. RIFKIN. 1991. Lipoprotein(a) inhibits the generation of transforming growth factor b: an endogenous inhibitor of smooth muscle cell migration. J. Cell Biol. **113:** 1439–1445.
53. GRAINGER, D.J., *et al.* 1994. Activation of transforming growth factor is inhibited in transgenic apolipoprotein(a) mice. Nature **370:** 460–462.
54. WILLIAMS, J.K., *et al.* 1993. Occlusive arterial thrombosis in cynomologous monkeys with varying plasma concentrations of lipoprotein(a). Arterioscler. Thromb. **13:** 548–554.
55. BIEMOND, B.J., *et al.* 1997. Apolipoprotein(a) attenuates endogenous fibrinolysis in vivo in a rabbit jugular vein thrombosis model. Circulation **96:** 1612–1615.

56. PALABRICA, T.M., et al. 1995. Antifibrinolytic acitivity of apolipoprotein(a) in vivo: human apolipoprotein(a) transgenic mice are resistant for tissue plasminogen activator-mediated thrombolysis. Nat. Med. **1:** 256–259.
57. BOONMARK, N.W., et al. 1997. Modification of apolipoprotein(a) lysine binding site reduces atherosclerosis in transgenic mice. J. Clin. Invest. **100:** 558–564.
58. KRAFT H.G., et al. 1989. Changes of genetic apolipoprotein phenotypes caused by liver transplantation. J. Clin. Invest **83:** 137–142.
59. KRONENBERG, F., et al. 1996. Lipoprotein(a) in Renal Disease. Am. J. Kidney Diss. **27:** 1–25.
60. SOULAT, T., et al. 1999. Evidence that modifications of Lp(a) in vivo inhibit plasmin formation on fibrin—a study with individual plasmas presenting natural variations of Lp(a). Thromb. Hæmost. **82:** 121–127.
61. SOULAT, T., et al. 2000. Effect of individual plasma lipoprotein(A) variations in vivo on its competition with plasminogen for fibrin and cell binding—an in vitro study using plasma from children with idiopathic nephrotic syndrome. Arterioscler. Thromb. Vasc. Biol. **20:** 575–584.
62. GAW, A., et al. 1994. Comparative analysis of the apo(a) gene, apo(a) glycoprotein, and plasma concentrations of Lp(a) in three ethnic groups. Evidence for non common "null" allele at the apo(a) locus. J. Clin. Invest. **93:** 2526–2534.
63. UTERMANN, G., et al. 1987. Lp(a) glycoprotein phenotypes: inheritance and relation to Lp(a)-lipoprotein concentations in plasma. J. Clin. Invest. **80:** 458–465.
64. MÖLGAARD, J., et al. 1992. Significant association between low molecular weight apolipoprotein (a) isoforms and intermittent claudication. Arterioscler. Thromb. **12:** 895–901.
65. SEED, M., et al. 1990. Relation of serum lipoprotein(a) and apolipoprotein(a) phenotype to coronary heart disease in patients with familial hypercholesterolemia. N. Engl. J. Med. **322:** 1494–1499.
66. SANDHOLZER, C., et al. 1992. Apo(a) isoforms predict risk for coronary heart disease. A study in six populations. Arterioscler. Thromb. **12:** 1214-1226.
67. HERVIO, L., et al. 1993. Does apolipoprotein (a) heterogeneity influence lipoprotein (a) effects on fibrinolysis? Blood **82:** 392–397.
68. KRONENBERG, F., et al. 1999. Role of lipoprotein(A) and apolipoprotein(A) phenotype in atherogenesis—prospective results from the Bruneck study. Circulation **100:** 1154–1160.

Genetic Manipulation of Fibrinogen and Fibrinolysis in Mice

JAY L. DEGEN, ANGELA F. DREW, JOSEPH S. PALUMBO,
KEITH W. KOMBRINCK, JORGE A. BEZERRA, MARY JO S. DANTON,
KENN HOLMBÄCK, AND THEODORE T. SUH

Children's Hospital Research Foundation and University of Cincinnati College of Medicine, Cincinnati, Ohio, USA

ABSTRACT: Vascular integrity is maintained by a sophisticated system of circulating and cell associated hemostatic factors that control local platelet deposition, the conversion of soluble fibrinogen to an insoluble fibrin polymer, and the dissolution of fibrin matrices. However, hemostatic factors are likely to be biologically more important than merely maintaining vascular patency and controlling blood loss. Specific hemostatic factors have been associated with a wide spectrum of physiological processes, including development, reproduction, tissue remodeling, wound repair, angiogenesis, and the inflammatory response. Similarly, it has been proposed that hemostatic factors are important determinants of a variety of pathological processes, including vessel wall disease, tumor dissemination, infectious disease, and inflammatory diseases of the joint, lung, and kidney. The development of gene targeted mice either lacking or expressing modified forms of selected hemostatic factors has provided a valuable opportunity to test prevailing hypotheses regarding the biological roles of key coagulation and fibrinolytic system components *in vivo*. Genetic analyses of fibrin(ogen) and its interacting factors in transgenic mice have proven to be particularly illuminating, often challenging long standing concepts. This review summarizes the key findings made in recent studies of gene targeted mice with single and combined deficits in fibrinogen and fibrinolytic factors. Studies illustrating the role and interplay of these factors in disease progression are highlighted.

KEYWORDS: Fibrogen-deficient mice; Gene targeting; Fibrinolysis; Fibrin; Hemostasis.

INTRODUCTION

The availability of gene targeted mice with single and combined deficits in coagulation and fibrinolytic factors, platelet function, and endothelial cell properties has provided a means to explore in greater detail the roles of selected hemostatic factors in development, hemostasis, reproduction, wound repair, inflammatory and immune responses, and disease. Although detailed studies of physiological and pathological processes in mice with specific hemostatic deficits remain relatively limited, initial

Address for correspondence: J. L. Degen, Ph.D., Children's Research Foundation, Children's Hospital Medical Center, IDR - NRB Room 2042, 3333 Burnet Avenue, Cincinnati, Ohio 45229-3039, USA. Voice: 513-636-4679;fax: 513-636-4317
degenjl@chmcc.org

findings have significantly improved our understanding of the roles of selected factors *in vivo* and occasionally challenged prevailing notions. Most of the major hemostasis related genes have been experimentally modified in mice, including tissue factor, factor VII, factor X, factor IX, factor VIII, vWF, factor XI, factor V, prothrombin, PAR1, PAR3, and fibrinogen.[1-18] Furthermore, many of the genes encoding the known suppressors and inhibitors of the coagulation system have been disrupted or altered, including tissue factor pathway inhibitor, protein C, and thrombomodulin.[19-22] As tools for understanding the impact of hemostatic factors on tissue repair, the inflammatory response, and disease pathophysiology, perhaps the most valuable of these mutant mice are those that are viable to adulthood. Disruption of the genes encoding five soluble hemostatic factors resulted in viable homozygous mutants, as defined by the ability to survive to sexual maturity: factor VIII-, IX-, XI-, vWF-, and fibrinogen-deficient mice. Mice with significant defects in platelet function were also found to be viable, including mice lacking thrombin receptor (e.g., PAR1 and PAR3), Gαq (a crucial G protein coupling extracellular signals to intracellular platelet activation pathways), and integrin (e.g., $\alpha_{IIb}\beta_3$) and non-integrin (e.g., P-selectin, GPIbα) adhesion molecules.[1,3,23-26] In addition, mice lacking the element of the fibrinogen γ chain that is recognized by the platelet integrin $\alpha_{IIb}\beta_3$ are viable.[18] Loss of any of the established fibrinolytic system components (e.g., tissue-type plasminogen activator [tPA], urokinase-type plasminogen activator [uPA], uPA receptor [uPAR], plasminogen activator inhibitor, plasminogen, and α_2 plasmin inhibitor) is compatible with development, survival to adulthood, and reproduction.[27-33] Taken together, there is an extraordinary library of viable mouse lines with deficits on both sides of the hemostatic balance that can now be exploited to define the roles and interplay of coagulation-, fibrinolytic-, platelet- and endothelial-factors in adult physiology and disease progression.

FIBRINOGEN: A KEY COAGULATION FACTOR WITH ROLES OUTSIDE OF HEMOSTASIS

Fibrin(ogen) is the target of two complex and opposing biochemical pathways, the coagulation and fibrinolytic cascades, that together preserve vascular integrity and maintain hemostatic balance.[34-38] The coagulation system includes more than a dozen cell-associated and soluble factors that initiate, promote, and ultimately restrain the conversion of prothrombin to thrombin, subsequent platelet activation, and fibrin formation. Fibrinogen is fundamentally important in hemostasis in that it functions: (1) as a dimeric bridging molecule, non-covalently linking adjacent platelets following agonist induced activation of fibrinogen receptors (e.g., $\alpha_{IIb}\beta_3$), and (2) as the primary building block for formation of fibrin matrices following thrombin cleavage of the fibrinogen Aα and Bβ chains. A carefully controlled fibrinolytic system also provides a means for the timely degradation and solubilization of fibrin matrices. Two major driving forces for detailed studies of fibrin(ogen) and other hemostatic factors are the high morbidity and mortality associated with vasoocclusive disease (e.g., myocardial infarction and stroke) and the identification of multiple congenital abnormalities leading to life threatening hemorrhagic and thrombic events. However, an additional incentive for detailed studies of fibrinogen has been to understand the impressive number of functional properties of fibrin(ogen) that are seemingly distinct from those necessary to maintain hemostasis. These properties, summarized below, strongly suggest that fibrin(ogen) plays a broader physiological role than merely controlling blood loss and a broader pathological role than merely participating in deleterious vasoocclusive events.

In addition to platelets, many types of cells specifically bind to fibrin(ogen) matrices, including endothelial cells, smooth muscle cells, keratinocytes, fibroblasts, leukocytes, and tumor cells.[39–44] Both integrin (e.g., $\alpha_v\beta_3$, $\alpha_M\beta_2$, CD11c/CD18, $\alpha_5\beta_1$, avb1) and non-integrin (e.g., I-CAM-1) receptors appear to contribute to these cell–fibrin(ogen) interactions.[44–47] In addition, various fibrin(ogen) degradation products are reported to have chemotactic, angiogenic, and immunosuppressive activities.[48–50] Finally, fibrin(ogen) appears to be important in tissue repair, leukocyte cell adhesion and transendothelial cell migration.[47,51,52] Therefore, in addition to supporting hemostatic plug formation, fibrino(gen) appears to function as a bridging molecule for many types of cell–cell adhesion events, and provides a critical provisional matrix at sites of injury, inflammation, or infection in which cells can proliferate, organize, and carry out specialized functions. If this hypothesis is correct, then the engagement of fibrin(ogen) via cell surface receptors may profoundly alter cellular behavior, including responses to other insoluble matrix components and soluble signaling molecules (e.g., chemokines, cytokines, and growth factors).

FIBRINOGEN-DEFICIENT MICE

The availability of gene targeted mice either lacking fibrinogen[17] or expressing modified forms of fibrinogen[18] has offered an opportunity to explore in greater detail the biological features and roles of fibrin(ogen) *in vivo*. Furthermore, crossbreeding studies of mouse lines with deficits in other hemostatic factors (e.g., plasminogen and vWF deficient mice) have begun to illuminate the interplay and relationship of fibrin(ogen) with other hemostatic factors *in vivo*.

Mice lacking the fibrinogen Aα chain (A$\alpha^{-/-}$ mice)[17] were shown to complete embryonic development with no evidence of fetal loss. Consistent with earlier *in vitro* studies showing that fibrinogen secretion is dependent on the synthesis and assembly of all three fibrinogen chains,[53] none of the component chains of fibrinogen (Aα, Bβ, or γ) were immunologically detectable in the circulation of either neonatal or adult A$\alpha^{-/-}$ mice. Whole blood analyses in control and fibrinogen deficient mice revealed no significant difference in platelet or other blood cell counts. Blood and plasma samples collected from A$\alpha^{-/-}$ mice failed to either clot or support platelet aggregation *in vitro*, but platelet function and thrombus formation *in vivo* have been well documented in fibrinogen deficient mice by using intravital microscopy (see below FIBRINOGEN, VWF, AND THROMBUS FORMATION). Overt bleeding events developed shortly after birth in a significant fraction of A$\alpha^{-/-}$ mice, most frequently in the peritoneal cavity, skin, and soft tissues around joints. Remarkably, newborns displaying signs of bleeding often controlled the loss of blood and survived the neonatal period. Juveniles and young adult A$\alpha^{-/-}$ mice were also predisposed to spontaneous fatal hemorrhagic events, but neonatal and long term survival was variable and highly dependent on genetic background. With a mixed CF-1/129 genetic background, the majority of fibrinogen deficient mice survive the neonatal period, but nearly half die between weaning age (about 21 days) and sexual maturity (about 50 days). In contrast, the majority of C57BL/6-inbred fibrinogen deficient mice die within the first few days after birth, but those surviving the neonatal period are likely to live well into adulthood. Interestingly, the development, health, and survival of mice with mutations in other hemostatic factors have also been shown to be dependent on genetic background, including tissue factor–[6] and tissue factor pathway inhibitor–deficient mice (Kombrinck *et al.*, unpublished results). In each of these cases, the identity of secondary factors that account for the distinct phenotypic consequences of a specific hemostatic deficit in mice of different genetic backgrounds remains an important unresolved question. However, it should be noted that subtle biochemical, anatomical, and/or behavioral differences could all be contributing factors to the survival of fibrinogen deficient mice.

FAILURE OF PREGNANCY IN FIBRINOGEN-DEFICIENT MICE

The periodic rupture of ovarian follicles in adult fibrinogen deficient females would be expected to result in local hemorrhagic events, but did not appear to significantly diminish life expectancy relative to males.[17] However, pregnancy uniformly resulted in fatal uterine bleeding (maternal source) in $A\alpha^{-/-}$ mice around the tenth day of gestation, even if the entire litter carried one normal fibrinogen $A\alpha$ chain allele. This is the time in gestation where embryonic trophoblasts invade the maternal vasculature in the placenta, implying that fibrinogen deficient mothers cannot tolerate the disruption of placental labyrinthine vessels that is associated with establishing the embryonic/maternal interface. Consistent with these findings, recent studies of mice expressing low levels of tissue factor suggest that maternal tissue factor may also be important for maintaining vascular integrity in the developing placenta.[54]

FIBRINOGEN IN TISSUE REPAIR PROCESSES

Microscopic analyses in $A\alpha^{-/-}$ mice suggested that fibrin(ogen) provides a provisional matrix that is important, but not strictly required, for the organization of cells at sites of injury.[17,55] Fibrinogen was found to be particularly important in the cellular organization and resolution of pooled blood "lakes" at sites of spontaneous subcapsular hematoma formation in the liver and kidney. The obliteration of these lesions in fibrinogen deficient mice to ultimately form a fibrotic scar was characterized by the progressive organization of cells around the lesion but not directly infiltrating the blood filled zone. Interestingly, the timing and general outcome of tissue repair following introduction of a full skin thickness incisional wound was remarkably similar in control and fibrinogen deficient animals.[55] Nevertheless, at early times after injury, the path of keratinocyte migration from the wound edge in fibrinogen deficient mice was clearly altered from the usual path across the wound face immediately above the newly forming granulation tissue and below the dried eschar.

Rather, proliferating and migrating keratinocytes tended to move down the wound edge and often appeared to be headed away from the wound center (Drew *et al.,* Blood, in press). This detour of keratinocytes in skin incisional wounds did not profoundly delay reepithelialization, but the tensile strength of newly reepithelialized skin wounds was found to be significantly less in fibrinogen deficient mice relative to control animals (Drew *et al.,* Blood, in press). More detailed studies of wound repair in fibrinogen deficient mice will be required to define more precisely the impact of fibrin(ogen) on cellular proliferation/migration, inflammatory response, angiogenesis, extracellular matrix deposition, and scar tissue maturation.

FIBRINOGEN, VWF, AND THROMBUS FORMATION

Perhaps the most surprising observations in the initial studies of fibrinogen deficient mice were that severe spontaneous bleeding events were not necessarily fatal, and that many types of surgical challenges were well tolerated.[17] One possible explanation for these findings is that thrombin generation (which is presumably not profoundly impaired in fibrinogen-deficient mice), and subsequent local thrombin mediated platelet activation and deposition, may be sufficient to control blood loss at sites of modest vascular breaks. Consistent with this hypothesis, the primary fibrinogen receptor on platelets, $\alpha_{IIb}\beta_3$, can bind multiple ligands, including vWF, fibronectin, vitronectin and thrombospondin.[56–58] Furthermore, at high shear, platelet adhesion to the vessel wall appears to be mediated primarily by GPIbα-dependent binding to vWF.[59–61] Therefore, despite the absence of fibrinogen mediated platelet adhesion/aggregation in standard *in vitro* assays,[17] appreciable residual capacity to form a platelet thrombus would be expected *in vivo*. Indeed, this concept has been recently confirmed. Intravital microscopy studies focusing on platelet thrombus formation in ferric chloride injured mesenteric arterioles of live mice revealed prominent platelet deposition/aggregation in fibrinogen deficient mice.[62] However, these thrombi were unstable and frequently embolized from the site of vascular injury. Indeed, blood flow through injured vessels was rapidly terminated in fibrinogen deficient mice, but flow stoppage was often the result of vasoocclusion downstream of the injured zone by embolized platelet aggregates. In the same study, vWF deficient mice were also shown to support platelet adhesion and stable thrombus formation, although initial platelet adhesion was delayed relative to controls. Most surprisingly, a combined deficit in both fibrinogen and vWF was also compatible with thrombus formation, although platelet deposition was both delayed and unstable.[62] Although the alternative ligand(s) supporting platelet thrombus formation in fibrinogen/vWF deficient mice have not been firmly identified, it seems likely that the general good health, ability to tolerate some bleeding events, and longevity of fibrinogen deficient mice is largely due to residual platelet function *in vivo*. This hypothesis can be further explored by crossing fibrinogen deficient mice with available (viable) mouse lines with severe deficits in platelet function.[23,24,26]

FIBRINOGEN γΔ5, A SITE-DIRECTED MUTANT LACKING THE $\alpha_{IIb}\beta_3$ BINDING MOTIF

Mouse gene targeting technology offers the opportunity to make either dramatic (e.g., complete deletion of coding sequence) or subtle (e.g, single nucleotide substitutions) changes within any selected gene. To establish an *in vivo* model system to examine the importance of $\alpha_{IIb}\beta_3$–fibrinogen interactions in platelet function, hemostasis and response to injury and disease, the last five residues (QAGDV) of mouse fibrinogen γ chain known to be essential for $\alpha_{IIb}\beta_3$ binding[63] were eliminated by gene targeting.[18] Mice that were homozygous for the modified γ chain gene ($\gamma^{\Delta5/\Delta5}$) displayed a generally normal hematological profile, including normal platelet count, plasma fibrinogen level, clotting time, and fibrin crosslinking. However, both γΔ5-fibrinogen binding to $\alpha_{IIb}\beta_3$ and platelet aggregation were defective. Nevertheless, another $\alpha_{IIb}\beta_3$-dependent process, clot retraction, was unaffected by the γΔ5 mutation. Unfortunately, it remains uncertain whether clot retraction in $\gamma^{\Delta5/\Delta5}$ mice is supported by another $\alpha_{IIb}\beta_3$ ligand intimately associated with γΔ5-fibrin (e.g., fibronectin, vWF) or supported by a second, fibrin specific $\alpha_{IIb}\beta_3$ binding element present in γΔ5-fibrin. Despite the preservation of clotting function, $\gamma^{\Delta5/\Delta5}$ mice were unable to control blood loss following some types of surgical challenges (e.g., nail bed injury) and occasionally developed intraabdominal bleeding events as neonates. Future comparative studies of wildtype, fibrinogen deficient ($A\alpha^{-/-}$) and $\gamma^{\Delta5/\gamma\Delta5}$ mice will be helpful in establishing the role the fibrinogen–platelet interaction in a variety of important biological processes, including leukocyte adhesion to activated endothelium, dissemination of tumor emboli, vessel wall disease, and sepsis.

FIBRINOGEN AαC AND AαE SPLICE VARIANTS SUPPORT CLOTTING FUNCTION *IN VIVO*

Genetic approaches have also been employed to express mouse fibrinogen Aα chain variants. Introduction of Aα chain transgenes into $A\alpha^{-/-}$ mice have been shown to restore hepatic expression of the Aα chain and restore circulating fibrinogen with the transgene encoded Aα chain derivative. This approach has been used to generate mice expressing exclusively either the AαC splice variant (the predominant Aα chain splice variant; 538 amino acids in mature mouse AαC polypeptide) or AαE splice variant (the minor Aα chain splice variant; 770 amino acids in the mature mouse AαE polypeptide) (our unpublished results). Although detailed studies of these animals remain to be performed, mice expressing solely the AαC- or AαE-fibrinogen retain clotting function and have no appreciable spontaneous pathologies. Furthermore, unlike $A\alpha^{-/-}$ mice, females expressing either fibrinogen splice variant can successfully carry litters to term.

PLASMINOGEN: KEY FIBRINOLYTIC FACTOR

In mammals, plasmin mediated proteolysis is controlled by two plasminogen activators, uPA and tPA, as well as several plasminogen- and PA-specific receptors and inhibitors.[64] One critical role of the plasminogen activation system is fibrinolysis. Plasminogen activation is potentiated by the binding of PA and plasminogen to fibrin; plasmin efficiently degrades and solubilizes fibrin matrices; and deficits in plasminogen activation lead to widespread terminal vessel thrombosis in mice.[29,30] However, the plasminogen activation system has been frequently proposed to play a broader physiological role than merely fibrin clearance. Many observations are consistent with this view, including: (1) the intricate pattern and complex regulation of PA expression *in vivo,* (2) the common association of plasminogen activation with tissue remodeling and cell migration events *in vivo,* and (3) the broad substrate specificity of plasmin *in vitro* (e.g., common matrix glycoproteins, latent growth factors, and protease zymogens). Nevertheless, based on the fact that plasminogen is not essential to complete embryonic development, survival to adulthood, or reproduction, any contribution of plasmin to tissue remodeling and cell migration outside of fibrinolysis must be unnecessary in many physiological settings and may be important only in selected contexts (see below).

Although not essential for developmental success, the consequences of plasminogen deficiency are severe.[29,30] Plasminogen deficient mice have been shown to develop widespread and progressive organ damage, overt conjunctival lesions (ligneous conjunctivitis), and rectal lesions (ulceration and prolapse). Furthermore, $Plg^{-/-}$ mice experience progressive wasting, display impaired wound/tissue repair, and have a dramatically shortened life expectancy.[29,65,66] Interestingly, a few "benefits" of plasminogen deficiency have also been recognized; $Plg^{-/-}$ mice appear to be resistant to neuronal degeneration induced by the glutamate analog, kainic acid, and the mice appear to be resistant to at least some types of bacterial pathogens.[67–72]

COMBINED PLASMINOGEN- AND FIBRINOGEN-DEFICITS IN MICE

The early studies of $Plg^{-/-}$ mice did not directly identify the substrates of plasmin that were relevant to the diverse spectrum of phenotypes observed in $Plg^{-/-}$ animals. Widespread fibrin deposition is a prominent feature of adult $Plg^{-/-}$ mice, but many of the pathological manifestations of plasminogen deficiency could be due to the absence of plasmin mediated proteolysis of non-fibrin substrates (e.g., fibronectin, laminin, latent growth factors, and protease zymogens). The generation and comparative analysis of mice with single and combined deficits in plasminogen and fibrinogen have been helpful in clarifying physiological or pathological roles of plasmin mediated proteolysis, including proposed roles outside of fibrinolysis.[55] The available findings, summarized below, illustrate the central physiological role of plasminogen in fibrinolysis, but both physiological and pathological roles of plasmin outside of fibrin clearance have been recognized.

Superimposing fibrinogen deficiency on mice lacking plasminogen was shown to alleviate essentially all of the *spontaneous* pathologies known to befall $Plg^{-/-}$ mice, including ligenous conjunctivitis, wasting, terminal vessel thrombosis, hepatic and pulmonary gastrointestinal lesions, and rectal ulceration and prolapse.[55,73] Furthermore, nearly all $Plg^{-/-}$ mice observed at weaning age (about 21 days) live to one-year of age if they lack circulating fibrinogen, whereas the median survival time for

Plg$^{-/-}$ mice expressing fibrinogen is about six months. Finally, the impaired (delayed) tissue repair observed in Plg$^{-/-}$ mice following skin incision,[55] corneal injury,[66,74] arterial challenge,[75] and sciatic nerve injury[76] was corrected by genetic removal of fibrinogen. Taken together, the data support the concept that a key physiological role of the plasminogen activation system is fibrinolysis. However, it should be noted that the phenotypic improvements observed in Plg$^{-/-}$ mice following removal of fibrinogen does not formally exclude the possibility that plasmin(ogen) participates in processes unrelated to fibrin(ogen) cleavage. Indeed, recent studies of experimentally challenged Plg$^{-/-}$ mice have revealed that plasmin mediated proteolysis of non-fibrin matrix components is important in several specific biological settings, including both physiological (e.g., hepatic tissue repair; see below) and pathological settings (e.g., neurodegeneration; see below).

FIBRIN(OGEN) INDEPENDENT ROLE OF PLASMINOGEN IN LIVER REPAIR

A role of plasmin(ogen) distinct from its role in fibrinolysis has now been documented in studies of liver repair following an acute injury with carbon tetrachloride (CCl_4).[77] Exposure of mice to CCl_4 results in the intoxication and death of hepatocytes adjacent to the central vein and the livers are grossly pale within two days of challenge. In normal mice, the necrotic zones are cleared and normal liver architecture is restored in 7 to 14 days. In contrast, the microscopic and gross appearance of the livers of both plasminogen and plasminogen activator deficient mice was abnormal, with no evidence of repair, even 30 days after CCl_4 exposure. The failure to restore normal liver architecture in Plg$^{-/-}$ mice was not due to ongoing liver injury or defects in the proliferative response, but rather, a failure of repair. Like most wound fields, fibrin was found to be a prominent and persistent feature of the damaged centrilobular zones in Plg$^{-/-}$ mice. Therefore, one hypothesis that might explain the failure to repair necrotic tissue in Plg$^{-/-}$ mice was that fibrin within the diseased tissue formed a mechanical impediment to the removal of necrotic debris in the absence of plasmin mediated proteolysis. Interestingly, genetically superimposing fibrin(ogen) deficiency did not correct the abnormal pattern of hepatic repair in Plg$^{-/-}$ mice. These data suggest that plasmin mediated proteolysis of one or more non-fibrin targets is critical in at least one physiological process, the repair of necrotic liver tissue. Possible fibrin(ogen)-independent roles of plasmin in other organ systems, the identity of the relevant non-fibrin targets, and the exact mechanistic role(s) of plasmin in the clearance and repair of necrotic tissue all remain to be established. A working hypothesis that is consistent with all the presently available experimental findings is that fibrin is an important target of plasmin within wound fields in general, but plasmin mediated proteolysis of additional targets is critical when the reparative process demands the clearance of necrotic tissue.

FIBRIN(OGEN) INDEPENDENT ROLE OF PLASMINOGEN IN NEURAL DEGENERATION

Plasmin mediated proteolysis of a non-fibrin target *in vivo* has also been established in a pathological context: excitotoxin (kainate) induced neurodegeneration in the hippocampus. Both plasminogen and tPA deficient mice are resistant to excitotoxin induced degeneration of the hippocampus, whereas control and fibrinogen deficient mice are sensitive.[67] Comparative studies of excitotoxin-induced neurodegeneration in mice with single and combined deficiencies in plasminogen and fibrinogen revealed that $Plg^{-/-}$ mice lacking fibrinogen maintain the same resistance to degenerative disease as $Plg^{-/-}$ mice expressing fibrinogen.[68] More detailed studies suggest that the non-fibrin target of plasmin relevant to neuronal cell death in the hippocampus is laminin.[78]

PLASMIN-INDEPENDENT FIBRINOLYSIS

Wound repair following a skin incision is greatly delayed in plasminogen deficient mice. Nevertheless, given sufficient time, $Plg^{-/-}$ mice do heal with an outcome that is similar to control animals. The successful clearance of fibrin rich matrices from wound fields in $Plg^{-/-}$ mice implies that at least one other protease system aside from the plasminogen activation system can contribute to the solubilization of fibrin *in vivo*. Notably, membrane type-1 matrix metalloprotease (MT1-MMP) was recently demonstrated to be essential for VEGF induced endothelial penetration through, but not across, fibrin gels.[79] Furthermore, several other MMPs have been shown to degrade fibrin *in vitro*.[80,81] Plasmin appears to be a component of a team of enzymes that participates in extracellular matrix proteolysis. Although plasmin could be viewed as a specialist in fibrin clearance, it is not the sole enzyme capable of solubilizing and clearing fibrin *in vivo*.

DISEASE PROGRESSION IN MICE LACKING FIBRINOGEN AND PLASMINOGEN

Hemostatic factors are consistently deployed at sites of tissue damage. Given that disease, by definition, involves tissue damage of some form or another, it follows that hemostatic factors are likely to participate in the pathogenesis of a wide variety of diseases. Consistent with this notion, fibrin deposition is a prominent feature in many diseased tissues, including atherosclerotic plaques, tumor stroma, fibrotic lung tissue, arthritic joints, and sites of infection. Studies of gene targeted mice strongly suggest that hemostatic factors are important determinants of disease progression, including vessel wall disease, metastatic disease, inflammatory disease in the lung, kidney, and joints, and bacterial infection.[67–72,82–88] Recent studies illustrating an important role of fibrinogen in vessel wall disease/vascular remodeling and tumor cell metastasis are highlighted below.

FIBRINOGEN AND ARTERIAL VASCULAR DISEASE

Fibrin(ogen) has long been considered to play a causative role in the development of vessel wall lesions for several reasons: (1) fibrin is a prominent component of both atherosclerotic and restenotic lesions, (2) the location and biological activity of fibrin and fibrin degradation products in vascular lesions have been well documented, (3) high fibrinogen levels are recognized as an independent risk factor in the development of cardiovascular disease, and (4) loss of fibrinolytic capacity (e.g., plasminogen deficiency) accelerates vascular lesion development in atherosclerosis susceptible apolipoprotein E-deficient (apo $E^{-/-}$) mice.[82] Consistent with the notion that fibrin contributes to local lesion development in vessel wall disease, in one well studied model of atherosclerosis, apo(a) transgenic mice, fibrinogen deficiency reduced lesion size.[83] Nevertheless, in a profound hyperlipidemia model, apo $E^{-/-}$ mice, a deficit in fibrinogen had only a minimal impact on the rate of atherosclerotic lesion formation.[89] In the context of injury induced vascular lesions, plasminogen plays a role in cell migration and hence repopulation of necrotic vascular lesions and neointimal development.[90] However, neointimal development was not affected by plasminogen deficiency in an inflammation mediated challenge (arterial cuff placement), but compensatory vascular remodeling was prevented.[75] Remodeling potential was restored by simultaneous fibrinogen deficiency, indicating that the role of plasmin in vascular remodeling involves fibrin, and is likely to be fibrinolysis. The accumulation of bands of fibrin in the adventitial layer of the $Plg^{-/-}$ and control arteries, and their persistence in $Plg^{-/-}$ arteries, may indicate that fibrin deposits impose a physical barrier to arterial expansion that cannot be resolved without plasmin mediated fibrinolysis. These studies indicate that both fibrin and fibrinolysis are intricately involved in the development of vascular lesions, but questions still remain regarding the mechanisms and interplay of hemostatic factors in vascular disease.

FIBRINOGEN IS A DETERMINANT OF THE METASTATIC POTENTIAL OF TUMOR CELLS

The hemostatic system has long been thought to play a role in malignant tumor growth and metastatic spread. Several studies have demonstrated a correlation in cancer patients between poor clinical outcome and abnormalities in various hemostatic parameters, including prothrombin time, partial thromboplastin time, fibrinogen levels, and circulating D-dimer.[91,92] It has been proposed that fibrin(ogen) plays a critical role in tumor stroma formation, possibly by providing a provisional matrix supporting tumor angiogenesis.[93] However, direct studies of transplantable tumor growth in fibrinogen deficient mice have revealed that fibrinogen is not essential for the growth and spontaneous metastasis of subcutaneous tumors.[94] Nevertheless, the metastatic potential of circulating tumor cells was shown to be significantly reduced in mice lacking fibrinogen relative to control mice expressing fibrinogen. Studies tracking the fate of circulating Lewis lung carcinoma cells demonstrated that fibrin(ogen) is important in sustained adherence and/or survival of tumor emboli in the lungs.[94] Further studies will be necessary to establish the role of other hemostatic factors and platelets in tumor cell dissemination and the mechanism(s) by which fibrin(ogen) and other hemostatic factors contribute to metastatic disease.

CONCLUSIONS

Studies of gene targeted mice have significantly contributed to our understanding of the roles of fibrin(ogen) and fibrinolytic enzymes in development, hemostasis, inflammatory response, wound repair, and disease pathogenesis. The viability of fibrinogen and plasminogen deficient mice demonstrated that dramatic swings in the hemostatic balance in either direction are compatible with development and survival to adulthood. However, it is clear that an opposing and balanced system of coagulation and fibrinolytic factors is critical to preserve vascular integrity, efficiently repair tissues, and maintain reproductive success. Studies of gene targeted mice also point to an important role of hemostatic factors in the inflammatory response and disease progression. Understanding the mechanisms by which fibrin(ogen) contributes to these processes will be a key area of future research. The availability of fibrinogen deficient mice will continue to be a useful tool in these studies, but the introduction of more subtle mutations into both fibrinogen and its interacting factors promises to be extremely informative, particularly if germline mutations can be made in both mice and other species.

REFERENCES

1. CONNOLLY, A.J., H. ISHIHARA, M.L. KAHN, et al. 1996. Role of the thrombin receptor in development and evidence for a second receptor. Nature **381:** 516–519.
2. COUGHLIN, S.R. 1998. Sol Sherry lecture in thrombosis: how thrombin "talks" to cells: molecular mechanisms and roles *in vivo*. Arterioscler. Thromb. Vasc. Biol. **18:** 514–518.
3. KAHN, M.L., Y.W. ZHENG, W. HUANG, et al. 1998. A dual thrombin receptor system for platelet activation. Nature **394:** 690–694.
4. BUGGE, T.H., Q. XIAO, K.W. KOMBRINCK, et al. 1996. Fatal embryonic bleeding events in mice lacking tissue factor, the cell-associated initiator of blood coagulation. Proc. Natl. Acad. Sci. U.S.A. **93:** 6258–6263.
5. TOOMEY, J.R., K.E. KRATZER, N.M. LASKY, et al. 1996. Targeted disruption of the murine tissue factor gene results in embryonic lethality. Blood **88:** 1583–1587.
6. TOOMEY, J.R., K.E. KRATZER, N.M. LASKY, & G. J. BROZE, JR. 1997. Effect of tissue factor deficiency on mouse and tumor development. Proc. Natl. Acad. Sci. U.S.A. **94:** 6922–6926.
7. CARMELIET, P., N. MACKMAN, L. MOONS, et al. 1996. Role of tissue factor in embryonic blood vessel development. Nature **383:** 73–75.
8. ROSEN, E.D., J.C. CHAN, E. IDUSOGIE, et al. 1997. Mice lacking factor VII develop normally but suffer fatal perinatal bleeding. Nature **390:** 290–294.
9. DEWERCHIN, M., Z. LIANG, L. MOONS, et al. 2000. Blood coagulation factor X deficiency causes partial embryonic lethality and fatal neonatal bleeding in mice. Thromb. Hæmost. **83:** 185–190.
10. DENIS, C., N. METHIA, P.S. FRENETTE, et al. 1998. A mouse model of severe von Willebrand disease: defects in hemostasis and thrombosis. Proc. Natl. Acad. Sci. U.S.A. **95:** 9524–9529.
11. WANG, L., M. ZOPPE, T.M. HACKENG, et al. 1997. A factor IX-deficient mouse model for hemophilia B gene therapy. Proc. Natl. Acad. Sci. U.S.A. **94:** 11563–11566.

12. BI, L., R. SARKAR, T. NAAS, et al. 1996. Further characterization of factor VIII-deficient mice created by gene targeting: RNA and protein studies. Blood **88:** 3446–3450.
13. GAILANI, D., N.M. LASKY, & G. J. BROZE, JR. 1997. A murine model of factor XI deficiency. Blood Coagul. Fibrinol. **8:** 134–144.
14. CUI, J., K.S. O'SHEA, A. PURKAYASTHA, et al. 1996. Fatal hæmorrhage and incomplete block to embryogenesis in mice lacking coagulation factor V. Nature **384:** 66–68.
15. SUN, W.Y., D.P. WITTE, J.L. DEGEN, et al. 1998. Prothrombin deficiency results in embryonic and neonatal lethality in mice. Proc. Natl. Acad. Sci. U.S.A. **95:** 7597–7602.
16. XUE, J., Q. WU, L.A. WESTFIELD, et al. 1998. Incomplete embryonic lethality and fatal neonatal hemorrhage caused by prothrombin deficiency in mice. Proc. Natl. Acad. Sci. U.S.A. **95:** 7603–7607.
17. SUH, T.T., K. HOLMBACK, N.J. JENSEN, et al. 1995. Resolution of spontaneous bleeding events but failure of pregnancy in fibrinogen-deficient mice. Genes Dev. **9:** 2020–2033.
18. HOLMBACK, K., M.J. DANTON, T.T. SUH, et al. 1996. Impaired platelet aggregation and sustained bleeding in mice lacking the fibrinogen motif bound by integrin alpha IIb beta 3. EMBO J. **15:** 5760–5771.
19. HEALY, A.M., H.B. RAYBURN, R.D. ROSENBERG & H. WEILER. 1995. Absence of the blood-clotting regulator thrombomodulin causes embryonic lethality in mice before development of a functional cardiovascular system. Proc. Natl. Acad. Sci. U.S.A **92:** 850–854.
20. HEALY, A.M., W.W. HANCOCK, P.D. CHRISTIE, et al. 1998. Intravascular coagulation activation in a murine model of thrombomodulin deficiency: effects of lesion size, age, and hypoxia on fibrin deposition. Blood **92:** 4188–4197.
21. JALBERT, L.R., E.D. ROSEN, L. MOONS, et al. 1998. Inactivation of the gene for anticoagulant protein C causes lethal perinatal consumptive coagulopathy in mice. J. Clin. Invest. **102:** 1481–1488.
22. HUANG, Z.F., D. HIGUCHI, N. LASKY, & G.J. BROZE, JR. 1997. Tissue factor pathway inhibitor gene disruption produces intrauterine lethality in mice. Blood **90:** 944–951.
23. OFFERMANNS, S., C.F. TOOMBS, Y.H. HU, & M. I. SIMON. 1997. Defective platelet activation in G alpha(q)-deficient mice. Nature **389:** 183–186.
24. HODIVALA-DILKE, K.M., K.P. MCHUGH, D.A. TSAKIRIS, et al. 1999. Beta3-integrin-deficient mice are a model for Glanzmann thrombasthenia showing placental defects and reduced survival. J. Clin. Invest. **103:** 229–238.
25. SUBRAMANIAM, M., P.S. FRENETTE, S. SAFFARIPOUR, et al. 1996. Defects in hemostasis in P-selectin-deficient mice. Blood **87:** 1238–1242.
26. WARE, J., S. RUSSELL & Z.M. RUGGERI. 2000. Generation and rescue of a murine model of platelet dysfunction: the Bernard-Soulier syndrome. Proc. Natl. Acad. Sci. U.S.A. **97:** 2803–2808.
27. CARMELIET, P., L. SCHOONJANS, L. KIECKENS, et al. 1994. Physiological consequences of loss of plasminogen activator gene function in mice. Nature **368:** 419–424.
28. CARMELIET, P., J.M. STASSEN, L. SCHOONJANS, et al. 1993. Plasminogen activator inhibitor-1 gene-deficient mice. II. Effects on hemostasis, thrombosis, and thrombolysis. J. Clin. Invest. **92:** 2756–2760.
29. BUGGE, T.H., M.J. FLICK, C.C. DAUGHERTY & J.L. DEGEN. 1995. Plasminogen deficiency causes severe thrombosis but is compatible with development and reproduction. Genes Dev. **9:** 794–807.
30. PLOPLIS, V.A., P. CARMELIET, S. VAZIRZADEH, et al. 1995. Effects of disruption of the plasminogen gene on thrombosis, growth, and health in mice. Circulation **92:** 2585–2593.
31. BUGGE, T.H., T.T. SUH, M.J. FLICK, et al. 1995. The receptor for urokinase-type plasminogen activator is not essential for mouse development or fertility. J. Biol. Chem. **270:** 16886–16894.
32. DEWERCHIN, M., A.V. NUFFELEN, G. WALLAYS, et al. 1996. Generation and characterization of urokinase receptor-deficient mice. J. Clin. Invest. **97:** 870–878.

33. LIJNEN, H.R., K. OKADA, O. MATSUO, *et al.* 1999. Alpha2-antiplasmin gene deficiency in mice is associated with enhanced fibrinolytic potential without overt bleeding. Blood **93:** 2274–2281.
34. DAVIE, E.W., K. FUJIKAWA & W. KISIEL. 1991. The coagulation cascade: initiation, maintenance, and regulation. Biochemistry **30:** 10363–10370.
35. ESMON, C.T. 1993. Cell mediated events that control blood coagulation and vascular injury. Annu. Rev. Cell Biol. **9:** 1–26.
36. LIJNEN, H.R. & D. COLLEN. 1995. Mechanisms of physiological fibrinolysis. Baillieres Clin. Hæmatol. **8:** 277–290.
37. BROZE, G.J., JR. 1995. Tissue factor pathway inhibitor and the revised theory of coagulation. Annu. Rev. Med. **46:** 103–112.
38. ROSENBERG, R.D. & W.C. AIRD. 1999. Vascular-bed–specific hemostasis and hypercoagulable states. N. Engl. J .Med. **340:** 1555–1564.
39. DEJANA, E., L.R. LANGUINO, N. POLENTARUTTI, *et al.* 1985. Interaction between fibrinogen and cultured endothelial cells. Induction of migration and specific binding. J. Clin. Invest. **75:** 11–18.
40. ALTIERI, D.C., P.M. MANNUCCI & A.M. CAPITANIO. 1986. Binding of fibrinogen to human monocytes. J. Clin. Invest. **78:** 968–976.
41. NAITO, M., C. FUNAKI, T. HAYASHI, *et al.* 1992. Substrate-bound fibrinogen, fibrin and other cell attachment-promoting proteins as a scaffold for cultured vascular smooth muscle cells. Atherosclerosis **96:** 227–234.
42. DONALDSON, D.J., J.T. MAHAN, D.L. AMRANI, *et al.* 1994. Further studies on the interaction of migrating keratinocytes with fibrinogen. Cell Adhes Commun **2:** 299–308.
43. BROWN, L.F., N. LANIR, J. MCDONAGH, *et al.* 1993. Fibroblast migration in fibrin gel matrices. Am. J. Pathol. **142:** 273–283.
44. UGAROVA, T.P., D.A. SOLOVJOV, L. ZHANG, *et al.* 1998. Identification of a novel recognition sequence for integrin alphaM beta2 within the gamma-chain of fibrinogen. J. Biol. Chem. **273:** 22519–22527.
45. KATAGIRI, Y., T. HIROYAMA, N. AKAMATSU, *et al.* 1995. Involvement of alpha v beta 3 integrin in mediating fibrin gel retraction. J. Biol. Chem. **270:** 1785–1790.
46. SUEHIRO, K., J. GAILIT & E.F. PLOW. 1997. Fibrinogen is a ligand for integrin alpha5beta1 on endothelial cells. J. Biol. Chem. **272:** 5360–5366.
47. LANGUINO, L.R., A. DUPERRAY, K.J. JOGANIC, *et al.* 1995. Regulation of leukocyte–endothelium interaction and leukocyte transendothelial migration by intercellular adhesion molecule 1-fibrinogen recognition. Proc. Natl. Acad. Sci. U.S.A. **92:** 1505–1509.
48. SKOGEN, W.F., R.M. SENIOR, G.L. GRIFFIN & G.D. WILNER. 1988. Fibrinogen-derived peptide B beta 1-42 is a multidomained neutrophil chemoattractant. Blood **71:** 1475–1479.
49. THOMPSON, W.D., E.B. SMITH, C.M. STIRK, *et al.* 1992. Angiogenic activity of fibrin degradation products is located in fibrin fragment E. J. Pathol. **168:** 47–53.
50. SPORN, L.A., L.A. BUNCE & C.W. FRANCIS. 1995. Cell proliferation on fibrin: modulation by fibrinopeptide cleavage. Blood **86:** 1802–1810.
51. TANG, L. & J.W. EATON. 1993. Fibrin(ogen) mediates acute inflammatory responses to biomaterials. J. Exp. Med. **178:** 2147–2156.
52. TANG, L., T.P. UGAROVA, E.F. PLOW & J.W. EATON. 1996. Molecular determinants of acute inflammatory responses to biomaterials. J. Clin. Invest. **97:** 1329–1334.
53. ROY, S.N., R. PROCYK, B.J. KUDRYK & C.M. REDMAN. 1991. Assembly and secretion of recombinant human fibrinogen. J. Biol. Chem. **266:** 4758–4763.
54. ERLICH, J., G.C. PARRY, C. FEARNS, *et al.* 1999. Tissue factor is required for uterine hemostasis and maintenance of the placental labyrinth during gestation. Proc. Natl. Acad. Sci. U.S.A. **96:** 8138–8143.
55. BUGGE, T.H., K.W. KOMBRINCK, M.J. FLICK, *et al.* 1996. Loss of fibrinogen rescues mice from the pleiotropic effects of plasminogen deficiency. Cell **87:** 709–719.
56. THIAGARAJAN, P. & K. L. KELLY. 1988. Exposure of binding sites for vitronectin on platelets following stimulation. J. Biol. Chem. **263:** 3035–3038.

57. LAWLER, J. & R.O. HYNES. 1986. The structure of human thrombospondin, an adhesive glycoprotein with multiple calcium-binding sites and homologies with several different proteins. J. Cell. Biol. **103:** 1635–1648.
58. PLOW, E. F. & M. H. GINSBERG. 1981. Specific and saturable binding of plasma fibronectin to thrombin-stimulated human platelets. J. Biol. Chem. **256:** 9477–9482.
59. RUGGERI, Z.M. 1997. Mechanisms initiating platelet thrombus formation. Thromb. Hæmost. **78:** 611–616.
60. RUGGERI, Z.M., J.A. DENT & E. SALDIVAR. 1999. Contribution of distinct adhesive interactions to platelet aggregation in flowing blood. Blood **94:** 172–178.
61. GOTO, S., Y. IKEDA, E. SALDIVAR & Z.M. RUGGERi. 1998. Distinct mechanisms of platelet aggregation as a consequence of different shearing flow conditions. J. Clin. Invest. **101:** 479–486.
62. NI, H., C.V. DENIS, S. SUBBARAO, et al. 2000. Persistence of platelet thrombus formation in arterioles of mice lacking both von Willebrand factor and fibrinogen. J. Clin. Invest. **106:** 385–392.
63. FARRELL, D.H. & P. THIAGARAJAN. 1994. Binding of recombinant fibrinogen mutants to platelets. J. Biol. Chem. **269:** 226–231.
64. COLLEN, D. 1999. The plasminogen (fibrinolytic) system. Thromb. Hæmost **82:** 259–270.
65. ROMER, J., T.H. BUGGE, C. PYKE, et al. 1996. Plasminogen and wound healing. Nat. Med. **2:** 287–92.
66. DREW, A.F., H.L. SCHIMAN, K.W. KOMBRINCK, et al. 2000. Persistent corneal haze after excimer laser photokeratectomy in plasminogen-deficient mice. Invest. Ophthalmol. Vis. Sci. **41:** 67–72.
67. TSIRKA, S.E., A.D. ROGOVE, T.H. BUGGE, et al. 1997. An extracellular proteolytic cascade promotes neuronal degeneration in the mouse hippocampus. J. Neurosci. **17:** 543–552.
68. TSIRKA, S.E., T.H. BUGGE, J.L. DEGEN & S. STRICKLAND. 1997. Neuronal death in the central nervous system demonstrates a non-fibrin substrate for plasmin. Proc. Natl. Acad. Sci. U.S.A. **94:** 9779–9781.
69. COLEMAN, J.L., J.A. GEBBIA, J. PIESMAN, et al. 1997. Plasminogen is required for efficient dissemination of B. burgdorferi in ticks and for enhancement of spirochetemia in mice. Cell **89:** 1111–1119.
70. GEBBIA, J.A., J.C. MONCO, J.L. DEGEN, et al. 1999. The plasminogen activation system enhances brain and heart invasion in murine relapsing fever borreliosis. J. Clin. Invest. **103:** 81–87.
71. COLEMAN, J.L. & J.L. BENACH. 2000. The generation of enzymatically active plasmin on the surface of spirochetes. Methods **21:** 133–141.
72. GOGUEN, J.D., T. BUGGE & J.L. DEGEN. 2000. Role of the pleiotropic effects of plasminogen deficiency in infection experiments with plasminogen-deficient mice. Methods **21:** 179–183.
73. DREW, A.F., A.H. KAUFMAN, K.W. KOMBRINCK, et al. 1998. Ligneous conjunctivitis in plasminogen-deficient mice. Blood **91:** 1616–1624.
74. KAO, W.W., C.W. KAO, A.H. KAUFMAN, et al. 1998. Healing of corneal epithelial defects in plasminogen- and fibrinogen-deficient mice. Invest. Ophthalmol. Vis. Sci. **39:** 502–508.
75. DREW, A.F., H.L. TUCKER, K.W. KOMBRINCK, et al. 2000. Plasminogen is a critical determinant of vascular remodeling in mice. Circ. Res. **87:** 133–139.
76. AKASSOGLOU, K., K.W. KOMBRINCK, J.L. DEGEN & S. STRICKLAND. 2000. Tissue plasminogen activator-mediated fibrinolysis protects against axonal degeneration and demyelination after sciatic nerve injury. J. Cell. Biol. **149:** 1157–1166.
77. BEZERRA, J.A., T.H. BUGGE, H. MELIN-ALDANA, et al. 1999. Plasminogen deficiency leads to impaired remodeling after a toxic injury to the liver. Proc. Natl. Acad. Sci. U.S.A. **96:** 15143–15148.
78. CHEN, Z. L. & S. STRICKLAND. 1997. Neuronal death in the hippocampus is promoted by plasmin-catalyzed degradation of laminin. Cell **91:** 917–925.
79. HIRAOKA, N., E. ALLEN, I.J. APEL, et al. 1998. Matrix metalloproteinases regulate neovascularization by acting as pericellular fibrinolysins. Cell **95:** 365–377.

80. BINI, A., Y. ITOH, B.J. KUDRYK & H. NAGASE. 1996. Degradation of cross-linked fibrin by matrix metalloproteinase 3 (stromelysin 1): hydrolysis of the gamma Gly 404-Ala 405 peptide bond. Biochemistry **35:** 13056–13063.
81. BINI, A., D. WU, J. SCHNUER & B. J. KUDRYK. 1999. Characterization of stromelysin 1 (MMP-3), matrilysin (MMP-7), and membrane type 1 matrix metalloproteinase (MT1-MMP) derived fibrin(ogen) fragments D-dimer and D-like monomer: NH_2-terminal sequences of late-stage digest fragments. Biochemistry **38:** 13928–13936.
82. XIAO, Q., M. J. DANTON, D. P. WITTE, et al. 1997. Plasminogen deficiency accelerates vessel wall disease in mice predisposed to atherosclerosis. Proc. Natl. Acad. Sci. U.S.A. **94:** 10335–10340.
83. LOU, X.J., N.W. BOONMARK, F.T. HORRIGAN, et al. 1998. Fibrinogen deficiency reduces vascular accumulation of apolipoprotein(a) and development of atherosclerosis in apolipoprotein(a) transgenic mice. Proc. Natl. Acad. Sci. U.S.A. **95:** 12591–12595.
84. BUGGE, T.H., K.W. KOMBRINCK, Q. XIAO, et al. 1997. Growth and dissemination of Lewis lung carcinoma in plasminogen-deficient mice. Blood **90:** 4522–4531.
85. BUGGE, T.H., L.R. LUND, K.W. KOMBRINCK, et al. 1998. Reduced metastasis of Polyoma virus middle T antigen-induced mammary cancer in plasminogen-deficient mice. Oncogene **16:** 3097–3104.
86. BUSSO, N., V. PECLAT, K. VAN NESS, et al. 1998. Exacerbation of antigen-induced arthritis in urokinase-deficient mice. J. Clin. Invest. 102: 41–50.
87. EITZMAN, D.T., R.D. MCCOY, X. ZHENG, et al. 1996. Bleomycin-induced pulmonary fibrosis in transgenic mice that either lack or overexpress the murine plasminogen activator inhibitor-1 gene. J. Clin. Invest. **97:** 232–237.
88. KITCHING, A.R., S.R. HOLDSWORTH, V.A. PLOPLIS, et al. 1997. Plasminogen and plasminogen activators protect against renal injury in crescentic glomerulonephritis. J. Exp. Med. **185:** 963–968.
89. XIAO, Q., M. J. DANTON, D. P. WITTE, et al. 1998. Fibrinogen deficiency is compatible with the development of atherosclerosis in mice. J. Clin. Invest. **101:** 1184–1194.
90. CARMELIET, P., L. MOONS, V. PLOPLIS, et al. 1997. Impaired arterial neointima formation in mice with disruption of the plasminogen gene. J. Clin. Invest. **99:** 200–208.
91. BUCCHERI, G., D. FERRIGNO, C. GINARDI & C. ZULIANI. 1997. Hæmostatic abnormalities in lung cancer: prognostic implications. Eur. J. Cancer **33:** 50–55.
92. WAYMAN, J., D. O'HANLON, N. HAYES, et al. 1997. Fibrinogen levels correlate with stage of disease in patients with oesophageal cancer. Br. J. Surg. **84:** 185–188.
93. DVORAK, H.F. 1986. Tumors: wounds that do not heal. Similarities between tumor stroma generation and wound healing. N. Engl. J. Med. **315:** 1650–1659.
94. PALUMBO, J.S., K.W. KOMBRINCK, A.F. DREW, et al. 2001. Fibrinogen is an important determinant of the metastatic potential of circulating tumor cells. Blood. **96:** 3302–3309.

Factor XIII: Structure, Activation, and Interactions with Fibrinogen and Fibrin

LASZLO LORAND

Department of Cell and Molecular Biology and the Feinberg Cardiovascular Research Institute, Northwestern University Medical School, Chicago, Illinois, USA

ABSTRACT: Fibrin stabilizing factor (factor XIII or FXIII) plays a critical role in the generation of a viable hemostatic plug. Following exposure to thrombin and calcium, the zymogen is activated to FXIIIa that, in turn, catalyzes the formation of N^ϵ(γ-glutamyl)lysine protein-to-protein side chain bridges within the clot network. Introduction of these covalent crosslinks greatly augments the viscoelastic storage modulus of the structure and its resistance to fibrinolytic enzymes. Analysis of the individual reaction steps and regulatory control mechanisms involved in clot stabilization enabled us to reconstruct the entire physiological process. This also serves as a guide for the differential diagnosis of the variety of molecular defects of fibrin stabilization.

KEYWORDS: Fibrin stabilization; Fibrinogen; Factor XIII; Structure; Activation; Interactions.

INTRODUCTION

No volume with the title Fibrinogen 2000 would be complete without discussing the role of the fibrin stabilizing factor that, although a relative newcomer, turns out to be a very important player in the final stages of blood coagulation. The significance of the factor for maturation of the fibrin network, to provide the clot with sufficient stiffness for wound closure and with the necessary resistance against thrombolytic enzymes, is now well recognized. Defects that interfere with the normal operation of the clot stabilizing system cause severe hemorrhagic diseases. On the other hand, the possibility of blocking these reactions in a selective and controlled manner opens up new therapeutic modalities for digesting thrombi at much lower concentrations of lytic agents than currently used.[1] At the turn of the new century, as we reflect and try to summarize our knowledge about clot stabilization, we should also take the opportunity to point out the many gaps that still exist and await to be answered.

Address for correspondence: Laszlo Lorand, Department of Cell and Molecular Biology, Northwestern University Medical School, Searle 4-555, 303 E. Chicago Avenue, Chicago, IL 60611-3008, USA. Voice: 312-503-0591; fax: 312-503-0590.

l-lorand@northwestern.edu

CLOT SOLUBILITIES

My own introduction to the field of fibrinogen/fibrin was as a final year medical student working in the Szent-Györgyi Institute of Biochemistry where in 1948 I began *A study on the solubility of fibrin clots in urea.*[2] Remarkably, at that time there were already three relevant—although quite confusing—articles in the literature. The most recent of these (in 1937), by Meissner and Wöhlisch,[3] reported that (as translated from the German) "Limbourg was the first to show (in 1889)[4] ... that fibrin could be dissolved in urea. This was confirmed by Spiro (in 1900)[5]". However, Meissner and Wöhlisch were unable to reproduce these results and wrote in conclusion that "according to our findings, fibrin cannot be solubilized in concentrated urea solutions". As I tried to shed light on these contradictions, I found that clots generated by the action of purified thrombin on purified fibrinogen were readily soluble in 5 M (30%) urea, and that removing the urea by dialysis allowed the fibrin to repolymerize into a clot.[2] This finding proved to be useful later for separating fibrin by repeated cycles of solubilization and reclotting.[6] Another very important observation was that the intrinsic viscosity of fibrin in urea was identical to that of the parent fibrinogen.[2] This indicated that thrombin treatment did not significantly alter the hydrodynamic properties of the protein particles which, of course, was fully borne out by the demonstration that thrombin caused only a minor alteration in the fibrinogen molecule with the limited proteolytic cleavage of the relatively small N-terminal fibrinopeptide moieties from the body of the large clotting protein:[7,8]

$$n \text{ fibrinogen} \xrightarrow[\text{fibrinopeptides}]{\text{thrombin}} n \text{ fibrin} \underset{\text{5M urea}}{\rightleftarrows} (\text{fibrin})_n \text{ clot}$$

In sharp contrast to the experiments with purified fibrinogen and thrombin, the clots obtained from recalcified citrated plasma samples could not be dissolved in urea.[2,9] This finding was in accord with a report by Robbins on the difference of solubilities of fibrin and plasma clots in acids.[10] However, results obtained with the solubilization of clots in acids could be misleading because of the activation of a circulating pepsin; the proteolytic degradation of fibrin, as opposed to solubilization, would confound the tests aimed at assessing the presence of fibrin stabilizing factor (factor XIII) in a patient's sample.[11] Hence, solubility in urea proved to be the more reliable diagnostic test for identifying the great variety of disorders of fibrin stabilization. In the absence of FXIII or in the presence of inhibitors that provide a total blockade of the action of this factor, 5M urea can dissociate all clots into monomeric fibrin. However, a urea-solubilized clot is not always the equivalent of monomeric fibrin. Partially crosslinked fibrin units—we do not know at present the limiting size or even if there is a sharp cut off for oligomers—can also be solubilized in urea. Therefore, urea-solubility of a patient's plasma clot does not necessarily signify the total absence or a complete blockade of factor XIII.

Plasma clots can be dissolved in urea only in the presence of a reducing agent (e.g., thioglycolic acid),[12] a mixture which, as we now know, breaks apart the three constituent chains of monomeric fibrin.[13]

FACTOR XIIIa AND TRANSGLUTAMINASE

Factor XIII needs to be activated by thrombin and Ca^{2+} to factor XIII_a in order to catalyze the covalent crosslinking of blood clots.[14] The reaction, which in vertebrates is superimposed on the non-covalent polymerization of fibrin, can be blocked selectively by small primary amines (H_2NR; e.g., glycine ethylester and dansylcadaverine) without interfering with the clotting time:[15]

$$(\text{fibrin})_n \text{ clot} \xrightarrow[H_2NR]{FXIII_a} \text{crosslinked (fibrin)}_n \text{ clot}$$

These findings lead directly to the conclusion that $FXIII_a$ was a transamidase of the transglutaminase family of enzymes[16] operating with a kinetic pathway of acylation and deacylation and producing N^ε(γ-glutamyl)lysine crosslinks.[15,17–19] The general features of the transamidating reaction, including the involvement of a cysteine active center thiol[20] assisted by a histidine residue,[21,22] are similar to what we know about the papain group of enzymes. Human fibrinogen and fibrin are good substrates for the Ca^{2+}-dependent transglutaminases present in many types of cells.[16] In fact, as exemplified by the clotting of lobster blood or that of the seminal vesicle secretion proteins in rodents in forming the copulation plug, the transglutaminase mediated direct polymerization of proteins may be regarded as the evolutionary prototype for clotting systems. In each of these instances, clotting itself is inhibited by the same primary amines that in human plasma would only block the stabilization of the preassembled fibrin units:[23,24]

$$\text{n protein} \xrightarrow[\substack{Ca^{2+} \\ NH_2R}]{\text{transglutaminase}} (\text{protein})_n \text{ polymer}$$

Although cellular transglutaminases (e.g., from red cells) are capable of crosslinking fibrin clots and can also polymerize fibrinogen,[16,25,26] the crosslinked product

$$\text{n fibrinogen} \xrightarrow[Ca^{2+}]{\text{transglutaminase}} \text{crosslinked (fibrinogen)}_n$$

does not generate a three-dimensional gel lattice resembling a fibrin clot.

Just as papain, transglutaminases are good hydrolytic enzymes and, just as the transglutaminases, the papain family of proteases also function well as transamidases. The difference lies in the fact that, in contrast to papain, $FXIII_a$ and the related transglutaminases show exquisite specificities for the amine-carrying second substrates; that is, the thiolester intermediate generated in the acylation step with the Gln substrate forms an extra Michaelis complex with the amine and can thus be saturated by the Lys containing second substrate. As such, aminolysis can take preference over hydrolysis.[21,27–30]

FACTOR XIII: STRUCTURE AND ACTIVATION

The FXIII zymogen is comprised of AB protomers, in which the A subunits possess the catalytic potential. Since measurements of molecular weights for the purified protein—although obtained at concentrations far above physiological—suggest an A_2B_2 assembly,[31] it is generally assumed that FXIII circulates as an A_2B_2 heterotetramer, but we do not really know. The amino acid sequence of the carrier B subunit shows it to be related to the family of proteins containing the characteristic small consensus repeats, or Sushi domains.[32] Unlike the A subunits in FXIII, the B subunits are heavily sialylated, which gives rise to an electrophoretic heterogeneity over and above the polymorphism seen in the primary sequences of the B protein.[33] The A subunits share sequence homologies with the transglutaminases,[34] the already mentioned group of Ca^{2+}-dependent enzymes that seemed to have evolved from the thiol proteases.[35] The papain-like catalytic triad of cysteine-histidine-aspartic acid is fully preserved in the crystallized recombinant rA_2 even though the structure examined was that of the inactive zymogen.[36] The crystallographic studies identified

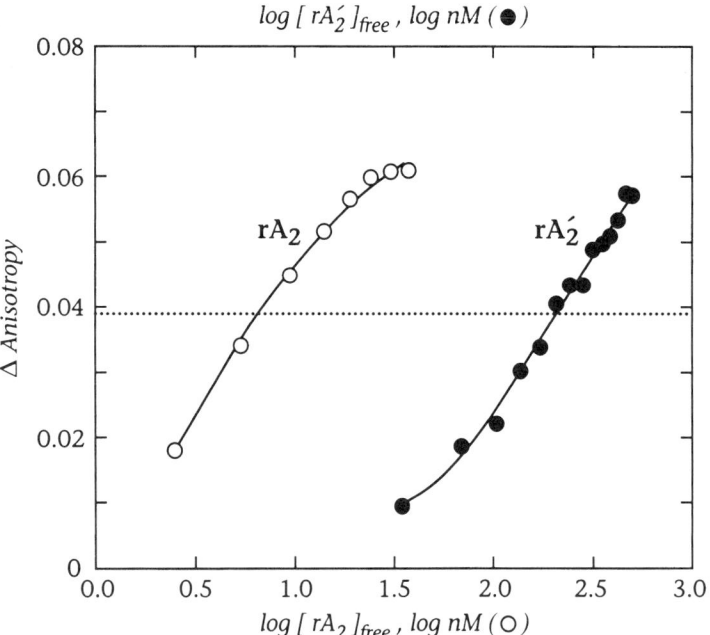

FIGURE 1. Activation of recombinant A_2 (rA_2) by thrombin (rA_2') weakens its association to the carrier B subunits (labelled with fluorescein: B_2^F). The *ordinate* shows increases in anisotropy from the starting B_2^F solution (19 nM in 2 ml of 75 mM Tris-Cl, pH 7.5, 0.15 M NaCl, and 0.5 mM EDTA). Aliquots of rA_2 (3 μl; ○, *bottom abscissa*) or thrombin-activated rA_2' (18 μl; ●, *top abscissa*) were injected and changes in anisotropy were recorded after 5 min.[40]

four distinct domains in the protein: an N-terminal β sandwich, the papain-like catalytic core, and a barrel 1 and a barrel 2 domain.[37]

Although FXIII can be activated under forced, non-physiological conditions without thrombin,[33,38] the far more facile physiological process requires the consecutive actions of thrombin and Ca^{2+}.[14] Thrombin initiates the conversion of the zymogen by hydrolyzing the peptide bond between Arg 37 and Gly 38 in the A subunits:[39]

$$A_2B_2 \xrightarrow{\text{thrombin}} A'_2 B_2 + 2 \text{ activation peptides}$$

The rationale for the limited proteolytic modification of A subunits became clear when it was shown that the cleavage of activation peptides significantly weakened the heterologous association of subunits (see FIGURE 1).[40] The Ca^{2+}-dependent dissociation:

$$A'_2 B_2 \xrightarrow{Ca^{2+}} A'_2 + B_2$$

requires a much higher concentration of Ca^{2+} than does the next step that leads to the unmasking of the active center cysteine thiols in the $A'_2 \rightarrow A^*_2$ transition. As such, the B subunits seem to act as a brake in the overall conversion of FXIII (A_2B_2) to XIIIa (A^*_2). [20,41]

Several important features of XIII activation, illustrated in FIGURE 2, still need to be resolved. In this diagram, the A_2B_2 protein is depicted with a C2 symmetry. However, without knowing the three dimensional structures of the B subunits and of the tetramer, this is obviously just an abstract illustration. We have no idea about the

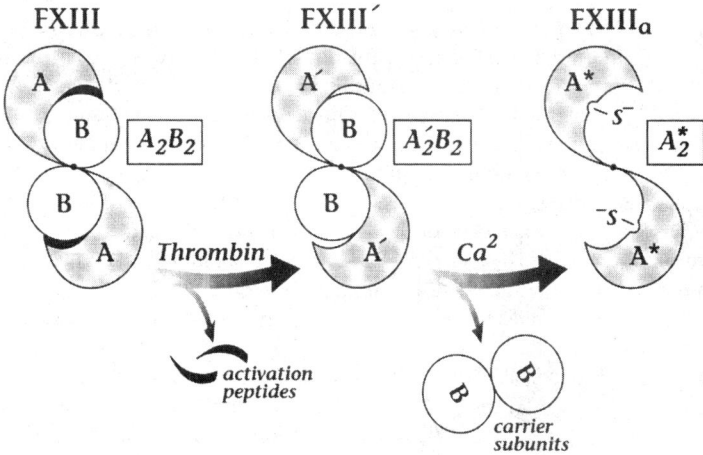

FIGURE 2. Conversion of the factor XIII zymogen to the XIIIa enzyme (A^*_2) occurs in two consecutive steps, respectively promoted by thrombin and Ca^{2+}. The thrombin-modified FXIII' allozymogen is still inactive. Dissociation from the carrier B subunits facilitates the conformational change necessary for unmasking the active center cysteine thiol in the catalytic A^*_2 subunits.

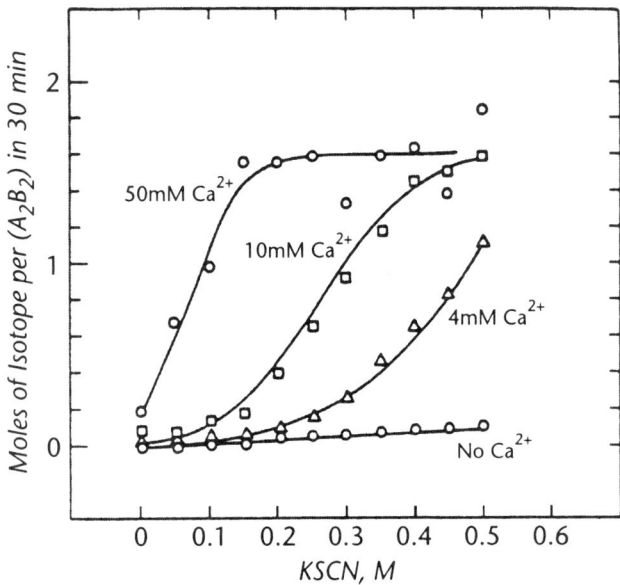

FIGURE 3. Thrombin-independent, non-physiological activation of factor XIII. The reaction mixtures ($\mu = 0.7$; pH 7.5, 37°C) contained various concentrations of the KSCN "chaotrope" (*abscissa*) and Ca^{2+}. Unmasking of the active sites in the A_2^o enzyme were titrated with iodo[1-^{14}C]acetamide (ordinate).[33]

side chains and interfaces involved in the heterologous A:B binding of the subunits. Studies on the thrombin-independent activation of $A_2B_2 \rightarrow A_2^o + B_2$ [33,38] allow only very general inferences to be made regarding the forces (electrostatic and hydrophobic) that hold the heterotetramer together. This mode of zymogen conversion requires a high ionic strength and is greatly enhanced in the presence of anions (*p*-toluenesulfonate ≥ thiocyanate ≥ iodide ≥ bromide) that are known to be particularly effective in disrupting the regular water lattice (see FIGURE 3). With small synthetic substrates, the steady-state kinetic constants of the A_2^o enzyme are indistinguishable from those of the thrombin-activated A_2^* enzyme.[33] Thus, the thrombin mediated release of the N-terminal activation peptide is not an obligatory requirement for the unmasking of the catalytic site; differences in the activities of A_2^o and A_2^* can be seen only with larger substrates.

In FIGURE 2, I have tried to illustrate the fact that the cleavage of activation peptides weakens the A:B interaction substantially (by about 2 kcal/mole), but this does not necessarily mean that the N-terminal peptide of A is actually in contact with B. A key issue would be to decide whether the activation peptides (ap) are released into the medium. The description of the thrombin reaction as $A_2B_2 \rightarrow A_2'B_2 + 2\,ap$ rests on data in which the fragments were separated from $A_2'B_2$ by alcohol precipitation.[39] Hence the question is whether, under native conditions, the N-terminal peptides, even though cleaved from the covalent backbone of the parent protein, would still remain with the A′ subunits, $A_2B_2 \rightarrow (A':ap)_2B_2$. Would the situation be similar to that of the S-peptide in subtilisin-treated RNAse (S-protein) or mimic the release of fibrinopeptides into the clot liquor from fibrinogen? The report that the structure of the thrombin plus Ca^{2+}-activated recombinant rA_2^* was identical to that of the rA_2 zymogen[42] would seemingly support the S peptide:S protein RNAse analogy. However, we can expect that the equilibrium governing the dissociation of ap from A would be quite different in the plasma milieu, and particularly in the presence of fibrin, than it would be under the conditions of very high A_2^* concentrations employed for crystallization. Moreover, the thrombin plus Ca^{2+}-treated crystallized material was clearly not the enzymatically active rA_2^* species, because its catalytic center still remained inaccessibly buried. If there was no structural difference between rA_2 and rA_2', how could one account for the substantial difference in the binding affinities of the virgin and thrombin-modified forms of the A subunits to the carrier B subunits (FIG. 1)?[40] How does the structure of free rA_2' differ from that of the same subunit when it is in the carrier bound $rA_2'B_2$ form?

Except for the fact that the active site cysteine thiol becomes accessible to titration when enzyme activity is generated in the

$$A_2' \xrightarrow{Ca^{2+}} A_2^*$$

step,[20] we really do not know any of the details of this critical conformational change on the pathway to FXIII activation. Given that the cysteine–histidine–aspartic acid construct for the papain-like catalytic triad is already in place in both the rA_2 and rA_2' zymogens,[36] we may assume that regardless of the nature of the conformational change, it would not alter the distances between the three catalytic residues, but would only open an access from the surface of the protein to the active site. Once a channel or a cleft is made available to bind the Gln side chain in the substrate for forming the thiolester intermediate with the active site cysteine, further conformational changes might occur to accommodate the Lys substrate. As far as Ca^{2+} binding is concerned, it is unlikely that the site which was identified in the X-ray study of the inactive rA_2 zymogen[37] would be the Ca^{2+} binding site necessary to be occupied by the cation for the unmasking of the active center cysteine in the $A_2' \rightarrow A_2^*$ conversion. In FIGURE 2, the dissociated forms of the subunits are shown as A_2^* and B_2, but we do not really know whether the homodimeric associations could be maintained in the plasma clotting environment. Nor do we know whether the free B_2 produced in the activation process is identical to the virgin B_2 as it originally exists in the A_2B_2 zymogen.

REGULATORY FUNCTIONS OF FIBRIN

Correct timing for the activation and operation of the FXIII system is of critical significance for efficient hemostasis. A key feature for synchronizing the clotting and crosslinking events is that thrombin plays the dual role of converting fibrinogen to fibrin and also initiating the process of activating FXIII. The fibrin substrate, itself, greatly accelerates the thrombin catalyzed cleavage of FXIII[43,44]

$$A_2B_2 \xrightarrow[\text{(fibrin)}]{\text{thrombin}} A_2'B_2 + 2ap$$

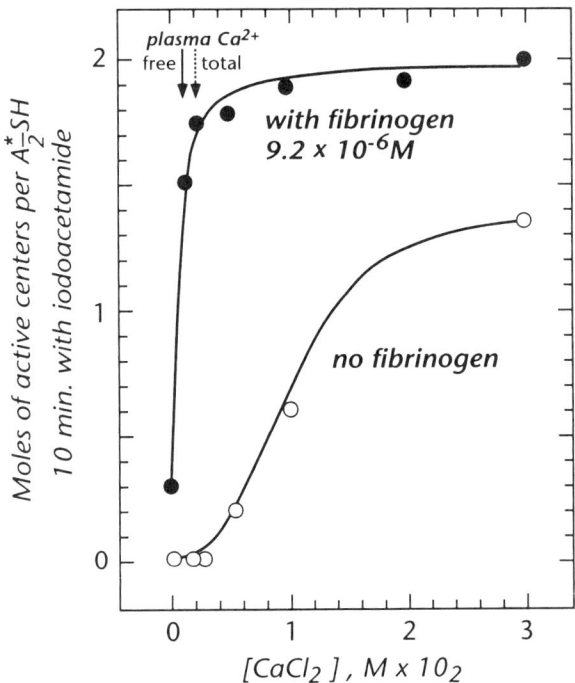

FIGURE 4. Calcium requirement for the unmasking of active center cysteine in A_2^* in the $A_2'B_2 \to A_2^* + B_2$ conversion, in the absence (○) and presence (●) of a physiological concentration of fibrinogen. Active sites were titrated with iodo[1-^{14}C]acetamide (*ordinate*).[20]

and enables the thrombin modified, but still inert, $A_2'B_2$ zymogen to dissociate and express enzyme activity in a concerted manner

$$A_2'B_2 \xrightarrow[\text{(fibrin)}]{Ca^{2+}} A_2^* + B_2$$

at the physiological 1.5 mM Ca^{2+} concentration in plasma, as is shown in FIGURE 4.[20] With fibrin assistance, the heterologous dissociation of $A_2'B_2$ subunits can actually proceed—albeit more slowly—even in the absence of Ca^{2+}.[20,40] These unique "feedforward" controls must have evolved to ensure that in the physiological sequence of events only fibrin, and not the parent fibrinogen molecule, would be the target for crosslinking by XIIIa. Based on these studies, we are now in a position to present a seemingly accurate outline for the entire physiological pathway of clot stabilization (see FIGURE 5), where the fibrin-dependent regulatory features are indicated by broken arrows.

Despite the disparity between the circulatory halflives of FXIII and of fibrinogen[45] there is good reason to think that almost all of the FXIII in the plasma phase is bound to fibrinogen. It has long been known that even highly purified fibrinogen preparations contain a form of FXIII that requires a thiol reducing agent (e.g., cysteine), in addition to thrombin and Ca^{2+}, to be activated.[46,47] Direct binding of FXIII to fibrinogen has also been demonstrated.[48] Siebenlist *et al.* made the interesting observation that the carrier B subunits contribute in a major way to the binding of FXIII to fibrinogen.[49] In this context, it would be important to find out if free B_2 would compete effectively against A_2B_2. These are critical issues, because it must be borne in mind that all the reactions depicted for the thrombin plus Ca^{2+}-ion promoted conversion of FXIII to XIIIa shown in FIGURE 2, occur in association with the clot matrix. The minority variants of fibrinogen with γ408–427 C-terminal extensions (γ') show the highest affinity for A_2B_2;[49,50] the homologous crosslinking of α chains is also faster for these molecules.[51] It is also likely that the FXIII zymogen interacts with the mobile mid-sections of the Aα chains because fragments representing Aα242–424 or Aα243–476 allow the XIII'→XIIIa conversion to proceed at the physiological 1.5 mM Ca^{2+} concentration in plasma when, in the absence of this fragment or of fibrin(ogen), a 10- to 20-fold higher concentration of Ca^{2+} would be required (see FIG. 4).[20] How do the various forms of XIII move among the different sites in fibrin? Given the quite strict timing that Glns of the γ chains react before the Glns of the α chains,[52,53] the question arises as to how A_2^* can accomplish this well ordered task. It would seem that each molecule of A_2^* would have to move from a γC site in one fibrin (perhaps from the extended γ' sites of the minor variants) to the γC sites in nearly a hundred other fibrin molecules before starting the reactions with the Glns in α chains. Here again, would each molecule of A_2^* have to finish the reaction with the two or three α Glns in a single fibrin unit before jumping to the next fibrin? Or would A_2^* react serially with, say, all the hundred or so αGln 328s before

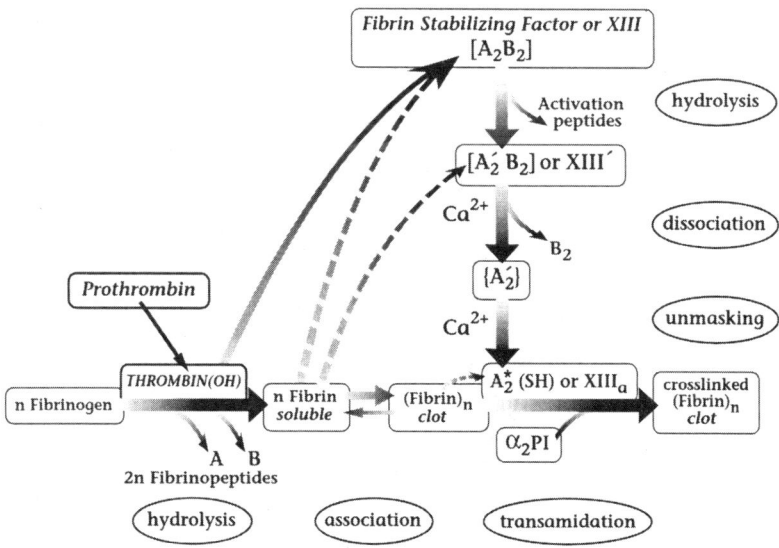

FIGURE 5. Reactions and controls of clot stabilization. In kinetic terms and in physiological significance, the substrate-level "feed-forward" regulation by fibrin, shown with *broken arrows,* matches the influence of factor VIII on thrombin generation.

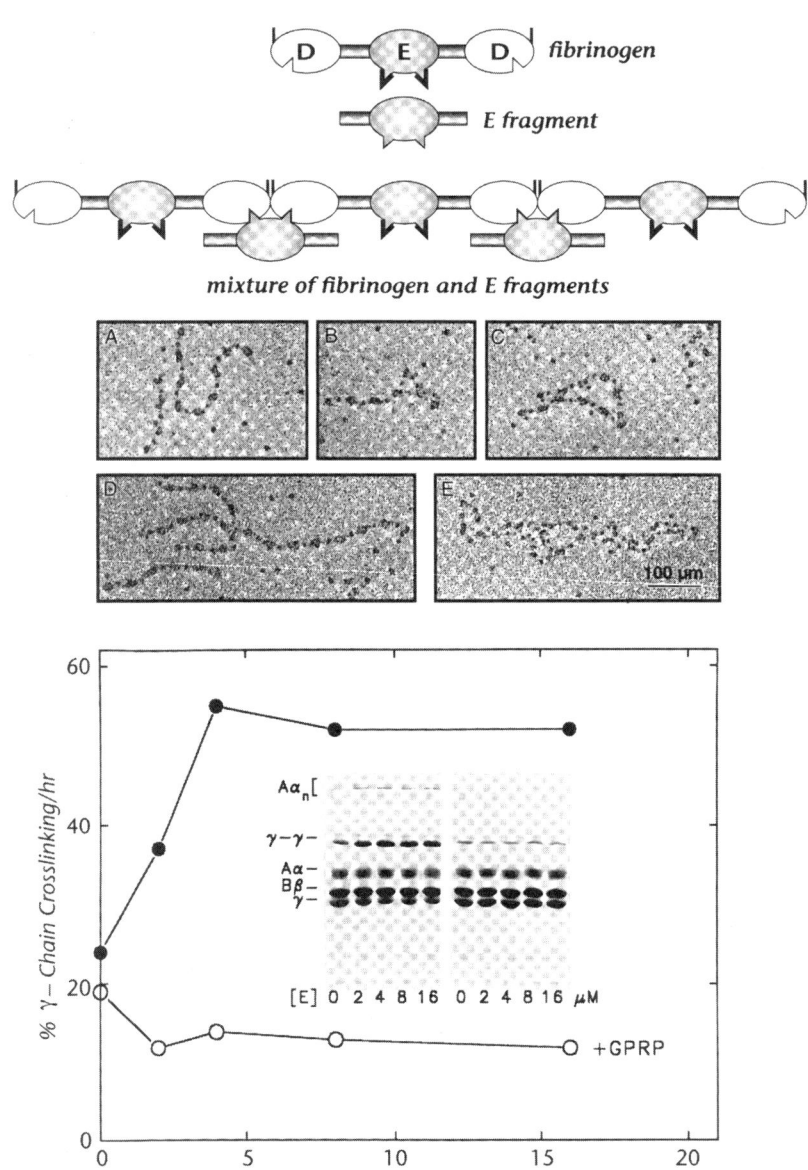

FIGURE 6. The assembly of fibrinogen molecules with the aid of fibrinopeptide-denuded E fragments accelerates the factor XIII-driven crosslinking of γ chains. GlyProArgPro blocks the formation of fibrinogen:E complexes and also abolishes the kinetic enhancement for crosslinking.[57,58]

engaging the same number of αGln 366s? Even though the clot is a nearly 100% hydrated medium, would solution kinetics explain such a highly ordered sequence of reactions? Would the A^* or A_2^* species be somehow propelled along the fibrin molecule and, if so, how would this processive movement be brought about; or would the enzyme, however transiently, be released into the clot liquor prior to participating in yet another cycle of crosslinking? What are the complementary binding sites in FXIII, XIII′, XIIIa, and in the fibrin substrate that mediate these interactions? Thus far, the only information we have is that saturating fibrin with A_2B_2 does not seem to affect the binding of a derivative of A_2^*, which indicates that A_2B_2 and A_2^* bind to separate non-interacting sites in fibrin.[54] Finally, how is the A_2^* action terminated so that it does not spread to proteins outside the immediate domain of the clot?

FXIIIa reacts about an order of magnitude faster with the γ chain Gln sites than with the α chain sites of fibrin even when possible differences of second substrate specificities are obviated by measuring the incorporation of an exogenous amine, such as glycine ester or dansylcadaverine.[52,53,55,56] Does this mean that the flanking sequences around the γ chain Gln sites react more favorably, perhaps by binding better to A^*, than the sequences around the enzyme-reactive Glns in the α chains? If that was the case, how would one explain that the tail end γC sequence, which is also present in the parent fibrinogen, shows a much weaker reactivity to XIIIa than the same sequence in fibrin? We find that the C-terminal γ chain segments when aligned end-to-end as in fibrin assemblies represent a far better substrate for A_2^* (or A^*) than these

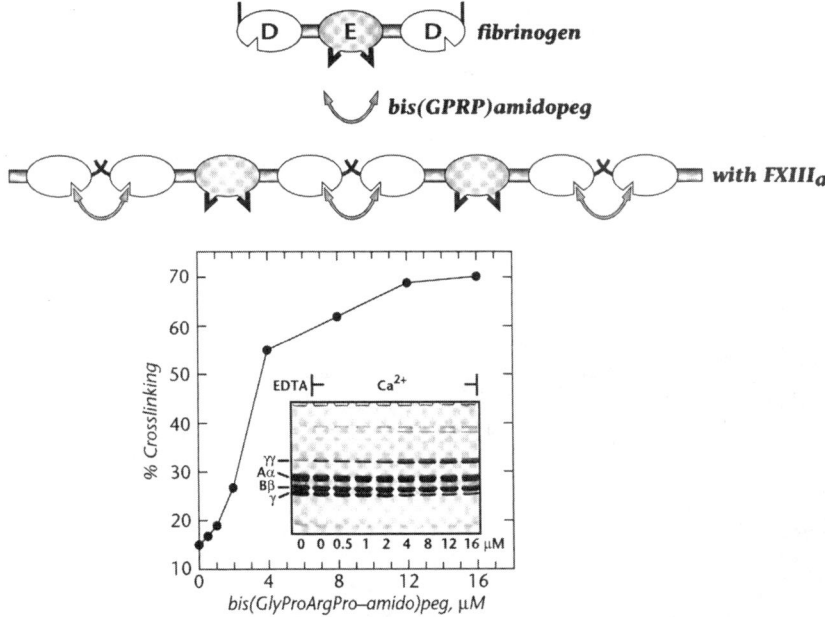

FIGURE 7. The synthetic, double-headed bis(GlyProArgPro) amidopolyethyleneglycol can substitute for E fragments (see FIG. 6) in promoting the factor XIIIa-catalyzed homologous crosslinking of γ chains with fibrinogen.[59]

same sites when they exist separately in the monomeric fibrinogen molecules in solution. The XIIIa reaction at the γC sites of fibrinogen is immediately enhanced when the parent fibrinogen molecules are allowed to form linear assemblies through complex formation with thrombin modified E fragments (see FIGURE 6)[57,58] or with a synthetic double-headed (GlyProArgPro)polyethyleneglycol as ligand (see FIGURE 7).[59] Thus, the construction of the proper landing site for the A_2^* enzyme appears to be an assembly-driven phenomenon.

THE CROSSLINKING REACTION

Though fibrin fibers become somewhat thinner and longer when stabilized by XIIIa, the reaction does not seem to affect clot morphology in a major way in regard to fiber and branch point densities.[60] Notwithstanding the morphological similarities, the properties of the network have been changed remarkably because the clot can no longer be dissolved in urea and, as is shown in FIGURE 8, there is a large (ca. five-fold) increase in viscoelastic storage modulus,[60,61] a measure of clot stiffness. In addition, the clots and thrombi acquire a high degree of resistance to digestion by thrombolytic agents (see FIGURE 9).[1,62–64] The latter effect is attributed to the XIIIa-mediated covalent attachment of a fraction of the α_2 plasmin inhibitor in plasma to the α chains of a few fibrin molecules.[65,66] In general, inhibitors of XIIIa-catalyzed crosslinking block all three features of thrombus maturation and act to increase clot solubility in urea, reduce clot stiffness and enhance fibrinolysis/thrombolysis.[1]

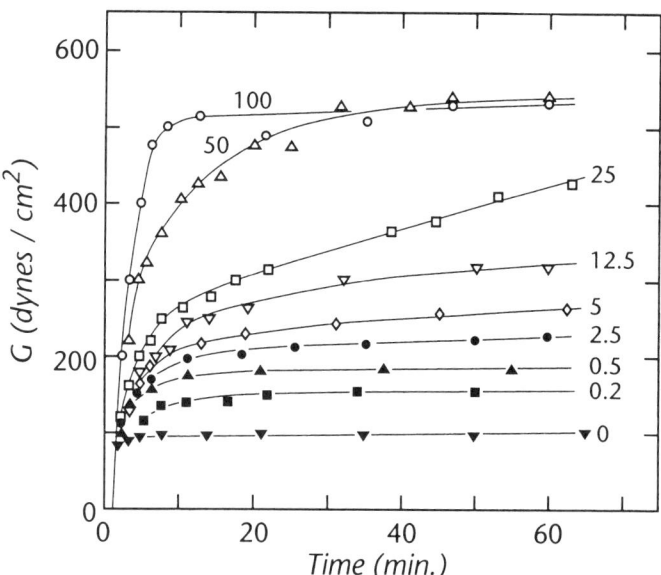

FIGURE 8. Supplementation of factor XIII-deficient plasma with the missing factor can normalize clot stiffness. Purified A_2B_2 [0.2 to 100×10^{-9} M] were added.[61]

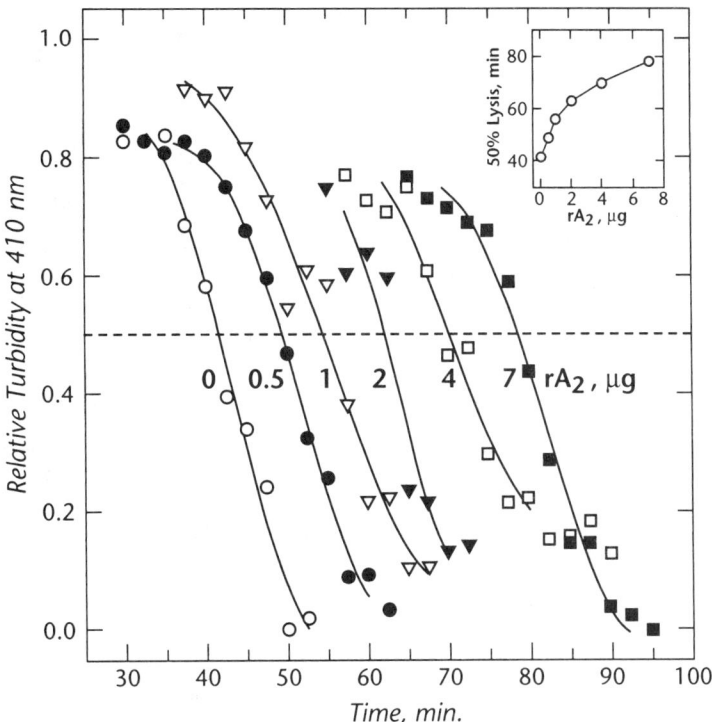

FIGURE 9. Failure of factor XIII-deficient clots to resist lysis can be corrected with supplementation of rA_2 (up to 7 μg/0.2 ml).[1]

The covalent crosslinking of fibrin units and attachments of other blood constituents (including α_2 plasmin inhibitor) to the clot matrix occurs through an amide exchange, that is, transamidation reaction between select Gln and Lys residues of neighboring protein molecules. As a result, a few strategically located N^ε(γ-glutamyl)lysine side chain bridges are formed. The end-to-end linear alignment of fibrin units is strengthened by two such crosslinks (ligations) with a reciprocal orientation very close to the C-termini of the γ chains.[67] Homologous crosslinking of the α chains with production of α_n is the next event in fibrin stabilization.[31,68] These are considered to be essential for generating urea insoluble clots, but we do not know what the precise relationship is between the type and the degree of crosslinking, on one hand, and the insolubility of the clot in urea, on the other. Crosslinks spanning more than two γ chains and several α chains, giving rise to homologous γ_2, γ_3, γ_4, and hybrid $\alpha_p\alpha_q$ crosslinked chain structures, are also thought to play important roles in clot stabilization.[25,69,70]

The human red cell transglutaminase crosslinks fibrinogen as well as fibrin by αγ and α_n type of intra- and intermolecular N^ε(γ-glutamyl)lysine bonds.[25,26] If this enzyme was released from trapped erythrocytes during the aging of thrombi, at

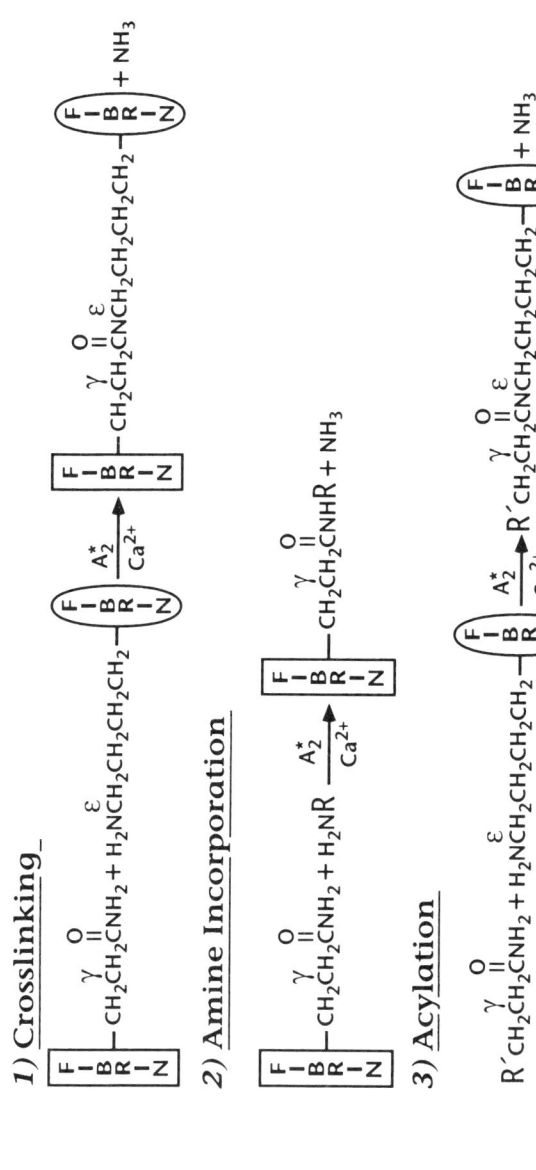

FIGURE 10. Competitive inhibitors block the factor XIIIa (A_2^*)-catalyzed crosslinking reactions (shown in *Scheme 1*). Serving as tracers, the inhibitors themselves become incorporated by the enzyme into the acceptor (*Scheme 2*) and donor (*Scheme 3*) crosslinking sites.

wound sites or in the fibrous caps of atherosclerotic plaques, its action would add to the complex pattern of crosslinking of the network.

The introduction of tracers for the FXIIIa-directed labelling of fibrin was essential not only for kinetic appraisal,[52,53,55,56] but also for localizing and identifying the Gln and Lys residues in fibrin, itself, and in its crosslinking protein partners that, in the absence of these compounds, would participate in forming N^ε(γ-glutamyl)lysine linkages. As is illustrated in FIGURE 10, primary amines (isotopic glycine esters and the fluorescent dansylcadaverine being the first examples) compete against the FXIIIa-reactive Lys donors in the deacylation step and label the Gln (i.e., acceptor) crosslinking sites of the protein substrate, whereas Gln containing peptides (dansyl-ε-aminocaproylAQQIV as the prototype) compete in the acylation step for forming the thiolester intermediate and label the Lys (i.e., donor) crosslinking sites.[71,72]

In the homologous crosslinking of the γ chains, γGln398 and 399 react with γLys406.[67,73] In the α chains, AαGln237, 328, and 366 were identified as acceptors.[74,75] Interestingly, the Lys donors are also concentrated into the flexible domains of the α chains in fibrin. With dansyl-ε-aminocaproylAQQIV as a tracer, we showed that AαLys 219, 427, 448, 539, 446, 580, 583, 601, and 606 were acylated by FXIIIa.[76] This work was later extended by using NQEQVSPLTLLK as a probe, which, in addition to labelling most of the same residues, identified some other lysines while missing AαLys583: AαLys 208, 219, 224, 418, 427, 429, 448, 508, 539, 556, 580, 601, and 606.[77] The overlap between these data is all the more impressive because they represent single time points for labelling under different experimental conditions. When the tryptic αC fragment was tested with dansyl-ε-aminocaproylAQQIV, we found AαLys418, 427, 429, 448, and 539 became acylated (P. Murthy, J. Wilson, and L. Lorand, unpublished results). In the FXIIIa-catalyzed reaction of a recombinant αC preparation Aα(221–610): AαLys 539, 556, 580, and 601 were shown to form crosslinks with at least three Gln residues in the same αC fragment: AαGln 221, 237, 328.[78] It is quite remarkable that the four Lys residues implicated in this crosslinking reaction are all part of the same set of lysines that were identified originally in the α chains of fibrin as FXIIIa targets with the dansyl-ε-aminocaproylAQQIV probe.[76] However, it is still somewhat of a puzzle as to why AαLys303 was not identified in any of these experiments as a potential amine donor. This is the residue that has been reported to be involved in the crosslinking of Gln2 of the α2 plasmin inhibitor to fibrinogen;[79] AαLys303 could not be modified even when the dodecapeptide: NQEQVSPLTLLK corresponding to the N-terminal segment of α_2PI was employed as a substrate[77] (see also Ref. 79a). In addition to Gln2, Gln21, 419, and 447 of α_2PI also seem to be involved in the FXIIIa-catalyzed crosslinking to fibrin.[79b] In the FXIIIa-mediated crosslinking of plasminogen activator inhibitor 2 (PAI-2), AαLys148, 230, and 413 of fibrin are involved, whereas guinea pig liver transglutaminase uses AαLys148, 176, 183, and 457.[79a]

Inter- and intramolecular hybrid A$\alpha \cdot \gamma$ and $\alpha \cdot \gamma$ types of crosslinking are the hallmarks of the reactions of the red cell transglutaminase with fibrinogen and fibrin. N^ε(γ-glutamyl)lysine sequences obtained from intramolecularly crosslinked fibrinogen showed[80] that both γGln398 and 399 were linked to AαLys413 and 418, that is, the transglutaminase utilized the same γGln acceptors and the same AαLys418 as FXIIIa, but also AαLys413 as an additional donor. As far as the intra-Aα chain crosslinks are concerned, transglutaminase was shown to catalyze the formation of multiple isopeptides between AαGln366 and AαLys413, 418, 421, and 448.

We still have a lot more to learn concerning the factor XIIIa- and the transglutaminase-driven incorporation of amines into the acceptor Gln residues and the acylation of donor Lys residues in fibrinogen and fibrin; these reactions need to be examined under a variety of kinetic conditions. The factor XIIIa and the transglutaminase-catalyzed processes are among the most important of all the physiological events in which fibrinogen and fibrin participate. From the data already available, it is quite clear that the XIIIa and the transglutaminase-dependent reactions occur at flexible regions in the γ and α or Aα chains of the protein, segments that are capable of assuming different conformations and that are not readily amenable to crystallographic examination. From the transglutaminase catalyzed intramolecular reaction of fibrinogen alone, we know that isopeptide bridging can occur between Aα(408–421) and γ(392–406), suggesting that these segments in two different chains of the protein are in sufficiently close contact to be involved in "zero distance" crosslinking.[80] Perhaps the most important lesson is that, even when focusing on fibrin and fibrinogen alone, the FXIIIa and transglutaminase catalyzed crosslinking reactions are far more complex than anyone would have predicted. Particularly with the reactions involving the α or Aα chains, multiple crosslinking possibilities exist. These will vary with different conditions and the possibilities will increase yet further with plasma proteins, blood and cellular constituents in the mixtures that can be covalently fused to the fibrin matrix. Because of the large excess of enzyme reactive donor sites, which do not seem to be required for fibrin-to-fibrin bonding, the Aα/α chains could serve as a nucleus for creating multibranched networks for the covalent attachments of proteins in wounds, in the fibrous caps of atherosclerotic plaques, in the endomysial lesions of celiac disease, and so forth. It is obvious that we have opened Pandora's box and that a tremendous amount of hard work is still left to be done in the twenty-first century! Fortunately, the path to these investigations, based on the chemistries in FIGURE 10, is quite clear.

MOLECULAR DEFECTS OF FIBRIN STABILIZATION

The outline presented in FIGURE 5 serves as the framework for classifying the molecular defects of fibrin stabilization. Most of these are hemorrhagic conditions with different degrees of severity, quite frequently fatal. Clotting and bleeding times may be normal, which indicates that immediate hemostasis, that is, the measure of the reversible assembly of fibrin molecules, proceeds at a normal rate. Delayed bleeding (oozing) occurs because the hemostatic plug fails on account of structural weakness (see FIG. 8) and because of its greater susceptibility to digestion by lytic enzymes (see FIG. 9). The disorders are usually recognized by the solubility of the patient's clot in 5 M urea,[2,9] as it was in the first report of such a case.[81] Further diagnostic methods may include an examination of the chain profile of fibrin[82] and measurements of the thrombin-dependent transamidase activity which can be generated in the patient's plasma.[83]

In the extreme forms of genetic deficiencies, the Class I disorders of fibrin stabilization, which comprise the majority of cases, virtually no thrombin dependent transamidase activity is detectable. This autosomal recessive condition[45,83,84] is caused by a failure of synthesis or the correct folding of either the A[85] or the B subunits.[86] Without the B, the A subunits do not seem to survive in the circulation.

With the development of quantitative amine incorporation assays for measuring factor XIIIa,[83] an approximately eightfold spread of activity was noted in the control group.[84] There is a Val34Leu polymorphism in the A subunits, present at various gene frequencies for the Leu allele in different ethnic groups (0.01 in Japan, 0.27 in Australia) which seems to account somewhat for the large range of FXIIIa values measured, because the mutation gives rise to an approximately twofold increase of FXIIIa activity.[87] Interestingly, there are also reports suggesting that the Val34Leu mutation might be protective against thrombotic disease[88–90] though increasing perhaps the risk for intracerebral hemorrhage.[91]

In the Class II defects of fibrin stabilization,[92] acquired inhibitors block one or more of the biochemical reactions in the conversion of factor XIII to XIIIa (Type I inhibitors), or prevent the A_2^* enzyme from carrying out its crosslinking functions on fibrin. Concerning the latter, distinction is drawn between Type II inhibitors which can interfere with the transamidase activity of XIIIa on test substrates other than fibrin (e.g., dimethylcasein-dansylcadaverine or putrescine) and Type III inhibitors that specifically block the γ or α chain crosslinking sites in fibrin. The severity of the Class II diseases (ca. 27% fatal) equals or exceeds that of the genetic deficiencies of Factor XIII. However, in contrast to the hereditary disorders, hemorrhagic manifestations become more prevalent with advancing age and there is usually no family history of bleeding. In the majority of the Class II disorders (about 68%), the acquired inhibitor was identified as an autoimmune antibody. However, in terms of target specificity, each patient's antibody seems to be quite unique thus far.

Further analysis of genetic mutations and identification of the epitopes recognized by the Type I and II antibodies could aid in resolving a number of outstanding structure–function related issues, such as the heterologous association of A_2 with B_2, interactions of A_2B_2 and of $A_2'B_2$ with fibrin, the Ca^{2+}-mediated dissociation of $A_2'B_2$ and others. Patient studies should also help in defining more precisely the relationship between XIIIa function and hemorrhagic phenotype.

ACKNOWLEDGMENTS

Support of the National Institutes of Health (HL-02212 and HL-16346) over the years is gratefully acknowledged.

REFERENCES

1. LORAND, L. 2000. Sol Sherry lecture in thrombosis. Research on clot stabilization provides clues for improving thrombolytic therapies. Arterioscler. Thromb. Vasc. Biol. **20:** 2–9.
2. LORAND, L. 1948. A study on the solubility of fibrin clots in urea. Hung. Acta Physiol. **1:** 192–196.
3. MEISSNER, I. & E. WÖHLISCH. 1937. Untersuchungen über die Einwirkung hydrotroper Substanzen auf das Fibrinogen und die Blutgerinnung. Biochem. Ziet. **293:** 133–141.
4. LIMBOURG, P. 1889. Ueber Lösung und Fällung von Eiweisskörpern durch Salze. Hoppe-Seyler's Zeit. Physiol. Chem. **13:** 450–463.

5. SPIRO, K. 1900. Ueber die Beeinflussung der Eiweisscoagulation durch stickstoffhaltige Substanzen. Hoppe-Seyler's Zeit. Physiol. Chem. **30:** 182–199.
6. LORAND, L. & W.R. MIDDLEBROOK. 1952. The action of thrombin on fibrinogen. Biochem. J. **52:** 196–199.
7. LORAND, L. 1951. "Fibrino-peptide": new aspects of the fibrinogen–fibrin transformation. Nature **167:** 992–994.
8. LORAND, L. 1952. Fibrino-peptide. Biochem. J. **52:** 200–203.
9. LAKI, K. & L. LORAND. 1948. On the solubility of fibrin clots. Science **108:** 280.
10. ROBBINS, K.C. 1944. A study on the conversion of fibrinogen to fibrin. Am. J. Physiol. **142:** 581–588.
11. IKEMORI, R., et al. 1975. Solubility of fibrin clots in monochloroacetic acid. A reflection of serum pepsinogen levels. Am. J. Clin. Pathol. **63:** 49–56.
12. LORAND, L. 1950. Fibrin clots. Nature **166:** 694–696.
13. IWANAGA, S., A. HENSCHEN & B. BLOMBACK. 1966. On the primary structure of human fibrinogen. I. Two-dimensional "finger prints" of tryptic digests of sulfitolyzed fibrinogen and fibrin. Acta Chem. Scand. **20:** 1183–1185.
14. LORAND, L. & K. KONISHI. 1964. Activation of the fibrin stabilizing factor of plasma by thrombin. Arch. Biochem. Biophys. **105:** 58–67.
15. LORAND, L., et al. 1968. Amine specificity in transpeptidation. Inhibition of fibrin crosslinking. Biochemistry **7:** 1214–1223.
16. BRUNER-LORAND, J., T. URAYAMA & L. LORAND. 1966. Transglutaminase as a blood clotting enzyme. Biochem. Biophys. Res. Commun. **23:** 828–834.
17. LORAND, L., et al. 1968. The transpeptidase system which crosslinks fibrin by γ-glutamyl-ε-lysine bonds. Biochem. Biophys. Res. Commun. **31:** 222–230.
18. MATACIC, S. & A.G. LOEWY. 1968. The identification of isopeptide crosslinks in insoluble fibrin. Biochem. Biophys. Res. Commun. **30:** 356–362.
19. PISANO, J.J., J.S. FINLAYSON & M.P. PEYTON. 1968. Crosslink in fibrin polymerized by factor XIII: ε(γ-glutamyl)lysine. Science **160:** 892–893.
20. CURTIS, C.G., et al. 1974. Calcium-dependent unmasking of active center cysteine during activation of fibrin stabilizing factor. Biochemistry **13:** 3774–3780.
21. PARAMESWARAN, K.N. & L. LORAND. 1981. New thioester substrates for fibrinoligase (coagulation factor XIII_a) and for transglutaminase. Transfer of the fluorescently labeled acyl group to amines and alcohols. Biochemistry **20:** 3703–3711.
22. MICANOVIC, R., et al. 1994. Role of histidine 373 in the catalytic activity of coagulation factor XIII. J. Biol. Chem. **269:** 9190–9194.
23. LORAND, L., et al. 1963. A new class of blood coagulation inhibitors. Arch. Biochem. Biophys. **102:** 171–179.
24. WILLIAMS-ASHMAN, H.G., et al. 1972. Transamidase reactions involved in the enzymic coagulation of semen: isolation of γ-glutamyl-ε-lysine dipeptide from clotted secretion protein of guinea pig seminal vesicle. Proc. Natl. Acad. Sci. U.S.A. **69:** 2322–2325.
25. MURTHY, S.N.P. & L. LORAND. 1990. Cross-linked $A\alpha \cdot \gamma$ chain hybrids serve as unique markers for fibrinogen polymerized by tissue transglutaminase. Proc. Natl. Acad. Sci. U.S.A. **87:** 9679–9682.
26. MURTHY, S.N.P., et al. 1991. Intramolecular crosslinking of monomeric fibrinogen by tissue transglutaminase. Proc. Natl. Acad. Sci. U.S.A. **88:** 10601–10604.
27. LORAND, L., et al. 1979. Specificity of guinea pig liver transglutaminase for amine substrates. Biochemistry **18:** 1756–1765.
28. CURTIS, C.G., et al. 1974. Kinetics of transamidating enzymes. Production of thiol in the reactions of thiol esters with fibrinoligase. Biochemistry **13:** 3257–3262.
29. STENBERG, P., et al. 1975. Transamidase kinetics. Amide formation in the enzymic reactions of thiol esters with amines. Biochem. J. **147:** 155–163.
30. LORAND, L. & S.M. CONRAD. 1984. Transglutaminases. Mol. Cell. Biochem. **58:** 9–35.
31. SCHWARTZ, M.L., et al. 1973. Human factor XIII from plasma and platelets. Molecular weights, subunit structures, proteolytic activation, and crosslinking of fibrinogen and fibrin. J. Biol. Chem. **248:** 1395–1407.
32. ICHINOSE, A., R.E. BOTTENUS & E.W. DAVIE. 1990. Structure of transglutaminases. J. Biol. Chem. **265:** 13411–13414.

33. LORAND, L., R.B. CREDO & T.J. JANUS. 1981. Factor XIII (fibrin stabilizing factor). Meth. Enzymol. **80:** 333–341.
34. ICHINOSE, A., *et al.* 1986. Amino acid sequence of the a subunit of human factor XIII. Biochemistry **25:** 6900–6906.
35. MAKAROVA, K.S., L. ARAVIND & E.V. KOONIN. 1999. A superfamily of archaeal, bacterial, and eukaryotic proteins homologous to animal transglutaminases. Protein Sci. **8:** 1714–1719.
36. PEDERSEN, L.C., *et al.* 1994. Transglutaminase factor XIII uses proteinase-like catalytic triad to crosslink macromolecules. Protein Sci. **3:** 1131–1135.
37. YEE, V.C., *et al.* 1994. Three-dimensional structure of a transglutaminase: human blood coagulation factor XIII. Biochemistry **91:** 7296–7300.
38. CREDO, R.B., C.G. CURTIS & L. LORAND. 1978. Ca^{2+}-related regulatory function of fibrinogen. Proc. Natl. Acad. Sci. U.S.A. **75:** 4234–4237.
39. TAKAGI, T. & R.F. DOOLITTLE. 1974. Amino acid sequence studies on factor XIII and the peptide released during its activation by thrombin. Biochemistry **13:** 750–756.
40. RADEK, J.T., *et al.* 1993. Association of the A subunits of recombinant placental factor XIII with the native carrier B subunits from human plasma. Biochemistry **32:** 3527–3534.
41. LORAND, L., *et al.* 1974. Dissociation of the subunit structure of fibrin stabilizing factor during activation of the zymogen. Biochem. Biophys. Res. Commun. **56:** 914–922.
42. YEE, V.C., *et al.* 1995. Structural evidence that the activation peptide is not released upon thrombin cleavage of factor XIII. Thromb. Res. **78:** 389–397.
43. JANUS, T.J., *et al.* 1983. Promotion of thrombin-catalyzed activation of factor XIII by fibrinogen. Biochemistry **22:** 6269–6272.
44. LEWIS, S.D., *et al.* 1985. Regulation of the formation of factor XIIIa by its fibrin substrates. Biochemistry **24:** 6772–6777.
45. LORAND, L., M.S. LOSOWSKY & K.J.M. MILOSZEWSKI. 1980. Human factor XIII: fibrin stabilizing factor. 1980. *In* Progress in Hemostasis and Thrombosis, Vol. 5. T.H. Spaet, Ed.: 245–290. Grune & Stratton, New York.
46. LORAND, L. & A. JACOBSEN. 1958. Studies on the polymerization of fibrin. The role of the globulin: fibrin-stabilizing factor. J. Biol. Chem. **230:** 421–434.
47. LORAND, L. 1961. Properties and significance of the fibrin stabilizing factor (FSF). *In* Progress in Coagulation. Thromb. Diath. Hæmorrh. Suppl. 1, Vol. VII. 238–248. Friedrich-Karl Schattauer-Verlag, Stuttgart.
48. GREENBERG, C.S. & M.A. SHUMAN. 1982. The zymogen forms of blood coagulation factor XIII bind specifically to fibrinogen. J. Biol. Chem. **257:** 6096–6101.
49. SIEBENLIST, K.R., D.A. MEH & M.W. MOSESSON. 1996. Plasma factor XIII binds specifically to fibrinogen molecules containing γ′ chains. Biochemistry **35:** 10448–10453.
50. MOADDEL, M., *et al.* 2000. Interactions of human fibrinogens with factor XIII: roles of calcium and the γ′ peptide. Biochemistry **39:** 6698–6705.
51. MOADDEL, M., L.A. FALLS & D.H. FARRELL. 2000. The role of γA/γ′ fibrinogen in plasma factor XIII activation. J. Biol. Chem. **41:** 32135–32140.
52. LORAND, L. & D. CHENOWETH. 1969. Intramolecular localization of the acceptor crosslinking sites in fibrin. Proc. Natl. Acad. Sci. U.S.A. **63:** 1247–1252.
53. LORAND, L., D. CHENOWETH & A. GRAY. 1972. Titration of the acceptor crosslinking sites in fibrin. Ann. N.Y. Acad. Sci. **202:** 155–171.
54. HORNYAK, T.J. & J.A. SHAFER. 1992. Interactions of factor XIII with fibrin as substrate and cofactor. Biochemistry **31:** 423–429.
55. LORAND, L. & H.H. ONG. 1966. Labeling of amine-acceptor cross-linking sites of fibrin by transpeptidation. Biochemistry **5:** 1747–1753.
56. LORAND, L. 1972. Fibrinoligase. The stabilizing factor system of blood plasma. Ann. N.Y. Acad. Sci. **202:** 6–30.
57. VEKLICH, Y., *et al.* 1998. The complementary aggregation sites of fibrin investigated through examination of polymers of fibrinogen with fragment E. Proc. Natl. Acad. Sci. U.S.A. **95:** 1438–1442.

58. SAMOKHIN, G.P. & L. LORAND. 1995. Contact with the N-termini in the central E domain enhances the reactivities of the distal D domains of fibrin to factor XIII_a. J. Biol. Chem. **270:** 21827–21832.
59. LORAND, L., K.N. PARAMESWARAN & S.N.P. MURTHY. 1998. A double-headed GlyProArgPro ligand mimics the function of the E domain of fibrin for promoting the end-to-end crosslinking of γ chains by Factor XIIIa. Proc. Natl. Acad. Sci. U.S.A. **95:** 537–541.
60. RYAN, E.A., et al. 1999. Structural origins of fibrin clot rheology. Biophys. J. **77:** 2813–2826.
61. SHEN, L. & L. LORAND. 1983. Contribution of fibrin stabilization to clot strength. Supplementation of factor XIII-deficient plasma with the purified zymogen. J. Clin. Invest. **71:** 1336–1341.
62. LORAND, L. & A. JACOBSEN. 1962. Accelerated lysis of blood clots. Nature **195:** 911–912.
63. BRUNER-LORAND, J., T.R.E. PILKINGTON & L. LORAND. 1966. Inhibitor of fibrin crosslinking: relevance for thrombolysis. Nature **210:** 1273–1274.
64. LORAND, L. & J.L.G. NILSSON. 1972. Molecular approach for designing inhibitors to enzymes involved in blood clotting. In Drug Design. E.J. Ariens, Ed.: 415–447. Academic Press, New York.
65. SAKATA, Y. & N. AOKI. 1982. Significance of crosslinking of α_2-plasmin inhibitor to fibrin in inhibition of fibrinolysis and hemostasis. J. Clin. Invest. **69:** 536–542.
66. ICHINOSE, A., T. TANAKI & N. AOKI. 1983. Factor XIII-mediated crosslinking of NH_2 terminal peptide of α_2-plasmin inhibitor to fibrin. FEBS Lett. **153:** 369–371.
67. CHEN, R. & R.F. DOOLITTLE. 1971. γ-γ Cross-linking sites in human and bovine fibrin. Biochemistry **10:** 4486–4491.
68. MCKEE, P.A., P. MATTOCK & R.L. HILL. 1970. Subunit structure of human fibrinogen, soluble fibrin and crosslinked insoluble fibrin. Proc. Natl. Acad. Sci. U.S.A. **66:** 738–744.
69. MOSESSON, M.W., et al. 1989. Identification of covalently linked trimeric and tetrameric D domains in crosslinked fibrin. Proc. Natl. Acad. Sci. U.S.A. **86:** 1113–1117.
70. MOSESSON, M.W., et al. 1993. Evidence for a second type of fibril branch point in fibrin polymer networks, the trimolecular function. Blood **82:** 1517–1521.
71. PARAMESWARAN, K.N., et al. 1990. Labeling of ε-lysine crosslinking sites in proteins with peptide substrates of factor XIIIa and transglutaminase. Proc. Natl. Acad. Sci. U.S.A **87:** 8472–8475.
72. LORAND, L., et al. 1992. Biotinylated peptides containing a factor XIIIa or tissue transglutaminase-reactive glutaminyl residue block protein cross-linking phenomena by becoming incorporated into amine donor sites. Bioconjug. Chem. **3:** 37–41.
73. PURVES, L., M. PURVES & W. BRANDT. 1987. Cleavage of fibrin-derived D-dimer into monomers by endopeptidase from puff adder venom (Bitis arietans) acting at crosslinked sites of the γ-chain. Sequence of carboxy-terminal cyanogen bromide γ-chain fragments. Biochemistry **26:** 4640–4646.
74. COTTRELL, B.A., et al. 1979. Amino acid sequence studies on the α chain of human fibrinogen. Exact location of cross-linking acceptor sites. Biochemistry **18:** 5405–5410.
75. FRETTO, L.J., et al. 1978. Localization of the α-chain cross-link acceptor sites of human fibrin. J. Biol. Chem. **253:** 2184–2195.
76. LORAND, L. 1991. Identification of ε-lysine cross-linking sites in the α chains of fibrin and the tissue transglutaminase carrying role of fibronectin in plasma. 20th Linderstrom-Lang Conference, Vingsted, Denmark.
77. SOBEL, J.H. & M.A. GAWINOWICZ. 1996. Identification of the α chain lysine donor sites involved in factor XIIIa fibrin cross-linking. J. Biol. Chem. **271:** 19288–19297.
78. MATSUKA, Y.V., et al. 1996. Factor XIIIa-catalyzed cross-linking of recombinant αC fragments of human fibrinogen. Biochemistry **35:** 5810–5816.
79. KIMURA, S. & N. AOKI. 1986. Cross-linking site in fibrinogen for α_2-plasmin inhibitor. J. Biol. Chem. **261:** 15591–15595.

79a. RITCHIE, H., et al. 2000. Cross-linking of plasminogen activator inhibitor 2 and α_2-antiplasmin to fibrin(ogen). J. Biol. Chem. **275:** 24915–24920.
79b. LEE, K.N., et al. 2000. Cross-linking of wild-type and mutant α_2-antiplasmins to fibrin by activated factor XIII and by a tissue transglutaminase. J. Biol. Chem. **275:** 37382–37389.
80. MURTHY, S.N.P., et al. 2000. Transglutaminase-catalyzed crosslinking of the Aα and γ constituent chains in fibrinogen. Proc. Natl. Acad. Sci. U.S.A. **97:** 44–48.
81. DUCKERT, F., E. JUNG & D.H. SCHMERLING. 1960. A hitherto undescribed congenital haemorrhagic diathesis probably due to fibrin stabilizing factor deficiency. Thromb. Diath. Hæmorrh. **5:** 179–186.
82. ROSENBERG, R.D., R.W. COLMAN & L. LORAND. 1974. A new haemorrhagic disorder with defective fibrin stabilization and cryofibrinogenemia. Brit. J. Hæmatol. **26:** 269–284.
83. LORAND, L., et al. 1969. Diagnostic and genetic studies on fibrin-stabilizing factor with a new assay based on amine incorporation. J. Clin. Invest. **48:** 1054–1064.
84. LORAND, L., et al. 1970. Inheritance of deficiency of fibrin-stabilizing factor (factor XIII). Am. J. Hum. Genet. **22:** 89–95.
85. MIKKOLA, H., et al. 1996. Four novel mutations in deficiency of coagulation factor XIII: consequences to expression and structure of the A-subunit. Blood **87:** 141–151.
86. IZUME, T., et al. 1996. Type 1 factor XIII deficiency is caused by a genetic defect of its b subunit: insertion of triplet AAC in exon III leads to premature termination in the second Sushi domain. Blood **87:** 2769–2774.
87. KANGSADALAMPAI, S. & P.G. BOARD. 1998. The Val34Leu polymorphism in the A subunit of coagulation factor XIII contributes to the large normal range in activity and demonstrates that the activation peptide plays a role in catalytic activity. Blood **92:** 2766–2770.
88. ANWAR, R., et al. 1999. Genotype/phenotype correlations for coagulation factor XIII: specific normal polymorphisms are associated with high or low factor XIII specific activity. Blood **93:** 897–905.
89. WARTIOVAARA, U., et al. 1999. Association of FXIII Val34Leu with decreased risk of myocardial infarction in Finnish males. Atherosclerosis **142:** 295–300.
90. ELBAZ, A., et al. 2000. The association between the Val34Leu polymorphism in the factor XIII gene and brain infarction. Blood **95:** 586–591.
91. CATTO, A., et al. 1998. Factor XIII Val34Leu: a novel association with primary intracerebral hemorrhage. Stroke **29:** 813–816.
92. LORAND, L. 1994. Acquired inhibitors of fibrin stabilization: a class of hemorrhagic disorders of diverse origins. *In* Anticoagulants, Physiologic, Pathologic and Pharmacologic. D. Green, Ed.: 169–191. CRC Press, Boca Raton.

The Structure and Function of the αC Domains of Fibrinogen

JOHN W. WEISEL[a] AND LEONID MEDVED[b]

[a]*Department of Cell and Developmental Biology, University of Pennsylvania School of Medicine, Philadelphia, Pennsylvania, USA*

[b]*Department of Biochemistry, The American Red Cross Holland Laboratory, 15601 Crabbs Branch Way, Rockville, Maryland, USA*

ABSTRACT: The αC domains have been localized on fibrinogen and fibrin. Several model systems have been developed to study their functions. Analysis of the amino acid sequence of the αC domains suggested that each is made up of a globular and an extended portion. Microcalorimetry confirmed this result and showed that the two αC domains interact intramolecularly. Electron microscopy of fibrinogen with a monoclonal antibody to the αC domains demonstrated that these regions normally interact with the central portion of the molecule. In the conversion from fibrinogen to fibrin there is a large scale conformational change, such that the αC domains dissociate from the central region and are available for intermolecular interaction. Experiments with highly purified and well characterized fragment X monomer, missing either one or both of the αC domains, indicate that intermolecular interactions between αC domains are important for the enhancement of lateral aggregation during fibrin polymerization. Isolated αC fragments polymerized at neutral pH and interacted with the αC domains of fibrin monomer to influence clot formation. Several dysfibrinogenemias in which there are amino acid substitutions in, or truncations of, the αC domains revealed that these changes can have dramatic effects on polymerization and clot structure. The polymerization of Aα251 recombinant fibrinogen, that contains Aα chains truncated at residue 251, was altered, as were the mechanical properties and the rate of fibrinolysis of the clots. Altogether, these results help to define the role of the αC domains in determining the structure and properties of clots.

KEYWORDS: Structure; Function; Fibrinogen; αC domain; Electron microscopy; Microcalorimetry.

INTRODUCTION

Each fibrinogen molecule contains two αC domains made up of the carboxy terminal two-thirds of the Aα chains, comprising about 25% of the mass of the molecule. Although this part of the molecule has been studied for many years, it is only recently that it has been localized on fibrinogen and fibrin, and that aspects of its function have been determined.

Address for correspondence: John W. Weisel, Ph.D., Department of Cell and Developmental Biology, University of Pennsylvania School of Medicine, Philadelphia, PA 19104-6058, USA. Voice: 215-898-3573; fax: 215-898-9871.
weisel@mail.cellbio.med.upenn.edu

STRUCTURE

Amino Acid Sequence

The αC domains are readily cleaved from fibrinogen by plasmin and other proteases and are degraded subsequently into smaller pieces.[1,2] Based on their high susceptibility to proteolysis, it was suggested that they are unordered in fibrinogen, forming "free-swimming appendages".[3] A number of sites whose cleavage results in generation of the early αC fragments were identified in the Aα200–253 region.[1,2] The largest αC fragments (molecular mass of 40 kDa) isolated from plasmin digests of human and bovine fibrinogen had amino termini at positions Aα220, 231, and 240,[4,5] suggesting that the former position can be regarded as a boundary between the αC domain and the bulk of the molecule. The αC domains would then consist of residues Aα220–610.

The sequence analysis of the αC domains revealed some unusual features. The polar and nonpolar residues are very unequally distributed along this region of the Aα chain. Its amino terminal half has a very low content of nonpolar residues and has an excess of Gly, Ser, Thr, and Pro,[1,6] which makes it very improbable that this part of the chain could form any compact structure. The carboxy terminal half is also rich in Gly, Ser, Thr, and Pro, but in addition this region contains a large number of amino acid residues with nonpolar side chains[1,6] that are usually involved in the formation of a compact core. In agreement with these observations, a computer analysis of the probability of formation of regular polypeptide chain conformations revealed that the α-helix and β-conformations are quite probable in the carboxy terminal 390–550 section of the αC domains, whereas the amino terminal 240–390 section appears to be predominantly in a random coil conformation.[7] Thus, although it was originally assumed that the αC domains are unordered,[3] the analysis just described suggests that, at least their carboxy terminal halves, can form compact, ordered structures.

Another unusual feature of the αC domain is the presence of tandem repeats in its amino terminal half. There are 10 tandem repeats of 13 amino acid residues in the Aα264–391 region of the human fibrinogen αC domain.[1,6] The number of repeats varies among species, as does the total number of amino acid residues in each repeat. However, most of the αC domain repeats contain a characteristic Trp residue and are rich in Gly, Ser, Thr, and Pro.[8,9] Such repeats are not unique for fibrinogen; various proteins with tandemly repeated proline rich sequences have been found in both procaryotes and eucaryotes.[10] It was suggested that such sequences adopt the extended helical conformation of the poly-L-proline type (PPII) that has unique features, reviewed by Williamson.[10] Briefly, it is a left handed, extended helix with three residues per turn; this structure is highly hydrated, since it is stabilized by hydrogen bonds with water molecules and often forms interdomain links or "sticky arms" involved in interactions with other proteins.[10] It is tempting to hypothesize that the amino terminal half of the αC domain is not unordered, as mentioned above, but is, at least partially, in a PPII conformation. The presence of the PPII regions in the αC domains would be consistent both with the unusually high degree of hydration of the fibrinogen molecule reported long ago (see Ref. 3 and references therein), but still unexplained, and with the difficulty of crystallization of intact fibrinogen,[11] since the extended PPII structures are hard to crystallize.[10] This hypothesis needs further testing.

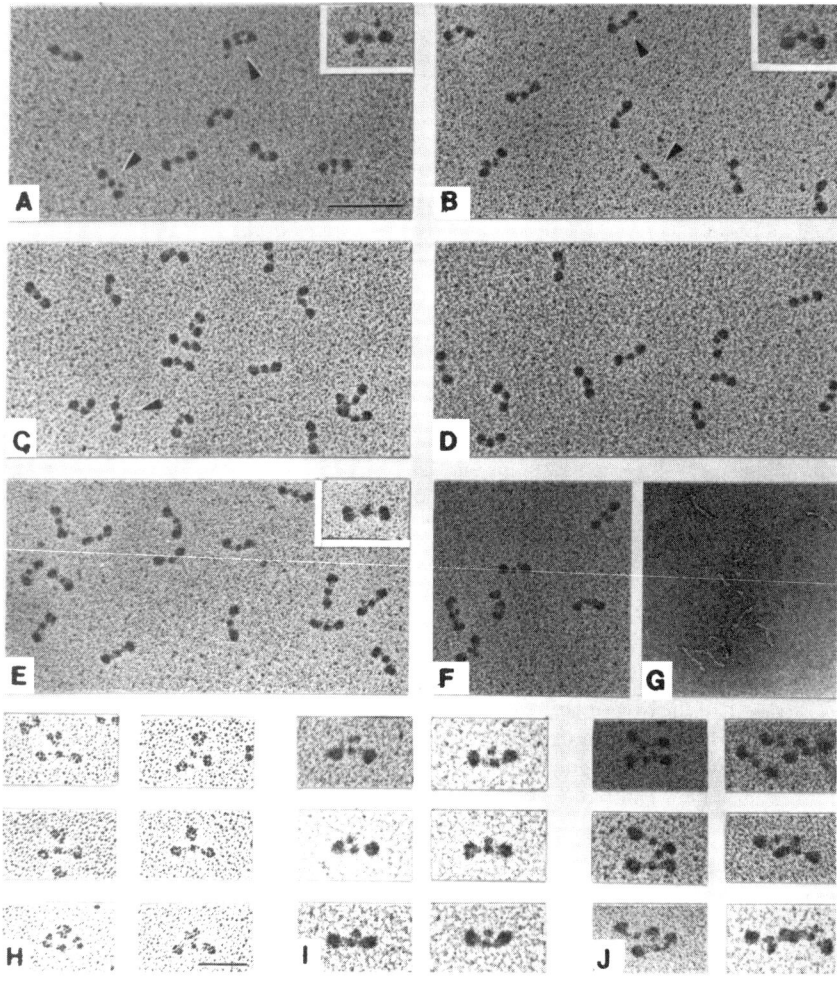

FIGURE 1. Electron microscopy of individual fibrinogen and fibrin molecules. All samples were rotary shadowed except those in *G*. **A.** Fibrinogen in dilute acetic acid; *arrows* point to examples of the most commonly observed species, containing one small projection from each end and the *inset* shows an example at higher magnification. **B.** Fibrin monomer in dilute acetic acid; *arrows* point to examples of the most commonly observed species, also containing one small projection from each end and the inset shows an example at higher magnification. **C.** Fragment X_1 monomer in dilute acetic acid; the *arrow* points to an example of a commonly observed species, with a single small projection from one end. **D.** Fragment X_2 monomer in dilute acetic acid; no molecules display additional mass. **E.** Fibrinogen at neutral pH; the *inset* shows at higher magnification the most commonly observed species, which is a trinodular structure with no additional nodules. However, occasionally some molecules display extra nodules. **F.** Fibrin monomer at neutral pH; most molecules have an additional large nodule adjacent to or near the central domain. (Legend continued on opposite page.)

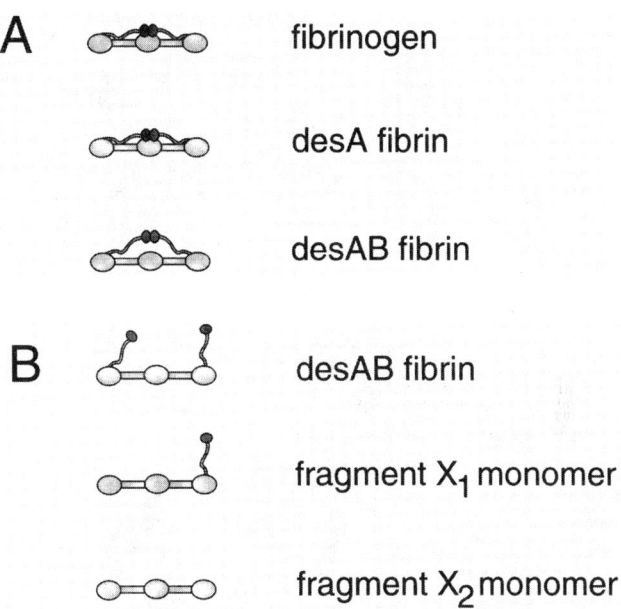

FIGURE 2. Schematic representations of the locations of the αC domains. **A.** Locations of the αC domains in fibrinogen and fibrin at neutral pH. Fibrinogen is shown at the *top*, with its αC domains interacting with the central region. In the *middle* is desA fibrin monomer, where most molecules have an appearance similar to fibrinogen. At the *bottom*, is desAB fibrin monomer, which has its αC domains interacting with each other, but not with the central region. **B.** Locations of the αC domains in fibrin monomer and fragment X monomer preparations in dilute acetic acid. At the *top* is fibrin monomer, with its αC domains free for intermolecular interactions; fragment X_1 monomer, missing one αC domain, and fragment X_2 monomer, missing both αC domains, are shown at the *bottom*.

FIGURE 1/continued. G. Negatively contrasted fibrin monomer at neutral pH; a subdivision of each end nodules into two domains is commonly seen, although the αC domains are not easily visualized with this technique. **H.** Fibrinogen with monoclonal antibody to the αC domains, which binds on or near the central domain of the molecule. **I.** Gallery of fibrin monomer molecules at neutral pH, showing the large nodule adjacent to the central region, consisting of the two αC domains interacting intramolecularly. **J.** Aggregates of fibrin monomer interacting intermolecularly via their αC domains. (Figure from Veklich, *et al.*Ref. 19)

Domain Structure as Determined by Microcalorimetry

The first evidence for the presence of a compact structure in the αC domains was obtained by differential scanning calorimetry, a method that makes it possible to measure the heat absorbed on denaturation or melting of compact, cooperative protein structures. When fibrinogen was heated in a calorimeter in various solution environments, it exhibited a complex heat absorption curve (endotherm) that reflected melting of a number of cooperative regions (domains) that have different stabilities. In particular, at pH 3.5, fibrinogen exhibited four well expressed heat absorption peaks, one of which was assigned to the melting of the structures removed at the early stage of proteolysis, that is, the αC domains.[12,13] This assignment was based on comparison of the endotherm of the intact fibrinogen with that of its fragment X (devoid of the αC domains), in which one of the heat absorption peaks was absent.

A detailed quantitative analysis of the heat absorption peak corresponding to the melting of the αC domains provided additional information about their structural organization.[7] First, the specific enthalpy of denaturation of the αC domains was lower than that for a typical globular protein, suggesting that only part of their residues are involved in the formation of the compact structure. This conclusion is in good agreement with the aforementioned theoretical analysis, which suggested that only the carboxy terminal half of each αC domain can form a compact structure. Second, comparison of the experimental enthalpy of denaturation of the αC domains with that calculated from the shape of the curve indicated that these domains must form two compact, cooperative units that strongly interact with each other, that is, they must be in direct contact, forming a single structural block. These structural features were further clarified by electron microscopy studies.

Localization of the αC Domains by Electron Microscopy

Early studies of individual fibrinogen molecules by electron microscopy of rotary shadowed preparations revealed the basic trinodular shape of the molecule, a fibrous protein about 45 nm in length, with nodules at either end and in the middle (see FIGURE 1E). In some of these images, a subdivision of the end nodule into two domains can be seen. The two domains at each end are often visualized more clearly in negatively contrasted preparations of fibrinogen or fibrin monomer (FIG. 1G). When electron microscope images of rotary shadowed fibrinogen molecules were examined in more detail, it was discovered that some molecules appeared to have extensions or appendages, that seemed to be related to the αC domains because they were not present in preparations missing this part of the molecule.[14–18] In particular, some fibrinogen molecules were found to have an additional fourth nodule adjacent to the central nodule (FIG. 1E).[14,18] Since this extra nodule was not observed in fragment X preparations devoid of αC domains (FIG. 1D), it was concluded that it is made up of the two carboxy terminal regions of the Aα chains.[14] However, these features were not seen in a large percentage of molecules,[18] indicating that the αC domains are not normally dissociated from the rest of the molecule. Where, then, are the αC domains normally located on fibrinogen and fibrin? The locations of the αC domains were determined by electron microscopy of fibrinogen in the presence of a monoclonal antibody to the carboxy terminal 150 amino acids of the Aα chain.[19] This antibody was seen to bind near the central region of fibrinogen (FIG. 1H), indicating that the αC domains of most molecules are not normally visible because they are on or near the central nodule of the molecule.

CHANGES IN CONFORMATION OF αC DOMAINS ON CONVERSION OF FIBRINOGEN TO FIBRIN

Electron microscopy of individual fibrin molecules has revealed striking changes in the αC domains that occur on cleavage of the fibrinopeptides A and B.[19] At pH 3.5, which prevents polymerization, fibrin monomer desAB and fibrinogen both show a projection terminating in a small globular domain from each end of most molecules (FIG. 1 A and B), whereas fragments of fibrinogen missing αC domains show no such projections (FIG. 1 D).[19] When fibrin monomer desAB is brought to neutral pH under conditions where polymerization is delayed, individual molecules are still visible showing the αC domains as a single additional nodule near the central region of the majority of molecules (FIG. 1 F and I). Moreover, clusters of molecules reveal some intermolecular associations via the αC domains (FIG. 1 J). At the same time, in fibrin monomer desA molecules, devoid of only fibrinopeptide A, at neutral pH, most of the αC domains, like those in fibrinogen, remain associated with the central region.[4] Therefore, on cleavage of both fibrinopeptides, the intramolecular interactions are weakened, so that the αC domains dissociate from the central region and are available for intermolecular interactions (see FIGURE 2 A).

FUNCTIONAL STUDIES

There have been many investigations of modified fibrinogen molecules missing or partially missing their αC domains. However, most previous studies were carried out on preparations that were either heterogeneous or were not well characterized. Here, we focus on more recent research, primarily from our laboratories.

Fragment X Monomer

The role of the αC domains in clot formation has been investigated by transmission and scanning electron microscopy, and by turbidity studies of clots made from preparations of fibrin monomer and proteolytically modified fibrin monomer molecules missing either one or both αC domains, designated fragment X_1 and X_2 (FIG. 1 C and D, and FIG. 2 B).[4,20] Although many studies had been carried out previously with various fragment X preparations, these were more homogeneous than those used in the past. Molecules with multiple cleavage sites were minimized by using very small amounts of proteolysis and then isolation of fractions containing molecules with different amounts of digestion. Another factor involved in producing homogeneous preparations was the use of clotting of the digests to remove nonclottable species. The resulting fragment X monomer was then polymerized by dilution into neutral pH buffer. These preparations formed gels on dilution to neutral pH, although the lag period associated with protofibril formation, maximum rate of turbidity increase associated with lateral aggregation, and final turbidity related to fiber thickness were all affected by the removal of the αC domains. In general, in preparations missing the αC domains the lag period was increased and the maximum rate was decreased in comparison with intact fibrin. These changes were more pronounced in dilute solutions.[4,20] Clots observed by either scanning or transmission electron microscopy were made up of a branched network of fibers, but the degree of order, lateral aggregation, and branching varied, indicating that the αC domain interactions affect both the rate of assembly and the final clot structure. These results suggest that the αC domains, that in fibrinogen interact intramolecularly, in fibrin desAB switch to intermolecular interactions to enhance both protofibril formation and lateral aggregation (see FIGURE 3A).

Sequential Cleavage of the Fibrinopeptides

Although polymerization of fibrin monomer described above is an important and commonly used approach, one should always keep in mind that it is not the same as

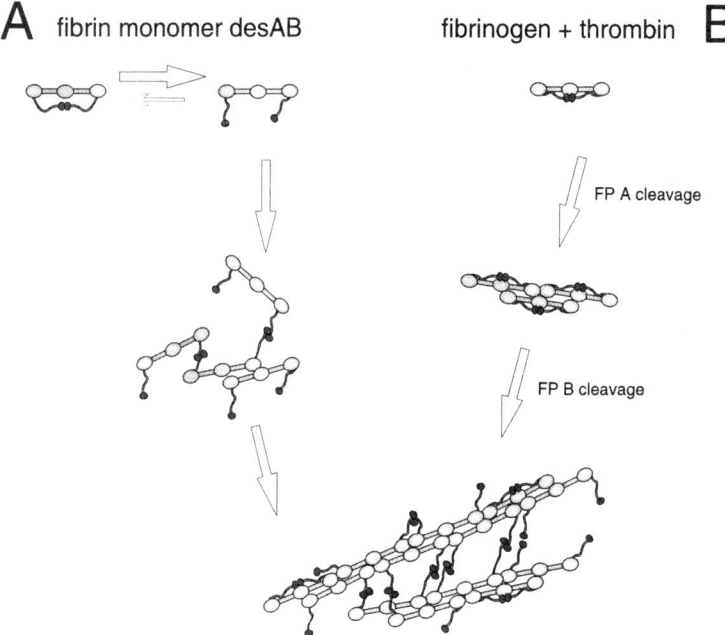

FIGURE 3. Schematic diagram of the αC domain switch from intramolecular to intermolecular interactions in polymerization. **A.** With fibrin monomer desAB, the αC domains are dissociated from each other and from the central domain. Thus, they are able to promote both protofibril formation and lateral aggregation via intermolecular interactions. **B.** With the sequential cleavage of the fibrinopeptides by thrombin, first the A fibrinopeptides are cleaved, leading to protofibril formation, but most of the αC domains remain associated with the central region. After protofibril formation and cleavage of the B fibrinopeptides, the αC domains dissociate from the central domain and are free to form intermolecular associations, promoting lateral aggregation by helping to bring protofibrils together. (Modified from Gorkun, et al. Ref. 4.)

polymerization initiated by thrombin. With fibrin monomer, the conformational changes associated with the conversion from fibrinogen to fibrin have already taken place, so that they cannot occur in the normal time sequence during polymerization. For example, in fibrin monomer desAB both fibrinopeptides A and B are removed and the αC domains have already been released from the central region (FIG. 2 B) with the consequence that they may interact intermolecularly during protofibril formation and lateral aggregation, affecting the rates of both steps. On the other hand, in the fibrinogen–thrombin system these fibrinopeptides are cleaved sequentially making the αC domains available for intermolecular interactions only after they dissociate from the central region on cleavage of the B fibrinopeptides (FIG. 2).

Thrombin cleaves the A fibrinopeptides from fibrinogen faster than B; cleavage of the B fibrinopeptides has been associated with enhanced lateral aggregation in the fibrin clot.[21,22] If thrombin is added to soluble fibrin desA during the lag period, increased lateral aggregation of fibers is observed by electron microscopy and the maximum turbidity is higher.[23] These results indicate that there must be a delay in the cleavage of the B fibrinopeptides for the enhancement of lateral aggregation. Changes that occur on cleavage of the B fibrinopeptides from molecules in a fiber, especially the release of the αC domains from the central domain, cause additional protofibrils to be added to a fiber. In conclusion, the usual delay in fibrinopeptide B cleavage appears to be necessary both for normal protofibril and fiber assembly, so that the changes that accompany removal of these peptides preferentially affect lateral aggregation rather than earlier steps of fibrin clot assembly. With respect to the αC domains, this means that in thrombin initiated polymerization they contribute mainly to the lateral aggregation of protofibrils (FIG. 3 B).

αC Fragments

The αC domains removed proteolytically from fibrinogen, called αC fragments, have been isolated and studied.[19] Electron microscopy of purified αC fragments at slightly acidic pH revealed individual globular structures, sometimes with a thin tail projecting from each one (see FIGURE 4 A and C). When these αC fragment preparations were brought to neutral pH, they polymerized to form oligomeric aggregates consisting of linear and branched structures with constant width (FIG. 4B).

The addition of αC fragments to fibrin monomer, followed by dilution to neutral pH to initiate polymerization, resulted in lower turbidity, longer lag period, and slower maximum rate of turbidity increase than controls without αC fragments.[19] Also, electron microscopy revealed complexes of αC fragments with the αC domains of fibrin monomer at neutral pH (FIG. 4D). It appears that αC fragments can bind to the αC domains of fibrin, competing with the normal αC domain interactions involved in polymerization. Therefore, these results again demonstrate that the αC domains act to enhance lateral aggregation and produce clots with thicker fibers.

Dysfibrinogenemias

The functional effects of modifications to the αC domains are revealed by several dysfibrinogenemias, one of the most striking of which is fibrinogen Dusart or Chapel Hill III, with the amino acid substitution of Aα 554arginine to reactive cysteine that becomes disulfide linked to albumin.[24–28] Polymerization of this mutant fibrinogen

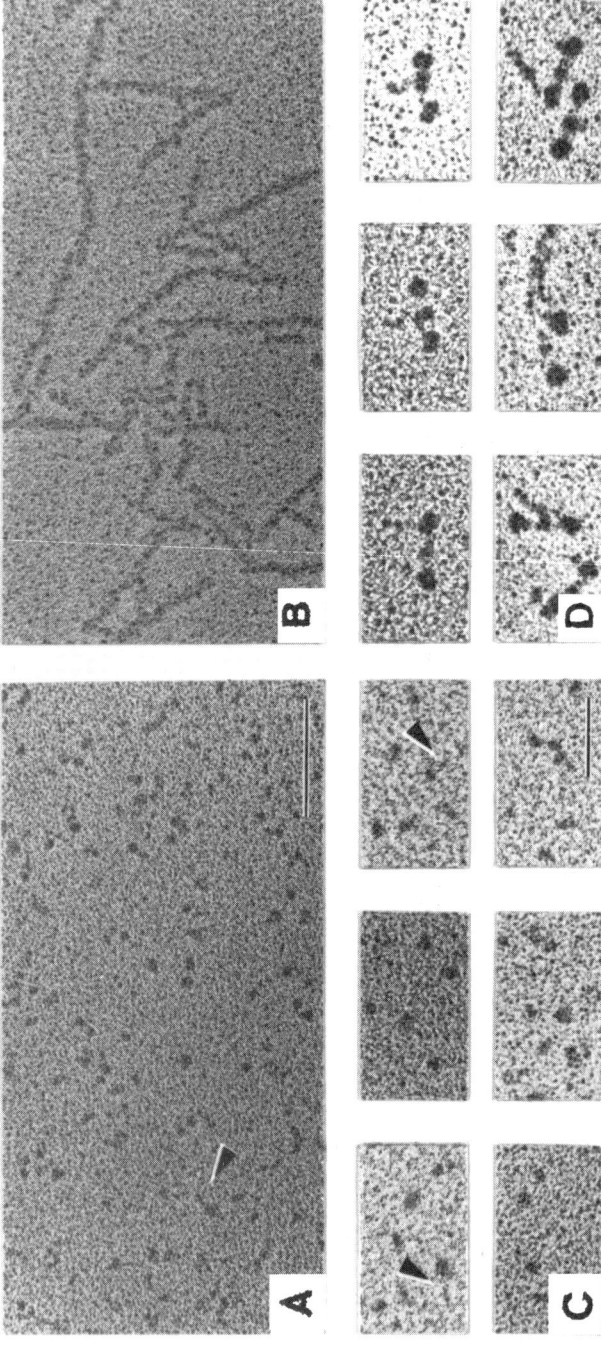

FIGURE 4. Electron microscopy of αC fragments and various aggregates. All specimens were rotary shadowed. **A.** Field of αC fragments at acidic pH; the αC fragments appear to be made up of a small nodule to which a thin appendage that is sometimes visible is attached. **B.** αC fragments at neutral pH form long thin polymers about twice the width of the individual particles. **C.** Gallery of individual and dimeric αC fragments at acidic pH. **D.** Individual and polymeric αC fragments interacting with fibrin monomers at neutral pH. (Figure from Veklich, *et al.* Ref. 19.)

on addition of thrombin is impaired, producing clots with decreased porosity and increased resistance to fibrinolysis, resulting in thrombotic complications and thromboembolism in the family members with this dysfibrinogenemia. Electron microscopy of rotary shadowed individual molecules revealed that, in contrast to control fibrinogen, most of the αC domains of fibrinogen or fibrin Dusart do not exhibit intramolecular or intermolecular interactions either with each other or with the central domains (see FIGURE 5).[29]

Scanning electron microscopy revealed that clots from fibrinogen Dusart were made up of thin fibers with many branch points and very small pore sizes (see FIGURE 6). Thus, the impairment of normal αC domain interactions that enhance

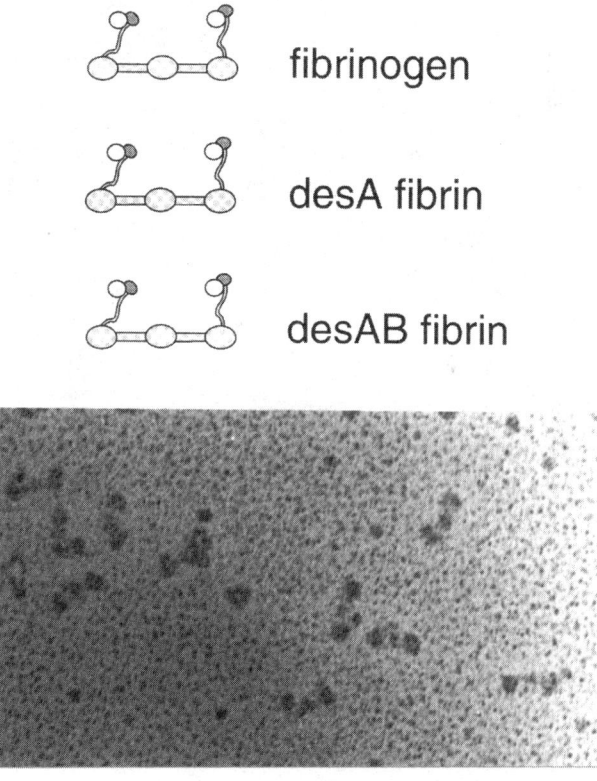

FIGURE 5. Fibrinogen Dusart molecules. Schematic representations are shown at the *top* and an example of an electron micrograph showing rotary shadowed molecules is shown at the *bottom*. The amino acid substitution (Aα 554 arginine to cysteine) and consequent albumin attachment prevent normal intramolecular and intermolecular αC domain interactions. Since this dysfibrinogenemia is heterozygous, about half of the molecules are normal in appearance. Fibrinogen Caracas II molecules are similar in appearance (not shown), but with extra glycosylation in the αC domains instead of the albumin. (Electron micrograph from Collet, *et al.* Ref. 29.)

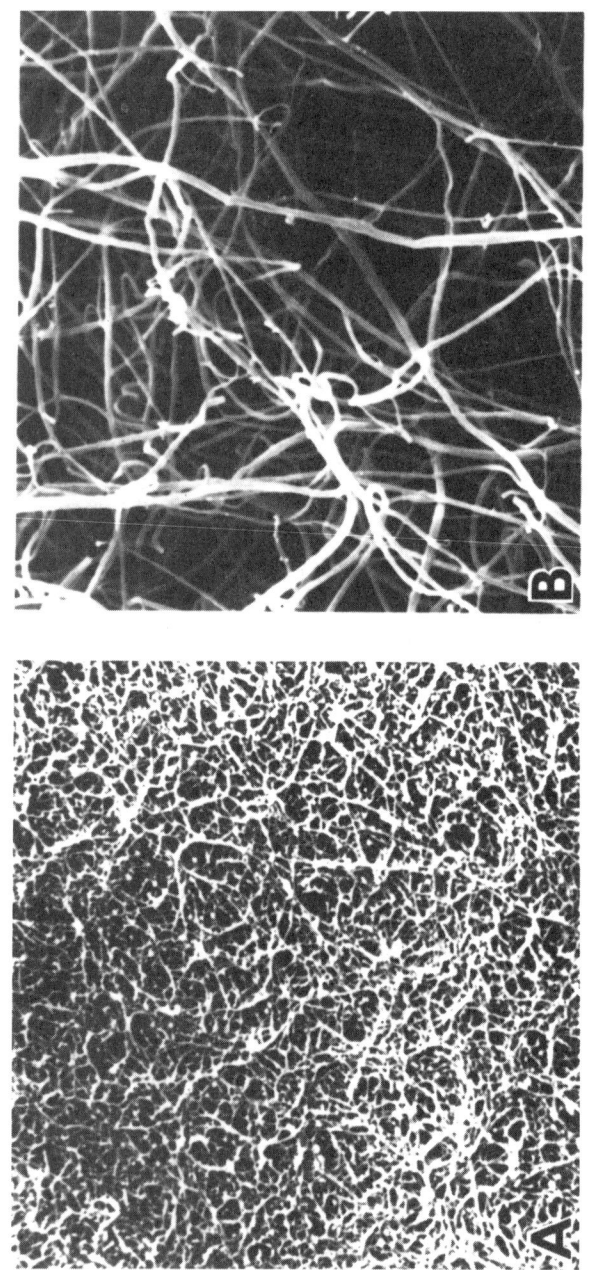

FIGURE 6. Effects of a mutation in the αC domains on clot structure. Scanning electron micrographs of clots from dysfibrinogenemia Dusart plasma (**A**) and control normal plasma (**B**). This mutation and consequent binding of albumin result in malfunctioning αC domains that do not enhance lateral aggregation, producing clots made up of very thin, highly branched fibers. (Electron micrographs from Collet, *et al.* Ref. 29.)

lateral aggregation leads to the formation of clots made up of very thin fibers. Because the association of protofibrils is greatly decreased, the number of branch points in these clots is also greatly increased.[30] The viscoelastic properties of Dusart fibrin clots measured with a torsion pendulum indicated a marked increase in stiffness, consistent with the structural observations, especially the greater number of branch points. Thus, this impairment of the normal function of the αC domains imposes striking effects on clot structure and function, with serious clinical consequences.

Fibrinogen Caracas II is an abnormal fibrinogen involving the mutation of Aα 434serine to N-glycosylated asparagine.[31] Some effects of this mutation on the ultrastructure of fibrinogen Caracas II molecules, fibers and clots were investigated by electron microscopy.[32] Electron microscopy of rotary shadowed individual molecules indicated that most of the αC domains of fibrinogen Caracas II do not interact with each other or with the central domain, in contrast to control fibrinogen, but similar to fibrinogen Dusart. Negatively contrasted Caracas II fibers were thinner and less ordered than control fibers, and many free fiber ends were observed. Scanning electron microscopy of whole clots revealed the presence of large pores bounded by local fiber networks made up of thin fibers. Permeation experiments also indicated that the average pore diameter was larger than that of control clots. The viscoelastic properties of the Caracas II clot, as measured by a torsion pendulum, were similar to those of control clots. Both the normal stiffness and increased permeability of the Caracas II clots are consistent with the observation that subjects with this dysfibrinogenemia are asymptomatic. In this case, the impairment of lateral aggregation again leads to clots made up of thinner fibers, but the abnormal molecules appear to "cap" the fibers, leading to many free fiber ends. As a result, the clots remain very permeable and are not abnormally stiff, giving a very different clinical picture than the Dusart syndrome.

There are now several reported dysfibrinogenemias in which the Aα chain is prematurely truncated, such as fibrinogens Marburg,[33] Milano III,[34] Lincoln,[35] and Otago.[36] Although the polymerization, clot structure, and other properties have not been reported in detail for all of these molecules, these fibrinogens all exhibit unusual polymerization properties, most of which are consistent with what has already been described.

Aα251 Recombinant Fibrinogen

Although the results from these and other dysfibrinogenemias are intriguing and give us many clues to the function of the αC domains, interpretation of the results is still somewhat confused by other accompanying changes to the fibrinogen and the fact that most of these cases are heterozygous. Studies using proteolytic fragments are perhaps more straightforward in interpretation, but any preparation of modified fibrinogen obtained by proteolysis will have some degree of heterogeneity, again making the results more difficult to interpret.

Another model for investigating the structure and function of the αC domains has recently been developed.[37] Aα251 recombinant fibrinogen contains Aα chains truncated at residue 251, but is otherwise identical to normal fibrinogen. The structural, physical, and biochemical properties of clots made from Aα251 fibrinogen were compared to those of normal clots to determine the function of the αC domains.[38] Both scanning electron microscopy and laser scanning confocal microscopy of clots made by adding thrombin to Aα251 fibrinogen revealed that the clots were normal in appearance, but the fibers were thinner and the densities of fibers and branch points were greater than those of control clots made from recombinant fibrinogen. Consistent with these results, the permeability of Aα251 fibrin clots was decreased to nearly half that of the control. Together, these results suggest that, in the formation of normal clots, the αC domains enhance lateral aggregation and produce thicker fibers.

In addition to these structural differences, there were greater differences between the viscoelastic or mechanical properties of Aα251 and control clots measured with a torsion pendulum.[38] Aα251 fibrin clots were only 37% as stiff as the control clots and showed more slippage of protofibrils, indicating that interactions between the αC domains in normal clots play a major role in determining the mechanical properties of the clot. With Factor XIIIa ligation, the rigidity of the control clots increased 2.5-fold, whereas that of the Aα251 fibrin clots increased 1.7-fold, providing information on the influence of different types of crosslinks. Crosslinking of the Aα251 fibrin clots, which would only have ligation of γ chains, yields stiffer clots, but the ligation of α chains in the control has a much greater effect on stiffness. Changes that were observed in the loss tangent, or irreversible deformation of the clot, demonstrate a considerable effect of the αC domain interactions on the slippage of protofibrils.

Fibrinolysis rates of the Aα251 fibrin clots at the microscopic level were determined by observing the lysis front over time by confocal microscopy of clots being digested with plasminogen activated by tissue type plasminogen activator (tPA).[38] Methods have been developed to label fibrin fibers with 5-nm colloidal gold beads, avoiding the fading that occurs with fluorescent labels and allowing quantitative measurements of clot structure as a function of time for fibrinolysis studies.[39,40] In these experiments, changes can be seen in individual fibers as they are digested; there is a sharp lysis front that moves across the clot with constant velocity as the fibers are cleaved. This velocity can be measured very accurately and used to characterize the rate of fibrinolysis. With this assay, the lysis front velocity was greater for Aα251 clots than for the controls.

Fibrinolysis rates, using tPA-activated plasminogen, were determined at the macroscopic level by measuring changes in turbidity of fibrin clots as a function of time,[38] since the turbidity decreases as the clots are digested. The rate of decrease of turbidity was lower for Aα251 clots than for the controls. These apparently contradictory results can be reconciled through an understanding of the differences between the two assays used. To calculate an approximate lysis front velocity from the turbidity measurements, the initial optical density must be taken into account. Both experiments indicate that lysis front propagation is faster for the less stiff Aα251 clots, which also have fewer intermolecular bonds and fewer plasmin cleavage sites. All of these results demonstrate that the αC domains play an important role in determining the structure and biophysical properties of clots and their fibrinolysis rate.

SUMMARY

The αC domains of fibrinogen interact with each other and are associated with the central region of the molecule, but there is a large conformational change on conversion to fibrin when the αC domains dissociate. They are then available for intermolecular interactions (FIG. 3B) that enhance lateral aggregation to produce clots made up of thicker fiber bundles and larger pores. Interactions between αC domains in fibrin have striking consequences for the mechanical properties and stability of the clot. Subsequent crosslinking of the αC domains in fibrin yields further important changes in the physical properties of clots. All of these effects are also significant for the rate of fibrinolysis of fibrin clots

ACKNOWLEDGMENTS

This work was supported by National Institutes of Health Grants HL30954 (to J.W.W.) and HL56051 (to L.M.). We would like to thank all of our collaborators and colleagues, especially Yuri Veklich, Oleg Gorkun, Jean-Philippe Collet, Sekar Nagaswami, John Woodhead, Jennifer Mullin, Susan Lord, Michio Matsuda, Agnes Henschen, Jeannette Soria, Claudine Soria, Peter Privalov, Vladimir Belitser, and Carmen Arocha-Piñango.

REFERENCES

1. DOOLITTLE, R.F., K.W. WATT, B.A. COTTRELL, et al. 1979. The amino acid sequence of the alpha-chain of human fibrinogen. Nature **280:** 464–468.
2. HENSCHEN, A. & J. MCDONAGH. 1986. Fibrinogen, fibrin and factor XIII. In Blood Coagulation. R.F.A. Zwall & H.C. Hemker, eds.: 171–234. Elsevier Science, Amsterdam.
3. DOOLITTLE, R.F. 1973. Structural aspects of the fibrinogen to fibrin conversion. Adv. Protein Chem. **27:** 1–109.
4. GORKUN, O.V., Y.I. VEKLICH, L.V. MEDVED', et al. 1994. Role of the αC domains in fibrin formation. Biochemistry **33:** 6986–6997.
5. RUDCHENKO, S., I. TRAKHT & J.H. SOBEL. 1996. Comparative structural and functional features of the human fibrinogen alphaC domain and the isolated alphaC fragment. J. Biol. Chem. **271:** 2523–2530.
6. HENSCHEN, A., F. LOTTSPEICH & B. HESSEL. 1979. Amino acid sequence of human fibrin. Preliminary note on the completion of the intermediate part of the alpha-chain sequence. Hoppe Seylers Z. Physiol. Chem. **360:** 1951–1956.
7. MEDVED', L.V., O.V. GORKUN & P.L. PRIVALOV. 1983. Structural organization of C-terminal parts of fibrinogen A alpha-chains. FEBS Lett. **160:** 291–295.
8. WANG, Y.Z., J.E. PATTERSON, C.Y. GRAY, et al. 1989. Complete sequence of the lamprey fibrinogen a chain. Biochem. **28:** 9801–9806.
9. MURAKAWA, M., T. OKAMURA, T. KAMURA, et al. 1993. Diversity of primary structures of the carboxy-terminal regions of mammalian fibrinogen Aα-chains. Characterization of the partial nucleotide and deduced amino acid sequences in five mammalian species; rhesus monkey, pig, dog, mouse and syrian hamster. Thromb. Hæmost. **69:** 351–360.
10. WILLIAMSON, M.P. 1994. The structure and function of proline-rich regions in proteins. Biochem. J. **297:** 249–260.

11. COHEN, C., J.W. WEISEL, G.N. PHILLIPS, et al. 1983. The structure of fibrinogen and fibrin: I. Electron microscopy and X-ray crystallography of fibrinogen. Ann. N.Y. Acad. Sci. **408:** 194–213.
12. MEDVED', L.V., E.I. TIKTOPULO, P.L. PRIVALOV, et al. 1980. Microcalorimetric studies of temperature transitions in fibrinogen and its proteolytic fragments. Mol. Biol. (U.S.S.R.) **14:** 835–842.
13. PRIVALOV, P.L. & L.V. MEDVED'. 1982. Domains in the fibrinogen molecule. J. Mol. Biol. **159:** 665–683.
14. ERICKSON, H.P. & W.E. FOWLER. 1983. Electron microscopy of fibrinogen, its plasmic fragments and small polymers. Ann. N.Y. Acad. Sci. **408:** 146–163.
15. PRICE, T.M., D.D. STRONG, M.L. RUDEE, et al. 1981. Shadow-cast electron microscopy of fibrinogen with antibody fragments bound to specific regions. Proc. Natl. Acad. Sci. U.S.A. **78:** 200–204.
16. MOSESSON, M.W., J. HAINFELD, J. WALL, et al. 1981. Identification and mass analysis of human fibrinogen molecules and their domains by scanning transmission electron microscopy. J. Mol. Biol. **153:** 695–718.
17. WALL, J., J. HAINFELD, R.H. HASCHEMEYER, et al. 1983. Analysis of human fibrinogen by scanning transmission electron microscopy. Ann. N.Y. Acad. Sci. **408:** 164–179.
18. WEISEL, J.W., C.V. STAUFFACHER, E. BULLITT, et al. 1985. A model for fibrinogen: domains and sequence. Science **230:** 1388–1391.
19. VEKLICH, Y.I., O.V. GORKUN, L.V. MEDVED', et al. 1993. Carboxyl terminal portions of the alpha chains of fibrinogen and fibrin: Localization by electron microscopy and the effects of isolated alphaC fragments on polymerization. J. Biol. Chem. **268:** 13577–13585.
20. MEDVED', L.V., O.V. GORKUN, V.F. MANYAKOV, et al. 1985. The role of fibrinogen alpha C-domains in the fibrin assembly process. FEBS Lett. **181:** 109–112.
21. BLOMBÄCK, B., B. HESSEL, D. HOGG, et al. 1978. A two-step fibrinogen-fibrin transition in blood coagulation. Nature **275:** 501–505.
22. WEISEL, J.W. 1986. Fibrin assembly. Lateral aggregation and the role of the two pairs of fibrinopeptides. Biophys. J. **50:** 1079–1093.
23. WEISEL, J.W., Y.I. VEKLICH & O.V. GORKUN. 1993. The sequence of cleavage of fibrinopeptides from fibrinogen is important for protofibril formation and enhancement of lateral aggregation in fibrin clots. J. Mol. Biol. **232:** 285–297.
24. SORIA, J., C. SORIA & J.P. CAEN. 1983. A new type of congenital dysfibrinogenemia with defective fibrin lysis: Dusart syndrome, possible relation to thrombosis. Brit. J. Hæmatol. **53:** 575–586.
25. KOOPMAN, J., F. HAVERKATE, J. GRIMBERGEN, et al. 1993. Molecular basis for fibrinogen Dusart (A alpha 554 Arg → Cys) and its association with abnormal fibrin polymerization and thrombophilia. J. Clin. Invest. **91:** 1637–1643.
26. COLLET, J.P., J. SORIA, M. MIRSHAHI, et al. 1993. Dusart syndrome: a new concept of the relationship between fibrin clot architecture and fibrin clot degradability: hypofibrinolysis related to an abnormal clot structure. Blood **82:** 2462–2469.
27. CARRELL, N., D.A. GABRIEL, P.M. BLATT, et al. 1983. Hereditary dysfibrinogenemia in a patient with thrombotic disease. Blood **62:** 439–47.
28. WADA, Y. & S. T. LORD. 1994. A correlation between thrombotic disease and a specific fibrinogen abnormality (A alpha 554 Arg → Cys) in two unrelated kindred, Dusart and Chapel Hill III. Blood **84:** 3709–3714.
29. COLLET, J.-P., J.L. WOODHEAD, J. SORIA, et al. 1996. Fibrinogen Dusart: electron microscopy of molecules, fibers and clots and viscoelastic measurements. Biophys. J. **70:** 500–510.
30. BARADET, T.C., J.C. HASELGROVE & J.W. WEISEL. 1995. Three-dimensional reconstruction of fibrin clot networks from stereoscopic intermediate voltage electron microscope images and analysis of branching. Biophys. J. **68:** 1551–1560.
31. MAEKAWA, H., K. YAMAZUMI, S. MURAMATSU, et al. 1991. An A alpha Ser-434 to N-glycosylated Asn substitution in a dysfibrinogen, fibrinogen Caracas II, characterized by impaired fibrin gel formation. J. Biol. Chem. **266:** 11575–11581.
32. WOODHEAD, J.L., C. NAGASWAMI, M. MATSUDA, et al. 1996. The ultrastructure of fibrinogen Caracas II molecules, fibers and clots. J. Biol. Chem. **271:** 4946–4953.

33. KOOPMAN, J., F. HAVERKATE, J. GRIMBERGEN, et al. 1992. Fibrinogen Marburg: a homozygous case of dysfibrinogenemia, lacking amino acids A alpha 461–610 (Lys 461 AAA→ stop TAA). Blood **80:** 1972–1979.
34. FURLAN, M., C. STEINMANN, M. JUNGO, et al. 1994. A frameshift mutation in Exon V of the A alpha-chain gene leading to truncated A alpha-chains in the homozygous dysfibrinogen Milano III. J. Biol. Chem. **269:** 33129–33134.
35. RIDGWAY, H.J., S.O. BRENNAN, S. GIBBONS, et al. 1996. Fibrinogen Lincoln: a new truncated alpha chain variant with delayed clotting. Brit. J. Hæmatol. **93:** 177–184.
36. RIDGWAY, H.J., S.O. BRENNAN, J.M. FAED, et al. 1997. Fibrinogen Otago: a major a chain truncation associated with severe hypofibrinogenemia and recurrent miscarriage. Brit. J. Hæmatol. **98:** 632–639.
37. GORKUN, O.V., A.H. HENSCHEN-EDMAN, L.F. PING, et al. 1998. Analysis of A alpha 251 fibrinogen: the alpha C domain has a role in polymerization, albeit more subtle than anticipated from the analogous proteolytic fragment X. Biochem. **37:** 15434–15441.
38. COLLET, J.-P., Y. VEKLICH, J.L. MULLIN, et al. 1999. The αC domains of fibrinogen affect the structure of the clot and its physical and biochemical properties. Thromb. Hæmost. Suppl. 692.
39. COLLET, J.-P., D. PARK, C. LESTY, et al. 2000. Influence of fibrin network conformation and fibrin fiber diameter on fibrinolysis speed: Dynamic and structural approaches by confocal microscopy. Arterioscler. Thromb. Vasc. Biol. **20:** 1354–1361.
40. WEISEL, J.W., Y. VEKLICH, J.-P. COLLET, et al. 1999. Structural studies of fibrinolysis by electron and light microscopy. Thromb. Hæmost. **82:** 277–282.

Fibrinogen αC Domains Contain Cryptic Plasminogen and tPA Binding Sites

GALINA TSURUPA AND LEONID MEDVED

Biochemistry Department, The Holland Laboratory,
American Red Cross, Rockville, Maryland, USA

> ABSTRACT: Surface plasmon resonance and ELISA experiments revealed that recombinant fibrinogen αC fragment (residues Aα221–610) corresponding to the αC domain binds tPA and plasminogen with high affinity. This binding was found to be Lys-dependent and occurred via independent binding sites. Study with truncated variants of the αC fragment located these sites in its COOH-terminal half. Binding of tPA and plasminogen to these sites stimulated activation of the latter whereas proteolytic degradation of the αC fragment reduced this effect substantially, suggesting the importance of the αC domains in regulation of fibrinolysis.
>
> KEYWORDS: Fibrinogen; αC domain; Plasminogen; tPA; Binding; Fibrinolysis.

The impaired fibrin-mediated activation of plasminogen by tissue plasminogen activator (tPA) caused by mutations in the fibrinogen αC domains[1–3] suggests the involvement of these domains in fibrinolysis. To test this suggestion we expressed in *E. coli* a recombinant αC fragment (Aα221–610) corresponding to the αC domain and studied its ability to stimulate activation of plasminogen by tPA and its interaction with both proteins.

When the activation of plasminogen by tPA was monitored by the cleavage of the chromogenic substrate S-2251 with newly formed plasmin, the reaction was much more efficient in the presence of the αC fragment. This result indicates that this fragment is an effective stimulator of plasminogen activation. It also suggests that the isolated αC fragment and presumably fibrinogen αC domains may contain plasminogen- and/or tPA-binding sites. To test this suggestion we studied interaction of the αC fragment with both proteins by surface plasmon resonance (SPR) and ELISA. In SPR experiments, when the αC fragment was immobilized on a sensor chip, both tPA and plasminogen bound to it in a dose-dependent manner with similar K_d values of 33 ± 5 nM and 32 ± 2 nM, respectively. The binding of both proteins was Lys-dependent since it was abolished by 50 mM ε-aminocaproic acid. Similar results were obtained by ELISA when increasing concentrations of tPA or plasminogen were added to the immobilized αC fragment. These results clearly indicate that the isolated αC domain contains tPA- and plasminogen-binding sites that are Lys-dependent. In a separate SPR experiment fibrinogen exhibited a negligible binding to immobilized tPA or plasminogen; in agreement the stimulating effect of

---

Address for correspondence: Galina Tsurupa, Ph.D., The Holland Laboratory, American Red Cross, 15601 Crabbs Branch Way, Rockville, MD 20855, USA. Voice: 301-738-0724; fax: 301-738-0794.

tsiouriog@usa.redcross.org

fibrinogen in a chromogenic substrate assay was negligible. This implies that the tPA and plasminogen binding sites are cryptic in fibrinogen and are exposed in the isolated αC domain.

To further localize tPA- and plasminogen-binding sites we expressed in *E. coli* the recombinant Aα221–391 and Aα392–610 fragments corresponding to the NH_2-terminal and COOH-terminal halves of the αC domain, respectively, and tested their effect on stimulation of plasminogen by tPA and their binding to these proteins. It was found that among the Aα221–391 and Aα392–610 fragments only the latter bound both tPA and plasminogen in ELISA and SPR experiments. In agreement, the COOH-terminal half of the αC domain was an active stimulator, whereas the NH_2-terminal fragment was not. These indicate that binding sites for both tPA and plasminogen are localized in the COOH-terminal part of the αC domain.

Since both plasminogen and tPA bind to the αC domain in a Lys-dependent manner with similar affinity one might suspect that both proteins bind to the same site(s) through their kringle domains. Alternatively, they may bind to different binding sites. To select between these alternatives we performed competition binding experiments. In SPR, in which the response signal is proportional to the bound mass, the amount of bound tPA and plasminogen was additive, even at saturating concentrations of both components (1 μM). In ELISA, when increasing concentrations of plasminogen were added to the immobilized αC fragment, the binding was unaffected by the presence of 10-fold molar excess of tPA and *vice versa*. These results indicate that tPA and plasminogen bind to the isolated αC domain via independent binding sites.

It is well established that αC domains are very sensitive to proteolysis and are removed from fibrin and degraded into a smaller pieces at the early stage of fibrinolysis.[4] To check whether the stimulating activity is preserved in these degradation products we digested the αC fragment with plasmin and tested the stimulating effect of the digest at different times. It was found that upon degradation of the αC fragment this effect was substantially reduced; the degree of the reduction correlated well with the accumulation of the degradation products. Thus, degradation of the αC domains abolishes their stimulating activity. The results presented above indicate that (1) the fibrinogen αC domains contain novel high affinity tPA and plasminogen binding sites that are cryptic in fibrinogen and are presumably exposed in fibrin; (2) binding of tPA and plasminogen to the αC domains results in increased activation of the latter, indicating the importance of these domains in fibrinolysis; (3) proteolytic inactivation of the stimulating effect of the αC fragment suggests that cleavage of the αC domains at the early stages of fibrin degradation may represent an *in vivo* mechanism regulating fibrinolysis. The latter also explains the physiological significance of high sensitivity of the αC domains to proteolysis.

ACKNOWLEDGMENT

This work was supported by National Institute of Health Grant HL-56051 (to L.M.).

REFERENCES

1. LIJNEN, H.R., J. SORIA, C. SORIA, *et al.* 1984. Dysfibrinogenemia (fibrinogen Dusard) associated with impaired fibrin-enhanced plasminogen activation. Thromb. Hæmost. **51:** 108–109.
2. KOOPMAN, J., F. HAVERKATE, J. GRIMBERGEN, *et al.* 1993. Molecular basis for fibrinogen Dusard (Aα 554 Arg$\rightarrow$Cys) and its association with abnormal fibrin polymerization and thrombophilia. J. Clin. Invest. **91:** 1637–1643.
3. SUGO, T., C. NAKAMIKAWA, M. TAKEBE, *et al.* 1998. Factor XIIIa cross-linking of the Marburg fibrin: formation of $\alpha_m \cdot \gamma_n$-heteromultimers and the α-chain-linked albumin·γ complex, and disturbed protofibril assembly resulting in acquisition of plasmin resistance relevant to thrombophila. Blood **91:** 3282–3288.
4. HENSCHEN, A. & J. MCDONAGH. 1986. Fibrinogen, fibrin and factor XIII. *In* Blood Coagulation. R.F.A. Zwaal & H.C. Hemker, Eds.: 171–241. Elsevier Science Publishers, Amsterdam.

Clot Lysis of Variant Recombinant Fibrinogens Confirms that Fiber Diameter is a Major Determinant of Lysis Rate

JENNIFER L. MULLIN,[a] SUSAN E. NORFOLK,[a] JOHN W. WEISEL,[b] AND SUSAN T. LORD[a,c]

[a]*Department of Chemistry, University of North Carolina at Chapel Hill, Chapel Hill, North Carolina, USA*

[b]*Department of Cell and Developmental Biology, University of Pennsylvania School of Medicine, Philadelphia, Pennsylvania, USA*

[c]*Department of Pathology and Laboratory Medicine, University of North Carolina at Chapel Hill, Chapel Hill, North Carolina, USA*

ABSTRACT: Previous studies have suggested that clots with thinner fiber diameter lyse at slower rates than clots with thicker fiber diameter. We examined lysis of fibrin clots formed from three variant fibrinogens, each with substitutions in the N-terminal region of the Bβ chain. When we measured lysis as the rate of decrease in turbidity at 350 nm, we found that the rate of lysis was slower than normal for clots with thinner fibers. We noted, however, that the time to complete lysis was the same for all clots. Thus, when the data were considered as the percent of lysis with time, we found that the curves were the same as normal. We suggest that a complete and accurate characterization of clot dissolution requires comparison of normalized lysis rates.

KEYWORDS: Lysis; Fiber diameter; Recombinant fibrinogen.

INTRODUCTION

Previous studies have shown a positive correlation between fibrin fiber diameter and fibrinolysis rate.[1–9] Specifically, clots containing thin fibers lyse at a slow rate, whereas clots containing thick fibers are lysed more quickly. Initial studies examined the effect of fibrin structure on fibrinolysis by varying the salt and thrombin concentrations,[1,8] and by determining the lysis rate of coarse versus tight clots. Since this time, studies of natural dysfibrinogens, such as fibrinogen Dusart,[4] and Chapel Hill III,[10] as well as disorders such as myeloma,[2] nephrotic disease,[5] and diabetes,[11] have also shown that thin fiber formation leads to impaired fibrinolysis. Together, these studies demonstrate that the diameter of fibers within fibrin clots, irrespective of the manner in which the altered clots are formed, is responsible for the rate of clot lysis.

Address for correspondence: Susan T. Lord, Ph.D., Department of Pathology and Laboratory Medicine, University of North Carolina at Chapel Hill, Chapel Hill, NC 27599-7525, USA. Voice: 919-966-3548; fax: 919-966-6718.

stl@med.unc.edu

In this work, we provide additional evidence that fiber diameter dictates fibrinolysis rate, by presenting the fibrinolysis of three recombinant fibrinogens, A'β, BβR14H and BβA68T. A'β fibrinogen contains an FpA-like substrate on the N-terminus of the β chain,[12] BβR14H has a His substituted for Arg at the thrombin cleavage site (position 14) of the Bβ chain,[13] and BβA68T is the recombinant counterpart of the dysfibrinogen Naples.[14] Although our work with these fibrinogens confirms that clots made of thinner fibers lyse at a slower rate than normal, we propose that lysis rate, as determined by the decrease in turbidity over time, is only one aspect of our generated lysis profiles. Specifically, we present the concept of plotting percent lysis versus time, which takes into account the turbidity of fibrin clots prior to their dissolution. Plotting the percentage of the clot lysed over time shifts the perspective from the rate of lysis of individual fibers to the overall lysis of the entire clot. This information is important because clots that are lysed at different rates may have similar percent lysis profiles, depending on their initial turbidity.

MATERIALS AND METHODS

Polymerization and Fibrinolysis

Normal, A'β, BβR14H, and BβA68T recombinant fibrinogens were all expressed and purified as described elsewhere.[15,16] BβR14H fibrinogen was synthesized using a mutated pMLP-β plasmid, by a similar protocol as BβA68T fibrinogen,[16] and will be published elsewhere. Polymerization of 90 μL fibrinogen (0.4 mg/mL) in 20 mM HEPES, pH 7.4, 150 mM NaCl, 1 mM $CaCl_2$ was initiated by the addition of 10 μL of human α-thrombin and immediate automixing by the instrument. The final thrombin concentration for normal, A'β and BβR14H fibrinogens was 0.4 U/mL (3.6 nM). BβA68T fibrinogen was polymerized with 10.8 U/mL (97 nM) thrombin, to normalize the rate of FpA release, as previously reported.[14] Polymerization was monitored in individual wells of 96-well plates at ambient temperatures on a SpectraMax 340PC for one hour as an increase in turbidity at 350 nm. On completion of polymerization, 100 μL of a tPA/plasminogen mixture, containing 10 nM tPA (Genentech) and 0.05 mg/mL glu-plasminogen, were overlaid onto each clot and fibrinolysis monitored as the decrease in turbidity at 350 nm for 150 minutes.

Scanning Electron Microscopy

Scanning electron microscopy was performed on these four fibrinogens as previously described,[12,14] and under conditions similar to those of polymerization. Fiber diameters were measured using Scion Image.

Analysis of Fibrinolysis Results

The absolute rate of fibrinolysis was determined as the slope of the lysis curve. The completion time represents the time at which the turbidity at 350 nm was zero. Percent lysis was calculated by the following formula: 100 × (max OD − measured OD)/max OD, where OD represents the turbidity value at 350 nm.

RESULTS AND DISCUSSION

Previous studies have shown that thrombin-catalyzed polymerization of A'β fibrinogen is similar to normal and results in clots with similar fiber diameter to normal,[12] whereas βA68T fibrin, under conditions of normalized FpA release, has impaired polymerization and forms thinner fibers than normal.[14] Thrombin-catalyzed polymerization of BβR14H fibrinogen is also impaired compared to normal, such that the final turbidity reached is approximately half that of normal. Fiber diameter analysis, from scanning electron micrographs, showed that βR14H fibrin clots contain thinner fibers than normal (fiber diameters: 102 ± 9 nm for normal, and 70 ± 8 nm for βR14H, $p < 0.001$). Fibrinolysis of clots formed with these variant fibrinogens was studied under conditions of physiologically relevant salt concentration and pH to determine if fiber diameter is the major determinant of lysis rate.

Fibrinolysis results showed that clots containing thinner fibers, made of βR14H and βA68T fibrins, lysed at a slower rate than normal. The lysis rate of βA68T fibrin was $1.4 \pm 0.5 \times 10^{-5} \text{sec}^{-1}$, compared to the normal rate of $2.3 \pm 0.4 \times 10^{-5} \text{sec}^{-1}$ ($p = 0.08$, $n = 3$). The lysis rate of βR14H fibrin was $1.6 \pm 0.1 \times 10^{-5} \text{sec}^{-1}$ ($n = 5$), compared to the normal rate of $2.5 \pm 0.2 \times 10^{-5} \text{sec}^{-1}$ ($n = 6$, $p < 0.001$). Additionally, the lysis rate of A'β fibrin, which forms clots containing fibers with similar fiber diameters to normal, was similar to normal ($2.5 \pm 0.3 \times 10^{-5} \text{sec}^{-1}$ for A'β and $2.9 \pm 0.9 \times 10^{-5} \text{sec}^{-1}$ for normal, $n = 12$, $p = 0.16$). In *all* cases, however, the lysis completion times of the variant fibrins were similar to normal, with complete lysis occurring after approximately 100 minutes under our conditions.

Our results agree with previous reports that conclude that fiber diameter correlates with lysis rate, such that clots made of thin fibers lyse more slowly than clots made of thick fibers.[1–9] Unexpectedly, however, the clots that appeared to lyse more slowly than normal did not require a longer time for complete dissolution of the clot. Moreover, plotting the lysis of A'β and βR14H variant fibrins as percent lysis over time showed that these percent lysis curves were identical to normal, despite the lower rate of fibrinolysis for βR14H seen when monitoring the decrease in turbidity at 350 nm over time. Thus, taking into account the difference in final turbidity on polymerization, and plotting percent lysis for these two fibrins shows that they lyse the same percentage of the clot over time as normal. For βA68T fibrin, however, the percent lysis curve compared to normal had a longer lag, and a steeper slope than normal, indicating a different pattern of lysis over time, although the completion time was the same as normal. This result suggests that the nature of the βA68T fibrin clot is different from normal in that it requires a longer time before efficient lysis of the clot begins.

From our results, we support the concept that fiber diameter is a major determinant of lysis rate, as characterized by turbidity decrease. In addition, we suggest that comparison of lysis rates measured by a decrease in turbidity may not provide a complete characterization of lysis, and that plotting percent lysis versus time may reveal additional important information.

ACKNOWLEDGMENTS

This work was funded by NIH Grant HL-31048 to S.T.L and NIH Grant HL-30954 to J.W.W. We acknowledge Kasim McLain and Li Fang Ping for their assistance in protein expression and purification. Cell medium for normal recombinant fibrinogen was from the Cell Culture Center (Minneapolis, MN).

REFERENCES

1. CARR, JR., M.E., & B.M. ALVING. 1995. Effect of fibrin structure on plasmin-mediated dissolution of plasma clots. Blood Coag. Fibrinol. **6:** 567–573.
2. CARR, JR., M.E., et al. 1996. Abnormal fibrin structure and inhibition of fibrinolysis in patients with multiple myeloma. J. Lab. Clin. Med. **128:** 83–88.
3. CARR, JR., M.E., et al. 1995. Effects of ionic and nonionic contrast media on clot structure, platelet function and thrombolysis mediated by tissue plasminogen activator in plasma clots. Hæmostasis **25:** 172–181.
4. COLLET, J.P., et al. 1993. Dusart syndrome: a new concept of the relationship between fibrin clot architecture and fibrin clot degradability: hypofibrinolysis related to an abnormal clot structure. Blood **82:** 2462–2469.
5. COLLET, J.-P., et al. 1999. Abnormal fibrin clot architecture in nephrotic patients is related to hypofibrinolysis: influence of plasma biochemical modifications. Thromb. Hæmostas. **82:** 1482–1489.
6. COLLET, J.-P., et al. 2000. Influence of fibrin network conformation and fibrin fiber diameter on fibrinolysis speed; dynamic and structural approaches by confocal microscopy. Arterioscleros. Thromb. Vasc. Biol. **20:** 1354–1361.
7. DHALL, D.P. & C.H. NAIR. 1994. Effects of gliclazide on fibrin network. J. Diabet. Compl. **8:** 231–234.
8. GABRIEL, D.A., et al. 1992. The effect of fibrin structure on fibrinolysis. J. Biol. Chem. **267:** 24259–24263.
9. VEKLICH, Y., et al. 1998. Structural studies of fibrinolysis by electron microscopy. Blood **92:** 4721–4729.
10. WADA, Y. & S.T. LORD. 1994. A correlation between thrombotic dsease and a specific fibrinogen abnormality (A alpha 554 Arg→Cys) in two unrelated kindred, Dusart and Chapel Hill III. Blood **84:** 3709–3714.
11. BROWNLEE, M., et al. 1983. Nonenzymatic glycosylation reduces the susceptibility of fibrin to degradation by plasmin. Diabetes **32:** 680–684.
12. MULLIN, J.L., et al. 2000. Recombinant fibrinogen studies reveal that thrombin specificity dictates order of fibrinopeptide release. J. Biol. Chem. **275:** 25239–25246.
13. MULLIN, J.L. 2000. An Expanded and Refined Model of Fibrin Polymerization Based on Studies with Variant Recombinant Fibrinogens. Ph.D. Dissertation, University of North Carolina, Chapel Hill.
14. MULLIN, J.L., et al. 2000. Decreased lateral aggregation of a variant recombinant fibrinogen provides insight into polymerization mechanism. Biochemistry **39:** 9843–9849.
15. BINNIE, C. G., et al. 1993. Characterization of purified recombinant fibrinogen: partial phosphorylation of fibrinopeptide A. Biochemistry **32:** 107–113.
16. LORD, S.T., et al. 1996. Strategy for recombinant multichain protein synthesis: fibrinogen B beta-chain variants as thrombin substrates. Biochemistry **35:** 2342–2348.

Crosslinking of α_2-Antiplasmin to Fibrin

KYUNG N. LEE, CHUNG S. LEE, WEON-CHAN TAE, KENNETH W. JACKSON, VICTORIA J. CHRISTIANSEN, AND PATRICK A. McKEE

William K. Warren Medical Research Institute and Department of Medicine, University of Oklahoma Health Sciences Center, Oklahoma City, Oklahoma, USA

ABSTRACT: Human α_2-antiplasmin (α_2AP) is the primary inhibitor of plasmin-mediated fibrinolysis and is an efficient substrate of activated factor XIII (FXIIa). Among 452 amino acid residues in α_2AP, Gln^2 is believed to be the sole FXIIIa-reactive site that participates in crosslinking α_2AP to fibrin. We studied the effect of mutating Gln^2 on the ability of FXIIIa to catalyze crosslinking of α_2AP to fibrin. By FXIIIa catalysis, [^{14}C]methylamine was incorporated into a Q2A-α_2AP mutant in which Gln^2 (Q) was replaced by Ala (A), thereby indicating that wildtype α_2AP has more than one FXIIIa-reactive site. To identify the FXIIIa-reactive sites in α_2AP, wildtype α_2AP and Q2A-α_2AP were labeled with 5-(biotinamido)pentylamine by FXIIIa. Each labeled α_2AP was digested with trypsin and applied to an avidin affinity column to capture labeled peptides. Edman sequencing and mass analysis of each labeled peptide showed that out of 35 Gln residues in wildtype α_2AP, four were labeled with the following order of efficiency: $Gln^2 > Gln^{21} > Gln^{419} > Gln^{447}$. Q2A-$\alpha_2$AP was also labeled at the three minor sites, $Gln^{21} > Gln^{419} > Gln^{447}$. Q2A-$\alpha_2$AP became crosslinked to fibirin(ogen) by FXIIIa catalysis at approximately one-tenth the rate of wt-α_2AP. These results demonstrate that α_2AP has one primary (Gln^2) and three minor substrate sites for FXIIIa and that the three minor sites identified in this study can also participate in crosslink formation between α_2AP and fibrin, but at a much lower efficiency than the Gln^2 site.

KEYWORDS: α_2-Antiplasmin; Crosslinking; Fibrin.

INTRODUCTION

Human α_2-antiplasmin (α_2AP) is synthesized primarily by the liver and circulates in blood as a single chain glycoprotein composed of 452 amino acid residues and 11–14% carbohydrate. α_2AP has three domains[1] that function to protect fibrin clots from plasmin mediated lysis as follows: (1) initially the Gln^2 residue in the N-terminal domain of α_2AP becomes crosslinked to each fibrin α chain at Lys^{303} by activated Factor XIII (FXIIIa); (2) then, the C-terminal domain of α_2AP interacts with the kringle structures of plasmin; and (3) the latter interaction causes Arg^{364} in the reactive site domain to align and form a covalent bond with the active site Ser in plasmin to give an inactive protease–inhibitor complex. Chemical modification[2] or mutation[3] of the reactive-site Arg^{364} in α_2AP results in complete loss of plasmin inhibitory activity; however, both forms compete effectively with native α_2AP for

---

Address for correspondence: Patrick A. McKee, M.D., Department of Medicine, University of Oklahoma Health Sciences Center, 941 Stanton L. Young Blvd., BSEB-321, Oklahoma City, OK 73104, USA. Voice: 405-271-5645; fax: 405-271-3229.
patrick-mckee@ouhsc.edu

becoming crosslinked to fibrin by FXIIIa catalysis. Lacking Arg^{364} reactivity to inhibit plasmin, fibrinolysis occurs significantly faster.[2,3]

Under catalysis by FXIIIa, only the Gln^2 site in α_2AP was believed to participate in crosslink formation with the α-chain of fibrin,[4] but recent mass spectrometric analyses of 5-(biotinamido)pentylamine (BPA)-labeled wt-α_2AP showed four such Gln substrate sites for FXIIIa.[5] This finding was expanded to determine whether mutating Gln^2→Ala (or Q2A in single letter code) would prevent α_2AP from crosslinking to fibrin. We report here the expression and purification of wt-α_2AP and Q2A-α_2AP; the identification and reactivity of respective Gln substrate sites; and the rates of wt-α_2AP and mutant Q2A-α_2AP crosslinking to fibrin by FXIIIa.

ISOLATION OF RECOMBINANT α_2AP

An *E. coli* expression vector containing a P_L promoter[6] and α_2AP cDNA was made, and with this construct, α_2AP was expressed as about 10% of total cellular proteins, with about 5% as soluble and 95% as insoluble inclusion body. To obtain active soluble α_2AP from the insoluble form, inclusion bodies were solubilized, refolded, and purified according to a published method.[7] Purified recombinant α_2AP showed a single band with a molecular mass of 51 Kda on SDS-PAGE analysis, which is the expected size for non-glycosylated α_2AP. Approximately 5 mg of active, purified α_2AP was obtained from each liter of culture. Purified recombinant α_2AP showed a single peak of 50,583 Da by electrospray mass spectrometry (ES-MS), which corresponds to its calculated molecular weight. The first 12 amino acids of its N-terminal sequence were determined to be MNQEQVSPLTLL, indicating an additional Met residue. Mutant α_2AP (Q2A-α_2AP) had an N-terminal amino acid sequence of MNAEQVSPLTLL in which Gln, the third amino acid from the N-terminus (equivalent to Gln^2 of native plasma α_2AP[8]), was replaced by Ala.

PLASMIN INHIBITION BY RECOMBINANT α_2AP

Plasmin inhibition activity was measured for wt-α_2AP and Q2A-α_2AP, and each was compared with that of native α_2AP. All three α_2AP proteins inhibited plasmin activity with the same rapidity. These observations were in accord with previous reports that non-glycosylated recombinant α_2AP,[3] α_1-antitrypsin,[9] and α_1-chymotrypsin[10] have activities equal to their glycosylated native counterparts. Despite our recombinant α_2AP having an additional Met residue as the N-terminus, it possessed full plasmin inhibition activity. Similarly, Q2A-α_2AP, prepared by replacing Gln^2 with Ala, inhibited plasmin as effectively as native α_2AP or wt-α_2AP. Hence, neither non-glycosylation, the extra N-terminal Met, or the Gln^2 mutation affected the inhibition of plasmin by α_2AP.

RECOMBINANT α_2AP AS FXIIIa SUBSTRATE

Both native and recombinant α_2AP proteins were assessed as FXIIIa substrates by comparing kinetic parameters of FXIIIa-catalyzed [^{14}C]methylamine incorporation

TABLE 1. Characterization of FXIIIa-catalyzed BPA-labeled tryptic peptides by ES-MS and Edman sequencing

Peptide	Amino Acid Sequence[a]	Observed Mass	Calculated Mass[b]
		Da	Da
T1	MN**Q**^{2}EQVSPLTLLK	1811.3	1500.8
T2	LGNQEPGG**Q**^{21}TALK	1623.1	1312.7
T31	DFL**Q**^{419}SLK	1160.9	850.4
T35	LVPPMEEDYP**Q**^{447}FGSPK	2144.7	1833.8

[a]The bold letter **Q** with number represents the BPA-labeled Gln residue. Number on the right shoulder of the letter **Q** refers to the position in the published sequence[8] in which the first extra Met residue at the N-terminus of recombinant α_2AP is not included.

[b]Calculated mass represents monoisotopic molecular weight and does not include the mass of incorporated BPA.

into α_2AP[11] at a saturating concentration of methylamine (2 mM) and selected concentrations of native α_2AP, wt-α_2AP, or Q2A-α_2AP. FXIIIa gave apparent K_m ($K_{m_{app}}$) values of 5.35 μM, 5.02 μM, and 39.5 μM, for native α_2AP, wt-α_2AP, and Q2A-α_2AP, respectively. Hence, FXIIIa showed full affinity toward wt-α_2AP, but a dramatic loss of affinity toward Q2A-α_2AP, and the kinetic efficiency (k_{cat}/K_m) of Q2A-α_2AP (22.2 min^{-1}/mM) was 35-fold lower than those of wt-α_2AP (780 min^{-1}/mM). Another amine substrate, BPA, could also be incorporated into both wt-α_2AP and Q2A-α_2AP by FXIIIa catalysis. These results showed that α_2AP has more than one Gln substrate site, but that Gln^2 is clearly the primary site for FXIIIa.

LOCALIZATION AND REACTIVITY OF Gln SUBSTRATE SITES IN α_2AP

To locate Gln substrate sites, α_2AP was labeled with BPA and then reduced, alkylated, and digested with trypsin. As a first step, BPA-labeled tryptic peptides were isolated from the tryptic digest by an avidin affinity chromatograpy. The labeled peptides were further purified by reversed-phase HPLC and subjected to ES-MS analysis. As is shown in TABLE 1, none of the four peptides (T1, T2, T31, or T35) isolated by avidin affinity chromatography and reversed-phase HPLC matched the calculated mass of any predicted tryptic peptide from wt-α_2AP. However, the observed mass of each tryptic peptide was consistent with that calculated, plus the addition of one BPA molecule, and this allowed each peptide that contained a potential Gln substrate site to be located: T1, residues 1–12; T2, residues 13–25; T31, 416–422; and T35, residues 437–452 (TABLE 1). To identify which of the two Gln residues in each of the T1 and T2 peptides is labeled with BPA, each peptide was subjected to Edman sequencing analysis. During Edman degradation of each of the two peptides, PTH-Gln^2 and PTH-Gln^{21} were not detected, indicating that these residues were labeled by FXIIIa catalysis. The other two peptides, T31 and T35, were also subjected to Edman degradation, and PTH-Gln^{419} and PTH-Gln^{447} were not detected, suggesting that these residues were modified.

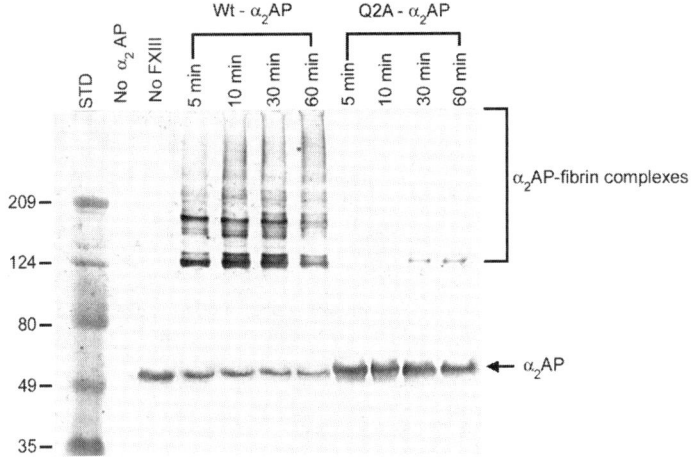

FIGURE 1. Time dependent crosslinking of wt-α_2AP and Q2A-α_2AP to fibrin with FXIIIa. FXIII, human fibrinogen, and α_2AP were clotted with thrombin and $CaCl_2$. After incubation for various times at 37°C, the clots were solubilized and subjected to SDS-PAGE. α_2AP-fibrin complexes were detected by immunoblot analysis. *Control lanes* without α_2AP (no α_2AP) or without either transglutaminase (no FXIII) are also shown.

To assess the reactivity of Gln substrate sites for FXIIIa, or the efficiency of Gln sites as FXIIIa substrates, liquid chromatography/mass spectrometry reconstructed ion current (RIC) analysis of tryptic digests of BPA-labeled α_2AP was performed to quantitate BPA-labeled and unlabeled peptides. To compare the reactivity of wt-α_2AP Gln substrate sites for FXIIIa, the percent modification was determined by division of the labeled peptide area by the combined areas of labeled and unlabeled forms for each peptide in the RIC analysis. The reactivities of the Gln^2 site in T1 peptide, Gln^{21} in T2 peptide, Gln^{419} in T31 peptide, and Gln^{447} in T35 peptide toward FXIIIa were found to be 92.9, 49.5, 37.5, and 16.9%, respectively, thereby demonstrating that Gln^2 is the major substrate site for FXIIIa, whereas Gln^{21}, Gln^{419}, and Gln^{447} are less reactive secondary sites.

CROSSLINKING wt-α_2AP OR Q2A-α_2AP INTO FIBRIN

To estimate the potential physiological relevance of Gln^2 being the primary site, and Gln^{21}, Gln^{419}, and Gln^{447} being minor sites in FXIIIa-catalyzed crosslinking of α_2AP to fibrin, physiological concentrations of fibrinogen (6 µM), FXIII (50 nM), and α_2AP (1 µM) were incubated with $CaCl_2$ and thrombin to induce fibrin formation. As is shown in FIGURE 1, wt-α_2AP was rapidly crosslinked to fibrin to form "α_2AP-fibrin complexes" within five minutes. In contrast, Q2A-α_2AP crosslinking slowly and appeared to start only at 30 min. However, Q2A-α_2AP retained full plasmin inhibition activity, equal to that of wt-α_2AP. Based on these results and those of

Weitz et al.,[12] Q2A-α_2AP, or an analogue, might have potential utility as a therapeutic agent to inhibit or reverse hyperplasminemic bleeding complications, simultaneously avoiding the inhibition of thrombolysis.

CONCLUSIONS

Of 452 amino acids in α_2AP, only Gln^2 is believed to be a FXIIIa reactive fibrin crosslinking site. With FXIIIa catalysis, however, [^{14}C]methylamine was also incorporated into Q2A-α_2AP, indicating wt-α_2AP has more than one Gln crosslink site. To identify FXIIIa-reactive sites in wt-α_2AP, or Q2A-α_2AP, each was labeled with 5-(biotinamido)pentylamine by FXIIIa catalysis. After each labeled α_2AP was digested by trypsin, amino acid sequence and mass analyses of each labeled peptide showed that four of 35 Gln residues were labeled with the following reactivities: Gln^2 > Gln^{21} > Gln^{419} > Gln^{447}. Q2A-α_2AP was also labeled at Gln^{21} > Gln^{419} > Gln^{447}, but became crosslinked to fibrin by FXIIIa at approximately one-tenth the rate for wt-α_2AP.

REFERENCES

1. MOROI, M. & N. AOKI. 1976. Isolation and characterization of α_2-plasmin inhibitor from human plasma. A novel proteinase inhibitor which inhibits activator-induced clot lysis. J. Biol. Chem. **251:** 5956–5965.
2. LEE, K.N., S.C. LEE, K.W. JACKSON, et al. 1998. Effect of phenylglyoxal-modified α_2-antiplasmin on urokinase-induced fibrinolysis. Thromb. Hæmost. **80:** 637–644.
3. LEE, K.N., W.C. TAE, K.W. JACKSON, et al. 1999. Characterization of wild-type and mutant α_2-antiplasmins: fibrinolysis enhancement by reactive site mutant. Blood **94:** 164–171.
4. TAMAKI, T. & N. AOKI. 1982. Cross-linking of α_2-plasmin inhibitor to fibrin catalyzed by activated fibrin-stabilizing factor. J. Biol. Chem. **257:** 14767–14772.
5. LEE, K.N., C.S. LEE, W.C. TAE, et al. 2000. Crosslinking of transglutaminase-reactive site mutant α_2-antiplasmin to fibrin. FASEB J. **14:** A1530.
6. MIESCHENDAHL, M., T. PETRI & U. HÄNGGI. 1986. A novel prophage independent Trp regulated lambda P$_L$ expression system. Bio/Tech. **4:** 802–808.
7. LIN, X.L., R.N. WONG & J. TANG. 1989. Synthesis, purification, and active site mutagenesis of recombinant porcine pepsinogen. J. Biol. Chem. **264:** 4482–4489.
8. LIJNEN, H.R., W.E. HOLMES, B. VAN HOEF, et al. 1987. Amino-acid sequence of human α_2-antiplasmin. Eur. J. Biochem. **166:** 565–574.
9. KWON, K.S., S. LEE & M.H. YU. 1995. Refolding of α_1-antitrypsin expressed as inclusion bodies in *Escherichia coli*: characterization of aggregation. Biochim. Biophys. Acta **1247:** 179–184.
10. RUBIN, H., Z.M. WANG, E.B. NICKBARG, et al. 1990. Cloning, expression, purification, and biological activity of recombinant native and variant human α_1-antichymotrypsins. J. Biol. Chem. **265:** 1199–1207.
11. LEE, K.N., L. FESUS, S.T. YANCEY, et al. 1985. Development of selective inhibitors of transglutaminase: phenylthiourea derivatives. J. Biol. Chem. **260:** 14689–14694.
12. WEITZ, J.I., B. LESLIE, J. HIRSH & P. KLEMENT. 1993. α_2-antiplasmin supplementation inhibits tissue plasminogen activator-induced fibrinogenolysis and bleeding with little effect on thrombolysis. J. Clin. Invest. **91:** 1343–1350.

Platelet–Fibrinogen Interactions

JOEL S. BENNETT

Hematology-Oncology Division, Department of Medicine,
University of Pennsylvania School of Medicine, Philadelphia, Pennsylvania, USA

ABSTRACT: Binding of fibrinogen to GPIIb-IIIa on agonist-stimulated platelets results in platelet aggregation, presumably by crosslinking adjacent activated platelets. Although unactivated platelets express numerous copies of GPIIb-IIIa on their surface, spontaneous, and potentially deleterious, platelet aggregation is prevented by tightly regulating the fibrinogen binding activity of GPIIb-IIIa. Preliminary evidence suggests that it is the submembranous actin or actin-associated proteins that constrains GPIIb-IIIa in a low affinity state and that relief of this constraint by initiating actin filament turnover enables GPIIb-IIIa to bind fibrinogen. Two regions of the fibrinogen α chain that contain an RGD motif, as well as the carboxyl-terminus of the fibrinogen γ chain, represent potential binding sites for GPIIb-IIIa in the fibrinogen molecule. However, ultrastructural studies using purified fibrinogen and GPIIb-IIIa, and studies using recombinant fibrinogen in which the RGD and relevant γ chain motifs were mutated indicate that sequences located at the carboxyl-terminal end of the γ chain mediates fibrinogen binding to GPIIb-IIIa. There is evidence that fibrinogen itself binds to regions in the amino terminal portions of both GPIIb and GPIIIa and that the sites interacting with the fibrinogen γ chain and with RGD-containing peptides are spatially distinct. Nonetheless, there appears to be allosteric linkage between these sites, accounting for the ability of RGD-containing peptides to inhibit platelet aggregation and arterial thrombosis.

KEYWORDS: GPIIb-IIIa; Fibrinogen receptors; Platelet aggregation; RGD peptides.

INTRODUCTION

Early *in vitro* studies indicated that in addition to calcium, a protein cofactor, fibrinogen, was required for ADP-stimulated platelet aggregation.[1,2] Moreover, the cutaneous bleeding time was often prolonged and *in vitro* platelet aggregation abnormal in patients with congenital afibrinogenemia, abnormalities that were corrected by fibrinogen infusions.[3] An explanation for these observations was provided when it was discovered that platelets express a specific receptor for fibrinogen, the function of which is regulated by agonist stimulation.[4,5] Thus, platelet aggregation requires the binding of soluble fibrinogen (and/or von Willebrand factor) to the fibrinogen receptor, glycoprotein IIb-IIIa (GPIIb-IIIa, αIIbβ3, CD41CD61).[4,6] GPI-Ib-IIIa, a calcium dependent heterodimer, the expression of which is restricted to platelets and cells of the megakaryocyte lineage,[7] is a member of the integrin family

---

Address for correspondence: Joel S. Bennett, M.D., Room 914, BRB II/III, 421 Curie Blvd., Philadelphia, PA 19104, USA. Voice: 215-573-3280; fax: 215-573-7039.
bennetts@mail.med.upenn.edu

of adhesion receptors[8] and contains a binding site for protein ligands such as fibrinogen and von Willebrand factor (vWf) that is exposed when platelets are stimulated by agonists.[9] Binding of either fibrinogen or von Willebrand factor to GPIIb-IIIa results in platelet aggregation, presumably by crosslinking adjacent activated platelets.[10] Nevertheless, by virtue of its high concentration in plasma, fibrinogen is the major physiologic GPIIb-IIIa ligand.

THE BIOCHEMISTRY OF FIBRINOGEN BINDING TO GPIIb-IIIa

Fibrinogen binds to GPIIb-IIIa on agonist-stimulated platelets with a dissociation constant (K_d) of approximately 100 nM, nearly 100-fold less than the concentration of fibrinogen in plasma, implying the fibrinogen binding site on GPIIb-IIIa is immediately occupied when platelets are activated in a plasma environment.[4] Because unactivated platelets express 50,000–100,000 copies of GPIIb-IIIa on their surface, spontaneous, and potentially deleterious, platelet aggregation is prevented by tightly regulating the fibrinogen binding activity of GPIIb-IIIa. However, the biochemical reactions that regulate the affinity of GPIIb-IIIa for fibrinogen are uncertain. It is likely that a product (or products) of these reactions eventually interacts with the cytoplasmic extensions of GPIIb and/or GPIIIa to induce a conformational change in GPIIb-IIIa to expose its ligand binding site.[10] Platelet stimulation is known to activate the phospholipases C (PLC) and A2 (PLA2) and the serine/threonine kinases protein kinase C (PKC), calcium-calmodulin dependent protein kinase, and myosin light chain kinase.[11,12] PLC activation is a G protein mediated process and generates diacylglycerol (DAG) and inositol 1,4,5-triphosphate (IP3) from phosphatidylinositol 4,5-bisphosphate. In turn, DAG activates PKC, whereas IP3 is responsible in part for the rise in intracellular calcium that accompanies platelet activation by most platelet agonists. Because phorbol esters are potent platelet agonists and activate PKC, it is likely that PKC plays an important role in inside-out signaling in platelets. Nevertheless, the ability of agonists to activate GPIIb-IIIa despite the presence of PKC inhibitors suggests that platelets contain both PKC dependent and PKC independent activation pathways.[13] Platelets also contain high concentrations of the protein tyrosine kinase Src, the Src-family members Fyn, Lyn, Hck, and Yes, and the protein kinase Syk.[12] Moreover, platelet activation results in the phosphorylation of a number of intracellular proteins on serine, threonine, and tyrosine residues. Which, if any, of these phosphorylated proteins is involved in GPIIb-IIIa activation is not known.

In addition to an extensive network of cytoplasmic actin filaments, platelets contain a submembranous cytoskeletal lattice composed of spectrin, talin, vinculin, and short actin filaments.[14,15] Thirty to 60 percent of the GPIIb-IIIa on the surface of unactivated platelets is associated with this lattice, although the structures mediating this association are not known.[14] It is noteworthy that following platelet stimulation, the actin filaments associated with the submembranous lattice are fragmented and GPIIb-IIIa is redistributed to a detergent insoluble cytoskeletal core.[16] We have observed that exposing unstimulated gel filtered human platelets to cytochalasin D and latrunculin A, drugs that impair actin polymerization, induces approximately 50% as much fibrinogen binding to GPIIb-IIIa as stimulating the platelets with ADP

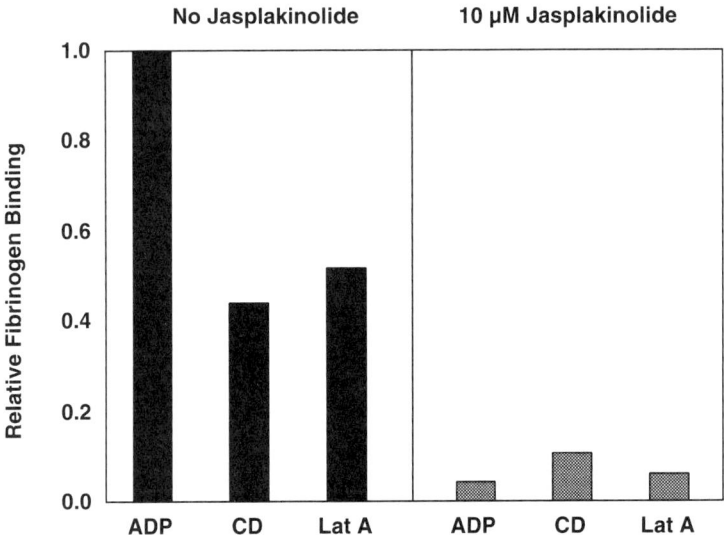

FIGURE 1. Effect of the actin filament stabilizing agent jasplakinolide on ADP-, cytochalasin D-, and latrunculin A-induced ^{125}I-labeled fibrinogen binding to GPIIb-IIIa on gel filtered human platelets. Platelets were preincubated at room temperature in the absence or presence of 10 μM jasplakinolide for 30 min. Fibrinogen binding was induced by 10 μM ADP, 1 μM cytochalasin D, or 1 μM latrunculin A. The data shown were normalized to the amount of fibrinogen binding induced by 10 μM ADP in the absence of jasplakinolide.

(see FIGURE 1). Conversely, incubating platelets with jasplakinolide, an agent that stabilizes actin filaments, prevents the cytochalasin D and latrunculin A effects and inhibits ADP stimulated fibrinogen binding as well. Accordingly, we have postulated that in unstimulated platelets, submembranous actin or actin associated proteins constrain GPIIb-IIIa in a low affinity state and that relief of this constraint by initiating actin filament turnover enables GPIIb-IIIa to bind ligands such as fibrinogen.

There is little actin filament turnover in unstimulated platelets.[18] Because neither cytochalasin D nor latrunculin A, by themselves, stimulate actin filament turnover, we postulated that agonist stimulated actin filament depolymerization was required for the cytochalasin and latrunculin effects. Consistent with this hypothesis, fibrinogen binding induced by these agents is inhibited by impairing platelet ATP metabolism with 2-deoxyglucose and sodium azide, or by exposing platelets to the inhibitory prostaglandin PGE_1. In addition, the ability of cytochalasin D and latrunculin A to induce fibrinogen binding to unstimulated platelets is prevented by either of the ADP scavengers, apyrase and creatine phosphate/creatine phosphokinase, suggesting that subthreshold concentrations of ADP provided the stimulus needed to initiate actin filament turnover.

ADP stimulates platelet aggregation by simultaneously binding to two purinergic receptors in the platelet membrane: P2Y1, a seven transmembrane domain receptor that activates phospholipase Cβ via the G protein Gαq and P2Y12, a receptor that inhibits platelet adenylyl cyclase via the G protein Gαi.[19] Accordingly, ADP stimulated platelet aggregation is inhibited by exposing platelets to either the P2Y1-specific inhibitor adenosine 3'phosphate 5'phosphate (A3P5P) or the P2Y12-specific inhibitor AR-C99096. Similarly, we found that both cytochalasin D- and latrunculin A-induced fibrinogen binding are inhibited by preincubating platelets with either A3P5P (IC_{50} 0.06 μM) or AR-C99096 (IC_{50} 1.2 nM). Thus, these data are consistent with the presence of subthreshold concentrations of ADP in the platelet suspension medium. Moreover, they suggest that the biochemical pathways responsible for ADP-stimulated platelet aggregation are also involved in the fibrinogen binding to GPIIb-IIIa induced by cytochalasin D and latrunculin A.

The biochemical reactions that follow ADP receptor stimulation and result in fibrinogen binding to GPIIb-IIIa are unknown. We have found that both cytochalasin D and latrunculin A-induced fibrinogen binding are inhibited by the phosphotidylinositol 3-kinase (PI 3-K) antagonists wortmannin and LY294002, by the protein kinase C (PKC) inhibitor bisindolylmaleimide I, and by the intracellular calcium chelators BAPTA-AM and EGTA-AM.[17] It is also noteworthy that the serine-threonine phosphatase inhibitor calyculin A is a potent inhibitor of the cytochalasin and latrunculin effects, whereas the protein tyrosine kinase inhibitor genistein is not. Thus, these observations suggest that pathways resulting in phosphorylation of membrane lipids and of serine and threonine residues, as well as increases in platelet cytosolic calcium ultimately release GPIIb-IIIa from cytoskeletal constraints. Platelets contain substantial amounts of the calcium activated actin filament severing protein gelsolin[20] and the low molecular weight actin depolymerizing protein cofilin, the activity of which is regulated by serine phosphorylation.[21] Whether either of these proteins is directly involved in GPIIb-IIIa activation remains to be determined.

IDENTIFICATION OF GPIIb-IIIa BINDING SITES IN FIBRINOGEN

Integrins bind to proteins and peptides containing the tetrapeptide sequence Arg-Gly-Asp-X, where X represents a variety of residues including serine, valine, phenylalanine, or alanine.[22] Thus, each of the proteins known to bind to GPIIb-IIIa, including fibrinogen, contains an Arg-Gly-Asp-X sequence, and peptides containing this sequence inhibit binding of these proteins to activated platelets.[23] Ligand binding to GPIIb-IIIa can also be inhibited by peptides corresponding to the carboxyl-terminal 10–15 amino acids of the fibrinogen γ chain and the γ chain pentadecapeptide crosslinked to albumin supports ADP-stimulated platelet aggregation like intact fibrinogen.[24] Thus, these data suggest the two regions of the fibrinogen α chain that contain an RGD motif, as well as the carboxyl-terminus of the fibrinogen γ chain, represent potential binding sites for GPIIb-IIIa in the fibrinogen molecule.

Electron Microscopy of Glycoprotein IIb-IIIa Bound to Fibrinogen

To address the location of the GPIIb-IIIa binding sites on fibrinogen, we took an ultrastructural approach using an electron microscope.[25] Intact GPIIb-IIIa heterodimers were purified from Triton X-100 extracts of human platelets by affinity chromatography using a monoclonal antibody that specifically recognizes GPIIIa.

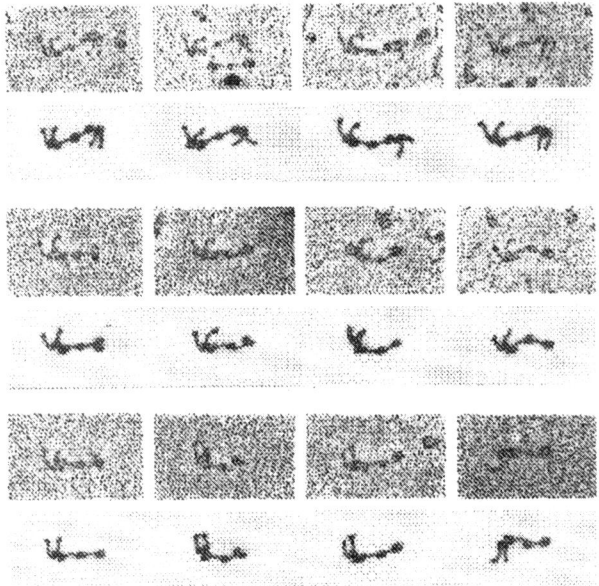

FIGURE 2. A gallery of electron microscope images of rotary shadowed preparations of fibrinogen bound to GPIIb-IIIa. *Tracings* of the shadowed molecules have been added to aid in the interpretation of the electron micrographs. (Adapted from Ref. 25, with permission.)

Following exchange of Triton X-100 with octyl-glucoside, the purified GPIIb-IIIa was incubated with purified human fibrinogen in the presence of 2 mM Ca^{2+}. The incubation mixture was then sprayed onto freshly cleaved mica, rotary shadowed with platinum/tungsten, and examined in an electron microscope at a magnification of 60,000×.

Similar to previous results,[26] we found that in the presence of 2 mM Ca^{2+} and detergent, GPIIb-IIIa consists of a 12 by 8 nm globular head and two 18 by 2 nm flexible tails extending from one side of the head (see FIGURE 2). In two thirds of the images, the tails were either parallel to each other or splayed apart, whereas in the remainder, the ends of the tails appeared to touch at their ends. Approximately 10% of the GPIIb-IIIa in the incubation mixture was bound to fibrinogen, either at one or both ends of the fibrinogen molecule. GPIIb-IIIa interacted with fibrinogen exclusively via its globular head. Moreover, the orientation of GPIIb-IIIa with respect to fibrinogen was specific, with lateral aspect of the head of GPIIb-IIIa bound to an end nodule of fibrinogen and with the tails of the GPIIb/IIIa extended laterally and away from the longitudinal axis of fibrinogen. Measurements based on the orientation of the tails indicated that the angle of GPIIb-IIIa with respect to the long axis of fibrinogen was approximately 98°. In the majority of images, the tails of GPIIb-IIIa bound to fibrinogen were parallel or splayed.

In 85% of the images, GPIIb-IIIa was associated with the distal ends of the fibrinogen molecule, whereas in the remaining images, possible binding sites involved a variety of other positions around the fibrinogen molecule. Because the carboxyl termini of the fibrinogen γ chains are located at the distal ends of fibrinogen,[27] these images indicate that fibrinogen binds to GPIIb-IIIa primarily via the γ chain sequences. Although the RGD motif is present at two places in the fibrinogen α chain, we did not observe statistically significant binding of GPIIb-IIIa at a location consistent with the position of either RGD motif.

The conclusion that a sequence motif located at the carboxyl-terminal end of the γ chain mediates fibrinogen binding to GPIIb-IIIa is supported by a substantial amount of additional data. Farrell et al. prepared fibrinogen mutants in which the RGD motifs in the α chain were converted to inactive RGE motifs and the four carboxyl terminal amino acids of the γ chain were replaced by the 20 amino acids of the carboxyl-terminus of the γ' chain.[28] They found that γ' containing fibrinogen was substantially less effective than γ containing fibrinogen in supporting platelet aggregation, whereas the ability of fibrinogen to support platelet aggregation was unaffected by the replacement of RGD by RGE. They also found that immobilized γ' fibrinogen was unable to support the platelet adhesion, whereas the RGD→RGE mutants were effective adhesive substrates.[29] Similarly, Hettasch et al. observed that

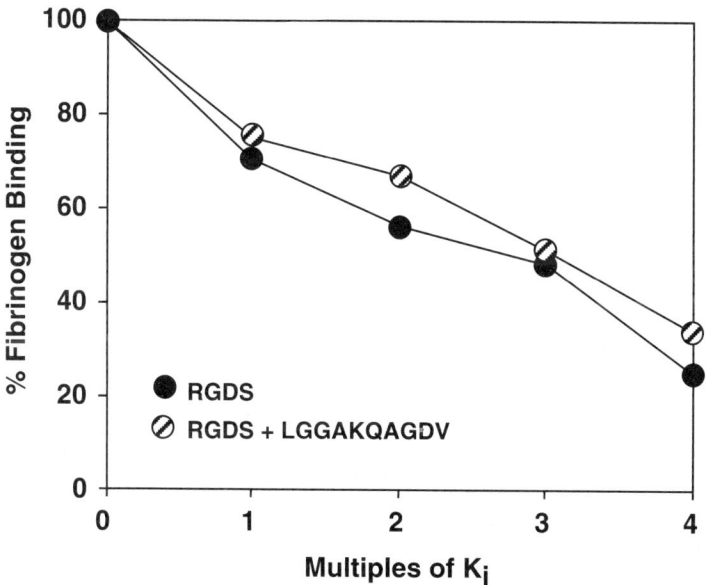

FIGURE 3. RGDS and LGGAKQAGDV binding to GPIIb-IIIa are mutually exclusive. To demonstrate that RGDS and LGGAKQAGDV binding to GPIIb-IIIa are mutually exclusive, the inhibitory effect of these peptides on fibrinogen binding was measured at concentrations that were multiples of their respective K_i values. For example, RGDS alone at a concentration corresponding to its K_i and RGDS and LGGAKAGDV together at concentrations corresponding to one-half of their respective K_i values, inhibited fibrinogen binding by 29% and 25%, respectively. (The data in the figure were adapted from Ref. 32, with permission).

γ chains lacking their usual carboxyl-terminal sequence of Ala-Gly-Asp-Val (AGDV) were unable to support platelet aggregation.[30] Furthermore, Cheresh *et al.* found that platelet adhesion to immobilized intact fibrinogen was unaffected by monoclonal antibodies recognizing its carboxyl- and amino-terminal RGD sequences.[31] By contrast, endothelial cell adhesion to fibrinogen, mediated by the vitronectin receptor αvβ3, was inhibited exclusively by the antibody recognizing the carboxyl terminal.

RGD and γ-Chain Peptide Binding to GPIIb-IIIa Are Mutually Exclusive

Despite the considerable evidence that fibrinogen binds to GPIIb-IIIa exclusively via the carboxyl terminus of the γ chains, clinical studies demonstrate that RGD containing peptides, or peptidomimetics based on the RGD sequence, effectively inhibit GPIIb-IIIa function *in vivo, ex vivo,* and *in vitro,* implying that the RGD containing regions of fibrinogen also interact with GPIIb-IIIa. To address these paradoxical observations, we measured fibrinogen binding to ADP stimulated platelets in the presence of one or both peptides and derived inhibition constants (K_i) from analyses of Dixon plots.[32] RGDS inhibited fibrinogen binding with a K_i of 16 μM, whereas the K_i for LGGAKQAGDV was 46 μM. When fibrinogen binding was measured in the presence of biochemically equivalent concentrations of both peptides, their inhibitory effects on fibrinogen binding were additive (see FIGURE 3). For example, RGDS, at a concentration equivalent to its K_i inhibited fibrinogen binding by 29%. However, when RGDS and the γ chain peptide were each present at a concentration equal to one half their respective K_i values, fibrinogen binding was inhibited by 25%. This behavior indicates that the interaction of RGDS and the γ chain peptide with GPIIb-IIIa are mutually exclusive, suggesting that either the peptides bind to the same site or to overlapping sites on GPIIb-IIIa, or that each peptide binds to GPIIb-IIIa at a distinct site and, as a consequence of binding, induces a change in the GPIIb-IIIa that precludes binding of the other peptide.

Other investigators have reported data supporting either of these mechanisms. Lam *et al.* reported that the γ chain peptide HHLGGAKQAGDV decreased the affinity of thrombin-activated GPIIb-IIIa for a peptide containing an RGDS motif and that GPIIb-IIIa adsorbed to an RGD or γ chain peptide affinity column could be eluted from either column with either GRGDSP or LGGAKQAGDV.[33] These observations suggest that either RGDS and LGGAKQAGDV bind to the same site on GPIIb-IIIa and the binding sites overlap. On the other hand, Santoro and Lawing, using photoaffinity labeling to crosslink RGDS and HHLGGAKQAGDV to GPIIb-IIIa, found that an RGDS derivative was incorporated into both GPIIb and GPIIIa, whereas HHLGGAKQAGDV was predominantly incorporated into GPIIb, suggesting that the binding sites for the peptides are not identical.[34]

LOCALIZATION OF THE FIBRINOGEN BINDING SITE IN GPIIb-IIIa

There is a substantial amount of data indicating the fibrinogen interacts with portions of both GPIIb and GPIIIa.

Identification of Binding Sites in GPIIIa

A region of GPIIIa encoded by exons C and D and encompassing amino acids 95–223, in particular a domain consisting of amino acids 164–202, has been implicated in ligand selection by GPIIb-IIIa.[35] Based on the results of chemical crosslinking experiments,[36] RGD peptides are generally thought to bind predominantly to GPIIIa in the vicinity of amino acids 109–171.[36] This is underscored by the presence of a naturally occurring mutation, Asp→Tyr at amino acid 119 that results in the expression of non-functional GPIIb-IIIa complexes on the platelet surface.[37] It is noteworthy that Asp119 resides at the amino terminal end of an array of oxygenated residues whose three dimensional structure may resemble that of the ligand binding I domains that are present in several integrin α subunits.[38] Conversion of neighboring Ser121 and Ser123 to Ala by *in vitro* mutagenesis has the same functional consequences as the Asp119→Tyr mutation, although substituting Ala for Asp126, Asp127, or Ser130 does not.[39] The ability of overlapping peptides corresponding to GPIIIa amino acids 211–222 to inhibit fibrinogen binding to purified GPIIb-IIIa and to bind to fibrinogen itself also suggest that this stretch of residues represents a portion of the fibrinogen binding site.[40,41] Again, two naturally occurring mutations in the midst of the sequence, Arg214→Trp and Arg214→Gln, result in the expression of non-functional GPIIb-IIIa.[42,43] However, in this case, disulfide bond reduction using dithiothreitol restores fibrinogen dependent platelet aggregation, suggesting that the mutation simply obscures the ligand binding site on GPIIb-IIIa by altering the folded conformation of the molecule.[44] There is also evidence that more distal regions of GPIIIa may be involved in fibrinogen binding. A naturally occurring GPIIIa mutation, Leu262→Pro, prevents GPIIb-IIIa binding to immobilized fibrinogen, although cells expressing the mutation are able to retract fibrin clots.[45] Finally, recombinant GPIIIa fragments encompassing amino acids 274–368 were found to bind to soluble and immobilized fibrinogen, although this binding was agonist-, RGD-, and calcium-independent.[46]

Identification of Binding Sites in GPIIb

The portion of GPIIb that interacts with ligands has been localized to the amino-terminal third of the molecule,[47] although the specific residues that define its ligand binding domain are uncertain. Chemical crosslinking experiments using a modified γ chain peptide indicate that the peptide binds to GPIIb in the vicinity of amino acids 294–314.[48] The physiologic significance of this observation is supported by the ability of a peptide corresponding to GPIIb residues 300–312 to inhibit platelet adhesion to fibrinogen.[49] However, more recent studies of both naturally occurring mutations and laboratory induced mutations involving amino acids 183, 184, 189, 190, 191, 193, and 224 suggest that these amino acids also interact with GPIIb-IIIa ligands.[50–52] In addition, a more proximal naturally occurring mutation Leu145→Pro mutation results in the expression GPIIb-IIIa that is unable to bind soluble and immobilized fibrinogen.[53] When viewed in the context of a β-propeller model of the GPIIb amino-terminus, these mutations would be juxtaposed on the upper surface of one quadrant of the propeller where they could constitute a portion of a ligand binding pocket.[53–55]

Fibrinogen binding to GPIIb-IIIa requires the presence of millimolar concentrations of divalent cations. The structural basis for this requirement is unknown.

Terbium luminescence spectroscopy suggests that ligand binding to GPIIb-IIIa is associated with the displacement of two of five tightly bound cations from the heterodimer.[65] One of the displaced cations may be derived from GPIIIa amino acids 118–131. Peptides containing this sequence bind terbium,[57] suggesting that the sequence could be a divalent cation binding site in the intact GPIIIa molecule. Peptides containing amino acids 118–131 also form stoichiometric complexes with RGD containing peptides with the concomitant displacement of a bound terbium,[56] suggesting the possibility that ligand binding to GPIIb-IIIa involves the formation of an initial ternary complex between GPIIIa, divalent cation, and ligand, with the subsequent displacement of the cation.[56] Although it is possible that the other displaced cation originated in GPIIb, there is no evidence that this is the case.

IDENTIFICATION OF ALLOSTERIC CHANGES IN GPIIb-IIIa THAT ACCOMPANY RGD BINDING

To resolve the apparent paradox that even though a sequence in the fibrinogen γ chain mediates fibrinogen binding to GPIIb-IIIa, molecules based on RGD motifs present in the α chain are potent GPIIb-IIIa antagonists, we studied the observation

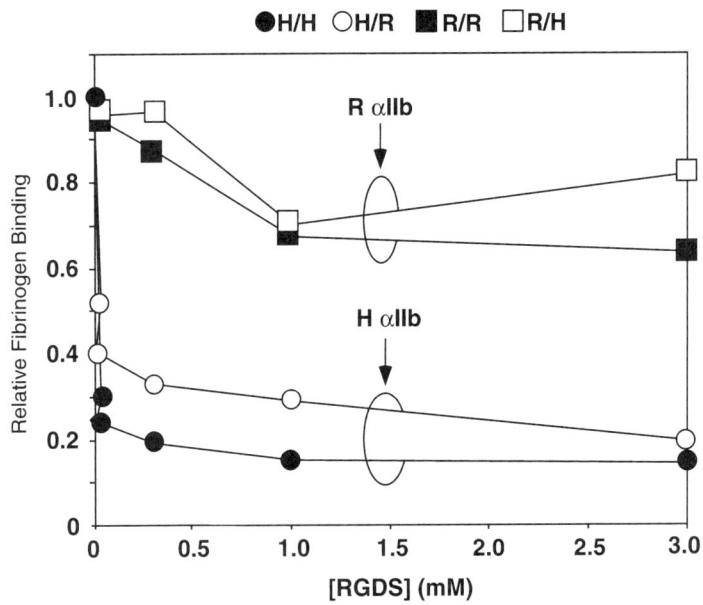

FIGURE 4. Inhibitory effect of RGDS on fibrinogen binding to human–rat GPIIb-IIIa hybrids. Hybrid GPIIb-IIIa heterodimers composed of human–human (*H/H*), human–rat (*H/R*), rat–human (*R/H*), or rat–rat (*R/R*) subunits were expressed in Chinese hamster ovary (CHO) cells. FITC labeled fibrinogen binding to the transfected cells was induced by 5 mM dithiothreitol and was measured using flow cytometry in the presence of increasing concentration of RGDS.

that rat platelets are relatively resistant to inhibition by RGD containing peptides.[58] A trivial explanation for this observation is that their resistance is simply due to a greater affinity for fibrinogen, but this was not the case because the K_d values of rat and human GPIIb-IIIa for fibrinogen were comparable. cDNAs for human and rat GPIIb-IIIa have been cloned.[59–62] Therefore, to identify sequences in the rat protein responsible for its resistance to RGD, we expressed GPIIb-IIIa complexes composed of human–human (H/H), human–rat (H/R), rat–human (R/H), or rat–rat (R/R) subunits in Chinese hamster ovary (CHO) cells and measured the ability of the tetrapeptide RGDS to inhibit FITC-labeled fibrinogen binding to GPIIb-IIIa using flow cytometry (see FIGURE 4). Recombinant GPIIb-IIIa expressed in CHO cells is constitutively inactive and cannot be activated by cellular agonists. To induce fibrinogen binding, we activated GPIIb-IIIa by exposing the CHO cells to 5 mM dithiothreitol,

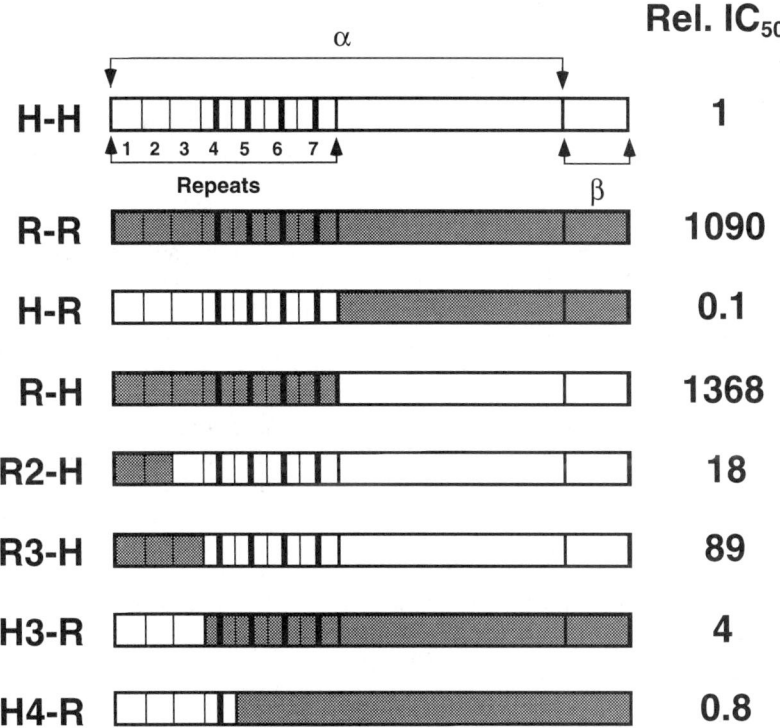

FIGURE 5. Effect of RGDS on fibrinogen binding to GPIIb-IIIa heterodimers composed of human GPIIIa and the indicated human–rat GPIIb chimeras. Data are expressed as the IC_{50} of RGDS inhibition relative to the IC_{50} for RGDS inhibition of fibrinogen binding to wildtype human GPIIb-IIIa. *H-H*, human GPIIb; *R-R*, rat GPIIb; *H-R*, human amino terminus–rat carboxyl terminus; *R-H*, rat amino terminus–human carboxyl terminus; *R2-H*, human GPIIb containing rat repeats 1 and 2; *R3-H*, human GPIIb containing rat repeats 1–3; *H3-R*, rat GPIIb containing human repeats 1–3; *H4-R*, rat GPIIb containing human repeats 1–4.

as was reported by Zucker and Masiello.[63] Surprisingly, despite the previous reports that RGD peptides bind to GPIIIa, we found H/H and H/R equally sensitive, and R/R and R/H equally resistant, to RGDS inhibition, indicating that the sequences determining RGD resistance are located in GPIIb. To further localize the relevant sequences, we coexpressed chimeric GPIIb subunits containing the amino-terminal half of GPIIb from one species and the carboxyl-terminal half of GPIIb from the other with human GPIIIa in CHO cells and measured the ability of RGDS to inhibit fibrinogen binding to the chimeric integrins (see FIGURE 5). We found that a chimera containing the amino-terminus of rat GPIIb and the carboxyl-terminus of human GPIIb was resistant to RGDS, whereas the opposite chimera was sensitive. Thus, these experiments indicate that the sequences determining the relative resistance of rat GPIIb-IIIa to RGD are located in the amino-terminal half of GPIIb.

The amino terminal half of αIIb is composed of seven tandem repeats of about 60 amino acids each and has been predicted to fold into a β-propeller configuration.[54] Accordingly, we measured RGDS inhibition of fibrinogen binding to GPIIb-IIIa in which various rat and human GPIIb had been exchanged (FIG. 5). We found that rat GPIIb-IIIa containing human repeats 3 and 4 was relatively sensitive to RGDS, whereas human GPIIb-IIIa containing the corresponding rat repeats was resistant. Our experiments suggest that either RGDS binds at or near GPIIb repeats 3 and 4, or that RGDS binds elsewhere in GPIIb-IIIa but affects the conformation of this portion of the GPIIb molecule.[64]

Hu *et al.* used plasmon resonance spectroscopy to suggest that GPIIb-IIIa contains two distinct ligand binding pockets, one for fibrinogen and a separate one for RGD-containing ligands.[65] Our experiments suggest that either RGDS binds at or near GPIIb repeats 3 and 4 or consistent with the report of Hu *et al.*,[65] RGDS binding elsewhere in GPIIb-IIIa, but affects the conformation of this portion of the GPIIb molecule. To differentiate between these possibilities, we measured RGDS stimulated binding of the GPIIIa-specific monoclonal antibody 10-783 to CHO cells expressing the H/H and R/H GPIIb-IIIa hybrids. Monoclonal antibody 10-783 recognizes ligand–induced binding sites in human GPIIIa and, therefore, has been termed a LIBS antibody. We found that RGDS concentrations that inhibited fibrinogen binding to human, but not rat, platelets also stimulated 10-783 binding to both H/H and R/H. Thus, these results indicate that RGDS indirectly impairs fibrinogen binding to GPIIb-IIIa by inducing an allosteric change in the third and fourth GPIIb repeats, thereby accounting for its ability to inhibit platelet aggregation. Moreover, they suggest that agonist stimulated conformational changes in these repeats may be critically involved in fibrinogen binding to platelets.

GPIIB-IIIa BINDING TO FIBRIN

Platelets are responsible for the retraction of fibrin clots.[66,67] The ability of GPIIb-IIIa antagonists to impair clot retraction indicates that GPIIb-IIIa interacts with fibrin as well as fibrinogen.[68] However, the GPIIb-IIIa binding sites in fibrin and fibrinogen appear to be substantially different. Thus, whereas fibrinogen lacking the four carboxyl-terminal γ chain residues is unable to support platelet aggregation, the ability to support clot retraction is not affected.[69] Similarly, substitution of each

RGD motif in the fibrinogen α chain with RGE was without effect.[70] Moreover, although clot retraction mediated by a triple fibrinogen mutant was somewhat delayed, it was eventually indistinguishable from that mediated by intact fibrinogen. Accordingly, the GPIIb-IIIa binding sites that mediate platelet adhesion to fibrin remain undefined. Nonetheless, platelet spreading on surfaces coated with fibrin lacking β chain residues 15–42 is substantially impaired,[71] suggesting that perhaps sequences located in the fibrinogen β chain are involved in GPIIb-IIIa binding to fibrin.

REFERENCES

1. McLean, J.R., R.E. Maxwell & D. Hertler. 1964. Fibrinogen and adenosine diphosphate-induced aggregation of platelets. Nature **202:** 605–606.
2. Cross, M.J. 1964. Effect of fibrinogen on the aggregation of platelets by adenosine diphosphate. Thromb. Diath. Hemorrh. **12:** 521–527.
3. Weiss, H.J. & J. Rogers. 1971. Fibrinogen and platelets in the primary arrest of bleeding. N. Engl. J. Med. **285:** 369–374.
4. Bennett, J.S. & G. Vilaire. 1979. Exposure of platelet fibrinogen receptors by ADP and epinephrine. J. Clin. Invest. **64:** 1393–1401.
5. Marguerie, G.A., E.F. Plow & T.S. Edgington. 1979. Human platelets possess an inducible and saturable receptor specific for fibrinogen. J. Biol. Chem. **254:** 5357–5363.
6. Ruggeri, Z.M., R. Bader & L. DeMarco. 1982. Glanzmann thrombasthenia; deficient binding of von Willebrand factor to thrombin-stimulated platelets. Proc. Natl. Acad. Sci. U.S.A. **79:** 6038–6041.
7. Sosnoski, D.M., B.S. Emanuel, A.L. Hawkins, *et al.* 1988. Chromosomal localization of the genes for the vitronectin and fibronectin receptors α subunits and for platelet glycoproteins IIb and IIIa. J. Clin. Invest. **81:** 1993–1998.
8. Hynes, R.O. 1987. Integrins: A family of cell surface receptors. Cell **48:** 549–554.
9. Phillips, D.R., I.F. Charo & R.M. Scarborough. 1991. GPIIb-IIIa: the responsive integrin. Cell **65:** 359–362.
10. Bennett, J.S. 1996. Structural biology of glycoprotein IIb-IIIa. Trends Cardiovasc.Med. **16:** 31–36.
11. Brass, L.F. 1995. Molecular basis for platelet activation. *In* Hematology: Basic Principles and Practice. R. Hoffman, E.J. Benz, Jr., S.J. Shattil, *et al.*, Eds.: 1536-1551. Churchill Livingstone, New York.
12. Shattil, S.J., H. Kashiwagi & N. Pampori. 1998. Integrin signaling: the platelet paradigm. Blood **91:** 2645–2657.
13. Paul, B.Z., J. Jin & S.P. Kunapuli. 1999. Molecular mechanism of thromboxane A(2)-induced platelet aggregation. Essential role for p2t(ac) and alpha(2a) receptors. J. Biol. Chem. **274:** 29108–29114.
14. Fox, J.E.B., L. Lipfert, E.A. Clark, *et al.* 1993. On the role of the platelet membrane skeleton in mediating signal transduction. J. Biol. Chem. **268:** 25973–25984.
15. Hartwig, J.H. & M. DeSisto. 1991. The cytoskeleton of the resting human blood platelet: structure of the membrane skeleton and its attachment to actin filaments. J. Cell Biol. **112:** 407–425.
16. Hartwig, J.H. 1992. Mechanisms of actin rearrangements mediating platelet activation. J. Cell Biol. **118:** 1421–1442.
17. Bennett, J.S., S. Zigmond, G. Vilaire, *et al.* 1999. The platelet cytoskeleton regulates the affinity of the integrin $\alpha_{IIb}\beta_3$ for fibrinogen. J. Biol. Chem. **274:** 25301–25307.
18. Fox, J.E.B. & D.R. Phillips. 1981. Inhibition of actin polymerization in blood platelets by cytochalasin. Nature **292:** 650–652.
19. Hollopeter, G., H.M. Jantzen, D. Vincent, *et al.* 2001. Identification of the ADP receptor targeted by antithrombotic drugs. Nature **409:** 202–207.

20. WITKE, W., A.H. SHARPE, J.H. HARTWIG, et al. 1995. Hemostatic, inflammatory, and fibroblast responses are blunted in mice lacking gelsolin. Cell **81:** 41–51.
21. DAVIDSON, M.M. & R.J. HASLAM. 1994. Dephosphorylation of cofilin in stimulated platelets: roles for a GTP-binding protein and Ca^{2+}. Biochem. J. **301:** 41–47.
22. PYTELA, R., M.D. PIERSCHBACHER, M.H. GINSBERG, et al. 1986. Platelet membrane glycoprotein IIb/IIIa: member of a family of arg-gly-asp-specific adhesion receptors. Science **231:** 1559–1562.
23. GARTNER, T.K. & J.S. BENNETT. 1985. The tetrapeptide analogue of the cell attachment site of fibronectin inhibits platelet aggregation and fibrinogen binding to activated platelets. J. Biol. Chem. **260:** 11891–11894.
24. KLOCZEWIAK, M., S. TIMMONS, T.J. LUKAS, et al. 1984. Platelet receptor recognition site on human fibrinogen. Synthesis and structure-function relationships of peptides corresponding to the carboxy-terminal segment of the g chain. Biochemistry **23:** 1767–1774.
25. WEISEL, J.W., C. NAGASWAMI, G. VILAIRE, et al. 1992. Examination of the platelet membrane glycoprotein IIb/IIIa complex and its interaction with fibrinogen and other ligands by electron microscopy. J. Biol. Chem. **267:** 16637–16643.
26. CARRELL, N.A., L.A. FITZGERALD, B. STEINER, et al. 1985. Structure of human platelet membrane glycoproteins IIb and IIIa as determined by electron microscopy. J. Biol. Chem. **260:** 1743–1749.
27. WEISEL, J.W., C.V. STAUFFACHER, E. BULLITT, et al. 1985. A model for fibrinogen: domains and sequence. Science **230:** 1388–1391.
28. FARRELL, D.H., P. THIAGARAJAN, D.W. CHUNG, et al. 1992. Role of fibrinogen α and γ chain sites in platelet aggregation. Proc. Natl. Acad. Sci. U.S.A. **89:** 10729–10732.
29. FARRELL, D.H. & P. THIAGARAJAN. 1994. Binding of recombinant fibrinogen mutants to platelets. J. Biol. Chem. **269:** 226–231.
30. HETTASCH, J.M., M.G. BOLYARD & S.T. LORD. 1992. The residues AGDV of recombinant γ chains of human fibrinogen must be carboxyl-terminal to support human platelet aggregation. Thromb. Hæmost. **68:** 701–706.
31. CHERESH, D.A., S.A. BERLINER, V. VINCENTE, et al. 1989. Recognition of distinct adhesive sites on fibrinogen by related integrins on platelets and endothelial cells. Cell **58:** 945–953.
32. BENNETT, J.S., S.J. SHATTIL, J.W. POWER, et al. 1988. Interaction of fibrinogen with its platelet receptor. Differential effects of α and γ chain fibrinogen peptides on the glycoprotein IIb-IIIa complex. J. Biol. Chem. **263:** 12948–12953.
33. LAM, S.C., E.F. PLOW, M.A. SMITH, et al. 1987. Evidence that arginyl-glycyl-aspartate peptides and fibrinogen gamma chain peptides share a common binding site on platelets. J. Biol. Chem. **262:** 947–950.
34. SANTORO, S.A. & W.J. LAWING. 1987. Competition for related but nonidentical binding sites on the glycoprotein IIb-IIIa complex by peptides derived from platelet adhesive proteins. Cell **48:** 867–873.
35. LIN, E.C.K., B.I. RATNIKOV, P.M. TSAI, et al. 1997. Identification of a region in the integrin $\beta 3$ subunit that confers ligand binding specificity. J. Biol. Chem. **272:** 23912–23920.
36. D'SOUZA, S.E., M.H. GINSBERG, T.A. BURKE, et al. 1988. Localization of an Arg-Gly-Asp recognition site within an integrin adhesion receptor. Science **242:** 91–93.
37. LOFTUS, J.C., T.E. O'TOOLE, E.F. PLOW, et al. 1990. A $\beta 3$ integrin mutation abolishes ligand binding and alters divalent cation-dependent conformation. Science **249:** 915–918.
38. TOZER, E.C., R.C. LIDDINGTON, M.J. SUTCLIFFE, et al. 1996. Ligand binding to integrin $\alpha IIb\beta 3$ is dependent on a MIDAS-like domain in the $\beta 3$ subunit. J. Biol. Chem. **271:** 21978–21684.
39. BAJT, M. & J. LOFTUS. 1994. Mutation of a ligand binding domain of beta 3 integrin. Integral role of oxygenated residues in alpha IIb beta 3 (GPIIb-IIIa) receptor function. J. Biol. Chem. **269:** 20913–20919.
40. CHARO, I.F., L. NANNIZZI, D.R. PHILLIPS, et al. 1991. Inhibition of fibrinogen binding to GPIIb-IIIa by a GP IIIa peptide. J. Biol. Chem. **266:** 1415–1421.

41. STEINER, B., A. TRZECIAK, G. PFENNINGER, et al. 1993. Peptides derived from a sequence within beta 3 integrin bind to platelet alpha IIb beta 3 (GPIIb-IIIa) and inhibit ligand binding. J. Biol. Chem. **268:** 6870–6873.
42. LANZA, F., A. STIERLE, D. FOURNIER, et al. 1992. A new variant of Glanzmann's thrombasthenia (Strasbourg I). Platelets with functionally defective glycoprotein IIb-IIIa complexes and a glycoprotein IIIa ^{214}Arg→^{214}Trp mutation. J. Clin. Invest. **89:** 1995–2004.
43. BAJT, M.L., M.H. GINSBERG, A.L. FRELINGER III, et al. 1992. A spontaneous mutation of integrin αIIb β3 (platelet glycoprotein IIb-IIIa) helps define a ligand binding site. J. Biol. Chem. **267:** 3789–3794.
44. KOUNS, W., B. STEINER, T. KUNICKI, et al. 1994. Activation of the fibrinogen binding site on platelets isolated from a patient with the Strasbourg I variant of Glanzmann's thrombasthenia. Blood **84:** 1108–1015.
45. WARD, C.M., A.S. KESTIN & P.J. NEWMAN. 2000. A Leu262Pro mutation in the integrin β3 subunit results in an $α_{IIb}β_3$ complex that binds fibrin but not fibrinogen. Blood **96:** 161–169.
46. ALEMANY, M., E. CONCORD, J. GARIN, et al. 1996. Sequence 274-368 in the β3 subunit of the integrin αIIbβ3 provides a ligand recognition and binding domain of the γ-chain of fibrinogen that is independent of platelet activation. Blood **87:** 562–601.
47. LOFTUS, J.C., C.E. HALLORAN, M.H. GINSBERG, et al. 1996. The amino-terminal one-third of $α_{IIb}$ defines the ligand recognition specificity of integrin $α_{IIb}β_3$. J. Biol. Chem. **271:** 2033–2039.
48. D'SOUZA, S.E., M.H. GINSBERG, T.A. BURKE, et al. 1990. The ligand binding site of the platelet integrin receptor GPIIb-IIIa is proximal to the second calcium binding domain of its α subunit. J. Biol. Chem. **265:** 3440–3446.
49. TAYLOR, D.B. &T.K. GARTNER. 1992. A peptide corresponding to GPIIb alpha 300–312, a presumptive fibrinogen gamma-chain binding site on the platelet integrin GPIIb/IIIa, inhibits the adhesion of platelets to at least four adhesive ligands. J. Biol. Chem. **267:** 11729–11733.
50. GRIMALDI, C.M., F. CHEN, C. WU, et al. 1998. Glycoprotein IIb Leu214Pro mutation produces Glanzman thrombasthenia with both quantitative and qualitative abnormalities in GPIIb/IIIa. Blood **91:** 1562–1571.
51. KAMATA, T., A. IRIE, M. TOKUHIRA, et al. 1996. Critical residues of integrin αIIb subunit for binding of αIIbβ3 (glycoprotein IIb-IIIa) to fibrinogen and ligand-mimetic antibodies (PAC-1, OP-G2, and LJ-CP3). J. Biol. Chem. **271:** 18610–18615.
52. TOZER, E.C., E.K. BAKER, M.H. GINSBERG, et al. 1999. A mutation in the α subunit of the platelet integrin αIIbβ3 identifies a novel region important for ligand binding. Blood **93:** 918–924.
53. BASANI, R.B., D.L. FRENCH, G. VILAIRE, et al. 2000. A naturally occurring mutation near the amino terminus of αIIb defines a new region involved in ligand binding to αIIbβ3. Blood **95:** 180–188.
54. SPRINGER, T.A. 1997. Folding of the N-terminal, ligand-binding region of integrin alpha-subunits into a beta-propeller domain. Proc. Natl. Acad. Sci. U.S.A. **94:** 65–72.
55. PUZON-MCLAUGHLIN, W., T. KAMATA & Y. TAKADA. 2000. Multiple discontinuous ligand-mimetic antibody binding sites define a ligand binding pocket in integrin αIIbβ3. J. Biol. Chem. **275:** 7795–7802.
56. D'SOUZA, S., T. HAAS, R. PIOTROWICZ, et al. 1994. Ligand and cation binding are dual functions of a discrete segment of the integrin β3 subunit: cation displacement is involved in ligand binding. Cell **79:** 659–667.
57. CIERNIEWSKI, C., T. HAAS, J. SMITH, et al. 1994. Characterization of cation-binding sequences in the platelet integrin GPIIb-IIIa (αIIbβ3) by terbium luminescence. Biochemistry **33:** 12238–12246.
58. HARFENIST, E.J., M.A. PACKHAM & J.F. MUSTARD. 1988. Effects of cell adhesion peptide, Arg-Gly-Asp-Ser, on responses of washed platelets from humans, rabbits, and rats. Blood **71:** 132–136.
59. PONCZ, M., R. EISMAN, R. HEIDENREICH, et al. 1987. Structure of the platelet membrane glycoprotein IIb. J. Biol. Chem. **262:** 8476–8482.

60. ZIMRIN, A.B., R. EISMAN, G. VILAIRE, *et al.* 1988. Structure of platelet glycoprotein IIIa. J. Clin. Invest. **81:** 1470–1475.
61. CIEUTAT, A.M., J.-P. ROSA, F. LETOURNEUR, *et al.* 1993. A comparative analysis of cDNA-derived sequences for rat and mouse beta 3 integrins (GPIIIA) with their human counterparts. Biochem. Biophys. Res. Commun. **193:** 771–778.
62. THORNTON, M.A. & M. PONCZ. 1999. Characterization of the murine platelet αIIb gene and encoded cDNA. Blood **94:** 3947–3950.
63. ZUCKER, M.B. & N.C. MASIELLO. 1984. Platelet aggregation caused by dithiothreitol. Thromb. Hæmostas. **51:** 119–124.
64. BASANI, R.B., G. D'ANDREA, N. MITRA, *et al.* 2001. RGD-containing peptides inhibit fibrinogen binding to platelet $\alpha_{IIb}\beta_3$ by inducing an allosteric change in the amino-terminal portion of α_{IIb}. J. Biol. Chem. In press.
65. HU, D.D., C.A. WHITE, S. PANZER-KNODLE, *et al.* 1999. A new model of dual interacting ligand binding sites on integrin $\alpha_{IIb}\beta_3$. J. Biol. Chem. **274:** 4633–4639.
66. COHEN, I., J.M. GERRARD & J.G. WHITE. 1982. Ultrastructure of clots during isometric contraction. J. Cell Biol. **93:** 775–787.
67. HANTGAN, R.R., R.G. TAYLOR & J.C. LEWIS. 1985. Platelets interact with fibrin only after activation. Blood **65:** 1299–1311.
68. GARTNER, K.T. & M.L. OGILVIE. 1988. Peptides and monoclonal antibodies which bind to platelet glycoproteins IIb and/or IIIa inhibit clot retraction. Thromb. Res. **49:** 43–53.
69. ROONEY, M.M., L.V. PARISE & S.T. LORD. 1996. Dissecting clot retraction and platelet aggregation. J. Biol. Chem. **271:** 8553–8555.
70. ROONEY, M.M., D.H. FARRELL, B.M. VAN HEMEL, *et al.* 1998. The contribution of the three hypothesized integrin-binding sites in fibrinogen to platelet-mediated clot retraction. Blood **92:** 2374–2381.
71. HAMAGUCHI, M., L.A. BUNCE, L.A. SPORN, *et al.* 1993. Spreading of platelets on fibrin is mediated by the amino terminus of the β chain including peptide β15–42. Blood **81:** 2348–2356.

Fibrin and Wound Healing

RICHARD A.F. CLARK

Department of Dermatology, SUNY at Stony Brook, Stony Brook, New York, USA

KEYWORDS: **Wounds; Fibrinogen; Fibronectin; Epidermal injury; Angiogenesis; Fibroplasia; Epidermal cells; Fibroblasts; Endothelial cells.**

INTRODUCTION

During the past two decades advances in the molecular and cellular biology of fibrin(ogen) have greatly expanded our comprehension of the role of fibrin in wound healing. Clearly, recent scientific breakthroughs in the understanding of these basic processes will lead to future therapeutic successes of fibrinogen preparations in wound care and tissue engineering. In this brief treatise the molecular and cellular biology of fibrin(ogen) are reviewed in the context of cutaneous wound repair. The reader is referred to *The Molecular and Cellular Biology of Wound Repair*[1] for a more detailed discussion of the many processes involved in wound healing.

INFLAMMATION

Severe tissue injury causes blood vessel disruption with concomitant extravasation of blood constituents. Blood coagulation and platelet aggregation generate a fibrin rich clot that plugs severed vessels and fills any discontinuity in the wounded tissue. Although the blood clot within vessel lumen reestablishes hemostasis, the clot within wound space provides a provisional matrix for cell migration.

Clearly, hemostasis is a major function of coagulation, however, blood clotting is also a part of the inflammatory response. For example, Hageman factor activation leads to generation of its fragments and bradykinin, potent vasoactive agents,[2] and to the initiation of classical and alternative complement cascades[3] with the resultant generation of the anaphylatoxins C3a and C5a. The anaphylatoxins directly increase blood vessel permeability and attract neutrophils and monocytes to sites of tissue injury.[4] In addition, these substances stimulate the release of other vasoactive mediators, such as histamine and leukotriene C4 and D4 from mast cells,[5] and the release of granule constituents and biologically active oxygen products from neutrophils and macrophages.[6]

The clot also provides a matrix scaffold for the recruitment of tissue cells to an injured site. Specifically, fibrin in conjunction with fibronectin act as a provisional matrix[7] for the influx of monocytes,[8,9] fibroblasts,[10–12] and endothelial cells.[13] These migrating cells use integrin receptors (see TABLE I),[14,15] that recognize fibrin,

---

Address for correspondence: Richard A.F. Clark, M.D., Department of Dermatology, SUNY at Stony Brook, Stony Brook, NY 11794-8165, USA. Voice: 613-444-3843; fax: 613-444-3844.

fibronectin, and vitronectin to interact with the clot matrix.[16–19] Since extracellular matrix molecules can provide signals for gene expression through integrin receptors,[20] the interaction of these tissue cells with the provisional matrix might be expected to alter cell phenotype and function. In fact, Damskey and Werb[21] have shown that a fibronectin-rich extracellular matrix can control the expression of collagenase (matrix metalloproteinase I, MMP-1) and our laboratory has recently found that fibronectin or fibrin matrix can modulate fibroblast response to cytokines[22] and endothelial cell expression of integrins.[23]

Proper clearance of the clot provisional matrix appears just as important as its deposition. The major proteolytic enzymes, plasminogen activators and plasmin, escape inactivation by fluid phase protease inhibitors, like plasminogen activator inhibitor and α_2-antiplasmin, through binding to the fibrin clot[24] and cell surfaces.[25] Although plasminogen activator and plasmin have the ability to degrade a wide variety of extracellular matrix proteins, a specific inhibitor of plasminogen activator (PAI-1) binds to the extracellular matrix[26] and limits matrix degradation to the microenvironment around cell surfaces. However, inadequate removal of the provisional matrix may lead to fibrosis.[27] For example, fibrin deposits and suppressed fibrinolysis are found in pulmonary fibrosis.[28,29] Furthermore, transgenic mice that

TABLE 1. Integrin superfamily

β1 Integrins	Ligands	αv Integrins	Ligands
$\alpha 1\beta 1$	fibrillar collagen, laminin-1	$\alpha v\beta 1$	fibronectin (RGD), vitronectin
$\alpha 2\beta 1$	fibrillar collagen, laminin-1	$\alpha v\beta 3$	vitronectin (RGD), fibronectin, fibrinogen, von Willebrand factor, thrombospondin, denatured collagen
$\alpha 3\beta 1$	fibronectin (RGD), laminin-5, entactin, denatured collagens		
$\alpha 4\beta 1$	fibronectin (LEDV), VCAM-1	$\alpha v\beta 5$	fibronectin (RGD), vitronectin
$\alpha 5\beta 1$	fibronectin (RGD)	$\alpha v\beta 6$	fibronectin, tenascin
$\alpha 6\beta 1$	laminin		
$\alpha 7\beta 1$	laminin	β2 Integrins	Ligands
$\alpha 8\beta 1$	fibronectin, vitronectin	$\alpha_M\beta 2$	ICAM-1, iC3b, fibrinogen, factor X
$\alpha 9\beta 1$	tenascin	$\alpha_L\beta 2$	ICAM-1, 2, and 3
Other ECM Integrins	Ligands	$\alpha_X\beta 2$	iC3b, fibrinogen
$\alpha IIb\beta 3$	Same as $\alpha v\beta 3$		
$\alpha 6\beta 4$	laminin		

overexpress plasminogen activator inhibitor accumulate significantly more collagen in their lungs after bleomycin injury than their normal littermates.[30] In addition, inadequate removal of fibrin may impede the normal wound healing processes. For example, transgenic mice that have the no plasminogen demonstrate a marked delay in cutaneous wound repair.[31] Interestingly, crossbreeding these mice with transgenic afibrinogenemic mice abrogates the delay in wound healing. Interestingly, afibrinogenemic mice do not have impaired wound healing as long as adequate hemostasis is maintained. This study strongly suggests that the major role of fibrin in wounds is hemostasis. However, once present in the wound, complex interactions must occur between fibrin and migrating leukocytes and tissue cells.

NEUTROPHILS

Neutrophils and monocytes begin to emigrate into injured tissue concurrently, but neutrophils arrive first in great numbers partly due to their abundance in the circulation. A variety of chemotactic factors attract both cell types to the site of injury.[32,33] General leukocyte chemoattractants include fibrinopeptides cleaved from fibrinogen by thrombin[34] and fibrin degradation products produced by plasmin degradation of fibrin.[35] As well as providing the stimulus for directed migration, chemotactic factors also increase CD11/CD18 expression on the neutrophil surface.[36] These heterodimeric complexes, in conjunction with L-selectin and sialyl Lewis factor χ, mediate adherence of neutrophils to blood vessel endothelium and thereby facilitate transmigration of leukocytes through the endothelium.[37] In addition, CD11c/CD18 mediates adhesion of neutrophils to fibrinogen through a domain at the N terminus of the A alpha chain.[38] Neutrophil activation by chemoattractants, including fibrinogen degradation products,[39] also stimulates release of neutrophil proteases. These enzymes facilitate cell penetration through blood vessel basement membranes as well as degradation of the fibrin clot.[40–42] Neutrophils at the wound site destroy contaminating bacteria via phagocytosis coupled with toxic oxygen radical generation and enzyme digestion.[43–45] One report[46] has demonstrated that these phagocytic and digestive processes can be modulated by the fibrinogen degradation products D and E.

MONOCYTES

Whether neutrophil infiltrates resolve or persist, monocyte accumulation continues, stimulated by selective monocyte chemoattractants, such as fragments of collagen,[47] elastin,[48] fibronectin,[49] enzymatically active thrombin,[50] TGF-β,[51] and fibrin(ogen) degradation peptides.[46] Similar to neutrophil recruitment, chemoattractants stimulate circulating monocytes to attach to the endothelium of blood vessels at the site of injury and to migrate through the blood vessel wall into the extracellular matrix.[52] Physicochemical properties of the three-dimensional fibrin matrix, as well as the presence of chemoattractants and fibronectin, control macrophage invasion of the wound clot.[8,9]

Binding of monocytes or macrophages to specific extracellular matrix proteins through integrin receptors stimulates extracellular matrix phagocytosis, and Fc- and C3b-mediated phagocytosis.[17] Fibrin interaction with the monocyte/macrophages can modulate these processes through the integrin receptor Mac-1 (CD11b/CD18; CR3).[53,54] Cultured macrophages, and presumably wound macrophages, release enzymes, such as collagenase, elastase, and plasminogen activator.[55] Fibrin fragment D-dimer has been reported to induce the secretion of interleukin-1, urokinase-type plasminogen activator, and plasminogen activator inhibitor-2 in a human promonocytic leukemia cell line, and may do so in human monocytes.[56] These proteases facilitate tissue debridement including the removal of fibrin itself. In fact, Mac-1 (CD11b/CD18) and the urokinase receptor (CD87) may form a functional unit on monocytic cells to digest fibrin(ogen).[57]

As well as promoting phagocytosis and debridement, adherence to extracellular matrix molecules also stimulates monocytes to undergo metamorphosis into inflammatory or reparative macrophages. Adherence induces selective mRNA expression of colony stimulating factor-1, a cytokine necessary for monocyte/macrophage survival; tumor necrosis factor-α (TNF-α), a potent inflammatory cytokine; and PDGF, a potent chemoattractant and mitogen for fibroblasts; as well as c-fos and c-jun, transactivating factors necessary for many activation signals.[58,59] mRNAs for other important macrophage cytokines are adherence-independent; for example, TGF-β is constitutively expressed; interleukin-1 (IL-1) mRNA is stimulated by bacterial endotoxin; and HLA-DR is stimulated by gamma-interferon (γ-IFN).[58] Thus, macrophages appear to play a pivotal role in the transition between wound inflammation and repair[60] and fibrin(ogen) can modulate macrophage activity and, thereby, the rate of this transition.

EPITHELIALIZATION

Reepithelialization of a wound begins within hours after injury. It is clear that rapid reestablishment of any epithelial barrier decreases victim morbidity and mortality. Epithelial cells from residual epithelial structures move quickly to dissect clot and damaged stroma from the wound space and repave the surface of viable tissue. The epithelial cells at the wound edge loose their apical-basal polarity and extend pseudopodia from their free baso-lateral sides into the wound.

If the epidermal basement membrane is damaged, provisional matrix, composed of fibronectin,[61] tenasin,[62] vitronectin,[63] and fibrinogen,[64] accumulates among stromal type I collagen bundles beneath the epidermis at the wound margin.[65] In contrast to normal epidermal cells, wound keratinocytes express integrin receptors for fibronectin, tenasin, and vitronectin,[66–69] which allows them to interact with these substrates. In addition, $\alpha 2\beta 1$ type 1 collagen receptors, that are normally disposed along the lateral sides of basal keratinocytes, redistribute to the basal membrane of wound keratinocytes that have come in contact with dermal type 1 collagen fibers. *In vitro* experiments suggest that these receptors are required for keratinocyte migration on type 1 collagen.[70]

The migrating wound epidermis does not simply transit over a wound surface coated with provisional matrix but rather dissects through the wound, separating desiccated eschar from viable tissue.[61] The path of dissection appears to be determined by the array of integrins that the migrating epidermal cells express on their cell membranes, as already described. Importantly, $\alpha v\beta 3$, the receptor for fibrinogen/fibrin[16] and denatured collagen,[71] is not expressed on keratinocytes *in vitro* or *in vivo*.[73] Hence, keratinocytes lack the capacity to interact with these matrix proteins.[74] Furthermore, epidermal cells fail to bind fibronectin in the presence of fibrin, but can bind in the presence of fibrinogen, whereas neither fibrin nor fibrinogen inhibit cell interaction with type 1 collagen.[74] Since the fibrinogen beneath the epidermis at the wound margin is not clotted,[64] the migrating wound epidermis avoids the fibrin/fibronectin-rich clot, migrating over a matrix composed of fibrinogen, fibronectin, and type 1 collagen.

Extracellular matrix degradation is clearly required for the dissection of migrating wound epidermis between the collagenous dermis the fibrin eschar[31] and probably depends on epidermal cell production of both collagenase[70,75] and plasminogen activator.[76] Plasminogen activator activates collagenase as well as plasminogen[77] and, therefore, facilitates the degradation of interstitial collagen as well as the fibrin eschar. These proteases would enzymatically digest the extracellular matrix in the plane of epidermal migration. Epidermal migration and dissection between viable and nonviable tissue ultimately results in sloughing the eschar.

As reepithelialization ensues, basement membrane proteins reappear in a very ordered sequence from the margin of the wound inward in a zipper-like fashion.[61] Epidermal cells revert to their normal phenotype, once again firmly attaching to reestablished basement membrane.

GRANULATION TISSUE

New stroma, often called granulation tissue, begins to invade the wound space approximately four days after injury. The name granulation tissue derives from the granular appearance of newly forming tissue when it is incised and visually examined. Numerous new capillaries endow the neostroma with its granular appearance. Macrophages, fibroblasts and blood vessels move into the wound space as a unit[78] that correlates well with the proposed biologic interdependence of these cells during tissue repair. The macrophages provide a continuing source of growth factors necessary to stimulate fibroplasia and angiogenesis, fibroblasts construct new extracellular matrix necessary to support cell ingrowth, and blood vessels carry oxygen and nutrients necessary to sustain cell metabolism.

Fibroplasia

Fibroblasts and the extracellular matrix that they synthesize are collectively known as fibroplasia. Growth factors, particularly PDGF[79] and TGF-β,[80] in concert with the clot matrix proteins fibrin and fibronectin,[81–83] presumably stimulate fibroblasts of the periwound tissue to proliferate, express appropriate integrin receptors and migrate into the wound space.

The early wound extracellular matrix was coined provisional matrix.[61] It is initially composed of plasma derived fibrin, fibronectin, and vitronectin and later composed of *in situ* produced hyaluronan and fibronectin. These provisional matrix constituents contribute to tissue formation by providing a scaffold or conduit for cell migration (fibronectin),[12] low impedance for cell invasion (hyaluronic acid),[84] a reservoir for cytokines (fibrinogen),[85] and direct signals to the cells through integrin receptors.[20]

Fibroblasts presumably require fibronectin for movement through the wound clot as they do for migration through fibrin matrices *in vitro*.[12] Nevertheless, fibroblasts can interact directly with fibrin and vitronectin at Arg-Gly-Asp-Ser (RGDS) sites via the integrin receptors $\alpha 3\beta 1$, $\alpha 5\beta 1$, $\alpha v\beta 1$, $\alpha v\beta 3$, and $\alpha v\beta 5$. The RGD-dependent, fibronectin receptors $\alpha 3\beta 1$ and $\alpha 5\beta 1$ are upregulated on periwound fibroblasts the day prior to moving into the wound clot and on early granulation tissue fibroblasts.[83] In contrast, $\alpha 1\beta 1$ and $\alpha 2\beta 1$ collagen receptors on these fibroblasts were either suppressed or did not appear to change appreciably.[83,86] Thus, periwound fibroblasts specifically upregulate integrins that can interaction with the provisional matrix just prior to their migration into the wound space.

Interestingly, fibrin and fibronectin begin to appear in the *periwound* environment two days *after* wounding (Greiling and Clark, unpublished observations). In the presence of PDGF these provisional matrix proteins are known to support an increase expression of their integrin receptors 24 hours later.[83] Thus, the appearance of provisional matrix proteins in the periwound stroma at day two may be causally related to the appearance of provisional matrix integrins on the fibroblasts at day three. Together, we believe that these phenomena are rate limiting for the commencement of granulation tissue formation.[12,13,83,86,87]

Fibroblast movement into a crosslinked fibrin blood clot or any tightly woven extracellular matrix may necessitate an active proteolytic system that can cleave a path for migration. A variety of fibroblast derived enzymes in conjunction with serum derived plasmin are potential candidates for this task, including plasminogen activator, interstitial collagenase-1 and -3 (MMP-1 and MMP-13, respectively), the 72-kDa gelatinase A (MMP-2), and stromelysin (MMP-3).[77,88] *In vitro* both plasminogen activator[12,89] and collagenase activity (Greiling and Clark, unpublished observations) are required for fibroblast movement from a collagen matrix into fibrin gel.

Once the fibroblasts have migrated into the wound they gradually switch their major function to collagen production.[90] After an abundant collagen matrix is deposited in the wound, fibroblasts cease collagen production. The process of fibroplasia creation from a fibrin clot context has been modeled *in vitro*.[91]

NEOVASCULARIZATION

Fibroplasia would halt if neovascularization failed to accompany the newly forming complex of fibroblasts and extracellular matrix. The process of new blood vessel formation is called angiogenesis.[92] Angiogenesis is a complex process that relies on an appropriate extracellular matrix in the wound bed as well as endothelial cell phenotype alteration, stimulated migration and mitogenic stimulation of endothelial cells.[92]

Soluble factors that may be responsible for wound angiogenesis include basic fibroblast growth factor (bFGF),[93] vascular endothelial growth factor (VEGF),[94] angiopoietin,[95] PDGF,[96] and many other.[97] Several isoforms of VEGF[98] and angiopoietins[99] have been identified that effect endothelial cell growth and angiogenesis differentially.

Besides growth factors and chemotactic factors, an appropriate extracellular matrix is also necessary for angiogenesis. Fibrin has been reported to induce angiogenesis directly.[100–102] If fibrin itself is sufficient to induce angiogenesis, it must perform at least two functions: (1) provide a three-dimensional matrix that supports cell migration, and (2) express selective chemotactic and/or chemokinetic activity such that endothelial cells migrate into fibrin clot. Given that bFGF and PDGF can bind to fibrin with a fairly high affinity[85] (Galanakis and Clark, unpublished observations), it is quite possible that one or more angiogenic (chemotactic) factors were present in the fibrin preparations previously found to induce angiogenesis directly. Recently we have developed an *in vitro* angiogenesis assay using human dermal microvascular endothelial cells that are cultured on microcarrier beads and suspended in an extracellular matrix. Our assay was derived from an assay previously described by Nehls and Drenckhahn.[103] When purified human fibrin is used as the extracellular matrix no angiogenesis occurs. However, either VEGF, VEGF-C, or bFGF added to this assay stimulant angiogenesis in a dose dependent manner (Feng, Clark, Galanakis, and Tonnesen, unpublished observations). Using a similar microcarrier based angiogenesis assay, Nehls and Herrmann demonstrated that fibrin structure plays an important role in bovine pulmonary artery endothelial cell migration and capillary morphogenesis. They showed that the degree of rigidity of fibrin gel strongly influences tube formation by bovine endothelial cells in response to bFGF or VEGF.[104]

In support of the finding that fibrin can support growth factor stimulated angiogenesis, angiogenesis in the chick chorioallantoic membrane is dependent on the expression of the $\alpha v \beta 3$ integrin that recognizes fibrin as well as fibronectin and vitronectin.[105] Furthermore, in porcine cutaneous wounds $\alpha v \beta 3$ is only expressed on capillary sprouts as they invade the fibrin clot.[13] *In vitro* studies in fact demonstrate that $\alpha v \beta 3$ can promote endothelial cell migration on provisional matrix proteins.[106] Tight regulation of $\alpha v \beta 3$ by fibrin matrix is supported by the finding that fibrin, but not collagen, gels induce $\alpha v \beta 3$ on cultured human microvascular endothelial cells.[23] Thus, endothelial cell expression of provisional matrix integrins is regulated by the provisional matrix proteins in a positive feedback fashion in a manner similar to the positive feedback between integrins and their ligands observed in fibroblasts.[22]

Within a day or two after removal of angiogenic stimuli, capillaries undergo regression as characterized by mitochondria swelling in endothelial cells, platelet adherence to the blood vessel wall, vascular stasis, and endothelial cell apoptosis and ingestion of by macrophages. Although $\alpha v \beta 3$ has been shown to regulate apoptosis of endothelial cells in culture and in tumors,[107] $\alpha v \beta 3$ is not present on wound endothelial cells as they undergo programmed cell death, indicating another pathway of apoptosis in healing wound blood vessels (Feng, Clark, and Tonnesen, unpublished observations). Thrombospondin appears to be a good candidate for this phenomenon.[108]

SUMMARY

Although hemostasis is the major role of fibrin in wound repair,[31] once the clot is present the wound cells must deal with it. The invasion and clearing of fibrin by these cells involves multiple complex processes that may go array XXX and delay wound repair. A good example, of the latter is leg ulcers. These chronic wounds contain a plethora of proteases that digest fibronectin and growth factors in the fibrin clot[109,110] resulting in a corrupt provisional matrix that no longer supports reepithelialization or granulation tissue formation. Every good wound care provider knows that these wounds will not heal unless the corrupt matrix is removed by vigorous debridement that stimulates the accumulation of a competent provisional matrix.

ACKNOWLEDGMENTS

The original data reported in this paper was supported by NIH grants AG 101143-14 and AR 42987-05 from the National Institute of Aging and the National Institute of Arthritis, Musculoskeletal and Skin Disease, respectively.

REFERENCES

1. CLARK, R.A.F. 1996. The Molecular and Cellular Biology of Wound Repair. Plenum Press, New York.
2. MULLER-ESTERL, N. 1989. Kininogens, kinins and kinships. Thromb. Hæm. **62:** 2–6.
3. GHEBREHIWET, B., M. SILVERBERG & A.P. KAPLAN. 1981. Activation of classic pathway of complement by Hageman factor fragment. J. Exp. Med. **153:** 665–676.
4. FERNANDEZ, H.N., P.M. HENSON, A. OTANI & T.E. HUGLI. 1978. Chemotactic response to human C3a and C5a anaphylatoxins. I. Evaluation of C3a and C5a leukotaxis *in vitro* and under simulated *in vivo* conditions. J. Immunol. **120:** 109–115.
5. STIMLER, N.P., BACH, M K, C.M. BLOOR & T.E. HUGLI. 1982. Release of leukotrienes from guinea pig lung stimulated by C5a des arg anaphylatoxin. J. Immunol. **128:** 2247–2257.
6. MCCARTHY, K. & P.M. HENSON. 1979. Induction of lysosomal enzyme secretion by macrophages in response to the purified complement fragments C5a and C5a des Arg. J. Immunol. **123:** 2511–2517.
7. YAMADA, K.M. & R.A.F. CLARK. 1996. Provisional matrix. *In* Molecular and Cellular Biology of Wound Repair. R.A.F. Clark, Ed.: 51–93. Plenum, New York.
8. CIANO, P.S., R.B. COLVIN, A.M. DVORAK, *et al.* 1986. Macrophage migration in fibrin gel matrices. Lab. Invest. **54:** 62–70.
9. LANIR, N., P.S. CIANO, L. VAN DE WATER, *et al.* 1988. Macrophage migration in fibrin gel matrices II. Effects of clotting factor XIII, fibronectin, and glycosaminoglycan content on cell migration. J. Immunol. **140:** 2340–2349.
10. KNOX, P., S. CROOKS & C.S. RIMMER. 1986. Role of fibronectin in the migration of fibroblasts into plasma clots. J. Cell Biol. **102:** 2318–2323.
11. BROWN, L.F., N. LANIR, J. MCDONAGH, *et al.* 1993. Fibroblast migration in fibrin gel matrices. Am. J. Pathol. **142:** 273–283.
12. GREILING, D. & R.A.F. CLARK. 1997. Fibronectin provides a conduit for fibroblast transmigration from a collagen gel into a fibrin gel. J. Cell Sci. **110:** 861–870.
13. CLARK, R.A.F., M.G. TONNESEN, J. GAILIT & D.A. CHERESH. 1996. Transient functional expression of αvβ3 on vascular cells during wound repair. Am. J. Path. **148:** 1407–1421.
14. HYNES, R.O. 1992. Integrins: versatility, modulation, and signaling in cell adhesion. Cell **69:** 11–25.

15. YAMADA, K.M., J. GAILIT & R.A.F. CLARK. 1996. Integrins in wound repair. In The Molecular and Cellular Biology of Wound Repair. R.A.F. Clark, Ed.: 311–338. Plenum, New York.
16. CHERESH, D.A., S.A. BERLINER, V. VICENTE & Z.M. RUGGERI. 1989. Recognition of distinct adhesive sites on fibrinogen by related integrins on platelets and endothelial cells. Cell **58:** 945–953.
17. BLYSTONE, S.D., I.L. GRAHAM, F.P. LINDBERG & E.J. BROWN. 1994. Integrin αvβ3 differentially regulates adhesive and phagocytic functions of the fibronectin receptor α5β1. J. Cell Biol. **127:** 1129–1137.
18. NEWMAN, D., R.A.F. CLARK, L. HONAKEN, et al. 1996. Human dermal microvascular endothelial cells use α5β1, as well as avb3, to interact with the major fibrinogen breakdown product E1. J. Invest. Dermtol. **106:** 823a.
19. GAILIT, J., C. CLARKE, D. NEWMAN, et al. 1997. Human fibroblasts bind directly to fibrinogen at RGD sites through integrin αvβ3. Exp. Cell Res. **232:** 118–126.
20. SCHWARTZ, M.A., M.D. SCHALLER & M.G. GINSBERG. 1995. Integrins: emerging paradigms of signal transduction. Annu. Rev. Cell Dev. Biol. **11:** 549–599.
21. HUHTALA, P., M.J. HUMPHRIES, J.B. MCCARTHY, et al. 1995. Cooperative signaling by α5β1 and α4β1 integrins regulates metalloproteinase gene expression in fibroblasts adhering to fibronectin. J. Cell Biol. **129:** 867–879.
22. XU, J., M.M. ZUTTER, S.A. SANTORO & R.A.F. CLARK. 1996. PDGF induction of α2 integrin gene expression is mediated by protein kinase C-ζ. J. Cell Biol. **134:** 1301–1311.
23. FENG, X., R.A.F. CLARK, D. GALANAKIS & M.G. TONNESEN. 1999. Fibrin and collagen differentially regulate human dermal microvascular endothelial cell integrins: stabilization of αvβ3 mRNA by fibrin. J. Invest. Dermatol. **113:** 913–919.
24. CASTELLINO, F.J., D.K. STRICKLAND, J.P. MORRIS, et al. 1983. Enhancement of the streptokinase-induced activation of human plasminogen by human fibrinogen and human fibrinogen fragment D1. Ann. N.Y. Acad. Sci. **408:** 595–601.
25. HAJJAR, K., A. JACOVINA & J. CHACKO. 1994. An endothelial cell receptor for plasminogen and tissue plasminogen activator: Identity with annexin II. J. Biol. Chem. **269:** 21191–21197.
26. SALONEN, E.-M., A. VAHERI, J. POLLANEN, et al. 1989. Interaction of plasminogen activator inhibitor (PAI-1) with vitronectin. J. Biol. Chem. **264:** 6339–6343.
27. OLMAN, M.A., N. MACKMAN, C.L. GLADSON, et al. 1995. Changes in procoagulant and fibrinolytic gene expression during bleomycin-induced lung injury in the mouse. J. Clin. Invest. **96:** 1621–1630.
28. CHAPMAN, H.A., C.L. ALLEN & O.L. STONE. 1986. Abnormalities in pathways of alveolar fibrin turnover among patients with interstitial lung disease. Am. Rev. Repir. Dis. **133:** 437–443.
29. KUHN, C.I., J. BOLDT, T.E.J. KING, et al. 1989. An immunohistochemical study of architectural remodeling and connective tissue synthesis in pulmonary fibrosis. Am. Rev. Respir. Dis. **140:** 1693–1703.
30. ELTZMAN, D.T., R.D. MCCOY, X. ZHENG, et al. 1996. Bleomycin-induced fibrosis in transgenic mice that either lack or overexpress the murine plasminogen activator inhibitor-1 gene. J. Clin. Invest. **97:** 232–237.
31. BUGGE, T.H., K.W. KOMBRINCK, M.J. FLICK, et al. 1996. Loss of fibrinogen rescues mice from the pleiotropic effects of plasminogen deficiency. Cell **87:** 709–719.
32. WILLIAMS, T.J. 1988. Factors that affect vessel reactivity and leukocyte emigration. In Molecular and Cellular Biology of Wound Repair. R.A.F. Clark & P.M. Henson, Ed.: 115–183. Plenum Press, New York.
33. BAGGIOLINI, M., B. DEWALD & B. MOSER. 1994. Interleukin-8 and related chemotactic cytokines-CXC and CC chemokines. Adv. Immunol. **55:** 97–179.
34. RICHARDSON, D.L., D.S. PEPPER & A.B. KAY. 1976. Chemotaxis for human monocytes by fibrinogen-derived peptides. Br. J. Hæmatol. **32:** 507–513.
35. GROSS, T.J., K.J. LEAVELL & M.W. PETERSON. 1997. CD11b/CD18 mediates the neutrophil chemotactic activity of fibrin degradation product D domain. Thromb. Hæmost. **77:** 894–900.

36. TONNESEN, M.G., D.C. ANDERSON, T.A. SPRINGER, et al. 1989. Adherence of neutrophils to cultured human microvascular endothelial cells. Stimulation by chemotactic peptides and lipid mediators and dependence upon the Mac-1, LFA-1, p150,95 glycoprotein family. J. Clin. Invest. **83:** 637–646.
37. FUHLBRIGGE, R.C., R. ALON, K.D. PURI, et al. 1996. Sialylated, fucosylated ligands for L-selectin expressed on leukocytes mediate tethering and rolling adhesions in physiologic flow conditions. J. Cell Biol. **135:** 837–848.
38. LOIKE, J.D., B. SODEIK, L. CAO, et al. 1991. CD11c/CD18 on neutrophils recognizes a domain at the N terminus of the A alpha chain of fibrinogen. Proc. Natl. Acad. Sci. U.S.A. **88:** 1044–1048.
39. WOJTECKA-LUKASIK, E. & S. MASLINSKI. 1992. Fibronectin and fibrinogen degradation products stimulate PMN-leukocyte and mast cell degranulation. J. Physiol. Pharmacol. **43:** 173–181.
40. PLOW, E.F. 1986. The contribution of leukocyte proteases to fibrinolysis. Blut. **53:** 1–9.
41. MACHOVICH, R., A. HIMER & W.G. OWEN. 1990. Neutrophil proteases in plasminogen activation. Blood Coagul. Fibrinolysis. **1:** 273–277.
42. ADAMS, S.A., S.L. KELLY, R.E. KIRSCH, et al. 1995. Role of neutrophil membrane proteases in fibrin degradation. Blood Coagul. Fibrinol. **6:** 693–702.
43. TONNESEN, M.G., G.S. WORTHEN & R.B.J. JOHNSTON. 1988. Neutrophil emigration, activation, and tissue damage. *In* Molecular and Cellular Biology of Wound Repair. R.A.F. Clark & P.M. Henson, Ed.: 149–183. Plenum Press, New York.
44. ELSBACH, P. & J. WEISS. 1992. Oxygen-independent antimicrobial systems of phagocytosis. *In* Inflammation: Basic Principles and Clinical Correlates. J.I. Gallin, I.M. Goldstein & R. Snyderman, Ed.: 603–636. Raven Press, Ltd, New York.
45. KLEBANOFF, S.J. 1992. Oxygen metabolites from phagocytes. In Inflammation: Basic Principles and Clinical Correlates. J.I. Gallin, I.M. Goldstein & R. Snyderman, Ed.: 541–601. Raven Press, Ltd, New York.
46. KAZURA, J.W., J.D. WENGER, R.A. SALATA, et al. 1989. Modulation of polymorphonuclear leukocyte microbicidal activity and oxidative metabolism by fibrinogen degradation products D and E. J. Clin. Invest. **83:** 1916–1924.
47. POSTLETHWAITE, A.E. & A.H. KANG. 1976. Collagen and collagen peptide-induced chemotaxis of human blood monocytes. J. Exp. Med. **143:** 1299–1307.
48. SENIOR, R.M., G.L. GRIFFIN & R.P. MECHAM. 1980. Chemotactic activity of elastin-derived peptides. J. Clin. Invest. **66:** 859–862.
49. CLARK, R.A.F., N.E. WIKNER, D.E. DOHERTY & D.A. NORRIS. 1988. Cryptic chemotactic activity of fibronectin for human monocytes resides in the 120 kDa fibroblastic cell-binding fragment. J. Biol. Chem. **263:** 12115–12123.
50. BAR-SHAVIT, R., M. BENEZRA, A. ELDOR, et al. 1990. Thrombin immobilized to extracellular matrix is a potent mitogen for vascular smooth muscle cells: nonenzymatic mode of action. Cell Regul. **1:** 453–463.
51. WAHL, S.M., D.A. HUNT, L.M. WAKEFIELD, et al. 1987. Transforming growth factor type β induces monocyte chemotaxis and growth factor production. Proc. Natl. Acad. Sci. U.S.A. **84:** 5788–5792.
52. DOHERTY, D.E., C. HASLET, M.G. TONNESEN & P.M. HENSON. 1987. Human monocyte adherence: a primary effect of chemotactic factors on the monocyte to stimulate adherence to human endothelium. J. Immunol. **138:** 1762–1771.
53. ALTIERI, D.C., F.R. AGBANYO, J. PLESCIA, et al. 1990. A unique recognition site mediates the interaction of fibrinogen with the leukocyte integrin Mac-1 (CD11b/CD18). J. Biol. Chem. **265:** 12119–12122.
54. TREZZINI, C., B. SCHUEPP, F.E. MALY & T.W. JUNGI. 1991. Evidence that exposure to fibrinogen or to antibodies directed against Mac-1 (CD11b/CD18; CR3) modulates human monocyte effector functions. Br. J. Hæmatol. **77:** 16–24.
55. ADAMS, D.O. & T.A. HAMILTON. 1992. Macrophages as destructive cells in host defense. *In* Inflammation: Basic Principles and Clinical Correlates. J.I. Gallin, I.M. Goldstein & R. Snyderman, Ed.: 637–662. Raven Press, New York.

56. HAMAGUCHI, M., Y. MORISHITA, I. TAKAHASHI, *et al.* 1991. FDP D-dimer induces the secretion of interleukin-1, urokinase-type plasminogen activator, and plasminogen activator inhibitor-2 in a human promonocytic leukemia cell line. Blood **77:** 94–100.
57. SIMON, D.I., N.K. RAO, H. XU, *et al.* 1996. Mac-1 (CD11b/CD18) and the urokinase receptor (CD87) form a functional unit on monocytic cells. Blood **88:** 3185–3194.
58. SHAW, R.J., D.E. DOHERTY, A.G. RITTER, *et al.* 1990. Adherence-dependent increase in human monocyte PDGF(B) mRNA is associated with increases in c-fos, c-jun, and EGF2 mRNA. J. Cell Biol. **111:** 2139–2148.
59. JULIANO, R.L. & S. HASKILL. 1992. Signal transduction from the extracellular matrix. J. Cell Biol. **120:** 577–585.
60. RICHES, D.W.H. 1996. Macrophage involvement in wound repair, remodeling and fibrosis. *In* The Molecular and Cellular Biology of Wound Repair. R.A.F. Clark, Ed.: 95–142. Plenum Press.
61. CLARK, R.A.F., J.M. LANIGAN, P. DELLAPELLE, *et al.* 1982. Fibronectin and fibrin(ogen) provide a provisional matrix for epidermal cell migration during wound reepithelialization. J. Invest. Dermatol. **79:** 264–269.
62. MACKIE, E.J., W. HALFTER & D. LIVERANI. 1988. Induction of tenascin in healing wounds. J. Cell Biol. **107:** 2757–2767.
63. CAVANI, A., G. ZAMBRUNO, A. MARCONI, *et al.* 1993. Distinctive integrin expression in the newly forming epidermis during wound healing in humans. J. Invest. Dermatol. **101:** 600–604.
64. CLARK, R.A.F., M.J. SPENCER, S.A. HERRICK, *et al.* 2001. Fibrinogen clears from epidermal and blood vessel matrix prior to healing of chronic wounds. J. Invest. Dermatol. Submitted.
65. ODLAND, G. & R. ROSS. 1968. Human wound repair. I. Epidermal regeneration. J. Cell Biol. **39:** 135–151.
66. CLARK, R.A.F. 1990. Fibronectin matrix deposition and fibronectin receptor expression in healing and normal skin. J. Invest. Dermatol. **94**(Suppl):128S–134S.
67. LARJAVA, H., T. SALO, K. HAAPASALMI, *et al.* 1993. Expression of integrins and basement membrane components by wound keratinocytes. J. Clin. Invest. **92:** 1425–1435.
68. JUHASZ, I., G.F. MURPHY, H.-C. YAN, *et al.* 1993. Regulation of extracellular matrix proteins and integrin cell substratum adhesion receptors on epithelium during cutaneous human wound healing *in vivo*. Am. J. Path. **143:** 1458–1469.
69. CLARK, R.A.F., G.S. ASHCROFT, M.-J. SPENCER, *et al.* 1996. Reepithelialization of normal human excisional wounds is associated with a switch from $\alpha v \beta 5$ to $\alpha v \beta 6$ integrins. Br. J. Dermatol. **135:** 46–51.
70. PILCHER, B.K., J.A. DUMIN, B.D. SUDBECK, *et al.* 1997. The activity of collagenase-1 is required for keratinocyte migration on a type I collagen matrix. J. Cell Biol. **137:** 1445–1457.
71. DAVIS, E.D. 1992. Affinity of integrins for damaged extracellular matrix: $\alpha v \beta 3$ binds to denatured collagen type I through RGD sites. Biochem. Biophys. Res. Comm. **182:** 1025–1031.
72. ADAMS, J.C. & F.M. WATT. 1991. Expression of $\beta 1$, $\beta 3$, $\beta 4$, and $\beta 5$ integrins by human epidermal keratinocytes and non-differentiating keratinocytes. J. Cell Biol. **115:** 829–841.
73. GAILIT, J., M.P. WELCH & R.A.F. CLARK. 1994. TGF-$\beta 1$ stimulates expression of keratinocyte integrins during re-epithelialization of cutaneous wounds. J. Invest. Derm. **103:** 221–227.
74. KUBO, M., L. VAN DE WATER, L.C. PLANTEFABER, *et al.* 2000. Fibrin(ogen) is an anti-adhesive for human keratinocytes: A possible mechanism for fibrin eschar slough during wound repair. J. Invest. Dermatol. In press.
75. WOODLEY, D.T., T. KALEBEC, A.J. BANES, *et al.* 1986. Adult human keratinocytes migrating over nonviable dermal collagen produce collagenolytic enzymes that degrade type I and type IV collagen. J. Invest. Dermatol. **86:** 418–423.
76. GRONDAHL-HANSEN, J., L.R. LUND, E. RALFKIAER, *et al.* 1988. Urokinase- and tissue-type plasminogen activators in keratinocytes during wound reepithelilaization *in vivo*. J. Invest. Dermatol. **90:** 790–795.

77. MIGNATTI, P., D.B. RIFKIN, H.G. WELGUS & W.C. PARKS. 1996. Proteinases and tissue remodeling. In The Molecular and Cellular Biology of Wound Repair. R.A.F. Clark, Ed.: 427–474. Plenum Press, New York.
78. HUNT, T.K. 1980. Wound Healing and Wound Infection: Theory and Surgical Practice. Appleton-Century-Crofts, New York.
79. HELDIN, C.-H. & B. WESTERMARK. 1996. Role of Platelet-derived growth factor *in vivo*. In The Molecular and Cellular Biology of Wound Repair. R.A.F. Clark, Ed.: 249–274. Plenum Press, New York.
80. ROBERTS, A.B. & M.B. SPORN. 1996. Transforming growth factor-β (TGF-β). In The Molecular and Cellular Biology of Wound Repair. R.A.F. Clark, Ed.: 275–310. Plenum Press, New York.
81. POSTLETHWAITE, A.E., J. KESKI-OJA, G. BALIAN & A. KANG. 1981. Induction of fibroblast chemotaxis by fibronectin. Location of the chemotactic region to a 140,000 molecular weight nongelatin binding fragment. J. Exp. Med. **153:** 494–499.
82. GRAY, A.J., J.E. BISHOP, J.T. REEVES & G.J. LAURENT. 1993. Aα and Bβ chains of fibrinogen stimulate proliferation of human fibroblasts. J. Cell Sci. **104:** 409–413.
83. XU, J. & R.A.F. CLARK. 1996. Extracellular matrix alters PDGF regulation of fibroblast integrins. J. Cell Biol. **132:** 239–249.
84. TOOLE, B.P. 1991. Proteoglycans and hyaluronan in morphogenesis and differentiation. In Cell Biology of the Extracellular Matrix. E.D. Hay, Ed.: 305–341. Plenum Press, New York.
85. SAHNI, A., T. ODRLJIN & C.W. FRANCIS. 1998. Binding of basic fibroblast growth factor to fibrinogen and fibrin. J. Biol Chem. **273:** 7554–7559.
86. GAILIT, J., J. XU, H. BUELLER & R.A.F. CLARK. 1996. Platelet-derived growth factor and inflammatory cytokines have differential effects on the expression of integrins $\alpha 1\beta 1$ and $\alpha 5\beta 1$ by human dermal fibroblasts *in vitro*. J. Cell Physiol. **169:** 281–289.
87. MCCLAIN, S.A., M. SIMON, E. JONES, *et al.* 1996. Mesenchymal cell activation is the rate limiting step of granulation tissue induction. Am. J. Path. **149:** 1257–1270.
88. VAALAMO, M., L. MATTILA, N. JOHANSSON, *et al.* 1997. Distinct populations of stromal cells express collagenase-3 (MMP-13) and collagenase-1 (MMP-1) in chronic ulcers but not in normally healing wounds. J. Invest. Dermatol. **109:** 96–101.
89. KNOX, P., S. CROOKS, M.C. SCAIFE & S. PATEL. 1987. Role of plasminogen, plasmin, and plasminogen activators in the migration of fibroblasts into plasma clots. J. Cell Physiol. **132:** 501–508.
90. WELCH, M.P., G.F. ODLAND & R.A.F. CLARK. 1990. Temporal relationships of F-actin bundle formation, collagen and fibronectin matrix assembly, and fibronectin receptor expression to wound contraction. J. Cell Biol. **110:** 133–145.
91. TUAN, T.L., A. SONG, S. CHANG, *et al.* 1996. In vitro fibroplasia: matrix contraction, cell growth, and collagen production of fibroblasts cultured in fibrin gels. Exp. Cell Res. **223:** 127–134.
92. MADRI, J.A., S. SANKAR & A.M. ROMANIC. 1996. Angiogenesis. In The Molecular and Cellular Biology of Wound Repair. R.A.F. Clark, Ed.: 355–372. Plenum Press, New York.
93. FOLKMAN, J. & M. KLAGSBRUN. 1987. Angiogenic factors. Science **235:** 442–448.
94. KECK, P.J., S.D. HAUSER, G. KRIVI *et al.* 1989. Vascular permeability factor, an endothelial cell mitogen related to PDGF. Science **246:** 1309–1313.
95. SURI, C., P.F. JONES, S. PATAN, *et al.* 1996. Requisite role of angiopoietin-1, a ligand for the TIE2 receptor, during embryonic angiogenesis [see comments]. Cell **87:** 1171–1180.
96. BATTEGAY, E.F., J. RUPP, L. IRUELA-ARISPE, *et al.* 1994. PDGF-BB modulates endothelial proliferation and angiogenesis *in vitro* via PDGF β-receptors. J. Cell Biol. **125:** 917–928.
97. CLARK, R.A.F. 2000. Wound repair: Lessons for tissue engineering. In Principles of Tissue Engineering. R.P. Lanza, R. Langer & W.J. Chick, Ed. Academic Press.
98. VEIKKOLA, T. & K. ALITALO. 2001. VEGFs, receptors and angiogenesis [In Process Citation]. Semin. Cancer Biol. **9:** 211–220.
99. DAVIS, S. & G.D. YANCOPOULOS. 1999. The angiopoietins: Yin and Yang in angiogenesis. Curr. Top Microbiol. Immunol. **237:** 173–185.

100. KNIGHTON, D.R., T.K. HUNT, K.K. THAKRAL & W.H. GOODSON. 1982. Role of platelets and fibrin in the healing sequence: an *in vivo* study of angiogenesis and collagen synthesis. Ann. Surg. **196:** 379–388.
101. JAKOB, W., J. ZIPPER & E.D. JENTZSCH. 1982. Is the formation of fibrin a necessary event for the initiation of angiogenic response in the chick chorioallantoic membrane? Exper. Pathol. **21:** 251–262.
102. DVORAK, H.F., V.S. HARVEY, P. ESTRELLA, *et al.* 1987. Fibrin containing gels induce angiogenesis. Implications for tumor stroma generation and wound healing. Lab. Invest. **57:** 673–686.
103. NEHLS, V. & D. DRENCKHAHN. 1995. A novel, microcarrier-based *in vitro* assay for rapid and reliable quantification of three-dimensional cell migration and angiogenesis. Microvasc. Res. **50:** 311–322.
104. NEHLS, V. & R. HERRMANN. 1996. The configuration of fibrin clots determines capillary morphogenesis and endothelial cell migration. Microvasc. Res. **51:** 347–364.
105. BROOKS, P.C., R.A.F. CLARK & D.A. CHERESH. 1994. Requirement of vascular integrin $\alpha v\beta 3$ for angiogenesis. Science **264:** 569–571.
106. LEAVESLEY, D.I., M.A. SCHWARTZ, M. ROSENFELD & D.A. CHERESH. 1993. Integrin $\beta 1$- and $\beta 3$-mediated endothelial cell migration is triggered through distinct signaling mechanisms. J. Cell Biol. **121:** 163–170.
107. BROOKS, P.C., A.M.P. MONTGOMERY, M. ROSENFELD, *et al.* 1994. Integrin $\alpha v\beta 3$ antagonists promote tumor regression by inducing apoptosis of angiogenic blood vessels. Cell **79:** 1157–1164.
108. DIPIETRO, L.A., N.N. NISSEN, R.L. GAMELLI, *et al.* 1996. Thrombospondin 1 synthesis and function in wound repair. Am. J. Pathol. **148:** 1851–1860.
109. WYSOCKI, A.B., L. STAIANO-COICO & F. GRINNELL. 1993. Wound fluid from chronic leg ulcers contains elevated levels of metalloproteinases MMP-2 and MMP-9. J. Invest. Dermatol. **101:** 64–68.
110. WYSOCKI, A.B. & F. GRINNELL. 1990. Fibronectin profiles in normal and chronic wound fluid. Lab. Invest. **63:** 825–831.

Recognition of Fibrinogen by Leukocyte Integrins

TATIANA P. UGAROVA AND VALENTIN P. YAKUBENKO

J.J. Jacobs Center for Thrombosis and Vascular Biology, Department of Molecular Cardiology, Cleveland Clinic Foundation, Cleveland, Ohio, USA

ABSTRACT: Numerous studies have provided evidence that fibrinogen plays a multifaceted role in the immune and inflammatory response. The ability of fibrinogen to participate in the inflammatory response depends on its specific interaction with leukocyte cell surface adhesion receptors, integrins. Two leukocyte integrins, $\alpha_M\beta_2$ (CD11b/CD18, Mac-1) and $\alpha_X\beta_2$ (CD11c/CD18, p150,95), are the main fibrinogen receptors expressed on neutrophils, monocytes, macrophages and several subsets of lymphocytes. The recognition site for $\alpha_M\beta_2$ has been previously mapped to the carboxyl-terminal globular γC domains (γ143–411) and two sequences, γ190–202 (P1) and γ377–395 (P2), were implicated as the putative binding sites. We now demonstrate that a second leukocyte integrin, $\alpha_X\beta_2$, which is highly homologous to $\alpha_M\beta_2$, mediates adhesion of the $\alpha_X\beta_2$-bearing cells to the D fragment and to the recombinant γ-module, γ143–411. Within the γC domain, $\alpha_X\beta_2$ may recognize P1 and P2 sequences since synthetic peptides duplicating these sequences effectively inhibits adhesion of the $\alpha_X\beta_2$-expressing cells to the D fragment. In addition, neutrophil inhibitory factor, NIF, a potent inhibitor of $\alpha_X\beta_2$, also inhibited $\alpha_X\beta_2$-mediated cell adhesion. These data suggest that recognition of the γC domain of fibrinogen by $\alpha_M\beta_2$ and $\alpha_X\beta_2$ may have common structural requirements.

KEYWORDS: Fibrinogen; Adhesion molecules; Integrin; $\alpha_M\beta_2$; Inflammation.

INTRODUCTION

During the past two decades there has been an increasing appreciation of the multifaceted role fibrinogen plays in the immune and inflammatory reactions. In general, inflammation is defined as the body's response to noxious or injurious stimuli that brings plasma molecules and leukocytes to sites of infection or tissue damage. This is accomplished through an increased vascular permeability caused by retraction of the endothelial cells and by enhanced migration of leukocytes across the local vascular endothelium and in the direction of the site of inflammation. Fibrinogen may participate in both of these aspects of inflammation. Leukocyte emigration from the blood through the endothelial barrier during the inflammatory reaction is currently viewed as an adhesion cascade that involves a coordinated function of a variety of adhesion receptors on the surface of leukocytes and the endothelial cells.[1,2] The abil-

---

Address for correspondence: T. Ugarova, Ph.D., CTVB, NB-50, Cleveland Clinic Foundation, 9500 Euclid Avenue, Cleveland, OH 44195, USA. Voice: 216-445-8209; fax: 216-445-8204.

ugarovt@ccf.org

ity of fibrinogen to bind to these receptors indicates the existence of a potential fibrinogen dependent pathway of leukocyte endothelium interaction *in vivo*. In this regard, recent studies demonstrated that simultaneous fibrinogen binding to leukocytes and endothelial cells did enhance adhesion of monocytes to endothelium by acting as a molecular bridge between the two cell types.[3,4] Plasma leakage is a second characteristic feature of inflammation. Within minutes after the onset of stimulation by inflammatory agonists, such as histamine and serotonin, there is a rapid formation of the gaps between endothelial cells that are normally joined by intercellular junctions.[5,6] The size of these openings ranges from 0.1 to 3 µm,[7] which permits the escape of plasma molecules, even as large as 0.045 µm fibrinogen, into the interstitial space. If the inflammatory reaction proceeds, fibrinogen can be converted to fibrin.[8] Fibrin(ogen) deposition in tissues has been demonstrated to occur during the course of many pathologies[9] and has been implicated as a factor that contributes to the progression of diseases. For example, local fibrin(ogen) deposition is a major component of atherosclerotic lesions.[10] However, little is known about the role and fate of fibrinogen that leaks into tissues during normally beneficial immune responses.

The link between fibrin(ogen) and cell mediated inflammatory reactions was first proposed by Colvin *et al.*,[11] who demonstrated that extensive deposition of fibrin(ogen) was essential for the development of induration, a central feature of classic delayed-type hypersensitivity. Later, this concept was supported by observations that patients with a congenital afibrinogenemia, a rare genetic defect in fibrinogen synthesis, lacked the indurated lesions.[12] Subsequent studies in many experimental disease models showed that systemic defibrinogenation of animals significantly reduced the magnitude and manifestation of inflammatory responses. For example, Wu *et al.*[13] demonstrated that defibrination of rats before the induction of nephrotoxic nephritis markedly reduced proteinuria, which serves as an index of the tissue damage caused by invading leukocytes. As an illustration of the same principle, transient removal of fibrinogen from circulation reduced the degree of intraabdominal abscess formation and significantly decreased the sustained joint inflammation in rheumatoid arthritis.[14,15] Perhaps the most express example of fibrinogen dependent inflammation was provided by Tang and Eaton.[16] Using a model of acute inflammatory response to experimental biomaterial implants, they showed that fibrin(ogen), spontaneously adsorbed on the surface of plastic disks implanted into mouse peritoneum, is responsible for leukocyte attraction and activation. Even more convincingly, hypofibrinogenemic mice failed to recruit leukocytes and this response could be restored by coating fibrinogen onto the surface of the implants. This study suggests that, in addition to its role as a helper in eliciting a competent inflammatory response, fibrinogen itself may trigger undesirable inflammatory reactions to foreign surfaces. Thus, accumulating experimental evidence underscores the accessory role of fibrin(ogen) in modulating inflammatory responses *in vivo*.

The mission of the inflammation is not to injure the organism but instead to return the affected tissue to a state of health. It is currently unresolved how the balance between local fibrin(ogen) deposition and dissolution may control inflammation. It is possible that small doses of fibrinogen leaking with plasma into the extracellular matrix during normal physiological inflammation can provide the optimal and transient recruitment and activation of leukocytes. In contrast, alterations in the hemostatic balance that result in formation of stable fibrin(ogen) deposits can induce a sustained leukocyte accumulation and their excessive activation, and, thus, contribute to progression of disease.

Other than the obvious link between fibrin(ogen) and leukocytes in the mounting of competent inflammatory response, our understanding of the molecular mechanisms underlying fibrinogen dependent leukocyte reactions is still in an embryonic state. Cumulatively, the available data would indicate that leukocyte *adhesion* to fibrinogen is a critical step required for both the localization of leukocytes to the sites of inflammation and for their activation. Consistent with the important role of adhesion, recent studies demonstrated that engagement by leukocytes of immobilized rather than soluble form of fibrin(ogen) may initiate a variety of intracellular signaling events which, in turn, regulate gene expression and induce synthesis of proinflammatory molecules. In this connection, adhesion of peripheral blood monocytes to immobilized fibrin(ogen) resulted in generation of active oxygen species,[17] upregulation of TNFα mRNA,[18] induction of Il-1β,[19] and metalloproteinase synthesis (Ugarova *et al.*, unpublished observation). In addition, adhesion to fibrinogen cooperated with endotoxin stimulation to induce tissue factor expression on monocytes.[20] Thus, leukocyte–fibrinogen interaction may contribute to inflammation by initiating the activation of signaling pathways that regulate specific leukocyte responses. At present, there is no clear understanding of how recognition of fibrinogen by leukocyte receptors is converted into specialized inflammatory responses. Central to understanding the mechanisms that control the development of inflammatory reactions is the molecular basis for fibrinogen recognition by leukocyte receptors.

LEUKOCYTE INTEGRINS

Adhesion of leukocytes to fibrinogen is mediated by cells surface receptors that belong to the integrin gene superfamily.[21] Integrins are heterodimers composed of two noncovalently associated α and β subunits. Each integrin subunit has a large extracellular domain, a transmembrane segment, and a short cytoplasmic tail. A distinctive feature of integrins is that their adhesive function is directly coupled with intracellular signaling and reorganization of actin cytoskeleton. Twenty-four integrins that have been identified to date were grouped into subfamilies based on shared β-chains. Members of the three major families of integrins, $β_1$ ($α_5β_1$), $β_2$ ($α_Mβ_2$ and $α_Xβ_2$), and $β_3$ ($α_Vβ_3$) are expressed on leukocytes and all can engage fibrinogen as a ligand. Although evidence that $α_5β_1$ on leukocytes can interact with fibrin(ogen) is somewhat circumstantial, recent data obtained with other cells directly confirmed that this integrin binds fibrinogen.[22–25] The interaction of fibrinogen with integrins has certain characteristics common for many integrin–ligand pairs. For example, cell adhesion to fibrinogen is cation-dependent; that is, it requires the presence Mg^{2+}, is enhanced by Mn^{2+}, and suppressed by Ca^{2+}. Stimulation by agonists activates integrins and renders them competent to bind soluble fibrinogen, whereas nonstimulated integrins can mediate adhesion to the immobilized fibrinogen. Finally, similar to other integrin ligands, fibrinogen is not the single ligand recognized by leukocyte integrins.

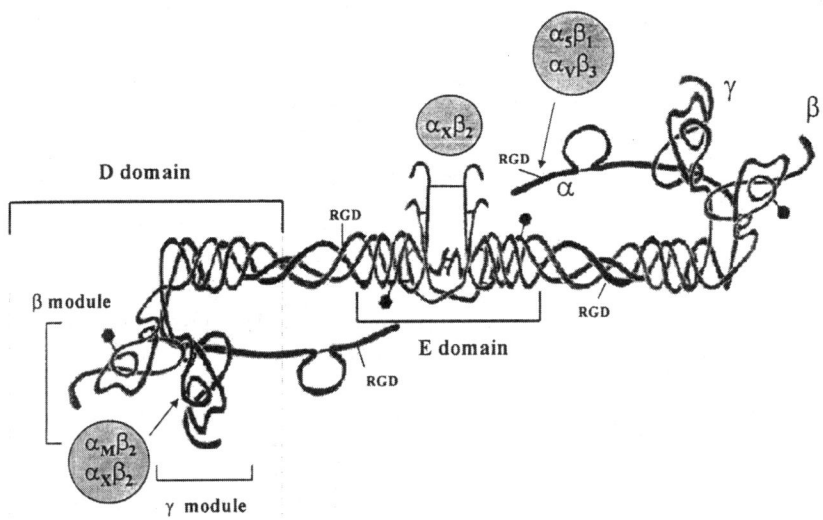

FIGURE 1. Schematic representation of the fibrinogen molecule. Aα, Bβ, and γ are the individual chains. The entire D domain with its constituent β- and γ-modules, and E domain are *boxed*. Integrins are positioned adjacent to the corresponding regions of fibrinogen with which they are known to interact.

Although the role of $\alpha_5\beta_1$ in leukocyte migration on fibrin matrices was clearly demonstrated,[26] it is via the β_2 integrins that fibrinogen is believed to mediate the inflammatory responses. Within the group of β_2 integrins, $\alpha_M\beta_2$, also known as Mac-1, is a key participant. The recently developed $\alpha_M\beta_2$-deficient mice were used to demonstrate that fibrinogen dependent inflammatory reactions were significantly subdued.[27] Specifically, $\alpha_M\beta_2$-deficient mice failed to accumulate phagocytes at the site of biomaterial implants, whereas migration of leukocytes was not affected. The role of the second integrin, $\alpha_X\beta_2$, which is enriched on macrophages and dendritic cells, is still poorly characterized.

Fibrinogen recognition by leukocyte integrins is mediated by adhesive sequences that reside in different domains of the molecule (see FIGURE 1). The interaction of fibrinogen with $\alpha_5\beta_1$ and $\alpha_V\beta_3$ is mediated by short RGD sequences that represent one of the major recognition systems for cell adhesion.[28] Among four RGD sites in dimeric fibrinogen, only those residing in the αC domains (Aα572–574) are exposed at the hydrated surface and can interact with cells, whereas two other RGD sequences at Aα95–97 are cryptic and not available in the intact soluble molecule for cell binding.[29] These latter two RGD sequences become exposed in immobilized fibrinogen and in fibrinogen bound to platelet integrin GPIIbIIIa.[29] However, the role of these sequences in cell adhesion is unclear.

The binding sites for $\alpha_M\beta_2$ and $\alpha_X\beta_2$ were shown to reside in the D and E domains.[30–32] The interaction of these integrins with fibrinogen does not depend on recognition of RGD as an adhesive signal, and overall recognition specificity of β_2 integrins is more complex than that of β_1 and β_3. It is clear now that $\alpha_M\beta_2$ and $\alpha_X\beta_2$

binding of fibrinogen cannot be accomplished by a single short amino acid sequence(s) and, likely, involves multiple contact residues scattered on the broad recognition interface. Therefore, the elucidation the architecture of the β_2 integrin–fibrinogen interface presents a major scientific challenge. This article addresses current understanding of the structural basis for fibrinogen recognition by β_2 integrins.

INTERACTION OF FIBRINOGEN WITH INTEGRIN $\alpha_M\beta_2$

The role of integrin $\alpha_M\beta_2$ as fibrinogen receptor was established in 1988–1989 when several groups independently described specific and saturable binding of soluble fibrinogen to myeloid cells.[33–36] The interaction was described as of moderate affinity with a K_d in the μM range and with the association of 40,000–150,000 fibrinogen molecules per cell. In addition, the binding of soluble fibrinogen to leukocytes was shown to increase after receptor activation with various agonists, including ADP and fMLP. Using defined proteolytic fragments of fibrinogen, Altieri et al.[30] demonstrated that the $\alpha_M\beta_2$-mediated interaction of soluble fibrinogen with leukocytes was mediated by the D domain of the molecule. A low molecular weight degradation product, D_{30}, generated by plasmin proteolysis of fibrinogen in the presence of 2 M urea, produced a dose dependent inhibition of fibrinogen binding to fMLP-stimulated neutrophils and THP-1 monocytoid cells. Since the binding of radiolabeled ^{125}I-D_{30} to the cells was inhibited by polyclonal antibodies against γ95–264, the binding site for $\alpha_M\beta_2$ was suggested to reside in the γ chain. Through the use of overlapping synthetic peptides spanning the constituent γ-chain of D_{30} (γ89–206), the peptide corresponding to γ190–202 (GWTVFQKRLDGSV) was identified as a mediator of ligand binding to $\alpha_M\beta_2$.[37] This peptide, designated P1, inhibited ^{125}I-fibrinogen binding to stimulated neutrophils and monocytes with IC$_{50}$ 8 μM and directly bound to the cells. A minimal core sequence was contained within γ195–202 since the peptide that duplicated this sequence retained some inhibitory activity. In addition, a variant peptide, in which Asp^{199} was either deleted or mutated to Glu or Asn, was substantially less potent in inhibiting fibrinogen binding to leukocytes,[37] suggesting that this residue may play a role in receptor recognition.

Despite general consensus that $\alpha_M\beta_2$ on leukocytes can function as a receptor for soluble fibrinogen, the functional significance of this interaction is not fully understood. It was speculated that local concentration of fibrinogen on the cell surface via $\alpha_M\beta_2$ could facilitate conversion to fibrin by the extrinsic tissue factor dependent pathway of coagulation that is assembled on the surface of activated monocytes.[38,39] However, $\alpha_M\beta_2$ is also a high affinity receptor for factor X,[40] which is activated during the extrinsic coagulation, and high concentrations of plasma fibrinogen do not compete with the binding of factor X to activated monocytes. More recently, simultaneous binding of fibrinogen to leukocytes and ICAM-1 on endothelial cells was suggested as a pathway of intercellular adhesion that helps to stabilize firm attachment of leukocytes to endothelium and to facilitate migration.[3,4,41] Nevertheless, the prerequisites to this interaction should be defined to account for the binding of fibrinogen to $\alpha_M\beta_2$ in the presence of a physiological plasma milieu. Therefore, the potential physiological relevance of this fibrinogen dependent intercellular bridging remains to be determined.

Since the majority of leukocyte responses are elicited by immobilized fibrinogen or fibrin, these forms of the molecule seem to be functionally relevant ligands for $\alpha_M\beta_2$. It is conceivable that the binding site(s) within soluble and immobilized fibrinogen, recognized by $\alpha_M\beta_2$, share common structural determinants, but are not completely identical. In this respect, previous *in vitro* studies demonstrated that P1 peptide inhibited adhesion of leukocytes to immobilized fibrinogen.[37] When immobilized on the surface of Mylar® disks and implanted into mouse peritoneum, P1 triggered recruitment of phagocytes.[42] However, the concentration of the peptide required to achieve 50% inhibition of adhesion was about 90 μM, or about 10-fold higher than that to inhibit the binding of soluble fibrinogen to leukocytes.[37] In addition, the recombinant γ-module (γ148–411) was also a more potent inhibitor of adhesion than P1.[43] Furthermore, selected mAbs directed against the α_M subunit differentially affected the interaction of leukocytes with soluble or immobilized fibrinogen. For example, whereas mAb OKM1 strongly inhibited the binding of soluble fibrinogen to leukocytes,[33,36] it did not inhibit adhesion of the cells to surface bound fibrinogen.[35,44] Taken together, these observations suggest that other fibrinogen sequences, most probably present in the γ-module, may contribute to the recognition of fibrinogen by $\alpha_M\beta_2$.

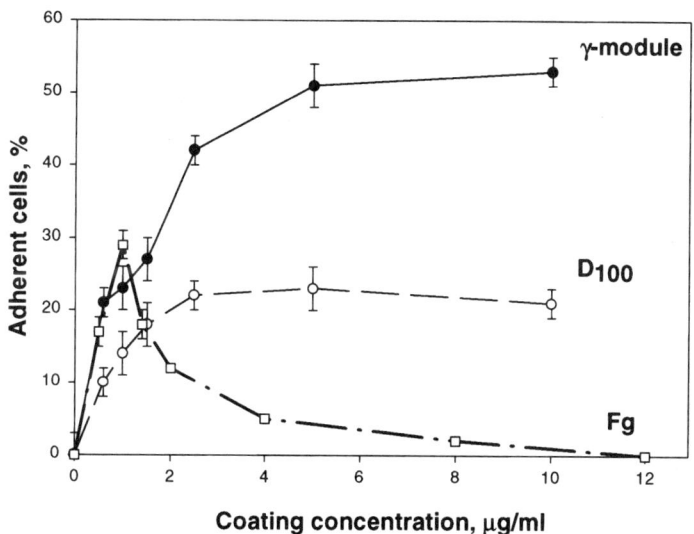

FIGURE 2. Adhesion of the $\alpha_M\beta_2$-transfected cells to immobilized fibrinogen, the D_{100} fragment and the recombinant γ-module. Aliquots of ^{51}Cr-labeled cells (5×10^4) in HBSS/HEPES, supplemented with 1 mM Ca^{2+} and 1 mM Mg^{2+}, were added to the wells of 48-well tissue culture plates coated with different concentrations of ligands. After 25 min at 37°C in a humidified atmosphere containing 5% CO_2, the nonadherent cells were removed by three washes with HBSS, the adherent cells were solubilized with 2% SDS, and their radioactivity was counted in a β counter. Data are expressed as a percent of adherent cells. *Abscissa* shows the concentration of proteins used for coating the wells.

To assess the capacity of different fibrinogen fragments to support the $\alpha_M\beta_2$-mediated adhesion, we have used 293 HEK cells stably transfected with $\alpha_M\beta_2$.[45] These cells bind fibrinogen exclusively via $\alpha_M\beta_2$, thereby offering an advantage over isolated leukocytes and their derivative cell lines, which express several fibrinogen receptors simultaneously. Furthermore, these cells do not require stimulation to adhere and, most importantly, adhesion in not accompanied by secretion of proteases that can degrade fibrinogen.[36] We have observed that fibrinogen and its derivatives support efficient adhesion of the $\alpha_M\beta_2$-transfected cells, although the pattern of adhesion varies. As shown in FIGURE 2, the transfected cells exhibited a dose-dependent and saturable adhesion to immobilized D_{100} fragment (M_r 100,000) and the recombinant γ-module (γ148–411) at coating concentrations up to 2–4 µg/ml. Above 5 µg/ml, the level of adhesion reached a plateau, and the attached cells were observed to spread extensively on these substrates. However, the extent of cell attachment was consistently lower on the D_{100} than on the γ-module. The pattern of cell adhesion to immobilized fibrinogen was quite different and anomalous. Whereas cell adhesion was dependent on fibrinogen coating concentrations at low doses (a maximum was reached at 1.0 µg/ml), further increases in fibrinogen concentration resulted in a precipitous decline in the adhesive activity of the coated surface. Thus, fibrinogen supported adhesion in a very narrow range of concentrations (not exceeding 10 µg/ml) and was non-adhesive at more than 10–30 µg/ml. Nevertheless, when the amounts of proteins bound to the surface were quantified using radiolabeled tracers, they were similar. Specifically, fibrinogen was immobilized onto plastic in a dose dependent and saturable manner. Thus, a nontrivial explanation is needed for

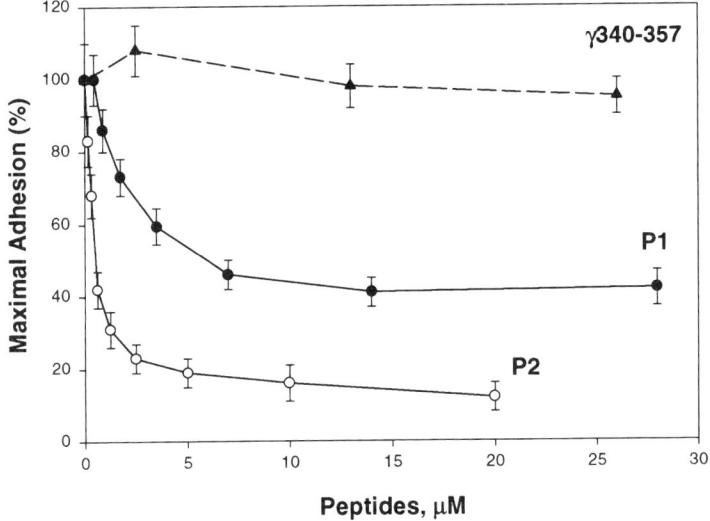

FIGURE 3. Inhibition of adhesion of the $\alpha_M\beta_2$-expressing cells to the D_{100} fragment by soluble P1 and P2 peptides. Increasing concentrations of peptides were preincubated with ^{51}Cr-labeled cells and added to the wells coated with D_{100}. (Reprinted from Ref. 44, with permission from *J. Biol. Chem.*)

the anomalous behavior of fibrinogen in adhesion assay. Since the above data clearly demonstrated the capability of the γ-module to support adhesion, we have focused upon the structural requirements for its recognition by β_2 integrins.

To characterize further the $\alpha_M\beta_2$-mediated adhesion, we tested the effect of synthetic peptide γ190–201 (P1). Soluble P1 produced a dose dependent inhibition of cell adhesion to the immobilized D_{100} fragment.[44] However, maximal inhibition by P1 did not exceed 60–65%, even at the highest concentration of P1 tested (130 µM) (see FIGURE 3). In addition, mutation of Asp^{199} in the recombinant γ-module, which was shown to be of major importance in the recognition of P1 by the receptor,[44] did not impair adhesion of the $\alpha_M\beta_2$-expressing cells. Thus, the results obtained with the $\alpha_M\beta_2$-transfected cells and with monocytoid cells[37] strengthened a notion that other sequences within γ148–411 contribute to the recognition by $\alpha_M\beta_2$. Initial insights into the functional significance of these sequences in receptor recognition was obtained from the analysis of the three-dimensional structure of the γ-module.[46,47] In this structure, the P1 sequence forms a β strand (residues 190–198) and a short connecting loop (residues 199–202). This β strand and loop are adjacent to a β strand formed by residues γ380–389 (see Ref. 44 for a ribbon model of the γ-module). The unusual feature of this latter β strand is that it originates from the COOH-terminal part of the γ-chain but folds back and inserts into the middle subdomain of the γ-module, next to the P1 β strand. Thus, although γ190–202 and γ380–390 are separated by 178 residues in terms of linear amino acid sequence, the specific folding of the γ-module brings these two regions in close proximity. This proximity led us to consider whether the sequences adjacent to γ380–389 contribute to receptor recognition. Accordingly, we synthesized a peptide γ377–395, designated P2, which

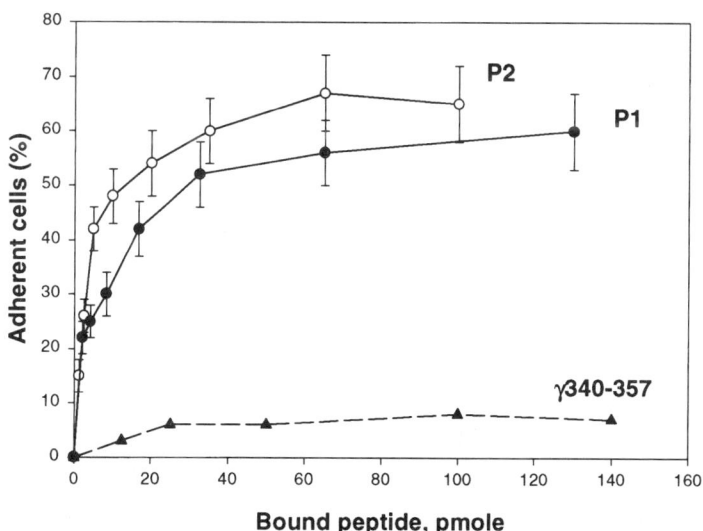

FIGURE 4. Adhesion of the $\alpha_M\beta_2$-expressing cells to P2 and P1 peptides. ^{51}Cr-labeled cells were added to wells coated with different concentrations of P2, P1, or control peptide (γ340–357). (From Ref. 44, with permission from *J. Biol. Chem.*)

duplicated β strand γ380–389 and its flanking residues and tested its effects on adhesion of $\alpha_M\beta_2$-expressing cells. As shown in FIGURE 3, this peptide inhibited adhesion of the $\alpha_M\beta_2$-expressing cells to the immobilized D_{100} in a dose dependent manner and was able to produce 80–90% inhibition at concentrations of 20–30 μM.[44] On a molar basis, P2 was 10–15-fold more potent than P1. Specificity of the effect of both P1 and P2 was confirmed by using peptides originating from the neighboring region of the γ-chain. Two control peptides, γ340–357 and γ350–375, did not inhibit adhesion of the cells to D_{100}.

We further found that, when immobilized onto plastic surface, P2 was in itself capable to support $\alpha_M\beta_2$-mediated adhesion (see FIGURE 4).[44] Comparison of P2 and P1 demonstrated that both peptides were almost equally efficacious adhesive substrates for the $\alpha_M\beta_2$-expressing cells. In addition, both peptides induced stable adhesion of THP-1 and U937 monocytoid cells and adhesion was only slightly enhanced by activation with PMA. Interestingly, although P1 in solution was a relatively poor inhibitor of cell adhesion, on immobilization, it supported adhesion as well as immobilized P2. It is possible that the low inhibition activity of P1 may arise from its inability to adopt a competent conformation for interaction with receptor. Similar to their behavior in inhibition analyses, several control peptides did not support adhesion. More supportive evidence that P1 and P2 contain specific structural information required for recognition by $\alpha_M\beta_2$ has been obtained from analysis of the adhesion promoting activity of the peptide CEIDGSGNGW, which duplicates sequence γ182–191 of the γ-chain. This sequence contains IDGS, which is almost identical to the COOH-terminal part of P1, ^{198}LDGS^{201}, and both these sequences

TABLE 1. Interaction of the $\alpha_M\beta_2$-expressing cells with various synthetic peptides[a]

Name of peptide and its position in the γ chain		Amino acid sequence	IC50, concentration of peptide required to produce 50% inhibition of adhesion to D_{100}, μM[b]	Adhesion to immobilized peptides, percentage of cells adhesion to P1
P1	(γ190–202)	GWTVFQKRLDGSV	5	100
P2	(γ377–395)	YSMKKTTMKIIPFNRLTIG	1.0	110
P2-N	(γ377–386)	YSMKKTTMKI	46	30
P2-C	(γ383–395)	TMKIIPFNRLTIG	1.5	85
	γ377–391	YSMKKTTMKIIPFNR	24	70
	γ340–354	HAGHLNGVYYQGGTY	no inhibition	16
	γ182–191	CEIDGSGNGW	no inhibition	10

[a]Most, but not all, of these data were reported in Reference 44.
[b]The data are the mean of five individual experiments performed with freshly propagated cells within 2–3 weeks. A gradual drift in the $\alpha_M\beta_2$-expressing cells was observed toward a more adhesive phenotype after several months in culture, which influenced the absolute concentrations of peptides required for inhibition of adhesion.

form loops exposed on the surface of the γ-module. The immobilized γ182–191 did not support cell adhesion suggesting that other residues are important.

Since 293 HEK cells express endogenously on their surface several integrins, including $\alpha_5\beta_1$ and $\alpha_V\beta_1$, that are known to interact with many extracellular matrix proteins, it was critical to substantiate that interaction of P2 with transfected cells was mediated by $\alpha_M\beta_2$. It was confirmed in the experiments with mock transfected 293 HEK cells which adhered poorly to immobilized P2 and P1 peptides. In addition, function blocking antibodies that recognize α_M and β_2 subunits of receptor produced 50–70% of adhesion inhibition.[44]

Because active sequences recognized by other integrins, including RGDS in fibrinogen and LDV in fibronectin, were narrowed down to several residues,[28] we sought to localize the active determinant(s) within 19-residue P2. As a first step in defining the minimal recognition sequences within P2, several overlapping peptides spanning the entire length of P2 were synthesized and tested (see TABLE 1). Analyses of synthetic peptides demonstrated that P2 may contain two adhesive sequences, one in its NH$_2$-terminal part, γ377–386 (designated P2-N), and a second in its COOH-terminal region, γ383–395 (designated P2-C). This conclusion was based on the observation that synthetic peptides duplicating these sequences were able to inhibit adhesion of the $\alpha_M\beta_2$-expressing cells and to directly support adhesion (TABLE 1).[44] However, adhesion promoting and inhibitory activity of the two peptides differed significantly; P2-N was about 65-fold less active than parental P2, and P2-C is almost as active as P2. We interpreted these results to mean that the 13-residue P2-C sequence represents a major $\alpha_M\beta_2$ recognition site and P2-N a minor one. Two other lines of evidence seem to prove that COOH-terminal part of P2-C is responsible for the majority of recognition by $\alpha_M\beta_2$. First, truncation of the COOH-terminal part of P2 at Arg^{391} resulted in significant loss of adhesive activity of peptide γ377–391, and second, mAb 4-2, directed against the γ392–406 inhibited adhesion of the $\alpha_M\beta_2$-expressing cells to immobilized P2-C.[44]

Identification of P2-C as a relatively short and potent peptide aided us in investigating the complementary binding site in $\alpha_M\beta_2$. In our initial analyses, we demonstrated that within the receptor, P2 and P1 specifically bind to the I-domain in the α_M subunit.[44] In subsequent studies, using mutational analyses of the recombinant I-domain and testing the binding of the generated mutants to P2-C, we mapped the binding site for P2-C in the α_M I-domain (Yakubenko et al., in press).

BINDING SITE(S) FOR INTEGRIN $\alpha_X\beta_2$ WITHIN THE γ-MODULE

In 1991, Loike et al.[31] demonstrated that stimulated neutrophils can adhere to immobilized fibrinogen via $\alpha_X\beta_2$. To identify the binding site for $\alpha_X\beta_2$ within fibrinogen, these authors examined adhesion of nonstimulated and TNFα activated neutrophils to surfaces coated with various fibrinogen fragments. In the absence of activation, both the D fragment and N-DSK supported cell adhesion. However, the D fragment was a more attractive substrate for neutrophils since about threefold more cells adhered to D than that to N-DSK. Stimulation of neutrophils with TNFα resulted in a fourfold increase of cell adhesion to N-DSK and slightly enhanced cell adhesion to the D fragment. Despite the documented fact that both fragments are capable of sustaining strong adhesion, only the ability of $\alpha_X\beta_2$ to recognize the E

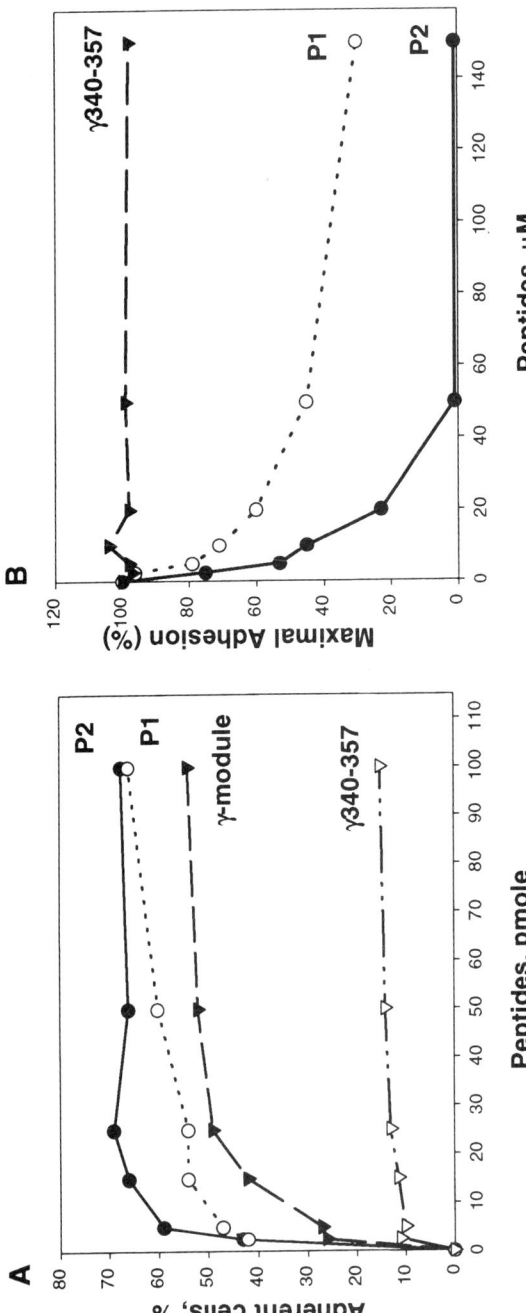

FIGURE 5. Interaction of the $\alpha_X\beta_2$-expressing cells with P2 and P1 peptides. **A**, $\alpha_X\beta_2$-expressing CHO cells were resuspended in HBSS/Hepes supplemented with 1 mM Ca^{2+} and Mg^{2+}. Aliquots (5×10^4) of the cells were added to the wells of tissues culture plates coated with different concentrations of the recombinant γ-module or peptides P2, P1, and γ340–357. After 25 min at 37°C, the nonadherent cells were removed by three washes with HBSS and the plates were placed overnight at −20°C. The cells were thawed and lysed with lysis buffer containing the fluorescent dye CyQuant GR. The amount of adherent cells was determined from the fluorescence and the reference curve which correlated the fluorescence to cell number. **B**, Inhibition of adhesion of the $\alpha_X\beta_2$-expressing cells by P2 and P1. The cells were incubated for 15 min with different concentrations of soluble P2, P1, and control peptides. Aliquots of the cells were added to the wells coated with D_{100} and the adhesion assay proceeded as above.

domain was cited in the subsequent literature, thus creating the impression that two β_2 integrins, $\alpha_M\beta_2$ and $\alpha_X\beta_2$, differ in their recognition specificity. Expression of $\alpha_X\beta_2$ on leukocytes is normally low and regulated by myeloid differentiation with large numbers found only on mature macrophages. Since neutrophils and blood monocytes express far more $\alpha_M\beta_2$ than $\alpha_X\beta_2$, this may explain why interaction of fibrinogen with $\alpha_X\beta_2$ captured less attention of investigators.

To substantiate that $\alpha_X\beta_2$ can interact with the D fragment and to identify sequences critical for this recognition, we have examined adhesion of the $\alpha_X\beta_2$-expressing cells to different fibrinogen derivatives. For these analyses, we have employed CHO cells transfected with cDNA for α_X and β_2 subunits. The cells readily adhered to the D fragments (not shown) and the recombinant γ-module (see FIGURE 5A) in a dose dependent and saturable manner, whereas mock-transfected cells did not exhibit appreciable affinity for these substrates (not shown). In parallel experiments, we confirmed that the E_1 and E_3 fragments also promoted adhesion of the $\alpha_X\beta_2$-transfected cells. Adhesion of the cells to the D_{100} fragment and γ-module was specific for $\alpha_X\beta_2$ as the mAb 3.9, directed against α_X, inhibited adhesion.

Since α_X and α_M subunits share 63% identity at the amino acid level, we hypothesized that α_X can recognize P1 and P2 sequences within the γ-module. To test the activity of P1 and P2, the peptides were immobilized onto tissue culture plates, and the $\alpha_X\beta_2$-expressing cells were added. As shown quantitatively in FIGURE 5A, both P1 and P2 supported efficient adhesion. As with $\alpha_M\beta_2$, P2 was a slightly more potent

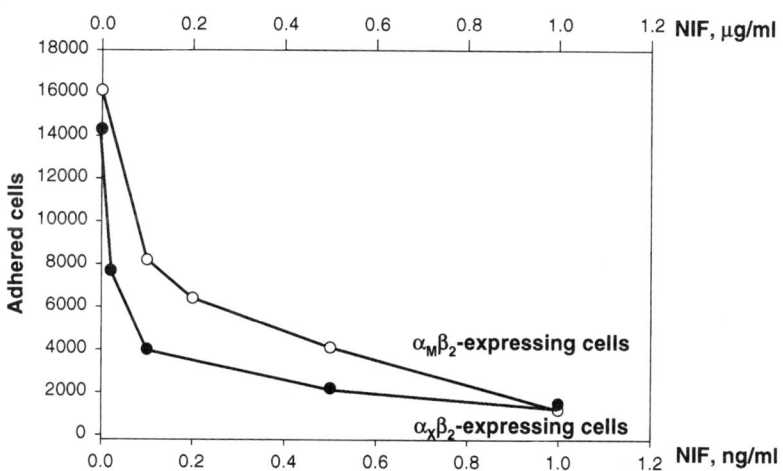

FIGURE 6. Inhibition of adhesion of the $\alpha_M\beta_2$- and $\alpha_X\beta_2$-expressing cells by NIF. Aliquots (5×10^4) of the Calcein M-labeled cells were preincubated for 15 min with different concentrations of NIF and then added to the wells with immobilized D_{100} fragment. After 25 min at 37°C in a humidified atmosphere containing 5% CO_2, the nonadherent cells were removed by washing with PBS and adherent cells were determined from the fluorescence corrected for nonspecific binding to PVD. Data are expressed as numbers of adherent cells. Note that the *bottom* and *top abscissæ* correspond to the range of NIF concentrations required to inhibit adhesion of the $\alpha_M\beta_2$ (○) and $\alpha_X\beta_2$ (●) expressing cells, respectively.

adhesive substrate than P1. In inhibition adhesion assays, P2 and P1 efficiently inhibited adhesion of the $\alpha_X\beta_2$-expressing cells to the immobilized D_{100} (FIG. 5B). On a molar basis, P2 was about 6-fold more potent inhibitor than P1; IC_{50} were 5 µM and 30 µM for P1 and P2, respectively. At a concentration 50 µM, P2 completely inhibited cell adhesion. Specificity of the interaction of $\alpha_X\beta_2$ with P2 and P1 was apparent, since control peptide duplicating the γ-chain sequence 340–357 did not inhibit adhesion and did not support direct adhesion. Both peptides in soluble form were also able to inhibit adhesion of the $\alpha_X\beta_2$-expressing cells to immobilized P1 and P2. Therefore, the ability of peptides to cross inhibit each others' function was similar to that observed with $\alpha_M\beta_2$-expressing cells.[44] Thus, these results suggested that within the α_X subunit, P1 and P2 can bind to the site(s) homologous to those in α_M. Previous results demonstrated that within the α_M subunit, P2 and P1 bind to the I-domain.[44] Accordingly, we tested whether P2 and P1 can directly bind to the α_XI-domain. Similar amounts of P2 and P1 were immobilized onto microtiter plate, and the binding of the recombinant α_XI-domain was assessed. Both P2 and P1, but not control peptide, bound the α_XI-domain in a dose dependent manner.

It has been shown that NIF (neutrophil inhibitory factor), a 41-kD glycoprotein isolated from canine hookworm,[48,49] is a potent inhibitor of $\alpha_M\beta_2$-mediated cell adhesion to fibrinogen.[44,45] Previous reports indicated that functional effects of NIF resulted from its specific binding to $\alpha_M\beta_2$ but not to other β_2 integrins. Since recognition specificity of $\alpha_X\beta_2$ and $\alpha_M\beta_2$ was similar, we tested the effect of NIF on adhesion of the $\alpha_X\beta_2$-expressing cells to fibrinogen derivatives. As shown in FIGURE 6, NIF effectively inhibited adhesion of the $\alpha_X\beta_2$-expressing CHO-cells to immobilized D_{100}; 50% inhibition was attained at 9.1 nM NIF. However, NIF was a more potent inhibitor of $\alpha_M\beta_2$-mediated adhesion with an IC_{50} of 2.4 pM. Nevertheless, since nanomolar concentrations of NIF were required in order to inhibit the adhesive function mediated by $\alpha_X\beta_2$, it can be viewed as an effective antagonist of this β_2 integrin as well. Taken together, these results suggest that the binding of $\alpha_M\beta_2$ and $\alpha_X\beta_2$ to the γ-module of fibrinogen has common structural requirements, with P1 and P2 as the likely recognition sites for both integrins.

RELATIONSHIP BETWEEN THE $\alpha_M\beta_2$-BINDING SITE AND FIBRIN POLYMERIZATION SITE IN THE γ-MODULE

Analysis of the three-dimensional structure of the γ-module indicated that among 32 residues in the P1, P2-N and P-C sequences, the side chains of 10 residues are exposed on the surface of the γ-module and, potentially, can be involved in receptor docking. The exposed residues in P2-N, Lys^{380} and Lys^{381}, and those in P1, Lys^{196}, Leu^{198}, and Asp^{199} are clustered on the single face of the γ-module (see FIGURE 7; see also Fig. 10 in Ref. 44 for details). Separated by the distance of about 30 Å, is an array of exposed residues in P2-C, including Thr^{389}, Asn^{390}, Arg^{391}, Leu^{392}, Thr^{393}, and Ile^{394}. This array is situated on the opposite side of the γ-module (the position of this cluster is marked by Asn^{390} in FIG. 7). Based on the analyses of atomic structure of the recognition sites seen in many known three-dimensional protein–protein complexes,[50] two comments can be made concerning the putative binding site for $\alpha_M\beta_2$ and $\alpha_X\beta_2$ in the γ-chain of fibrinogen. First, the binding pocket seems to represent a wide surface that can engage multiple contact residues in distant

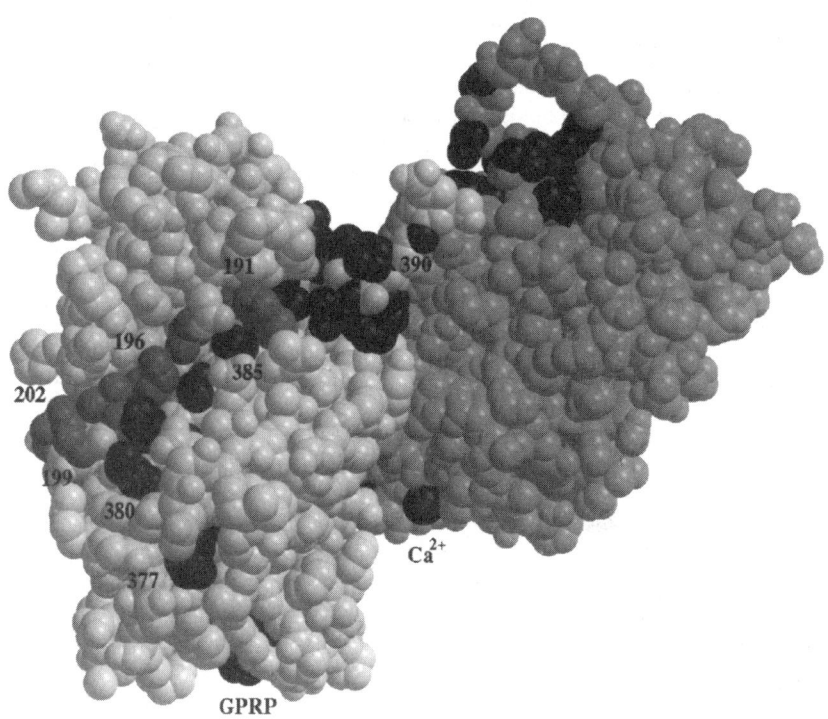

FIGURE 7. Relationship between γ190–202 (P1), γ377–395 (P2) sequences and the fibrin polymerization site. The space-filling model of the γ–γ dimer is based upon its crystal structure,[47] PDB identifier 1FZB. Two γ-modules are colored in different shades of *grey*. The exposed residues in the P1 and P2 sequences are shown in *dark grey* and *black*, respectively. Several residues within both sequences are labeled to trace the direction of the chain in one of the γ-modules. Only the COOH-terminal part of P2 can be viewed in the second γ-module. The COOH-terminal parts of P2 in the two abutting γ-modules border a crevice formed upon end-to-end association. GPRP ligand and the calcium atom reside on the opposite sides of γ–γ.

sequences. The distance between the residues in P2-N and the COOH-terminal loop of P1 is about 10 Å, and about 34 Å separate Lys^{380} in P2-N from Asn^{390} in P2-C. The residues in P1 and P2 lie in the middle of a large concave surface that can potentially be involved in recognition by receptor. If so, then the interface area formed between the γ-module and receptor would be compatible with the "standard-size" interfaces of 1600 (± 400) Å^2 reported for many complexes.[50] Second, most active determinants in P1 and P2 were found in the segments that form loops connecting the adjacent β strands γ190–198 and γ380–389 in the three-dimensional structure of the γ-module. These results corroborate the findings in many ligand–receptor systems, demonstrating that residues in the loops frequently mediate molecular recognition.

Of particular import in considering the structural organization of the interface area between two partners, the γ-module and the I-domain of the receptor, is the relationship between the integrin binding site and the fibrin polymerization site. In this respect, three major recognition clusters in P1 and P2 are remote from the binding cavity for the GPRP[47] (FIG. 7). Furthermore, the Ca^{2+}-binding site, which is functionally connected with GPR-binding pocket, lies on the opposite side of the γ-module. Therefore, it is likely that formation of the "D_2E" complex in protofibrils, as a result of ligation of GPR in the E domain by the γ-module, should not affect the accessibility of the integrin binding site. The COOH-terminal part of the P2-C cluster (γ389–395) resides in close proximity to the γ–γ interface formed by the end-to-end association of joined γ-modules on DD-E formation. Since accessibility measurements indicated that about 750 $Å^2$ of surface becomes inaccessible to solvent due to γ–γ association,[47] the possibility that the integrin binding site could be masked within this interface should be considered. As shown in FIGURE 7, the majority of exposed residues in P2-C are not lost in the γ–γ interface. This excludes only two residues, Phe^{389} and Thr^{393}, that become completely and partially, respectively, buried in the crevice formed between the ends of associated nodules. Thus, protofibril formation seems not to occlude access to integrin recognition sequences. It should be noted, however, that despite the capacity of this model to attractively accommodate an increasing number of observations, more convincing evidence should be obtained from mutational analyses and ultimately from crystallization of a cocomplex between the γ-module and the I-domain. Finally, the critical and as yet unanswered question is whether the binding site(s) for integrins $α_Mβ_2$ and $α_Xβ_2$ remain accessible after protofibrils coalesce in fully developed fibrin.

ACKNOWLEDGMENTS

This work was supported by NIH grant HL 63991. The authors wish to thank Dr. Vivien Yee for help with generating FIGURE 7 and for stimulating discussions. We thank Dr. D. Golenbock for kindly providing the $α_Xβ_2$-expressing CHO cells and Dr. E. Plow for NIF. Dr. D. Solovjov was instrumental in conducting several experiments. Tim Burke provided helpful comments on this manuscript.

REFERENCES

1. CARLOS, T.M. & J.M. HARLAN. 1994. Leukocyte-endothelial adhesion molecules. Blood **84:** 2068–2101.
2. SPRINGER, T.A. 1994. Traffic signals for lymphocyte recirculation and leukocyte emigration: the multistep paradigm. Cell **76:** 301–314.
3. LANGUINO, L.R., J. PLESCIA, A. DUPERRAY, et al. 1993. Fibrinogen mediates leukocyte adhesion to vascular endothelium through an ICAM-1-dependent pathway. Cell **73:** 1423–1434.
4. LANGUINO, L.R., A. DUPERRAY, K.J. JOGANIC, et al. 1995. Regulation of leukocyte-endothelium interaction and leukocyte transendothelial migration by intercellular adhesion molecule 1-fibrinogen recognition. Proc. Natl. Acad. Sci. U.S.A. **92:** 1505–1509.

5. McDonald, D.M. 1994. Endothelial gaps and permeability of venules in rat tracheas exposed to inflammatory stimuli. Am. J. Physiol. Lung Cell. Mol. Physiol. **10:** L61–L83.
6. Michel, C.C. & C.R. Neal. 1999. Openings through endothelial cells associated with increased microvascular permeability. Microcirc. **6:** 45–54.
7. McDonald, D.M., G. Thurston & P. Baluk. 1999. Endothelial gaps as sites for plasma leakage in inflammation. Microcirc. **6:** 7–22.
8. Dvorak, H.F., D.R. Senger, A.M. Dvorak, et al. 1985. Regulation of extravascular coagulation by microvascular permeability. Science **227:** 1059–1061.
9. Robbins, S., R. Cotran & V. Kumar. 1984. Pathologic Basis of Disease. W.B. Saunders, Philadelphia.
10. Bini, A., J.J. Fenoglio, Jr., R. Mesa-Tejada, et al. 1989. Identification and distribution of fibrinogen, fibrin, and fibrin(ogen) degradation products in atherosclerosis. Use of monoclonal antibodies. Arteriosclerosis **9:** 109–121.
11. Colvin, R.B., R.A. Johnson, M.C. Mihm, Jr., et al. 1973. Role of the clotting system in cell-mediated hypersensitivity I. Fibrin deposition in delayed skin reactions in man. J. Exp. Med. **138:** 686.
12. Colvin, R.B., M.W. Mosesson & H.F. Dvorak. 1979. Delayed-type hypersensitivity skin reactions in congenital afibrinogenemia lack fibrin deposition and induration. J. Clin. Invest. **63:** 1302–1306.
13. Wu, X., M.H. Helfrich, M.A. Horton, et al. 1994. Fibrinogen mediates platelet-polymorphonuclear leukocyte cooperation during immune-complex glomerulonephritis in rats. J. Clin. Invest. **94:** 928.
14. McRitchie, D.I., M.J. Girotti, M.F.X. Glynn, et al. 1991. Effect of systematic fibrinogen depletion on intraabdominal abscess formation. J. Lab. Clin. Med. **118:** 48.
15. Busso, N., V. Peclat, K. Van Ness, et al. 1998. Exacerbation of antigen-induced arthritis in urokinase-deficient mice. J. Clin. Invest. **102:** 41–50.
16. Tang, L. & J.W. Eaton. 1993. Fibrin(ogen) mediates acute inflammatory responses to biomaterials. J. Exp. Med. **178:** 2147–2156.
17. Trezzini, C., T.W. Jungi, F.E. Maly, et al. 1989. Low-affinity interaction of fibrinogen carboxy-γ terminus with human monocytes induces an oxidative burst and modulates effector functions. Biochem. Biophys. Res. Commun. **1:** 7–13.
18. Fan, S.-T. & T.S. Edgington. 1993. Integrin regulation of leukocyte inflammatory functions: CD11b/CD18 enhancement of the tumor necrosis factor-α responses of monocytes. J. Immunol. **150:** 2972–2980.
19. Perez, R.L. & J. Roman. 1995. Fibrin enhances the expression of IL-1 beta by human peripheral blood mononuclear cells. Implications in pulmonary inflammation. J. Immunol. **154:** 1879–1887.
20. Fan, S.-T. & T.S. Edgington. 1991. Coupling of the adhesive receptor CD11b/CD18 to functional enhancement of effector macrophage tissue factor response. J. Clin. Invest. **87:** 50–57.
21. Hynes, R.O. 1992. Integrins: versatility, modulation, and signaling in cell adhesion. Cell **69:** 11–25.
22. Suehiro, K., J. Gailit & E.F. Plow. 1997. Fibrinogen is a ligand for integrin $\alpha_5\beta_1$ on endothelial cells. J. Biol. Chem. **272:** 5360–5366.
23. Farrell, D.H. & H.A. Al-Mondhiry. 1997. Human fibroblast adhesion to fibrinogen. Biochemistry **36:** 1123–1128.
24. Miettinen, H., J.M. Gripentrog & A.J. Jesaitis. 1998. Chemotaxis of Chinese hamster ovary cells expressing the human neutrophil formyl peptide receptor: role of signal transduction molecules and $\alpha_5\beta_1$ integrin. J. Cell Sci. **111:** 1921–1928.
25. Asakura, S., K. Niwa, T. Tomozawa, et al. 1997. Fibroblasts spread on immobilized fibrin monomer by mobilizing a b1-class integrin, together with a vitronectin receptor $\alpha_V\beta_3$ on their surface. J. Biol. Chem. **272:** 8824–8829.
26. Loike, J.D., L. Cao, S. Budhu, et al. 1999. Differential regulation of β_1 integrins by chemoattractants regulates neutrophil migration through fibrin. J. Cell Biol. **144:** 1047–1056.
27. Lu, H., C.W. Smith, J. Perrard, et al. 1997. LFA-1 is sufficient in mediating neutrophil emigration in Mac-1 deficient mice. J. Clin. Invest. **99:** 1340–1350.

28. RUOSLAHTI, E. 1996. RGD and other recognition sequences for integrins. Annu. Rev. Cell Biol. **12:** 697–715.
29. UGAROVA, T.P., A.Z. BUDZYNSKI, S.J. SHATTIL, *et al.* 1993. Conformational changes in fibrinogen elicited by its interaction with platelet membrane glycoprotein GPIIb-IIIa. J. Biol. Chem. **268:** 21080–21087.
30. ALTIERI, D.C., F.R. AGBANYO, J. PLESCIA, *et al.* 1990. A unique recognition site mediates the interaction of fibrinogen with the leukocyte integrin Mac-1 (CDIIb/CD18). J. Biol. Chem. **265:** 12119–12122.
31. LOIKE, J.D., B. SODEIK, L. CAO, *et al.* 1991. CD11c/CD18 on neutrophils recognizes a domain at the N terminus of the Aα chain of fibrinogen. Proc. Natl. Acad. Sci. U.S.A. **88:** 1044–1048.
32. NHAM, S.U. 1999. Characteristics of fibrinogen binding to the domain of CD11c, an α subunit of p150,95. Biochem. Biophys. Res. Commun. **264:** 630–634.
33. ALTIERI, D.C., R. BADER, P.M. MANNUCCI, *et al.* 1988. Oligospecificity of the cellular adhesion receptor MAC-1 encompasses an inducible recognition specificity for fibrinogen. J. Cell Biol. **107:** 1893–1900.
34. TREZZINI, C., T.W. JUNGI, P. KUHNERT, *et al.* 1988. Fibrinogen association with human monocytes: evidence for constitutive expression of fibrinogen receptors and for involvement of Mac-1 (CD18, CR3) in the binding. Biochemical and Biophysical Research Communications **156:** 477–484.
35. WRIGHT, S.D., J.I. WEITZ, A.J. HUANG, *et al.* 1988. Complement receptor type three (CD11b/CD18) of human polymorphonuclear leukocytes recognizes fibrinogen. Proc. Natl. Acad. Sci. U.S.A. **85:** 7734–7738.
36. GUSTAFSON, E.J., H. LUKASIEWICZ, Y.T. WACHTFOGEL, *et al.* 1989. High molecular weight kininogen inhibits fibrinogen binding to cytoadhesins of neutrophils and platelets. J. Cell Biol. **109:** 377–387.
37. ALTIERI, D.C., J. PLESCIA & E.F. PLOW. 1993. The structural motif glycine 190-valine 202 of the fibrinogen gamma chain interacts with CD11b/CD18 integrin amb2, Mac-1) and promotes leukocyte adhesion. J. Biol. Chem. **268:** 1847–1853.
38. ALTIERI, D.C. 1993. Coagulation assembly on leukocytes in transmembrane signaling and cell adhesion. Blood **81:** 569–579.
39. ALTIERI, D.C. 1995. Leukocyte interaction with protein cascades in blood coagulation. Curr. Opin. Hematol. **2:** 41–46.
40. ALTIERI, D.C. & T.S. EDGINGTON. 1988. The saturable high affinity association of factor X to ADP- stimulated monocytes defines a novel function of the Mac-1 receptor. J. Biol. Chem. **263:** 7007–7015.
41. SRIRAMARAO, P., L.R. LANGUINO & D.C. ALTIERI. 1996. Fibrinogen mediates leukocyte-endothelium bridging *in vivo* at low shear forces. Blood **88:** 3416–3423.
42. TANG, L., T.P. UGAROVA, E.F. PLOW, *et al.* 1996. Molecular determinants of acute inflammatory responses to biomaterials. J. Clin. Invest. **97:** 1329–1334.
43. MEDVED, L., S. LITVINOVICH, T. UGAROVA, *et al.* 1997. Domain structure and functional activity of the recombinant human fibrinogen y-module (γ148–411). Biochemistry **36:** 4685–4693.
44. UGAROVA, T.P., D.A. SOLOVJOV, L. ZHANG, *et al.* 1998. Identification of a novel recognition sequence for integrin $\alpha_M\beta_2$ within the gamma-chain of fibrinogen. J. Biol. Chem. **273:** 22519–22527.
45. ZHANG, L. & E.F. PLOW. 1996. Overlapping, but not identical sites, are involved in the recognition of C3bi, NIF, and adhesive ligands by the $\alpha_M\beta_2$ integrins. J. Biol. Chem. **271:** 18211–18216.
46. YEE, V.C., K.P. PRATT, H.C.F. COTE, *et al.* 1997. Crystal structure of 30 kDa carboxyl terminal fragment from the y chain of human fibrinogen. Structure **5:** 125–138.
47. SPRAGGON, G., S.J. EVERSE & R.F. DOOLITTLE. 1997. Crystal structures of fragment D from human fibrinogen and its crosslinked counterpart from fibrin. Nature **389:** 455–462.
48. RIEU, P., T. UEDA, I. HARUTA, *et al.* 1994. The A-domain of β_2 integrin CR3 (CD11b/CD18) is a receptor for the hookworm-derived neutrophil adhesion inhibitor NIF. J. Cell Biol. **127:** 2081–2091.

49. MUCHOWSKI, P.J., L. ZHANG, E.R. CHANG, et al. 1994. Functional interaction between the integrin antagonist neutrophil inhibitory factor and the I domain of CD11b/CD18. J. Biol. Chem. **269:** 26419–26423.
50. LO CONTE, L., C. CHOTHIA & J. JANIN. 1999. The atomic structure of protein-protein recognition sites. J. Mol. Biol. **285:** 2177–2198.

Interaction of Fibrin with VE-Cadherin

JOSÉ MARTINEZ, ANDRÉS FERBER, TAMI L. BACH,
AND CHRISTOPHER H. YAEN

Cardeza Foundation for Hematologic Research, Jefferson Medical College of Thomas Jefferson University, Philadelphia, Pennsylvania, USA

ABSTRACT: The conversion of fibrinogen into fibrin and the association of fibrin(ogen) with activated platelets play a fundamental role in hemostasis because their interaction with the injured vessel prevents blood extravasation. Platelet aggregates and fibrin also participate in the occlusion of the vascular lumen in pathological conditions. Fibrin II also promotes the formation of new blood vessels, for example, during wound healing and tumor growth. Using an *in vitro* assay, we have studied the mechanism by which fibrin II induces formation of capillaries. Generation of fibrin II on top of an endothelial cell monolayer rapidly rearranged the ECs into a capillary network. In contrast, neither fibrin I nor fibrin 325 induced these morphogenetic changes, indicating that exposure of the N-terminal peptide β15–42 is involved in this process. Binding studies, using the N-terminal fragment of fibrin (NDSK II), showed that NDSK II binds to EC with high affinity, but neither NDSK nor NDSK325 bound specifically. Binding of NDSK II to endothelial cells was blocked with an antibody to VE-cadherin. Direct association of NDSK II and VE-cadherin was also demonstrated in a VE-cadherin antibody capture assay. NDSK II bound specifically with the captured VE-cadherin but NDSK or NDSK 325 did not associate with VE-cadherin. Moreover, fibrin II associated with EC VE-cadherin and this interaction triggered the formation of capillary-like structures. A better understanding of the cellular responses to fibrin, identification of the fibrin binding site within VE-cadherin and the intracellular signaling that follows this interaction, could yield important information that may translate into better control of the angiogenic process.

KEYWORDS: Fibrin; Interaction with VE-cadherin; Association of fibrinogen; Angiogenic process.

INTRODUCTION

The formation of fibrin is a central event in blood coagulation because following vascular injury, the generation of fibrin and its interaction with the vessel wall plays a major role in hemostasis. In a different clinical setting, formation of fibrin inside of the lumen of blood vessels leads to organ damage such as myocardial or brain infarction. In addition, the generation of fibrin also intervenes in other biological processes, for example, formation of new blood vessels during tissue repair in wound healing or in tumor growth. Other studies have also demonstrated that fibrin, similar to some extracellular matrix proteins, promotes the formation of capillary-like structures. Our

---

Address for correspondence: José Martinez, M.D., Cardeza Foundation for Hematologic Research, 1015 Walnut Street, Philadelphia, PA 19107-5099, USA. Voice: 215-955-8458; fax: 215-923-3836.
Jose.Martinez@mail.tju.edu

laboratory has investigated the interactions of fibrin with endothelial cells, interactions that are responsible for the assembly of endothelial cells into capillary-like structures.

Interaction of Fibrin II with Endothelial Cells Induces the Formation of Capillaries

The interaction of fibrin with the vessel wall induces a series of biological changes leading to remodeling of the blood vessels and surrounding tissues. Fibrin interaction with EC causes a release of von Willebrand factor that further interacts with EC and also with platelets, promoting the hemostatic process.[1] Fibrin also serves as a provisional extracellular matrix protein, providing a scaffold to which endothelial cells can migrate, proliferate, and differentiate into capillaries.[2,3] Under appropriate conditions, such as during wound healing and tumor growth, the EC immersed in fibrin undergo major morphogenetic changes leading to the formation of new capillaries.[4–6] *In vitro* angiogenic assays have also demonstrated that blood vessel rings immersed in fibrin form capillary sprouts.[3] In other experimental systems,

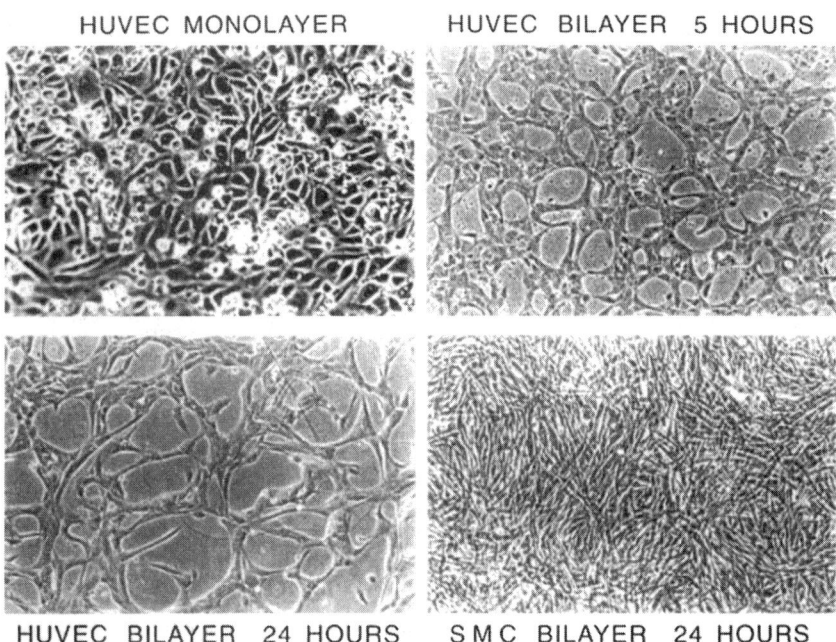

FIGURE 1. Effect of Fibrin II gels on endothelial cell capillary-like tube formation. Human umbilical vein endothelial cells were allowed to become confluent in a fibrin II gel (**top left**), before the addition of the overlying the fibrin II apical gel the EC formed a monolayer. Four hours following the addition of the apical fibrin II gel, a net-like array of tubes were seen (**top right**), and by 24 hours, the tube-like structures were fully developed (**bottom left**). Arterial smooth muscle cells failed to form capillary tube-like structures in response to an apical fibrin II gel (**bottom right**). (Reproduced from Ref. 8.)

isolated EC cultured in a fibrin gel and in the presence of growth factors and media containing serum were able to migrate, forming capillary-like structures inside the fibrin gel.[2,7] To characterize the interactions of fibrin with the endothelial cells responsible for these morphogenetic changes, our laboratory has developed a simple *in vitro* assay in which human EC cultured in a fibrin II gel form a dense monolayer. These cells rapidly rearrange to form a network of capillary tubes on generation of a second fibrin II gel overlaying the endothelial cells (see FIGURE 1).[8] The capillary network that is formed in a culture media devoid of serum can be seen as early as five hours after generating the fibrin apical gel, and the net continues to develop for another 24 hours.[8] The net of capillaries developed by EC sandwiched between the fibrin II gels contain lumens formed by the intercellular adhesion of EC, as visualized by electron microscopy[8] (see FIGURE 2), and degradation of fibrin was not necessary for the formation of the capillary net.

The Sequence β15–42 of Fibrin II Is Necessary for Endothelial Cell Capillary Tube Formation

The structural properties of fibrin responsible for the formation of capillary structures were also analyzed by comparing the effect of fibrin II, fibrin I, and fibrin 325. Fibrin II was prepared by treating fibrinogen with thrombin that releases fibrinopeptides A and B. Fibrin I was generated by treating fibrinogen with the enzyme Atroxin that cleaves only fibrinopeptide A. Fibrin 325 was generated by treating fibrinogen with protease III from *Crotalus atrox,* which cleaves segment β1–42 from fibrinogen. This derivative, named fibrinogen 325, was then treated with thrombin converting fibrinogen 325 into fibrin 325.[9] The molecular structures of the different fibrins, analyzed by SDS-PAGE, are presented in FIGURE 3. To determine whether the morphogenetic changes of the EC monolayer interacting with the double fibrin gel were due to the effect of basal versus apical fibrin, we cultured the EC in fibrin I followed by formation of fibrin II gel on top of the monolayer. Formation of capillary tubes was seen within 4–6 hours and maximal effect was seen at 24 hours.[8] In contrast, when HUVEC were cultured in fibrin II and a fibrin I gel was generated apically, no distinct tubular structures were seen and most of the endothelial cells formed aggregates (not shown). The results were more striking when an EC was cultured in fibrin II and a preformed fibrin 325 gel was placed on top of the monolayer. In this case, the cell monolayer kept the cobblestone appearance without formation of capillary structures (see FIGURE 4). In contrast, when EC monolayer was formed in a fibrin 325 gel and fibrin II was generated apically, capillary structures were seen. In addition, in this experimental system, the formation of capillaries was not inhibited by antiintegrin antibodies directed against $\alpha_v\beta_3$ or by the peptide RGDS (see TABLE 1). However, when peptide β15–42 at a concentration of 1 mM was added to the medium prior to the placement of fibrin II gels on top of the monolayer, a reduction of the number and length of tubular forms, by approximately 40%, was seen.[8] Moreover, the tubes formed in presence of the β15–42 peptide also showed a moderate increase in their diameters. In contrast, the N-terminal peptide of fibrin β15–18 (GHRP) did not have an effect (TABLE 1).[8] It is known that integrins play a major role in angiogenesis and inhibition of $\alpha_v\beta_3$ interferes with EC migration and proliferation that results in inhibition of the angiogenic response.[10,11] However, it is likely that in our experimental system the EC sandwiched between two fibrin gels rearrange into

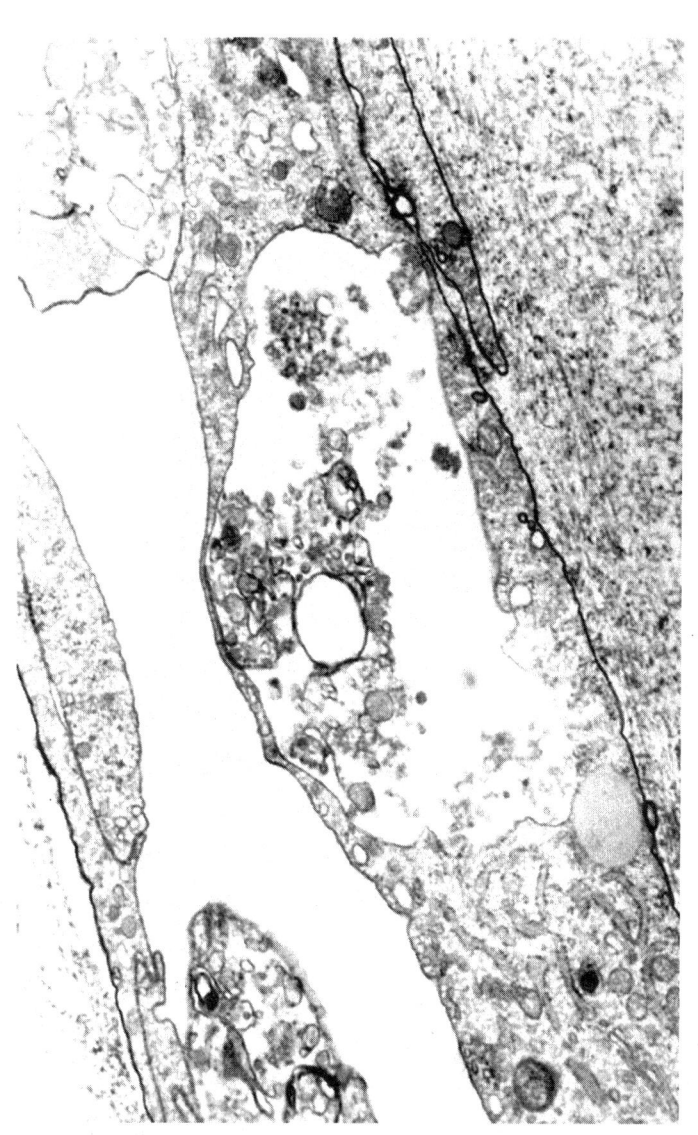

FIGURE 2. Transmission electron micrograph of capillary tubes formed between fibrin II gels that manifest intercellular and intracellular lumina. Endothelial cell capillary tube has been sectioned along its longitudinal axis, reveals both an intercellular lumen formed from interdigitation of several endothelial cell processes as well as a putative intracellular lumen present within the cytoplasm of a single cell. Both lumina are filled with intercellular debris. Intercellular junctions between endothelial cells can be seen. Overlying and underlying fibrin gels are shown. ×10,000. (Reproduced from Ref. 8.)

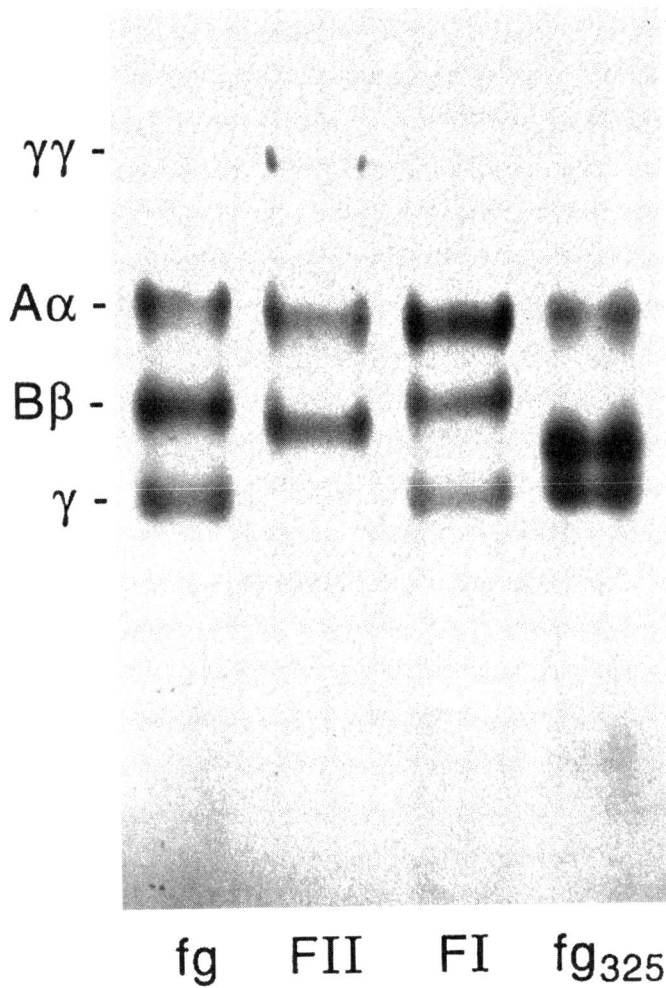

FIGURE 3. Electrophoretic mobility of various fibrin derivatives. Fibrin II (*FII*) and fibrin I (*FI*) were prepared from fibrinogen (*fg*) by treatment with thrombin and Atroxin, respectively. Fibrinogen 325 (*fg325*) was prepared by treatment with protease III. The α chains of FII and FI migrate faster than the Aα chain of fg, due to the cleavage of fibrinopeptide A. The β chain of FII migrates faster than the Bβ chain of fg, due to the cleavage of fibrinopeptide B. The β chain of fg325 migrates just above the γ chain due to the absence of the Bβ1–42. (Reproduced from Ref. 8.)

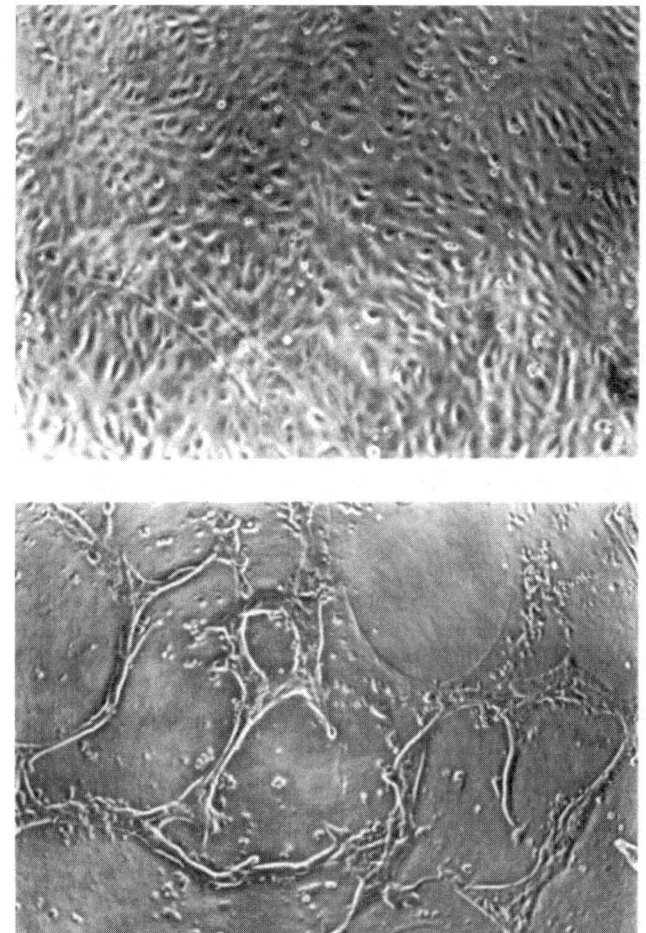

FIGURE 4. Effect of fibrin II on capillary tube formation. Interaction of overlying fibrin II gel with the apical surface of the cell monolayer optimally promotes capillary tube formation. HUVEC monolayers were established on fibrin II gels and overlaid with preformed fibrin-325 gels or were established on fibrin -325 gels and overlaid with preformed fibrin II gels. With fibrin-325 on top, the cells remained as a monolayer (**left**), and with fibrin II on top the cells formed capillary tubes (**right**). (Reproduced from Ref. 8.)

TABLE 1. Effects of RGDS or 7E3 and β15–42 on fibrin II-induced capillary tube formation

Treatment	Capillary tube parameters		
	Number	Length μm	Width μm
Control	26 ± 7	122 ± 24	25 ± 6
RGDS 150 μm	25 ± 7	117 ± 19	26 ± 7
RGDS 500 μm	23 ± 9	102 ± 21	22 ± 3
7E3 10 μg/ml	27 ± 7	124 ± 17	25 ± 6
Control	12.7 ± 4.3	101.0 ± 24.4	7.6 ± 1.6
β15–42	7 ± 1.4	66.7 ± 14.8	12.8 ± 2.4

capillary structures without a need for cell migration. Our studies demonstrate the ability of fibrin II to induce the formation of capillaries, and that this cellular response is not observed with fibrins I and 325 indicates that the interaction of fibrin II with the apical surface of EC induces the formation of capillaries. Moreover, these results also indicate that the exposure of segment β15–42 in fibrin II is responsible for this biological effect.[8]

Search for a β15–42 Receptor in Endothelial Cells

The studies outlined above demonstrate that thrombin mediates the exposure of new biochemical determinants in the N-terminal region of fibrin, as opposed to fibrinogen, that interact with the apical surface of endothelial cells, an association that induces the differentiation of the endothelial cells monolayer into capillary structures. The terminal fragment β15–42 of fibrin appears to possess most of these biological properties.

The fibrinogen molecule exhibits multiple associations with endothelial cells, for example, fibrin(ogen) associates with the integrin $\alpha_v\beta_3$ via the peptide RGD present in the C-terminal region of the Aα chain.[12] RGD peptide 572–574 of the Aα chain also recognizes the integrin $\alpha_5\beta_1$, present in endothelial cells.[13] In addition, fibrin(ogen) binds to other non-integrin cell adhesive surface receptors such as ICAM-1, via γ-chain 117–133,[14,15] and this interaction modulates the inflammatory response by attracting myelomonocytic differentiated cells to the vessel wall in areas of inflammation.[14–16] Fibrin(ogen) also binds with other cells besides EC, and these interactions may also promote the attachment to the vessel wall of cells with bound fibrinogen. The most extensively studied is the platelet integrin GPIIb/IIIa that, following activation, recognizes the C-terminal γ-chain dodecapeptide,[17–19] and this sequence appears to be specific for the interaction with GPIIb/IIIa that induces platelet aggregation. In contrast with EC integrin $\alpha_v\beta_3$, GPIIb/IIIa does not appear to ligate fibrinogen RGD peptide Aα 572–574.[20] Binding of fibrinogen to GPIIB/IIIa plays a major role in the formation of the platelet thrombus that seals the vessel wall during the hemostatic process. Under pathological conditions these platelet aggregates may cause occlusion of the vascular lumen causing organ damage. Other nucleated blood cells possess a specific family of integrins, formed by the β2

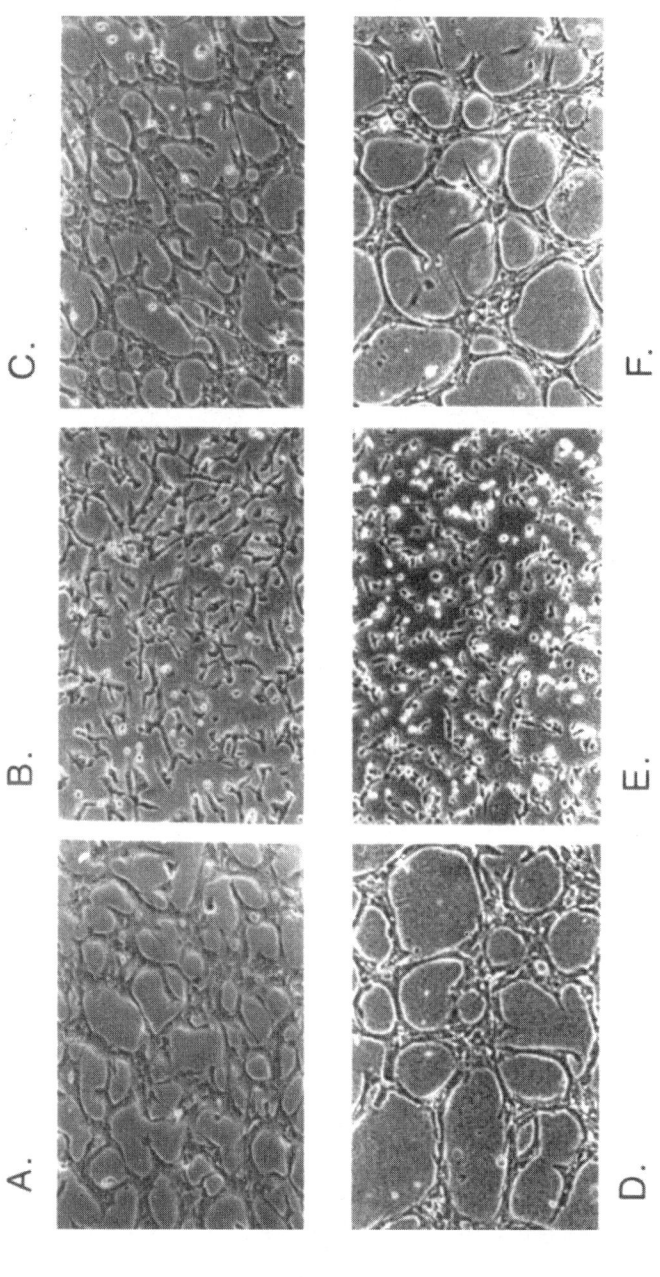

FIGURE 5. Effect of anticadherin antibodies on the capillary tube formation in fibrin II sandwiches. Confluent human umbilical vein endothelial cells sandwiched between fibrin II gels differentiate into capillary tube-like structures, assessed by phase contrast microscopy, at either 2 (**A**, **B**, and **C**) or 24 (**D**, **E**, and **F**) hours after formation of the overlaying fibrin gel. (**A** and **D**) control tubes in the absence of antibody; (**B** and **E**) in presence of VE-cadherin antibody; (**C** and **F**) tubes in the presence of an N-cadherin antibody. Original magnification 100 ×. (Reproduced from Ref. 24.)

family, which is composed of different α subunits linked to β2 and is the common subunit to this integrin family. A particular integrin of the β2 family, $\alpha_M\beta_2$ (Mac-1), ligates fibrinogen and this interaction is mediated through the cooperative effect of two regions of the γ chain, peptides γ190–202, and γ 377–395.[21,22] Activated myelomonocytic cells bind fibrinogen through Mac-1 integrin and the bound fibrinogen can form intercellular bridges by associating activated white cells with ICAM-1 of EC, and this mechanism accounts, at least in part, for the attachment and transendothelial migration of white cells.[15,16,21]

The finding that β15–42 is involved in fibrin binding with the apical portion of the EC suggested that a novel receptor was responsible for this interaction. Other investigators have demonstrated that immobilized β15–42 peptide associated with a surface EC protein of 130 KD, and this surface protein was eluted from the column with soluble β15–42 peptide.[23] The eluted EC protein appeared to be distinct from EC surface receptors of the integrin and Ig superfamilies.[23] Our studies also demonstrated that when EC are sandwiched between two fibrin or collagen gels they differentiate into capillary-like structures and this process was not inhibited by antibodies directed against integrin $\alpha_v\beta_3$, PECAM-1 or against the extracellular domain of N-cadherin.[24] However, an antibody that recognizes an epitope within domains 1–2 of VE-cadherin (Transduction Lab, Lexington, KY) prevented the formation of capillaries in double fibrin or collagen gel assays (see FIGURE 5).[24] Moreover, the VE-cadherin antibody did not cause a major morphological disruption of the cell monolayer, suggesting that the interaction of fibrin with a component of the EC and/or the homophilic association of VE-cadherin play a fundamental role in the morphogenetic changes seen in these *in vitro* assays.

Our experimental approach to identify the fibrin II EC receptor was based on binding studies testing the N-terminal disulfide knot fragments of fibrinogen (NDSK), fibrin II (NDSK II), and fibrin 325 (NDSK 325) for their ability to specifically interact with EC. Since fibrin is insoluble and this property precludes its use in binding studies, we elected to perform binding studies with the soluble NDSK fragments. This choice was based on previous studies indicating that the specific interaction of fibrin with EC was present in the N-terminal fragment of fibrin II that contains β15–42 in terminal position. The N-terminal fibrinogen fragment (NDSK) was isolated after cleavage of fibrinogen with cyanogen bromide followed by separation of the N-terminal fragment by column chromatography. The chain composition of these fragments can be seen in TABLE 2. Binding studies were done, after labeling NDSK with ^{125}I, by treating the labelled fragment with thrombin and then incubating ^{125}I-NDSK II with HUVEC. NDSK II bound to the cells was separated from the unbound by centrifugation.[25] Time course experiments showed that NDSK

TABLE 2. Generation and biochemical structure of NDSK derivatives

Starting material	Cleaving enzyme	Peptide released	Resulting material
NDSK	Thrombin	FPA + FPB	NDSK II
NDSK	Atroxin	FPA	NDSK I
NDSK	Protease III	Bβ1–42	NDSK 325
NDSK II	Protease III	FPA + Bβ1–42	Thrombin-cleaved NDSK 325

II binding to HUVEC, in suspension or in cell monolayer, reached maximum binding after ten minutes of incubation and the binding was cation independent (see FIGURE 6). Saturation isotherms demonstrated that the association of NDSK II with HUVEC reached a maximum at a concentration of 20 nM, and about 55% of the binding was specific. Ligand affinity was high, on the order of 7 nM Kd, with about 25,000 molecules of ligand bound per cell at saturation (see FIGURE 7).[25] In contrast with NDSK II neither NDSK or its derivative thrombin treated NDSK 325 bound specifically to EC (see FIGURE 8), indicating that the presence of β15–42 in a terminal position is indispensable for binding of NDSK II to EC. This finding was corroborated by studies demonstrating inhibition of binding of NDSK II to endothelial cells by peptide β15–42 crosslinked to albumin[25] (see FIGURE 9). To elucidate the component of the EC that interacts with NDSK II, we tested several antibodies to EC surface receptors for their ability to inhibit NDSK II binding. Two monoclonal antibodies against $\alpha_v\beta_3$ (7E3 and LM 609) and the peptide RGDS failed to interfere with NDSK II binding.[25] As shown in FIGURE 10, only an antibody that, according to the manufacturer, recognizes an epitope within amino acids 26 to 156 of VE-cadherin interfered with the EC association of NDSK II (FIG. 10). To further characterize the association of NDSK II with EC, a VE-cadherin antibody capture assay was developed. In this system, plastic dishes were coated with a pan-cadherin polyclonal antibody directed against the C-terminal region of this family of molecules, and EC lysates were incubated with the pan-cadherin coated dishes. After blocking with milk powder proteins, labeled NDSK II was added to the system and after incubation the bound radioactive NDSK II was counted (see FIGURE 11). Non-specific binding was

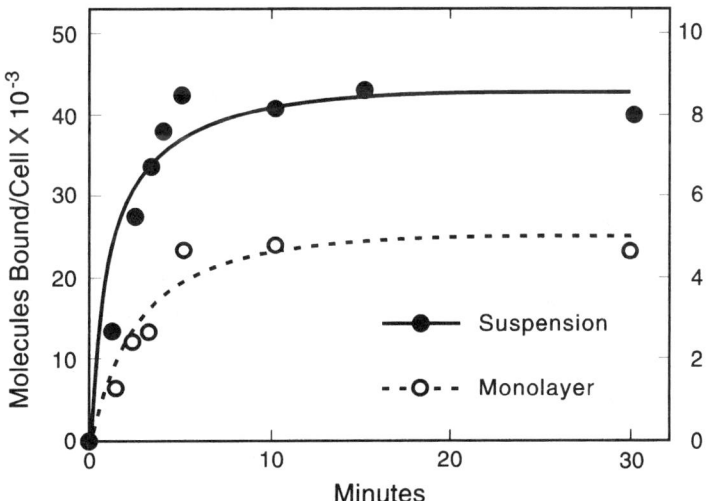

FIGURE 6. Time course of ^{125}I-NDSK II total binding to endothelial cells. Human umbilical vein endothelial cells (HUVEC) in suspension or in monolayer were incubated with 20 nM NDSK II. The *ordinate* to the *left* represents the scale of total ^{125}I-NDSK II binding to HUVEC suspensions (●); the ordinate to the right represents the scale of total ^{125}I-NDSK II binding to HUVEC in monolayer (○). (Reproduced from Ref. 25.)

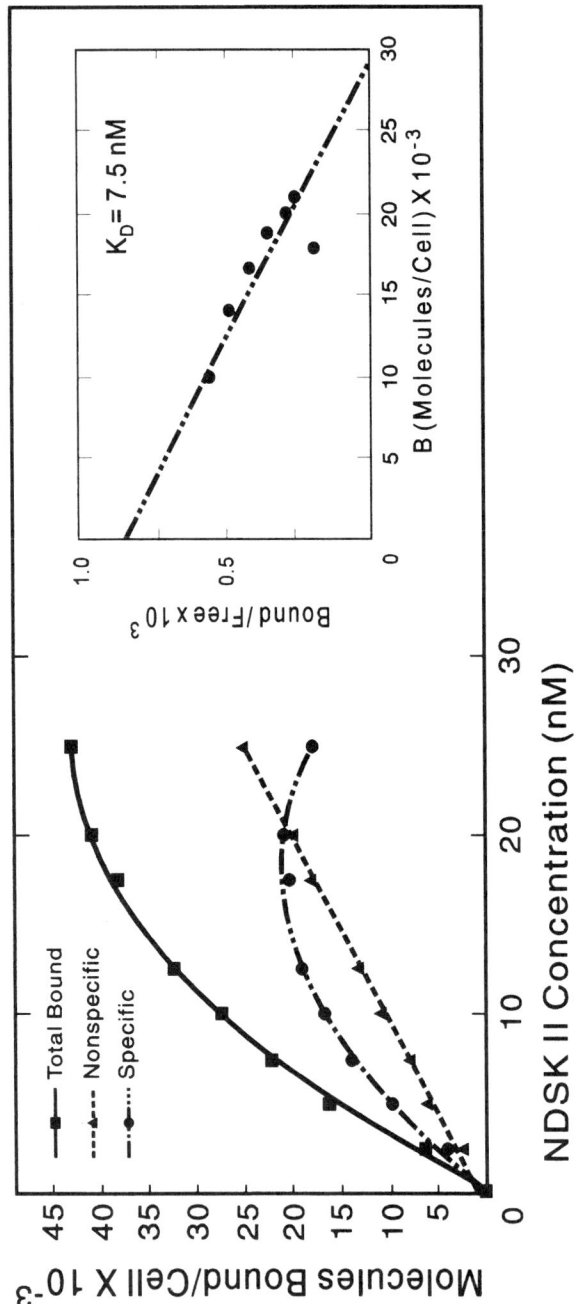

FIGURE 7. Saturation isotherm and Scatchard analysis of ^{125}I-NDSK II binding to HUVEC. Increasing concentrations of ^{125}I-NDSK II were incubated without or with addition of 100-fold molar excess unlabeled NDSK II. Specific values (●) were calculated by subtracting nonspecific values (▲), measured as the residual binding in the presence of 100-fold molar excess of unlabeled NDSK II at each concentration, from the total binding (■). **Inset.** Scatchard plot of NDSK II binding to HUVEC. Analysis of the results using a nonlinear curve fitting program, LIGAND, defined a one-site binding isotherm for NDSK II with a K_d of 7.5 nM and B_{max} = 29,000 molecules/cell. (Reproduced from Ref. 25.)

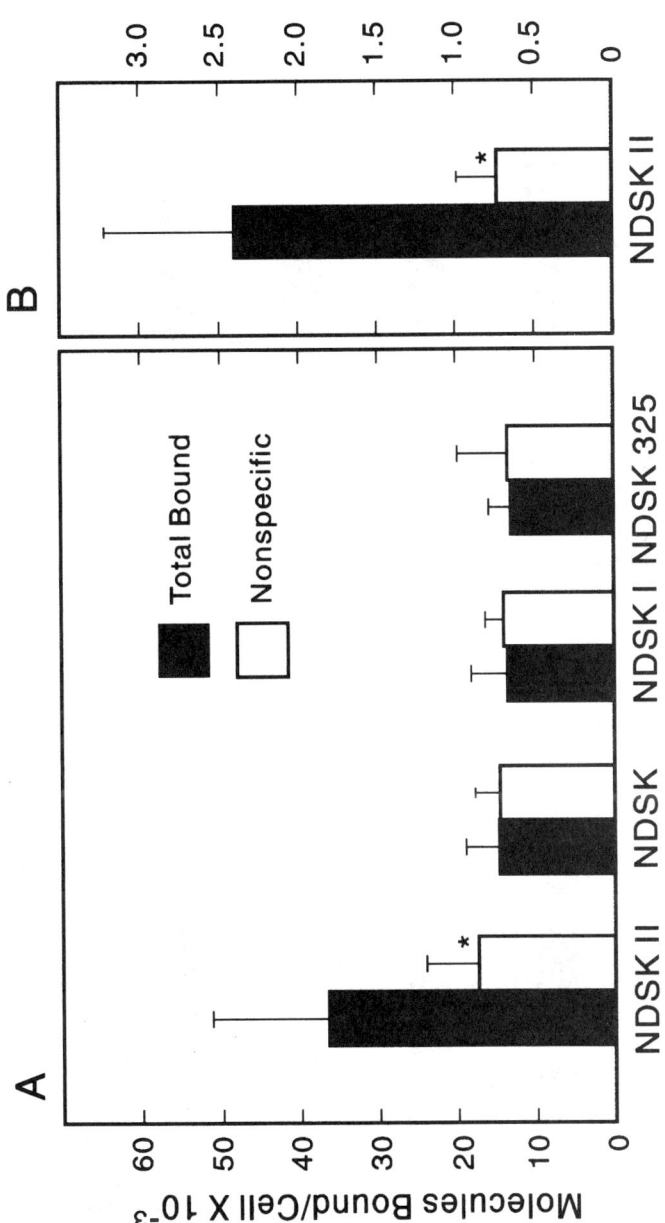

FIGURE 8. Specific binding of various ^{125}I-NDSK derivatives to HUVEC. Specific binding of each ^{125}I-NDSK derivative was determined by incubation with HUVEC in suspension (**A**) or in monolayer (**B**), with (*nonspecific*) or without (*total bound*) a prior 20-min incubation with a 100-fold molar excess of each respective unlabeled NDSK fragment. NDSKs were added at a concentration of 20 nM. NDSK II, thrombin treated; NDSK, untreated; NDSK I, Atroxin-treated; NDSK 325, protease III treated. Values for total and nonspecific binding represent the means ± S.D. of duplicate measurements from three to six separate binding experiments; * $p < 0.05$. (Reproduced from Ref. 25.)

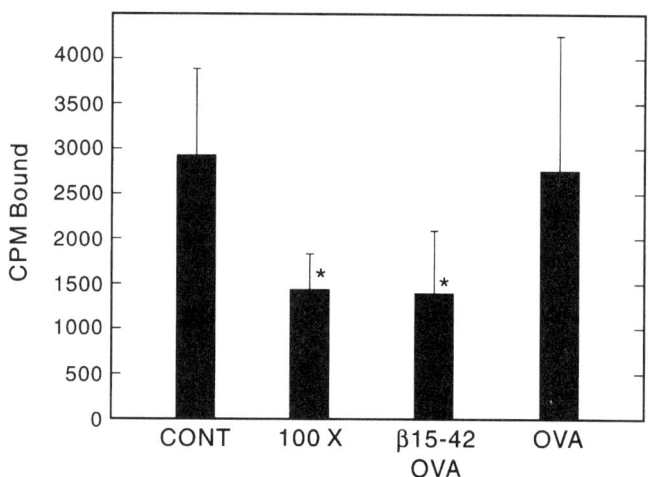

FIGURE 9. Inhibition of ^{125}I-NDSK II binding to HUVEC by β15–42-ovalbumin conjugate. HUVEC were preincubated with either unlabeled NDSK II, (2 μM) β15–42 peptide coupled to ovalbumin (25 micrograms/ml, or ovalbumin alone (25 μg/ml). ^{125}I-NDSK II (20 nM) added in the absence of inhibitor (*CONT*), unlabeled NDSK II (*100X*), β15–42 coupled to ovalbumin (β15–42-OVA), and ovalbumin alone (OVA). Means ± S.D. of duplicate measurements from four experiments. ∗ $p < 0.05$. (Reproduced from Ref. 25.)

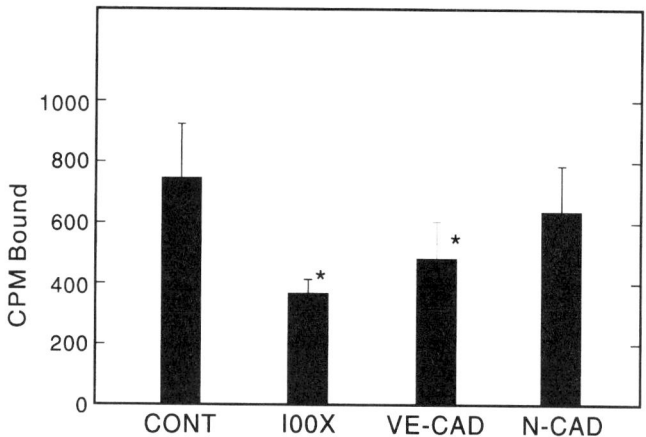

FIGURE 10. Effect of anticadherin monoclonal antibody on ^{125}I-NDSK II binding to HUVEC. HUVEC suspensions were preincubated with unlabeled NDSK II or monoclonal antibodies against VE-cadherin or N-cadherin prior to the addition of ^{125}I-NDSK II. ^{125}I-NDSK II in the absence of antibody (*CONT*), unlabeled NDSK II (*100X*), anti-VE-cadherin antibody (*VE-CAD*), anti-N-cadherin (*N-CAD*). Means ± S.D. of duplicate measurements from four experiments. ∗ $p < 0.05$. (Reproduced from Ref. 25.)

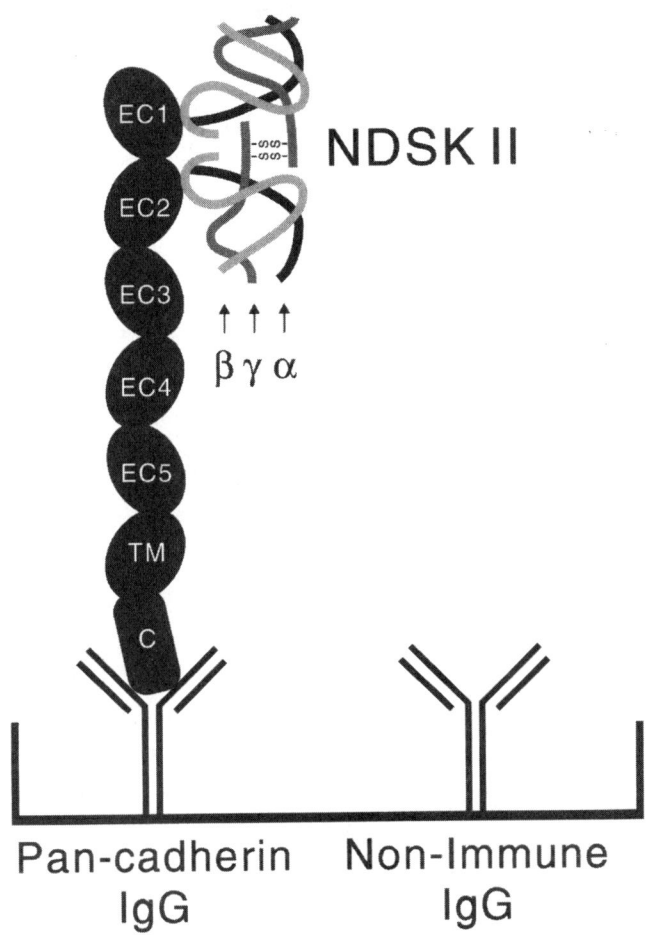

FIGURE 11. Schematic representation of modified antigen capture assay. Multiwell dishes were pretreated with a polyclonal rabbit anti-pan-cadherin antibody, known to interact with the intracellular domain of known cadherins. Lysates from HUVEC were then added to the antibody treated wells and allowed to interact with the anti-pan-cadherin antibody. Some wells were then treated with NDSK II unlabeled to measure nonspecific binding, followed by addition of ^{125}I-NDSK II. A non-immune polyclonal rabbit IgG was used to measure specificity of antibody–antigen complex.

measured by counting the radioactivity of NDSK II bound to VE-cadherin in presence of 100-fold unlabeled ligand. Analysis by Western blot of the captured component from the EC lysate revealed a 130-KD band, similar to that of EC lysates (see FIGURE 12 A, lanes 1 and 2), whereas no band was seen when the EC lysate was captures with a non-immune rabbit serum (FIG. 12 A, lane 3). Saturation of the captured VE-cadherin with labeled NDSK II, about 20 nM, was similar to the maximum binding obtained with EC (FIG. 12 B), indicating that the captured VE-cadherin has an

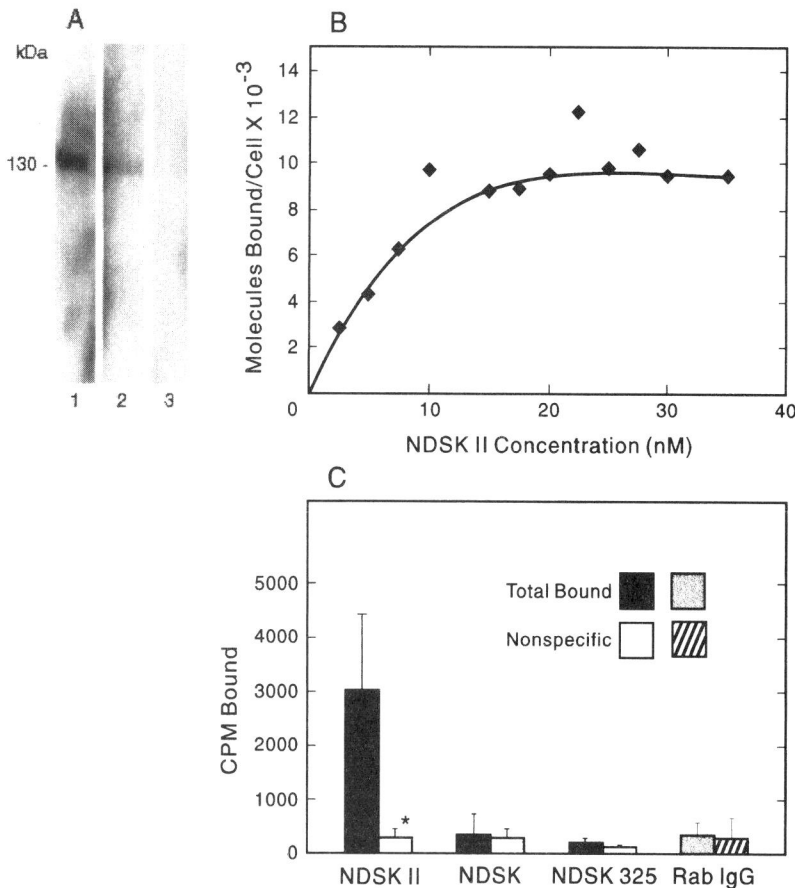

FIGURE 12. Saturation isotherm and specificity of ^{125}I-NDSK II binding to capture HUVEC cadherins. (**A**) Western blot of anti-pan-cadherin and non-immune IgG treated wells following addition of HUVEC lysates. *Lane 1,* control HUVEC lysate; *lane 2,* anti-pan cadherin IgG immunocaptured material; *lane 3,* non-immune rabbit IgG immunocaptured material. HUVEC lysates were applied to the anti-pan-cadherin coated wells, followed by the addition of increasing concentrations of ^{125}I-NDSK II (**B**). HUVEC lysates were incubated on antibody treated wells, followed by the addition of ^{125}I-NDSK II, NDSK, and NDSK 325 (**C**). To determinate nonspecific binding, preincubation with 100-fold molar excess of unlabeled NDSK II (non-specific) was performed prior to the addition of ^{125}I-NDSK II. Values represent the means ± S.D. of duplicate measurements from three–five experiments. Polyclonal pan-cadherin (*black bar, open bar*) or rabbit IgG (*dotted bar, hatched bar*). ∗ $p < 0.05$. (Reproduced from Ref. 25.)

affinity similar to that expressed by EC. Binding in this system was highly specific, with about 90% of NDSK II binding inhibited in presence of 100-fold excess cold ligand.[25] (FIG. 12 C). In contrast, NDSK and NDSK 325 binding were negligible (FIG. 12 C). Specificity of association of NDSK II with VE-cadherin was also indicated by studies showing that EC devoid of VE-cadherin, but expressing N-cadherin, did not bind NDSK II (not shown). These studies have identified a novel interaction of fibrin with VE-cadherin, a surface adhesive protein that is mainly localized at the intercellular junctions,[26,27] and the interaction fibrin-VE cadherin appear to be mediated by the N-terminal segment β15–42 of fibrin II. The association of the N-terminal region of fibrin, fragment NDSK II, with VE-cadherin exhibits unique properties; for example, cadherins participate mainly in homophilic associations, where cadherin dimers from one cell interact with dimers from the opposite cell causing cell-cell association in a reaction that is calcium dependent.[28–30] In contrast, the association of NDSK II with VE-cadherin is heterophilic and calcium independent.[25]

Other investigators have recently demonstrated that fibrinogen with cleaved fibrinopeptide B and exposed β15–42 binds to saphenous vein EC and this association upregulates the synthesis of ICAM-1.[31] The peptide β15–42 also interacts with EC, but the binding has low affinity, about 0.18 μM K_d.[31] (see FIGURE 13). Nevertheless, β15–42 peptide when added at high concentration was able to partially block the up regulation of ICAM-1 by fibrin or by an antibody to VE-cadherin. These results are also consistent with the presence of a specific receptor in endothelial cells that associate with exposed β15–42 of fibrin(ogen) and this receptor has been identified as VE-cadherin.

The studies presented above indicate that the interaction of the N-terminal domain of fibrin II with EC elicits a series of different cellular responses. EC attach and spread on fibrin II[32] and the exposure of EC to fibrin triggers the release of von Willebrand factor.[1] The association of fibrin II with the apical region of the EC induces the rearrangement of the EC into capillary structures, and these effects appear to be mediated by the interaction of β15–42 of fibrin II with VE-cadherin.[8,24,25] In addition, the binding of fibrin II with VE-cadherin also upregulates the synthesis of ICAM-1,[31] a proinflammatory adhesive protein that attracts white cells to the vessel wall.[14,15] A summary of the biological effects on EC by β15–42 of fibrin II is presented in TABLE 3. Thus, it seems that the N-terminal region of the β-chain, β15–42, induces several biological responses in the endothelium, but whether β15–42 peptide expresses full biological activity is unclear. It is possible that other sequences within the N-terminal domain of the β chain that also include all or some sequences of β15–42 may express higher biological activity. For example, peptide GHRP (β15–18) did not inhibit formation of capillaries,[8] suggesting that this sequence which is involved in fibrin polymerization,[9,33,34] does not influence the binding of β15–42 to VE-cadherin. Other studies also showed that peptide β19–26, a sequence that contains charged amino acids, was not able to elute the 130-kD protein bound to segment β15–42.[23] Nevertheless, several studies have shown that the functional activity expressed by the peptide β15–42 is rather weak, as demonstrated by the high concentrations (close to 1 mM) required to inhibit the functional responses of fibrin II.[8,23,31] It is likely that the monomeric form of β15–42 expresses low activity because a specific conformation of the dimer, as present in the N-terminal region of fibrin II, is necessary for its full functional activity. Thus, it is possible that a β-chain peptide that extends beyond Bβ1–42 and

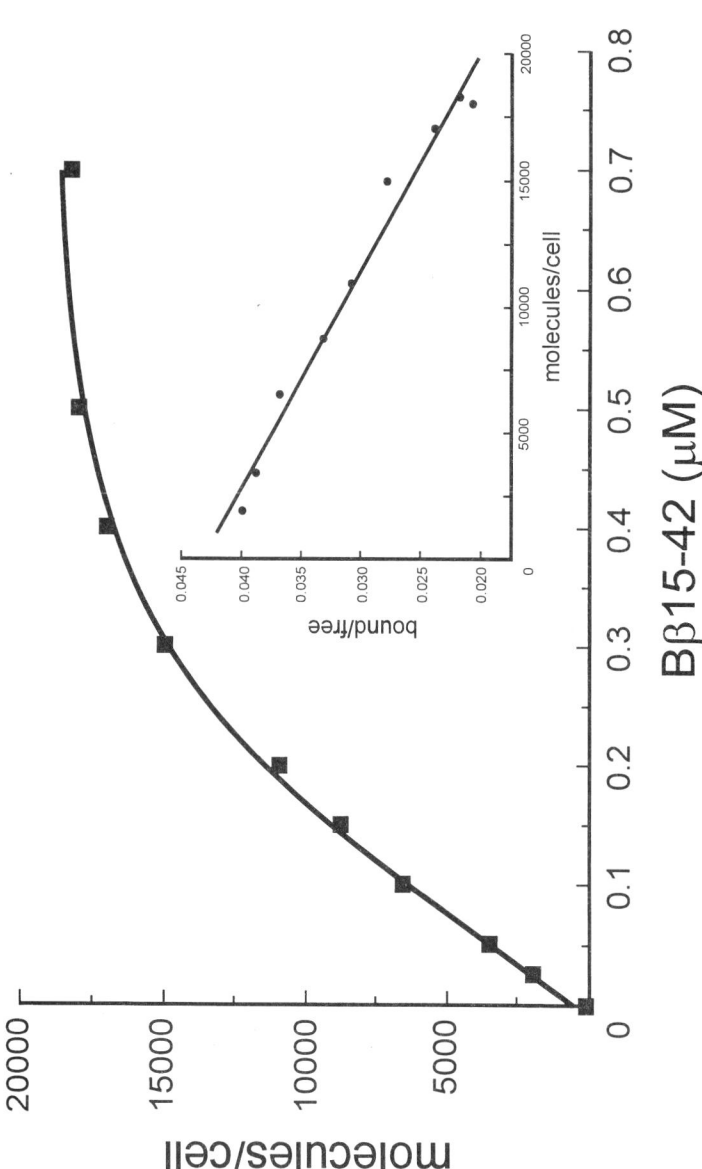

FIGURE 13. Binding of [^{125}I]Bβ15–42 to HUVECs. The specific binding of [^{125}I] Bβ15–42 to HUVECs is shown in the Scatchard plot **inset**. Binding studies were performed at 4°C for 2.5 hours, each well containing 18,000 to 20,000 endothelial cells. The data shown are the averages of triplicate measurements at each experimental point. The apparent dissociation constant (K_d) was 0.18 μM!. (Reproduced from Ref. 31.)

TABLE 3. Biological responses of endothelial cells to fibrin II mediated by β15–42

Release of von Willebrand factor[1]
Endothelial cell spreading and cytoskeletal reorganization[32]
Induction of capillary tube formation, via binding to VE-cadherin[8,24]
Upregulating of ICAM-1[31]

includes Cys Bβ65, one that is known to play an important role in fibrinogen dimerization by interacting with Aα Cys 36,[35,36] may be necessary for the proper conformation and biological activity of the N-terminal domain of the β chain. The high affinity binding of NDSKII to endothelial cells when compared to that of β15–42[25] is also consistent with this hypothesis. The finding that crosslinking of β15–42 to other proteins, such as ovalbumin, seems to increase its functional properties[23,25] and suggests that this modification stabilizes its conformation.

Identification of VE-cadherin Binding Site of Fibrin II

VE-cadherin, similar to other cadherins, is composed of three main regions, an extracellular domain of 546 amino acids, followed by a single transmembrane domain terminating with an intracellular domain of 144 amino acids.[26] The intracellular domain serves for the interaction with catenins that participates in cell–cell adhesion and in signal transduction. The extracellular domain is involved in homophilic and heterophilic associations. Distinguishing between sites involved in homophilic versus heterophilic associations (i.e., the fibrin binding site) may lead to a new understanding of important biological processes, such as wound healing and tumor growth. Preliminary data performed with CHO cells transfected with wildtype VE-cadherin showed specific interaction with fibrin and when these cells grow to confluency they form intercellular adhesions via VE-cadherin similar to those present in endothelial cells. Deletion and mutation studies involving the extracellular domain should identify the regions of VE-cadherin that are involved in homophilic cell–cell interactions and possibly distinguish this domain from a distinct region that may mediate the binding of fibrin II. Identification of these regions of VE-cadherin may lead to the development of new agents that can intervene in certain pathological processes such as neovascularization during tumor growth and metastasis.

ACKNOWLEDGMENTS

We thank A.S. Likens for the graphics and R.M. Silvano for preparation of the manuscript. All figures are reproduced with permission from J. Cell. Biol., Exp. Cell Res., J. Biol. Chem., or Arterioscler. Thromb. Vasc. Biol.

REFERENCES

1. RIBES, J.A., C.W. FRANCIS & D.D. WAGNER. 1987. Fibrin induces release of von Willebrand factor from endothelial cells. J. Clin. Invest. **79:** 117–123.

2. MONTESANO, R., M.S. PEPPER, J.-D. VASSALLI & L. ORCI. 1987. Phorbol ester induces cultured endothelial cells to invade a fibrin matrix in the presence of fibrinolytic inhibitors. J. Cell. Physiol. **132:** 509–516.
3. NICOSIA, R.F. & A. OTTINETTI. 1990. Modulation of microvascular growth and morphogenesis by reconstituted basement membrane gel in three-dimensional cultures of rat aorta; a comparative study of angiogenesis in matrigel, collagen, fibrin, and plasma clot. In Vitro Cell. Dev. Biol. **26:** 119–128.
4. THOMPSON, W.D., S.J. MCNALLY, N. GANESALINGAM, et al. 1996. Wound healing, fibrin and angiogenesis. In Molecular, Cellular, and Clinical Aspects of Angiogenesis. M.E. Maragoudakis, Ed.: 161–172. Plenum Press, New York.
5. NEHLS, V. & R. HERRMANN. 1996. The configuration of fibrin clots determines capillary morphogenesis and endothelial cell migration. Microvasc. Res. **51:** 347–364.
6. DVORAK, H.F. 1986. Tumors: wounds that do not heal. Similarities between tumor stroma generation and wound healing. N. Engl. J. Med. **325:** 1650–1659.
7. GAMBLE, J.R., L.J. MATTHIAS, G. MEYER, et al. 1993. Regulation of in vitro capillary tube formation by anti-integrin antibodies. J. Cell. Biol. **121:** 931–943.
8. CHALUPOWICZ, D.G., Z.A. CHOWDHURY, T.L. BACH, et al. 1995. Fibrin II induces endothelial cell capillary tube formation. J. Cell Biol. **130:** 207–215.
9. PANDYA, B.V., J.L. GABRIEL, J. O'BRIEN & A.Z. BUDZYNSKI. 1991. Polymerization site in the β chain of fibrin: mapping of the Bβ1–55 sequence. Biochemistry **30:** 162–168.
10. STRÖMBLAD, S. & D.A. CHERESH. 1996. Cell adhesion and angiogenesis. Trends Cell Biol. **6:** 462–467.
11. BROOKS, P.C., R.A.F. CLARK & D.A. CHERESH. 1994. Requirement of vascular integrin $\alpha_v\beta_3$ for angiogenesis. Science **264:** 569–571.
12. CHERESH, D.A., S. A. BERLINER, V. VICENTE & Z.M. RUGGERI. 1989. Recognition of distinct adhesive sites on fibrinogen by related integrins on platelets and endothelial cells. Cell **58:** 945–953.
13. SUEHIRO, K., J. GALIT & E.F. PLOW. 1997. Fibrinogen is a ligand for integrin $\alpha_5\beta_1$ on endothelial cells. J. Biol. Chem. **272:** 5360–5366.
14. ALTIERI, D.C., A. DUPERRAY, J. PLESCIA, et al. 1995. Structural recognition of a novel fibrinogen γ chain sequence (117-133) by intercellular adhesion molecule-1 mediates leukocyte-endothelium interaction. J. Biol. Chem. **270:** 696–699.
15. LANGUINO, L.R., A. DUPERRAY, K.J. JOGANIC, et al. 1995. Regulation of leukocyte-endothelium interaction and leukocyte transendothelial migration by intercellular adhesion molecule 1-fibrinogen recognition. Proc. Natl. Acad. Sci. U.S.A. **92:** 1505–1509.
16. COLLER, B.S. 1999. Binding of Abciximab to $\alpha_v\beta_3$ and activated $\alpha_M\beta_2$ receptors: with a review of platelet–leukocyte interactions. Thromb. Hæmost. **82:** 326–336.
17. PHILLIPS, D.R., CHARO, I.F., L.V. PARISE & L.A. FITZGERALD. 1988. The platelet membrane glycoprotein IIb-IIIa complex. Blood **71:** 831–843.
18. BENNETT, J.S. & G. VILAIRE. 1979. Exposure of platelet fibrinogen receptors by ADP and epinephrine. J. Clin. Invest. **64:** 1393–1401.
19. SHATTIL, S.J. 1998. Integrin signaling: The platelet function. 1998. Blood **91:** 2645–2657.
20. FARRELL, D.H. & R. THIAGARAJAN. 1994. Binding of recombinant fibrinogen mutants to platelets. J. Biol. Chem. **269:** 226–231.
21. ALTIERI, D.C., J. PLESCIA & E. F. PLOW. 1993. The structural motif glycine 190-valine 202 of the fibrinogen γ chain interacts with CD 11b/CD 18 integrin ($\alpha_M\beta_2$, Mac-1) and promotes leukocyte adhesion. J. Biol. Chem. **268:** 1847–1853.
22. UGAROVA, T.P., D.A. SOLOVJOV, L. ZHANG, et al. 1998. Identification of a novel recognition sequence for integrin $\alpha_M\beta_2$ within the γ-chain of fibrinogen. J. Biol. Chem. **273:** 22519–22527.
23. ERBAN, J.K. & D.D. WAGNER. 1992. A 130-kDa protein on endothelial cells binds to amino acids 15-42 of the Bβ chain of fibrinogen. J. Biol. Chem. **267:** 2451–2458.
24. BACH, T.L., C. BARSIGIAN, D.G. CHALUPOWICZ, et al. 1998. VE-cadherin mediates capillary tube formation in fibrin and collagen gels. Exp. Cell Res. **238:** 324–334.

25. BACH, T.L., C. BARSIGIAN, C.H. YAEN & J. MARTINEZ. 1998. Endothelial cell VE-cadherin functions as a receptor for the β15–42 sequence of fibrin. J. Biol. Chem. **272:** 30719–30728.
26. DEJANA, E. 1996. Perspectives Series. Cell adhesion in vascular biology: endothelial adherens junctions: implications in the control of vascular permeability and angiogenesis. J. Clin. Invest. **98:** 1949–1952.
27. NAVARRO, P., L. RUCO & E. DEJANA. 1998. Differential localization of VE- and N-cadherins in human endothelial cells: VE-cadherin competes with N-cadherin for junctional localization. J. Cell Biol. **140:** 1475–1484.
28. TAKEICHI, M. 1991. Cadherin cell adhesion receptors as a morphogenetic regulator. Science **251:** 1451–1455.
29. GUMBINER, B.M. 1996. Cell adhesion: the molecular basis of tissue architecture and morphogenesis. Cell **84:** 345–357.
30. GUMBINER, B.M. 2000. Regulation of cadherin adhesive activity. J. Cell Biol. **148:** 399–403.
31. HARLEY, S.L., J. STURGE & J.T. POWELL. 2000. Regulation by fibrinogen and its products of intercellular adhesioin molecule-1 expression in human saphenous vein endothelial cells. Arterioscler. Thromb. Vasc. Biol. **20:** 652–658.
32. BUNCE, L.A., L.A. SPORN & C.W. FRANCIS. 1992. Endothelial cell spreading on fibrin requires fibrinopeptide B cleavage and amino acid residues 15–42 of the β chain. J. Clin. Invest. **89:** 842–850.
33. LAUDANO, A.P. & R.F. DOOLITTLE. 1978. Synthetic peptide derivatives that bind to fibrinogen and prevent the polymerization of fibrin monomers. Proc. Natl. Acad. Sci. U.S.A. **75:** 3085–3089.
34. SIEBENLIST, K.R., J.P. DIORIO, A.Z. BUDZYNSKI & M.W. MOSESSON. 1990. The polymerization and thrombin-binding properties of des-(Bβ1–42)-fibrin. J. Biol. Chem. **265:** 18650–18655.
35. ZHANG, J-Z. & C.M. REDMAN. 1994. Role of interchain disulfide bonds on the assembly and secretion of human fibrinogen. J. Biol. Chem. **269:** 652–658.
36. HUANG, S., Z. CAO & E.W. DAVIE. 1993. The role of amino-terminal disulfide bonds in the structure and assembly of human fibrinogen. Biochem. Biophys. Res. Commun. **190:** 488–495.

Tumors and Fibrinogen

The Role of Fibrinogen as an Extracellular Matrix Protein

P.J. SIMPSON-HAIDARIS[a,b,c] AND BRIAN RYBARCZYK[a,b]

Departments of [a]Medicine, [b]Pathology, and [c]Microbiology & Immunology, University of Rochester School of Medicine and Dentistry, Rochester, New York, USA

ABSTRACT: The progression of a tumor from benign and localized to invasive and metastatic growth is the major cause of poor clinical outcome in cancer patients. Much like in a healing wound, the deposition of fibrin(ogen), along with other adhesive glycoproteins, into the extracellular matrix (ECM) serves as a scaffold to support binding of growth factors and to promote the cellular responses of adhesion, proliferation, and migration during angiogenesis and tumor cell growth. Inappropriate synthesis and deposition of ECM constituents is linked to altered regulation of cell proliferation, leading to tumor cell growth and malignant transformation. Fibrin deposition occurs within the stroma of a majority of tumor types. In contrast, abundant FBG, not fibrin, is present within the stroma of breast cancers. It is thought to originate from exudation of plasma FBG and subsequent deposition into the tumor stroma and not endogenous synthesis and secretion of FBG by breast tumor cells. However, we show that MCF-7 human breast cancer cells synthesize and secrete FBG polypeptides, suggesting that the origin of FBG in the stroma of breast carcinoma may be due to endogenous synthesis and deposition. Moreover, FBG assembles into ECM as conformationally altered FBG, not as fibrin. Studies in our laboratory demonstrate that FBG alters the ability of breast cancer cells to migrate. Together, the results of studies from our laboratory, as well as the laboratories of others, indicate that the presence of fibrin(ogen) within the tumor stroma likely affects the progression of tumor cell growth and metastasis. This review focuses on FBG within tumors and its relationship with other tumor constituents, ultimately focusing on the role of FBG in breast cancer.

KEYWORDS: Tumors; Fibrinogen; Extracellular matrix protein; Breast cancer; Angiogensis; Fibrin; Cell migration.

FIBRIN(OGEN) AND CANCER

The association between coagulation factors and malignancy was recognized more than a century ago.[1] More recently, Costantini and Zacharski[2] reviewed the evidence addressing the significance of blood coagulation activation in fibrin deposition in tumor tissues. The techniques of immunofluorescence, immunohistochemical staining, and immunoelectron microscopy have been employed to study fibrin(ogen) within tumors and its relationship with other tumor constituents.

Address for correspondence: P.J. Simpson-Haidaris, University of Rochester School of Medicine and Dentistry, Rochester, NY 14620, USA. Voice: 716-275-8267; fax: 716-473-4314.
pj_simpsonhaidaris@urmc.rochester.edu

Monoclonal antibodies (MoAb) against fibrin(ogen) have been powerful tools in the analysis of tumor composition. Three such MoAb used are 18C6, which recognizes the N-terminal β_{1-21} sequence of fibrinogen (FBG),[3] T2G1, which recognizes the conformational epitope β_{15-21} once FBG is converted to fibrin by thrombin cleavage,[4] and GC4, which recognizes plasmin D fragment or D-dimer from crosslinked fibrin but not intact fibrin or FBG.[5] Because of recent findings in our laboratory showing that the fibrin specific epitope β_{15-21} is exposed once FBG is incorporated into ECM[6] (see EXTRAHEPATIC FBG PRODUCTION), caution must be taken when interpreting immunohistochemical staining for fibrin(ogen) within tumor stroma. Nonetheless, the results of studies using immunohistochemical staining with MoAb 18C6, T2G1, and GC4, together with antibodies against coagulation factors and fibrinolytic components, have revealed a heterogeneous pattern of fibrin(ogen) deposition in various tumor types.[2]

Both FBG and fibrin have been localized to the tumor–host cell interface.[7,8] Fibrin is abundant in different types of tumors[9–13] such as primary brain lesions[14] and prostate cancer.[15] Tumor cell associated fibrin deposition is also found in small cell carcinoma of the lung, renal carcinoma and malignant melanoma in association with activated coagulation factors such as Factor XIIIa or Xa, indicative of thrombin generation at the primary tumor site. Positive fibrin staining is frequently found in focal sites at the interface of the tumor cells and surrounding stroma, but not in the ECM of normal host cells. In addition, abundant deposition of FBG, that is, 18C6 positive material, is found predominantly within the extracellular tumor stroma. In contrast, using immunohistochemical staining with MoAbs 18C6 and T2G1, Costantini et al.[7] reported that abundant FBG, not fibrin, is present within the tissue stroma in breast cancer, although fibrin deposition has been localized to the tumor-normal tissue interface.[16] In addition, FBG, not fibrin, deposition is a feature of mesothelioma,[17] colon cancer,[18] and lymphoma.[19] Fibrin surrounding tumors may protect them from infiltrating inflammatory cells by acting as a barrier, thus preventing inflammatory reactions directed towards the tumor cells.[13,16]

EXTRACELLULAR MATRIX AND CANCER

Many biological activities require specific interactions with ECM proteins including development, morphogenesis, growth, and tumor progression. Thus, there exists a dynamic relationship between cells and the ECM they produce.[20] ECM is composed of a number of different structures and molecules; the most common are fibronectin (FN), FBG, laminin, collagen, vitronectin, and thrombospondin. Although these proteins differ in primary structure, they retain similar functional motifs that contribute to their adhesive properties for both cells and other proteins, and the ability to organize into fibrillar structures. Matrix provides structure and elasticity for tissues, compartmentalization of different cell types, and growth factor sequestration.[21,22] The ECM not only influences growth factor synthesis and degradation, but also sequesters these growth factors to limit their diffusion and degradation.[21] For example, FBG can bind growth factors such as fibroblast growth factor (FGF)-2[23] and vascular endothelial cell growth factor (VEGF).[24] Thus, the ECM controls cell proliferation by controlling the availability of growth factors and cytokines in tumor stroma, which in turn is dependent on the ability of a cell to degrade this matrix to release the sequestered growth factors.

Loss of cell surface FN is associated with oncogenic transformation and is correlated with an increase in tumorigenicity, particularly with respect to metastatic potential.[25,26] Reasons for loss of cell surface associated FN include reduced rates of FN biosynthesis, increased turnover, and decreased ability to bind FN. The loss of FN involves alterations in cell adhesion, which contributes to its role in invasion and metastasis. Immunostaining of breast cancer tissue sections for FN is a potential prognostic marker. Positive FN is associated with low metastatic potential, whereas a lack of FN at the invasive edge suggests a more invasive tumor cell phenotype.[27,28]

FBG resembles FN in that FBG also contains many important functional binding regions, including heparin binding domains (HBD) on the N-terminal Bβ chain,[29,30] platelet recognition sites ($\gamma_{400-411}$),[31] and ICAM-1 binding sites ($\gamma_{117-133}$).[32] Other important sites on FBG include the arginine-glycine-aspartate (RGD) sites at Aα_{95-98} and A$\alpha_{572-575}$ that are necessary for integrin ligation.[33] RGD sites on fibrin(ogen) promote adhesion of endothelial cells to provisional matrix through specific integrins such as $\alpha_v\beta_3$;[34] engagement of $\alpha_v\beta_3$ supports angiogenesis that is critical for tumor growth and dissemination. The FBG RGD sites also bind to the classic FN receptor, $\alpha_5\beta_1$.[35] FBG fragments function as chemotactic factors for inflammatory cell infiltration;[36] the Bβ_{1-42} fragment generated by plasmin cleavage serves as a neutrophil chemoattractant.[37] In addition, FBG is capable of binding inflammatory cells through the CD11/CD18 family of leukocyte integrins.[38] During hemostasis, FN becomes crosslinked into the fibrin clot. A close association of FN and FBG is maintained in the ECM as well. We have shown that FBG deposition in the ECM is dependent on deposition of a FN matrix (discussed below), suggesting a link between the levels of FN and FBG in the ECM during tumorigenesis.

EXTRAHEPATIC FBG PRODUCTION

Recent evidence suggests that FBG is expressed and synthesized by cell types other than hepatocytes. FBG is synthesized and secreted from alveolar epithelium of *Pneumocystis carinii* infected lungs (see FIGURE 1A)[39] and from established extrahepatic epithelial cell lines including human uterine cervical carcinoma cells,[40] intestinal carcinoma cells,[41] and alveolar adenocarcinoma cells (FIG. 1B and C).[42] These extrahepatic epithelial cells synthesize FBG composed of Aα, Bβ, and γ chain polypeptides with the characteristic M_r of 340 kDa. Hepatocytes secrete mainly intact FBG; although, occasionally, very small amounts of intermediate chain complexes, unassembled fragments, or individual chains are observed.[43] However, bovine granulosa cells appear to synthesize and secrete only β and γ chains of FBG.[44] Furthermore, studies of COS cells transfected with FBG cDNAs show that FBG assembly is not required for its secretion and that intermediates such as α–γ complexes are secreted along with fully assembled FBG.[45]

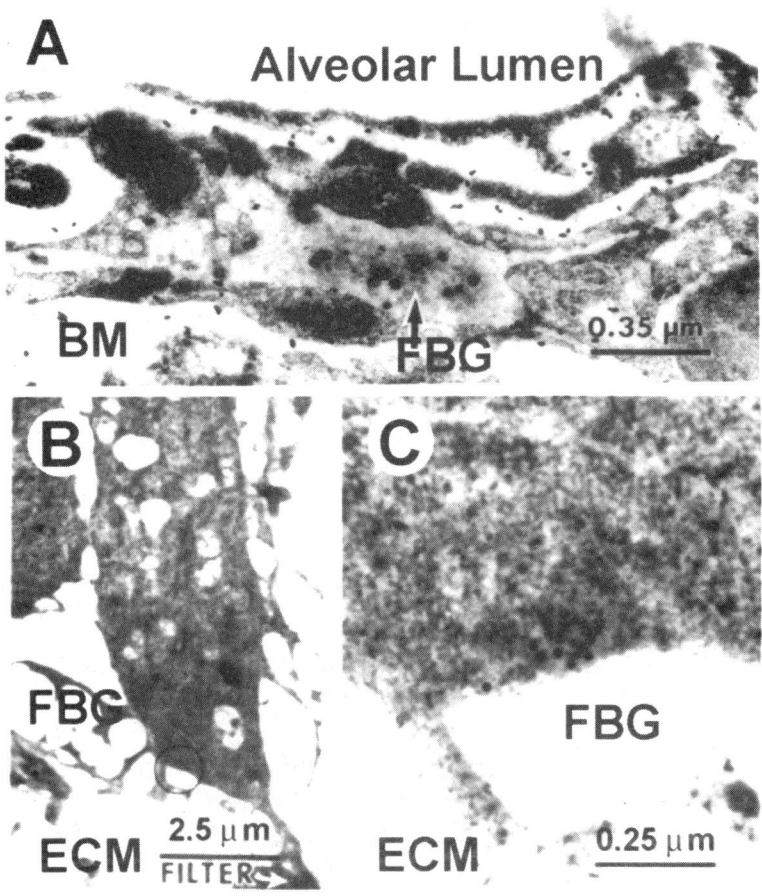

FIGURE 1. Fibrinogen produced by lung epithelial cells is secreted basolaterally. Immunoelectron microscopy was performed on lung sections from *Pneumocystis carinii* infected mice (**A**) or the human lung adenocarcinoma A549 cell line cultured on Transwell-Col™ filters (coated with collagen type I) and induced with interleukin-6 and dexamethasone (**B** and **C**). Primary monospecific antiserum was raised in rabbits against highly purified mouse FBG for use in panel **A**. Both the antimouse FBG and antihuman FBG were purified over homologous species FN-Sepharose affinity columns to remove contaminating antibodies to FN; the IgG fractions were affinity purified. FBG-specific staining was detected using Protein A-gold (20 nm) and the grids were counter stained with aqueous uranyl acetate and lead citrate. Positive staining for FBG was found intracellular in secretory vesicles at or fusing with the basolateral plasma membrane in lung alveolar epithelial cells in infected lung sections (**A**, *arrow*) and cultured A549 cells (**B** and **C**). Panel **C** is an enlargement of the *circled area* in panel **B**. BM, basement membrane; FBG, fibrinogen, ECM; extracellular matrix; filter, a portion of the Transwell-Col™ filters which represents the basolateral face of the A549 cell. (Panel **A** was reprinted with permission from Ref. 39 and panels **B** and **C** reprinted with permission from Ref. 48.)

FIBRONECTIN AND FIBRINOGEN ASSEMBLY INTO EXTRACELLULAR MATRIX

FN matrix assembly is a cell-mediated process that begins with protomeric FN binding to the cell surface mediated through the N-terminal 70-kDa region, inducing a conformational change that enhances FN polymerization with other FN protomers to become an insoluble matrix.[46,47] FN is known to interact with collagen (gelatin) and fibrin.[26] We have shown that FBG colocalizes with FN in ECM.[48] Using a fibrin specific antibody, T2G1,[4] indirect immunofluorescent staining of FBG incorporated into fibroblast matrix shows positive staining for the fibrin specific epitope, $B\beta_{15-21}$ (see FIGURE 2). Further analysis of matrix FBG showed that FBG assembles into matrix as an intact molecule, not as monomeric or polymeric fibrin, suggesting that cryptic "fibrin" specific epitopes are exposed upon FBG incorporation into matrix.[6] We have also shown that the incorporation of FBG into matrix is dependent upon metabolically active cells (Pereira and Simpson-Haidaris, this volume) and heparan sulfate proteoglycans. Current studies also suggest that assembly of a FN matrix is required for deposition of FBG into fibroblast ECM (Pereira et al., manuscript submitted).

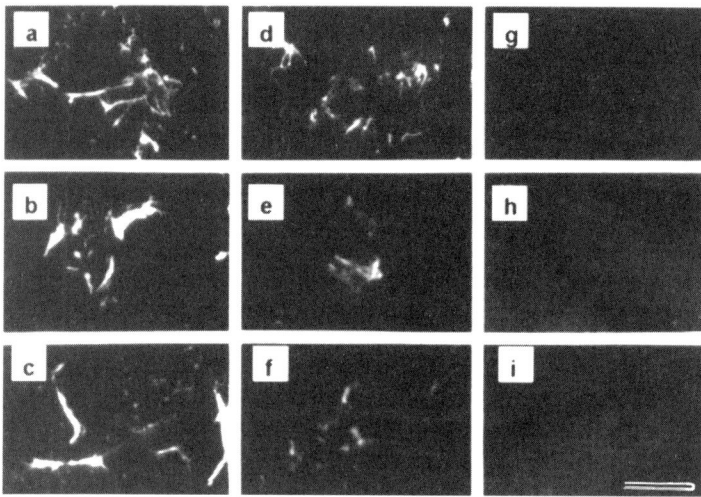

FIGURE 2. Exposure of FBG and fibrin specific epitopes on FBG deposited in the ECM of fibroblasts. Immunofluorescent staining was performed using MoAbs for FPA (RGD3, panels **a**, **d**, and **g**), FPB (18C6, panels **b**, **e**, and **h**) and fibrin β15–21 (T2G1, panels **c**, **f**, and **i**). HFF were overlaid with conditioned media from A549 cells (**a–c**), with complete media containing purified plasma FBG (**d–f**) or complete media with no added ligand (**g–i**). The *bar* (panel **i**) represents 10 μm. (Figure was reprinted with permission from Ref. 6.)

BREAST CANCER EPIDEMIOLOGY

The incidence of breast cancer is currently nearly one million cases worldwide. The American Cancer Society estimated that nearly 184,200 new cases of invasive breast cancer and approximately 41,200 deaths from this disease would occur in the United States in the year 2000 (www.cancer.org). Breast cancer is the second leading cause of cancer deaths among women in the United States. Women in North America and Northern Europe have the highest risk for breast cancer; women in Africa and Asia have the lowest risk.[49] Risks for breast cancer include age, obesity, smoking, alcohol consumption, and long term postmenopausal estrogen replacement therapy. The most pronounced risk factor is a family history of breast cancer.[50] The molecular basis of breast cancer is only beginning to be understood and has lead to the critical investigation of specific genes involved in promoting this disease. Mutations in the BRCA1 and BRCA2 genes are correlated with a 50–85% risk of breast cancer, ovarian cancer, or both.[51] All breast cancers have some type of somatic genetic alteration including abnormalities in genes such as p53, bcl-2, and c-myc. Genetic alterations combined with altered cell growth, influenced by steroid hormones, growth factors, and cytokines, promote unique pathologies during the progression of breast cancer.

NORMAL VERSUS TUMORIGENIC BREAST MORPHOLOGY

The origins of breast tumorigenesis are attributed to genetic alterations in the breast epithelial cell compartments. Neighboring fibroblasts displaying abnormal phenotypes produce an altered stroma that may influence the progression of tumorigenesis in the breast epithelia. On tumorigenic transformation, particularly in the breast, a stromal response termed desmoplasia occurs. Desmoplasia occurs due to quantitative changes in synthesis and degradation of ECM proteins. An imbalance between these two processes may result in breast tumor stroma containing increased amounts of collagen, FN, proteoglycans,[52] and, perhaps, FBG, as discussed in this paper. Thus, tumor stroma may be considered "wound healing gone awry" and tumors themselves may be considered "wounds that do not heal".[53] The source of some of these matrix proteins is thought to be from leaky blood vessels[10] or inappropriate synthesis by the tumor cells. Some consequences of inappropriate or modified synthesis of ECM molecules in the context of tumorigenesis include changes in proliferation, induction of angiogenesis, altered cell adhesion and cell migration, and ultimately tumor metastasis. Cancer implies aberrant cell growth. A combination of autocrine and paracrine growth factors and cytokines and ECM molecules leads to degregulation of cell cycle control, which results in increased cell proliferation and malignant transformation.[52]

Metastasis is the migration of tumor cells from a primary tumor to a secondary site. Metastasis is a multistep process involving multiple interactions of tumor cells with normal cells and interstitial tissue constituents. The process of metastasis involves a series of migratory and invasive events that allow cells to pass through connective tissues and vascular environments. Thus, cells must differentiate from their original adhesive phenotype, degrade matrix, migrate through vascular barriers, enter the circulatory system, evade host defense mechanisms, and invade distant organs.[54] The most common organs for breast cancer metastasis include brain, bone, and lung.[55] Morbidity is not usually due to the establishment of the primary tumor but rather results from metastatic disease.[56] The metastatic potential of a cell is dependent upon its ability to destroy basement membrane and become motile.[57] Not only does a breast tumor cell require adhesion receptors but it also requires enzymes specific for degradation of ECM proteins.[58] Once in the vasculature, the tumor cells migrate into distant organs and establish secondary sites of tumor growth, which alters the normal functioning of the organ. The major difference between normal cell functioning and one of a pathological nature is a difference in cell cycle regulation.[59]

ANGIOGENESIS

Neovascularization, the formation of new blood vessels, is essential for tumor growth beyond a certain size.[8,13] Angiogenesis depends on growth factors such as VEGF, matrix proteins, and cell surface receptors involved in cell migration and proliferation.[60–62] ECM components can promote the process of angiogenesis as well as inhibit it.[63] Endothelial cell migration and establishment of blood vessels depends on matrix proteins for cell attachment and detachment that further trigger downstream intracellular signaling events.[64] Endothelial cells contain multiple cell surface receptors important for angiogenesis. Most important of these cell surface receptors is the integrin $\alpha_v\beta_3$, which has specificity for vitronectin and fibrin(ogen).[34] An increase in $\alpha_v\beta_3$ integrin expression is also an important prognostic marker of breast cancer[65] and contributes to breast tumor growth due to its involvement in angiogenesis.[66] Martinez and colleagues[67,68] have shown that the β_{15-42} sequence of fibrin interacts with a component of the apical cell surface of endothelial cells and that this interaction plays a fundamental role in the induction of endothelial capillary tube formation. This component is likely to be VE-cadherin, which is localized at areas of cell–cell contact where it functions to maintain the tubular architecture.

INTEGRINS AND BREAST CANCER

The integrin superfamily is a class of cell surface, transmembrane receptors that mediate cell–matrix and cell–cell adhesion interactions. These interactions are mediated by heterodimeric combinations of α and β integrin subunits. Various combinations of α and β subunits give rise to differential specificities for ECM ligands.[69] Integrins are functionally important during development, morphogenesis, wound healing, angiogenesis, and oncogenesis.[70] Ligands for integrins include FN, FBG, collagen, vitronectin, laminin, and thrombospondin. Integrins act as a bridge between the outside of the cell and intracellular proteins. Most integrins interact with ECM molecules via RGD sequences on the matrix proteins. Thus, the binding of a ligand to its corresponding integrin leads to conformational changes in the integrin structure that transduces intracellular signals, resulting in changes in cellular behavior including proliferation, adhesion, and migration.[71] Recent studies have focused on patterns of integrin expression in different types of breast cancers. The $\alpha_2\beta_1$ (collagen/laminin receptor) is highly expressed on ductule epithelium of normal breast tissue and in benign fibroadenomas.[72] Moderate decreases of $\alpha_5\beta_1$ and $\alpha_v\beta_3$ integrin expression were observed in mammary tumors,[73–75] however, increased β_1 integrin expression was found to be associated with enhanced tumor aggressiveness, invasiveness, and metastatic potential correlating with decreased patient survival.[76] Furthermore, increased $\alpha_v\beta_3$ integrin levels are positively correlated with the ability of the cell to adhere and migrate, thus increasing metastatic potential.[77] Blockage of β_1 and β_4 integrins in a three-dimensional culture assay resulted in reversion of mammary epithelial cells from a malignant to a normal phenotype.[78] Together, integrins play an essential role in the regulation and progression of breast tumorigenesis.

EXTRACELLULAR MATRIX AND BREAST CANCER

The ECM is required for normal functional differentiation of mammary epithelia.[63] However, the mechanisms by which ECM constituents affect differentiation and mammary gland function are only beginning to be understood. The intricate variety of ECM-mediated events and the remarkable degree of plasticity of matrix structure and composition complicate these types of studies. Recent data suggest that the ECM and ECM-receptors might participate in the control of most, if not all, of the successive stages of breast tumors, from appearance to progression and metastasis (reviewed in Ref. 63). As a matrix protein, FBG could potentially influence cell behavior through FBG protein–protein and protein–cell interactions. The novel finding that FBG is a matrix protein can be extrapolated to the context of tumorigenesis. The studies below describe FBG production by breast cancer cells and the potential role of FBG on tumor cell migration.

FOCUS OF RESEARCH REVIEW

The basis for the research described in this review stemmed from studies in our laboratory concerning extrahepatic FBG production in a lung epithelial cell line, A549.[42] Since these cells synthesize FBG, we hypothesized that other extrahepatic cells may also synthesize FBG. In this review, we discuss data from studies that examined the mechanisms of assembly of FBG into ECM. Because FBG, not fibrin, is a major constituent in breast cancer,[7] we hypothesized that FBG could be synthesized and secreted from breast epithelial cells and that FBG, in general, affects the metastatic potential of breast cancer cells. Herein, we present a review of data showing FBG gene expression, protein production and cell surface and matrix assembly of endogenous FBG by the breast adenocarcinoma cell line, MCF-7. We also examine the role of FBG in cell migration of the more invasive breast carcinoma cell line, MDA-MB-231, and the involvement of integrins during this process. Together, these data analyze the importance and potential involvement of FBG during nonmalignant breast cell matrix deposition, adenocarcinoma breast cell synthesis of FBG, and FBG effects during cell migration of malignant breast cells.

FIBRINOGEN ASSEMBLY, SECRETION, AND DEPOSITION INTO ECM BY MCF-7 HUMAN BREAST CARCINOMA CELLS

A hallmark of breast carcinoma is the deposition of FBG without subsequent conversion to fibrin in the tumor stroma. Recently, we studied the ability of the MCF-7 human breast cancer epithelial cell line to synthesize, secrete, and deposit FBG into the ECM. MCF-7 cells produced low levels of intact FBG; however, abundant levels of FBG intermediate complexes or degraded Aα, Bβ, and γ chain polypeptides were observed (see FIGURE 3). Most of the Bβ chain was degraded to fragments of 46–48 kDa, the majority of which was missing the N-terminus. RT-PCR analysis indicated that only γ chain mRNA was present in detectable steady state levels, although Southern hybridization revealed that the FBG Aα, Bβ, and γ chain genes were intact in MCF-7 cells (not shown). Indirect immunofluorescence staining using PoAb anti-FBG and MoAbs for FBG A$\alpha_{566-580}$ (134B29), Bβ_{1-21} (18C6), and γ_{63-78} (J88B) chains confirmed positive extracellular staining for all three FBG chains (see FIGURE 4 A–D). These results suggest that MCF-7 cell FBG binds to the cell surface in a punctate pattern, reminiscent of receptor binding.

Because plasma FBG assembles into the ECM of heterologous cells such as lung epithelial cells and fibroblasts,[6] we wanted to determine whether exogenous plasma FBG incorporates into breast epithelial cell ECM. Medium containing Oregon

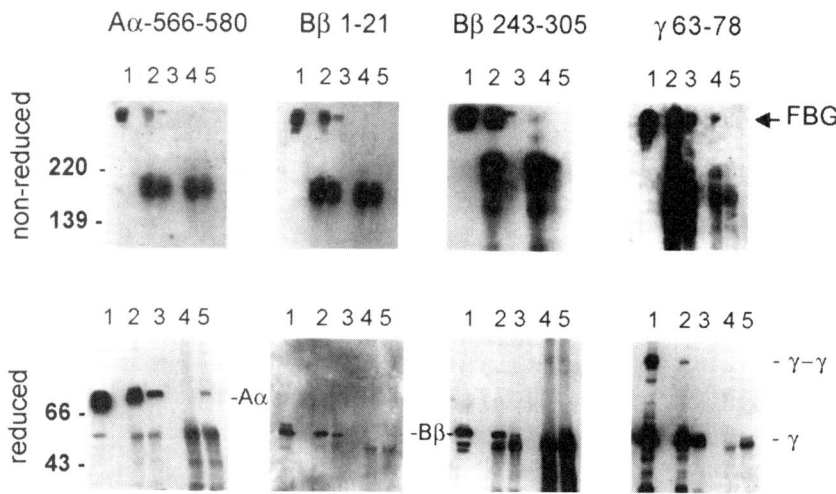

FIGURE 3. Western blot analysis of MCF-7 FBG. HepG2 and MCF-7 cells were two to three days past confluency. Accumulated FBG was immunoprecipitated and analyzed by SDS-PAGE under non-reducing (**A**) and reducing (**B**) conditions, followed by Western blot with FBG chain specific MoAb: MoAb134B29, anti-FBG A$\alpha_{566-580}$ (a gift from Dr. Z. Ruggeri); MoAb 18C6, anti-FBG Bβ_{1-21}; MoAb D73H, anti-FBG B$\beta_{243-305}$; MoAb J88B, anti-FBG γ_{63-78}. The MoAb name and its chain specificity are denoted above the panels. The lanes on each gel correspond to: (1) purified plasma FBG, (2) HepG2 medium, (3) HepG2 cell lysate, (4) MCF-7 medium, and (5) MCF-7 cell lysate. *Arrow* indicates intact FBG at 340 kDa. (Figure reprinted with permission from Ref. 79.)

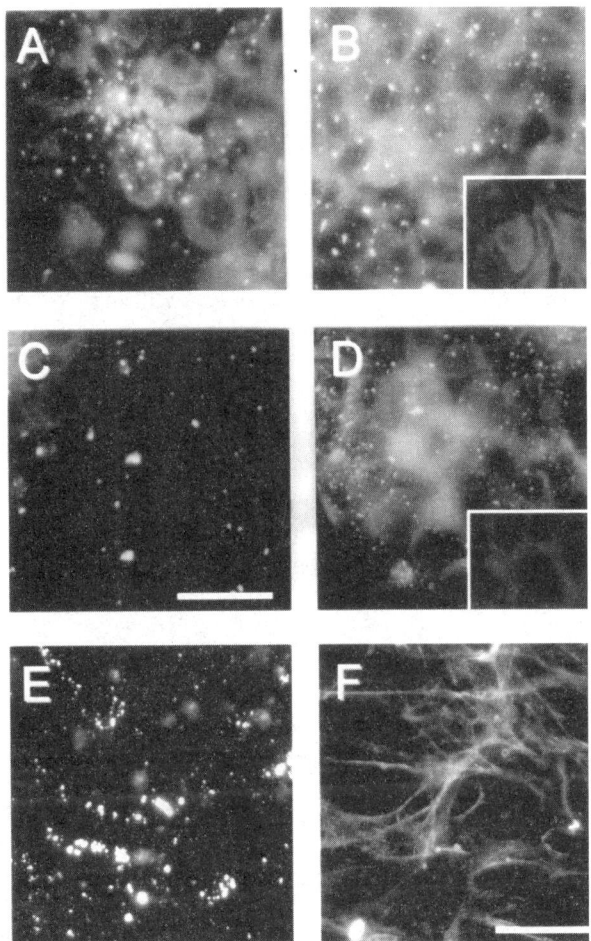

FIGURE 4. Immunodetection of FBG in the ECM of MCF-7 and HFF cells. MCF-7 cells were immunostained with anti-FBG PoAb or FBG-specific MoAb: **(A)** Aα-RGD, **(B)** 18C6, **(C)** J88B, or **(D)** anti-FBG PoAb. *Inset* in panel **B** shows staining with rabbit antimouse polyvalent Ig alone; *inset* in panel **C** shows staining with goat antirabbit IgG alone. The *bar* in panel **C** represents 25 μm. Purified plasma FBG labeled with Oregon Green™ was added to MCF-7 **(E)** and HFF **(F)** cells and incubated for 24 h. The *bar* in panel **F** represents 15 μm. HFF cells assemble plasma FBG into a mature fibrillar matrix **(F)**, whereas, MCF-7 cells show only punctuate cell surface binding of both endogenously produced FBG **(A–D)** and plasma FBG, **(E)**.

Green™ labeled-purified plasma FBG was overlaid onto MCF-7 cells (FIG. 4E). As a control, HFF were also overlaid with the FBG-Oregon Green (FIG. 4F). The results showed the expected fibrillar pattern of plasma FBG in the ECM of HFF (FIG. 4F). In contrast, FBG-Oregon Green appeared in a punctuate pattern of staining (FIG. 4E) indicative of FBG binding to the MCF-7 cell surface. Together, these results indicate a defect in the synthesis and processing of endogenous FBG by MCF-7 cells, as well as a defect in the ability of MCF-7 breast carcinoma cells to assemble exogenously added plasma FBG into mature matrix fibrils.[79] The data in this report show that the intracellular assembly of intact, dimeric FBG is limited in MCF-7 cells. Analysis of the chain composition indicates that most of the FBG secreted by MCF-7 cells is intact γ chain with partially degraded Aα and Bβ chains. The absence of intact Bβ chain likely hinders the formation of fully assembled FBG, because these NH_2-terminal regions are known to be important in disulfide bond formation between two FBG half-molecules.[80] The loss of FBG Bβ chain N-terminal peptides may contribute to the lack of intact FBG assembly in MCF-7 cells, which may further affect its ability to assemble FBG into a fibrillar ECM. These data suggest that endogenous synthesis and secretion of FBG is, at least in part, the source of FBG deposition in the ECM of breast cell carcinomas and may signify the production of a specialized ECM molecule that, in turn, affects cellular processes that modulate the progression of breast cancer.

FIBRINOGEN EFFECTS ON MDA-MB-231 BREAST CARCINOMA CELL MIGRATION

Cell motility is a continuum of sequential events in which a cell extends pseudopodia, forms nascent attachments, assembles and contracts the cytoskeleton, and disengages distal adhesions as it translocates forward. Substratum contact is mediated by integrin receptors that trigger cell movement. During migration, cells use matrix proteins as substrates upon which to adhere and migrate. Adhesive glycoproteins that modulate cell migration include thrombospondin, laminin, FN, and FBG.[81] Because MCF-7 cells synthesize and secrete FBG, we evaluated two other breast epithelial cell lines, the nonmalignant HBL-100 cell line and the invasive malignant MDA-MB-231 cell line, to determine whether endogenous FBG synthesis was a generalized phenomenon of mammary epithelia. Although neither of these cell lines produced detectable levels of FBG related polypeptides, they showed a differing ability to assembly FBG into the ECM. HBL-100 cells assembled a modest, but mature FBG matrix that colocalized with FN matrix fibrils. In contrast, MDA-MB-231 cells supported only cell surface binding of FBG in a manner similar to that described for the less invasive malignant cell MCF-7 cells (not shown).

A wound scrape model was used to assess the influence of FBG on the migration of the invasive and moderately metastatic breast cell line, MDA-MB-231. Once the cells reached confluence, the cells were wounded down the middle of the coverslip by scraping with a sterile pipette tip. FBG or FBG-Oregon Green at 50 µg/ml was added to wounded cells and incubated for an additional 18 h. For pretreatment conditions, the cells were cultured as described except 24 h prior to wounding, fresh medium was added in the presence of purified plasma FBG 50 µg/ml. When FBG was added at the time of wounding (see FIGURE 5, +FBG and ATOW), MDA-MB-231 cells did not achieve wound closure in the 18 h time period as compared to untreated control cells (FIG. 5, Control and ATOW); approximately 40–60% fewer cells were observed in the wound space in the presence of FBG as compared to cells that were not treated with FBG at the time of wounding. In contrast, when MDA-MB-231 cells were pretreated with FBG 24 h prior to wounding, wound closure was achieved during the 18 h time period and similar numbers of cells migrated into the wound space in the presence of FBG as compared to control cells not treated with FBG (FIG. 5, Pretreated+FBG compared to Pretreated Control). Thus, depending on when FBG is introduced into the system in relation to induction of migration, the results showed that FBG matrix either inhibits (added at time of wounding) or supports (added 24 h prior to wounding) cell migration. FBG stimulation of MDA-MB-231 migration depended on $\alpha_5\beta_1$ and $\alpha_v\beta_3$ integrins (not shown). These results implicate FBG and $\alpha_5\beta_1/\alpha_v\beta_3$ integrins in modulating MDA-MB-231 cell migration during breast cancer metastasis and support our hypothesis that once FBG becomes incorporated into ECM, exposing new epitopes, FBG influences matrix composition and provides a surface/conduit for migration. The exact mechanism of the influence of FBG on cell migration remains unclear, but most likely involves signaling events that alter cell attachment by modulating focal adhesion plaque formation, integrin engagement, and/or actin cytoskeleton rearrangement.[82,83]

Cytoskeletal involvement in FN matrix assembly epitomizes the interactions between extracellular proteins and intracellular events such as actin cytoskeleton

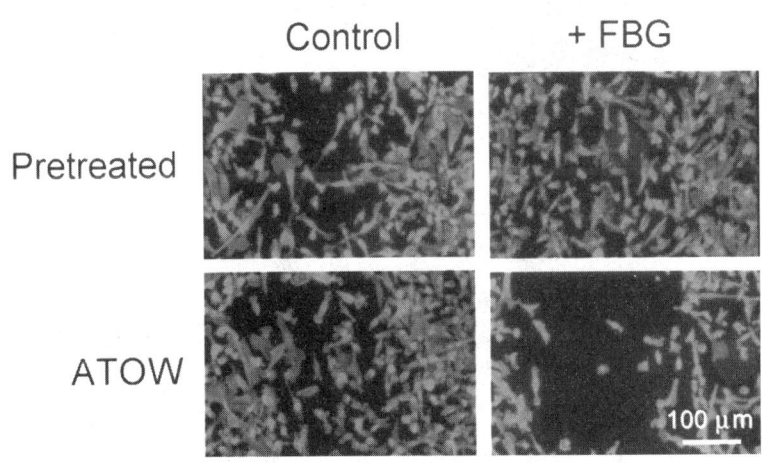

FIGURE 5. FBG modulation of MDA-MB-231 cell migration using a wound injury model. MDA-MB-231 cells were grown on glass coverslips until confluent. Cells were washed, wounded, and fresh medium was added at time of wounding (*ATOW*) with (*+FBG*) or without (*Control*) 50 µg/ml FBG for 18 h. Cells were also pretreated with FBG for 24 h (*Pretreated*), then wounded in the presence (*+FBG*) or absence (*Control*) of FBG. Cells were fixed, permeabilized, stained with phalloidin-Texas Red. Random fields were counted and normalized to control conditions to determine the percent of cells that migrated into the wound space.

changes, focal adhesion plaque formation, and initiation of signal transduction cascades.[84,85] Actin architecture and cell morphology play important roles in the "inside-out" regulation of matrix assembly.[85–87] Using immunofluorescent staining techniques, fibroblast cells show a parallel array of actin stress fibers and the fibroblast cells grow in a contact inhibited monolayer (see FIGURE 6). In contrast, MDA-MB-231 cells display disorganized actin architecture, are heterogeneous in shape, and do not grow in an organized, contact inhibited monolayer (FIG. 6). The nonmalignant HBL-100 and less invasive MCF-7 cells showed intermediate levels of disorganization of the actin cytoskeleton consistent with the adherent or migratory phenotype of these cells (FIG. 6). The lack of a homogeneous morphology, due to actin architecture, may lead to unorganized surface integrin presentation and thus may contribute to the lack of significant matrix deposition by the more invasive tumor cells.

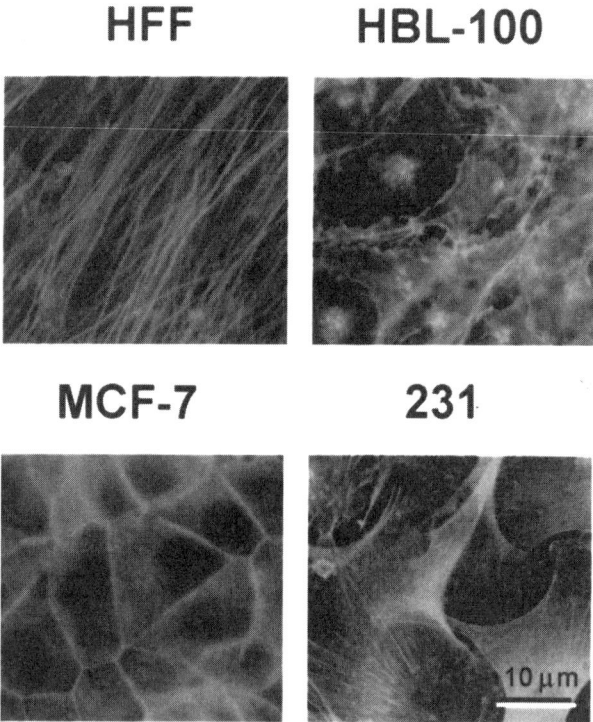

FIGURE 6. The actin cytoskeleton is grossly disorganized in malignant breast carcinoma cells. Confluent HFF, HBL-100, MCF-7, and MDA-MB-231 cells were stained for actin filaments with Texas Red-phalloidin. HFF show parallel arrays of actin filament, indicative of adherent and spread cells; whereas, the actin cytoskeleton and cell morphology of the breast epithelial cells becomes increasingly disorganized depending on the degree of malignancy exhibited by each cell line from nonmalignant to most malignant: HBL-100<MCF-7<MDA-MB-231.

The FBG inhibition/stimulation events seen with MDA-MB-231 cells are diagramed in FIGURE 7. An as yet to be determined "critical point" of FBG cell/matrix binding in the time continuum is hypothesized to consist of events that are sufficient to stimulate a nonmotile cell to migrate. Potential mechanisms responsible for such changes in cell migration might involve integrin–cytoskeletal interactions, focal adhesion kinase (FAK) dependent pathways, and activation of Rho GTPases.[88] Focal adhesion plaques are tightly regulated cellular structures that participate in cell migration. Components of focal adhesions include integrins, FAK, paxillin, vinculin, and small GTPases, including Rho and Rac.[83] Once matrix proteins engage cell surface receptors, intracellular signals are transduced, phosphorylation events occur, ultimately leading to changes in gene expression and protein synthesis. FAK, a key focal adhesion component, is overexpressed in colon, prostate, and breast cancer.[89] Thus, FAK may play a role in tumor cell migration, invasion, and metastasis.

Evidence provided in the present studies highlight the potential involvement of FBG in the context of biological processes of matrix assembly and cell migration in normal and pathological breast epithelium. Due to the multifunctional domain composition of FBG and its dimeric structure, FBG is capable of binding numerous cell

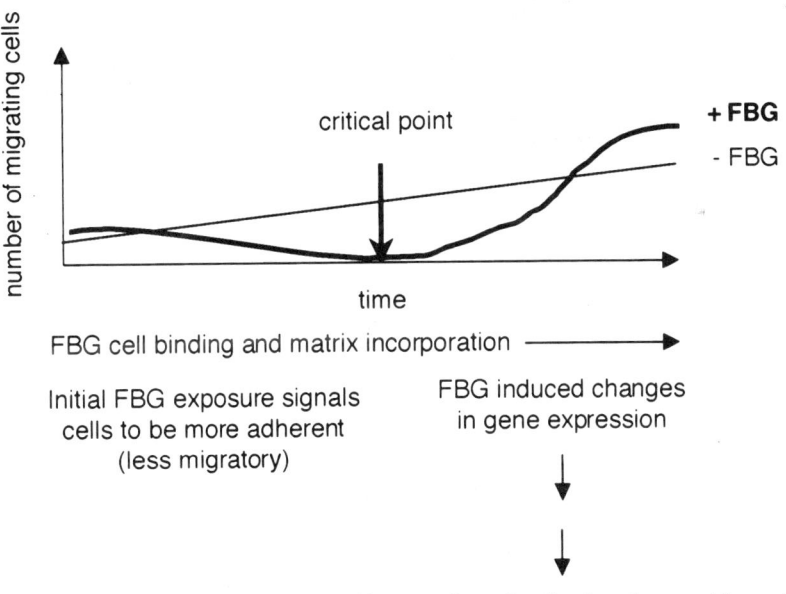

FIGURE 7. Proposed time continuum of FBG effects on cell migration. Schematic diagram depicting the effects of FBG during wound repair as measured by modulation of cell migration. FBG added at the time of cell injury induces a cellular response to inhibit cell migration and support enhanced cell adhesion. However, once a critical point is reached and FBG is cell/matrix bound, cell migration increases up to or slightly beyond basal levels. The data suggest that the critical point is reached when cells shift from a nonmotile, adhesive phenotype to a more loosely attached migratory phenotype.

types, matrix components, and growth factors acting as a bridge between cells, providing traction for cell adhesion and motility, and altering cell morphology and invasive potential. Although the studies presented were performed with cultured cells, fibrin(ogen) has been localized in numerous tumor types including breast cancer.[7,8,10,13,53] The presence of FBG in breast ECM suggests that FBG plays specialized roles in the process of desmoplasia leading to alterations in cellular behavior during malignant breast cell growth, motility, and invasion.

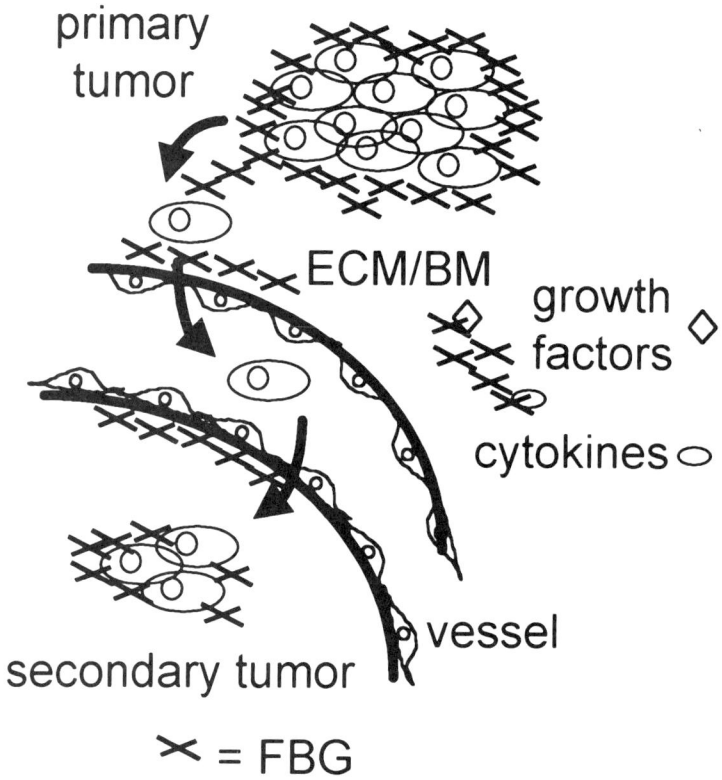

FIGURE 8. Schematic of the role of FBG in breast cancer progression and metastasis. FBG potentially participates in the processes of tumor growth, tumor invasion into blood vessels, binding to tumor cells, extravasation out of vessels, cell migration, and establishment of secondary tumors. FBG is characterized as protumorigenic, since it binds growth factors such as VEGF, promoting endothelial cell proliferation, interacts with other matrix proteins providing a physical protective barrier, and serves as a scaffold for cell migration due to integrin binding sites. On the other hand, FBG also serves as an antitumorigenic agent in that FBG has been shown to inhibit cell motility due to its adhesive properties and acting as a chemoattractant for inflammatory cell infiltration in the immune response against oncogenic transformation. Ultimately, the microenvironment, cell–cell interactions, cell–matrix interactions, and signaling pathways shift the balance between these pro- and antitumorigenic properties depending on the disease context and mechanisms involved.

The functions of FBG in tumor cell growth include regulating inflammatory cell recruitment, providing structure to the tumor stroma, and inducing angiogenesis.[90] In transplantable carcinomas, fibrin(ogen) deposition occurs within minutes of tumor injection resulting in the formation of a three-dimensional gel encasing tumor cells or tumor cell aggregates.[11] This is illustrated (see FIGURE 8) in human breast cell carcinomas, where tumor masses grow in discrete structural nodules surrounded by stromal components. The fibrin(ogen) surrounding breast tumor masses can also serve as a provisional matrix with which macrophages, fibroblasts, and endothelial cells interact during inflammatory processes, matrix remodeling, and angiogenesis, respectively.[10,12,53] Angiogenesis is critical for a tumor to growth beyond a certain size.[8,13] Once blood vessels are established, tumor cells have access to necessary growth factors and nutrients such as oxygen. In turn, the highly neovascularized tumor mass ultimately serves as a conduit for disseminating tumor cells to other organs. The binding and localization of growth factors such as FGF-2 and VEGF by FBG, or other matrix constituents, localizes growth factors to tumor sites and subsequently influence neovascularization, which induces tumor growth. The proteolytic cleavage of matrix proteins releases bound cytokines, chemokines, and growth factors into the microenvironment. Thus, ECM proteins such as FBG may control the bioavailability and accessibility of sequestered tumor promoting factors.

In summary, FBG represents a dynamic, multifunctional protein that influences a multitude of cellular processes during breast tumorigenesis. In the future, FBG may serve as a target in therapeutic strategies for localizing antitumorigenic agents to inhibit angiogenic and metastatic processes.

REFERENCES

1. TROSSEAU, A. 1865. Phlegmasia alba dolens. Clinique medicale de l'Hotel-Dieu de Paris. (New Sydenham Society, London.) **3**: 94.
2. COSTANTINI, V. & L.R. ZACHARSKI. 1993. Fibrin and cancer. Thromb. Hæmost. **69**: 406–414.
3. KUDRYK, B., A. ROHOZA, M. AHADI, et al. 1983. A monoclonal antibody with ability to distinguish between NH_2-terminal fragments derived from fibrinogen and fibrin. Mol. Immunol. **20**: 1191–1200.
4. KUDRYK, B., A. ROHOZA, M. AHADI, et al. 1984. Specificity of a monoclonal antibody for the NH_2-terminal region of fibrin. Mol. Immunol. **21**: 89–94.
5. KAPLAN, K.L., A. BINI, J. FENOGLIO, JR. & B. KUDRYK. 1990. Fibrin and the vessel wall. Adv. Exp. Med. Biol. **281**: 313–318.
6. GUADIZ, G., L.A. SPORN & P.J. SIMPSON-HAIDARIS. 1997. Thrombin cleavage-independent deposition of fibrinogen in extracellular matrices. Blood **90**: 2644–2653.
7. COSTANTINI, V., L.R. ZACHARSKI, V.A. MEMOLI, et al. 1991. Fibrinogen deposition without thrombin generation in primary human breast cancer tissue. Cancer Res. **51**: 349–353.
8. BROWN, L.F., L. VAN DE WATER, V.S. HARVEY & H.F. DVORAK. 1988. Fibrinogen influx and accumulation of cross-linked fibrin in healing wounds and in tumor stroma. Am. J. Pathol. **130**: 455–465.
9. CLARK, R.A., J.M. LANIGAN, P. DELLAPELLE, et al. 1982. Fibronectin and fibrin provide a provisional matrix for epidermal cell migration during wound reepithelialization. J. Invest. Dermatol. **79**: 264–269.
10. NAGY, J.A., L.F. BROWN, D.R. SENGER, et al. 1989. Pathogenesis of tumor stroma generation: a critical role for leaky blood vessels and fibrin deposition. Biochim. Biophys. Acta. **948**: 305–326.

11. DVORAK, H.F., J.A. NAGY, B. BERSE, et al. 1992. Vascular permeability factor, fibrin, and the pathogenesis of tumor stroma formation. Ann. N.Y. Acad. Sci. **667:** 101–111.
12. DVORAK, H.F., V.S. HARVEY, P. ESTRELLA, et al. 1987. Fibrin containing gels induce angiogenesis. Implications for tumor stroma generation and wound healing. Lab. Invest. **57:** 673–686.
13. DVORAK, H.F., D.R. SENGER & A.M. DVORAK. 1983. Fibrin as a component of the tumor stroma: origins and biological significance. Cancer Metastasis Reviews **2:** 41–73.
14. BARDOS, H., P. MOLNAR, G. CSECSEI & R. ADANY. 1996. Fibrin deposition in primary and metastatic human brain tumours. Blood Coag. Fibrinoly. **7:** 536–548.
15. WOJTUKIEWICZ, M.Z., L.R. ZACHARSKI, V.A. MEMOLI, et al. 1991. Fibrin formation on vessel walls in hyperplastic and malignant prostate tissue. Cancer **67:** 1377–1383.
16. DVORAK, H.F., G.R. DICKERSIN, A.M. DVORAK, et al. 1981. Human breast carcinoma: fibrin deposits and desmoplasia. Inflammatory cell type and distribution. Microvasculature and infarction. J. Natl. Cancer Inst. **67:** 335–345.
17. WOJTUKIEWICZ, M.Z., L.R. ZACHARSKI, V.A. MEMOLI, et al. 1989. Absence of components of coagulation and fibrinolysis pathways *in situ* in mesothelioma. Thromb. Res. **55:** 279–284.
18. WOJTUKIEWICZ, M.Z., L.R. ZACHARSKI, V.A. MEMOLI, et al. 1989. Indirect activation of blood coagulation in colon cancer. Thromb. Hæmost. **62:** 1062–1066.
19. COSTANTINI, V., L.R. ZACHARSKI, V.A. MEMOLI, et al. 1992. Fibrinogen deposition and macrophage-associated fibrin formation in malignant and nonmalignant lymphoid tissue. J. Lab. Clin. Med. **119:** 124–131.
20. SAGE, E.H. & P. BORNSTEIN. 1991. Extracellular proteins that modulate cell-matrix interactions. SPARC, tenascin, and thrombospondin. J. Biol. Chem. **266:** 14831–14834.
21. CLARK, R.A.F. 1996. Wound repair. *In* The Molecular and Cellular Biology of Wound Repair, edit. 2. R.A.F. Clark, Ed. Plenum Press, New York.
22. MOSHER, D.F., J. SOTTILE, C. WU & J.A. MCDONALD. 1992. Assembly of extracellular matrix. Curr. Opin. Cell Biol. **4:** 810–818.
23. SAHNI, A., T. ODRLJIN & C.W. FRANCIS. 1998. Binding of basic fibroblast growth factor to fibrinogen and fibrin. J. Biol. Chem. **273:** 7554–7559.
24. SAHNI, A. & C.W. FRANCIS. 2000. Vascular endothelial growth factor binds to fibrinogen and fibrin and stimulates endothelial cell proliferation. Blood **96:** 3772–3778.
25. VAHERI, A. & D.F. MOSHER. 1978. High molecular weight, cell surface-associated glycoprotein (fibronectin) lost in malignant transformation. Biochim. Biophys. Acta. **516:** 1–25.
26. HYNES, R.O., A.T. DESTREE, M.E. PERKINS & D.D. WAGNER. 1979. Cell surface fibronectin and oncogenic transformation. J. Supramol. Struct. **11:** 95–104.
27. CHRISTENSEN, L., M. NIELSEN, J. ANDERSEN & I. CLEMMENSEN. 1988. Stromal fibronectin staining pattern and metastasizing ability of human breast carcinoma. Cancer Res. **48:** 6227–6233.
28. CHRISTENSEN, L. 1990. Fibronectin: a discrimination marker between small invasive carcinomas and benign proliferative lesions of the breast. APMIS **98:** 615–623.
29. ODRLJIN, T.M., C.W. FRANCIS, L.A. SPORN, et al. 1996. Heparin-binding domain of fibrin mediates its binding to endothelial cells. Arterioscler. Thromb. Vasc. Biol. **16:** 1544–1551.
30. ODRLJIN, T.M., J.R. SHAINOFF, S.O. LAWRENCE & P.J. SIMPSON-HAIDARIS. 1996. Thrombin cleavage enhances exposure of a heparin binding domain in the N-terminus of the fibrin beta chain. Blood **88:** 2050–2061.
31. HANTGEN, R.R., C.W. FRANCIS & V.J. MARDER. 1994. Fibrinogen structure and physiology. *In* Hemostasis and Thrombosis: Basic Principles and Clinical Practice, 3rd. edit. R.W. Colman, V.J. Marder & E.W. Salzman, Eds. J.B. Lippincott Company, Philadelphia.
32. ALTIERI, D.C., A. DUPERRAY, J. PLESCIA, et al. 1995. Structural recognition of a novel fibrinogen gamma chain sequence (117–133) by intercellular adhesion molecule-1 mediates leukocyte-endothelium interaction. J. Biol. Chem. **270:** 696–699.

33. RUOSLAHTI, E. 1996. RGD and other recognition sequences for integrins. Annu. Rev. Cell Dev. Biol. **12:** 697–715.
34. CHERESH, D.A. 1987. Human endothelial cells synthesize and express an Arg-Gly-Asp-directed adhesion receptor involved in attachment to fibrinogen and von Willebrand factor. Proc. Natl. Acad. Sci. U.S.A. **84:** 6471–6475.
35. SUEHIRO, K., J. GAILIT & E.F. PLOW. 1997. Fibrinogen is a ligand for integrin alpha5beta1 on endothelial cells. J. Biol. Chem. **272:** 5360–5366.
36. SUEISHI, K., S. NANNO & K. TANAKA. 1981. Permeability enhancing and chemotactic activities of lower molecular weight degradation products of human fibrinogen. Thromb. Hæmost. **45:** 90–94.
37. SKOGEN, W.F., R.M. SENIOR, G.L. GRIFFIN & G.D. WILNER. 1988. Fibrinogen-derived peptide Bβ1–42 is a multidomained neutrophil chemoattractant. Blood **71:** 1475–1479.
38. ALTIERI, D.C., J. PLESCIA & E.F. PLOW. 1993. The structural motif glycine 190-valine 202 of the fibrinogen gamma chain interacts with CD11b/CD18 integrin (αMβ2, Mac-1) and promotes leukocyte adhesion. J. Biol. Chem. **268:** 1847–1853.
39. SIMPSON-HAIDARIS, P.J., M.A. COURTNEY, T.W. WRIGHT, *et al.* 1998. Induction of fibrinogen expression in the lung epithelium during pneumonia. Infect. Immun. **66:** 4431–4439.
40. LEE, S.Y., K.P. LEE & J.W. LIM. 1996. Identification and biosynthesis of fibrinogen in human uterine cervix carcinoma cells. Thromb. Hæmost. **75:** 466–470.
41. MOLMENTI, E.P., T. ZIAMBARAS & D.H. PERLMUTTER. 1993. Evidence for an acute phase response in human intestinal epithelial cells. J. Biol. Chem. **268:** 14116–14124.
42. SIMPSON-HAIDARIS, P.J. 1997. Induction of fibrinogen biosynthesis and secretion from cultured pulmonary epithelial cells. Blood **89:** 873–882.
43. AMRANI, D.L., P.W. PLANT, J. PINDYCK, *et al.* 1983. Structural analysis of fibrinogen synthesized by cultured chicken hepatocytes in the presence or absence of dexamethasone. Biochim. Biophys. Acta. **743:** 394–400.
44. PARROTT, J.A., P.D. WHALEY & M.K. SKINNER. 1993. Extrahepatic expression of fibrinogen by granulosa cells: potential role in ovulation. Endocrinology **133:** 1645–1649.
45. HARTWIG, R. & K.J. DANISHEFSKY. 1991. Studies on the assembly and secretion of fibrinogen. J. Biol. Chem. **266:** 6578–6585.
46. MAGNUSSON, M.K. & D.F. MOSHER. 1998. Fibronectin: structure, assembly, and cardiovascular implications. Arterioscler. Thromb. Vasc. Biol. **18:** 1363–1370.
47. MOSHER, D.F., F.J. FOGERTY, M.A. CHERNOUSOV & E.L. BARRY. 1991. Assembly of fibronectin into extracellular matrix. Ann. N.Y. Acad. Sci. **614:** 167–180.
48. GUADIZ, G., L.A. SPORN, R.A. GOSS, *et al.* 1997. Polarized secretion of fibrinogen by lung epithelial cells. Am. J. Respir. Cell Mol. Biol. **17:** 60–69.
49. KELSEY, J.L. & L. BERNSTEIN. 1996. Epidemiology and prevention of breast cancer. Annu. Rev. Public Health. **17:** 47–67.
50. ANDERSON, D.E. 1992. Familial versus sporadic breast cancer. Cancer **70:** 1740–1746.
51. CASEY, G. 1997. The BRCA1 and BRCA2 breast cancer genes. Curr. Opin. Oncol. **9:** 88–93.
52. RONNOV-JESSEN, L., O.W. PETERSEN & M.J. BISSELL. 1996. Cellular changes involved in conversion of normal to malignant breast: importance of the stromal reaction. Physiol. Rev. **76:** 69–125.
53. DVORAK, H.F. 1986. Tumors: wounds that do not heal. Similarities between tumor stroma generation and wound healing. N. Engl. J. Med. **315:** 1650–1659.
54. AHMAD, A. & I.R. HART. 1997. Mechanisms of metastasis. Crit. Rev. Oncol. Hematol. **26:** 163–173.
55. PAULI, B.U., H.G. AUGUSTIN-VOSS, M.E. EL-SABBAN, *et al.* 1990. Organ-preference of metastasis. The role of endothelial cell adhesion molecules. Cancer Metastas. Rev. **9:** 175–189.
56. NICOLSON, G.L. & A.S. MOUSTAFA. 1998. Metastasis-Associated genes and metastatic tumor progression. *In Vivo* **12:** 579–588.

57. LIOTTA, L.A., C.N. RAO & U.M. WEWER. 1986. Biochemical interactions of tumor cells with the basement membrane. Annu. Rev. Biochem. **55:** 1037–1057.
58. CROWE, D.L. & C.F. SHULER. 1999. Regulation of tumor cell invasion by extracellular matrix. Histol. Histopathol. **14:** 665–671.
59. STETLER-STEVENSON, W.G., S. AZNAVOORIAN & L.A. LIOTTA. 1993. Tumor cell interactions with the extracellular matrix during invasion and metastasis. Annu. Rev. Cell Biol. **9:** 541–573.
60. BROWN, L.F., S.M. OLBRICHT, B. BERSE, et al. 1995. Overexpression of vascular permeability factor (VPF/VEGF) and its endothelial cell receptors in delayed hypersensitivity skin reactions. J. Immunol. **154:** 2801–2807.
61. CLAFFEY, K.P. & G.S. ROBINSON. 1996. Regulation of VEGF/VPF expression in tumor cells: consequences for tumor growth and metastasis. Cancer Metastas. Rev. **15:** 165–176.
62. SOLDI, R., S. MITOLA, M. STRASLY, et al. 1999. Role of alphavbeta3 integrin in the activation of vascular endothelial growth factor receptor-2. EMBO J. **18:** 882–892.
63. LOCHTER, A. & M.J. BISSELL. 1995. Involvement of extracellular matrix constituents in breast cancer. Semin. Cancer Biol. **6:** 165–173.
64. ELICEIRI, B.P. & D.A. CHERESH. 1999. The role of alphav integrins during angiogenesis: insights into potential mechanisms of action and clinical development. J. Clin. Invest. **103:** 1227–1230.
65. GASPARINI, G., P.C. BROOKS, E. BIGANZOLI, et al. 1998. Vascular integrin alpha(v)beta3: a new prognostic indicator in breast cancer. Clin. Cancer Res. **4:** 2625–2634.
66. BROOKS, P.C., S. STROMBLAD, R. KLEMKE, et al. 1995. Antiintegrin alpha v beta 3 blocks human breast cancer growth and angiogenesis in human skin. J. Clin. Invest. **96:** 1815–1822.
67. CHALUPOWICZ, D.G., Z.A. CHOWDHURY, T.L. BACH, et al. 1995. Fibrin II induces endothelial cell capillary tube formation. J. Cell Biol. **130:** 207–215.
68. BACH, T.L., C. BARSIGIAN, D.G. CHALUPOWICZ, et al. 1998. VE-Cadherin mediates endothelial cell capillary tube formation in fibrin and collagen gels. Exp. Cell Res. **238:** 324–334.
69. HYNES, R.O. 1992. Integrins: versatility, modulation, and signaling in cell adhesion. Cell **69:** 11–25.
70. ALBELDA, S.M. & C.A. BUCK. 1990. Integrins and other cell adhesion molecules. FASEB J. **4:** 2868–2880.
71. APLIN, A.E., A. HOWE, S.K. ALAHARI & R.L. JULIANO. 1998. Signal transduction and signal modulation by cell adhesion receptors: the role of integrins, cadherins, immunoglobulin-cell adhesion molecules, and selectins. Pharmacol. Rev. **50:** 197–263.
72. HOWLETT, A.R., N. BAILEY, C. DAMSKY, et al. 1995. Cellular growth and survival are mediated by beta 1 integrins in normal human breast epithelium but not in breast carcinoma. J. Cell Sci. **108:** 1945–1957.
73. GREEN, L.J., A.P. MOULD & M.J. HUMPHRIES. 1998. The integrin beta subunit. Int. J. Biochem. Cell Biol. **30:** 179–184.
74. ZUTTER, M.M., G. MAZOUJIAN & S.A. SANTORO. 1990. Decreased expression of integrin adhesive protein receptors in adenocarcinoma of the breast. Am. J. Pathol. **137:** 863–870.
75. ZUTTER, M.M., H.R. KRIGMAN & S.A. SANTORO. 1993. Altered integrin expression in adenocarcinoma of the breast. Analysis by *in situ* hybridization. Am. J. Pathol. **142:** 1439–1448.
76. FRIEDRICHS, K., P. RUIZ, F. FRANKE, et al. 1995. High expression level of alpha 6 integrin in human breast carcinoma is correlated with reduced survival. Cancer Res. **55:** 901–906.
77. LIAPIS, H., A. FLATH & S. KITAZAWA. 1996. Integrin alpha V beta 3 expression by bone-residing breast cancer metastases. Diagn. Mol. Pathol. **5:** 127–135.
78. WEAVER, V.M., O.W. PETERSEN, F. WANG, et al. 1997. Reversion of the malignant phenotype of human breast cells in three-dimensional culture and *in vivo* by integrin blocking antibodies. J. Cell Biol. **137:** 231–245.

79. RYBARCZYK, B.J. & P.J. SIMPSON-HAIDARIS. 2000. Fibrinogen assembly, secretion, and deposition into extracellular matrix by MCF-7 human breast carcinoma cells. Cancer Res. **60:** 2033–2039.
80. ZHANG, J.Z. & C.M. REDMAN. 1996. Assembly and secretion of fibrinogen. Involvement of amino-terminal domains in dimer formation. J. Biol. Chem. **271:** 12674–12680.
81. DONALDSON, D.J., J.T. MAHAN, D. AMRANI & J. HAWIGER. 1989. Fibrinogen-mediated epidermal cell migration: structural correlates for fibrinogen function. J. Cell Sci. **94:** 101–108.
82. MIYAMOTO, S., B.Z. KATZ, R.M. LAFRENIE & K.M. YAMADA. 1998. Fibronectin and integrins in cell adhesion, signaling, and morphogenesis. Ann. N.Y. Acad. Sci. **857:** 119–129.
83. CARY, L.A. & J.L. GUAN. 1999. Focal adhesion kinase in integrin-mediated signaling. Frontiers in Bioscience **4:** D102–113.
84. WU, C. 1997. Roles of integrins in fibronectin matrix assembly. Histol. Histopathol. **12:** 233–240.
85. SECHLER, J.L. & J.E. SCHWARZBAUER. 1997. Coordinated regulation of fibronectin fibril assembly and actin stress fiber formation. Cell Adhes. Commun. **4:** 413–424.
86. APLIN, A.E. & R.L. JULIANO. 1999. Integrin and cytoskeletal regulation of growth factor signaling to the MAP kinase pathway. J. Cell Sci. **112:** 695–706.
87. CHRISTOPHER, R.A., A.P. KOWALCZYK & P.J. MCKEOWN-LONGO. 1997. Localization of fibronectin matrix assembly sites on fibroblasts and endothelial cells. J. Cell Sci. **110:** 569–581.
88. SCHLAEPFER, D.D. & T. HUNTER. 1998. Integrin signalling and tyrosine phosphorylation: just the FAKs? Trends Cell Biol. **8:** 151–157.
89. KORNBERG, L.J. 1998. Focal adhesion kinase and its potential involvement in tumor invasion and metastasis. Head Neck **20:** 745–752.
90. BROWN, L.F., A.M. DVORAK & H.F. DVORAK. 1989. Leaky vessels, fibrin deposition, and fibrosis: a sequence of events common to solid tumors and to many other types of disease. Am. Rev. Respir. Dis. **140:** 1104–1107.

Role of Fibrin Matrix in Angiogenesis

VICTOR W.M. VAN HINSBERGH,[a,b] ANNEMIE COLLEN,[b] AND PIETER KOOLWIJK[b]

[a]*Gaubius Laboratory TNO-PG, Leiden, the Netherlands*

[b]*Department of Physiology, Institute for Cardiovascular Research, Vrije Universiteit, Amsterdam, the Netherlands.*

ABSTRACT: Angiogenesis, the formation of new blood vessels from existing vessels, plays an important role during development. In the adult, it is limited to the female reproductive system and to tissue repair and pathological conditions. Repair associated angiogenesis is usually accompanied by the presence of inflammatory cells, vascular leakage, and fibrin deposition. The temporary fibrin matrix acts, not only as a sealing matrix, but also as a scaffolding for invading leukocytes and endothelial cells during tissue repair. We have used a three-dimensional fibrin matrix to study the outgrowth of human microvascular endothelial cells in capillary-like tubular structures. This process is induced by the simultaneous addition of an angiogenic growth factor (bFGF or VEGF) and the cytokine TNFα, and is enhanced by hypoxia. It involves proteolytic activities, in particular cell bound urokinase/plasmin and matrix metalloproteinase activities. Modulation of the fibrin structure markedly affects the extent and stability of capillary tube formation *in vitro*. Preparation of fibrin at different pH (7.0–7.8) or crosslinking of the fibrin matrix induces differences in fibrin matrix rigidity and structure. This is accompanied by a change in capillary ingrowth. Heparins, in particular low molecular weight heparins, modulate the fibrin structure and by this action affect angiogenesis *in vitro*. A mutant fibrinogen$_{Nieuwegein}$, which lacks the terminal part of the Aα chain of fibrin harboring an RGD sequence and the transglutaminase sequence, provided additional evidence that the structure of fibrin is an important determinant for angiogenesis. These findings may have impact on improving wound healing and on influencing angiogenesis in malignancies with a fibrinous stroma.

KEYWORDS: Fibrin matrix; Angiogenesis; Blood vessels; Wound healing; Fibrinous stoma.

INTRODUCTION

An essential part of the healing process is the restoration of an adequate blood supply. This is achieved by the formation of new microvessels from existing vessels, a process called angiogenesis.[1,2] Angiogenesis is often followed by the remodeling of small arterial blood vessels into larger vessels, indicated as arteriogenesis,[3] that is needed for proper perfusion of the extended vascular bed. Angiogenesis is indispensable for the

---

Address for correspondence: Victor W.M. van Hinsbergh, Ph.D., Gaubius Laboratory TNO Prevention and Health Zernikedreef 9 2333 CK Leiden, the Netherlands. Fax: +31-71-518 1904.
vwm.vanhinsbergh@pg.tno.nl

proper development of the embryo and the fetus. Much has been learned about developmental angiogenesis from studies on mice in which specific genes had been deleted, in particular with respect to the involvement of growth factors and their receptors and the role of specific cell–matrix interactions.[4]

In adult organisms the vascular tree is fully developed, and angiogenesis is normally limited to the female reproductive system. However, the formation of new blood vessels becomes essential in adults during tissue repair after severe wounding or inflammation. Furthermore, it is associated with a number of pathological conditions, such as malignancies, diabetic retinopathy, and rheumatoid arthritis.[5–7] In repair associated or pathological angiogenesis the formation of new microvessels is usually accompanied by the occurrence of inflammatory cells, vascular leakage, and the deposition of fibrin and vitronectin in a fibrinous exudate.[8,9] Indeed activation of the clotting system not only limits blood loss, but it provides a number of factors that stimulate angiogenesis including the temporary fibrin matrix. This fibrin matrix provides a suitable scaffolding for invading inflammatory, endothelial, and other tissue cells during the healing process.

MECHANISM OF ANGIOGENESIS

The process of angiogenesis is generally thought to involve initial activation and subsequent liberation of endothelial cells by degradation of their basement membrane.[10] This is followed by the migration of endothelial cells into the interstitial extracellular matrix, which is accompanied by the proliferation of endothelial cells. Subsequently, lumen formation occurs and the newly formed endothelial tubes develop a basement membrane. Finally, the vessels become stabilized, a process in

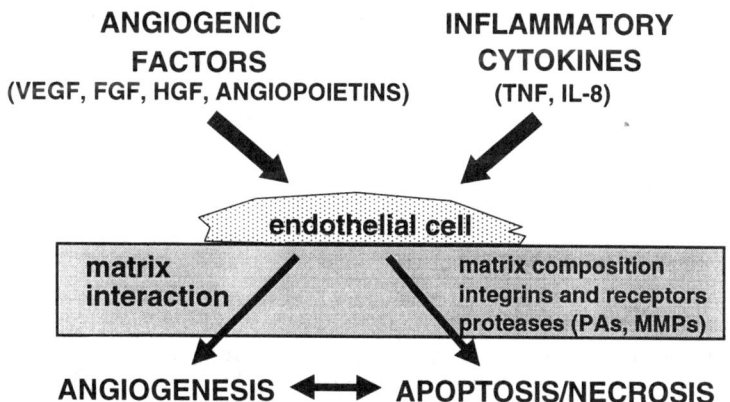

FIGURE 1. Angiogenic growth factors usually in interplay with other cytokines induce endothelial cells to go into the angiogenesis process if a supportive matrix interaction exists. The matrix interaction is regulated by the matrix composition, the expression of cellular receptors, and the expression of matrix remodelling proteases. This response is modulated or prevented by angiogenesis inhibitors.

which specific cytokines and pericytes are thought to play a role.[11,12] The initiation and progression of angiogenesis is controlled by angiogenic factors, often in interplay with cytokines. Only when these factors act on endothelial cells in a suitable matrix environment does angiogenesis proceed (see FIGURE 1). Otherwise the cells are refractory to stimulation, or they respond but subsequently go into apoptosis and die. Cellular receptors, in particular integrins, recognize specific matrix interactions sites. They not only facilitate cell adhesion, but also transfer biochemical signals into the cell as a response to this interaction.[3] These biochemical signals provide the cell with information about its position, influencing the responsiveness of the cell to growth factors and cytokines.[14] Angiogenic growth factors and cytokines not only influence endothelial cell proliferation, but also affect the production of matrix modifying proteases.[9,10] These proteases are involved in cell migration and invasion by acting on cell–matrix interactions, and they contribute to the formation of a lumen of the new vascular structures.

ANGIOGENESIS IN A THREE-DIMENSIONAL FIBRIN MATRIX *IN VITRO*

Better understanding of the mechanisms that contribute to the invasion and tube formation in a fibrin matrix was obtained by a model, in which human microvascular endothelial cells form capillary-like tubular structures in a three-dimensional fibrin matrix[14] (modified from Ref. 15). With the exception of endometrial endothelial cells that directly respond to angiogenic growth factors, human microvascular endothelial cells have to be exposed simultaneously to both an angiogenic growth factor, such as VEGF-A, bFGF, aFGF, or HGF, and the cytokine TNFα (see FIGURE 2A and B).[14] The onset of tubular structures becomes visible after two to three days and these grow into complex networks during the subsequent five to ten days. The structures are real tubules and have an open lumen (FIG. 2C). The extent of tube formation can be altered by steroid hormones[16] and by the oxygen tension.[17] When the oxygen tension is reduced from ambient oxygen (20%) to a hypoxic condition (1% oxygen) the extent of tube formation is increased (FIG. 2D).

Electron microscopy examination of the invading capillary-like structures made it clear that the fibrin structure adjacent to the invading cells was partly degraded, suggesting that proteolytic processes are involved in cell invasion. Cell bound u-PA and plasmin activities played a major role in endothelial cell invasion in a fibrin matrix.[14,18] The formation of tubular structures was completely prevented by inhibitors of the activation of plasmin (ε-aminocaproic acid) or plasmin activity (aprotinin) and by the presence of anti-u-PA antibodies.[14,18,19] Anti-t-PA antibodies had no effect. Prevention of u-PA interaction with its cellular receptor by a blocking monoclonal antibody against the u-PA receptors (mAb H-2) also inhibited this process, suggesting that the action of u-PA occurred at the u-PAR.[20] Because the amino terminal fragment of u-PA could not act as a substitute for u-PA, it is unlikely that signal transduction through the u-PA receptor was responsible. It is more probable that the u-PAR acted as a docking site for cell bound u-PA activity.

When u-PA and u-PA receptor are required for neovascularization in fibrin, one expects that these proteins are expressed in newly formed microvessels. Indeed, both

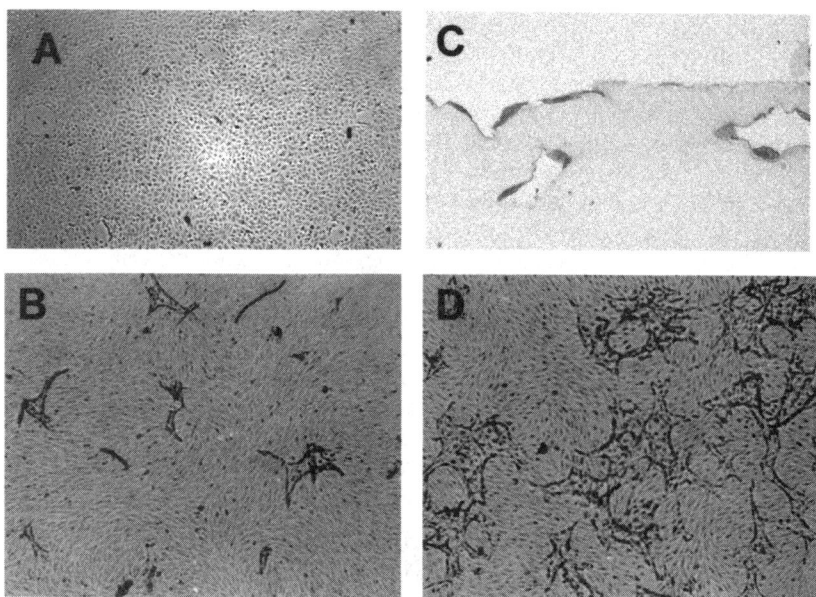

FIGURE 2. Human foreskin microvascular endothelial cells seeded on top of a three-dimensional fibrin matrix remain quiescent when stimulated by bFGF or TNFα in a manner comparable to non-stimulated conditions (**A**). When bFGF and TNFα are added simultaneously, they form capillary-like tubular structures (**B**), a process that is enhanced when the cells grow under hypoxic condition at 1% oxygen atmosphere (**D**). Panel **C** shows a cross-section through these structures. (Modified from Ref. 17.)

u-PA and u-PA receptor antigens were present in these new vascular structures, whereas the endothelial cells that remained on top of the fibrin matrix did not express these proteins.[20] These findings were confirmed *in vivo* in the neovessels of a mural thrombus.[20]

INVOLVEMENT OF UPA AND MMPS IN THE NEOVASCULARIZATION OF A PLASMA CLOT

In vivo a fibrinous exudate is not encountered as a pure fibrin matrix, but as a plasma clot that also contains vitronectin, fibronectin and other proteins. It is also often intermingled with the collagen fibers of the interstitial matrix. Because of the more complex matrix and cell–matrix interaction sites, it is likely that the requirement of cell bound proteases is also more complex. This is indeed the case. When human microvascular endothelial cells were seeded on top of a plasma clot and subsequently stimulated with bFGF and TNFα, we observed that both the u-PA/plasmin system and matrix metalloproteinase activity contributed to invasion and tube formation by endothelial cells. Tube formation was partly inhibited by aprotinin and partly by the metalloproteinase inhibitor BB94. These inhibitors acted additively. These data point to the involvement of both protease systems in the neovascularization of a plasma clot. In this context it is of interest to note that TNFα induces both u-PA and a number of MMPs in human endothelial cells.[21–23] Thus, the endothelial cell is enabled to execute a coordinated expression of several types of proteases that can modulate its extracellular matrix.[10] Cooperative action between the plasminogen activator/plasmin system and matrix metalloproteinases has been suggested by many experiments. Plasmin can contribute to the activation of several MMPs *in vitro*.[24] Furthermore, in plasminogen deficient conditions MMPs can even act as fibrinolysins.[25] Cooperative interaction between the plasminogen activator/plasmin system and matrix metalloproteinases is also suggested in the tissue repair. Although data on u-PA and plasminogen deficient mice indicate that developmental angiogenesis proceeds in the absence of a functional u-PA/plasmin cascade,[26–28] the u-PA system can enforce the action of matrix metalloproteinases when increased demand on neovascularization occurs after tissue damage (cf. Ref.29).

CONTRIBUTION OF THE FIBRIN STRUCTURE TO ANGIOGENESIS *IN VITRO*

Cell–matrix interactions determine whether or not a stimulated endothelial cell will invade the matrix successfully. In addition to matrix remodeling proteases and cellular matrix receptors, in particular integrins, the structure of the matrix itself is an important determinant in the angiogenic process. On the one hand it provides the binding sites for cellular integrins and as such it acts as a scaffold for invading cells. On the other hand, the structure of the matrix also determines the rate and extent of the proteolytic degradation of this matrix by invading cells. To understand how the matrix contributes to angiogenesis, we focused on the structure of the fibrin matrix itself. We used fibrin matrices *in vitro* with altered structures induced by different physical conditions during coagulation, by the addition of heparins during coagulation, and by using a mutant fibrinogen with a truncated Aα chain.

In the first group of experiments the structure of the fibrin matrix was altered by polymerizing of fibrin at various pH ranging from 7.0 to 7.8.[30] At pH 7.0 the fibrin matrix consists of relatively thick fibers and is malleable and turbid. At pH 7.8 the fibrin matrix is a dense network of fine fibers, which is rigid and translucent. The pH 7.0 fibrin matrix is more susceptible to proteolysis by plasmin than the rigid pH 7.8 fibrin matrix.[31] A fibrin matrix formed at pH 7.4 has intermediate properties. After these matrices were formed, they were equilibrated in serum containing culture medium at pH 7.4, and subsequently human microvascular endothelial cells were seeded on top of these matrices. The ingrowth of capillary like tubular structures induced by bFGF/TNFα or VEGF/TNFα was considerably faster in the more malleable matrix than in the rigid one (see FIGURE 3B). However, after prolonged incubation the structures in the more rigid matrix were more stable, because the malleable fibrin 7.0 matrix lysed to such an extent that it lost its scaffolding function. Thus, the structure of the fibrin matrix appears to be an important determinant in the extent of neovascularization and possibly in the stability of the newly formed

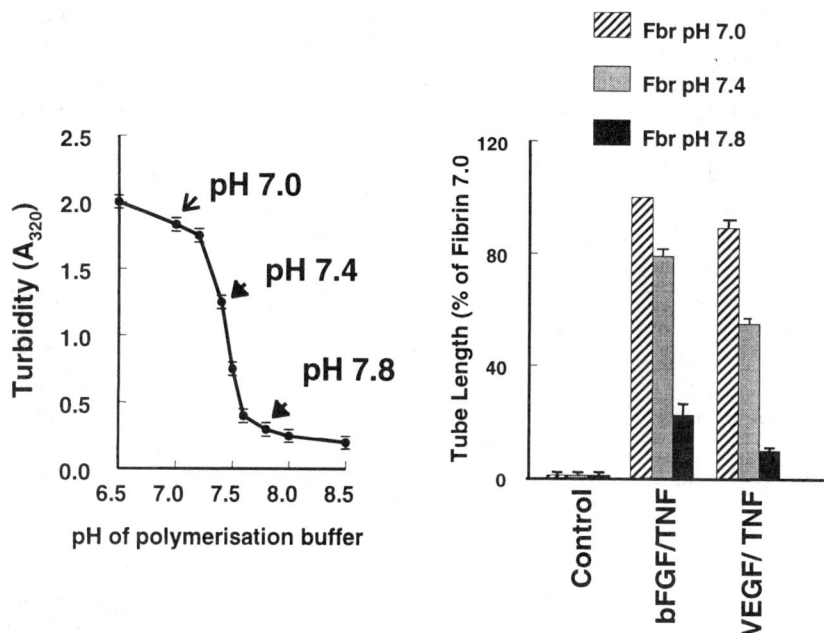

FIGURE 3. When fibrin is polymerized at different pH the structure is affected. This results in a proteolytically sensitive, more turbid and malleable fibrin matrix at pH 7.0. When these matrices were equilibrated with serum containing medium at pH 7.4, they affected the subsequent ingrowth of capillary like tubular structures by human microvascular endothelial cells to a considerable extent. (Modified from Ref. 19.)

vessels.[19,32] Stabilization of the fibrin matrix is also achieved by crosslinking the fibrin matrix, either by plasma factor XIII[33,34] or by tissue transglutaminase.[35] *In vivo* the addition of tissue transglutaminase and the subsequent crosslinking of the fibrin matrix improves wound healing.[36] The improved healing process is probably related to the stabilization of the matrix, which is needed to support the invading cells for a prolonged period.

In a second group of experiments the fibrin structure was altered by the addition of unfractionated heparin (UFH) and low molecular weight heparin (LMWH).[37] The presence of LMWH during polymerization made the matrices more rigid, whereas UFH did the opposite. Parallel to the altered structure, the invasion of the fibrin matrix by bFGF/TNFα- or VEGF/TNFα-stimulated human microvascular endothelial cells was changed; that is, the ingrowth of vascular structures was reduced in the fibrin matrices that were prepared in the presence of LMWH (see FIGURE 4). Because UFH and LMWH had identical effects on the proliferation of these endothelial cells, it is likely that indeed the altered structure of fibrin was responsible for the effect.[37] The finding that LMWH reduced the extent of angiogenesis may help to explain the clinical observation that the survival of tumor patients treated with LMWH is higher than that of patients treated with UFH.[38]

Further information regarding the contribution of the fibrin matrix to the process of angiogenesis was obtained with a mutant fibrinogen in which the Aα-chain was truncated. In this fibrinogen mutant, called fibrinogen$_{Nieuwegein}$, a homozygous insertion of a single nucleotide (C) in codon Aα453 (Pro remained unaltered) introduced a stop codon at position 454.[39] As a consequence the parent fibrinogen with only a deletion of the carboxy-terminal segment Aα454–610 was obtained (see FIGURE 5A). Because of this deletion the RGD sequence at Aα572–574 and the crosslinking site for tissue transglutaminase were absent. Furthermore, the deletion caused an unpaired cysteine at Aα442, which generated fibrinogen–albumin complexes. This resulted in a marked alteration of the fibrin matrix, that became a finer and more rigid network than control fibrin (FIG. 5B and C).

The truncation of fibrinogen$_{Nieuwegein}$ had no effect on the adherence of endothelial cells to fibrin suggesting that the RGD-sequence at Aα572–574 was dispensable for endothelial adherence to fibrin.[39] An important role of this sequence was previously suggested.[40,41] The formation of tubular structures, however, was markedly impaired.[39] This appeared mainly to be due to the altered fibrin structure, which was

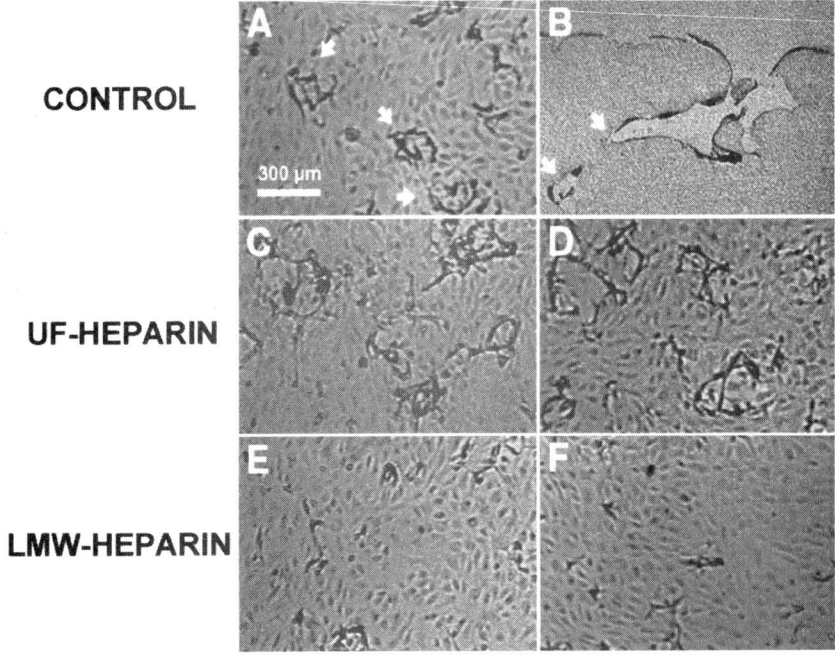

FIGURE 4. Low molecular weight (LMW) heparin and unfractionated (UF) heparin affect fibrin polymerization and structure. This causes a reduced the ingrowth of endothelial cells and the formation of capillary like tubular structures in fibrin matrices that had been polymerized in the presence of LMW-heparin. Panel **B** shows a cross section through a tubular structure in the control matrix. (Reproduced from Ref. 37, with permission.)

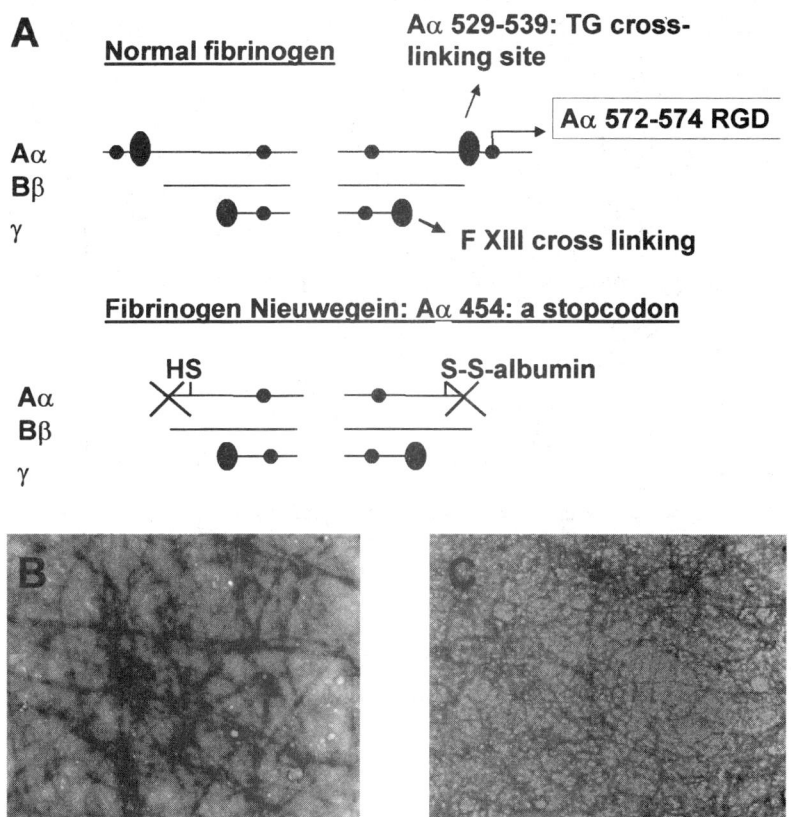

FIGURE 5. A. The mutant fibrinogen$_{Nieuwegein}$ has a truncated Aα chain at codon Aα454. This causes the loss of an RGD sequence at Aα572–574, the loss of the crosslinking site for tissue transglutaminase, and the generation of a free sulfydryl group that binds albumin. **B** and **C**, the structure of fibrinogen$_{Nieuwegein}$ as visualized by transmission electron microscopy.

more rigid and resistant to plasmin degradation than the normal fibrin matrix formed at pH 7.4. Indeed lowering the pH during polymerization, which made the matrix more malleable, partially increased tube formation. A reduction of the sensitivity of the fibrin matrix to plasmin is also obtained by crosslinking either by plasma derived factor XIII or by tissue transglutaminase. In this context it is of interest that fibrinogen$_{Nieuwegein}$ lacks the tissue transglutaminase crosslinking site. Whereas factor XIII further decreased the ingrowth of microvascular endothelial cells in control fibrin and fibrin$_{Nieuwegein}$, tissue transglutaminase did this only in control fibrin (see FIGURE 6). The more dense structure of fibrin$_{Nieuwegein}$ may compensate for the inability of tissue transglutaminase to crosslink the fibrin matrix.

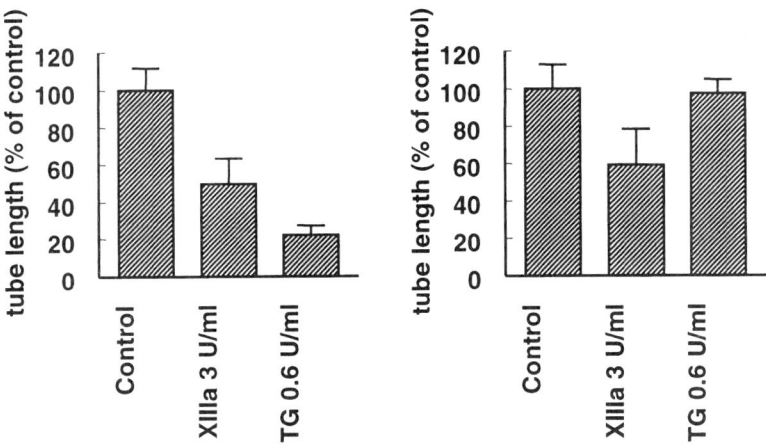

FIGURE 6. The effect of fibrin crosslinking by plasma factor XIII and tissue transglutaminase on capillary like tube formation by human microvascular endothelial cells on fibrin matrices of control fibrin and fibrinogen$_{\text{Nieuwegein}}$. (Data modified from Ref. 39.)

CONCLUSION

In this short survey we have focused on the specific role of the fibrin matrix in angiogenesis. We demonstrated that not only fibrin does act as a scaffold for invading endothelial cells, but also that the spatial structure of fibrin is an important determinant in the progression of angiogenesis *in vitro*. This is partially related to its altered sensitivity to proteases, such as plasmin and MMPs. In addition, the spatial organization of sites involved in cell matrix interaction may also facilitate or retard the ingrowth of endothelial cells. A similar phenomenon has been described for the spatial organization of type I collagen matrices.[42] Incorporation of vitronectin and other plasma proteins probably further improves the facilitation of cell invasion by the fibrin matrix. Indeed, our data on a plasma clot indicate that additional factors become involved in this more complex fibrin matrix.

In repair associated angiogenesis angiogenic growth factors, usually in interaction with inflammatory cytokines, activate endothelial cells to participate in the angiogenic response (see FIGURE 7). This response, also called the angiogenic switch,[43] is only successful in the proper environment. The inflammatory process and at least a major angiogenic factor VEGF cause vascular leakage, by which a fibrinous exudate is formed.[44] This fibrinous exudate provides additional scaffolding for the invasion process and as such accelerates neovascularization. Our data show that the structure of the fibrin matrix, additional molecules that are mixed in the fibrinous exudate, and proteases synthesized by and acting on the surface of endothelial and other invading cells, determine the extent of cell invasion and, thus, contribute to the success of the angiogenic process.

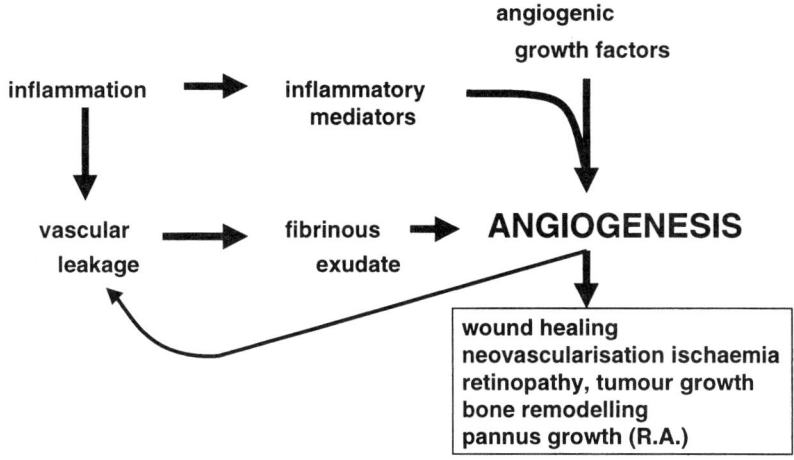

FIGURE 7. Schematic presentation of factors that contribute to repair associated angiogenesis.

The observation that the structure of fibrin affects the rate and extent of neovascularization as well as the stability of the new vessels may have an impact on the treatment of poorly healing wounds. Furthermore, because a fibrinous exudate is found not only in healing wounds, but also in many pathological conditions, such as the stroma surrounding many solid tumors, better understanding of the role of fibrin in angiogenesis may contribute to a better management of the angiogenic process, that is, the inhibition of unwanted angiogenesis and the stimulation of angiogenesis in ischemic tissues.

REFERENCES

1. FOLKMAN, J. 1995. Angiogenesis in cancer, vascular, rheumatoid and other disease. Nature Med. **1:** 27–31.
2. CARMELIET, P. & R.K. JAIN. 2000. Angiogenesis in cancer and other diseases. Nature **407:** 249–257.
3. SCHAPER, W. & I. BUSCHMANN. 1999. Arteriogenesis, the good and bad of it. Cardiovasc. Res. **43:** 835–837.
4. CARMELIET, P. & D. COLLEN. 2000. Transgenic mouse models in angiogenesis and cardiovascular disease. J. Pathol. **190:** 387–405.
5. FOLKMAN, J. 1995. Tumor angiogenesis. *In* The Molecular Basis of Cancer. J. Mendelsohn, P.M. Howley, M.A. Isreal & L.A. Liotta, Eds.: 206–232. W.B. Saunders, Philadelphia.
6. FERRIS, F.L., M.D. DAVIS & L.M. AIELLO. 1999. Treatment of diabetic retinopathy. N. Engl. J. Med. **341:** 667–678.
7. COLVILLE-NASH, P.R. & D.L. SCOTT. 1992. Angiogenesis and rheumatoid arthritis—pathogenic and therapeutic implications. Ann. Rheum. Dis. **51:** 919–925.
8. DVORAK, H.F., S. HARVEY, P. ESTRELLA, *et al.* 1987. Fibrin containing gels induce angiogenesis. Implications for tumor stroma generation and wound healing. Lab. Invest. **57:** 673–686.

9. VAN HINSBERGH, V.W.M., P. KOOLWIJK & R. HANEMAAIJER. 1997. Role of fibrin and plasminogen activators in repair-associated angiogenesis: *in vitro* studies with human endothelial cells. Reg. Angiogen. EXS. **79:** 391–411.
10. PEPPER, M.S. 1997. Manipulating angiogenesis—from basic science to the bedside. Arterioscler. Thromb. Vasc. Biol. **17:** 605–619.
11. YANCOPOULOS, G.D., M. KLAGSBRUN & J. FOLKMAN. 1998. Vasculogenesis, angiogenesis, and growth factors: Ephrins enter the fray at the border. Cell **93:** 661–664.
12. BENJAMIN, L.E., D. GOLIJANIN, A. ITIN, *et al.* 1999. Selective ablation of immature blood vessels in established human tumors follows vascular endothelial growth factor withdrawal. J. Clin. Invest. **103:** 159–165.
13. INGBER, D.E. 1990. Fibronectin controls capillary endothelial cell growth by modulating cell shape. Proc. Natl. Acad. Sci. U.S.A. **87:** 3579–3583.
14. KOOLWIJK, P., M.G.M. VAN ERCK, W.J.A. DE VREE, *et al.* 1996. Cooperative effect of TNFα, bFGF, and VEGF on the formation of tubular structures of human microvascular endothelial cells in a fibrin matrix. Role of urokinase activity. J. Cell Biol. **132:** 1177–1188.
15. MONTESANO, R. 1992. Regulation of angiogenesis *in vitro*. Eur. J. Clin. Invest. **22:** 504–515.
16. LANSINK, M., P. KOOLWIJK, V. VAN HINSBERGH, *et al.* 1998. Effect of steroid hormones and retinoids on the formation of capillary-like tubular structures of human microvascular endothelial cells in fibrin matrices is related to urokinase expression. Blood **92:** 927–938.
17. KROON, M.E., P. KOOLWIJK, B. VAN DER VECHT, *et al.* 2000. Urokinase receptor expression on human microvascular endothelial cells is increased by hypoxia: implications for capillary-like tube formation in a fibrin matrix. Blood **96:** 2775–2783.
18. PEPPER, M.S., D. BELIN, R. MONTESANO, *et al.* 1990. Transforming growth factor-β-1 modulates basic fibroblast growth factor induced proteolytic and angiogenic properties of endothelial cells *in vitro*. J. Cell Biol. **111:** 743–755.
19. COLLEN, A., P. KOOLWIJK, M.E. KROON, *et al.* 1998. Influence of fibrin structure on the formation and maintenance of capillary-like tubules by human microvascular endothelial cells. Angiogenesis **2:** 153–165.
20. KROON, M.E., P. KOOLWIJK, H. VAN GOOR, *et al.* 1999. Role and localization of urokinase receptor in the formation of new microvascular structures in fibrin matrices. Am. J. Pathol. **154:** 1731–1742.
21. VAN HINSBERGH, V.W.M., E.A. VAN DEN BERG, W. FIERS, *et al.* 1990. Tumor necrosis factor induces the production of urokinase-type plasminogen activator by human endothelial cells. Blood **75:** 1991–1998.
22. HANEMAAIJER, R., P. KOOLWIJK, L. LECLERCQ, *et al.* 1993. Regulation of matrix metalloproteinase expression in human vein and microvascular endothelial cells—effects of tumour necrosis factor-alpha, interleukin-1 and phorbol ester. Biochem. J. **296:** 803–809.
23. HANEMAAIJER, R., T. SORSA, Y.T. KONTTINEN, *et al.* 1997. Matrix metalloproteinase-8 is expressed in rheumatoid synovial fibroblasts and endothelial cells - regulation by tumor necrosis factor-alpha and doxycycline. J. Biol. Chem. **272:** 31504–31509.
24. MURPHY, G., H. STANTON, S. COWELL, *et al.* 1999. Mechanisms for pro matrix metalloproteinase activation. Apmis. **107:** 38–44.
25. HIRAOKA, N., E. ALLEN, I.J. APEL, *et al.* 1998. Matrix metalloproteinases regulate neovascularization by acting as pericellular fibrinolysins. Cell **95:** 365–377.
26. CARMELIET, P., L. SCHOONJANS, L. KIECKENS, *et al.* 1994. Physiological consequences of loss of plasminogen activator gene function in mice. Nature **368:** 419–424.
27. BUGGE, T.H., M.J. FLICK, C.C. DAUGHERTY, *et al.* 1995. Plasminogen deficiency causes severe thrombosis but is compatible with development and reproduction. Genes Dev. **9:** 794–807.
28. PLOPLIS, V.A., P. CARMELIET, S. VAZIRZADEH, *et al.* 1995. Effects of disruption of the plasminogen gene on thrombosis, growth, and health in mice. Circulation **92:** 2585–2593.

29. HEYMANS, S., A. LUTTUN, D. NUYENS, et al. 1999. Inhibition of plasminogen activators or matrix metalloproteinases prevents cardiac rupture but impairs therapeutic angiogenesis and causes cardiac failure. Nature Med. **5:** 1135–1142.
30. FERRY, J.D. & P.R. MORRISON. 1947. Preparation and properties of serum and plasma proteins. VIII. The conversion of human fibrinogen to fibrin under various conditions. J. Am. Chem. Soc. **69:** 388–400.
31. CARR, M.E.J., D.A. GABRIEL & J. MCDONAGH. 1987. Influence of factor XIII and fibronectin on fiber size and density in thrombin-induced fibrin gels. J. Lab. Clin. Med. **110:** 747–752.
32. NEHLS, V. & R. HERRMANN. 1996. The configuration of fibrin clots determines capillary morphogenesis and endothelial cell migration. Microvasc. Res. **51:** 347–364.
33. MATACIC, S. & A.G. LOEWY. 1966. Transglutaminase activity of the fibrin crosslinking enzyme. Biochem. Biophys. Res. Commun. **24:** 858–866.
34. PISANO, J.J., J.S. FINLAYSON, & M.P. PEYTON. 1968. Cross-link in fibrin polymerized by factor 13: epsilon-(gamma-glutamyl)lysine. Science **160:** 892–893.
35. MITKEVICH, O.V., J.H. SOBEL, J.R. SHAINOFF, et al. 1996. Monoclonal antibody directed to a fibrinogen A alpha #529-539 epitope inhibits alpha-chain crosslinking by transglutaminases. Blood Coagul. Fibrinol. **7:** 85–92.
36. HAROON, Z.A., J.M. HETTASCH, T.S. LAI, et al. 1999. Tissue transglutaminase is expressed, active, and directly involved in rat dermal wound healing and angiogenesis. FASEB J. **13:** 1787–1795.
37. COLLEN, A., S.M. SMORENBURG, E. PETERS, et al. 2000. Unfractionated and low molecular weight heparin affect fibrin structure and angiogenesis in vitro. Cancer Res. **60:** 6196–6200.
38. SIRAGUSA, S., B. COSMI, F. PIOVELLA, et al. 1996. Low-molecular-weight heparins and unfractionated heparin in the treatment of patients with acute venous thromboembolism: results of a meta-analysis. Am. J. Med. **100:** 269–277.
39. COLLEN, A., A. MAAS, T. KOOISTRA, et al. 2001. Aberrant fibrin formation and crosslinking of fibrinogen Nieuwegein, a variant with a shortened Aα-chain, alters endothelial capillary tube formation. Blood. **97:** 973–980.
40. BYZOVA, T.V., R. RABBANI, S.E. D'SOUZA, et al. 1998. Role of integrin alpha(v)beta3 in vascular biology. Thromb. Hæmost. **80:** 726–734.
41. THIAGARAJAN, P., A.J. RIPPON & D.H. FARRELL. 1996. Alternative adhesion sites in human fibrinogen for vascular endothelial cells. Biochem. **35:** 4169–4175.
42. VERNON, R.B., S.L. LARA, C.J. DRAKE, et al. 1995. Organized type I collagen influences endothelial patterns during "spontaneous angiogenesis in vitro": planar cultures as models of vascular development. In Vitro Cell Dev. Biol.—Animal **31:** 120–131.
43. HANAHAN, D. & J. FOLKMAN. 1996. Patterns and emerging mechanisms of the angiogenic switch during tumorigenesis. Cell **86:** 353–364.
44. DVORAK, H.F., L.F. BROWN, M. DETMAR, et al. 1995. Vascular permeability factor/vascular endothelial growth factor, microvascular hyperpermeability, and angiogenesis. Am. J. Pathol. **146:** 1029–1039.

Fibrinogen Modulates Gene Expression in Wounded Fibroblasts

MARIAN PEREIRA[a,b] AND PATRICIA J. SIMPSON-HAIDARIS[a,b,c]

Departments of [a]Medicine-Vascular Medicine Unit,
[b]Pathology and Laboratory Medicine, [c]Microbiology and Immunology,
University of Rochester School of Medicine and Dentistry, Rochester, New York, USA

> ABSTRACT: Fibrinogen (FBG) has long been regarded as serving essentially a hemostatic role by its conversion from a soluble, plasma protein to an insoluble fibrin gel. However, several extrahepatic sites of FBG biosynthesis have been identified. Indeed, we have demonstrated that both lung epithelial cell derived and plasma FBG assemble into the extracellular matrix (ECM) of epithelial cells and fibroblasts. In this report, we determined that FBG assembly into the ECM is a cell dependent step that occurs in the absence of *de novo* protein synthesis. Using an *in vitro* model of wound repair, we examined the role of FBG in modulating gene expression. Data collected from cDNA array analysis indicated that FBG downregulates steady state levels of fibronectin mRNA, whereas cyclin D1 mRNA levels were upregulated in fibroblasts. Taken together, these data suggest that FBG may function independently of hemostasis in cellular adhesive interactions to modulate cellular signaling processes during wound repair.
>
> KEYWORDS: Fibrinogen; Gene expression; Wounded fibroblasts; cDNA array; Extracellular matrix; Fibronectin.

Although fibrinogen (FBG) is typically described as a hepatically derived plasma protein, extrahepatic synthesis of intact FBG also occurs. Recent *in vitro* studies have detected extrahepatic gene expression of FBG in epithelial cells from intestine[1] and lung.[2] FBG expression is upregulated five- to ten-fold in lung epithelial cells (A549) following induction with dexamethasone and IL-6,[2] the proinflammatory mediators of FBG gene expression during the acute phase response to inflammation. Furthermore, the majority of the FBG secreted by A549 cells is directed basolaterally, where it becomes incorporated into the extracellular matrix independent of either thrombin or plasmin cleavage.[3,4] In addition to the finding that matrix FBG colocalizes with fibronectin (FN) and laminin fibrils, studies by Guadiz *et al.*[3] also revealed that upon its deposition into the matrix, FBG displays an epitope on the Bβ chain (β_{15-42}), normally only accessible after FBG has been proteolytically converted to fibrin. Because fibrin(ogen) functions in inflammation and wound repair by providing an ECM support for cellular responses involved in restoration of normal tissue architecture, we hypothesize that matrix FBG triggers signalling cascades to modulate gene expression by injured cells.

Address for correspondence: P.J. Simpson-Haidaris, Ph.D., Department of Medicine, P.O. Box 610, University of Rochester, 601 Elmwood Ave., Rochester, NY 14642, USA. Voice: 716-275-8267; fax: 716-473-4314.
pj_simpsonhaidaris@urmc.rochester.edu

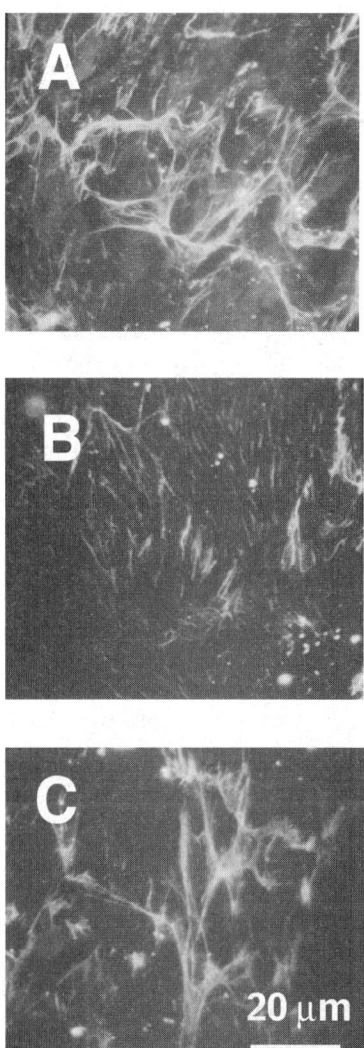

FIGURE 1. Effect of NaN$_3$ on FBG incorporation into ECM. HFF were cultured on glass coverslips for seven days, pretreated with NaN$_3$ (0.1%) for 30 min, and incubated with FBG-Oregon Green in the absence (**A**) or presence (**B** and **C**) of NaN$_3$ for 24 h. Some coverslips were washed to remove residual NaN$_3$ and incubated in the presence of FBG-Oregon Green for an additional 24 h (**C**).

To determine whether the incorporation of FBG into the ECM is a cell-mediated process as has been shown for FN, human foreskin fibroblasts (HFF) were metabolically arrested with sodium azide (NaN_3) and assayed for matrix deposition of FBG. FBG incorporation into the matrix in the presence of 0.1% NaN_3 (see FIGURE 1 B) was negligible as compared to control conditions (FIG. 1 A). To ascertain whether HFF could assemble FBG into the ECM after being metabolically arrested, a set of HFF cultures were washed to remove NaN_3, and incubated in the presence of only FBG-Oregon Green for another 24 hours. In the absence of NaN_3, HFF were again able to assemble exogenous FBG into matrix fibrils (FIG. 1 C). Taken together, these data suggest that FBG assembly into the ECM requires the presence of metabolically active cells.

To determine if the *de novo* protein synthesis is required for FBG incorporation into the ECM, HFF were treated with cyclohexamide (CHX), a protein synthesis inhibitor, and assayed for FBG assembly into the matrix. Although the pattern of FBG matrix fibrils in the presence of CHX (see FIGURE 2 B) differs from that of the control condition (FIG. 1 A), FBG is nonetheless assembled into the matrix in the absence of *de novo* protein synthesis. Therefore, even though newly synthesized proteins may be required for higher order or complex matrix fibril formation, the data suggest that the factors necessary for FBG incorporation into the ECM are functional in the absence of new protein synthesis.

From studies (Lawrence *et al.*, manuscript in preparation) performed in a wound scrape model of cell injury, we have shown that a preestablished FBG matrix induces ^{3}H-thymidine uptake in HFF as compared to control HFF that were not treated with FBG. To address whether matrix FBG plays a role in inducing cellular responses during the wound repair process, a cDNA expression array profile was performed on mRNA obtained from HFF that were treated as described in FIGURE 3. In the presence of matrix FBG, gene expression of cyclin D1 was upregulated, whereas that of

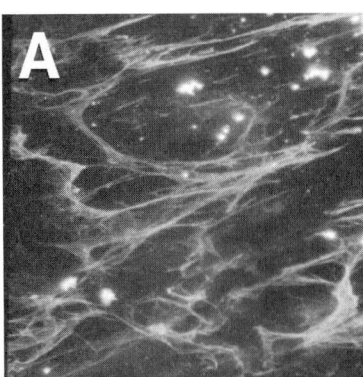 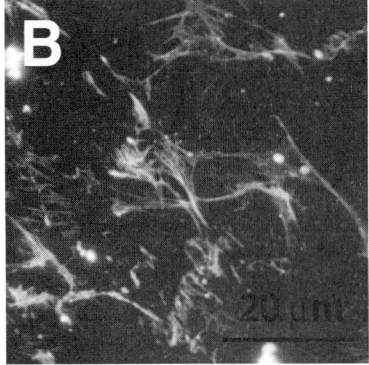

FIGURE 2. Effect of CHX on FBG incorporation into ECM. HFF were cultured on glass coverslips for seven days, pretreated with CHX (10 μg/ml) for two hours, and incubated with FBG-Oregon Green in the absence (**A**) or presence (**B**) of CHX for an additional 24 h.

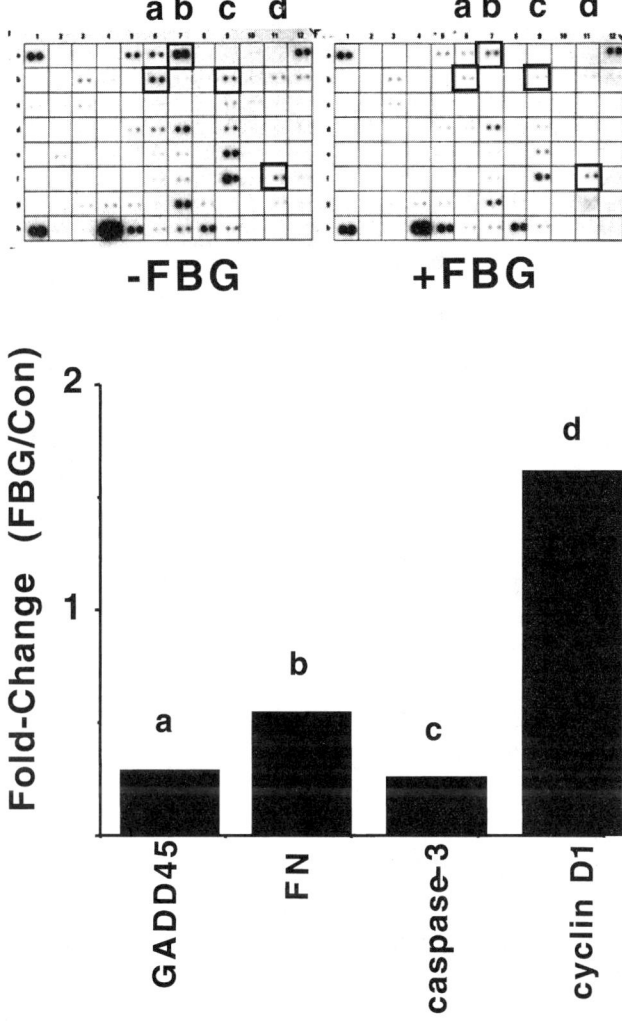

FIGURE 3. cDNA array analysis of HFF wound scrape model. HFF were cultured on 100 mm dishes and used in assay six days after seeding. HFF were untreated or pretreated for 24 h with FBG (40 μg/ml), wounded by scraping dish once with a 2-cm wide plastic cell scraper, and total RNA isolated 16 h after wounding. Radiolabeled cDNA was prepared from poly (A^+) mRNA and hybridized to the Clonetech Atlas cDNA expression. Blots were exposed to X-ray film, analyzed by densitometry and gene expression was normalized to GAPDH values. Data is presented as the change in gene expression in response to FBG treatment relative to untreated levels.

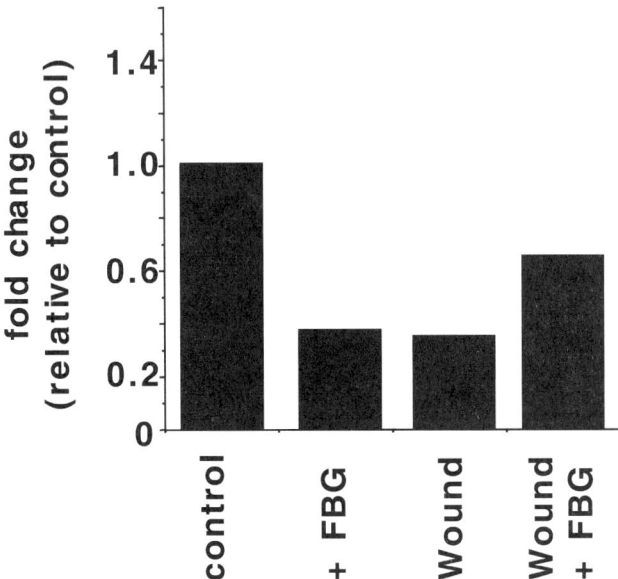

FIGURE 4. Effect of FBG on FN synthesis in fibroblasts. HFF were untreated (*control* and *wound*) or pretreated (*+FBG* and *wound +FBG*) with FBG (40 μg/ml) for 24 h, wounded (*wound, wound+FBG*) and concomitantly metabolically labeled with 40μCi/ml ^{35}S-methionine + cysteine. After 16 h, cells were lysed[6] with the addition of 100U/ml aprotinin. FN was immunoprecipitated from cell lysate using affinity purified rabbit antihuman FN IgG and analyzed by SDS-PAGE and fluorography. Data is presented as the change in FN protein levels in response to FBG treatment and/or wounding relative to untreated and non-wounded (*control*) condition.

FN, growth-arrest and DNA damage inducible 45A (GADD45A), and caspase-3 was downregulated. In the context of wound repair, both the upregulation of cyclin D1 and the downregulation of GADD45A serve to induce cellular proliferation. The transcriptional downregulation of caspase-3 in the presence of matrix FBG during wound repair may serve to protect components of the cytoskeleton and nucleus as well as proteins involved in signaling pathways that are possible targets of caspase-3; activation of the procaspase to its active form as well as inhibitors of apoptosis proteins are other important means of regulating caspase activity. The downregulation of FN synthesis in the presence of matrix FBG may promote cell migration into the wound space, since high concentrations of FN cause focal adhesion formation and inhibition of cell movement.[5] The downregulation of FN gene expression is supported by preliminary data showing a decrease in FN biosynthesis in the presence of matrix FBG during wound repair (see FIGURE 4) and Northern blot analysis (not shown). Taken together these data support a functional role for FBG as an ECM protein that regulates cellular processes involved in wound healing.

REFERENCES

1. MOLMENTI, E.P., T. ZIAMBARAS & D.H. PERLMUTTER. 1993. Evidence for an acute phase response in human intestinal epithelial cells. J. Biol. Chem. **268:** 14116–14124.
2. HAIDARIS, P.J. 1997. Induction of fibrinogen biosynthesis and secretion from cultured pulmonary epithelial cells. Blood **89:** 873–882.
3. GUADIZ, G., L.A. SPORN & P.J. SIMPSON-HAIDARIS. 1997. Thrombin cleavage-independent deposition of fibrinogen in extracellular matrices. Blood **90:** 2644–2653.
4. GUADIZ, G., L.A. SPORN, R.A. GOSS, et al. 1997. Polarized secretion of fibrinogen by lung epithelial cells. Am. J. Respir. Cell Mol. Biol. **17:** 60–69.
5. DIMILLA, P.A., J.A. STONE, J.A. QUINN, et al. 1993. Maximal migration of human smooth muscle cells on fibronectin and type IV collagen occurs at an intermediate attachment strength. J. Cell Biol. **122:** 729–737.
6. SPORN, L.A., V.J. MARDER & D.D. WAGNER. 1989. Differing polarity of the constitutive and regulated secretory pathways for von Willebrand factor in endothelial cells. J. Cell Biol. **108:** 1283–1289.

Mutations on Fibrinogen (γ316–322) Are Associated with Reduction in Platelet Adhesion Under Flow Conditions

JASPER A. REMIJN,[a] KARIM C. LOUNES,[b] KELLY A. HOGAN,[b] SUSAN T. LORD,[b] DENNIS K. GALANAKIS,[c] JAN J. SIXMA,[a] AND PHILIP G. DE GROOT[a]

[a]*Thrombosis and Hæmostasis Laboratory, Department of Hæmatology, University Medical Center, Utrecht, the Netherlands*

[b]*Department of Pathology and Laboratory Medicine, University of North Carolina at Chapel Hill, Chapel Hill, North Carolina, USA*

[c]*State University of New York at Stony Brook, Stony Brook, New York, USA*

> ABSTRACT: In this paper we report on studies of platelet adhesion to several fibrinogen γ chain variants under physiological flow conditions. Reduced platelet adhesion was found to patient dysfibrinogen Vlissingen and its recombinant form (deletion of γ319–320). Furthermore, substitutions of the amino acids 318, 320, or both in the recombinant fibrinogen γ chain showed a strong decrease in platelet adhesion under flow conditions in our perfusion system. Antibodies raised against peptides covering these sequences inhibited platelet adhesion completely, which suggested that the γ316–322 sequence could be involved in platelet adhesion in flowing blood.
>
> KEYWORDS: Fibrinogen γ chain; Platelet adhesion; Flow conditions.

INTRODUCTION

Fibrinogen plays a major role in hemostatic plug formation, not only as the precursor of fibrin formation and as a mediator of platelet–platelet interaction, but also as an adhesive molecule for platelets. Resting, unstimulated platelets do not interact with fibrinogen in the fluid phase. Following activation of platelets, glycoprotein IIbIIIa ($\alpha_{IIb}\beta_3$) undergoes a conformational change and provides a binding site for fibrinogen.[1,2] When fibrinogen is immobilized on a surface, however, platelets adhere to it without any proceeding activation.[3,4] Fibrinogen binds to platelets through multivalent interactions with $\alpha_{IIb}\beta_3$ integrins present on the surface of platelets. Three sites on fibrinogen have been postulated to bind to $\alpha_{IIb}\beta_3$ during the interaction with platelets, two Arg-Gly-Asp (RGD) sites at positions 95–97 and 572–574 of the α chain and the carboxyterminal dodecapeptide of the γ chain (γ400–411). Experiments with antibodies and genetically engineered fibrinogen variants lacking the RGD sequences in the α chain have demonstrated that the α chain RGD sequences are not

---

Address for correspondence: Dr. J.A. Remijn, Thrombosis and Hæmostasis Laboratory, Department of Hæmatology, University Medical Center, Utrecht, P.O.Box 85500, 3508 GA Utrecht, the Netherlands. Voice: + 31 30 250.6498; fax: + 31 30 251.1893.

J.A.Remijn@lab.azu.nl

required for $\alpha_{IIb}\beta_3$ binding to soluble and surface bound fibrinogen or for platelet aggregation.[5–8] Instead, genetic variants of fibrinogen with either an extension of the γ chain or truncation of the γ chain show that the C-terminal sequence of the γ chain is an important determinant on fibrinogen for platelet binding and aggregation.[9] These studies, however, do not exclude the possibility that other domains of fibrinogen may also play a crucial role in the interaction with platelets, assuming that the γ chain dodecapeptide sequence is necessary but not sufficient to mediate irreversible platelet attachment.[10] In order to obtain more insight into the involvement of other domains of fibrinogen in platelet adhesion in flowing blood, we studied platelet adhesion under flow conditions to dysfibrinogen Vlissingen with a known deletion in the γ chain (Δ γ319–320) purified from patient plasma and to several recombinant fibrinogen γ chain variants.

PERFUSION STUDIES

To study platelet adhesion under flow conditions, perfusions were performed with a single pass perfusion chamber under non-pulsatile flow conditions using a modified small perfusion chamber.[11] Coverslips were incubated with 100 µg/ml fibrinogen γ chain variants in PBS for one hour at room temperature and coverslips were blocked for one hour with 1% human albumin solution in PBS. Citrate anticoagulated blood, prewarmed to 37°C for ten minutes, was drawn for five minutes through the perfusion chamber by a syringe placed in an infusion pump. After perfusion the coverslips were rinsed in Hepes buffered saline, fixed in 0.5% glutaraldehyde in PBS, dehydrated in methanol, and stained with May-Grünwald-Giemsa. Evaluation of platelet deposition was performed using a computer assisted analysis.

PLATELET ADHESION TO FIBRINOGEN γ CHAIN VARIANTS UNDER FLOW CONDITIONS

Results obtained from whole blood perfusion experiments demonstrated reduced platelet adhesion to patient dysfibrinogen Vlissingen (Δ γ319–320). These results were obtained both at physiological shear rates of 300 sec^{-1} and 1,600 sec^{-1} representing veins and arteries. Perfusion experiments using the recombinant fibrinogen γ chain variants: γ318 D→A, γ320 D→A, 318,320 D→A, and deletion of γ319–320 (Δ γ319–320) also showed reduced platelet adhesion under flow conditions at shear rates of 300 sec^{-1} and 1,600 sec^{-1}. To exclude the possibility that the lack of platelet adhesion was due to improper coating of the dysfibrinogens, an ELISA was set up to measure the amount of fibrinogen coated to the surface. Comparable amounts of fibrinogen were present at all fibrinogen preparations.

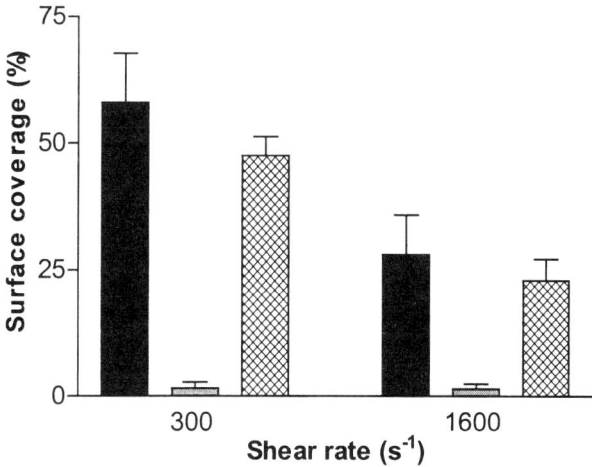

FIGURE 1. Effect of antipeptide F(ab)$_2$ fragments on platelet adhesion to fibrinogen under flow conditions (300 sec^{-1} and 1,600 sec^{-1}). Coverslips were coated with fibrinogen and incubated with F(ab)$_2$ fragments directed against γ308–322 and γ316–333 and F(ab)$_2$ fragments directed against ED-1 fibronectin as a control. Results are expressed as percentage of the surface covered with platelets. Data are mean ± SEM of results obtained in three separate perfusion experiments; $p < 0.001$ for fibrinogen preincubated with the antipeptide F(ab)$_2$ fragments versus fibrinogen preincubated with control anti-ED1 F(ab)$_2$ fragment. ■, Fg; ▨, Fg + F(ab)$_2$ γ308–333; ▩, Fg + F(ab)$_2$ ED-1;

EFFECT OF ANTIPEPTIDE ANTIBODIES ON PLATELET ADHESION UNDER FLOW CONDITIONS

To further examine the participation of the γ318–320 region of the γ chain on fibrinogen in platelet adhesion, F(ab)$_2$ fragments of the antibodies directed against peptides covering the sequences γ308–322 and γ316–333 were prepared. Coverslips coated with fibrinogen were preincubated with F(ab)$_2$ fragments (100 µg/ml). Preincubation of fibrinogen with either the F(ab)$_2$ fragments directed against γ308–322 or γ316–333 separately did not significantly inhibit platelet adhesion. However, when the coverslips coated with fibrinogen were incubated together with both antibodies, platelet adhesion was inhibited by more than 90% at both shear rates 300 sec^{-1} and 1,600 sec^{-1} (see FIGURE 1). Incubation with a control F(ab)2 fragment directed against fibronectin did not influence platelet adhesion.

DISCUSSION

The study reported here demonstrated that platelet adhesion to surface bound fibrinogen in flowing blood requires the involvement of more than one site in the γ chain of fibrinogen. We found that the sequence around γ316–322 is also essential for platelet interaction with immobilized fibrinogen. Deletion of the amino acids at position 319 and 320 resulted in reduced platelet adhesion in the patient dysfibrinogen and the recombinant γ chain variant. Furthermore, substitutions of amino acids 318, 320, or both in the recombinant fibrinogen γ chain showed a strong decrease in platelet adhesion under physiological flow conditions. To exclude the possibility that these mutations resulted from a conformational change in the molecule, antibodies were raised against linear peptides covering these sequences. When the antibodies were added together, platelet adhesion was totally absent, indicating that the loss of adhesiveness was due to loss of a binding site for platelets, although we can not exclude completely a pertubation in the structure of fibrinogen.

Interestingly, the antibodies did not inhibit adhesion when incubated separately with fibrinogen. The antibodies were raised against the sequences γ308–322 and γ316–333, respectively. Neither antibody recognized the peptide sequence used for the generation of the other antibody, indicating that the antibodies did not recognize the overlapping sequence of the peptides. Thus, the first antibody recognized the sequence γ308–316, whereas the second antibody recognized the sequence γ322–333. Clearly, neither sequence, individually, is involved in the recognition by platelets. Only the combination of antibodies results in sufficient steric hindrance to prevent platelet adhesion. This reasoning points to the sequence γ316–322 as the crucial domain. Our data indicate that the γ316–322 sequence in fibrinogen is involved in platelet adhesion to fibrinogen under flow conditions.

ACKNOWLEDGMENT

J.A. Remijn was supported by Grant 95.169 of the Netherlands Heart Foundation.

REFERENCES

1. BENNETT, J.S., *et al.* 1979. Exposure of platelet fibrinogen receptors by ADP and epinephrine. J. Clin. Invest. **64:** 1393–1401.
2. MARGUERIE, G.A., *et al.* 1980. Interaction of fibrinogen with its platelet receptor as part of a multistep reaction in ADP-induced platelet aggregation. J. Biol. Chem. **255:** 154–161.
3. SAVAGE, B., *et al.* 1991. Selective recognition of adhesive sites in surface-bound fibrinogen by glycoprotein IIb-IIIa on nonactivated platelets. J. Biol. Chem. **266:** 11227–11233.
4. KIEFFER, N., *et al.* 1991. Adhesive properties of the beta 3 integrins: comparison of GP IIb-IIIa and the vitronectin receptor individually expressed in human melanoma cells. J. Cell Biol. **113:** 451–461.
5. HANTGAN, R.R., *et al.* 1995. Evidence that fibrin alpha-chain RGDX sequences are not required for platelet adhesion in flowing whole blood. Blood **86:** 1001–1009.
6. FARRELL, D.H., *et al.* 1992. Role of fibrinogen alpha and gamma chain sites in platelet aggregation. Proc. Natl. Acad. Sci. U.S.A. **89:** 10729–10732.
7. FARRELL, D.H., *et al.* 1994. Binding of recombinant fibrinogen mutants to platelets. J. Biol. Chem. **269:** 226–231.
8. HOLMBACK, K., *et al.* 1996. Impaired platelet aggregation and sustained bleeding in mice lacking the fibrinogen motif bound by integrin alpha IIb beta 3. EMBO J. **15:** 5760–5771.

9. ZAIDI, T.N., *et al.* 1996. Adhesion of platelets to surface-bound fibrinogen under flow. Blood **88:** 2967–2972.
10. CHEN, C.S., *et al.* 1988. Fibrin(ogen) peptide B beta 15-42 inhibits platelet aggregation and fibrinogen binding to activated platelets. Biochemistry **27:** 6121–6126.
11. SIXMA, J.J., *et al.* 1998. A new perfusion chamber to detect platelet adhesion using a small volume of blood. Thromb. Res. **92:** S43–S46.

A New Model for *in Vitro* Clot Formation that Considers the Mode of the Fibrin(ogen) Contacts to Platelets and the Arrangement of the Platelet Cytoskeleton

EBERHARD MORGENSTERN,[a] MATTHIAS DAUB,[a] AND ROLF DIERICHS[b]

[a]*Medical Biology, Faculty of Medicine, Saarland University, Homburg, Germany*

[b]*Institute of Anatomy, University of Muenster, Germany*

ABSTRACT: The constitution of platelet–fibrin(ogen) contacts, the separation of the platelets initially aggregated, and the rearrangement of the platelet cytoskeleton during clot formation (0.5 to 60 minutes after thrombin stimulation) were investigated using ultrastructural and immunocytochemical techniques. After aggregation, fibrin polymerizing within focal contacts and from degranulating secretory granules contributed to the fibers. The initially formed focal contacts with fibrin obviously persisted during clot formation. The physiological branching of the fibers enabled separation of platelets. The contact associated cytoskeleton formed a constricting and fiber internalizing sphere, but later stress fiber like bundles. As retraction progressed, the cytoskeleton changed to stress fiber connecting focal contacts with fibers. A model of clot formation *in vitro* is presented that reflects both the contributions of platelets (fibrin fiber internalization and retraction) and of fibers (branching) enabling the retraction.

KEYWORDS: Clot retraction; Platelet cytoskeleton; Platelet–fibrin(ogen) interaction.

INTRODUCTION

A previous study concerned the interaction of internalizing platelets with fibrin(ogen) during the first 10 minutes of clot formation.[1] The side-to-side binding mode of fibrin fibers on platelets has been demonstrated in retracted clots 30–60 minutes after thrombin addition.[2] The study reported here deals with the entire process of clot formation, taking into account the mode and the fate of the platelet–fibrin contacts, the separation of platelets after aggregation, and the retraction performed by the action of the contractile cytoskeleton in platelets.

Address for correspondence: Dr. Eberhard Morgenstern, Medizinische Biologie, Saarland Universitaet, D-66421 Homburg, Saar, Germany. Voice: +49 6841 16 6252; fax: +49 6841 16 6256.
em@rz.uni-saarland.de

MATERIALS AND METHODS

Thrombin induced platelet clots were prepared as described elsewhere.[1] Clot formation was stopped after 30 seconds, 1, 5, 15, 30, and 60 minutes by glutaraldehyde fixation. After postfixation with OsO_4 the clots were embedded in araldite. For immunolabeling, the clots were fixed with 4% paraformaldehyde and 0.1% glutaraldehyde at 4°C, dehydrated with ethanol by progressively lowering the temperature, and embedded at $-30°C$ in Lowicryl HM20 (Chemische Werke Lowi, Waldkraiburg, Germany). Ultrathin sections were labeled with rabbit antihuman fibrinogen IgG (Dakopatts, Denmark) or rabbit antihuman platelet myosin IgG (a gift from Prof. Dr. D. Drenckhahn, Institute of Anatomy, University Wuerzburg, Germany). After washing, a secondary goat antirabbit 6 nm gold antibody (Aurion, Wageningen, Netherlands) was used.

RESULTS

During the first 60 seconds, the degranulating platelets formed aggregates with focal contacts mediated by fibrinogen or polymerizing fibrin (see FIGURE 1). Invaginations of the plasma membrane filled with polymerizing fibrin were observed. Polymerizing fibrin also covered the luminal surfaces of the membrane of the degranulating alpha-granules. Five minutes after thrombin addition, the clot reduced its volume slightly. Branching fibrin fibers can be observed between the separating platelets (see FIGURE 2). A side-to-side binding of fibrin fibers, located in invaginations of the platelet membrane could be detected on serial sections (see FIGURES 3 and 4). The more or less degranulated platelets formed a centralized, fiber internalizing contractile cytoskeleton in which myosin was present (FIG. 4). After 15 minutes, the clots reduced their volume to 50%. Within the clots the platelets were found to be separated and located distant from each other. The platelets interacted with

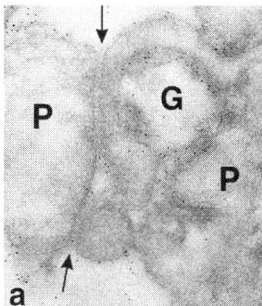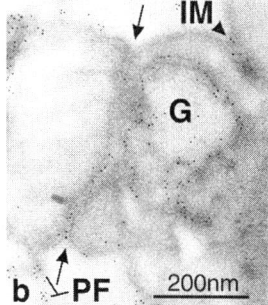

FIGURE 1. Contact of two aggregating platelets 60 sec after thrombin addition. Gold labels show fibrinogen or polymerizing fibrin (*PF* in **b**) within the contact (*arrows*) and on the surface of the platelets (*P*), in membrane invaginations (*IM*), and on the surface of alpha-granules in exocytosis (*G*).

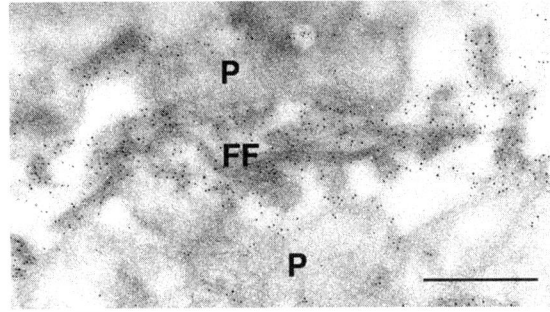

FIGURE 2. Contact of two aggregating platelets five minutes after thrombin addition. Gold labels show branching fibrin (*FF*) between two platelets (*P*).

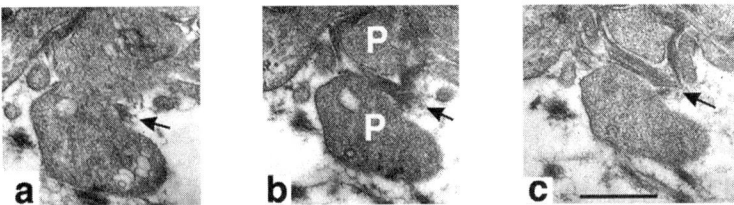

FIGURE 3. Binding of fibrin fibers on the platelet surface five minutes after thrombin addition. The fibers indicated by *arrows* seem to bind end-to-side on the platelet surface in sections **a** and **b**. However, in section **c** the side-to-side binding mode can be clearly seen.

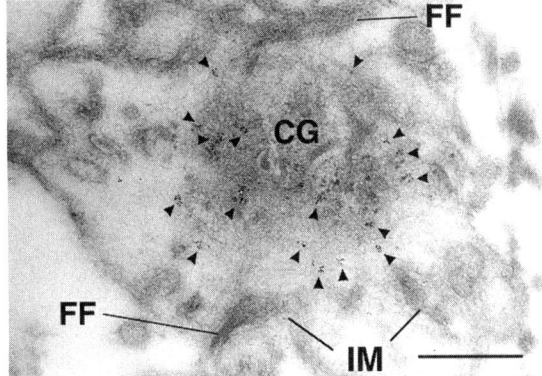

FIGURE 4. Myosin distribution in a platelet five minutes after thrombin addition. Labels indicated by *arrowheads* are located in the constricting sphere of the contractile gel (*CG*). Fibrin fibers (*FF*) are located within membrane invaginations (*IM*).

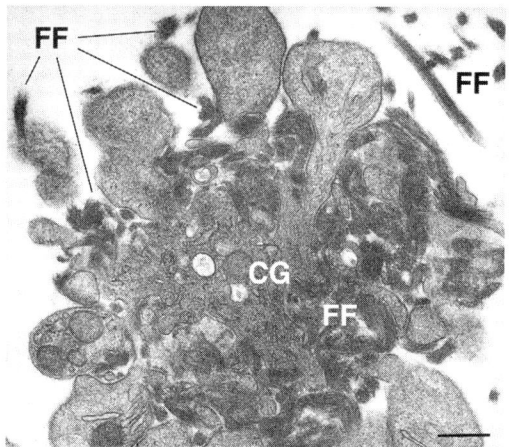

FIGURE 5. A separated platelet interacting with fibrin fibers 15 min after thrombin addition. Fibrin fibers (*FF*) are located in deep membrane invaginations. The contractile cytoskeleton (*CG*) is changing from the constricting sphere into filament bundles connecting the fiber contacts, as examined on 33 serial sections of this platelet (not shown).

thicker fibrin fibers, stretching between the platelets (see FIGURE 5). The focal contacts were located in membrane invaginations associated with the contractile gel (FIG. 5). Examination of serial sections showed that this contractile gel changed during this period from a constricting sphere into stress fiber like filament bundles connecting the fibrin contacts. Later, the platelets again came into close contact and were surrounded by fiber masses. The focal contacts of the platelets with side-to-side binding fibers were connected by stress fibers, inserted end-to-side in the plasma

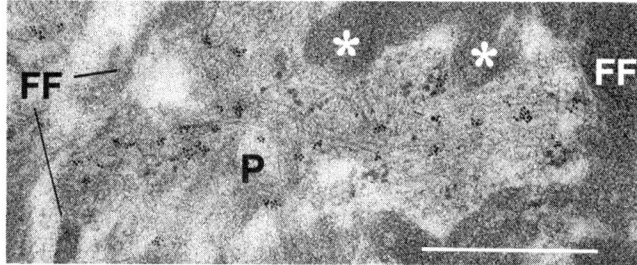

FIGURE 6. Myosin distribution in a platelet 30 min after thrombin addition. Labels are seen on the microfilament bundles of the contractile cytoskeleton connecting, from right to left, the contacts with fibrin fibers (*FF*). The fibers (*FF*) are located not only in membrane invaginations but also on the extended surface of the platelets. A portion of the fibers is attached to the platelet surface and is not connected with filament bundles inserting end-to-side in the plasma membrane (*asterisks*).

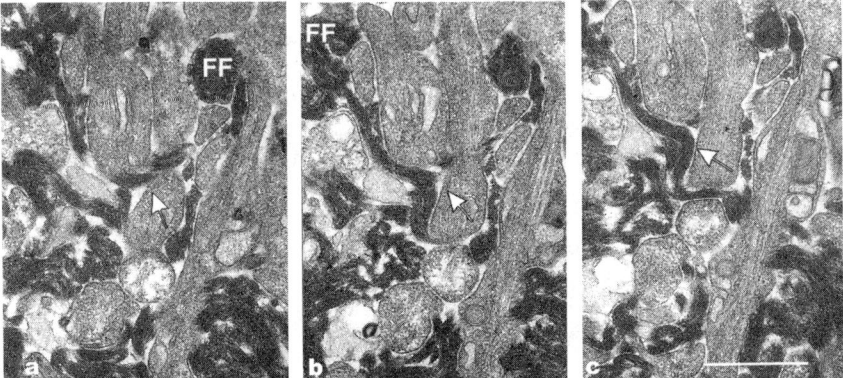

FIGURE 7. Binding of fibrin fibers on the platelet surface 60 min after thrombin addition. The fibers indicated by *arrows* appear to bind end-to-side on the platelet surface in section **a**. However, in sections **b** and **c** the side-to-side binding mode can be clearly recognized. The fibers (*FF* in **a** and **b**) are not located in membrane invaginations.

membrane (see FIGURES 6 and 7). The focal contacts were also located on extended surface areas of the platelets, not only in membrane invaginations (FIGS. 6 and 7). Myosin was located within the stress fibers throughout their length up to the focal contacts (FIG. 6). A portion of the fibers was not connected with filament bundles inserting end-to-side in the plasma membrane but attached to the platelet surface (FIG. 6).

DISCUSSION AND CONCLUSIONS

From a plethora of suggestions aimed at elucidating the clot formation, two models are based on detailed morphology studies. The first model reflects the generation of stress fiber like bundles in platelets during fibrin retraction under isometric conditions.[3] Internalization of fibrin fibers is considered in another model.[4] Neither model reflects in detail on the binding mode of fibrin fibers to the platelet surface, nor the separation of the initially aggregated platelets, nor the rearrangement of the contractile cytoskeleton from a constricting sphere to stress fiber like filament bundles. The model proposed in the schematic drawing shown in FIGURE 8 separates the thrombin induced formation of an *in vitro* clot into four, fairly well distinguishable, steps: aggregation, separation, retraction, and stabilization by generating tension. Results from an examination of serial sections concerning the fibrin contacts suggest that the initially formed focal contacts with fibrin (FIG. 8A) persist during clot formation. Polymerizing fibrin within focal contacts and from degranulating secretory granules contribute to the fibers bound on platelets, as is also suggested by biochemical findings.[5] The fibers bind all the time in a side-to-side mode on the platelet surface (FIG. 8B–D) in agreement with previous findings.[6] Separation of the platelets, a result of a physiological branching of the growing fibers,[7–9] enables the platelets to retract fibers (FIG. 8B). Initially, the contact associated cytoskeleton forms a constricting and fiber internalizing sphere (FIG. 8A–C). Due to increasing tension within the clot, the cytoskeleton changes to stress fiber-like filament bundles connecting focal contacts with fibers (FIG. 8D). The stress fibers insert in an end-to-side mode within the focal contacts, but at the end of retraction fibers are attached at the platelet surface, even where the stress fibers did not insert in this mode. Because in later phases of retraction a novel binding site on fibrin was shown to be involved,[10] our results support the hypothesis that in late phases of clot retraction fibrin fibers may bind on the surface of platelets anew.

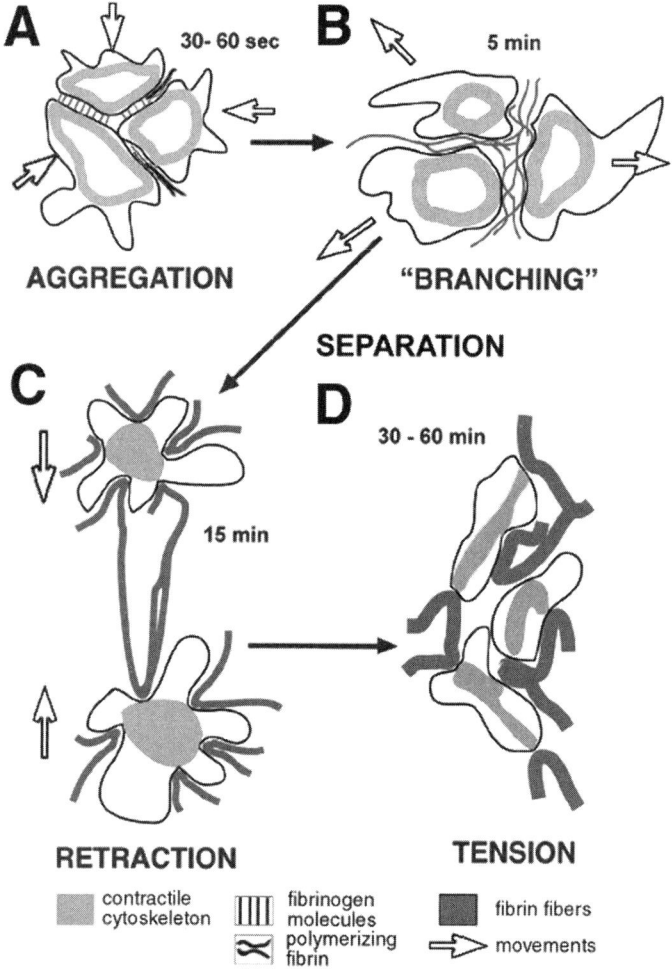

FIGURE 8. Schematic drawing of proposed model.

REFERENCES

1. MORGENSTERN, E., et al. 1990. Ultrastructure of the interaction between human platelets and polymerizing fibrin within the first minutes of clot formation. Blood Coagul. Fibrinolys. **1:** 543–546.
2. MORGENSTERN, E., et al. 1984. Platelets and fibrin strands during clot retraction. Thromb. Res. **33:** 617–623.
3. COHEN, I., et al. 1982. Ultrastructure of clots during isometric contraction. J. Cell. Biol. **93:** 775–787.
4. LEISTIKOW, E.A. 1996. Platelet internalization in early thrombogenesis. Sem. Thromb. Hemost. **22:** 289–294.
5. LEGRAND, C., et al. 1989. Studies on the mechanism of expression of secreted fibrinogen on the surface of activated human platelets. Blood **73:** 1226–1234.
6. LEWIS, J.C., et al. 1988. Orientation and specificity of fibrin protofibril binding to ADP-stimulated platelets. Blood **72:** 1992–2000.
7. BURCHARD, W. & M. MÜLLER. 1980. Statistics of branched polymers composed of rod substructures: A model for fibrin before clot formation. Int. J. Biol. Macromol. **2:** 225–234.
8. BARADET, T.C., et al. 1995. Three-dimensional reconstruction of fibrin clot networks from stereoscopic intermediate voltage electron microscope. Images and analysis of branching. Biophys. J. **168:** 1551–1560.
9. BARK, N., et al. 1999. The incipient stage in thrombin-induced fibrin polymerization detected by FCS at the single molecule level. Biochem. Biophys. Res. Commun. **260:** 35–41.
10. ROONEY, M.M., et al. 1998. The contribution of the three hypothesized integrin-binding sites in fibrinogen to platelet-mediated clot retraction. Blood **82:** 2374–2381.

Mutation of W215 Compromises Thrombin Cleavage of Fibrinogen, but Not of PAR1 or Protein C

YOUHNA M. AYALA, DANIELE AROSIO, AND ENRICO DI CERA

Department of Biochemistry and Molecular Biophysics,
Washington University School of Medicine, St. Louis, Missouri, USA

ABSTRACT: W215 is a highly conserved residue that shapes the S3 and S4 specificity sites of thrombin. Replacement of W215 with Phe produces modest effects on thrombin function, whereas the W215Y replacement significantly compromises the amidolytic activity toward synthetic and natural substrates. Replacement of W215 with Ala reduces fibrinogen and PAR4 cleavage 500-fold and 280-fold, respectively. On the other hand, the mutant decreases protein C activation and PAR1 cleavage only threefold and 25-fold, respectively. The W215A mutant cleaves PAR1 with a specificity constant more than 13-fold greater than that of fibrinogen and protein C, and 800-fold greater than PAR4. This is the first thrombin derivative to be described that functions as an almost exclusive activator of PAR1. The environment of W215 influences differentially three physiologically important interactions of thrombin, a feature that should assist in the separate study of each of these functions *in vivo*.

KEYWORDS: W215 mutation; Thrombin cleavage; Fibrinogen; PAR1; Protein C.

INTRODUCTION

Thrombin plays two important and opposing roles in the blood.[1] It acts as a procoagulant when it cleaves fibrinogen and promotes the formation of a fibrin clot. This action is reinforced and amplified by activation of the transglutaminase factor XIII and inhibition of fibrinolysis; activation of factors V, VIII, and XI upstream in the coagulation cascade; and cleavage of the platelet receptors PAR1 and PAR4 that leads to platelet activation and aggregation. In addition to its procoagulant function, thrombin also functions as an anticoagulant when it cleaves and activates protein C with the assistance of the endothelial receptor thrombomodulin. Activated protein C initiates a natural anticoagulant pathway by cleaving and inactivating factors Va and VIIIa, two essential cofactors of the intrinsic pathway, thereby downregulating both the amplification and progression of the coagulation cascade. The multiple roles of thrombin in hemostasis raise the possibility that individual functions, such as cleavage of fibrinogen, PAR1, PAR4, or protein C, can be dissociated.

Address for correspondence: Enrico Di Cera, Department of Biochemistry and Molecular Biophysics, Washington University School of Medicine, Box 8231, St. Louis, MO 63110, USA. Voice: 314-362-4185; fax: 314-747-5354.
enrico@biochem.wustl.edu

Residue W215 is absolutely conserved in thrombins from hagfish to human[2] and is also highly conserved in serine proteases, regardless of their specific function or species.[3] In this paper we show that mutation of W215 has profound effects on substrate recognition by thrombin, and that it differentially affects cleavage of fibrinogen, PAR1, PAR4, and protein C.

RESULTS AND DISCUSSION

Whereas the W215F substitution does not affect significantly the value of the specificity constant, $s = k_{cat}/K_m$, for the hydrolysis of the chromogenic substrate FPR, the W215Y and W215A mutants exhibit specificity constants perturbed by 20- to 100-fold (see TABLE 1). The aromatic nature of residue 215 is considered essential for high affinity substrate binding. Indeed ablation of this property in the W215A mutant produces a significant increase in K_m, with a smaller effect on k_{cat} in FPR hydrolysis. The presence of the hydroxyl group on the aromatic ring of Tyr also reduces substrate binding relative to the Phe derivative, mainly through an effect on K_m, suggesting that the polar group of the Tyr is not well accommodated in the highly hydrophobic environment around W215 in the wildtype.[3,4]

Mutation of W215 profoundly affects the ability of thrombin to release fibrinopeptide A from fibrinogen (TABLE 1). This property is compromised in all mutants. A comparison with the cleavage of FPR reveals a similar loss of specificity (10-fold) of W215Y relative to W215F, again suggesting that the presence of the hydroxyl group on the aromatic ring of Tyr is detrimental to substrate binding. However, the loss of specificity in the W215A mutant relative to wildtype is more pronounced in

TABLE 1. Specificity of wildtype and mutant thrombins of residue W215

	Wildtype	W215F	W215Y	W215A
FPR k_{cat}/K_m ($\mu M^{-1} sec^{-1}$)[a]	88 ± 4	44 ± 1	5.4 ± 0.4	0.91 ± 0.01
k_{cat} (sec^{-1})	56 ± 3	77 ± 8	120 ± 10	43 ± 2
K_m (μM)	0.63 ± 0.04	1.75 ± 0.09	22 ± 1	47 ± 2
FPA k_{cat}/K_m ($\mu M^{-1} sec^{-1}$)	17 ± 1	2.3 ± 0.1	0.19 ± 0.01	0.034 ± 0.002
FpB k_{cat}/K_m ($\mu M^{-1} sec^{-1}$)	8.1 ± 0.5	1.4 ± 0.1	0.08 ± 0.01	0.053 ± 0.003
Protein C k_{cat}/K_m ($\mu M^{-1} sec^{-1}$)[b]	0.22 ± 0.01	0.13 ± 0.01	0.032 ± 0.008	0.075 ± 0.006
PAR1 k_{cat}/K_m ($\mu M^{-1} sec^{-1}$)	30 ± 1	8.0 ± 0.3	1.2 ± 0.1	1.0 ± 0.1
PAR4 k_{cat}/K_m ($mM^{-1} sec^{-1}$)	336 ± 0.08	N.D.	N.D.	1.2 ± 0.1

NOTE: Experimental conditions were 5 mM Tris, 0.1% PEG, 145 mM NaCl, pH 7.4, 37°C, unless otherwise specified.
[a]Experimental conditions: 5 mM Tris, 0.1% PEG, 200 mM NaCl, pH 8.0, 25°C.
[b]In the presence of 100 nM rabbit thrombomodulin and 5 mM $CaCl_2$. The chromogenic substrate FPR was synthesized by solid phase and purified by HPLC. Progress curves of the release of p-NA following the hydrolysis of chromogenic substrate were measured at 405 nm as a function of substrate concentration and analyzed to extract the values of k_{cat}/K_m and k_{cat}, after proper correction for product inhibition.

the case of fibrinogen relative to FPR. When the aromatic side chain of Trp is removed altogether in the Ala substitution, the decrease in specificity toward fibrinogen reaches 500-fold.

An interesting feature of the role of W215 in the interaction of thrombin with fibrinogen is that the W215A mutant releases fibrinopeptide A from fibrinogen and fibrinopetide B from fibrin with similar kinetics and rate constants. It is conceivable that steric constraints in the S3 and S4 sites of thrombin oppose the release of fibrinopeptide B directly from fibrinogen and that these. constraints are removed with the W215A substitution.

The drastic perturbation of substrate binding seen in the case of fibrinogen is not matched by protein C in the presence of thrombomodulin. This substrate experiences only a modest loss of specificity, as the value of s decreases threefold in the W215A mutant. We conclude that the environment of W215 of thrombin is not significantly involved in binding protein C.

Mutation of W215 affects cleavage of PAR1 to an extent intermediate to those of fibrinogen and protein C. For PAR1, replacement of Trp with Phe and Tyr progressively reduces specificity, but the Ala mutant behaves like the Tyr mutant. The aromatic ring of Phe preserves the interaction with PAR1, but introduction of an hydroxyl group in the ring, or removal of the ring altogether, produce similar loss of cleavage of PAR1. As a result of the replacement of Trp with Ala at position 215, a differential effect on substrate recognition is observed. W215A exhibits severely compromised clotting activity (500-fold), with moderate loss of PAR1 cleavage (26-fold), and insignificant loss of protein C cleavage (threefold). Although the difference between PAR1 and PAR4 specificities in wildtype reaches 90-fold, W215A has 800-fold preference for PAR1. Hence, mutation of W215 with Ala affords a significantly different perturbation of important thrombin functions mediated by the cleavage of fibrinogen leading to clot formation, cleavage of PAR1 and PAR4 leading to platelet aggregation, and cleavage of protein C. Whereas the wildtype cleaves fibrinogen and PAR1 with comparable specificity constants, the W215A mutant preferentially cleaves PAR1 (TABLE 1). Mutation of W215 to Ala results in a change of specificity from fibrinogen to PAR1 and practically obliterates cleavage of PAR4.

REFERENCES

1. DI CERA, E., Q.D. DANG & Y.M. AYALA. 1997. Molecular mechanisms of thrombin function. Cell. Mol. Life Sci. **15:** 701–730.
2. BANFIELD, D.K. & R.T.A. MACGILLIVRAY. 1992. Partial characterization of vertebrate prothrombin cDNAs: amplification and sequence analysis of the B chain of thrombin from nine different species. Proc. Natl. Acad. Sci. U.S.A. **89:** 2779–2783.
3. LESK, A.M. & W.D. FORDHAM. 1996. Conservation and variability in the structures of serine proteinases of the chymotrypsin family. J. Mol. Biol. **258:** 501–537.
4. BODE, W., D. TURK & A. KARSHIKOV. 1992. The refined 1.9-Å X-ray crystal structure of D-Phe-Pro-Arg chloromethyl ketone-inhibited human alpha-thrombin: structure analysis, overall structure, electrostatic properties, detailed active-site geometry, and structure–function relationships. Protein Sci. **1:** 426–471.

Platelets in Suspension Require Preactivation to Adhere to Immobilized Fibrinogen

ARNAUD BONNEFOY,[a] QINGDE LIU,[b] W. GRAY JEROME,[c] CHANTAL LEGRAND,[a] AND MONY M. FROJMOVIC[b]

[a]*Unité 353 INSERM, Institut d'Hématologie, Hôpital St Louis, Paris, France*

[b]*Physiology Department, McGill University, Montreal, Quebec, Canada*

[c]*Pathology, Wake Forest University, North Carolina, USA*

ABSTRACT: Previous studies using whole blood perfusion through flow chambers have suggested that unactivated platelets can adhere to surface immobilized fibrinogen (Fg). However, the red blood cells needed for surface delivery of the platelets may activate platelets by released adenosine diphosphate (ADP). Our studies of coaggregation of unactivated or ADP-activated platelets with Fg-coated latex beads in flowing suspensions show that only preactivated platelets can adhere to Fg-coated surfaces.

KEYWORDS: Platelet adhesion; Activation; Immobilized fibrinogen.

Fibrinogen (Fg) in solution does not bind to GPIIbIIIa on resting (unactivated) platelets, but rather does bind to activated receptors (GPIIbIIIa*), thereby promoting platelet aggregation.[1] The literature suggests that resting platelets can adhere to surface-immobilized Fg, presumably via unactivated GPIIbIIIa.[2–4] However, the state of activation of the platelets was not directly evaluated and, furthermore, flow experiments generally require use of whole blood, where a fraction of platelets may be readily activated by ADP released from red blood cells.[3,4] We, therefore, compared the adhesion of resting versus ADP activated platelets to Fg immobilized on polystyrene latex beads (Fg beads) or on purified GPIIbIIIa-beads, in flowing suspensions (Couette) and that can be studied free from red blood cells[5,6] (shown schematically in FIGURE 1). The state of activation of our platelets was evaluated in parallel by flow cytometry, measuring soluble FITC–Fg binding to the activated platelets.[1]

Electron microscopy studies show that Fg is horizontally elongated when adsorbed directly onto polystyrene latex beads or onto beads bearing covalently attached pure GPIIbIIIa receptors (see FIGURE 2). The accessibility of the AGDV residues (position 408–411 on the γ chains) of the Fg coated on the beads was assessed using FITC-labelled MoAb 4A5, directed against the γ chain carboxyl terminus domain of the Fg. The stoichiometry of 4A5/Fg was about 1.55 when the Fg density on the surface of the beads was 1–10% of maximum, and about 0.70 for maximal (100% = 2,882 Fg/μm^2), indicating that about one AGDV site is accessible per Fg

---

Address for correspondence: Mony M. Frojmovic, Ph.D., Physiology Department, McIntyre Medical Building, McGill University, 3655 Drummond # 1137, Montreal, Quebec, Canada. Voice: 514-398-4326; fax: 514-398-5372.

mony@med.mcgill.ca

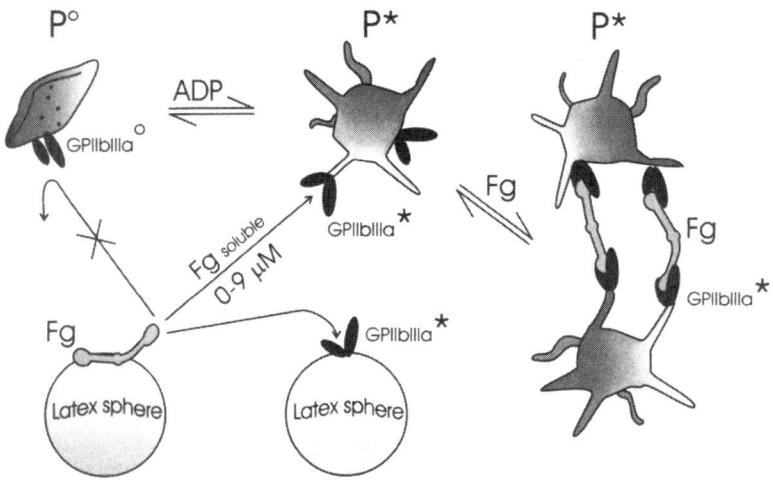

FIGURE 1. Coaggregation of fibrinogen (Fg) coated latex spheres with resting (unactivated) (P°) or ADP-activated (P*) platelets, or activated GPIIbIIIa-latex spheres. The P° and P* contain, respectively, resting (GPIIbIIIa°) and activated (GPIIbIIIa*) receptors. Soluble Fg was added at 0–9 μM with Fg-spheres in competition assays.

over the entire range of 1–100% surface coverage, and can participate in cross bridging to GPIIbIIIa* receptors.

The efficiency of platelet adhesion to Fg beads was compared for ADP activated versus resting platelets. Washed platelet preparations[6] containing detectable levels of activated platelets by flow cytometry were not used in these studies. The effects of shear rate (100–2000 sec^{-1}), Fg density on the beads (24–2,882 Fg/μm^2), concentration of ADP used to activate the platelets (0.1–100 μM), and presence of soluble Fg (1.1–9 μM) were assessed. We found no detectable adhesion of resting platelets to Fg beads, even under static conditions. The apparent efficiency of platelet adhesion to Fg beads readily correlated with the proportion of platelets "quantally" activated by doses of ADP.[7] That is, only ADP activated platelets adhere to Fg beads, with a maximal adhesion efficiency of 6–10% at shear rates of 100–300 sec^{-1}, decreasing with increasing shear rates up to 2,000 sec^{-1}. The adhesion efficiency of ADP activated platelets decreased threefold when the Fg density on the beads was decreased from 100% to 1% and only twofold in presence of a physiological concentration of soluble Fg (9 μM). Similar results were obtained when replacing ADP activated platelets by RGD activated GPIIbIIIa-beads.[6]

Our results suggest that a platelet adheres to surface bound Fg only if it is already preactivated in suspension. We suggest that previous studies were associated with adhesion of *preactivated* rather than *resting* platelets. In the widely used studies using whole blood perfused in flow through chamber models, we calculate that only 1% of all platelets need be in an activated state to yield a "carpet" of adherent platelets after a few minutes of flow. In contrast, our method uses diluted platelet suspensions (5–20×10^3/μL) where even about 5% "contamination" by activated platelets would be just below detectability.

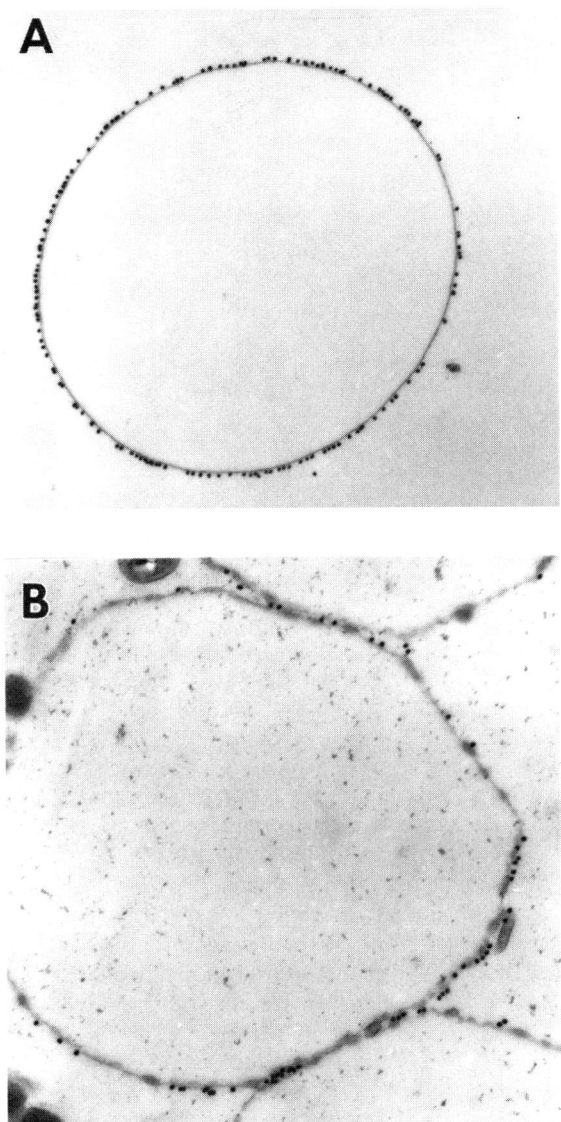

FIGURE 2. Surface organization of fibrinogen on latex spheres by intermediate voltage electron microscopy (15,400×). The Fg beads or GPIIbIIIa beads (4.5 μm diameter) first coated with soluble Fg were stained with the antibody 4A5, and then with gold anti-mouse antibody.

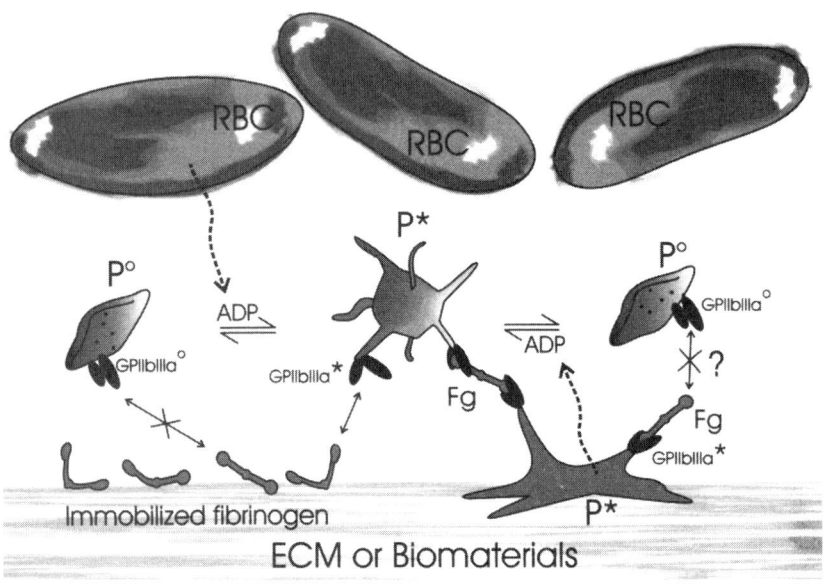

FIGURE 3. Schematic representation of capture of activated (P*) and not resting (P°) platelets by Fg immobilized on biomaterials or on extracellular matrix (ECM), or on activated GPIIbIIIa* receptors on P* in suspension or adherent on ECM. The activator, adenosine diphosphate (ADP), may be "released" from red blood cells (RBC) or secreting P*, and then activate P° to P*. It is hypothesized from our data that neither surface nor receptor immobilized Fg will recognize GPIIbIIIa° on P°, but only the activated GPIIbIIIa* on P*.

We propose that activated rather than resting platelets play a central role in both platelet adhesion to fibrinogen deposited on biomaterials or on the damaged vessel wall, as well as Fg presented on activated GPIIbIIIa receptors on activated platelets in platelet recruitment in the growing thrombus (shown schematically in FIGURE 3). Such activation could readily occur at sites of damage containing local concentrations of activators such ADP or thrombin, as shown in FIGURE 3 for ADP locally secreted from red blood cells in shear, or from secreting platelets, where ADP may be sustained on the activated platelet surface by continuous secretion from dense granules.

ACKNOWLEDGMENTS

We gratefully thank Professors Theo van de Ven (Chemistry, McGill) and Harry Goldsmith (Medicine, McGill) for helpful discussions and suggestions; Gary Matsueda, (Princeton University, NJ) for the 4A5 monoclonal antibody; the Medical Research Council of Canada and Heart and Stroke Foundation of Quebec for research support. Arnaud Bonnefoy was a recipient for salary support from Sanofi-Thrombose, from the International Council for Canadian Studies, and from the Heart and Stroke Foundation of Canada, with travel money from the Quebec-France exchange program of FRSQ-INSERM, the latter supporting exchange between the Montreal and Paris laboratories.

REFERENCES

1. FROJMOVIC, M.M., T. WONG & T. VAN DE VEN. 1991. Dynamic measurements of the platelet membrane glycoprotein IIb–IIIa receptor for fibrinogen by flow cytometry. Biophys. J. **59**: 815–827.
2. GARTNER, T.K., D.L. AMRANI, J.M. DERRICK, et al. 1993. Characterization of adhesion of "resting" and stimulated platelets to fibrinogen and its fragments. Thromb. Res. **71**: 47–60.
3. SAVAGE, B., E. SALDIVAR & Z.M. RUGGERI. 1996. Initiation of platelet adhesion by arrest onto fibrinogen or translocation on von Willebrand factor. Cell **84**: 289–297.
4. ZAIDI, T., L.V. MCINTIRE, D.H. FARRELL & P. THIAGARAJAN. 1996. Adhesion of platelets to surface-bound fibrinogen under flow. Blood **88**: 2967–2972.
5. XIA, Z. & M.M. FROJMOVIC. 1994. Aggregation efficiency of activated normal and fixed platelets in a simple shear field: Effect of shear and fibrinogen occupancy. Biophys. J. **66**: 2190–2201.
6. BONNEFOY, A., Q. LIU, C. LEGRAND & M.M. FROJMOVIC. 2000. Efficiency of platelet adhesion to fibrinogen depends on both cell activation and flow. Biophys. J. **78**: 2834–2843.
7. FROJMOVIC, M.M., R.F. MOONEY & T. WONG. 1994. Dynamic measurements of the platelet membrane GPIIbIIIa receptor expression and fibrinogen binding. II. Quantal activation parallels platelet capture in stir-associated microaggregation. Biophys. J. **67**: 2069–2075.

Platelet Adhesion to Fibrinogen Coated at Various Densities

MARKÉTA JIROUŠKOVÁ AND BARRY S. COLLER

Mount Sinai School of Medicine, New York, New York 10029, USA

ABSTRACT: Platelet adhesion to low-density coated fibrinogen induces greater protein tyrosine phosphorylation of SYK and FAK than adhesion to high-density coated fibrinogen, and leads to activation of integrin αIIbβ3 on the luminal side of adherent platelets.

KEYWORDS: Platelet adhesion; Fibrinogen; Protein tyrosine phosphorylation; SYK; FAK; Integrin activation.

Integrin αIIbβ3 is the major platelet membrane protein that mediates platelet interactions with fibrinogen. Unactivated platelets do no bind fibrinogen in solution, but adhere to immobilized fibrinogen under both static and low shear rate conditions.[1] We have previously reported differences in platelet morphology following adhesion to fibrinogen immobilized at various densities.[2] In the study reported here, we analyzed protein tyrosine phosphorylation, activation of SYK and FAK kinases, and activation of integrin αIIbβ3 in platelets following adhesion to fibrinogen adsorbed at different surface densities.

Gel filtered platelets were allowed to adhere for one hour to microtiter plates precoated with 3 µg/ml fibrinogen (low-density) or 100 µg/ml fibrinogen (high-density). With the aid of monoclonal antibodies, surface and intracellular proteins of adherent platelets were detected with fluorescent microscopy. When signal transduction was studied, adherent platelets were lysed and the protein lysates were analyzed by immunoblotting with antibodies specific for phosphorylated tyrosines. In addition, lysates were immunoprecipitated with antibodies specific for FAK and SYK kinases, and their phosphorylation was assessed by immunoblotting.[3]

Platelets adherent to low and high-density coated fibrinogen underwent equal increases in P-selectin expression. Focal adhesion formation, as determined by anti-vinculin antibody, was greater on platelets spread on low-density coated fibrinogen. Binding of antibody to β3 (7H2) revealed differences in localization of β3 on the surface of adherent platelets. On platelets adherent to low-density coated fibrinogen, 7H2 bound to the entire surface and stained the central granulomere heavily. When platelets were adherent to high-density coated fibrinogen, 7H2 also bound to the whole surface, but the staining was stronger around the edge of the spread platelets. Antibody PAC1, which binds only to activated αIIbβ3, bound strongly to platelets spread on low-density coated fibrinogen, whereas there was almost no binding to

Address for correspondence: Markéta Jiroušková, Ph.D., Mount Sinai School of Medicine, Box 1118, One Gustave L. Levy Place, New York, NY 10029, USA. Voice: 212-241-8928; fax: 212-876-5844.

marketa_jirouskova@mssm.edu

platelets adherent to high-density coated fibrinogen. However, PAC1 bound to platelets adherent to high-density coated fibrinogen if the αIIbβ3 receptors were activated with antibodies D3 or AP5. Tyrosine phosphorylation of proteins in platelets adherent to low-density coated fibrinogen was greater than that in platelets adherent to high-density coated fibrinogen. The intensity of antiphosphotyrosine antibody binding to a 100-kDa protein band was 55% less in platelets adherent to high versus low-density coated fibrinogen. A lower level of tyrosine phosphorylation of FAK was observed by immunoblotting after immunoprecipitation from platelets adherent to high-density coated fibrinogen and compared to platelets adherent to low-density coated fibrinogen. Tyrosine phosporylation of SYK was readily observed in platelets on low-density but not on high-density coated fibrinogen.

We conclude that, under static conditions, adhesion to low-density coated fibrinogen induces greater tyrosine phosporylation of SYK and FAK kinases, as well as other proteins, than adhesion to high-density coated fibrinogen, and platelets adherent to low-density coated fibrinogen have more extensive activation of luminal αIIbβ3 receptors.

REFERENCES

1. SAVAGE, B., E. BOTTINI & Z.M. RUGGERI. 1995. Interaction of integrin alpha IIb beta 3 with multiple fibrinogen domains during platelet adhesion. J. Biol. Chem. **270:** 28812–28817.
2. COLLER, B.S., J.L. KUTOK, L.E. SCUDDER, *et al.* 1993. Studies of activated GPIIb/IIIa receptors on the luminal surface of adherent platelets. Paradoxical loss of luminal receptors when platelets adhere to high density fibrinogen. J. Clin. Invest. **92:** 2796–2806.
3. HAIMOVICH, B., N. KANESHIKI & P. JI. 1996. Protein kinase C regulates tyrosine phosphorylation of pp125FAK in platelets adherent to fibrinogen. Blood **87:** 152–161.

Attenuation of Neointima Formation Following Arterial Injury in PAI-1 Deficient Mice

VICTORIA A. PLOPLIS AND FRANCIS J. CASTELLINO

W.M. Keck Center for Transgene Research and the Department of Chemistry and Biochemistry, University of Notre Dame, Notre Dame, Indiana, USA

ABSTRACT: Atherosclerosis is a chronic inflammatory disease in which the fibrinolytic system has been implicated as playing a major role. In order to directly assess the physiological impact an imbalanced fibrinolytic system has on both early and late stages of this disease, mice deficient for PAI-1 ($PAI\text{-}1^{-/-}$) were used in a model of vascular injury/repair and compared to wildtype mice (*WT*). Copper-containing cuffs were placed around the carotid arteries of these mice and the injured arteries were removed at either 7 or 21 days for histological analyses. At both times after injury, fibrin was prevalent in *WT* arteries, whereas only diffuse in $PAI\text{-}1^{-/-}$ arteries. At 21 days after injury, a prominent, multilayered neointima was evident in *WT* arteries, with no evidence of a neointima in $PAI\text{-}1^{-/-}$ arteries. Results from this study directly confirm the involvement of the fibrinolytic system in vascular repair processes following injury and indicate that fibrin could potentially play a role in lesion formation by stimulating smooth muscle cell proliferation, collagen synthesis, and intracellular cholesterol accumulation.

KEYWORDS: Neointima attenuation; Arterial injury; PAI-1 deficient mice.

INTRODUCTION

The fibrinolytic system has been implicated in playing a major role in the development and progression of atherosclerosis, a chronic inflammatory disease. Clinical studies have indicated that high plasma levels of fibrinogen and PAI-1 correlate with an increased risk of cardiovascular disease. Other studies have indicated that insoluble fibrin may affect smooth muscle cell proliferation, collagen deposition, and cholesterol accumulation and thus promote atherosclerotic lesion formation.

The development of mice deficient for components of the fibrinolytic system provides a means to directly assess the physiological impact of an imbalanced fibrinolytic system on a number of inflammation based diseases. Use of these mice in such disease models has identified a role for urokinase (uPA) and plasminogen (Pg) in the inflammatory response.[1–3]

Since the role of genetic factors in inflammation can be investigated effectively in gene targeted animals, these animals are extremely valuable for studying vascular injury and repair. In the study we report here, a copper induced model of inflammation has been adapted to *WT* mice,[4] and further studied in mice deficient in the PAI-1 gene ($PAI\text{-}1^{-/-}$).

Address for correspondence: Victoria A. Ploplis, Department of Chemistry and Biochemistry, Nieuwland Science Hall, University of Notre Dame, Notre Dame, IN 46556, USA. Voice: 219-631-4017; fax: 219-631-4048.
ploplis.3@nd.edu

METHODS

The left carotid arteries of anesthetized male and female C57BL/6J wild type (*WT*) and *PAI-1*$^{-/-}$ mice (8–12 weeks old) were surgically exposed and a copper–silicone cuff, 1–1.5 mm in length, was placed around the periphery of the artery proximal to the bifurcation. The surgical site was then closed with a 6-0 nylon suture and the mice allowed to recover. At 7 and 21 days postimplantation, the left carotid artery was reexposed to remove the cuffed artery. The contralateral right artery served as a negative control.

A number of histological (H&E, Masson's trichrome, Verhoeff von Giesson and Oil Red O) and immunohistological (smooth muscle cell α-actin, von Willebrand factor, fibrin(ogen), PAI-1 and BrdU) analyses were performed to assess qualitative and quantitative changes in the vessel wall following oxidative injury.

RESULTS AND DISCUSSION

Elevated plasma levels of fibrinogen, or PAI-1, are associated with an increased risk of cardiovascular disease. Other studies have shown that insoluble fibrin can promote atherosclerosis by affecting smooth muscle cell proliferation, collagen deposition, and cholesterol accumulation. A direct *in vivo* analysis of the physiological impact of an imbalanced fibrinolytic system on both early and late stages of this disease was made using *PAI-1*$^{-/-}$ mice, in a model of vascular injury and repair, and the resulting phenotype compared to that of *WT* mice.

At seven days after implantation of the cuff, intimal, medial, and adventitial compartments were significantly increased in size in arteries from *WT* mice relative to arteries from *PAI-1*$^{-/-}$ mice (see TABLE 1). Additionally, fibrin deposits were significantly enhanced in the medial and adventitial compartments of *WT* mice. This was not observed in injured arteries from *PAI-1*$^{-/-}$ mice.

At 21 days after implantation of the cuff, arteries from *WT* mice revealed an enlarged multilayered neointima, which was significantly larger than that observed in *PAI-1*$^{-/-}$ arteries. Fibrin deposits in the injured *WT* arteries remained unresolved,

TABLE 1. Morphometric analyses of carotid arteries from *WT* and *PAI-1*$^{-/-}$ mice

	DAY 7 (mm^2)			
	Lumen	Intima	Media	Adventitia
WT (n = 6)	0.057 ± 0.006	0.011 ± 0.001	0.046 ± 0.004	0.108 ± 0.009
PAI-1$^{-/-}$ (n = 6)	0.059 ± 0.008	0.004 ± 0.0004	0.020 ± 0.001	0.075 ± 0.008
p values	0.846	< 0.001	0.003	0.021
	DAY 21 (mm^2)			
	Lumen	Intima	Media	Adventitia
WT (n = 6)	0.031 ± 0.009	0.051 ± 0.008	0.071 ± 0.007	0.092 ± 0.007
PAI-1$^{-/-}$ (n = 6)	0.023 ± 0.005	0.007 ± 0.002	0.024 ± 0.004	0.182 ± 0.020
p values	0.436	< 0.0001	< 0.0001	< 0.0015

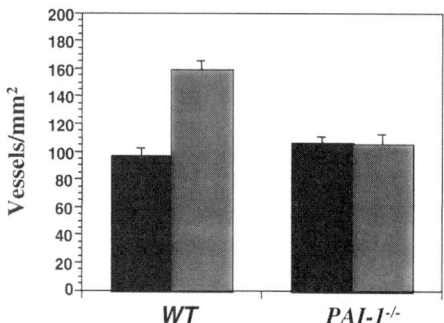

FIGURE 1. Vessel counts per mm² area of the adventitial compartment of arteries from WT and PAI-1⁻/⁻ mice seven days (*black bars*) and 21 days (*gray bars*) after perivascular placement of the cuff. WT versus PAI-1⁻/⁻ day 21, $p = 0.0057$. (Reprinted from Ref. 5, with permission from the American Journal of Pathology.)

primarily in the adventitial compartment, whereas injured arteries from PAI-1⁻/⁻ mice demonstrated little, if any, fibrin in the vessel wall. Enhanced PAI-1 expression and fat deposition were seen in the arterial wall of WT mice. No fat deposits were observed in PAI-1⁻/⁻ carotid arteries. At 21 days, neovascularization in the adventitia was more evident in injured arteries from WT mice relative to PAI-1⁻/⁻ mice (see FIGURE 1). Neovascularization of the adventitial compartment in PAI-1⁻/⁻ arteries was relatively unchanged in arteries at day 21 compared to those analyzed at day seven (FIG. 1).

This study demonstrates that copper induced injury of murine carotid arteries results in accelerated progression of pathological events associated with the development of atherosclerotic like lesions. Furthermore, results of this investigation emphasize the involvement of the fibrinolytic system in vascular repair processes following injury and indicate that alterations in the fibrinolytic balance in the vessel wall have a profound effect on the development and progression of vascular lesion formation.

REFERENCES

1. GYETKO, M.R., *et al.* 1996. Urokinase is required for the pulmonary inflammatory response to Cryptococcus neoformans. A murine transgenic model. J. Clin. Invest. **15:** 18–26.
2. MOONS, L., *et al.* 1998. Reduced transplant arteriosclerosis in plasminogen-deficient mice. J. Clin. Invest. **102:** 1788–1797.
3. PLOPLIS, V.A., *et al.* 1998. Plasminogen deficiency differentially affects recruitment of inflammatory cell populations in mice. Blood **91:** 2005–2009.
4. VOLKER, W., *et al.* 1997. Copper-induced inflammatory reactions of rat carotid arteries mimic restenosis/arteriosclerosis-like neointima formation. Atherosclerosis **130:** 29–36.
5. PLOPLIS, V.A., I. CORNELISSEN, M. SANDOVAL-COOPER, *et al.* 2001. Remodeling of the vessel wall after copper-induced injury is highly attenuated in mice with a total deficiency of plasminogen activator inhibitor-1. Am. J. Pathol. **158:** 107–117.

Transcriptional Control Mechanism of Fibrinogen Gene Expression

GERALD M. FULLER AND ZHIXIN ZHANG[a]

University of Alabama at Birmingham, Birmingham, Alabama, USA

ABSTRACT: Although fibrinogen genes are expressed constitutively in hepatocytes, their transcription can be greatly increased during inflammatory stress. Extensive studies have focused on the cytokine mediated transcriptional regulation of fibrinogen genes. It is clear that interleukin-6 (IL-6) and its family of cytokines are the major inducers of fibrinogen gene expression. Functional analyses of all three fibrinogen promoters for human and rat all demonstrate that the conserved CTGGGAA motifs within the proximal promoter of each fibrinogen gene are the IL-6 responsive elements. Exploration of the rat γ fibrinogen gene demonstrated that the IL-6 activated transcription factor, STAT3, binds to the CTGGGAA motif and is required for the IL-6 mediated upregulation of this gene. IL-6 mediated fibrinogen production can be significantly elevated by glucocorticoid treatment. The synergistic effect of glucocorticoids and IL-6 relies on the functional interaction between STAT3 and glucocorticoid receptor. In addition to the upregulation signals for fibrinogen gene expression during inflammatory stress, other signaling also downregulates the expression of fibrinogen genes. For example, the proinflammatory cytokine IL-1β exerts inhibitory function on IL-6 mediated fibrinogen gene expression. Given the fact that elevated levels of fibrinogen in blood correlate with increased risk for cardiovascular disease, there is strong motivation to explore the molecular mechanisms that control fibrinogen expression, especially those signals that may downmodulate expression and thus provide novel approaches to controlling fibrinogen levels.

KEYWORDS: Transcription control mechanism; Fibrinogen; Gene expression.

INTRODUCTION

The fibrinogen genes are clustered within a 65-kb region on chromosome 4 (4q23–q32) in human.[1] One unique feature of fibrinogen gene family is that the three genes are transcribed in a tightly coordinated manner, such that, in responding to an upregulation signal, the transcription of all three genes increase to the same extent.[2,3] Inspection of the regulatory regions of fibrinogen genes shows that, although there are common transcription binding motifs for all three genes, there are also unique motifs for each gene.[4] In this review, we explore details of how transcription of the rat γ fibrinogen is regulated. We believe that a detailed understanding of

---

Address for correspondence: Gerald M. Fuller, Ph.D., Department of Cell Biology, BHSB 680, University of Alabama at Birmingham, Birmingham, AL 35294, USA. Voice: 205-934-7596.
GMFULLER@uab.edu

[a]Present address: Zhixin Zhang, Ph.D., WTI378 Division of Clinical and Developmental Immunology, University of Alabama at Birmingham, Birmingham, AL 35294, USA.

control mechanisms for one fibrinogen gene will likely shed light on how this entire gene family is regulated at the transcriptional level.

Organization of the Fibrinogen Genes

Within the fibrinogen gene locus, the γ and Aα genes reside on the same DNA strand and are transcribed in the same direction, whereas, the Bβ is on the opposite DNA stand and transcribed in the reverse order.[1] Interestingly, there is considerable variation in the number of exons for the human fibrinogen genes, for example the human Aα gene has eight exons, the Bβ gene has at least five exons, and the γ gene has seven exons.[5] There is only one gene for each fibrinogen subunit, however different isoforms of Aα and γ fibrinogen subunits occur due to alternate mRNA splicing events.[6–8]

The Promoter Regions of the Fibrinogen Genes

The DNA sequences of the promoter regions for rat fibrinogen genes have been reported.[4,9] The principal regulatory regions of fibrinogen genes reside within the first kilobase upstream from the transcription initiation site. A schematic depicting identified binding elements in each promoter of the rat fibrinogen genes is shown in FIGURE 1. Comparison of the nucleotide sequences among the three genes does not show strong homology, although there are a number of sequence similarities among the three genes.[4] Importantly, these sequences represent binding motifs for specific transcription factors that may participate in the highly coordinated expression of the three genes. For example, all three promoters contain a CAAT-enhancing binding element (C/EBP) and CTGGGAA motif(s) that are known as IL-6 responsive elements (type I and type II IL-6RE, respectively).[51] Both Aα and Bβ promoters contain a tissue specific motif for hepatic nuclear factor-1 (HNF-1), which is required for expression in hepatocytes.[10–13] There is a clearly identified TATA-box in the γ gene promoter, however such a sequence is not as well defined in Aα and Bβ genes. An element known as the adenovirus major late promoter element (MLP) is found

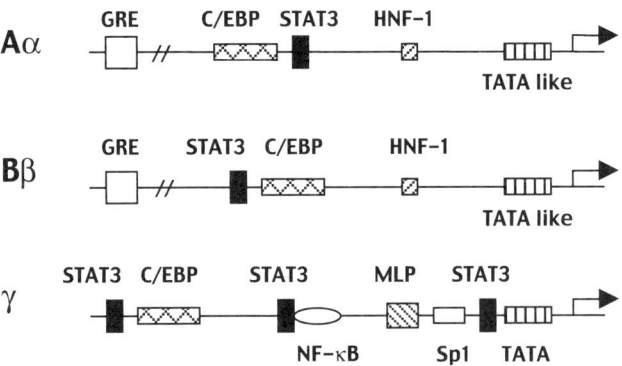

FIGURE 1. The promoter regions and transcriptional regulatory elements of rat Aα, Bβ, and γ fibrinogen genes.

in the γ gene and is required for its basal transcription.[9] Interestingly, the γ promoter has three CTGGGAA elements that are involved in upregulating gene transcription during an acute inflammatory response.[14] Glucocorticoid response elements have been proposed for both Aα and Bβ genes, but none has been identified to date.

Basal Expression of Fibrinogen Genes

Fibrinogen is produced almost exclusively in the liver. The circulating concentration of fibrinogen is stringently maintained at a level between 2.5–3.0 mg/ml.[15] The circulating concentration of fibrinogen appears to be maintained by precise regulation of its production and degradation. It is not clear if the circulating levels of fibrinogen itself affect its production. Evidence has been presented indicating that fibrin split products may indirectly influence fibrinogen gene transcription via stimulating IL-6 production. Hepatic tissue specific expression of fibrinogen genes is controlled mainly by a liver enriched transcription factor, HNF-1.[10–12] The CAAT enhancer binding protein (C/EBP) has also been implicated in the basal expression of fibrinogen genes.[16] The expression and secretion of "clottable" fibrinogen has been reported from a lung fibroblastic cell line (A549) on stimulation with the proinflammatory cytokine IL-6.[17] This finding indicates that non-hepatic cells also have the capacity to produce fibrinogen during injury, although the level of fibrinogen produced by these cells is very low and its physiological relevance is not clearly understood. It has been reported that overexpression of the Bβ fibrinogen gene in human HepG2 cells results in an increase in fibrinogen secretion, suggesting an overall increase in transcription of the other two genes.[18] Similarly, by elevating the expression of any one of the three fibrinogen genes, transcription of other subunits appeared to increased.[19] These findings suggested the notion that there is a tightly linked coordination of transcription for fibrinogen genes. However, when Aα fibrinogen gene was disrupted in mice such that no Aα mRNA was produced, expression of the other two fibrinogen genes were unaffected.[20] These observations argue that regulation of transcription does not occur via sensing the levels or fidelity of fibrinogen mRNAs.

Fibrinogen Gene Expression During the Acute Phase Response

Nuclear run-on studies demonstrated that transcription of fibrinogen genes are coordinately regulated. When hepatocytes were stimulated with IL-6, transcription of all three genes increased simultaneously, and following removal of IL-6, they coordinately returned to normal level.[2] Furthermore, quantification of steady state mRNA levels demonstrated that the halflife of each fibrinogen transcript is very similar (about 9.0 h).[3] Fibrinogen gene expression is elevated during an acute inflammatory response. Both *in vitro* studies using cultured hepatocytes or hepatoma cells and *in vivo* studies using animals showed that IL-6 is the major inducer of fibrinogen gene expression.[2,16,21] The IL-6 response of fibrinogen genes can be modulated by other acute phase response mediators, such as glucocorticoids, IL-1, and TNFα. It is well recognized that glucocorticoids synergize with IL-6 to induce fibrinogen genes expression,[2] the molecular interplay has been recognized recently.[39] The effect of IL-1β on fibrinogen gene expression was controversial. For example, IL-1β inhibited the IL-6 response in hepatocyte cultures, however injection of IL-1β into animals stimulated fibrinogen expression.[22] Subsequently it was found that IL-1β induces

IL-6 expression, and that the increase in fibrinogen production *in vivo* was the result of the IL-6 signal.[22,50] When either IL-1β or TNFα was coadministered to animals they inhibited the IL-6 signal. The molecular basis for this signaling paradox has now been resolved.[52]

Because human Bβ fibrinogen was considered to be the "nucleating chain" for fibrinogen assembly, the Bβ gene was the first to be chosen by several groups in order to investigate how IL-6 controls fibrinogen expression.[13,16,21] Functional assays using Bβ fibrinogen promoter driving a luciferase reporter gene revealed several regulatory elements, including a CTGGGAA motif (within a 150-bp region), a C/EBP site, and an HNF-1 site, all of which are required to achieve a full IL-6 response.[13,16,21] An intact HNF-1 element is required for Bβ fibrinogen IL-6 response within the native promoter context. However, when the CTGGGAA motif was moved nearer to the transcription start site, deletion of HNF-1 had no effect on the IL-6 response. This result indicated that the HNF-1 site was important to the basal promoter function rather than IL-6 signaling.[13] The C/EBP site on the Bβ fibrinogen gene provided only a slight enhancement of the IL-6 response, but it was critical for the basal level expression and maximal IL-6 induction.[16] The CTGGGAA motif (150 bp to -144 bp) was essential for the IL-6 response because mutation of this region completely abolished IL-6 response for the Bβ gene. Electrophoretic mobility shift assays (EMSA) showed that several protein complexes associated with this promoter region, however, none of these protein/DNA complexes were affected by IL-6 treatment.[16] For the rat Aα fibrinogen gene, functional analyses defined the CTGGGAA motif (-300 bp) as the IL-6RE. The binding of STAT3 to its DNA element could not be demonstrated, instead, a non-STAT3 protein associated with this region. How it activates the Aα gene has not been established.[23] Although several detailed studies of the Aα and the Bβ promoter regions all demonstrated that the consensus sequence CTGGGAA is a functional motif for the IL-6 mediated signal. How IL-6 initiated signaling events control the expression of these two genes remains unclear. As we discuss below, the role of IL-6/STAT3 in controlling the rat γ fibrinogen gene is more clearly defined.

TRANSCRIPTIONAL REGULATION OF THE RAT γ FIBRINOGEN GENE

The rat γ gene is the first one within the fibrinogen locus,[1] and as such its expression may exert influence on the expression of the downstream Aα and Bβ genes (see FIGURE 2). The tightly coordinated transcription of fibrinogen gene family may begin with the first gene in the locus, the γ gene.

IL-6 Controls Rat γ Fibrinogen Gene Transcription Through Activation of STAT3

It had been shown previously that three important elements, including a Sp1 site, a CAAT enhancer binding protein site, and a major late promoter (MLP) transcription factor binding site, were essential for basal transcription of the rat γ gene.[9,24] Inspection of the rat γ gene promoter region showed three putative STAT3 binding motifs (FIG. 1). These elements were designated as Sites I (-296 bp to -291 bp), Site

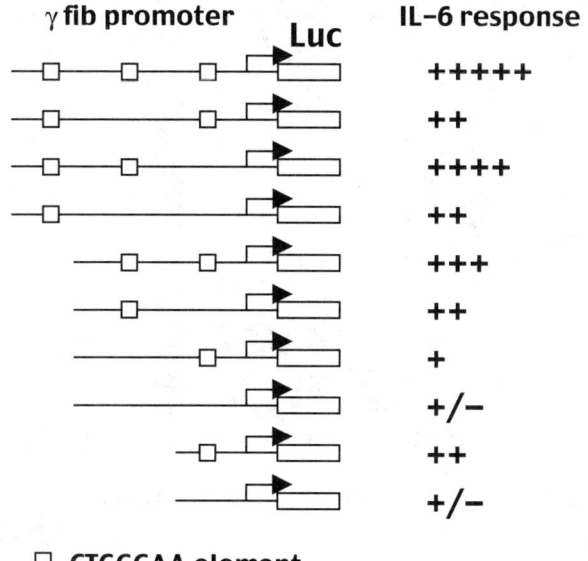

FIGURE 2. Function of the three CTGGGAA elements in the rat γ fibrinogen gene promoter. The IL-6 response is presented as relative fold induction. (Summarized from Figure 3 in Ref. 14.)

II (−152 bp to −146 bp), and Site III (−43 bp to −37 bp). Their importance in regulating the expression of the γ gene on IL-6 induction was shown by deletion mapping of the rat γ fibrinogen promoter (−1466 to +54) and selective deletion or mutation of each individual CTGGGAA motif (see FIGURE 3).[14] Although all three binding motifs were required for maximal induction of the gene, Site II showed the most potency in activating the rat γ fibrinogen promoter driving reporter gene. EMSA results demonstrated that IL-6-activated STAT3 associated with all three CTGGGAA regions.[14] In addition, using a stable cell line overexpressing wildtype STAT3, we were able to show that the IL-6 response of rat γ fibrinogen was elevated up to 40-fold (see FIGURE 4). Based on these results, we concluded that IL-6 controlled rat γ fibrinogen gene expression through activation of transcription factor STAT3.

Molecular Interaction Between STAT3 and Glucocorticoid Receptor Synergistically Upregulates Rat γ Fibrinogen Gene Expression

Glucocorticoids are important mediators of the immune system and modulate the biological activities of inflammatory cytokines such as IL-1 and IL-6.[2,25,27] Subsequent to binding the latent intracellular glucocorticoid receptor (GR), the ligand–receptor complex translocates into the nucleus, where it activates transcription of downstream genes through interaction with the glucocorticoid responsive element (GRE).[28,29] Recent studies have also shown that the GR can modulate other signaling events through interaction with transcription factors such as AP-1 (Jun or

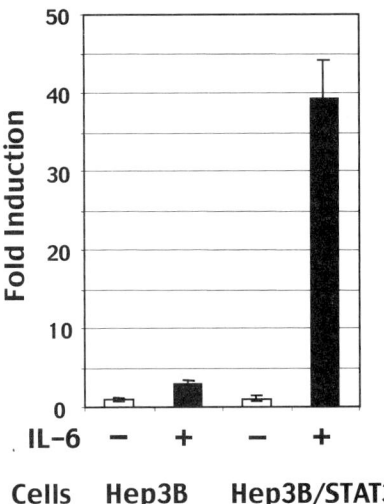

FIGURE 3. Overexpression of STAT3 elevated IL-6 response on rat γ fibrinogen gene. The reporter constructs containing the rat γ fibrinogen gene promoter (−300 bp to +54 bp) driving the luciferase gene were transfected into wildtype Hep3B cells or Hep3B cells stably transfected with pRc/CMV-STAT3. The results are presented as fold induction of luciferase activity over control levels from triplicate experiments. *Error bars* indicate standard deviation.

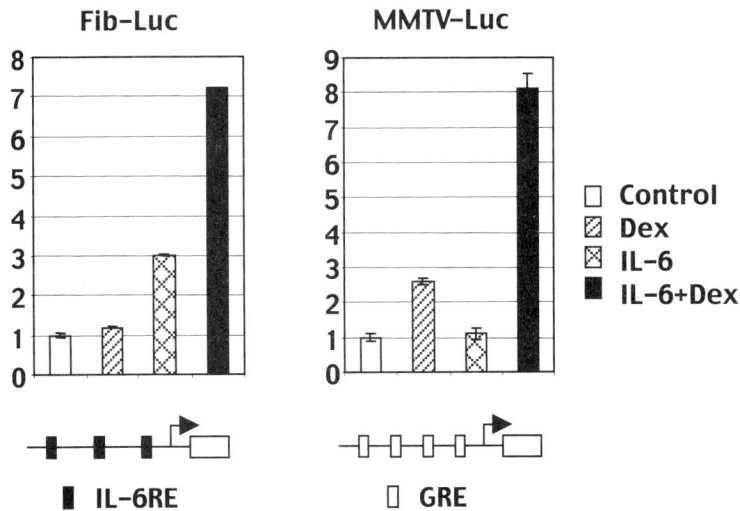

FIGURE 4. Mutual synergism between IL-6 and glucocorticoid signaling. Transient transfections were performed in H4IIE rat hepatoma cells using reporter constructs Fib-Luc and MMTV-Luc. (Extracted from Figure 1 in Ref. 39.)

Fos),[30–32] NF-κB (p65),[33,34] and STAT5[35] or nuclear cofactors, such as NcoR,[36] CBP/300, and pCAF.[37,38] These findings provide another level of gene regulation based upon protein–protein interaction between transcriptional regulators.

Several reports suggest that glucocorticoids synergize with IL-6 signaling in controlling fibrinogen gene expression,[2,25] however the molecular mechanism was unknown. Using the rat γ fibrinogen gene reporter constructs, we first demonstrated that glucocorticoid/IL-6 synergism is not dependent on any glucocorticoid responsive element on the γ fibrinogen gene promoter region (Zhang and Fuller, unpublished observation). The functional synergism between IL-6 and glucocorticoids was further analyzed using two different types of reporter constructs. One construct (Fib-Luc) is the γ fibrinogen promoter (−300 to +54 bp) containing three IL-6REs linked to the luciferase reporter, which has no GRE. The second construct (MMTV-Luc) is mouse mammary tumor virus LTR region containing four functional glucocorticoid binding elements linked to the luciferase reporter, which has no STAT3 binding motif. Using these two constructs, we observed that there is a mutual synergism between glucocorticoid signaling and IL-6 signaling. As shown in FIGURE 5, glucocorticoid treatment enhances the IL-6 response on Fib-Luc construct. IL-6 stimulation also enhances the glucocorticoid response on the MMTV-Luc construct (FIG. 5). A detailed analysis using a reconstitution system further indicated that IL-6 activated STAT3 and ligand bound glucocorticoid receptor are essential for functional synergism. Finally, coimmunoprecipitation experiments provide the evidence that STAT3 interacts with GR.[39] Thus, for the GRE containing reporter construct, STAT3 acts as a transcription coactivator for GR, in return, GR acts as a transcription coactivator for IL-6 activated STAT3 in enhancing IL-6 mediated rat γ fibrinogen gene expression.

IL-1β Inhibits IL-6 Mediated Rat γ Fibrinogen Gene Expression

The initial proinflammatory cytokines produced in an acute inflammatory response are tumor necrosis factor α (TNFα) and interleukin-1β (IL-1β).[40–42] These two cytokines not only affect the production of a number of hepatic derived acute phase proteins, but are also responsible for the production of IL-6, the other major

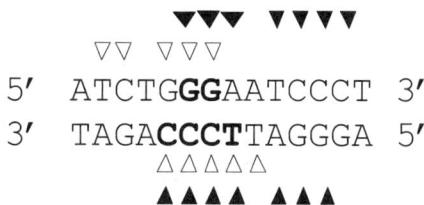

FIGURE 5. STAT3 and NF-κB compete for DNA contact on rat γ fibrinogen gene site II region. The STAT3 contacting bases (△) and NF-κB contacting bases (▲) are shown based on the three-dimensional structures of STAT3β homodimer or NF-κB p50 homodimer binding to its own consensus binding sites.[52] The directly competing bases are shown in *bold*.

cytokine involved in regulating hepatic genes. TNFα and IL-1β both activate transcription factor complexes designated as nuclear factor κB (NF-κB).[42] Upon IL-1β binding to its receptor on the hepatocyte surface, a cascade of signaling events results in the release of NF-κB subunits from the complex known as IκB. NF-κB factors translocate to the nucleus and interact with the κB binding motif (GGGAATTCCC).[43] The IL-6 signal is mediated by two different sets of transcription factors, C/EBP and STAT3.[44–46] In most instances, IL-β and IL-6 act cooperatively and upregulate several acute phase responsive genes, such as serum amyloid A (SAA),[33,47] C-reactive protein (CRP),[48] and α1-acid glycoprotein.[49] In these cases, the coordinated activation of NF-κB and C/EBP explains the molecular mechanism.[47] However, the IL-6 mediated expression of other acute phase responsive genes, such as fibrinogen, α1-antichymotrypsinogen, α2-macroglobulin, and thiostatin are inhibited by IL-1β costimulation.[22,50] A number of mechanisms for this paradoxical response have been proposed, including the suggestion that IL-1β downregulates the expression of IL-6 receptor gp80 subunit.[51] Careful inspection of the DNA sequence of the binding element for STAT3 in the rat γ fibrinogen promoter revealed that it was overlapped by a NF-κB binding element (see FIGURE 6). This observation suggested the possibility that the IL-1β activated NF-κB, occurring prior to the initiation of IL-6 signaling during acute phase response, and this might bind to the overlapping STAT3/NF-κB motif, thereby preventing STAT3 from binding to the CTGGGAA region and activating the downstream γ fibrinogen gene. This hypothesis was tested in EMSA studies using a DNA probe corresponding to the site II region of rat γ fibrinogen promoter. The results showed that, indeed, NF-κB competed with STAT3 for binding to the site II region, which is important for the IL-6 response of this gene.[52] Binding of NF-κB alone to the κB site within the γ fibrinogen gene did not transactivate this gene. Instead, binding of NF-κB to the rat γ fibrinogen gene promoter region blocks STAT3 binding to the overlapping binding site. Thus, NF-κB

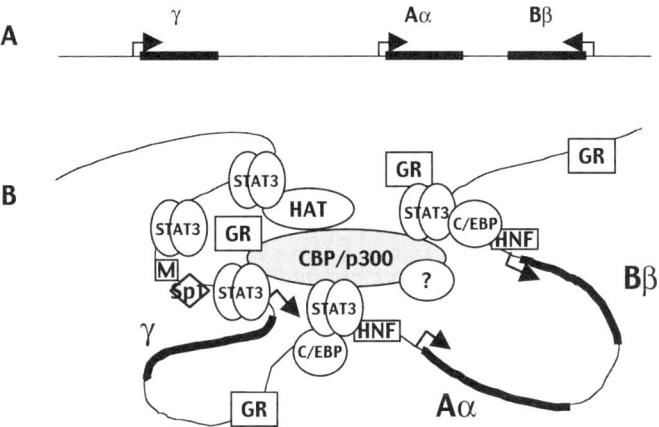

FIGURE 6. A. The relative position and transcription orientation for Aα, Bβ, and γ fibrinogen genes. **B.** Model for fibrinogen gene transcription regulation.

acts as an inhibitor for STAT3 mediated transcription. These observations provide an unrecognized function for NF-κB in regulating fibrinogen gene expression.

SUMMARY

Using the rat γ fibrinogen gene as a model system, we have dissected the functional interplay among three signaling cascades: IL-6/STAT3, IL-1β/NF-κB, and glucocorticoid/GR, and provided information on how the combinatorial signaling events control the transcription of rat γ fibrinogen. The schematic shown in FIGURE 6 depicts a possible model for understanding the transcriptional regulation of all three fibrinogen genes as a coordinated event. The promoter of rat γ fibrinogen possesses three weak STAT3 binding sites that collectively act as functional IL-6RE. STAT3 associates with all three CTGGGAA regions and transactivates γ fibrinogen gene. GR interacts with STAT3 and functions as a costimulator, thus enhancing the STAT3 effect. The STAT3/GR complexes may also attract other cofactors to achieve a full IL-6/glucocorticoid synergism. It is known that both GR and STAT3 can interact with nuclear factors such as CBP/P300.[53,54] CBP/P300 are the nuclear integrators for several signaling pathways and they act as a transcriptional coactivators for AP-1, GR, NF-κB, p53, STAT1, and STAT3.[55] Importantly, CBP/P300 exerts intrinsic histoneacetyltransferase activity[56] and can recruit other histoneacetyltransferases to remodel chromatin structure for transcriptional events. STAT3 and GR binding to the γ fibrinogen promoter region may recruit these nuclear co-factors to the rat γ fibrinogen promoter region. Given the fact that the γ fibrinogen gene is positioned first in the fibrinogen locus, its activation might help to open the chromosomal structure of the fibrinogen gene locus and facilitate transcription of the Aα and Bβ genes.

REFERENCE

1. KANT, J. 1985. Organization and evolution of the human fibrinogen gene locus on the chromosome 4. Proc. Natl. Acad. Sci. U.S.A. **82:** 2344.
2. OTTO, J.M., et al. 1987. The coordidated regulation of fibrinogen gene transcription by hepatocyte-stimulating factor and dexamethasone. J. Cell Biol. **105:** 1067–1072.
3. NESBITT, J.E. & G.M. FULLER. 1991. Transcription and translation are required for fibrinogen mRNA degradation in hepatocytes. Biochim. Biophys. Acta **1089:** 88–94.
4. FOWLKES, D.M., et al. 1984. Potential basis for regulation of the coordinately expressed fibrinogen genes: homology in the 5′ flanking regions. Proc. Natl. Acad. Sci. U.S.A. **81:** 2313–2316.
5. CRABTREE, G.R. 1987. The molecular biology of fibrinogen. *In* The Molecular Basis of Blood Diseases. 631–657.
6. CRABTREE, G.R. & J.A. KANT. 1982. Organization of the rat γ-fibrinogen gene: Alternative mRNA splice patterns produce the γA and γB (γ′) chains of fibrinogen. Cell **31:** 159–166.
7. CHUNG, D.W., et al. 1983. Characterization of a complementary deoxyribonucleic acid coding for the γ chain of human fibrinogen. Biochem. **22:** 3250–3256.
8. CHUNG, D.W. & E.W. DAVIE. 1984. Gamma and gamma′ chains of human fibrinogen are produced by alternative mRNA processing. Biochem. **23:** 4232–4236.
9. MORGAN, J.G., et al. 1988. Sp1, a CAAT-binding factor, and the adenovirus major late promoter transcription factor interact with functional regions of the gamma-fibrinogen promoter. Mol. Cell Biol. **8:** 2628–2637.

10. BAUMHUETER, S., et al. 1988. A variant nuclear protein in dedifferentiated hepatoma cell binds to the same functional sequences in the β fibrinogen gene promoter as HNF-1. EMBO J. **7:** 2485–2483.
11. BAUMHUETER, S., et al. 1990. HNF-1 shares three sequence motifs with the POU domain proteins and is indetical to LF-B1 and APF. Genes Dev. **4:** 372–379.
12. COURTOIS, G., et al. 1987. Interaction of a liver-specific nuclear factor with the fibrinogen and α_1-antitrypsin promoters. Science **238:** 688–692.
13. DALMON, J., et al. 1993. The human β fibrinogen promoter contains a hepatocyte nuclear factor 1-dependent interleukin-6-responsive element. Mol. Cell Biol. **13:** 1183–1193.
14. ZHANG, Z., et al. 1995. Characterization of the IL-6 responsive elements in the γ fibrinogen gene promoter. J. Biol. Chem. **270:** 24287–24291.
15. FULLER, G.M. 1993. Fibrinogen: a multifunctional acute phase protein. *In* Acute Phase Proteins: Molecular Biology, Biochemistry, and Clinical Applications. A. Mackiewicz, I. Kushner & H. Baumann, Eds.: 170–181. CRC Press, Boca Raton.
16. ANDERSON, G.M., et al. 1993. Functional characterization of promoter elements involved in regulation of human Bβ-fibrinogen expression. J. Biol. Chem. **268:** 22650–22655.
17. HAIDARIS, P.J.S. 1997. Induction of fibrinogen biosynthesis and secretion from cultured pulmonary epithelial cells. Blood **89:** 873–882.
18. ROY, S.N., et al. 1990. Regulation of fibrinogen assembly. J. Biol. Chem. **265:** 6389–6393.
19. ROY, S., et al. 1994. Overexpression of any fibrinogen chain by HepG2 cells specifically elevates the expression of the other two chains. J. Biol. Chem. **269:** 691–695.
20. SUH, T.T., et al. 1995. Resolution of spontaneous bleeding events but failure of pregnancy in fibrinogen-deficient mice. Genes Dev. **9:** 2020–2033.
21. HUBER, P., et al. 1990. Human beta fibrinogen gene extression, Upstream sequences involved in the its tissue specific expression and its dexamethasone and interleukin-6 stimulation. J. Biol. Chem. **265:** 5695–5701.
22. SIPE, J.D., et al. 1991. IL-1 receptor antagonist simultaneously inhibits SAA and stimulates fibrinogen synthesis *in vivo* and *in vitro*: a proposed mechanism of action. Cytokine **3:** 497–502.
23. LIU, Z. & G.M. FULLER. 1995. Detection of a novel transcription factor for the Aα fibrinogen gene in response to interleukin-6. J. Biol. Chem. **270:** 7580–7586.
24. CHODOSH, L.A., et al. 1987. The adenovirus major late transcription factor activates the rat γ fibrinogen promoter. Science **238:** 684–687.
25. BAUMANN, H., et al. 1990. Interaction of cytokine and glucocorticoid-response elements of acute phase plasma protein genes. Importance of glucocorticoid receptor level and cell type for regultion of the elements from rat alpha-1 acid glycoprotein and beta-fibrinogen gene. J. Biol. Chem. **265:** 20390–20399.
26. AUPHAN, N., et al. 1995. Immunosuppression by glucocorticoids: inhibition of NF-κB activity through induction of IκB synthesis. Science **270:** 286–290.
27. SCHEINMAN, R.I., et al. 1995. Role of transcriptional activation of IκBα in mediation of immunosuppression by glucocorticoids. Science **270:** 283–286.
28. BEATO, M. 1989. Gene regulation by steriod hormones. Cell **56:** 335–344.
29. BEATO, M., et al. 1995. Steroid hormone receptors: many actors in search of a plot. Cell **83:** 851–857.
30. JONAT, C., et al. 1990. Antitumor promotion and antiinflammation: down-modulaiton of AP-1 (Fos/Jun) activity by glucocorticoid hormone. Cell **62:** 1189–1204.
31. SCHULE, R., et al. 1990. Functional antagonism between oncoprotein c-Jun and the glucocorticoid receptor. Cell **62:** 1217–1226.
32. YANG-YEN, H., et al. 1990. Transcriptional interference between c-Jun and the glucocorticoid receptor: mutual inhibition of DNA binding due to direct protein–protein interaction. Cell **62:** 1205–1215.
33. RAY, A. & K.E. PERFONTAINE. 1994. Physical association and functional antagonism between the p65 subunit of transcription factor NF-κB and the glucocorticoid receptor. Proc. Natl. Acad. Sci. U.S.A. **91:** 752–756.

34. CALDENHOVEN, E., *et al.* 1995. Negative cross-talk between RelA and the glucocorticoid receptor: a possible mechanism for the antiinflammatory action of glucocorticoids. Mol. Endo. **9:** 401–412.
35. STOCKLIN, E., *et al.* 1996. Functional interactions between Stat5 and the glucocorticoid receptor. Nature **383:** 726–728.
36. CHEN, J.D. & R.M. EVANS. 1995. A transcriptional co-repressor that interacts with nuclear hormone receptors. Nature **377:** 454–457.
37. CHAKRAVARTI, D., *et al.* 1996. Role of CBP/P300 in nuclear receptor signaling. Nature **383:** 99–103.
38. KAMEI, Y., *et al.* 1996. A CBP integrator complex mediates transcriptional activation and AP-1 inhibition by nuclear receptors. Cell **85:** 403–414.
39. ZHANG, Z., *et al.* 1997. STAT3 acts as a co-activator of glucocorticoid receptor signaling. J. Biol. Chem. **272:** 30607–30610.
40. DINARELLO, C.A. 1984. Induction of the acute phase reactants by interleukin-1. Adv. Inflam. Res. **8:** 223–225.
41. DINARELLO, C.A. 1988. Biology of Interleukin-1. FASEB J. **2:** 108–115.
42. DINARELLO, C.A. 1994. The interleukin-1 family: 10 years of discovery. FASEB J. **8:** 1314–1325.
43. THANOS, D. & T. MAINIATIS. 1995. NF-κB: a lesson in family values. Cell **80:** 529–532.
44. AKIRA, S., *et al.* 1990. A nuclear factor for IL-6 expression (NF-IL-6) is a member of C/EBP family. EMBO J. **9:** 1897–1906.
45. AKIRA, S., *et al.* 1994. Molecular cloning of APRF, a novel IFN-stimulated gene factor 3 p91-related transcription factor involved in the gp130-mediated signaling pathway. Cell **77:** 63–71.
46. ZHONG, Z., *et al.* 1994. Stat3: a STAT family member activated by tyrosine phosphorylation in response to epidermal growth factor and interlerkin-6. Science **264:** 95–98.
47. RAY, A., *et al.* 1995. Concerted participation of NF-κB and C/EBP heteromer in lipopolysaccharide induction of serum amyloid A gene expression in liver. J. Biol. Chem. **270:** 7365–7374.
48. ZHANG, D., *et al.* 1996. STAT3 participates in transcriptional activation of the C-reactive protein gene by interleukin-6. J. Biol. Chem. **271:** 9503–9509.
49. ALAM, T., *et al.* 1993. *Trans*-activation of the α1-acid glycoprotein gene acute phase responsive element by multiple isoforms of C/EBP and glucocorticoid receptor. J. Biol. Chem. **268:** 15681–15688.
50. CONTI, P., *et al.* 1995. The down-regulation of IL-6-stimulated fibrinogen steady state mRNA and protein levels by human recombinant IL-1 is not PGE2-dependent: Effects of IL-1 receptor antagonixt (IL-1RA). Mol. Cell Biochem. **142:** 171–178.
51. BAUMANN, H., *et al.* 1988. Phorbol ester modulates interleukin 6- and interleukin 1-regulated expression of acute phase plasma proteins in hepatoma cells. J. Biol. Chem. **263:** 17390–17396.
52. ZHANG, Z. & G.M. FULLER. 2000. IL-1β inhibits IL-6-mediated rat γ fibrinogen gene expression. Blood **96:** 3466–3472.
53. PAULSON, M., *et al.* 1999. STAT protein transactivation domains recruit p300/CBP through widely divergent sequences. J. Biol. Chem. **274:** 25343–25349.
54. NAKASHIMA, K., *et al.* 1999. Synergistic signaling in fetal brain by STAT3-Smad1 complex bridged by p300. Science **284:** 479–482.
55. GOODMAN, R.H. & S. SMOLIK. 2000. CBP/p300 in cell growth, transformaiton, and development. Genes & Dev. **14:** 1553–1577.
56. OGRYZKO, V.V., *et al.* 1996. The transcriptional coactivators p300 and CBP are histone acetyltransferases. Cell **87:** 953–959.

Fibrinogen Biosynthesis

Assembly, Intracellular Degradation, and Association with Lipid Synthesis and Secretion

COLVIN M. REDMAN AND HUI XIA

Lindsley F. Kimball Research Institute of the New York Blood Center, New York, New York, USA

ABSTRACT: Plasma fibrinogen is synthesized primarily in hepatocytes and assembly of the three component chains (Aα, Bβ, and γ) into its final form as a six-chain dimer (Aα, Bβ, γ)$_2$ occurs rapidly in the lumen of the endoplasmic reticulum (ER). Assembly takes place in a stepwise manner with single chains interacting with each other to form Aα–γ and Bβ–γ complexes. The two-chain complexes then acquire another chain to form half-molecules (Aα, Bβ, γ)$_1$, which in a final step are linked to form the six-chain (Aα, Bβ, γ)$_2$ complex. As with other secreted glycoproteins, N-linked glycosylation of Bβ and γ chains commences in the ER and is completed in Golgi organelles. Sulfation and phosphorylation occur at post-ER stages of the secretory process. Since some ER chaperones coisolate with nascent fibrinogen chains they have been implicated in assisting chain assembly. Studies with recombinant systems, using deletion and substitution mutants, indicate that initial chain assembly depends on hydrophobic interactions present in the C-terminal half of the coil–coil domains and that inter- and intra-disulfide bonds that stabilize fibrinogen are needed to complete chain assembly. Not all the chains that are synthesized are assembled into fibrinogen and the unassembled chains are not secreted. HepG2 cells contain surplus Aα and γ chains that accumulate as free γ chains and as an Aα–γ complex. Aα–γ is degraded by lysosomes whereas the γ chain is degraded by the proteasome–ubiquitin system. Studies with expression of single chains by COS cells confirm that γ and Bβ are hydrolyzed by proteasomes and indicate that Aα is degraded partially both by lysosomes and proteasomes. The role of surplus chains in regulating fibrinogen assembly is not understood but overexpression of any one chain, elicited by transfection of HepG2 cells, results in the upregulation of the other two genes, increased fibrinogen synthesis and secretion, and maintenance of surplus intracellular Aα and γ chains. HepG2 cells, programmed in this manner to increase basal fibrinogen expression, have higher HMG-CoA reductase mRNA levels, enhanced cholesterol and cholesterol ester synthesis, and increased secretion of apolipoprotein B (apoB). Overexpression of basal levels of fibrinogen does not affect synthesis of other acute phase proteins. Enhanced secretion of apoB is due to diminished degradation of nascent apoB by proteasomes and not to increased expression. Increased secretion of apoB is associated with increased basal expression of fibrinogen and is not affected when fibrinogen expression is stimulated by interleukin-6. In HepG2 cells, a feedback mechanism exists and extracellular sterols specifically downregulate expression of the three fibrinogen genes. These studies link, at the cellular level, basal fibrinogen expression with lipid metabolism.

KEYWORDS: Fibrinogen assembly; Intracellular degradation; Apolipoprotein B secretion; Cholesterol.

Address for correspondence: C.M. Redman, The New York Blood Center, 310 East 67 Street, New York, NY 10021, USA. Voice: 212-570-3059; fax: 212-879-0243.
credman@nybc.org

INTRODUCTION

Fibrinogen is mainly expressed in hepatocytes and is secreted into the circulation to attain wide ranging normal values of 200 to 300 mg per 100 ml of plasma. Its level can be greatly increased, two- to 10-fold, as part of the acute phase response following tissue injury, infections, and inflammation. Fibrinogen is a dimer, composed of two identical sets, each set containing three different polypeptides, Aα (MW 66,000), Bβ (MW 52,000), and γ (MW 46,500).[1] The fibrinogen hexamer is assembled in the endoplasmic reticulum (ER) and is stabilized by 29 inter- and intrachain disulfide bonds. A number of other posttranslational modifications also occur intracellularly, including N-glycosylation of the Bβ and γ chains,[2] phosphorylation of the Aα chain,[3] tyrosine sulfation of the γ' variant,[4] and partial proteolytic processing of 15 carboxy-terminal amino acids of the Aα chain.[5] Under normal circumstances, fibrinogen is secreted into the blood only as a fully assembled hexamer and there are no indications of single fibrinogen chains or other precursor forms.

Since fibrinogen is composed of different polypeptide subunits and correct assembly is normally a prerequisite for secretion, it serves as a useful model for studying the mechanism by which multichain proteins are assembled. Fibrinogen assembly requires several coordinated steps. These steps include translation of each of the chains, independent translocation into the lumen of the ER, interactions of the nascent chains with chaperone proteins that assist in the assembly and folding processes, and quality control mechanisms that distinguish properly assembled fibrinogen destined for secretion from unassembled forms that are degraded intracellularly.

Studies on the mechanisms of fibrinogen biosynthesis and secretion have been aided by *in vitro* studies using HepG2 cells, and recombinant systems with mammalian cells, although previous *in vivo* animal studies also yielded valuable preliminary information. This article reviews the intracellular assembly of fibrinogen and the degradation of non-secreted forms and more recent studies that indicate an association of increased constitutive expression of fibrinogen in HepG2 cells, with enhanced lipid synthesis and apolipoprotein B (apoB) secretion.

BIOSYNTHESIS AND ASSEMBLY OF HUMAN FIBRINOGEN

Biosynthesis

Fibrinogen is mainly synthesized in the liver by hepatic parenchymal cells, although small amounts are expressed in non-hepatic tissues. Each fibrinogen chain is encoded by a separate gene and, in liver, the levels of the three different mRNAs are approximately equal, both in normal and acute phase conditions.[6–11] The nascent fibrinogen chains, each on separate polysomes, are cotranslationally directed towards the lumen of the ER where an ordered and sequential series of steps lead to the formation of fibrinogen. The process of chain assembly is rapid, a six-chain molecule being assembled in less than five minutes and complete assembly occurring in the ER.[3,12,13] In comparison to other proteins involved in coagulation, fibrinogen is expressed at relatively high levels under normal basal conditions and its expression is greatly upregulated in response to proinflammatory agents.[14,15]

The intracellular levels of the three fibrinogen chains vary in the various systems studied. In rats, the levels of the three fibrinogen mRNAs in liver are nearly equal, both under normal conditions and following induction of the acute phase response.[6] HepG2 cells, a human hepatocellular carcinoma, contain surplus Aα and γ chains, present as an Aα–γ complex and as free γ chains.[12] Surplus Aα and γ chains may result from a combination of unequal synthesis an degradation rates. In HepG2 cells, the initial rate of synthesis of Aα and γ chains is slightly greater than that of Bβ and, as is discussed below, surplus chains are degraded by different cellular mechanisms. *In vivo* studies with rabbits injected with labeled amino acid, indicate a similar mechanism to HepG2 cells, with radioactivity initially appearing in Bβ, followed by Aα and γ.[16] On the other hand, primary chicken hepatocytes, maintained under hormone deficient conditions, contain more Bβ than Aα, which correlates with a reduced Aα mRNA level, and have surplus γ chains.[15,17] Rat hepatocytes, are similar to chicken hepatocytes, also containing more β than Aα and surplus γ chains.[18] By contrast, an *in vivo* study with dogs only detected nascent fully formed fibrinogen in the ER, after a 15 minute pulse, and did not detect other fibrinogen forms.[3]

Sequential Steps in the Assembly of Fibrinogen Chains

The intracellular assembly of fibrinogen has been studied in HepG2 cells, in stably transfected COS, and in baby hamster kidney cell (BHK) lines expressing combinations of the three fibrinogen chains.[12,13,19–21] Taken together these studies demonstrate that fibrinogen is assembled in a stepwise manner. In the first step, two-chain complexes, α–γ and Bβ–γ, are formed and little, if any, Aα–Bβ is detected. In the second step, a third chain is affixed to the two-chain intermediates to produce a three-chain half-molecule, $(A\alpha–B\beta–\gamma)_1$. In a final step the two half-molecules are joined at their N-termini to form the dimeric hexamer $(A\alpha–B\beta–\gamma)_2$ (see FIGURE 1).

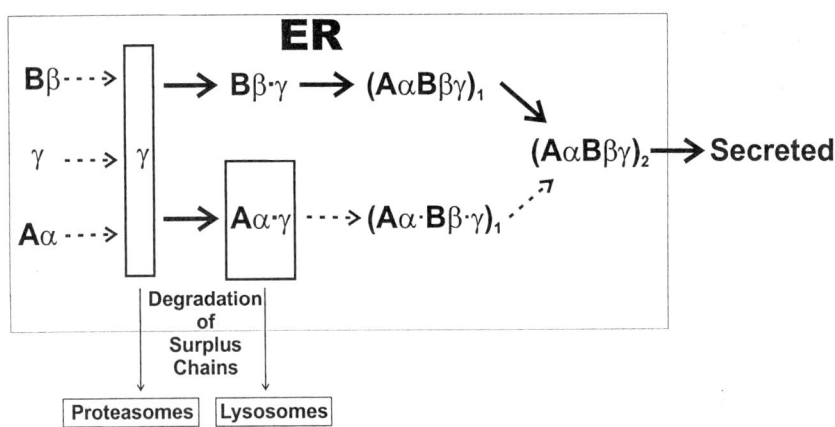

FIGURE 1. Diagram illustrating the steps in fibrinogen assembly. The boxes enclosing γ and Aα–γ indicate that they accumulate as intracellular pools in HepG2 cells. Pulse chase experiments indicate that that the path marked by *solid arrows* is the principal pathway although the path marked by *dotted arrows* also occurs.

After the individual chains have been translated and two-chain precursors formed, the assembly process does not require further protein synthesis[22] and is probably driven by the unique primary sequences of the chains, assisted by a number of molecular chaperones probably present as a macromolecular complex in the ER.

The β and γ fibrinogen chains are glycoproteins, with N-linked sugars. Carbohydrate processing, such as occurs with other secreted glycoproteins, commences in the ER and is finalized in Golgi organelles.[2,23,24] Phosphorylation occurs in a post-ER cellular compartment and nearly all of the newly secreted fibrinogen is phosphorylated.[3] Since a significant percent of plasma fibrinogen is not phosphorylated this indicates that dephosphorylation occurs in the circulation.

Although the stepwise assembly of fibrinogen chains leading to intact fibrinogen is well supported by experimental evidence from different systems, there are still several unresolved issues. Although α–γ complexes are formed and are prevalent in most systems studied, the extent to which they participate as intermediate precursors in the assembly process is not yet clear. Pulse chase experiments with HepG2 cells do not show a definitive precursor–product relation between Aα–γ, (Aα–Bβ–γ)$_1$ and (Aα–Bβ–γ)$_2$ whereas another two-chain complex, Bβ–γ, does. Nascent Aα–γ complex accumulates within the cell and appears to act as a pool of residual chains rather than as an intermediate precursor.[12] In a steady state situation α–γ complex, together with free γ chains and fully formed fibrinogen, are present in HepG2 cells and in transfected cells expressing all three fibrinogen chains. There is evidence, however, that Aα–γ can act as a precursor to fibrinogen, since BHK cell lines expressing Aα–γ, when fused with cells only expressing Bβ, are capable of forming and secreting fibrinogen.[22] Another unresolved issue concerns the first step in the assembly process, when nascent single chains interact with another chain to form two-chain complexes. Incubation of HepG2 cells for two to three minutes with radioactive L-methionine allows the detection of incomplete radioactive fibrinogen chains, present on polysomes, disulfide linked to preexisting proteins present in the ER. The molecular weight of the radioactive complex, which contains the incompletely synthesized chains, is between 102,000 and 113,000, as determined by non-reduced SDS-PAGE, and the radioactivity is quickly "chased" when cells are incubated with non-radioactive L-methionine. That the radioactive proteins present in the complex are incomplete polypeptide chains is supported by the fact that in the presence of cycloheximide the radioactive proteins are not chased.[12] Originally we suggested that the nascent incomplete chains assembled with preexisting pools of Aα and γ chains, but this first step in fibrinogen assembly is not yet clear. Given the role that ER resident chaperones are now known to play in the folding, assembly, and processing of proteins,[25] it is likely that fibrinogen assembly takes place on chaperones close to its entry point into the ER. A complex set of chaperones may act both to place the various chains together and to assist in all of the subsequent interactions. As is discussed below the chaperone complex may also be involved in distinguishing and separating properly assembled chains destined for secretion from other forms that are degraded intracellularly.

Biosynthesis of Fibrinogen$_{420}$

A subclass of fibrinogen contains a large globular extension of 236 amino acids at the C-terminus of the Aα chain that is homologous to the C-terminal domains of the Bβ and γ chains. The fibrinogen subclass containing an extended Aα chain, termed α_E, is called fibrinogen$_{420}$ or Fib$_{420}$.[26,27] Although both fibrinogen and Fib$_{420}$ share common Bβ and γ chains, and are synthesized simultaneously, Fib$_{420}$ is assembled and secreted as a homodimer (α_E, Bβ, γ)$_2$ and not as a heterodimer. The ability to discriminate between Aα and αE chains in the assembly process is maintained in a recombinant COS cell system that has been programmed to coexpress both fibrinogen and Fib$_{420}$.[28] The mechanism that favors formation of the homodimer in the face of a large stoichiometric difference in the amounts of Aα and αE is not yet well understood, but may involve steric considerations in the assembly process and/or rapid removal, by proteolysis, of heterodimers.

Primary Structures Required for Assembly

Expression of substitution and deletion mutants in recombinant systems has proved useful in identifying the primary structures that are necessary for fibrinogen assembly and secretion. These studies indicate that hydrophobic interactions in the coiled-coil region are the driving force in promoting chain interactions, although both inter- and intrachain disulfide bonds also play important roles in assembly. Coexpression of normal Aα and γ chains with deletion mutants of the Bβ chain in COS cells first indicated that the intact coiled-coil domain is necessary for fibrinogen assembly and secretion.[29,30] Deletion of the N-terminal 72 amino acids of Bβ that precede the coil-coil region, or of the C-terminal 208–461 amino acids that follow the coiled-coil region, resulted in chain assembly and secretion of a mutant fibrinogen containing wildtype Aα and γ together with mutant Bβ chains. By contrast, deletion of Bβ amino acids 1–93, a deletion that encroaches into the N-terminal domain of the coil-coil, allowed formation of heteromeric complexes, but BβΔ1–93 was not incorporated into a secreted product. A high molecular weight form that was secreted was composed only of wildtype Aα and γ chains and did not contain BβΔ1–93. By contrast, BβΔ1–80, that retains the intact coiled-coil region, but lacks the N-terminal disulfide ring that includes BβCys76 and BβCys80, is assembled into a three-strand half-molecule that is secreted. This indicated that a complete coiled-coil sequence is required for assembly of a three-strand complex that is competent to be secreted. It also indicated that intact disulfide rings flanking the N-terminal side of the coiled-coil region are needed for dimerization. Xu *et al.*[31] demonstrated that deletion of the first half of the coiled-coil domain of each the three chains allowed two-chain and three-chain half-molecules to be formed, but disallowed formation of the six-chain dimer. On the other hand, deletion of the distal second half of the coiled-coil region of each of the chains prevented the formation of two-chain molecules, and thus further disallowed assembly into the hexamer. These experiments show that the initial two-strand interactions occur at the distal half of the coiled–coil region and that a complete first half of the coiled-coil region is necessary to accommodate three-strand formation.

The disulfide bonds that stabilize fibrinogen also play critical roles in chain assembly. Although substituting the cysteine residues (AαCys28, γCys8 and 9), which are involved in formation of the symmetrical disulfide bonds between the two half-molecules, with serine do not affect dimer formation,[32] Huang *et al.*[33] showed that if in addition the AαCys36 and BβCys65 linkage is disrupted dimerization is prevented. AαCys36 and BβCys65, by themselves, can hold the two half-molecules together indicating that this disulfide bond bridges the half-molecules. Other interchain disulfide bonds that are not involved in directly linking the two half-molecules also participate in dimerization. Dimerization is also prevented when the disulfide rings that flank the proximal end of the coiled-coil region (BβCys76-AαCys49, BβCys80-γCys19, and γCys23-AαCys45) are disrupted by site directed mutagenesis. Since this occurs in the presence of all the cysteine residues known to link the half-molecules, it suggests that the N-terminal disulfide rings that flank the coiled-coil region must be formed before dimerization occurs. On the other hand, disruption of the disulfide rings at the proximal end of the coiled-coil region inhibits chain assembly, with some six-chain fibrinogen formed but not secreted, indicating that these disulfide rings are not required for chain assembly but participate in the process.[34]

The fibrinogen intrachain disulfide loops are conserved in different species, suggesting that they may play roles in maintaining structures that allow biological functions. These intrachain disulfide bonds also affect fibrinogen chain assembly and secretion.[34,35] Each half-molecule has six intrachain disulfide loops in the carboxy terminal region. The Aα chain has one (Cys442-Cys472), Bβ has three (Cys201-Cys286, Cys211-Cys240, and Cys394-Cys339), and γ has two (Cys153-Cys182 and Cys326-Cys339). Preventing formation of the disulfide loops closest to the coiled-coil region in the Bβ and γ chains (i.e., BβCys201-Cys286, BβCys211-Cys240, and γCys153-Cys182) markedly inhibits the formation of two-, three-, or six-chain fibrinogen molecules, although some Aα–γ and Bβ–γ complexes are formed when BβCys201–286 is disrupted. This indicates that it is not sufficient to only have the correct primary coiled-coil sequences and adjacent disulfide rings for chain assembly; the neighboring C-terminal disulfide rings are also required. These disulfide rings may be necessary for proper alignment and placement of the chains into the coiled-coil configuration. The more distal disulfide rings in the Aα and γ chains appear not to be necessary for chain assembly. Disruption of AαCys442-Cys472 does not affect fibrinogen assembly or secretion. γCys326-Cys339 is not necessary for chain assembly, but may be needed for correct folding since its disruption prevents secretion of the six-chain product. FIGURE 2 summarizes the domains that actively participate in fibrinogen chain assembly.

Role of Chaperones on Chain Assembly

Secreted proteins, like fibrinogen, enter the ER and are assembled and folded within this organelle. To support this process efficiently the ER maintains an oxidizing environment, it contains chaperones and glycosylation enzymes that assist protein folding, chain assembly, and disulfide bond formation.[25] The ER also maintains a quality control mechanism, again using chaperones, ensuring that only correctly assembled and folded proteins exit the ER. Misfolded or misassembled proteins are retained and ultimately degraded. (For a review see Ref. 36.) The evidence that chaperones are involved in fibrinogen chain assembly derives mainly from experiments that coisolate fibrinogen precursors together with known chaperones that are present in the ER. In addition, *in vitro* studies with a cell free systems demonstrate that an oxidizing environment is not sufficient and that fibrinogen chain assembly requires the presence of components of the ER. Nascent fibrinogen chains have been found associated with BiP (GRP78), a resident ER chaperone, in several recombinant

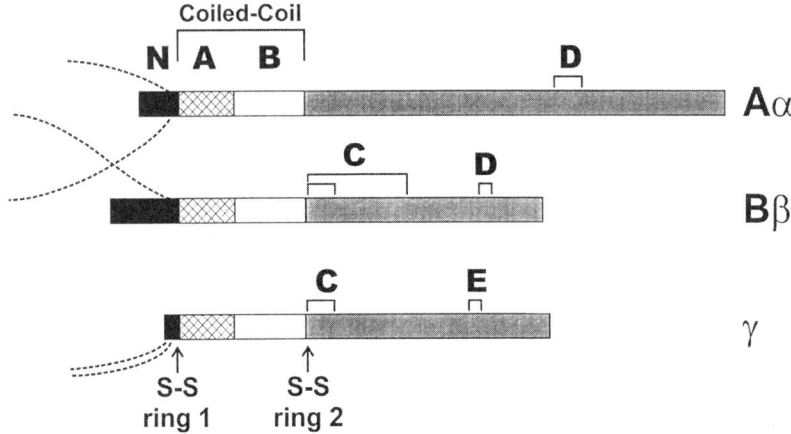

FIGURE 2. Sequences involved in fibrinogen chain assembly. The figure shows a fibrinogen half-molecule. The *dotted lines* represent disulfide bonds that link it to the other half-molecule. *N* depicts the amino-terminal domains of each of the chains. The coiled-coil region is separated into a proximal half (*A*) and a distal half (*B*). The disulfide rings that flank the N-terminal side of the coiled-coil region is termed the S-S ring 1, and that at the C-terminal end of the coiled-coil is named S-S ring 2. The Bβ and γ disulfide loops closest to the coiled-coil are marked *C*. The disulfide loops at the C-terminal regions are labeled *D* and *E*. The effects of deletion or disruption of the various regions are as follows: N, A, and S-S ring 1 prevent dimer formation. B and C prevent or markedly inhibit initial formation of two-chain complexes. S-S ring 2 inhibits all chain assembly. E allows chain assembly, but fibrinogen is not secreted. D has no affect on chain assembly or secretion.

systems expressing fibrinogen, and in HepG2 cells and chicken hepatocytes.[20,21,37,38] In an *in vitro* system composed of rabbit reticulocyte lysate and dog pancreas microsomes, antibodies to Bip, Grp94, protein disulfide isomerase and calnexin coisolated the chaperones together with nascent fibrinogen complexes.[39] Thus, although it is apparent that chaperones are involved in fibrinogen assembly, the mechanisms by which these chaperones coordinate and assist chain assembly remains to be determined.

INTRACELLULAR DEGRADATION OF FIBRINOGEN CHAINS

Although the majority of fibrinogen synthesized by HepG2 cells, approximately 77%, is secreted under normal conditions, a substantial amount is retained and eventually degraded.[12] In addition to fibrinogen, HepG2 cells accumulate Aα–γ complex and some free γ chains. Various cellular mechanisms are used in degrading the surplus fibrinogen chains that are not secreted (see TABLE 1). There are two main proteolytic systems in mammals, lysosomes and proteasomes. Lysosomes contain acid hydrolases and mostly degrade macromolecules that enter cells by the endocytic pathway, but they are also involved in processing internalized plasma membrane receptors and in degrading proteins that have exited the ER and are marked not to be secreted. Proteasomes, on the other hand, are part of the ER-associated degradation (ERAD) process that eliminates proteins that are not properly folded or assembled in the ER. Proteins that fail to pass quality control are retrotranslocated from the ER to the cytosol, where ubiquitin conjugating enzymes target them for degradation by proteasomes.[36]

In HepG2 cells, degradation of the Aα–γ complex is markedly diminished by compounds that inhibit the acid hydrolases present in lysosomes. Thus, chloroquine, NH_4Cl, and leupeptin all inhibit the rate at which Aα–γ is degraded. By contrast, these reagents do not affect the degradation of free γ chains.[37] Degradation of free γ chain is, however, inhibited by MG132, a compound that specifically inhibits proteolysis by proteasomes. In HepG2 cells, ubiquinated fibrinogen chains were detected, confirming the role of proteasomes in fibrinogen chain degradation.[40]

To determine the mode of degradation of individual fibrinogen chains, COS cells that were programmed to express single fibrinogen chains were studied. Leupeptin, an inhibitor of lysosome enzymes, does not affect the degradation of Bβ or γ chains, but MG132, a proteasome inhibitor, prevents degradation. This indicates that free Bβ and γ chains are degraded by proteasomes and not lysosomes. Surprisingly, free Aα chains appear to be degraded by both lysosomes and proteasomes. Inhibitors of both lysosome and proteasome enzymes partially inhibit the degradation of free Aα chains expressed by COS cells.[37,40] The reason why Aα chains are degraded by dual paths is not known, but time courses of degradation indicate that leupeptin mostly inhibits Aα degradation at late times, 3–8 hours after synthesis, whereas proteasome inhibitors act within the first three hours.

Time course studies that measure rates of degradation of the individual chains expressed by COS cells indicate that γ chains are degraded much slower than Aα and Bβ chains. The halflife of Aα and Bβ chains is about one and a half hours, that of γ chain exceeds three hours.[40] Recent unpublished studies from our laboratory also indicate that, in HepG2 cells, nascent glycosylated Bβ chain associates with Sec61p much more quickly than nascent γ chains. Since Sec61p is a protein component of the translocon that is involved in retrotranslocation and ERAD, this, together with the noted longer halflife of free γ chains, suggests that unassembled γ chains are

TABLE 1. Intracellular degradation of surplus fibrinogen chains

Cells Expressing Fibrinogen	Intracellular Fibrinogen	Proteolytic Inhibitor	Degradation Site
HepG2	Aα–γ	chloroquine, NH_4Cl, leupeptin	lysosomes
HepG2	free γ	MG132	proteasomes
COS	free Aα	leupeptin, MG132	lysosomes and proteasomes
COS	free Bβ	lactacystin, MG132	proteasomes
COS	free γ	MG132	proteasomes

NOTE: HepG2 cells or COS cells expressing single fibrinogen chains were radiolabeled with L-[^{35}S]methionine and chase incubated for various periods of time in the presence and absence of the specific proteolytic inhibitors.[37,40] MG132 and lactacystin are specific inhibitors of proteasomes and chloroquine, NH_4Cl and leupeptin inhibit the acid hydrolases present in lysosomes.

transported from the ER to proteasomes at a lower rate than the other fibrinogen chains. The unequal rates of degradation may contribute to the accumulation of surplus γ chains in hepatocytes.

Surplus Aα and γ Chains are Maintained when Fibrinogen is Over Expressed

Fibrinogen expression is regulated at two levels; the basal or constitutive level and during the acute phase response when stimulated by interleukin-6 (IL-6). When fibrinogen is expressed at basal levels, in HepG2 cells, there is steady state pool of surplus Aα and γ chains, present as Aα–γ and free γ chains. This condition remains when fibrinogen expression is stimulated with IL-6 since there is a coordinated upregulation of expression of all three fibrinogen genes. Overexpression of the basal rate of fibrinogen expression can be accomplished in HepG2 cells by transfection of cDNA encoding a single fibrinogen chain and selection of cell clones that overexpress and secrete increased levels of fibrinogen. In this situation, in which a single fibrinogen chain is overexpressed, the pools of surplus Aα and γ chains are maintained and the cells compensate by upregulating the expression of the other two chains. Originally this was accomplished by transfecting HepG2 cells with Bβ cDNA.[41] Similar results are obtained by transfection with Aα or γ cDNA, although there is less overexpression of fibrinogen with the latter two cDNAs.[42] Overexpression of fibrinogen by this procedure does not affect the synthesis and secretion of other acute phase proteins. The mechanism by which this occurs is not understood, although nuclear run-on experiments demonstrated that increased expression of the fibrinogen chains is due to increased transcription.[41,42]

Although *in vivo* fibrinogen expression is mostly regulated as part of the acute phase response and basal levels are considered to be stable, there are genetic circumstances wen basal fibrinogen expression is increased. Two common polymorphisms, Bβ-455G/A and –854G/A, show increased basal rates of transcription when transfected into HepG2 cells and associated *in vivo* with higher levels of plasma fibrinogen.[43]

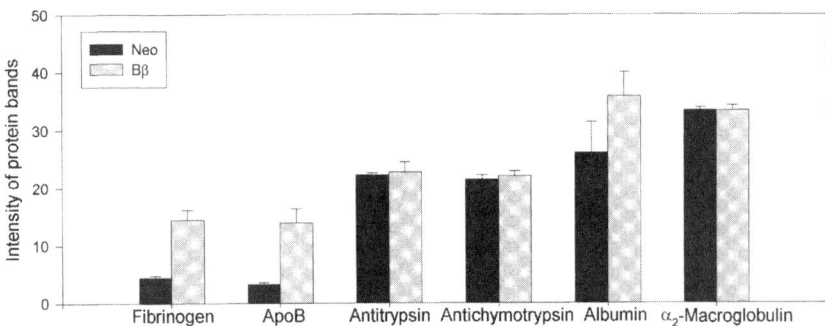

FIGURE 3. Association of increased apoB secretion and overexpression of fibrinogen. The secretion of fibrinogen, apoB, albumin, and three representative acute phase proteins is shown. *Neo* is a control HepG2 cell and *Bβ* is a HepG2 cell that overexpresses fibrinogen. (Reproduced from Ref. 44, with permission from Academic Press.)

Although in these cases increased basal expression is caused by genetic mutations in the promoter region of the Bβ gene, it suggests that basal rates of fibrinogen *in vivo* may also be influenced by increased expression of a single chain.

ASSOCIATION OF INCREASED BASAL FIBRINOGEN EXPRESSION, ENHANCED LIPID SYNTHESIS, AND APOLIPOPROTEIN B SECRETION

In determining the cellular parameters that are affected in HepG2 cells that are programmed to overexpress basal levels of fibrinogen elicited by transfection with Bβ cDNA, we determined that, although expression and secretion of the other acute phase proteins are unaffected, there is a fourfold increase in secretion of apoB (see FIGURE 3). Increased apoB secretion is independent of the fibrinogen cDNA used to overexpress fibrinogen, and similar results are obtained when Aα, Bβ, or γ cDNA are transfected. ApoB is a large (540-kDA) protein that is the principal structural protein in human LDL and VLDL. ApoB assembles lipids in the liver and intestines, and the resulting lipoprotein particles are secreted into the circulation. In plasma, LDL and VLDL are the major carriers of cholesterol and triglycerides.[45,46] Epidemiological studies show that elevated levels of plasma fibrinogen and LDL-cholesterol are correlated with coronary artery disease.[47–49] The reasons for this association are not known but studies with transgenic mice that overexpress fibrinogen indicate that elevated levels of fibrinogen, by itself, may not cause coronary artery disease.[50]

Overexpression of Fibrinogen is Associated with Increased apoB Secretion Due to Diminished Intracellular Degradation

Enhanced secretion of apoB is due to diminished intracellular proteolysis of nascent apoB and not to increased expression. Overexpression of fibrinogen in HepG2 cells does not affect apoB expression as measured by Northern blot analysis and by the initial rate of synthesis determined by the incorporation of L-[^{35}S]methionine. Although both control HepG2 cells and cells overexpressing fibrinogen synthesize equal amounts of apoB, there is a large difference in the amount of apoB secreted into the incubation medium. Quantification of the amount of apoB synthesized at the end of a 15-minute pulse and that secreted into the medium and retained by the cells after a two-hour chase demonstrated that, as described by others, the majority of apoB is degraded intracellularly and not secreted. HepG2 cells that overexpress fibrinogen and secrete increased amounts of apoB degrade less nascent apoB. In control HepG2 cells, 86% of apoB is degraded within the cell, and in cells overexpressing fibrinogen 78% is degraded.

Degradation of apoB, like that of individual fibrinogen chains, is via the proteasome–ubiquitin pathway, and treatment of both control cells and cells overexpressing fibrinogen with MG132, a specific inhibitor of proteasomes, resulted in increased secretion of apoB, indicating that inhibition of degradation by itself is sufficient to increase apoB secretion.[40]

Overexpression of Fibrinogen is Associated with Increased HMG-CoA Reductase mRNA Levels and Increased Synthesis of Cholesterol and Cholesterol Esters

HMG-CoA reductase is a key enzyme that is sensitive to feedback mechanisms regulating cholesterol synthesis via the mevalonate biosynthetic pathway. Its mRNA level is an index of the level of cholesterol biosynthesis.[51] Northern blot analysis showed that HepG2 cells overexpressing fibrinogen had threefold greater levels of HMG-CoA reductase mRNA than control cells. Compared with control cells, HepG2 cells overexpressing fibrinogen synthesized three times more cholesterol and 1.6 times more cholesterol esters, as measured by determining incorporation of [^{3}H]acetate into lipids. There was, however, no significant difference in the synthesis of triglycerides and phospholipids (see FIGURE 4). ApoB secretion is regulated post-translationally by balancing the amount that is targeted for secretion with the amount that is degraded intracellularly by proteasomes. Assembly of lipids by nascent apoB protects it from intracellular degradation.[52–55] However, the role of free and esterified cholesterol in influencing the degradation of nascent apoB in hepatocytes is not clear. Although it is reported that apoB secretion increases, both *in vitro* and *in vivo*, with higher cholesterol levels[56–58] there are also studies with contrary results.[59–61]

The sterol responsive element protein (SREBP) family of transcription factors regulates the uptake and synthesis of cholesterol. These transcription factors, which are part of larger membrane proteins, emanate from the ER following sterol regulated proteolysis.[62] Although the SREBPs are essential factors in activation of HMG-CoA reductase expression, there is no increase in the amounts of these factors in HepG2 cells overexpressing fibrinogen, suggesting that other coregulatory factors may also participate.

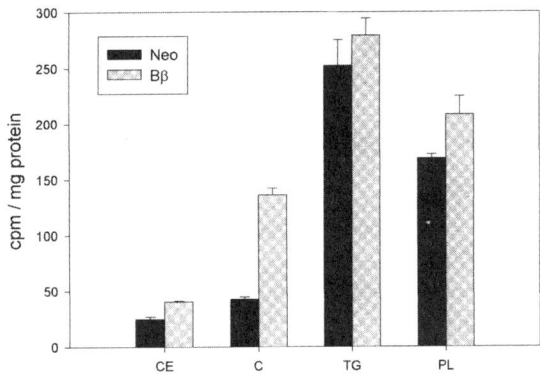

FIGURE 4. Increased cholesterol biosynthesis in HepG2 cells overexpressing fibrinogen. Control HepG2 cells (*Neo*) and cells overexpressing fibrinogen (Bβ) were labeled with [^{3}H]-acetate and their incorporation into cholesterol esters (*CE*), cholesterol (*C*), triglycerides (*TG*), and phospholipids (*PL*) determined. (Reproduced from Ref. 44, with permission from Academic Press.)

Increased apoB Secretion is Independent of the Acute Phase Response

Il-6 stimulates the coordinate expression of the three fibrinogen genes in HepG2 cells emulating the acute phase response. Treatment of both control cells and cells overexpressing fibrinogen with IL-6 causes increased expression of fibrinogen in both cell types. Il-6, however, has no effect on the secretion of apoB in either control cells or in cells overexpressing fibrinogen.[44] This, together with the fact that increased fibrinogen expression does not affect the synthesis of other acute phase proteins such as α_1-antitrpsin and anti-chymotrypsin, α_2-macroglobulin, or C-reactive protein, demonstrates that increased apoB secretion is independent of the acute phase response. Thus, increased apoB secretion and cholesterol synthesis is only linked to increased level of basal fibrinogen expression, as elicited experimentally by transfection of single fibrinogen cDNAs and selection of cells that overexpress fibrinogen.

Extracellular Sterols Downregulate Fibrinogen Expression

In addition to linking increased basal expression of fibrinogen with cholesterol biosynthesis and apoB secretion, HepG2 cells also have a feedback mechanism by which extracellular sterols downregulate the basal expression of fibrinogen. As determined by separate procedures (Northern blot analysis, initial rate of synthesis, and level of secretion) sterols (mixtures of cholesterol and 25-hydroxycholesterol) downregulate the expression of Aα, Bβ, and γ fibrinogen, both in normal HepG2 cells and in HepG2 cells overexpressing fibrinogen. Feedback regulation is dose dependent and responds to the extracellular levels of 25-hydroxycholesterol. Feedback inhibition is limited to the basal expression of fibrinogen and does not affect the stimulation by IL-6 (Xia and Redman, unpublished). These studies indicate that sterol and basal fibrinogen expression are balanced at the cellular level.

Significance of the Association of Basal Fibrinogen Expression, Cholesterol Metabolism, and apoB Secretion

The association of increased cholesterol biosynthesis and apoB secretion with increased basal expression of fibrinogen is not likely to be due to a selection of a special cell line or to a transfection artifact. Eleven different HepG2 clones, all of which expressed and secreted more fibrinogen than the control cell lines that were transfected with empty vectors, had enhanced apoB secretion. In addition, since not all cells that were originally transfected with Bβ cDNA overexpressed fibrinogen, those cells that did not overexpress fibrinogen were also analyzed and found not to secrete greater amounts of apoB.[44] Further evidence against the view that increased apoB secretion is due to a transfection artifact is that others have transfected HepG2 cells with an expression vector containing another cDNA, a minigene for apolipoprotein (a), and those stable transfected cells did not exhibit increased apoB secretion.[63]

The association of increased, basal fibrinogen expression, elicited by transfection of a single fibrinogen cDNA and increased cholesterol synthesis and apoB secretion, is quite specific since the secretion of other plasma proteins is unaffected. In addition, the feedback mechanism by sterols that downregulates basal fibrinogen expression occurs in both normal HepG2 cells and in HepG2 cells overexpressing fibrinogen, further indicating that the association is independent of the transfection procedure.

As previously discussed, both the association of basal fibrinogen expression and increased lipid metabolism and the feedback inhibition of fibrinogen expression by sterols are independent of the acute phase response. The acute phase response is an immediate reaction to tissue damage or to infection, whereas basal expression of fibrinogen is an inherent function that probably remains constant and may only vary in persons with genetic differences in fibrinogen genes. Although the role of genetic factors in determining plasma fibrinogen levels is still being debated,[64–67] it has been demonstrated that two unrelated polymorphisms in the Bβ fibrinogen gene promoter exhibit increased expression when transfected in HepG2 cells.[43] These Bβ polymorphisms are associated with increased fibrinogen plasma concentrations in men and may be compared to the HepG2 system that overexpresses Bβ chain.

Increased basal expression by the introduction of extra copies of a single fibrinogen transcript is not a physiological process, but it serves to expose and perhaps magnify existing cellular pathways that are not apparent under normal circumstances. Since the studies are performed with cells in culture that are removed from extracellular influences, the associations noted in isolated cells may be modified *in vivo*.

Epidemiological studies indicate that elevated levels of plasma fibrinogen and LDL-cholesterol are independent risk factors in coronary artery disease, but there is a positive association between increased fibrinogen plasma concentration and increased lipid levels. Given that this positive association occurs *in vivo*, the cellular association of basal fibrinogen expression and cholesterol synthesis, although perhaps exaggerated in HepG2 cells, may be of physiological significance.

REFERENCES

1. BLOMBACK, B. 1996. Fibrinogen and fibrin—proteins with complex roles in hemostasis and thrombosis. Thromb. Res. **83:** 1–75.
2. NICKERSON, J.M. & G.M. FULLER. 1981. Modification of fibrinogen chains during synthesis: glycosylation of B beta and gamma chains. Biochemistry **20:** 2818–2821.
3. KUDRYK, B., M. OKADA, C.M. REDMAN, *et al.* 1982. Biosynthesis of dog fibrinogen. Characterization of nascent fibrinogen in the rough endoplasmic reticulum. Eur. J. Biochem. **125:** 673–682.
4. FARRELL, D.H., E.R. MULVIHILL, S.M. HUANG, *et al.* 1991. Recombinant human fibrinogen and sulfation of the gamma′ chain. Biochemistry **30:** 9414–9420.
5. FARRELL, D.H., S. HUANG & E.W. DAVIE. 1993. Processing of the carboxyl 15-amino acid extension in the alpha-chain of fibrinogen. J. Biol. Chem. **268:** 10351–10355.
6. CRABTREE, G.R. & J.A. KANT. 1982. Coordinate accumulation of the mRNAs for the alpha, beta, and gamma chains of rat fibrinogen following defibrination. J. Biol. Chem. **257:** 7277–7279.
7. CRABTREE, G.R. & J.A. KANT. 1981. Molecular cloning of cDNA for the alpha, beta, and gamma chains of rat fibrinogen. A family of coordinately regulated genes. J. Biol. Chem. **256:** 9718–9723.
8. KANT, J.A., S.T. LORD & G.R. CRABTREE. 1983. Partial mRNA sequences for human A alpha, B beta, and gamma fibrinogen chains: evolutionary and functional implications. Proc. Natl. Acad. Sci. U.S.A. **80:** 3953–3957.
9. KANT, J.A. & G.R. CRABTREE. 1983. The rat fibrinogen genes. Linkage of the A alpha and gamma chain genes. J. Biol. Chem. **258:** 4666–4667.
10. CHUNG, D.W., M.W. RIXON, B.G. QUE, *et al.* 1983. Cloning of fibrinogen genes and their cDNA. Ann. N.Y. Acad. Sci. **408:** 449–456.
11. CHUNG, D.W., J.E. HARRIS & E.W. DAVIE. 1990. Nucleotide sequences of the three genes coding for human fibrinogen. Adv. Exper. Med. Biol. **281:** 39–48.

12. YU, S., B. SHER, B. KUDRYK, et al. 1984. Fibrinogen precursors. Order of assembly of fibrinogen chains. J. Biol. Chem. **259:** 10574–10581.
13. YU, S., B. SHER, B. KUDRYK, et al. 1983. Intracellular assembly of human fibrinogen. J. Biol. Chem. **258:** 13407–13410.
14. MARINKOVIC, S., G.P. JAHREIS, G.G. WONG, et al. 1989. IL-6 modulates the synthesis of a specific set of acute phase plasma proteins in vivo. J. Immunol. **142:** 808–812.
15. GRIENINGER, G., P.W. PLANT, T.J. LIANG, et al. 1983. Hormonal regulation of fibrinogen synthesis in cultured hepatocytes. Ann. N.Y. Acad. Sci. **408:** 469–489.
16. ALVING, B.M., S.I. CHUNG, G. MURANO, et al. 1982. Rabbit fibrinogen: time course of constituent chain production in vivo. Archiv. Biochem. Biophys. **217:** 1–9.
17. PLANT, P.W. & G. GRIENINGER. 1986. Noncoordinate synthesis of the fibrinogen subunits in hepatocytes cultured under hormone-deficient conditions. J. Biol. Chem. **261:** 2331–2336.
18. HIROSE, S., K. ODA & Y. IKEHARA. 1988. Biosynthesis, assembly and secretion of fibrinogen in cultured rat hepatocytes. Biochem. J. **251:** 373–377.
19. ROY, S.N., R. PROCYK, B.J. KUDRYK, et al. 1991. Assembly and secretion of recombinant human fibrinogen. J. Biol. Chem. **266:** 4758–4763.
20. HARTWIG, R. & K.J. DANISHEFSKY. 1991. Studies on the assembly and secretion of fibrinogen. J. Biol. Chem. **266:** 6578–6585.
21. HUANG, S., E.R. MULVIHILL, D.H. FARRELL, et al. 1993. Biosynthesis of human fibrinogen. Subunit interactions and potential intermediates in the assembly. J. Biol. Chem. **268:** 8919–8926.
22. HUANG, S., Z. CAO, D.W. CHUNG, et al. 1996. The role of betagamma and alphagamma complexes in the assembly of human fibrinogen. J. Biol. Chem. **271:** 27942–27947.
23. HENSCHEN, A., F. LOTTSPEICH, M. KEHL, et al. 1983. Covalent structure of fibrinogen. Ann. N.Y. Acad. Sci. **408:** 28–43.
24. BLOMBACK, B., N.J. GRONDAHL, B. HESSEL, et al. 1973. Primary structure of human fibrinogen and fibrin. II. Structural studies on NH_2-terminal part of gamma chain. J. Biol. Chem. **248:** 5806–5820.
25. ELLGAARD, L., M. MOLINARI & A. HELENIUS. 1999. Setting the standards: quality control in the secretory pathway. Science **286:** 1882–1888.
26. FU, Y., L. WEISSBACH, P.W. PLANT, et al. 1992. Carboxy-terminal-extended variant of the human fibrinogen alpha subunit: a novel exon conferring marked homology to beta and gamma subunits. Biochemistry **31:** 11968–11972.
27. FU, Y. & G. GRIENINGER. 1994. Fib420: a normal human variant of fibrinogen with two extended alpha chains. Proc. Natl. Acad. Sci. U.S.A. **91:** 2625–2628.
28. FU, Y., J.Z. ZHANG, C.M. REDMAN, et al. 1998. Formation of the human fibrinogen subclass fib420: disulfide bonds and glycosylation in its unique (alphaE chain) domains. Blood **92:** 3302–3308.
29. ZHANG, J.Z. & C.M. REDMAN. 1996. Assembly and secretion of fibrinogen—involvement of amino-terminal domains in dimer formation. J. Biol. Chem. **271:** 12674–12680.
30. ZHANG, J.Z. & C.M. REDMAN. 1992. Identification of B beta chain domains involved in human fibrinogen assembly. J. Biol. Chem. **267:** 21727–21732.
31. XU, W., D.W. CHUNG & E.W. DAVIE. 1996. The assembly of human fibrinogen. The role of the amino-terminal and coiled-coil regions of the three chains in the formation of the alphagamma and betagamma heterodimers and alphabetagamma halfmolecules. J. Biol. Chem. **271:** 27948–27953.
32. ZHANG, J.Z., B. KUDRYK & C.M. REDMAN. 1993. Symmetrical disulfide bonds are not necessary for assembly and secretion of human fibrinogen. J. Biol. Chem. **268:** 11278–11282.
33. HUANG, S., Z. CAO & E.W. DAVIE. 1993. The role of amino-terminal disulfide bonds in the structure and assembly of human fibrinogen. Biochem. Biophys. Res. Commun. **190:** 488–495.
34. ZHANG, J.Z. & C.M. REDMAN. 1994. Role of interchain disulfide bonds on the assembly and secretion of human fibrinogen. J. Biol. Chem. **269:** 652–658.
35. ZHANG, J.Z. & C. REDMAN. 1996. Fibrinogen assembly and secretion—role of intrachain disulfide loops. J. Biol. Chem. **271:** 30083–30088.

36. BONIFACINO, J.S. & A.M. WEISSMAN. 1998. Ubiquitin and the control of protein fate in the secretory and endocytic pathways. Annu. Rev. Cell Dev. Biol. **14:** 19–57.
37. ROY, S., S. YU, D. BANERJEE, *et al.* 1992. Assembly and secretion of fibrinogen. Degradation of individual chains. J. Biol. Chem. **267:** 23151–23158.
38. DANISHEFSKY, K., R. HARTWIG, D. BANERJEE, *et al.* 1990. Intracellular fate of fibrinogen B beta chain expressed in COS cells. Biochim. Biophys. Acta **1048:** 202–208.
39. ROY, S., A. SUN & C. REDMAN. 1996. *In vitro* assembly of the component chains of fibrinogen requires endoplasmic reticulum factors. J. Biol. Chem. **271:** 24544–24550.
40. XIA, H. & C. REDMAN. 1999. The degradation of nascent fibrinogen chains is mediated by the ubiquitin proteasome pathway. Biochem. Biophys. Res. Commun. **261:** 590–597.
41. ROY, S.N., G. MUKHOPADHYAY & C.M. REDMAN. 1990. Regulation of fibrinogen assembly. Transfection of Hep G2 cells with B beta cDNA specifically enhances synthesis of the three component chains of fibrinogen. J. Biol. Chem. **265:** 6389–6393.
42. ROY, S., O. OVERTON & C. REDMAN. 1994. Overexpression of any fibrinogen chain by Hep G2 cells specifically elevates the expression of the other two chains. J. Biol. Chem. **269:** 691–695.
43. 'T HOOFT, F.M., S.J. VON BAHR, A. SILVEIRA, *et al.* 1999. Two common, functional polymorphisms in the promoter region of the beta-fibrinogen gene contribute to regulation of plasma fibrinogen concentration. Arterioscler. Thromb. Vasc. Biol. **19:** 3063–3070.
44. XIA, H. & C. REDMAN. 2000. Enhanced secretion of ApoB by transfected HepG2 cells overexpressing fibrinogen. Biochem. Biophys. Res. Commun. **273:** 377–384.
45. BOREN, J., A. WHITE, M. WETTESTEN, *et al.* 1991. The molecular mechanism for the assembly and secretion of ApoB-100-containing lipoproteins. Progress Lipid Res. **30:** 205–218.
46. DIXON, J.L. & H.N. GINSBERG. 1993. Regulation of hepatic secretion of apolipoprotein B-containing lipoproteins: information obtained from cultured liver cells. J. Lipid Res. **34:** 167–179.
47. ERNST, E. & K.L. RESCH. 1993. Fibrinogen as a cardiovascular risk factor: a meta-analysis and review of the literature. Ann. Int. Med. **118:** 956–963.
48. DANESH, J., R. COLLINS, P. APPLEBY, *et al.* 1998. Association of fibrinogen, C-reactive protein, albumin, or leukocyte count with coronary heart disease—meta-analyses of prospective studies. J. Amer. Med. Assoc. **279:** 1477–1482.
49. WELTY, F.K., M.A. MITTLEMAN, P.W. WILSON, *et al.* 1997. Hypobetalipoproteinemia is associated with low levels of hemostatic risk factors in the Framingham offspring population. Circulation **95:** 825–830.
50. LORD, S.T. & A.A. GULLEDGE. 1999. Exploring a risk factor: diet-induced atherosclerosis in transgenic mice with elevated plasma fibrinogen. Thromb. Hæmost. Suppl. 521–521.
51. GOLDSTEIN, J.L. & M.S. BROWN. 1990. Regulation of the mevalonate pathway. Nature **343:** 425–430.
52. WANG, S., R.S. MCLEOD, D.A. GORDON. *et al.* 1996. The microsomal triglyceride transfer protein facilitates assembly and secretion of apolipoprotein B-containing lipoproteins and decreases cotranslational degradation of apolipoprotein B in transfected COS-7 cells. J. Biol. Chem. **271:** 14124–14133.
53. DU, E.Z., J.F. FLEMING, S.L. WANG, *et al.* 1999. Translocation-arrested apolipoprotein B evades proteasome degradation via a sterol-sensitive block in ubiquitin conjugation. J. Biol. Chem. **274:** 1856–1862.
54. YEUNG, S.J., S.H. CHEN & L. CHAN. 1996. Ubiquitin-proteasome pathway mediates intracellular degradation of apolipoprotein B. Biochemistry **35:** 13843–13848.
55. FISHER, E.A., M. ZHOU, D.M. MITCHELL, *et al.* 1997. The degradation of apolipoprotein B100 is mediated by the ubiquitin–proteasome pathway and involves heat shock protein 70. J. Biol. Chem. **272:** 20427–20434.

56. MUSANTI, R., L. GIORGINI, P.P. LOVISOLO, et al. 1996. Inhibition of acyl-CoA: cholesterol acyltransferase decreases apolipoprotein B-100-containing lipoprotein secretion from HepG2 cells. J. Lipid Res. **37:** 1–14.
57. CIANFLONE, K.M., Z. YASRUEL, M.A. RODRIGUEZ, et al. 1990. Regulation of apoB secretion from HepG2 cells: evidence for a critical role for cholesteryl ester synthesis in the response to a fatty acid challenge. J. Lipid Res. **31:** 2045–2055.
58. WILCOX, L.J., P.H. BARRETT & M.W. HUFF. 1999. Differential regulation of apolipoprotein B secretion from HepG2 cells by two HMG-CoA reductase inhibitors, atorvastatin and simvastatin. J. Lipid Res. **40:** 1078–1089.
59. HUFF, M.W., D.E. TELFORD, P.H. BARRETT, et al. 1994. Inhibition of hepatic ACAT decreases ApoB secretion in miniature pigs fed a cholesterol-free diet. Arterioscler. Thromb. **14:** 1498–1508.
60. BURNETT, J.R., L.J. WILCOX, D.E. TELFORD, et al. 1997. Inhibition of HMG-CoA reductase by atorvastatin decreases both VLDL and LDL apolipoprotein B production in miniature pigs. Arterioscler. Thromb. Vasc. Biol. **17:** 2589–2600.
61. WATTS, G.F., R.P. NAOUMOVA, J.M. KELLY, et al. 1997. Inhibition of cholesterogenesis decreases hepatic secretion of apoB-100 in normolipidemic subjects. Am. J. Physiol. **273:** E462–E470.
62. BROWN, M.S. & J.L. GOLDSTEIN. 1997. The SREBP pathway: regulation of cholesterol metabolism by proteolysis of a membrane-bound transcription factor. Cell **89:** 331–340.
63. BONEN, D.K., A.M.L. HAUSMAN, C. HADJIAGAPIOU, et al. 1997. Expression of recombinant apolipoprotein(a) in HepG2 cells. Evidence for intracellular assembly of lipoprotein(a). J. Biol. Chem. **272:** 5659–5667.
64. HAMSTEN, A., L. ISELIUS, U. DE FAIRE, et al. 1987. Genetic and cultural inheritance of plasma fibrinogen concentration. Lancet **2:** 988–991.
65. BARA, L., V. NICAUD, L. TIRET, et al. 1994. Expression of a paternal history of premature myocardial infarction on fibrinogen, factor VIIC and PAI-1 in European offspring—the EARS study. European Atherosclerosis Research Study Group. Thromb. Hæmost. **71:** 434–440.
66. BERG, K. & P. KIERULF. 1989. DNA polymorphisms at fibrinogen loci and plasma fibrinogen concentration. Clin. Genet. **36:** 229–235.
67. REED, T., R.P. TRACY & R.R. FABSITZ. 1994. Minimal genetic influences on plasma fibrinogen level in adult males in the NHLBI twin study. Clin. Genet. **45:** 71–77.

Fibrinogen Gene Mutations Accounting for Congenital Afibrinogenemia

MARGUERITE NEERMAN-ARBEZ

Division of Medical Genetics, University Medical School, Geneva, Switzerland

ABSTRACT: This article reviews recent progress made in understanding the molecular basis of congenital afibrinogenemia, an autosomal recessive coagulation disorder characterized by the complete absence of detectable fibrinogen. We have identified the first causative mutations for this disorder in a nonconsanguineous Swiss family; these were homozygous deletions of approximately 11 kb of the fibrinogen alpha chain (*FGA*) gene. Haplotype data implied that the deletions occurred on distinct ancestral chromosomes, suggesting that this region may be susceptible to deletion by a common mechanism. All the deletions were identical to the base pair, and probably resulted from non-homologous (illegitimate) recombination. In a subsequent study of 13 unrelated patients with congenital afibrinogenemia we analyzed the *FGA* gene in order to identify the causative mutations, and to determine the prevalence of the 11-kb *FGA* deletion. Although this deletion was found in an additional unrelated patient, the most common mutation was at the donor splice site of *FGA* intron 4 (IVS4+1 G>T). Three frameshift mutations, two nonsense mutations, and one other splice site mutation were also characterized. Other studies identified one further *FGA* nonsense mutation, two *FGB* missense mutations, and one *FGG* nonsense mutation, all in homozygosity in a single patient. In conclusion, the majority of patients have truncating mutations in the *FGA* gene although, intuitively, all three fibrinogen genes could be predicted to be equally implicated. These results will facilitate molecular diagnosis of the disorder, permit prenatal diagnosis for families who so desire, and pave the way for new therapeutic approaches such as gene therapy.

KEYWORDS: Afibrinogenemia; Fibrinogen; Mutations; Large deletion; Non-homologous recombination.

INTRODUCTION

The intrinsic and extrinsic coagulation pathways involve a complex cascade of reactions leading ultimately to the cleavage of fibrinogen to fibrin and formation of a fibrin clot. Abnormal activities of the enzymes or cofactors of this pathway are predicted or known to lead to various forms of coagulation disorders, the most common and best documented of these being hemophilia A and hemophilia B, due to mutations in the *F8C* gene and the *F9* gene, respectively. Mutations have also been identified in deficiency of factors V, VII, X, XI, XII, XIII, and II, in combined Factor V–Factor VIII deficiency, and in hypo- and dysfibrinogenemia.

Address for correspondence: Marguerite Neerman-Arbez, Ph.D., Division of Medical Genetics, University Medical School, Geneva, Switzerland. Voice: +41 22 702 5809; fax: +41 22 702 5706.

Marguerite.Arbez@medecine.unige.ch

Congenital afibrinogenemia (Mendelian Inheritance in Man No. 202400) is a rare, autosomal recessive disorder characterized by the complete absence of fibrinogen, the precursor that is processed by thrombin to form fibrin, the major protein constituent of the blood clot. The disease was originally described in 1920 by Rabe and Solomon[1] with an estimated prevalence of around 1 in 1,000,000.[2,3] In populations where consanguineous marriages are common, however, the prevalence of afibrinogenemia, as for other autosomal recessive coagulation disorders is increased.[4]

Although functional assays of clot formation are markedly prolonged, the coagulation defect is surprisingly no more severe than in the hemophilias A and B, varying from severe to mild. This is due, at least in part, to the presence of functional vWF, which allows platelet aggregation and adhesion even in the absence of fibrin. Umbilical cord hemorrhage is often the first sign of the disorder; gum bleeding, epistaxis, menorrhagia, gastrointestinal bleeding, and hemarthrosis occur with varying intensity, and spontaneous intracerebral bleeding and splenic rupture can occur throughout life.[5] Patients respond well to fibrinogen replacement therapy. The genetic defect was proposed to be either at the level of fibrinogen synthesis, fibrinogen degradation, or since the three fibrinogen mRNAs are coordinately regulated, at the level of a common fibrinogen gene regulator molecule. Because the halflife of infused fibrinogen is essentially normal, the fibrinogen degradation hypothesis was excluded but the responsible gene remained elusive until recently.[6]

The identification of the precise genetic defect of coagulation disorders is of value (1) to improve the differential diagnosis, (2) to permit early testing of other at risk individuals, (3) to understand the correlation (if any) between genotype and clinical phenotype, (4) to assist in therapeutic choices, and (5) as an essential prerequisite for the development of new specific treatments, such as gene therapy.

IDENTIFICATION OF THE FIRST GENETIC DEFECT RESPONSIBLE FOR CONGENITAL AFIBRINOGENEMIA

Fibrinogen is produced predominantly in the hepatocyte from three homologous polypeptide chains, Aα, Bβ, and γ that assemble to form the hexameric structure (AαB$\beta\gamma$)$_2$. Fibrin is produced by proteolytic cleavage of the fibrinogen alpha and beta chains by thrombin, thus releasing fibrinopeptides A and B, allowing polymerization to occur. The three genes coding for fibrinogen gamma (*FGG*), alpha (*FGA*), and beta (*FGB*) are clustered in a region of approximately 50 kb on chromosome 4q28–q31.[7] Early studies using Southern blotting suggested that no gross structural changes of the fibrinogen genes were present in afibrinogenemia patients.[8] In contrast, in various forms of dysfibrinogenemia, numerous missense mutations have been found in the syntenic genes for the gamma, alpha, and beta subunits of fibrinogen, approximately 300 abnormal fibrinogens have been reported, with some 83 structural defects identified.[2]

We studied the fibrinogen gene region in a Swiss family with two pairs of affected brothers by microsatellite analysis, PCR amplification, and Southern blotting.[6] Despite intensive genealogical investigation, no consanguinity was identified (see FIGURE 1). The diagnosis of afibrinogenemia for all the affected individuals was made in the neonatal period, after bleeding from the umbilical cord. All four patients

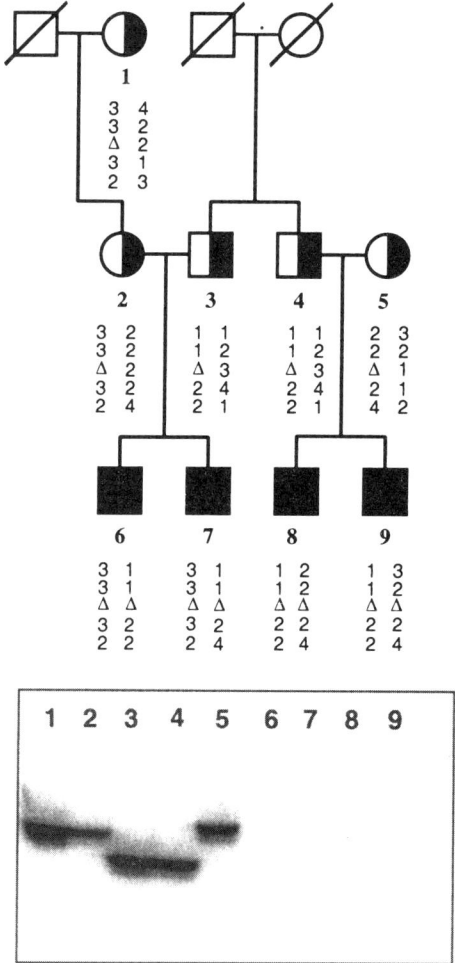

FIGURE 1. Top panel: pedigree of the Swiss family with congenital afibrinogenemia, together with haplotype data obtained for five microsatellite markers (D4S1625, D4S2962, FGAi3, D4S2934, and D4S1629) covering a region of approximately 20 centimorgans around the fibrinogen locus. Δ: deleted. **Bottom panel**: PAGE analysis of polymorphic marker FGAi3, showing hemizygosity in carriers (*lanes 1–5*) and absence in patients (*lanes 6–9*). (Reprinted from Ref. 6).

(now aged 30–40 years) were first treated with plasma derived fibrinogen preparations only after trauma or before operative surgery, a strategy replaced in early adulthood by a prophylactic protocol consisting of a fibrinogen infusion every two to four weeks. Antifibrinogen antibodies have never been detected in any of the patients. All coagulation tests performed (activated partial thromboplastin time, prothrombin time, thrombin time, reptilase time, and fibrinogen measured either by functional or

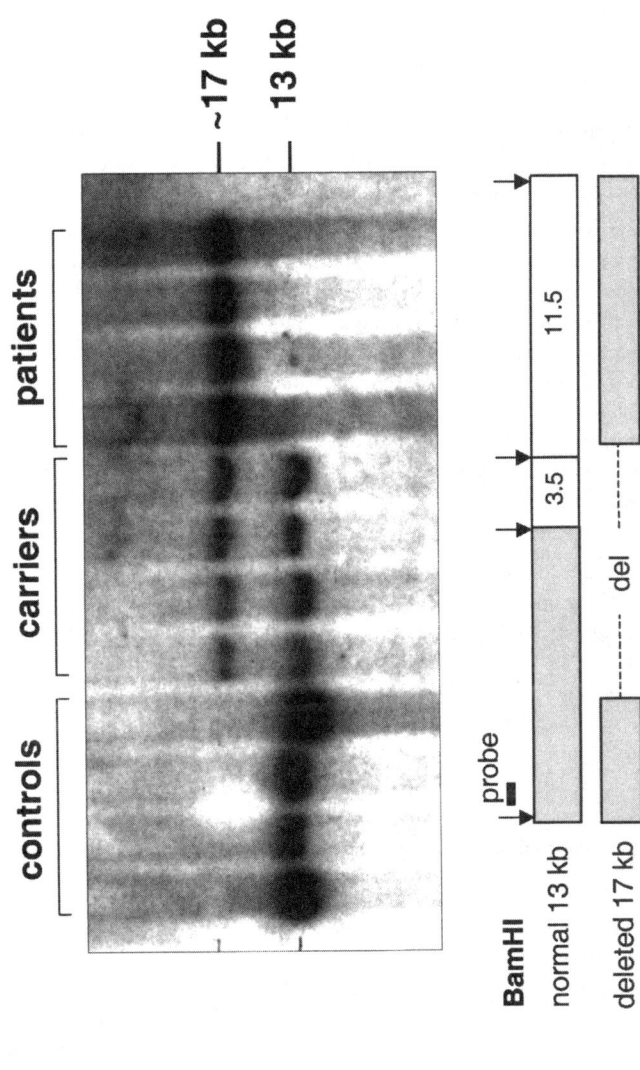

FIGURE 2. BamHI Southern blot analysis of the *FGA* deletion. For control samples the probe detects a 13-kb band. The deletion in heterozygous samples (*carriers*) and in homozygous patients appears as a band of increased size because two BamHI sites are eliminated.

immunological assays) were compatible with the complete absence of active fibrinogen, the plasma being incoagulable.

FIGURE 1 shows the family tree, together with haplotype data obtained for five microsatellite markers (D4S1625, D4S2962, FGAi3, D4S2934, and D4S1629) covering a region of approximately 20 centimorgans around the fibrinogen locus. One of these, FGAi3, a (TCTT)n polymorphic marker situated in intron 3 of the *FGA* gene, was deleted in all four affected individuals and hemizygous in the obligate carriers (FIG. 1). Flanking microsatellite markers revealed that the deletions had occurred on three different ancestral chromosomes, with the haplotypes arbitrarily numbered 3-3-del-3-2, 2-2-del-2-4, and 1-1-del-2-2 (the last being shared by the two fathers of the affected patients). This implied that homozygous deletion of the *FGA* gene is responsible for the congenital afibrinogenemia in this family, and that this region of the fibrinogen locus is susceptible to deletion by a common mechanism.

PCR amplification of portions of the three fibrinogen genes revealed that the *FGG* and *FGB* genes were intact in the affected individuals. Exon 1 was the only *FGA* exon that could be amplified from patient DNA, thus placing the centromeric deletion breakpoint in *FGA* intron 1. In order to estimate the size of the deletions, and to establish if they were, in fact, identical, we performed BamHI Southern blot analysis. Using a probe in the 5' region of *FGA,* a normal band of 13 kb was detected for the control samples (see FIGURE 2) as well as for the obligate carriers. In addition, a larger band (approximately 17 kb) was detected in these same carriers. Only the larger fragment was observed in the affected individuals. This abnormal fragment is the result of a deletion of approximately 11 kb, which includes two BamH1 sites and so generates a band of paradoxically increased size.

We concluded that the genetic defect in this family was a recurrent deletion of approximately 11 kilobases of DNA that eliminates the majority of the *FGA* gene and so leads to an absence of fibrinogen. These results showed that afibrinogenemia is caused by a defect in fibrinogen synthesis and demonstrated unequivocally that humans, like mice,[9] may be born without any functional fibrinogen. This identification of the first genetic defect for the disorder justified the analysis of the three fibrinogen genes in all patients with congenital afibrinogenemia.

A DELETION HOT SPOT IN THE FIBRINOGEN GENE CLUSTER?

The three afibrinogenemia alleles of the family were found on different haplotypes with closely linked markers, implying recurrence of the mutation, but the deletions were apparently identical within the limits of resolution of the Southern blot. Although the centromeric breakpoints of all deletions were known to occur within *FGA* intron 1, the exact sequence of the deletion junctions was unknown since the telomeric end of the deletions, as determined by the size of the deletion measured in Southern blots (FIG. 2), was situated in the previously uncharacterized *FGA–FGB* intergenic region. The cloning and partial sequencing of this portion of chromosome 4q (Genbank accession number AF145725) was therefore an essential prerequisite to the sequencing of the deletion junctions.[10]

FIGURE 3 shows the sequence of the afibrinogenemia deletion junction, with the corresponding normal sequences from *FGA* intron 1 (the centromeric breakpoint) and from the *FGA–FGB* intergenic sequence (the telomeric breakpoint). This deletion junction sequence was identical in all the affected individuals and in their heterozygous parents.[10] The breakpoint is situated in a stretch of four Ts located in both *FGA* intron 1 and on the same DNA strand in the *FGA–FGB* intergenic sequence. The regions bordering the deletion breakpoint contain a number of sequence elements that have been implicated in the generation of human gene deletions. First, the breakpoint is located in a 7-bp direct repeat, AACTTTT. Second, there is an imperfect inverted copy of this repeat (6 out of 7 nucleotides) immediately telomeric of the deletion junction (FIG. 3). Third, the breakpoints are surrounded by further inverted repeats: analysis with the computer program mfold[11] identified one centromeric and two telomeric predicted stem–loop structures. The recurrent deletion responsible for congenital afibrinogenemia in four affected members of the non-consanguineous Swiss family reported is most likely due to non-homologous recombination between sites with limited homology[12] that has been implicated in various rearrangements, such as deletions, insertions, inversions, duplications, and translocations. Repetitive genomic elements, such as Alu sequences, have been shown to be responsible for deletions in delta^0-beta^0-thalassemia[13] and familial hypercholesterolemia.[14] Recombination between non-repetitive, homologous sequences of several kilobases have also been described, for example with the *F8* gene inversions in hemophilia A[15] and alpha-globin gene deletions in alpha-thalassemias.[16]

In the case of the afibrinogenemia deletion described here, no single long region of homology appears to be involved, but a number of short homologous sequences are present. The position of the breakpoint strongly suggests that the deletion is mediated by the two direct AACTTTT repeats. Meiotic recombination leading to deletion could have been facilitated by the various inverted repeats around the breakpoints as a 7-bp repeat alone is unlikely to be sufficiently long to support crossing over. Slipped mispairing during DNA replication could also have led to the deletion; this mechanism is supported by the fact that only one direct repeat remains in the deleted allele, although an 11-kb deletion is large for such a mechanism. The loss of one thymidine at the deletion junction is unusual in large deletions and may be caused by an error of DNA polymerase β, which is responsible for repairing single stranded nicks and is known to be particularly error prone in polypyrimidine tracts.[12] The single base deletion would, therefore, also support the slipped mispairing model,

```
centromeric        ACTTGTCCATAATAAGCAGAACTTTTAGTGTTAGTACAGTTTTGCTGAA
deletion junction  ACTTGTCCATAATAAGCAGAACTTT-GCAAATGTTTCATAGGTGAAAAG
telomeric          CTTTGTGAACAAGTTGCGTAACTTTTGCAAATGTTTCATAGGTGAAAAG
                                     *** ***|**** ***
```

FIGURE 3. Sequence of the afibrinogenemia deletion junction. The normal centromeric sequence is from *FGA* intron 1 (*top line*) and the normal telomeric sequence (Genbank accession number AF145725) from the *FGA–FGB* intergenic region (*bottom line*). Homologies are indicated by *grey shading*. The 7-bp direct repeat is shown **bold** and the inverted repeat is indicated by *asterisks*. The deletion junction has a deletion of one T.

in which a single stranded loop between direct repeats is excised and the nick is incorrectly repaired by DNA polymerase β.

We screened more than 460 control DNA samples isolated from unrelated individuals for presence of the deletion. No deletion carriers were identified using this method, indicating that although this mutation may be found repetitively in afibrinogenemia patients, it is very rare in the general population.[10]

THE MAJORITY OF PATIENTS WITH AFIBRINOGENEMIA HAVE MUTATIONS IN THE *FGA* GENE

We studied 13 additional unrelated patients with congenital afibrinogenemia in order to determine the prevalence of the original *FGA* deletion in the patient population and to identify other causative mutations.[17] The *FGA* gene encodes two different transcripts that are generated by alternative splicing at the 3' end. The major α isoform, representing 99% of mRNAs is encoded by *FGA* exons 1–5 whereas an extended variant αE is produced by the addition of coding sequences from exon 6.[18] We chose to screen for mutations by PCR amplification of the entire coding regions and intron–exon junctions of the α isoform, followed by DNA sequencing. A total of seven novel mutations were identified, accounting for 22/26 (85%) of disease alleles in this study[17] (see also TABLE 1).

Deletion of 11 kb

We had predicted that the originally reported 11-kb *FGA* deletion would be a relatively common recurrent mutation and we, therefore, wished to determine its prevalence in an extended patient collection. The deletion was excluded in some patients, because of heterozygosity for the *FGAi3* microsatellite, which is situated within the deletion in intron 3, and in compound heterozygotes for two different mutations. For all other patients, Southern blotting was performed. One patient of American origin was found to be a heterozygous carrier of the deletion, confirming its recurrent nature.

Splice Site Mutations

The most common mutation was a donor splice mutation in intron 4, IVS4+1 G>T, representing 14/26 alleles in our patient population (54%). Haplotype data suggest that this mutation is also recurrent, or a very ancient mutation since the IVS4+1 mutation was found on at least eight discrete haplotypes.[17] Although it proved impossible to determine the effect of this mutation in the leukocyte RNA of patients (the level of illegitimate expression being too low to detect), the fact that these patients have clinical afibrinogenemia, as opposed to severe hypofibrinogenemia, implies that essentially no normal splicing occurs at this site. Computer analysis of the region around the IVS4 donor site with the program Spliceview[19] predicted, in the normal sequence, two distinct donor sites with nearly equal scores, the "physiological" one and a second four bases downstream, the usage of which would result in a frameshift mutation leading to premature termination. We are currently studying the splicing of the IVS4 +1 mutant in transfected COS-7 cells (Attanasio *et al.,* Blood, in press). The mutation leads to the production of close to 100% abnormally spliced molecules with, in particular, 83% of mRNA transcripts spliced at the predicted downstream donor site, leading to a 4-base pair frameshift. Two other cryptic donor splice sites situated in exon 4 are also used.

In one patient who was heterozygous for the IVS4+1 mutation, the only other change identified by sequencing of the *FGA* gene was an A>G mutation in the intron 1 donor site (IVS1+3 A>G). Spliceview[19] analysis of the normal sequence around the IVS1 donor site predicts a single site; however, in the mutant IVS1+3 G sequence, no donor site was identified. Furthermore, although the consensus donor splice site has either A or G at this position,[20] eleven disease causing splice site +3 A>G mutations have been reported to date; in ten of these, aberrant splicing was proven.[21] Analysis of this variant in transfected COS-7 cells is currently being performed.

TABLE 1. List of all causative afibrinogenemia mutations reported to date

Mutation	Type	Reference	Alternative Name	Percent Alleles
FGA				
11 kb-deletion	deletion	6, 17	n.a.	8.3%
34 ins C	frameshift	17	g.3-4 ins C	2.8%
IVS1 +3 A>G	splice-site	17	n.a.	2.8%
3121 del AA	frameshift	17	g.3091-3092 del AA	5.5%
Aa 149Arg> stop	nonsense	23	R168X	5.5%
IVS4 +1 G>T	splice-site	17	n.a.	38.9%
G316X	nonsense	17	n.a.	2.8%
W334X	nonsense	17	n.a.	2.8%
4329 del C	frameshift	17	g.4300 del C	2.8%
FGB				
L353R	missense	28	L383R	5.5%
G400D	missense	28	G430D	5.5%
FGG				
E231X	nonsense	29	E257X	5.5%

NOTE: The mutations are ordered according to their 5' > 3' position in the respective gene. The mutations are named according to the original publication, but for some an alternative name is proposed that complies with human mutation nomenclature guidelines[22] as discussed in the text, with amino acids numbered from the initiator ATG codon (this adds 19 amino acids for the alpha chain, 30 for beta, and 26 for gamma, based on Swiss-Prot accession numbers P02671, P02675, and P02679). For mutations described at the DNA level, the g. in front of nucleotide number indicates that the numbering is based on the genomic sequence (with the A in the initiator ATG codon counted as nucleotide 1). The allele percentage is calculated out of the total of 18 unrelated afibrinogenemia families documented in References 6, 17, 23, 28, and 29. To simplify, all patients are considered to have two disease alleles, although in reality patients from consanguineous families have two copies of the same allele. n.a., not applicable.

Nonsense Mutations

Two nonsense mutations were identified: G316X and W334X, both in exon 5. The numbering of the amino acids is according to the Genbank *FGA* sequence (M64982) and Swiss-Prot P02671 with the first ATG encoding amino acid number 1. This numbering, which complies with guidelines for human gene mutation nomenclature,[22] may lead to some confusion since other authors number the amino acids according to the mature Aα chain, that is, missing the first 19 amino acids corresponding to the signal peptide. This nomenclature was used when abnormal fibrinogen chains responsible for the dominant disease dysfibrinogenemia were characterized by amino acid sequencing of the secreted polypeptide lacking the signal peptide, before the DNA sequences of the fibrinogen genes were known. For example, we have identified the same nonsense mutation (Aα 149 Arg > stop) described by Fellowes *et al.*[23] in a consanguineous Norwegian patient and in an unrelated French patient (Neerman-Arbez *et al.,* Hum. Genet., in press). We, however, name this mutation R168X. In an effort to clarify this point, TABLE 1 lists all the afibrinogenemia mutations published in the literature to date with both their original identification, as well as the preferred nomenclature for human mutation description,[22] where the first ATG codon encodes amino acid number 1. In all cases the description of a mutation leading to an amino acid change should also be described at the DNA level.

Frameshift Mutations

Three *FGA* frameshift mutations were identified.[17] These were originally numbered according to Genbank #M64982, where the A in the first ATG codon is nucleotide 31: 34insC, a one base pair insertion in the second codon of exon 1; 3121delAA (in exon 4); and 4329delC, in exon 5. However, in order to comply to human mutation nomenclature standards, these should ideally be renumbered using the A in the first ATG codon as nucleotide 1. The frameshift mutations are, thus, better described as g.3-4insC (exon 1); g.3091-3092delAA (exon 4), and g.4300delC (exon 5), where the g. indicates that the reference sequence is genomic. These frameshift mutations are predicted to encode truncated proteins due to premature introduction of in frame stop codons.

All the mutations described so far are predicted to lead to premature FGA termination. Four such mutations have been previously described in *FGA* exon 5, in association with either dysfibrinogenemia or severe hypofibrinogenemia. Three of these are situated closer to the carboxy-terminus than the mutations described.[24–26] Fibrinogen Otago,[27] however, was associated with a homozygous frameshift mutation leading to truncation of fibrinogen Aα at amino acid 270 (or amino acid 289 counting from the initiator Methionine), which is proximal to three of the afibrinogenemia mutations we describe. In our patients, these mutations were present in compound heterozygosity associated with more amino-terminal truncation mutations. It is possible that the more distal truncation mutations are functionally less severe and permit the secretion of minute quantities of fibrinogen which, in homozygosity or in compound heterozygosity with less severe *FGA* mutations, result in hypofibrinogenemia rather than afibrinogenemia. If this were the case, we would expect to find some of the mutations we have described here in patients with less severe fibrinogen disorders, in association with less severe mutations.

NON *FGA* MUTATIONS RESPONSIBLE FOR CONGENITAL AFIBRINOGENEMIA

We were unable to find *FGA* mutations in two patients. Both these patients are non-Caucasian, consanguineous, and are homozygous for the three polymorphic markers used. It is likely that they have mutations in the closely linked *FGG* or *FGB* genes, which were not analyzed. Recently, two missense mutations in *FGB* have been shown by Duga *et al.* to cause afibrinogenemia in a patient from Italy (L353R, in exon 7) and one from Iran (G400D, in exon 8).[28] The mutations both introduce charged amino acids at conserved hydrophobic sites in the C-terminal global portion of the fibrinogen beta chain. To date, these are the only missense mutations found to cause afibrinogenemia (see TABLE 1 and FIGURE 4). Since this disease is the most extreme form of fibrinogen deficiency as compared to hypofibrinogenemia or dysfibrinogenemia, null mutations would be expected to be the most common, if not the only, causative mutations identified. Expression of the two *FGB* missense mutations in COS-1 cells followed by pulse–chase experiments revealed that the afibrinogenemia was caused by deficient secretion of the fibrinogen hexameric complex.[28] A single nonsense mutation in *FGG* exon 7 has also been identified in a Japanese patient with afibrinogenemia.[29] TABLE 1 recapitulates all afibrinogenemia mutations reported in the literature to date. The relative distributions of mutations between the three fibrinogen genes are shown in FIGURE 4.

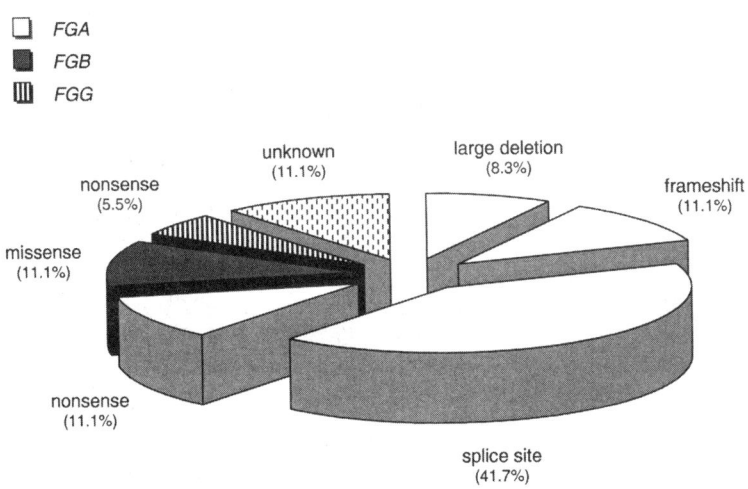

FIGURE 4. Relative distributions of afibrinogenemia causative mutations between the three fibrinogen genes: *FGA* mutations are in *white,* *FGB* mutations in *grey,* and *FGG* mutations in *striped.* The unknown mutations are from two consanguineous patients from Reference 17. Percentages are calculated as for TABLE 1 out of the total of 18 unrelated afibrinogenemia families documented in References 6, 17, 23, 28, and 29. The type of mutation nonsense, frameshift, splice-site, or missense is indicated.

WHY SO MANY MUTATIONS IN *FGA*?

It is not clear why, at least for patients of Caucasian origin, the causative mutations for afibrinogenemia are predominantly found in the *FGA* gene.[17] Because of extensive allelic heterogeneity, and because several different mutations appear to be recurrent, this is not simply the consequence of a founder effect. The *FGA* mRNA (approximately 2.2 kb for the major Aα transcript, Genbank accession number M64982) is larger than the *FGB* and *FGG* mRNAs (approximately 1.6 kb for Genbank M64983.1 and 1.3 kb for Genbank M10014, respectively) but this difference is clearly not sufficient to explain the excessively high proportion of *FGA* mutations (more than 70% of alleles, see FIG. 4). The analysis of patient populations from other ethnic groups, such as the large sample collection from Iranian patients[4,5] will reveal if this interesting finding is true for all afibrinogenemia patient groups.

WHAT ARE THE IMPLICATIONS FOR THE TREATMENT OF AFIBRINOGENEMIA?

The identification of the gene (or in this case genes) mutated in a human disease is of great importance for several obvious reasons. First, the precise molecular diagnosis of the disorder is possible, and may reveal correlations between the patient genotype and the clinical presentation. Second, it allows prenatal diagnosis for interested families. The identification of an affected fetus during gestation, for example, could prevent the often life threatening cord hemorrhage observed at birth, by fibrinogen replacement at the appropriate time. Third, gene therapy is now achievable. For very rare disorders such as afibrinogenemia, gene therapy is rarely considered feasible from a financial point of view. Furthermore, there exists a treatment for this disease, consisting of purified fibrinogen preparations, that is considered effective, although not perfect. However, the finding that congenital afibrinogenemia is caused by the mutation of a single small gene (in the majority of cases, *FGA,* but in some patients, *FGB* or *FGG*) leads to obvious possibilities for gene therapy, by delivery of the normal gene or cDNA to the liver. Such a protocol would also be of potential benefit to patients with hypofibrinogenemia, increasing the potential patient base. Finally, there exists a knockout mouse model for afibrinogenemia, which by remarkable coincidence was generated by the disruption of exon 1 of the *FGA* gene.[9] These mice have no detectable liver *FGA* mRNA nor circulating fibrinogen although the *FGG* and *FGB* mRNA levels in the liver are normal. These mice represent the perfect model for the development of a gene therapy protocol for congenital afibrinogenemia.

CONCLUSION

During only two years the field has evolved from knowing virtually nothing about the molecular basis of the bleeding disorder congenital afibrinogenemia to the characterization of numerous causative mutations for this disease. At least in patients of Caucasian origin, the great majority of cases are due to truncating mutations in the *FGA* gene, although intuitively all three fibrinogen genes could be predicted to be equally implicated. These results allow a precise molecular diagnosis of the disorder, permit prenatal diagnosis for families who so desire, and pave the way for new therapeutic approaches such as gene therapy.

[NOTE ADDED IN PROOF: Recently, additional causative mutations for congenital afribrinogenemia have been identified in *FGG* and *FGA*.[30,31,32]]

ACKNOWLEDGMENTS

This work is supported by Swiss National Science Foundation Grants 31-55898.98 and 31-59399.99. I would like to thank my collaborators Dr. Michael Morris, Dr. Philippe de Moerloose, and Prof. Stylianos Antonarakis for their helpful suggestions and comments on this review.

REFERENCES

1. RABE, F. & E. SALOMON. 1920. Ueber-faserstoffmangel im Blute bei einem Falle von Hämophilie. Arch. Int. Med. **95:** 2–14.
2. MARTINEZ, J. 1997. Congenital dysfibrinogenemia. Curr. Opin. Hemat. **4:** 357–365.
3. CRABTREE, G.R. 1983. The molecular biology of fibrinogen. *In* The Molecular Basis of Blood Diseases, Chapter 17. Stamatoyannopoulos, *et al.*, Eds. Saunders. Philadelphia.
4. PEYVANDI, F. & P.M. MANNUCCI. 1999. Rare coagulation disorders. Thromb. Hæmost. **82:** 1207–1214.
5. LAK, M., *et al.* 1999. Bleeding and thrombosis in 55 patients with inherited afibrinogenaemia. Br. J. Hæmatol. **107:** 204–206.
6. NEERMAN-ARBEZ, M., *et al.* 1999. Deletion of the Fibrinogen alpha-chain gene (*FGA*) causes congenital afibrinogenemia. J. Clin. Invest. **103:** 215–218.
7. KANT, J.A., *et al.* 1985. Evolution and organization of the fibrinogen locus on chromosome 4: gene duplication accompanied by transposition and inversion. Proc. Natl. Acad. Sci. U.S.A. **82:** 2344–2348.
8. UZAN, G., *et al.* 1984. Analysis of fibrinogen genes in patients with congenital afibrinogenemia. Biochem. Biophys. Res. Commun. **120:** 376–383.
9. SUH, T.T., *et al.* 1995. Resolution of spontaneous bleeding events but failure of pregnancy in fibrinogen-deficient mice. Genes Dev. **9:** 2020–2033.
10. NEERMAN-ARBEZ, M., *et al.* 1999. The 11kb *FGA* Deletion responsible for congenital afibrinogenemia is mediated by a short direct repeat in the fibrinogen gene cluster. Eur. J. Hum. Genet. **7:** 897–902.
11. SANTALUCIA, JR., J. 1998. A unified view of polymer, dumbbell, and oligonucleotide DNA nearest-neighbor thermodynamics. Proc. Natl. Acad. Sci. U.S.A. **95:** 1460–1465.
12. COOPER, D.N. & M. KRAWCZAK. 1993. Human Gene Mutation. BIOS Scientific, Oxford.
13. OTTOLENGHI, S. & B. GIGLIONI. 1982. The deletion in a type of delta 0-beta 0-thalassaemia begins in an inverted AluI repeat. Nature **300:** 770–771.
14. LEHRMAN, M.A., *et al.* 1985. Mutation in LDL receptor: Alu-Alu recombination deletes exons encoding transmembrane and cytoplasmic domains. Science **227:** 140–146.
15. LAKICH, D., *et al.* 1993. Inversions disrupting the factor VIII gene are a common cause of severe haemophilia A. Nat. Genet. **5:** 236–241.
16. PHILLIPS, J.A., *et al.* 1980. Unequal crossing-over: a common basis of single alpha-globin genes in Asians and American blacks with hemoglobin-H disease. Blood **55:** 1066–1069.

17. NEERMAN-ARBEZ, M., *et al.* 2000. Mutations in the fibrinogen Aα gene account for the majority of cases of congenital afibrinogenemia. Blood **96:** 149–152.
18. FU, Y., *et al.* 1992. Carboxy-terminal-extended variant of the human fibrinogen alpha subunit: a novel exon conferring marked homology to beta and gamma subunits. Biochemistry **31:** 11968–11972.
19. ROGOZIN, I.B. & L. MILANESI. 1997. Analysis of donor splice signals in different organisms. J. Mol. Evol. **45:** 50–59.
20. KRAWCZAK, M., J. REISS & D.N. COOPER. 1992. The mutational spectrum of single base-pair substitutions in mRNA splice junctions of human genes: causes and consequences. Hum. Genet. **90:** 41–54.
21. OHNO, K., *et al.* 1999. Congenital end-plate acetylcholinesterase deficiency caused by a nonsense mutation and an A→G splice-donor-site mutation at position +3 of the collagen-like-tail-subunit gene (COLQ): how does G at position +3 result in aberrant splicing? Am. J. Hum. Genet. **65:** 635–644.
22. ANTONARAKIS, S.E. 1998. Recommendations for a nomenclature system for human gene mutations. Nomenclature Working Group. Hum. Mutat. **11:** 1–3.
23. FELLOWES, A.P., *et al.* 2000. Homozygous truncation of the fibrinogen Aalpha chain within the coiled coil causes congenital afibrinogenemia. Blood **96:** 773–775.
24. FURLAN, M., *et al.* 1994. A frameshift mutation in Exon V of the A alpha-chain gene leading to truncated A alpha-chains in the homozygous dysfibrinogen Milano III. J. Biol. Chem. **269:** 33129–33134.
25. KOOPMAN, J., *et al.* 1992. Fibrinogen Marburg: a homozygous case of dysfibrinogenemia, lacking amino acids A alpha 461-610 (Lys 461 AAA→stop TAA). Blood **80:** 1972–1979.
26. RIDGWAY, H.J., *et al.* 1996. Fibrinogen Lincoln: a new truncated alpha chain variant with delayed clotting. Br. J. Hæmatol. **93:** 177–184.
27. RIDGWAY, H.J., *et al.* 1997. Fibrinogen Otago: a major alpha chain truncation associated with severe hypofibrinogenaemia and recurrent miscarriage. Br. J. Hæmatol. **98:** 632–639.
28. DUGA, S., *et al.* 2000. Missense mutations in the human β fibrinogen gene cause congenital afibrinogenemia by impairing fibrinogen secretion. Blood **95:** 1336–1341.
29. IIDA, H., *et al.* 2000. A case of congenital afibrinogenemia: fibrinogen Hakata, a novel nonsense mutation of the fibrinogen γ-chain gene. Thromb. Hæmost. **84:** 49–53.
30. ASSELTA, R., *et al.* 2000. Afibrinogenemia: first identification of a splicing mutation in the fibrinogen gamma gene leading to a major gamma chain truncation. Blood **96:** 2496–2500.
31. MARGAGLIONE, M., *et al.* 2000. A G-to-A mutation in IVS-3 of the human gamma fibrinogen gene causing afibrinogenemia due to abnormal RNA splicing. Blood **96:** 2501–2505.
32. NEERMAN-ARBEZ, M., *et al.* 2001. Molecular analysis of the fibrinogen gene cluster in 16 patients with congenital afibrinogenemia: novel truncating mutations in the *FGA* and *FGG* genes. Hum. Gen. In press.

Effects of Diet, Drugs, and Genes on Plasma Fibrinogen Levels

MONIEK P.M. DE MAAT

Gaubius Laboratory TNO-PG, Leiden, the Netherlands

ABSTRACT: Plasma levels of fibrinogen have been identified as independent risk predictors of cardiovascular disease. This has greatly increased interest in the regulation of plasma fibrinogen levels. Many demographic and environmental factors are known to affect fibrinogen levels, such as diet, use of several drugs, age, smoking, body mass, gender, physical exercise, race, and season. Additionally, it is also known that genetic factors determine the fibrinogen levels, and also that they determine the response of fibrinogen levels to environmental factors. Estimates, based on twin studies, suggest that 30–50% of the plasma fibrinogen level is genetically determined. The effect of dietary components on plasma fibrinogen levels is modest. Several components have been identified as factors that influence fibrinogen levels. Among those are fish oil, other lipids, and fibers. Dietary components that were expected to have an effect on fibrinogen, but for which no association was observed are black and green tea. Several drugs are known to influence fibrinogen levels, the most studied of which are platelet aggregation inhibiting drugs, such as ticlopidine, and the lipid lowering fibric acid derivatives (fibrates). Both types of drugs decreased the plasma fibrinogen level by about 10%, and bezafibrate lowers fibrinogen even more in patients with diabetes. No clear effect was observed for the HMG-CoA reductase inhibitors (statins). In the Bezalip study, fibrinogen levels decreased in patients treated with bezafibrate, but this had no clear effect on the risk of cardiovascular disease. This suggests that several mechanisms influence the fibrinogen level and that these mechanisms may contribute differently to cardiovascular disease. Several variations in the fibrinogen genes have been described and especially variations in the promoter region of the fibrinogen β-gene are interesting, because the synthesis of the fibrinogen Bβ chain is considered to be the rate limiting step in the fibrinogen biosynthesis. In many studies the fibrinogen β-gene polymorphisms (–455G/A, –148C/T, and BclI) are found to be associated with the plasma levels of fibrinogen. However, they are not associated with the risk of cardiovascular events, although in several studies an association with the severity and progression of atherosclerosis has been reported. It has also been observed frequently that the fibrinogen β-gene promoter polymorphisms are associated with the response of fibrinogen levels to environmental factors, such as exercise and trauma. In conclusion, plasma fibrinogen levels are regulated by an interesting and complex interplay between environmental and genetic factors.

KEYWORDS: Diet effects; Drug effects; Gene effects; Plasma fibrinogen levels.

Address for correspondence: Moniek P.M. de Maat, Ph.D., Gaubius Laboratory TNO-PG, P.O. Box 2215, 2301 CE Leiden, the Netherlands. Voice: +31 71 5181502; fax: +31 71 5181904.

mpm.demaat@pg.tno.nl

PLASMA LEVELS OF FIBRINOGEN AND RISK OF CARDIOVASCULAR DISEASE

The Northwick Park Heart Study was the first large scale prospective longitudinal study that described an association between plasma fibrinogen levels and cardiovascular disease risk.[1] Since then, a large number of other epidemiological studies confirmed that the plasma fibrinogen level is an independent risk indicator for cardiovascular disease.[2] There is still a lot of discussion about the mechanism that underlies the observed association. Fibrinogen is an acute phase protein, and thus plasma fibrinogen levels may reflect the inflammatory status of the vascular wall and higher plasma fibrinogen levels may reflect a higher degree of atherosclerosis.[3,4] Additionally, there are also many mechanisms known in which fibrinogen causally contributes to the severity and progression of cardiovascular disease.[5-8] It is, therefore, most plausible that the relationship between fibrinogen and cardiovascular disease is a combination of reflecting underlying inflammatory processes and a causal contribution. The possible causal role of fibrinogen has greatly increased interest in fibrinogen and the ways in which its level is regulated.

REGULATION OF PLASMA FIBRINOGEN LEVELS

Many demographic and environmental factors are known to affect fibrinogen levels, such as diet, use of several drugs, age, smoking, alcohol consumption, body mass, gender, physical exercise, race, and season.[9] The effect of some of these environmental factors is largely through an effect on the acute phase reaction.

It is also known that genetic factors determine the fibrinogen levels, and the response of fibrinogen levels to environmental factors. Estimates, based on twin studies, suggest that 30–50% of the plasma fibrinogen level is genetically determined.[10-12]

The regulation of fibrinogen synthesis is for a large part at the transcriptional level, but also at the level of mRNA stability, translation, posttranslational modifications, such as glycosylation and phorphorylation, and assembly of the fibrinogen chains.[13,14] The plasma fibrinogen levels are also determined by degradation of fibrinogen in the circulation and the clearance rate.[15,16]

EFFECT OF DIET ON FIBRINOGEN LEVELS

The effect of dietary components on plasma fibrinogen levels is modest (see TABLE 1). Several components have been identified that have a modest effect on plasma fibrinogen levels. Among those are fish oil, other lipids, and fibers (TABLE 1). Only moderate alcohol consumption was found to have a consistent and clear fibrinogen lowering effect.[9,17-21]

These weak associations are surprising, since several dietary components affect mechanisms that regulate the fibrinogen synthesis. For example, dietary antioxidants such as those present in tea,[22-24] onions, fish oil, and parsley, would be expected to protect from processes brought about by oxidative damage, such as an upregulation of genes of inflammatory components. Furthermore, an effect of free fatty acids (FFA) on plasma fibrinogen levels would be expected, since FFA induce fibrinogen synthesis in perfused liver systems[25] and infusion of FFA in rabbits stimulates fibrinogen synthesis. This latter effect could be nullified by injection of defatted albumin.[26]

In patients with high lipid levels, the plasma fibrinogen levels are increased.[27] Therefore, it would be expected that lowering the lipid content of the diet results in lowering the serum lipid levels and, as a result, lowering the plasma fibrinogen levels.[28-32]

Unsaturated fatty acids are relatively insensitive to β-oxidation and this has been suggested to result in the lowering of the fibrinogen levels. The underlying mechanism may be a reduction of the acute phase reaction because the unsaturated fatty acids reduce the oxidant stress, or it may be a result of the decreased aggregation of platelets. Results of dietary studies with fish oil or n-3 or n-3 fatty acids are not consistent.[33-37] It has been suggested that the favorable effect of fish oil preparations depends on their antioxidant vitamin E content.[38]

One explanation for the fact that only minimal and often contradicting effects of dietary components on fibrinogen levels are observed may be that the study design often is far from optimal. The intraindividual variation in fibrinogen levels is quite high (12% of the total variance),[39] which warrants large study populations or multiple sampling. However, many studies have been performed in small groups, often no control groups were used, and many different assays for fibrinogen have been employed.

TABLE 1. Described effects of diet on plasma fibrinogen levels

Detary Component	Effect	References
Fiber	in cross sectional studies no effect	74–76
Alcohol	in cross sectional studies a decrease for moderate consumption and increase for high consumption; in experimental studies a moderate consumption gives a decrease	9,17,19,77–80
Energy content	no effect in obese subjects	81–83
n-3 Fatty acids/fish oil	decrease or no effect	34,84
Lipid	patients with high lipid levels have high fibrinogen; in experimental studies no effects	28,29,31,32,75,76
Vegetarian diet	no effect	85,86
Antioxidants	no effect	24,76
Tea	no effect	22–24
Garlic	no effect	87
Pectin	no effect	88,89

EFFECT OF DRUGS ON FIBRINOGEN

No drugs are known that have a specific effect on plasma fibrinogen levels, but several drugs influence plasma fibrinogen levels as an additional effect (see TABLE 2).[40–45] The most studied drugs are platelet aggregation inhibiting drugs such as ticlopidine and the lipid lowering fibrates. Both types of drugs decrease the plasma fibrinogen level by about 10%, and bezafibrate lowers fibrinogen levels even more in patients with diabetes.[40,46,47] No clear effect was observed for the HMG-CoA reductase inhibitors (statins).[48] In the BIP study, fibrinogen levels decreased in patients treated with bezafibrate, but this had no clear effect on the risk of cardiovascular disease.[49] This suggests that not all mechanisms that influence plasma fibrinogen levels are important for cardiovascular disease.

Fibric acid derivatives (fibrates) have an effect on plasma fibrinogen levels because they (1) activate PPARα resulting in inhibition of C/EBPβ,[50] (2) inhibit the increase of cAMP induced by adrenaline resulting in a reduction of IL6 production, (3) inhibit lipolysis by reduction of free fatty acids (FFA), (4) inhibit acetyl CoA carboxylase affecting the hepatic secretion efficiency, and (5) inhibit the release of factors from endothelium to affect monocyte function. The effect of fibrates on

TABLE 2. Effects of drugs on plasma fibrinogen levels

Drug	Effect	References
Ticlopidine	10% decrease in healthy volunteers and patients with CAD	40,41,43,90–93
Vasodilators (β-adrenergic antagonists)	modest reduction of fibrinogen in patients with stress or pathologic adrenaline excess	43
Fibrates	15% in hypercholesterolemia and 25% in NIDDM comparable results with bezafibrate, fenofibrate, ciprofibrate, and clofibrate; no effect with Gemfibrozil	43,94–99
Statins	effects on fibrinogen varies between and within the different statins	98,100–108
Pentoxifylline:	decreases fibrinogen	41
Ca-antagonists	no effect in patients with CAD	43
Defibrotide	20% decrease in patients with PAD	43
Stanozolol	40% decrease in healthy volunteers and 30% decrease in patients with vascular and inflammatory disorders	43,109
Antiinflammatory drugs (prednisolone, nabumetone)	47% reduction in patients with inflammatory disorders; 15% reduction in healthy volunteers	43
Aspirin	no effect on plasma level known	110–113

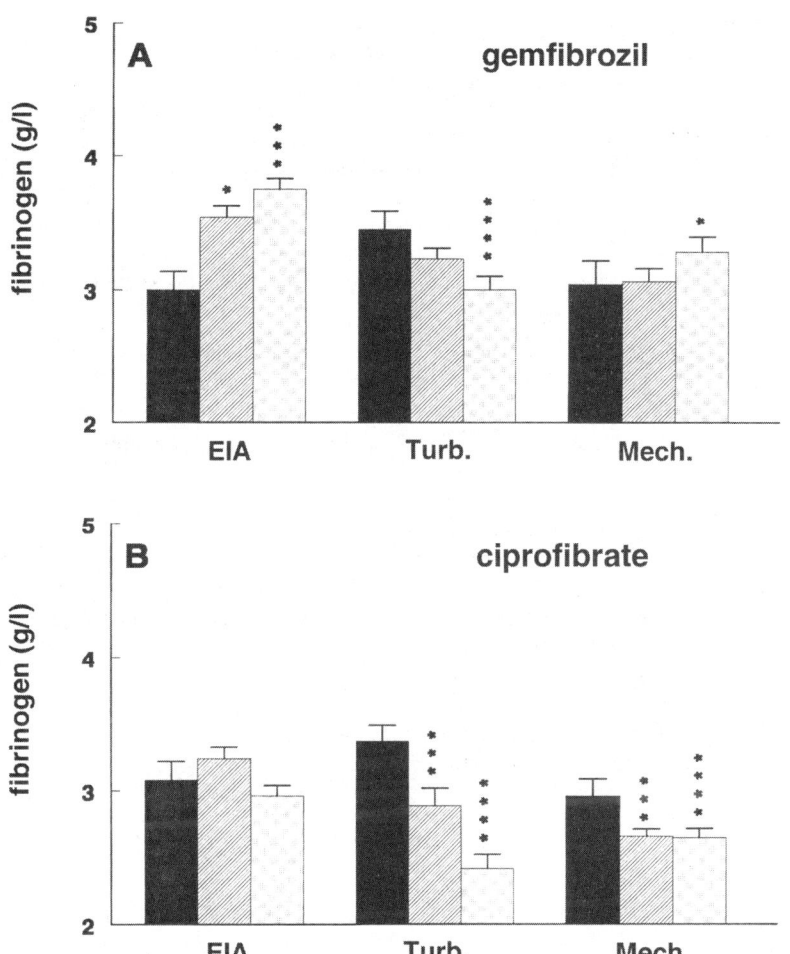

FIGURE 1. Effects of gemfibrozil (**A**) and ciprofibrate (**B**) on plasma fibrinogen levels at baseline (*black box*), after six weeks of fibrate treatment (*hatched box*) and after twelve weeks of fibrate treatment (*dotted box*). *Bars* represent means ± SEM ($n = 48$ and $n = 41$ for gemfibrozil and ciprofibrate, respectively). *Turb.*, funcional fibrinogen levels determined by the Clauss assay with a turbidity based end point; *Mech.*, functional fibrinogen levels determined by the Clauss assay with a mechanical end point; *EIA*, enzyme immuno assay measuring HMW + LMW fibrinogen anigen levels. * $p < 0.05$; *** $p < 0.001$; **** $p < 0.0001$ and #$p = 0.1$ using the Student's paired *t*-test. (Reproduced from Ref. 52, with permission)

fibrinogen does not seem to be associated with a general effect on inflammation because CRP levels are not lowered.[51]

One further reason for the variation that is observed in different studies is the characteristics of the fibrinogen assay used. For example, in a study on hyperlipidemic patients who were randomly treated with gemfibrozil or ciprofibrate, the apparent effect of the drug on the plasma fibrinogen level was different when the HMW+LMW fibrinogen EIA, the Clauss method with a turbidimetric endpoint, and the Clauss method with a mechanical endpoint were used.[52] This also suggests that these drugs may affect not only the synthesis rate but also the quality of the fibrinogen (see FIGURE 1).[52]

GENETIC VARIATION AND PLASMA FIBRINOGEN LEVELS

Several variations in the fibrinogen genes have been described.[51,53,54] Especially interesting are variations in the promoter region of the fibrinogen β-gene, because the synthesis of the fibrinogen Bβ-chain is considered to be the rate limiting step in the fibrinogen biosynthesis. In many studies the fibrinogen β-gene polymorphisms (−455G/A, −148C/T, and BclI) are found to be associated with the plasma levels of fibrinogen.[11,54–58] However, they are usually not associated with the risk of cardiovascular events[59–62] although in several studies an association with the severity and progression of atherosclerosis has been observed.[55,63] A relationship between the BclI-polymorphism, 3′ of the fibrinogen β-gene, and risk of myocardial infarction has been described in the presence of a second risk factor, or trigger, such as a Helicobacter pylori infection.[64] It has also frequently been observed that the fibrinogen β-gene promoter polymorphisms are associated with the response of fibrinogen levels to environmental factors, such as exercise and trauma.[65–67] The −148C/T polymorphism may be functional because it is located very close to an interleukin-6 responsive element[14] and it has also been suggested that the −455G/A polymorphism is functional because there is differential binding of nuclear factors to the alleles.[68,69]

INTERACTION BETWEEN DIET, GENES, AND FIBRINOGEN LEVELS

It has already been shown in clinical and population studies that variations in the promoter region of fibrinogen genes are associated with transcription regulation by environmental factors.[56,65,66] Since dietary components and drugs may have an effect through the same regulatory mechanisms, it may be anticipated that there is also an interaction between the effect of diet, drugs, fibrinogen genes and plasma fibrinogen levels.

Not much is known yet about the interaction between fibrinogen, diet, and genes. Dietary studies in mouse strains recently showed that increases in inflammation markers after an atherogenic diet depend on the genetic background,[70,71] indicating that modulation of inflammation factors by diet is genetically determined. Part of the genetic contribution may be explained by variations in the fibrinogen genes that are sensitive to regulation by inflammatory mediators.

In one study on 28 subjects with impaired glucose tolerance the effects of a monounsaturated, fat enriched diet and a diet recommended by the National Cholesterol Education Program (NCEP) on plasma fibrinogen levels were compared. A difference in fibrinogen levels was observed but this difference was independent of the genotype (see TABLE 3).[29] However, these results should be interpreted with caution because this was a very small study. In another study the relationship between glycemic control and fibrinogen gene polymorphisms was estimated (TABLE 3).[72,73] The decrease of fibrinogen levels during treatment was greatest in patients with the −455AA genotype, but these patients also had the highest pretreatment levels. It is not clear from this study if the genetic variation is associated with increased fibrinogen levels during a period of bad glycemic control or if it really is hyperresponsiveness to the insulin treatment. However, the results suggest that the fibrinogen gene polymorphisms partly explain the variation seen in the relation between plasma fibrinogen levels and the level of glycemic control.

We studied whether the effects of ticlopidine, ciprofibrate, and gemfibrozil on plasma fibrinogen levels are associated with the −455G/A genotype (TABLE 3).[40] We observed no association, but the same caution with respect to sample size applies here.

In the Regress study (a study in 885 male patients with symptomatic cardiovascular disease) we examined the relationship of the plasma fibrinogen level and the −455G/A polymorphism with progression of atherosclerosis during a two-year follow up period.[63] Patients were randomized to the lipid lowering HMG-CoA reductase inhibitor pravastatin or to placebo. Patients with the −455AA genotype in the placebo group had the strongest progression of atherosclerosis as measured by mean segment diameter (MSD) and minimal obstruction diameter (MOD), measures of generalized and focal atherosclerosis, respectively. Interestingly, in the patients with the −455AA genotype in the pravastatin treated group, the progression was the lowest, suggesting that carriers of the −455AA genotype hyper-respond to pravastatin

TABLE 3. Interaction between diet, drugs and fibrinogen genes

Treatment	Patient Group	Effect	References
Glycemic control	25 patients with NIDDM	effect largest in −455AA	72,73
Monounsaturated fat-enriched vs. recommended diet	28 subjects with impaired glucose tolerance	higher levels in heterozygotes, no difference in effect of diet	29
Ticlopidine	healthy volunteers and patients with stable angina	fibrinogen decrease, but no effect of −455G/A polymorphism	40
Gemfibrozil and ciprofibrate	healthy volunteers and patients with hyperlipidemia	fibrinogen decrease, but no effect of −455G/A polymorphism	46
Pravastatin	patients with CAD	−455G/A polymorphism associated with effect drug on progression of atherosclerosis	63

treatment. The plasma fibrinogen levels in this study were not associated with progression. The results from the Regress study suggest that a hyper-response to stimuli may be more important in predicting progression of atherosclerosis, than the fibrinogen level itself.

Plasma fibrinogen levels have been identified as a cardiovascular risk in many studies. Research on prevention/intervention and life style determinants is now imminent and diet and drugs are important candidates for study. Although there are not many effects known from experimental studies, there are many mechanisms by which effects of dietary components and drugs, possibly modified by polymorphisms in fibrinogen genes, may contribute to the development of cardiovascular disease through their effects on plasma fibrinogen levels.

In conclusion, plasma fibrinogen levels are regulated by an interesting and complex interplay between environmental and genetic factors.

REFERENCES

1. MEADE, T.W., S. MELLOWS, M. BROZOVIC, et al. 1986. Hæmostatic function and ischaemic heart disease: principal results of the Northwick Park Heart Study. Lancet **2:** 533–537.
2. ERNST, E. & W. KOENIG. 1997. Fibrinogen and cardiovascular risk. Vasc. Med. **2:** 115–125.
3. THOMPSON, S.G., J. KIENAST, S.D. PYKE, et al. 1995. Hemostatic factors and the risk of myocardial infarction or sudden death in patients with angina pectoris. European Concerted Action on Thrombosis and Disabilities Angina Pectoris Study Group (see comments). N. Engl. J. Med. **332:** 635–641.
4. BAUMANN, H. & J. GAULDIE. 1994. The acute phase response. Immunol. Today **15:** 74–80.
5. BLOMBACK, B. & M. OKADA. 1982. Fibrin gel structure and clotting time. Thromb. Res. **25:** 51–70.
6. FATAH, K., A. HAMSTEN, B. BLOMBACK, et al. 1992. Fibrin gel network characteristics and coronary heart disease: relations to plasma fibrinogen concentration, acute phaes protein, serum lipoproteins and coronary atherosclerosis. Thromb. Hæmost. **68:** 130–135.
7. GABRIEL, D.A., K. MUGA & E.M. BOOTHROYD. 1992. The effect of fibrin structure on fibrinolysis. J. Biol. Chem. **267:** 24259–24263.
8. NAITO, M., C. FUNAKI, T. HAYASHI, et al. 1992. Substrate-bound fibrinogen, fibrin and other cell attachment- promoting proteins as a scaffold for cultured vascular smooth muscle cells. Atherosclerosis **96:** 227–234.
9. KROBOT, K., H.W. HENSE, P. CREMER, et al. 1992. Determinants of plasma fibrinogen: relation to body weight, waist-to-hip ratio, smoking, alcohol, age, and sex. Results from the second MONICA Augsburg survey 1989–1990. Arterioscler. Thromb. **12:** 780–788.
10. HAMSTEN, A., L. ISELIUS, U. DE FAIRE, et al. 1987. Genetic and cultural inheritance of plasma fibrinogen concentration. Lancet **2:** 988–991.
11. HUMPHRIES, S.E., M. COOK, M. DUBOWITZ, et al. 1987. Role of genetic variation at the fibrinogen locus in determination of plasma fibrinogen concentrations. Lancet **1:** 1452–1455.
12. PANKOW, J.S., A.R. FOLSOM, M.A. PROVINCE, et al. 1998. Segregation analysis of plasminogen activator inhibitor-1 and fibrinogen levels in the NHLBI family heart study. Arterioscler. Thromb. Vasc. Biol. **18:** 1559–1567.
13. HENSCHEN-EDMAN, A.H. 1999. On the identification of beneficial and detrimental molecular forms of fibrinogen. Hæmostasis **29:** 179–186.

14. ANDERSON, G.M., A.R. SHAW & J.A. SHAFER. 1993. Functional characterization of promoter elements involved in regulation of human B beta-fibrinogen expression. Evidence for binding of novel activator and repressor proteins. J. Biol. Chem. **268:** 22650–22655.
15. PRINCEN, H.M.G., H.J. MOSHAGE, J.J. EMEIS, *et al.* 1985. Fibrinogen fragments X, Y, D and E increase levels of plasma fibrinogen and liver mRNAs coding for fibrinogen polypeptides in rats. Thromb. Hæmost. **53:** 212–215.
16. VERSCHUUR, M., M. BEKKERS, M. VAN ERCK, *et al.* 2000. Influence of plasma triglyceride and plasma cholesterol levels on the clearance rate of fibrinogen. Ann. N.Y. Acad. Sci. **936:** this volume.
17. DIMMITT, S.B., V. RAKIC, I.B. PUDDEY, *et al.* 1998. The effects of alcohol on coagulation and fibrinolytic factors: a controlled trial. Blood Coagulat. Fibrinol. **9:** 39–45.
18. MARQUES VIDAL, P., J.P. CAMBOU, V. NICAUD, *et al.* 1995. Cardiovascular risk factors and alcohol consumption in France and Northern Ireland. Atherosclerosis **115:** 225–232.
19. WANG, Z.Q., T.H. BARKER & G.M. FULLER. 1999. Alcohol at moderate levels decreases fibrinogen expression *in vivo* and *in vitro*. Alcohol. Clin. Exp. Res. **23:** 1927–1932.
20. SIERKSMA, A. 2000. Effect of moderate alcohol consumption on fibrinogen levels in healthy volunteers is discordant with effects on C-Reactive Protein. Ann. N.Y. Acad. Sci. **936:** this volume.
21. MARCKMANN, P. 1995. Diet, blood coagulation and fibrinolysis. Dan. Med. Bull. **42:** 410–425.
22. VINSON, J.A. & Y.A. DABBAGH. 1998. Effect of green and black tea supplementation on lipids, lipid oxidation and fibrinogen in the hamster: mechanisms for the epidemiological benefits of tea drinking. FEBS **433:** 44–46.
23. VORSTER, H., J. JERLING, W. OOSTHUIZEN, *et al.* 1996. Tea drinking and hæmostasis: a randomized, placebo-controlled, crossover study in free-living subjects. Hæmostasis. **26:** 58–64.
24. DE MAAT, M.P.M., H. PIJL, C. KLUFT, *et al.* 2000. Consumption of black and green tea has no effect on inflammation, hæmostasis and endothelial markers in smoking halthy individuals. Eur. J. Clin. Nutr. **54:** 757–763.
25. PICKART, L.R. & M.M. THALER. 1980. Fatty acids, fibrinogen and blood flow: a general mechanism for hyperfibrinogenemia and its pathologic consequences. Med. Hypotheses. **6:** 545–557.
26. PICKART, L. 1981. Free fatty acidemia as an inducer of systemic hyperfibrinogenemia and fibrinolytic inhibition. Inflammation **5:** 61–70.
27. ANDERSEN, P. 1992. Hypercoagulability and reduced fibrinolysis in hyperlipidemia: relationship to the metabolic cardiovascular syndrome. J. Cardiovasc. Pharmacol. **20**(Suppl. 8): S29–S31.
28. MARCKMANN, P., B. SANDSTROM & J. JESPERSEN. 1992. Fasting blood coagulation and fibrinolysis of young adults unchanged by reduction in dietary fat content. Arterioscler. Thromb. **12:** 201–205.
29. NISKANEN, L., U.S. SCHWAB, E.S. SARKKINEN, *et al.* 1997. Effects of dietary fat modification on fibrinogen, factor VII, and plasminogen activator inhibitor-1 activity in subjects with impaired glucose tolerance. Metabolism **46:** 666–672.
30. NYDAHL, M.C., I.B. GUSTAFSSON & B. VESSBY. 1994. Lipid-lowering diets enriched with monounsaturated or polyunsaturated fatty acids but low in saturated fatty acids have similar effects on serum lipid concentrations in hyperlipidemic patients. Am. J. Clin. Nutr. **59:** 115–122.
31. OKAZAKI, M., H. ZHANG, Y. YOSHIDA, *et al.* 1994. Correlation between plasma fibrinogen and serum lipids in rats with hyperlipidemia induced by cholesterol free-high fructose or high cholesterol diet. J. Nutr. Sci. Vitaminol. (Tokyo) **40:** 479–489.
32. SANDERS, T.A.B., F.R. OAKLEY, G.J. MILLER, *et al.* 1997. Influence of n-6 versus n-3 polyunsaturated fatty acids in diets low in saturated fatty acids on plasma lipoproteins and hemostatic factors. Arterioscler. Thromb. Vasc. Biol. **17:** 3449–3460.
33. KNAPP, H.R. 1997. Dietary fatty acids in human thrombosis and hemostasis. Am. J. Clin. Nutr. **65:** 1687S–1698S.

34. GARCIA CLOSAS, R., L. SERRA MAJEM & R. SEGURA. 1993. Fish consumption, omega-3 fatty acids and the Mediterranean diet. Eur. J. Clin. Nutr. **47**(Suppl. 1): S85–90.
35. SAYNOR, R. & T. GILLOTT. 1992. Changes in blood lipids and fibrinogen with a note on safety in a long term study on the effects of n-3 fatty acids in subjects receiving fish oil supplements and followed for seven years. Lipids **27**: 533–538.
36. SEMPLICINI, A. & R. VALLE. 1994. Fish oils and their possible role in the treatment of cardiovascular diseases. Pharmacol. Ther. **61**: 385–397.
37. SHAHAR, E., A.R. FOLSOM, K.K. WU, et al. 1993. Associations of fish intake and dietary n-3 polyunsaturated fatty acids with a hypocoagulable profile. The Atherosclerosis Risk in Communities (ARIC) Study. Arterioscler. Thromb. **13**: 1205–1212.
38. HAGLUND, O., R. LUOSTARINEN, R. WALLIN, et al. 1991. The effects of fish oil on triglycerides, cholesterol, fibrinogen and malondialdehyde in humans supplemented with vitamin E. J. Nutr. **121**: 165–169.
39. DE MAAT, M.P.M., A.C.W. DE BART, B.C. HENNIS, et al. 1996. Interindividual and intraindividual variability in plasma fibrinogen, TPA antigen, PAI activity, and CRP in healthy, young volunteers and patients with angina pectoris. Arterioscler. Thromb. Vasc. Biol. **16**: 1156–1162.
40. DE MAAT, M.P.M., A.E.R. ARNOLD, S. VAN BUUREN, et al. 1995. Modulation of plasma fibrinogen levels by ticlopidine in healthy volunteers and patients with stable angina pectoris. Thromb. Hæmost. **73**: 1040–1040.
41. DI MINNO, G. & M. MANCINI. 1992. Drugs affecting plasma fibrinogen levels. Cardiovasc. Drugs Ther. **6**: 25–27.
42. ERNST, E. 1992. Lowering the plasma fibrinogen concentration with drugs. Clin. Pharm. **11**: 968–971.
43. HANDLEY, D.A. & T.E. HUGHES. 1997. Pharmacological approaches and strategies for therapeutic modulation of fibrinogen. Thromb. Res. **87**: 1–36.
44. ERNST, E. & K.L. RESCH. 1995. Therapeutic interventions to lower plasma fibrinogen concentration. Eur. Heart J. **16**(Suppl. A): 47–52.
45. MARGAGLIONE, M., E. GRANDONE, F.P. MANCINI, et al. 1996. Drugs affecting plasma fibrinogen levels. Implications for new antithrombotic strategies. Prog. Drug Res. **46**: 169–181.
46. DE MAAT, M.P.M., H.C. KNIPSCHEER, J.J.P. KASTELEIN & C. KLUFT. 1994. Modulation of plasma fibrinogen levels by ciprofibrate and gemfibrozil in primary hyperlipidaemia. Blood Coagul. Fibrinol. **5**: 17–17.
47. NIORT, G., M. CASSADER, R. GAMBINO, et al. 1992. Comparison of the effects of bezafibrate and acipimox on the lipid pattern and plasma fibrinogen in hyperlipidaemic type 2 (non-insulin-dependent) diabetic patients. Diabete. Metab. **18**: 221–228.
48. HAVERKATE, F., J. KOOPMAN & M.P.M. DE MAAT. 1998. Statins and fibrinogen. Lancet **351**: 1430–1431.
49. BEHAR, S. 1999. Lowering fibrinogen levels: clinical update. BIP Study Group. Bezafibrate Infarction Prevention. Blood Coagul. Fibrinol. **10**(Suppl. 1): S41–S43.
50. KOCKX, M., P.P. GERVOIS, P. POULAIN, et al. 1999. Fibrates suppress fibrinogen gene expression in rodents via activation of the peroxisome proliferator-activated receptor-alpha. Blood **93**: 2991–2998.
51. DE MAAT, M.P.M., H.C. KNIPSCHEER, J.J.P. KASTELEIN, et al. 1997. Modulation of plasma fibrinogen levels by ciprofibrate and gemfibrozil in primary hyperlipidaemia. Thromb. Hæmost. **77**: 75–79.
52. KOCKX, M., M.P.M. DE MAAT, H.C. KNIPSCHEER, et al. 1997. Effects of gemfibrozil and ciprofibrate on plasma levels of tissue-type plasminogen activator, plasminogen activator inhibitor-1 and fibrinogen in hyperlipidaemic patients. Thromb. Hæmost. **78**: 1167–1172.
53. BAUMANN, R.E. & A.H. HENSCHEN. 1994. Linkage disequilibrium relationships among four polymorphisms within the human fibrinogen gene cluster. Hum. Genet. **94**: 165–170.
54. THOMAS, A.E., F.R. GREEN & S.E. HUMPHRIES. 1996. Association of genetic variation at the beta-fibrinogen gene locus and plasma fibrinogen levels; interaction between allele frequency of the G/A-455 polymorphism, age and smoking. Clin. Genet. **50**: 184–190.

55. BEHAGUE, I., O. POIRIER, V. NICAUD, et al. 1996. Beta Fibrinogen gene polymorphisms are associated with plasma fibrinogen and coronary artery disease in patients with myocardial infarction: The ECTIM study. Circulation **93:** 440–449.
56. GREEN, F., A. HAMSTEN, M. BLOMBACK, et al. 1993. The role of beta-fibrinogen genotype in determining plasma fibrinogen levels in young survivors of myocardial infarction and healthy controls from Sweden. Thromb. Hæmost. **70:** 915–920.
57. MARGAGLIONE, M., G. CAPPUCCI, D. COLAIZZO, et al. 1998. Fibrinogen plasma levels in an apparently healthy general population—relation to environmental and genetic determinants. Thromb. Hæmost. **80:** 805–810.
58. SCARABIN, P.Y., L. BARA, S. RICARD, et al. 1993. Genetic variation at the beta-fibrinogen locus in relation to plasma fibrinogen concentrations and risk of myocardial infarction. The ECTIM Study. Arterioscler. Thromb. **13:** 886–891.
59. VAN DER BOM, J.G., M.P.M. DE MAAT, M.L. BOTS, et al. 1998. Elevated plasma fibrinogen—cause or consequence of cardiovascular disease? Arterioscler. Thromb. Vasc. Biol. **18:** 621–625.
60. TYBJAERG-HANSEN, A., B. AGERHOLM-LARSEN, S.E. HUMPHRIES, et al. 1997. A common mutation (g(-455)→a) in the beta-fibrinogen promoter is an independent predictor of plasma fibrinogen, but not of ischemic heart disease—a study of 9,127 individuals based on the Copenhagen city heart study. J. Clin. Invest. **99:** 3034–3039.
61. SCHMIDT, H., R. SCHMIDT, K. NIEDERKORN, et al. 1998. Beta-fibrinogen gene polymorphism (c-148→t) is associated with carotid atherosclerosis—results of the Austrian stroke prevention study. Arterioscler. Thromb. Vasc. Biol. **18:** 487–492.
62. Lane, D.A. & P.J. Grant. 2000. Role of hemostatic gene polymorphisms in venous and arterial thrombotic disease. Blood **95:** 1517–1532.
63. DE MAAT, M.P.M., J.J.P. KASTELEIN, J.W. JUKEMA. et al. 1998. -455G/A polymorphism of the beta-fibrinogen gene is associated with the progression of coronary atherosclerosis in symptomatic men. Proposed role for an acute-phase reaction pattern of fibrinogen. Arterioscler. Thromb. Vasc. Biol. **18:** 265–271.
64. ZITO, F., A. DI CASTELNUOVO, C. AMORE, et al. 1997. Bcl I polymorphism in the fibrinogen beta-chain gene is associated with the risk of familial myocardial infarction by increasing plasma fibrinogen levels—a case-control study in a sample of GISSI-2 patients. Arterioscler. Thromb. Vasc. Biol. **17:** 3489–3494.
65. MONTGOMERY, H.E., P. CLARKSON, O.M. NWOSE, et al. 1996. The acute rise in plasma fibrinogen concentration with exercise is influenced by the G-(453)-A polymorphism of the beta- fibrinogen gene. Arterioscler. Thromb. Vasc. Biol. **16:** 386–391.
66. FERRER-ANTUNES, C., M.P.M. DE MAAT, A. PALMEIRO, et al. 1998. Association between polymorphisms in the fibrinogen alpha- and beta-genes on the post-trauma fibrinogen increase. Thromb. Res. **92:** 207–212.
67. GARDEMANN, A., O. SCHWARTZ, W. HABERBOSCH, et al. 1997. Positive association of the beta fibrinogen H1/H2 gene variation to basal fibrinogen levels and to the increase in fibrinogen concentration during acute phase reaction but not to coronary artery disease and myocardial infarction. Thromb. Hæmost. **77:** 1120–1126.
68. BROWN, E.T. & G.M. FULLER. 1998. Detection of a complex that associates with the B beta fibrinogen G(–455)A polymorphism. Blood **92:** 3286–3293.
69. GREEN, F., A. LANE & S. HUMPHRIES. 1995. The effect of genetic variation in the β-fibrinogen 5′-flanking region on nuclear protein binding, transcriptional activity and plasma fibrinogen level. Thromb. Hæmost. **73:** 1224–1224.
70. LIAO, F., A. ANDALIBI, F.C. DEBEER, et al. 1993. Genetic control of inflammatory gene induction and NF-kappa B-like transcription factor activation in response to an atherogenic diet in mice. J. Clin. Invest. **91:** 2572–2579.
71. REZAEE, F., A. MAAS, J.H. VERHEIJEN, et al. 2000. Effect of genetic background and diet on plasma fibrinogen in mice. Possible relation with susceptibility for atherosclerosis. Ann. N.Y. Acad. Sci. **936:** this volume.
72. CERIELLO, A., F. MERCURI, D. FABBRO, et al. 1998. Effect of intensive glycaemic control on fibrinogen plasma concentrations in patients with Type II diabetes mellitus. Relation with beta-fibrinogen genotype. Diabetologia **41:** 1270–1273.

73. MERCURI, F., D. FABBRO, R. GIACOMELLO, et al. 1998. Effect of intensive glycemic control on fibrinogen plasma levels in diabetic subjects—relationship with beta-fibrinogen genotype. Molecular and Cell Biology of Type 2 Diabetes. 216.
74. DJOUSSE, L., R.C. ELLISON, Y.Q. ZHANG, et al. 1998. Relation between dietary fiber consumption and fibrinogen and plasminogen activator inhibitor type 1: the national heart, lung, and blood institute family heart study. Am. J. Clin. Nutr. **68:** 568–575.
75. MARCKMANN, P., B. SANDSTROM & J. JESPERSEN. 1993. Dietary effects of circadian fluctuation in human blood coagulation factor VII and fibrinolysis. Atherosclerosis **101:** 225–234.
76. JAMES, S., H.H. VORSTER, C.S. VENTER, et al. 2000. Nutritional status influences plasma fibrinogen concentration: evidence from the THUSA survey. Thromb. Res. **98:** 383–394.
77. VOLPI, E., P. LUCIDI, G. CRUCIANI, et al. 1998. Moderate and large doses of ethanol differentially affect hepatic protein metabolism in humans. J. Nutr. **128:** 198–203.
78. FOLSOM, A.R. 1995. Epidemiology of fibrinogen. Eur. Heart J. **16**(Suppl. A): 21–23.
79. ELWOOD, P.C., A.D. BESWICK, J.R. O'BRIEN, et al. 1993. Inter-relationships between haemostatic tests and the effects of some dietary determinants in the Caerphilly cohort of older men. Blood Coagul. Fibrinolysis. **4:** 529–536.
80. ROGERS, S., J.W.G. YARNELL & A.M. FEHILY. 1988. Nutritional determinants of haemostatic factors in the Caerphily Study. Eur. J. Clin. Nutr. **42:** 197–205.
81. POGGI, M., G. PALARETI, R. BIAGI, et al. 1994. Prolonged very low calorie diet in highly obese subjects reduces plasma viscosity and red cell aggregation but not fibrinogen. Int. J. Obes. Relat. Metab. Disord. **18:** 490–496.
82. SLABBER, M., H.C. BARNARD, J.M. KUYL, et al. 1992. Effect of a short-term very low calorie diet on plasma lipids, fibrinogen, and factor VII in obese subjects. Clin. Biochem. **25:** 334–335.
83. SVENDSEN, O.L., C. HASSAGER, C. CHRISTIANSEN, et al. 1996. Plasminogen activator inhibitor-1, tissue-type plasminogen activator, and fibrinogen—effect of dieting with or without exercise in overweight postmenopausal women. Arterioscler. Thromb. Vasc. Biol. **16:** 381–385.
84. HEINRICH, J., U. WAHRBURG, H. MARTIN, et al. 1990. The effect of diets, rich in mono- or polyunsaturated fatty acids, on lipid metabolism and haemostasis. Fibrinolysis **4:** 76–78.
85. HOSTMARK, A.T., E. LYSTAD, O.D. VELLAR, et al. 1993. Reduced plasma fibrinogen, serum peroxides, lipids, and apolipoproteins after a 3-week vegetarian diet. Plant Foods Hum. Nutr. **43:** 55–61.
86. PAN, W.H., C.J. CHIN, C.T. SHEU, et al. 1993. Hemostatic factors and blood lipids in young Buddhist vegetarians and omnivores. Am. J. Clin. Nutr. **58:** 354–359.
87. JANSSEN, K., R.P. MENSINK, F.J. COX, et al. 1998. Effects of the flavonoids quercetin and apigenin on hemostasis in healthy volunteers: results from an *in vitro* and a dietary supplement study. Am. J. Clin. Nutr. **67:** 255–262.
88. VELDMAN, F.J., C.H. NAIR, H.H. VORSTER, et al. 1997. Dietary pectin influences fibrin network structure in hypercholesterolaemic subjects. Thromb. Res. **86:** 183–196.
89. VELDMAN, F.J., C.H. NAIR, H.H. VORSTER, et al. 1999. Possible mechanisms through which dietary pectin influences fibrin network architecture in hypercholesterolaemic subjects. Thromb. Res. **93:** 253–264.
90. BOUDREAUX, M.K., C. JEFFERS, D. LIPSCOMB, et al. 1992. The effect of ticlopidine on canine platelets, fibrinogen, and antithrombin III activity. J. Vet. Pharmacol. Ther. **15:** 1–9.
91. DROSTE, D.W., H.J. SIEMENS, M. SONNE, et al. 1996. Hemostaseologic and hematologic parameters with aspirin and ticlopidine treatment in patients with cerebrovascular disease: a cross-over study. J. Cardiovasc. Pharmacol. **28:** 591–594.
92. GRYGLEWSKI, R.J., R. KORBUT, J. SWIES, et al. 1996. Thrombolytic action of ticlopidine: possible mechanisms. Eur. J. Pharmacol. **308:** 61–67.
93. MAZOYER, E., L. RIPOLL, M.R. BOISSEAU, et al. 1994. How does ticlopidine treatment lower plasma fibrinogen? Thromb. Res. **75:** 361–370.
94. CHADAREVIAN, R., E. BRUCKERT & G. TURPIN. 1998. Lipid-hemostasis interactions: a new approach to cardiovascular risk factors. Presse Med. **27:** 1437–1439.

95. HERNANDEZ-MIJARES, A., I. LLUCH, E. VIZCARRA, et al. 2000. Ciprofibrate effects on carbohydrate and lipid metabolism in type 2 diabetes mellitus subjects. Nutr. Metab. Cardiovasc. Dis. **10:** 1–6.
96. KOCKX, M., H.M.G. PRINCEN & T. KOOISTRA. 1998. Fibrate-modulated expression of fibrinogen, plasminogen activator inhibitor-1 and apolipoprotein a-1 in cultured cynomolgus monkey hepatocytes—role of the peroxisome proliferator-activated receptor-alpha. Thromb. Hæmost. **80:** 942–948.
97. MILLER, D.B. & J.D. SPENCE. 1998. Clinical pharmacokinetics of fibric acid derivatives (fibrates). Clin. Pharmacokinet. **34:** 155–162.
98. ZAMBRANA, J.I., F. VELASCO, P. CASTRO, et al. 1997. Comparison of bezafibrate versus lovastatin for lowering plasma insulin, fibrinogen, and plasminogen activator inhibitor-1 concentrations in hyperlipemic heart transplant patients. Am. J. Cardiol. **80:** 836–840.
99. HERBERT, J.M., A. BERNAT & L. CHATENETDUCHENE. 1999. Effect of ciprofibrate on fibrinogen synthesis *in vitro* on hepatoma cells and *in vivo* in genetically obese zucker rats. Blood Coagulat. Fibrinol. **10:** 239–244.
100. ATHYROS, V.G., A.A. PAPAGEORGIOU, H.A. HATZIKONSTANDINOU, et al. 1998. Effect of atorvastatin versus simvastatin on lipid profile and plasma fibrinogen in patients with hypercholesterolaemia—a pilot, randomised, double-blind, dose-titrating study. Clin. Drug Invest. **16:** 219–227.
101. BERTOLOTTO, A., S. BANDINELLI, L. RUOCCO, et al. 1999. More on the effect of atorvastatin on plasma fibrinogen levels in primary hypercholesterolemia. Atherosclerosis **143:** 455–457.
102. BRANCHI, A., A. ROVELLINI, D. SOMMARIVA, et al. 1993. Effect of three fibrate derivatives and of two HMG-CoA reductase inhibitors on plasma fibrinogen level in patients with primary hypercholesterolemia. Thromb. Hæmost. **70:** 241–243.
103. DANGAS, G., J.J. BADIMON, D.A. SMITH, et al. 1999. Pravastatin therapy in hyperlipidemia: effects on thrombus formation and the systemic hemostatic profile. J. Amer. Coll. Cardiol. **33:** 1294–1304.
104. FARRER, M., P.H. WINOCOUR, K. EVANS, et al. 1994. Simvastatin in non-insulin–dependent diabetes mellitus: effect on serum lipids, lipoproteins and hæmostatic measures. Diabetes Res. Clin. Pract. **23:** 111–119.
105. SIRTORI, C.R. & S. COLLI. 1993. Influences of lipid-modifying agents on hemostasis. Cardiovasc. Drugs Ther. **7:** 817–823.
106. TSUDA, Y., K. SATOH, T. TAKAHASHI, et al. 1993. Effect of medication with pravastatin sodium on hemorheological parameters in patients with hyperlipoproteinemia. Int. Angiol. **12:** 360–364.
107. TSUDA, Y., K. SATOH, M. KITADAI, et al. 1996. Effects of pravastatin sodium and simvastatin on plasma fibrinogen level and blood rheology in type II hyperlipoproteinemia. Atherosclerosis **122:** 225–233.
108. WADA, H., Y. MORI, T. KANEKO, et al. 1992. Hypercoagulable state in patients with hypercholesterolemia: effects of pravastatin. Clin. Ther. **14:** 829–834.
109. PRESTON, F.E., R.G. MALIA, M. GREAVES, et al. 1982. The effect of stanozolol on thrombotic risk factors in healthy individuals. Hæmostasis. **11:** 56–56.
110. IBBOTSON, S.H., A.M. LAYTON, J.A. DAVIES, et al. 1995. The effect of aspirin on hæmostatic activity in the treatment of chronic venous leg ulceration. Br. J. Dermatol. **132:** 422–426.
111. KRUSIUS, T., M. HIRVONEN, E. VAHTERA, et al. 1994. Short-term aspirin treatment does not reduce plasma fibrinogen concentration in young healthy adults. Thromb. Res. **75:** 653–656.
112. MA, J., C.H. HENNEKENS, P.M. RIDKER, et al. 1999. A prospective study of fibrinogen and risk of myocardial infarction in the Physicians' Health Study. J. Am. Coll. Cardiol. **33:** 1347–1352.
113. TOHGI, H., H. TAKAHASHI, M. KASHIWAYA, et al. 1994. Effect of plasma fibrinogen concentration on the inhibition of platelet aggregation after ticlopidine compared with aspirin. Stroke **25:** 2017–2021.

Genetic and Immunological Characterization of Fibrinogen Inclusion Bodies in Patients with Hepatic Fibrinogen Storage and Liver Disease

DANIELA MEDICINA,[a] GIOVANNA FABBRETTI,[a] STEPHEN O. BRENNAN,[b] PETER M. GEORGE,[b] BO KUDRYK,[c] AND FRANCESCO CALLEA[a]

[a]*Molecular Pathology Laboratories, Spedali Civili Brescia, Italy*

[b]*Christchurch University, New Zealand*

[c]*Blood Transfusion Center, New York, USA*

ABSTRACT: Fibrinogen storage in liver cells can occur under three different morphological inclusions. Type I contain all three fibrinogen chains (Aα, Bβ, and γ) as well as D and E fragments, whereas type II and III lack Bβ as well as D and E fragments. Patients with type I inclusions carry a point mutation (γ284 Gly-Arg). The mutation is not present in patients with type II and III inclusions. These results appear to suggest that the three various phenotypic expressions (i.e., morphological variants) reflect different genetical abnormalities of fibrinogen.

KEYWORDS: Fibrinogen; Liver disease; Immunohistochemistry; Hypofibrinogenemia.

INTRODUCTION

Selective and exclusive retention of fibrinogen within the RER of hepatocytes occurs in three different morphological variants with well characterized light and electron microscopic features.[1] Type I appears as round inclusions with an irregular outline or elongated somewhat acicular appearance (fiber-glass-like) in histological section stained with hematoxylin-eosin. Under an electron microscope (E.M.), the inclusions correspond to tubular structures arranged in curved bundles in a finger print fashion within dilated cisternae of the RER. Type II inclusions appear as large single eosinophilic inclusions filling up the entire cytoplasm of the hepatocytes. Under and E.M. the inclusions correspond to dilated cisternae of the RER contain fragmented or fibrillar material. Type III appears as round small single or multiple deeply eosinophilic or glassy inclusions. Under an E.M. the material is found within the cisternae of the RER and displays a tubular appearance (somewhat reminiscent finger prints) in the central part and a fluffy or fragmented fibrillar appearance in the periphery.

Address for correspondence: Francesco Callea, M.D., Ph.D., Department of Pathology, Spedali Civili of Brescia, Brescia 25123, Italy. Voice: +39-030-3995378; fax +39-030-3995377.
callea@bshosp.osp.unibs.it

Recently it has been shown that Type I inclusions identify a mutant variant of the fibrinogen molecule designated Fibrinogen Brescia. The mutation is located at the level of the exon VIII of the γ gene (γ284 Gly-Arg) and results in hereditary hypofibrinogenemia with hepatic storage of the mutant protein.

Additional characteristics of Fibrinogen Brescia have been reported in detail,[2] almost all the circulating protein is functionally normal, no γ chain is detectable in the plasma, the storage predisposes the development of cirrhosis.

AIM OF THE STUDY

The objectives of the study we report here were: (1) to characterize the nature of the three variants of fibrinogen inclusions from the immunological viewpoint; and (2) sequence the DNA of the three genes corresponding to the various inclusions.

MATERIAL OF THE STUDY

Four groups of patients were included in the study:

Group 1: Comprising 16 member of the Fibrinogen Brescia Family with type I inclusions.

Group 2: Comprising 10 patients with chronic liver disease (CALD) and type II fibrinogen inclusions.

Group 3: Comprising 10 patients with CALD and type III fibrinogen inclusions.

Group 4: Control group, comprising 20 patients with CALD and no fibrinogen inclusions. Biological samples from these groups consisted of liver tissue specimens (a total of 41 patients) as well as blood and plasma (from all patients).

METHODOLOGY OF THE STUDY

The methodology included:
1. Morphological investigations at both light and EM level.
2. Immunohistochemistry on serial sections by using polyclonal antifibrinogen antibodies (Dakopath, Denmark, dilution 1:500), as well as monoclonal antibodies against Aα, Bβ, γ chains, E and D fragments, the latter having been tested in previous work.[3,4] Polyclonal antibodies against other secretory proteins, a.o. alpha-1-antitrypsin, were used.
3. Biochemical investigations on the structure and the function of the protein.
4. Molecular biology techniques, including DNA amplification and sequencing and mutagenic PCR.

RESULTS

The morphological results confirmed previous observations related to the three types of inclusions.[1] The content of the three inclusions did react exclusively and

TABLE 1. Immunohistochemical results with the fibrinogen inclusions in tissue sections

	Polyclonal Abs	Monoclonal Antibodies				
		Aα	β	γ	D	E
Type I	++++	++	++	+++	++	++
Type II	+++	++	–	++	–	–
Type III	+++	++	–	++	–	–
Controls	–	–	–	–	–	–

TABLE 2. Molecular biology and fibrinogen plasma levels results associated with the three inclusion types

	Mutation (γ 284 Gly-Arg)	FBR Plasma Levels	Functional Tests
Type I	present (6/13)	low (6/13)	normal (13)
Type II	absent	low (4/10)	normal (10)
Type III	absent	low (2/10)	normal (10)
Controls	absent	normal (20/20)	normal (20)

selectively with the polyclonal antifibrinogen antibodies and was negative for all other tested proteins.

The immunohistochemical results obtained with the various monoclonal antibodies (MoAbs) are summarized in TABLE 1. The molecular biology results are summarized in TABLE 2. Table 2 shows that only type I inclusions are associated with the mutation γ284 Gly-Arg and that low fibrinogen plasma levels are consistently associated with the mutation.

DISCUSSION

The results from this study appear to indicate that the newly discovered mutation (γ284 Gly-Arg) is associated with type I fibrinogen inclusions in liver cells and identifies a new disease, hereditary hypofibrinogenemia, that is analogous to congenital alpha-1-antitrypsin deficiency.[1] The molecular abnormality leads to misfolding of the mutant protein, fibrille formation as well as intracellular aggregation ("conformational disease", according to the Carrel's concept[5]), within the RER ("endoplasmic reticulum storage disease", according to Callea's concept[6]).

Type I inclusions appear to contain all three fibrinogen chains (Aα, Bβ, and γ), whereas both Type II and III lack the Bβ as well as D and E fragments. From the data generated in this study it would seem that fibrinogen storage in liver cells represents a new cause of CALD. Moreover, fibrinogen storage disease represents the first example of a fibrinogen molecular abnormality resulting in a defective export of the protein without functional derangement of the circulating plasma protein.

Further investigations are expected to clarify whether the various phenotypic expressions (i.e., morphological variants) reflect different genetical abnormalities of fibrinogen.

REFERENCES

1. CALLEA, F., M. BRISIGOTTI, G. FABBRETTI, et al. 1992. Hepatic endoplasmic reticulum storage diseases. Review. Liver **12**(6): 357–362.
2. BRENNAN, S.O., J. WYATT, D. MEDICINA, et al. 2000. Fibrinogen Brescia: hepatic endoplasmic reticulum storage and hypofibrinogenemia due to a 284 Gly-Arg mutation. Am. J. Pathol. **157:** 189–196.
3. BINI, A., R. MESA-TEJADA, J.J. FENOGLIO, JR., et al. 1989. Himmunohistochemical characterization of fibrin(ogen)-related antigens in human tissues using monoclonal antibodies. Lab. Invest. **60**(6): 814–821.
4. KUDRYK, B.J., A. BINI, S.F. ROSEBROUGH, et al. 1991. Fibrinogen-fibrin: preparation and use of monoclonal antibodies as diagnostics. Biotechnology **19:** 281–313.
5. CARRELL, R.W. & D.A. LOMAS. 1997. Conformational disease. Lancet **350**(9071): 134–138.
6. CALLEA, F., R. DE VOS, J. PINACKAT, et al. 1987. Hereditary hypofibrinogenaemia with hepatic storage of fibrinogen: a new endoplasmic reticulum storage disease. *In* Fibrinogen 2. Biochemistry, Physiology and Clinical Relevance. G.D.O. Lowe, Ed.: 75–78. Amsterdam, Elsevier.

Hypofibrinogenemia Associated with a Heterozygous C→T Nucleotide Substitution at Position −1138 BP of the 5′-Flanking Region of the Fibrinogen Aα-Chain Gene

NOBUO OKUMURA,[a] FUMIKO TERASAWA,[a] OSAMU YONEKAWA,[b] ETSUKO HAMADA,[c] AND HIROSHI KANEKO[d]

[a]*Laboratory of Clinical Chemistry, Department of Medical Technology, School of Allied Medical Sciences, Shinshu University, Matsumoto, Japan*

[b]*Clinical Laboratory, Seirei Hamamatsu Hospital, Hamamatsu, Japan*

[c]*Central Clinical Laboratory, Hamamatsu University Hospital, Hamamatsu, Japan*

[d]*Matsuda Surgical Hospital, Hamamatsu, Japan*

ABSTRACT: We found a novel genetic abnormality, heterozygous C→T nucleotide substitution at position −1138 bp in the 5′-flanking region of the fibrinogen Aα gene, in patients with hypofibrinogenemia. Luciferase reporter assay using the pGL3-basic vector and CHO cells indicates that the transcriptional activity of a vector incorporated with −1138T was reduced to one-third that of a vector incorporated with −1138C. These results suggest that the region adjacent to the −1138C bp of the 5′-flanking region of the fibrinogen Aα gene is one of the most crucial sites for the transcription of the fibrinogen Aα gene.

KEYWORDS: Hypofibrinogenemia; Nucleotide substitution; Fibrinogen Aα chain gene; 5′-Flanking region.

INTRODUCTION

Although congenital afibrinogenemia was originally described in 1920, the genetic abnormalities are still not fully understood. In the last two years, 14 abnormalities of the fibrinogen Aα-, Bβ-, and γ-chain gene have been reported in congenital afibrinogenemias or hypofibrinogenemias.[1–5] In this report, we demonstrated the hypofibrinogenemia associated with a novel genetic abnormality, heterozygous C→T nucleotide substitution at position −1138 bp in the 5′-flanking region of the fibrinogen Aα gene that led to reduced transcriptional activity of luciferase.

Address for correspondence: Dr. Nobuo Okumura, Laboratory of Clinical Chemistry, Department of Medical Technology, School of Allied Medical Sciences, Shinshu University, 3-1-1 Asahi, Matsumoto 390-8621, Japan. Voice: +81 263 37 2393; fax: +81 263 37 2370.
nobuoku@gipac.shinshu-u.ac.jp

MATERIALS AND METHODS

Patient Identification. The propositus was an 81-year-old man with arteriosclerosis obliterans of iliac artery substituted with an artificial vessel.

Coagulation Screening Tests. Prothrombin time (PT), activated partial thromboplastin time (APTT), the fibrinogen concentration, determined by thrombin time method, and that by the immunologic method were measured as coagulation screening tests.

Polymerase Chain Reaction (PCR) and DNA Sequence Analysis. To amplify all exons and exon–intron boundaries of the fibrinogen Aα-, Bβ-, and γ-chain gene, the DNA was amplified by PCR and directly sequenced as described elsewhere.[2] To analyze 5'-flanking region of fibrinogen Aα-, Bβ-, and γ-chain gene, additional PCR primers were designed and PCR amplification and DNA sequence analysis were performed as described above.

To examine the frequency of the nucleotide "T" at −1138 bp of the 5'-flanking region of the fibrinogen Aα gene, allele specific PCR was performed on healthy volunteers. In addition, so as to exclude the possibility of the propositus as heterozygous large DNA deletion, long range PCR amplification of Aα-, Bβ-, and γ-chain genes were performed for 30 cycles using the pairs of primer designed in both first and last exon of each gene and Takara LA Taq DNA polymerase (Takara, Tokyo, Japan).

Construction of Vector for Luciferase (Luc) Reporter Assay. The DNA fragment of −1499 to +171 bp from the propositus fibrinogen Aα gene was amplified by PCR and cloned into the pCR2.1 plasmid vector (TA Cloning System, In Vitrogen, San Diego, CA). We used pGL3-basic vector and pGL3-promoter vector as Luc reporter assay (both from Promega, Madison, WI). First, the −1321 to 32 bp fragment of Aα gene, was inserted into the pGL3-basic vector. Second, to obtain the −1193 to −1118 bp 5'-flanking region of Aα-chain gene, cloned pCR2.1 plasmid vector was altered to the *Kpn I* restriction enzyme site by oligonucleotide directed mutagenesis using the Transformer™ site directed mutagenesis kit (Clontech Laboratories, Palo Alto, CA), purified mutated plasmid was digested by *Kpn I,* and the digested fragment was inserted into the pGL3-promoter vector.

Transfection and Luc Assay. Each of the cloned pGL3-basic vectors and pGL3-promoter vectors was cotransfected with pRL-TK (Promega) producing Renilla Luc, which compensates for the difference in transfection efficiency, into CHO cells or HepG2 cells, using TransFast™ Transfection Reagent (Promega). After two days incubation, the Luc activity of transfected cells was measured using Dual-Luciferase Reporter Assay System (Promega) as a reagent and MiniLumat LB9506 (Berthold, Badwildbad, Germany) as a luminometer.

RESULTS

Although the propositus showed no bleeding or thrombotic tendency, coagulation screening tests revealed a lower than normal plasma fibrinogen concentration (1.50 to 3.00 g/L), which was determined by both the thrombin time method (1.01 g/L) and the immunologic method (0.84 g/L). The plasma FDP-D-dimer value was 18.4 µg/ml, which was much higher than normal (less than 0.7 µg/ml). However, the

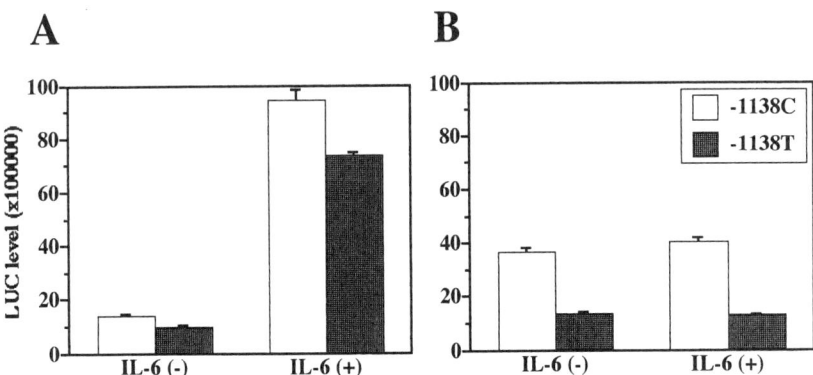

FIGURE 1. Luciferase reporter assay using the pGL3-basic vector inserted Aα chain gene fragment of −1321 to 32 bp of 5′-flanking region. pGL3-basic vector having −1138C or −1138T was cotransfected with pRL-TK plasmid into HepG2 cells (**A**) or CHO cells (**B**) and Luc activity was measured and compensated as described in MATERIALS AND METHODS.

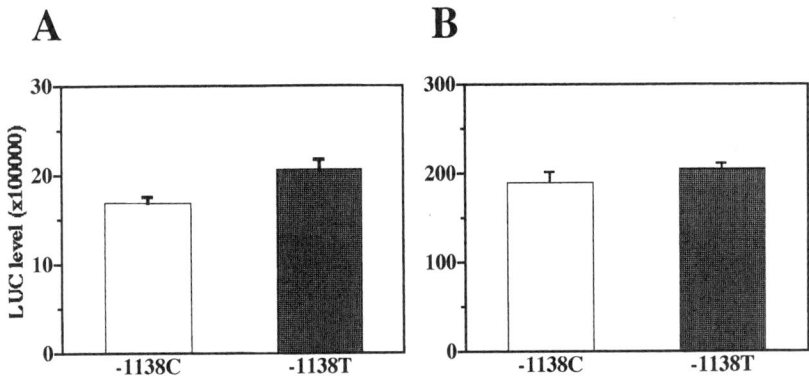

FIGURE 2. Luciferase reporter assay using the pGL3-promoter vector inserted Aα chain gene fragment of −1193 to −1118 bp of 5′-flanking region. pGL3-promoter vector having −1138C or −1138T was cotransfected with pRL-TK plasmid into HepG2 cells (**A**) or CHO cells (**B**) and Luc activity was measured and compensated as described in MATERIALS AND METHODS.

possibility of decreased synthesis of fibrinogen such as liver cirrhosis was excluded by normal values of APTT, serum albumin, and serum acylcholinesterase.

Direct sequencing of a PCR-amplified Aα-chain gene segment showed a novel C to T nucleotide substitution at position −1138 bp of the 5′-flanking region. No other mutations were found in the 5′-flanking region, the entire coding region or the exon–intron boundaries of the Aα-, Bβ-, and γ-chain genes due to the hypofibrinogenemia. Unfortunately, we were not able to analyze other family members because none of 60 healthy individuals had C by allele specific PCR analysis. Long range PCR amplification of Aα-, Bβ-, and γ-chain genes might exclude the possibility of the propositus as heterozygous large DNA deletion of one of three fibrinogen genes.

Luc reporter assay using pGL3-basic vector incorporated with −1321 to 32-bp fragment of Aα-chain gene indicated that Luc activity of CHO cells transfected with −1138T type vector was 37.2% and 33.2% of that of CHO cells transfected with −1138C wildtype vector in the absence or presence of interleukine-6 (IL-6), respectively (see FIGURE 1B). However, Luc activity of HepG2 cells transfected with −1138T type vector was 70.2% and 78.1% of that of HepG2 cells transfected with −1138C wildtype vector in the absence or presence of IL-6, respectively. Moreover, Luc activity in the presence of IL-6 was 6.6- and 7.4-fold that in the absence of IL-6 in HepG2 cells transfected with −1138T and −1138C type vector, respectively (FIG. 1A). On the other hand, Luc reporter assay using pGL3-promoter vector incorporated with −1193 to −1118-bp fragment of Aα-chain gene showed that Luc activity of CHO cells and HepG2 cells transfected with −1138T type vector was 7.7% and 22.2% higher than that transfected with −1138C wildtype vector, respectively (see FIGURE 2).

DISCUSSION

A novel −1138C→T nucleotide substitution in the 5′-flanking region of the fibrinogen Aα chain gene was found on the propositus of hypofibrinogenemia in Hamamatsu, Japan. However, there is no possibility of decreased synthesis of fibrinogen and no other mutations were found in the 5′-flanking region, the entire coding region or the exon-intron boundaries of the Aα-, Bβ-, and γ-chain genes due to the hypofibrinogenemia. We speculated that the elevated FDP-D dimer level was due to hypercoagulation and fibrinolysis on the surface of the artificial vessel. In addition, C.-H. Hu, et al.[6] demonstrated that the region from −1133 to −1393 bp of the 5′-flanking region of the fibrinogen Aα chain gene contained a positive enhancer element to the hepatocyte nuclear factor-1 (HNF-1) binding site (from −47 to −59 bp), interleukine-6 (IL-6) responsive element (from −122 to −127), and CCAAT/enhancer binding protein (C/EBP) site (from −134 to −142).

Therefore, to demonstrate the reduced enhancer activity of nucleotide T at position −1138 of the fibrinogen Aα gene directly, we inserted propositus aberrant and wildtype gene fragment into Luc reporter assay vectors and determined Luc activity of vector transfected cell lines. Luc activity of CHO cells transfected with pGL3-basic vector inserted the fibrinogen Aα gene including −1138T was one third that of the vector including −1138C, both in the absence or presence of IL-6. Moreover, in the presence of IL-6, Luc activity of HepG2 cells transfected with pGL3-basic vector inserted the fibrinogen Aα gene including −1138T were about seven-folds higher than that in the absence of IL-6. These results suggest that −1138C is important enhancer activity to the promoter region of HNF-1 binding site, C/EBP site, or novel one. Luc activity of CHO cells transfected with pGL3-promoter vector inserted the positive element of the 5′-flanking region of the fibrinogen Aα gene, −1193 to −1118 including −1138T, was slightly higher than that transfected with pGL3-promoter vector inserted −1138C. This result suggests that −1138T has an enhancer activity to the promoter of SV40, but not to the fibrinogen Aα gene promoters.

In conclusion, our case of hypofibrinogenemia suggests that the region adjacent to the −1138 bp of the 5′-flanking region of the fibrinogen Aα chain gene is one of the most important sites for transcription of the Aα chain gene, and a decrease of its mRNA following a decrease of the polypeptide induces a decreased level of fibrinogen synthesis in hepatocytes. Since this activity of −1138T in HepG2 cells was about twofold higher than that in CHO cells, we speculate that the regulation of human fibrinogen synthesis in normal hepatocytes is significantly different than that in the HepG2 cell line.

REFERENCES

1. NEERMAN-ARBEZ, M., *et al.* 1999. Deletion of the fibrinogen alpha-chain gene (FGA) causes congenital afibrinogenemia. J. Clin. Invest. **103:** 215–218.
2. TERASAWA, F., *et al.* 1999. Hypofibrinogenemia associated with a heterozygous missense mutation γ153Cys to Arg (Matsumoto IV): *in vitro* expression demonstrates defective secretion of the variant fibrinogen. Blood **94:** 4122–4131.
3. DUGA, S., *et al.* 2000. Missense mutations in the human β fibrinogen gene cause congenital afibrinogenemia by impairing fibrinogen secretion. Blood **95:** 1336–1341.
4. BRENNAN, S.O., *et al.* 2000. Hypofibrinogenemia in an individual with 2 coding (γ82A→G and Bβ235P→L) and 2 noncoding mutations. Blood **95:** 1709–1713.
5. NEERMAN-ARBEZ, M., *et al.* 2000. Mutations in the fibrinogen Aα gene account for the majority of cases of congenital afibrinogenemia. Blood **96:** 149–152.
6. HU, C-H., *et al.* 1995. Characterization of the 5′-flanking region of the gene for the α chain of human fibrinogen. J. Biol. Chem. **270:** 28342–28349.

Modified Clotting Properties of Fibrinogen in the Presence of Acetylsalicylic Acid in a Purified System

SHU HE,[a] MARGARETA BLOMBÄCK,[a] GINA YOO,[b]
RAKHI SINHA,[b] AND AGNES H. HENSCHEN-EDMAN[b]

[a]*Coagulation Research, Department of Surgical Sciences, Karolinska Institutet, Stockholm, Sweden*

[b]*Department of Molecular Biology and Biochemistry, University of California, Irvine, California, USA*

ABSTRACT: To assess how treatment with acetylsalicylic acid (ASA) alters the fibrin network structure, clotting was initiated in purified fibrinogen incubated with ASA by adding thrombin. Clotting time and maximum absorbance of the fibrin aggregation curve were used to demonstrate the potential of fibrin generation. The results showed that the clotting properties of fibrinogen decreased and that the affinity of plasminogen to fibrin or thrombin inhibition by antithrombin increased if plasminogen or antithrombin, respectively, were present in the reaction system. The effect of ASA varied in a dose dependent manner. It was concluded that ASA may directly or indirectly confer positive or negative effects on the stability of the fibrin clot and that the balance between these effects may be regulated by the ASA dose.

KEYWORDS: Fibrinogen clotting; Acetylsalicylic acid.

The fibrin clot is derived from fibrinogen via cleavage by thrombin. It is believed that blood flow through a fibrin clot and, accordingly, transport of fibrinolysis promoting components is more efficient when the network has a more porous structure. Thus, increases in gel porosity may be beneficial to clot dissolution.

Our previous studies showed that tighter fibrin gels were obtained with plasma samples derived from patients with cardiovascular disease or diabetes than those derived from healthy controls.[1,2] We also found that treatment with acetylsalicylic acid (ASA) increased the permeability of the fibrin network. A greater effect was observed for a daily intake of 75 mg than for 320 mg.[3] These findings may explain the good prevention of arterial thrombosis by ASA therapy, especially when administered in low dosages.

A quantitative decrease in crosslinked fibrin, the architectural material of the network, a depressed thrombin function, and/or a quantitative increase in plasminogen lead to increased gel porosity with higher permeability.[4,5] In the present study, we were interested in assessing whether ASA modifies the clotting properties of

Address for correspondence: Shu He, M.D. Ph.D., Coagulation Research, L2-05, Karolinska Hospital, S-17176, Stockholm, Sweden. Fax: 46+8+312438.
he_shu@yahoo.com

fibrinogen, the affinity of plasminogen to fibrinogen or fibrin, and the thrombin inhibition by antithrombin in a purified system.

METHODS AND RESULTS

Human fibrinogen (2 mg/ml) free from plasminogen, fibronectin, and factor XIII was incubated with ASA (0, 0.16, 0.48, 1.44, 4.32, and 12.96 mmol/l) in triethanolamine-acetate buffer (pH 7.4) at 37°C for 24 h. The concentrations of ASA and its degradation product, salicylic acid, were assayed at several times with an HPLC method. In this purified system, the halflife of ASA was found to be about 15 h. After incubation, the solutions were dialyzed overnight to remove remaining ASA and its hydrolysis products.

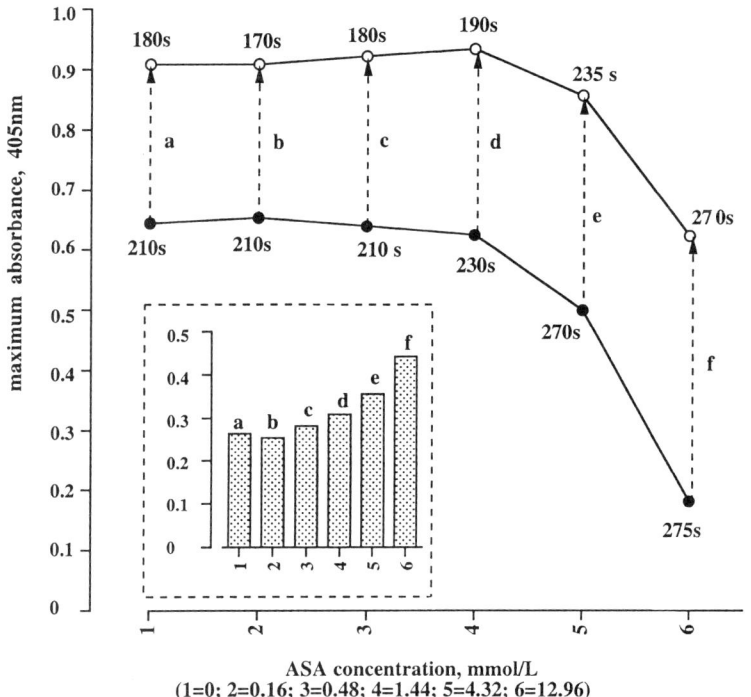

FIGURE 1. Modification of fibrinogen clotting properties, expressed as maximum absorbance of the fibrin aggregation curve, resulting from the dose dependent effect of ASA in the absence or presence of plasminogen. *Curve A* (Fbg-ASA), fibrinogen incubated with ASA of different concentrations. *Curve B* (Fbg-ASA) + (plg-ASA), Lys-plasminogen incubated with ASA solutions at different concentrations, and then added to the fibrinogen solutions preincubated with ASA of the corresponding concentrations. *Bars a–f,* differences in the maximum absorbance values between points on curves *A* and *B*. The clotting times in seconds are indicated next to the corresponding absorbance values. ●, A: (fbg-ASA); ○, B: (fbg-ASA) + (plg-ASA).

Clotting was initiated in Tris-HCl buffer (pH 7.4) at 37°C with final concentrations of the fibrinogen (1.5 mg/ml) incubated with ASA, factor XIII 0.1mg/ml, $CaCl_2$ 20mmol/L, and thrombin 0.2 IU/ml. A fibrin aggregation curve, measured as turbidity, was plotted from the absorbance values at 405 nm recorded every 20 sec. during 60 minutes. In this simple fibrinogen–thrombin system, a maximum absorbance of the fibrin aggregation curve, as well as the clotting times, were used to demonstrate the overall information of fibrin generation. In the fibrinogen solutions incubated

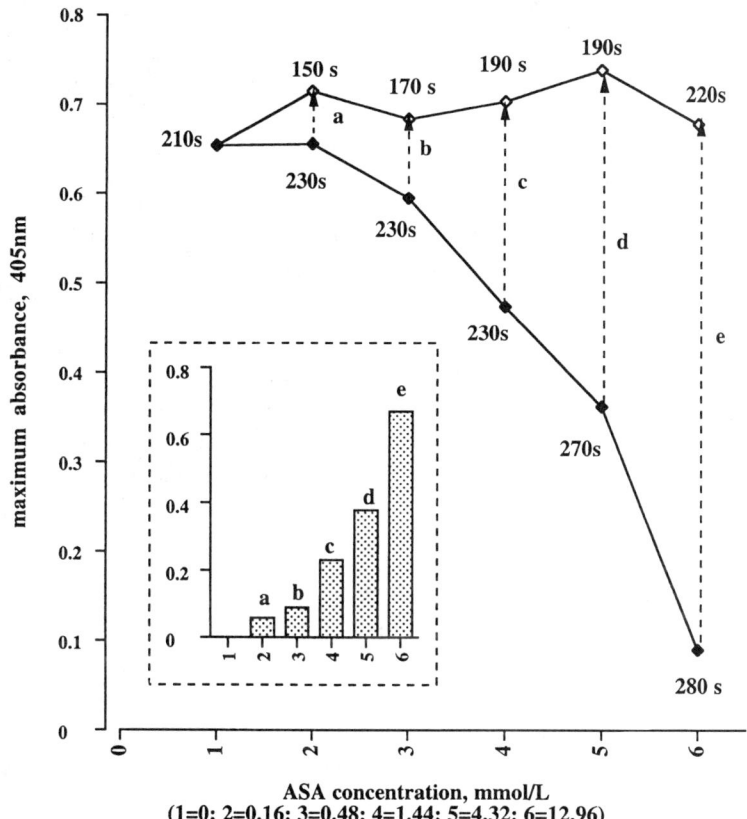

FIGURE 2. Modifications of fibrinogen clotting properties, expressed as maximum absorbance of the fibrin aggregation curve, resulting from the dose dependent effect of ASA on the antithrombin–heparin reaction. *Curve A* (Fbg-ASA) + (AT-heparin), fibrinogen incubated with ASA of different concentrations; heparin and AT incubated without ASA. *Curve B* (Fbg-ASA) + (AT-ASA-heparin), heparin incubated with AT previously dissolved in ASA solutions at different concentrations. The resulting mixtures were then incubated with thrombin and subsequently added to the fibrinogen solutions preincubated with ASA at the corresponding concentrations. *Bars a–e*, difference in the maximum absorbance values between points on curves A and B. The clotting times in seconds are indicated next to the corresponding absorbance values. ♦, A: (fbg-ASA) + (AT-heparin); ◊, B: (fbg-ASA) + (AT-ASA-heparin).

with ASA, the maximum absorbance decreased from 0.644 at an ASA concentration of 0 mmol/L, to 0.499 at 4.32 mmol/L, and then to 0.18 at 12.96 mmol/L (see FIGURE 1, curve A). The clotting times increased, i.e. from 210 seconds, to 270 seconds, and 275 seconds, respectively.

In order to estimate the modification of crosslinking, the clots obtained from the fibrinogen incubated with ASA were washed with NaCl (0.15 mol/L) and then with urea (9 mol/L) to remove the soluble proteins, including soluble fibrin oligomers. The washed clots were dissolved in an alkaline urea solution (40% urea in 0.8% NaOH) and measured at 282 nm in a spectrophotometer. The amounts of insoluble protein in the fibrin clots decreased with the increase in ASA concentration in parallel with the changes of the clotting properties of fibrinogen indicated by Curve A in FIGURE 1, indicating an ASA dose dependent effect on crosslinking fibrin formation (data not shown).

To assess the effect of modification of plasminogen by ASA, Lys-plasminogen (8 mg/ml) was incubated with ASA (0, 0.16, 0.48, 1.44, 4.32, or 12.96 mmol/l) in triethanolamine-acetate buffer at 37°C for two hours. It was then added (to give 0.2 mg/ml) into the fibrinogen solutions that had been preincubated with ASA at the corresponding concentration. We previously observed that when a fibrin aggregation curve is made in a purified fibrinogen–thrombin system,[5] higher absorbance values, measured as turbidity, were obtained when more plasminogen was present. This was also the case in the present experiments, but the increase in the maximal absorbance values was greater at the higher levels of ASA, (FIG. 1, Curve B and Insert). This demonstrates that fibrin formation is altered by the interaction between fibrinogen and plasminogen at presence of ASA of different concentrations.

In order to obtain evidence for the effect of antithrombin (AT) modification, unfractionated heparin (0.16 IU/ml) was added either to untreated antithrombin (0.08 IU/ml) or to antithrombin (0.08 IU/ml) previously dissolved in ASA solutions (0, 0.16, 0.48, 1.44, 4.32, or 12.96 mmol/L), and incubated at 37°C for two hours. The resulting mixtures were then incubated with thrombin (36 IU/ml) at room temperature for one hour, and subsequently added into the fibrinogen solutions (to give an expected concentration of 0.4 IU/ml) producing fibrin aggregation curves (see FIGURE 2, Curve A for untreated AT and Curve B for treated AT). The fibrinogen in both groups had been preincubated with ASA at the same concentrations as in the above experiments. The pre-incubation with ASA increased the values of maximum absorbance in an ASA dose dependent manner, indicating that the treatment of AT interferes with its interaction with heparin.

DISCUSSION

The inhibition of platelet aggregation by ASA induced acetylation of a serine residue in cyclooxygenase has been extensively documented. However, in other proteins, the acetylation site is primarily the side chain of lysine residues. In the present *in vitro* study, we showed that ASA treatment has a dose dependent influence on the clotting properties of fibrinogen. The main observed effect of ASA was that it impairs the ability of the fibrin monomers to polymerize, which may be explained by the fact that acetylation leads to changes in charge distributions and, possibly, conformation. In addition, ASA treatment resulted in a decreased stability of the clots as evidenced by their higher solubility.

When fibrin aggregation curves were made in the purified fibrinogen–thrombin system with plasminogen (no plasminogen activators), the increased turbidity was related to increased fibrin gel porosity.[5] Accordingly, the ASA dose dependent enhancements in the maximum absorbance, shown by Curve B in FIGURE 1, may be due to an increased binding of plasminogen to fibrin/fibrinogen caused by ASA. This implies that ASA treatment may indirectly lead to an increase in the proteolytic effect of plasmin(ogen) by contributing to the permeability of fibrin network.[5]

The lysine residues in AT are well documented as being involved in the heparin–AT reaction.[6] In the present study, AT treated with ASA was incubated with heparin, and subsequently incubated with thrombin. Due to partial acetylation of AT, leading to impaired heparin binding, the thrombin inhibition by AT was depressed. As a consequence, the thrombin activity was enhanced, causing an increase in clotting of fibrinogen in an ASA dose dependent manner (FIG. 2).

Our findings indicate that ASA modifies the clotting properties of fibrinogen and other proteins in plasma in several ways, altering the structure of the resulting fibrin network. Both a decrease in fibrinogen clotting properties and an increase in plasminogen binding may render the fibrin gel more porous. However, the decline in thrombin inhibition caused by the interference by ASA with the AT–heparin interaction may counteract the favorable influence on the gel porosity. In order to prove this inference, studies of the fibrin gel permeability in the purified system are required. In the plasma of ASA users, not only the proteins analyzed in the present study but also most other proteins will be acetylated. The corresponding modifications may directly or indirectly confer positive or negative effects on the fibrin gel porosity. The balance between the positive and negative effects may be regulated by the ASA dosage *in vivo*.

ACKNOWLEDGMENTS

This investigation was supported by grants from Torsten and Ragnar Söderberg's Foundation, Sweden. We thank Dr. Birger Blombäck for the constructive discussions.

REFERENCES

1. FATAH, K., *et al.* 1996. Proneness to formation of tight and rigid fibrin gel structures in men with myocardial infarction at a young age. Thromb. Hæmost. **76:** 535–540.
2. JÖRNESKOG, G., *et al.* 1996. Altered properties of the fibrin gel structure in patients with IDDM. Diabetologia **39:** 1519–1523.
3. WILLIAMS, S., *et al.* 1998. Better increase in fibrin gel porosity by low dose than intermediate dose acetylsalicylic acid. Eur. Hear. J. **19:** 1666–1672.
4. BLOMBÄCK, B., *et al.* 1994. Fibrin in human plasma: gel architectures governed by rate and nature of fibrinogen activation. Thromb. Res. **75:** 521–538.
5. FATAH-ARDALANI, K., *et al.* 2000. More porous final gel structure obtained by interaction with Lys-plasminogen than with Glu-plasminogen. Blood Coag. Fibrinol. **11:** 1–8.
6. VILLANUEVA, G.B. & N. ALLEN. 1986. Acetylation of antithrombin III by aspirin. Semin. Thromb. Hæmost. **12:** 213–215.

Identification and Characterization of Five New Fibrinogen Gene Polymorphisms

ANDREW P. FELLOWES, STEPHEN O. BRENNAN, AND PETER M. GEORGE

Molecular Pathology Laboratory, Canterbury Health Laboratories, Christchurch, New Zealand

ABSTRACT: It is clear that plasma fibrinogen levels are strongly influenced by genetic factors. To date 14 polymorphic sites have been identified within the fibrinogen gene cluster, mainly by restriction fragment length polymorphism (RFLP) and single-stranded conformation polymorphism (SSCP) analyses. Since elevated plasma fibrinogen is an independent risk factor for cardiovascular disease, these and other polymorphisms are of practical interest in defining haplotypes that correlate with fibrinogen levels. Here, DNA sequencing of fibrinogen genes from four patients led to the identification of 17 variations from the published sequence. Nine of these occurred in all chromosomes sequenced and were considered to be errors in the published data. Of the remaining eight, five represented novel variations, three having been previously described. The population frequency of the five novel variations, together with six known polymorphisms, was estimated by genotyping 50 normal individuals at each locus. The five new variations were all found at polymorphic frequencies in this group. Two of these new polymorphisms, Bβ intron 2 and Bβ codon 159, belong to the Bβ linkage group defined by Behague *et al.*,[1] since their rare alleles occurred in complete concordance with the rare alleles of Bβ *Mnl* I and Bβ *Bcl* I. Calculation of pairwise linkage disequilibrium coefficients showed that the three remaining novel polymorphisms, Aα *Dde* I, Bβ *Hinf* I, and γ intron 9 exhibited linkage equilibrium with respect to all other loci examined, including nearby polymorphisms that are themselves in strong linkage disequilibrium. This data indicates that these polymorphisms occur randomly with respect to background haplotype, and suggests that they are mutational hot spots.

KEYWORDS: Alpha-fibrinogen; Beta-fibrinogen; Gamma-fibrinogen; Polymorphisms; Haplotype; Linkage disequilibrium.

INTRODUCTION

Fibrinogen biosynthesis increases dramatically after general trauma or stress, and the magnitude of this response is influenced by genetic factors.[2] To date 14 polymorphisms have been identified within the fibrinogen gene cluster (see FIGURE 1) and these are of practical interest in defining markers that correlate with fibrinogen levels and cardiovascular risk. Since strong linkage disequilibrium has been observed, particularly in the Bβ fibrinogen gene,[1] some population studies may be observing

---

Address for correspondence: Andrew P. Fellowes, Molecular Pathology Laboratory, Canterbury Health Laboratories, Christchurch, New Zealand. Voice: (64 3) 364 0551; fax: (64 3) 364 0545.

andrew.fellowes@chmeds.ac.nz

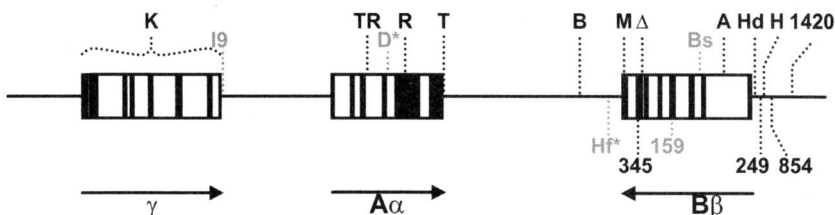

FIGURE 1. Schematic representation of the human fibrinogen gene cluster on 4q28 showing the location of common polymorphisms. Each gene is depicted as an *open box* with exons represented as *filled regions*. Arrows indicate the direction of transcription. Polymorphic sites are denoted by K (*Kpn* I, precise location unknown);[20] I9 (γ intron 9 C→T); TR (tetranucleotide repeat);[21] D* (*Dde* I); R (*Rsa* I);[7] T (*Taq* I);[8] B (*Bcl* I);[8] Hf* (*Hinf* I); M (*Mnl* I);[7] 345 (Bβ345Y→Y);[1] Δ (Bβ intron 6 4-bp insertion/deletion);[1] 159 (Bβ159S→S); Bs (*BsmA* I); A (*Ava* II);[8] Hd (*Hind* III);[8] 249 (Bβ–249 C→T);[1] H (*Hae* III);[8] 854 (Bβ–854 G→A);[1] 1420 (Bβ–1420 G→A).[1] Polymorphisms in *gray* were identified in this study, those followed by an *asterisk* may be mutation hot spots.

linkage between the polymorphism studied and other functional determinants of fibrinogen output.[3–5]

In the study described here, we characterized several novel sequence variations noted during the course of investigations into the molecular basis of new fibrinogen variants. We demonstrate that nine of these are probably due to errors in the published sequence, whereas five are novel polymorphisms. Two novel polymorphisms are completely associated with a previously identified haplotype, however the remaining three exhibited linkage equilibrium with all existing polymorphisms. Given the low rate of recombination seen in the fibrinogen gene cluster, these may represent mutational hot spots.

METHODS

Variations from the published Genbank sequence entries M64982 (Aα), M64983 (Bβ), and M10014 (γ) were originally identified by cycle sequencing of PCR fragments amplified from every fibrinogen exon from four unrelated individuals. Subsequent genotyping of 50 unrelated control subjects was performed by PCR and restriction enzyme digestion as described elsewhere.[6–8] The Aα *Dde* I polymorphism was amplified with the primers 3137α (acagcatatccagcttctgca) and 3283α (gtagacactcagtgcataactat) and digested directly with *Dde* I. The γ intron 9 polymorphism was amplified with the primers 7819γ (gatttgtagaaaaattactgtta) and 7952γ (ccattgaaggctaaatgtcc) and digested directly with *Hpa* I.

The degree of allelic association was estimated by calculating the linkage disequilibrium coefficient (D) for selected pairs of loci:

$$D = P_{AB} - p_A p_B$$

where P_{AB} is the observed frequency of the AB haplotype (the most common allele at each of loci A and B), p_A is the frequency of the common allele at locus A, and p_B is the frequency of the common allele at locus B.[9]

RESULTS

Nine of the variations identified occurred in all eight patient chromosomes sequenced and were, therefore, considered to represent errors in the published sequence (see TABLE 1). One of these differences, γ796 T→A, has been identified previously[10] and has recently been redescribed.[11] In that study, a total of five variations were noted, all of which were identified in our study. In agreement with our interpretation, these variations were classified by the authors as errors in the Genbank entry.

We found an additional eight heterozygous variations (TABLE 1). Five of these were new and their population frequency was estimated by genotyping 50 control

TABLE 1. Genbank entry variations[a]

Location	Variation	Other Details	Frequency	Designation
Aα exon 1	32 A→G	−11L→L	8/8	error
Aα exon 1	35 T→C	−10V→V	8/8	error
Aα intron 1	272–273 ins A		8/8	error
Aα intron 1	274–275 ins C		8/8	error
Aα intron 3	1918 T→G		8/8	error
Aα intron 4	3198–3200 GAG→AGA		8/8	error
Aα intron 4	3212 T→C	*Dde* I	10/100	polymorphism
Bβ intron 2	3980 G→C	*BsmA* I	20/100	polymorphism
Bβ exon 4	5580 C→T	159S→S	20/100	polymorphism
Bβ intron 6	7404–7405 del GA		8/8	error
Bβ intron 6	7515–7519 del GTTT		20/100	polymorphism
Bβ exon 7	7594 C→T	345Y→Y	—	polymorphism
Bβ exon 8	8520 G→A	448R→K, *Mnl* I	20/100	polymorphism
Bβ 3′ UTR	8986 A→C	*Hinf* I	32/100	polymorphism
γ intron 1	206–7 TC→CT		8/8	error
γ exon 4	796 T→A	88I→K	8/8	error
γ intron 9	7843 C→T		22/100	polymorphism

[a]Variations from the Genbank entries M64982 (Aα), M64983 (Bβ), and M10014 (γ) identified in four sequenced fibrinogen patients, and their frequency in 50 control individuals. Novel polymorphisms are *shaded*. Nucleotide numbering corresponds to the distance from the major site of transcription initiation for each gene,[22–24] whereas amino acid numbering corresponds to residues in the mature protein.

individuals at each locus. In addition, the genotypes for six known polymorphisms (Aα *Rsa* I,[7] Aα *Taq* I,[8] Bβ *Hind* III,[8] Bβ intron 6 insertion/deletion,[1] Bβ *Mnl* I,[7] and Bβ *Bcl* I[8]) were also determined in this group.

Three of the polymorphisms found in our patients have been described previously. The Bβ intron 6 insertion/deletion, Bβ codon 345 polymorphism, and Bβ *Mnl* I polymorphism were reported as being in complete linkage disequilibrium in the ECTIM study.[1] In the present study, these three polymorphisms were identified in one of the four original individuals. Analysis of 50 controls confirmed their complete association and established a rare allele frequency of 0.20, very close to that established previously. The Bβ codon 345 polymorphism was not genotyped in this group, however, because its complete association with Bβ *Mnl* I seems well established.

Two of the new polymorphisms identified during this study also appear to belong to this linkage group. In the original patient, and in all 50 control individuals, the common alleles of the Bβ *Bsm*A I and Bβ codon 159 polymorphisms always occurred together with the common alleles of the Bβ intron 6 insertion/deletion and Bβ *Mnl* I. These polymorphisms were also in complete association with Bβ *Hind* III, and in very strong association with Bβ *Bcl* I, with only one individual showing recombination between Bβ *Mnl* I and Bβ *Bcl* I. This is reflected in a high positive D value for this pair of loci in TABLE 2, and confirms earlier findings of strong linkage disequilibrium across the Bβ gene.[1]

The Aα *Dde* I polymorphism exhibits linkage equilibrium with all other loci examined including the nearby Aα *Rsa* I, and Aα *Taq* I polymorphisms, which are themselves in strong linkage disequilibrium (TABLE 2). The same is true of the Bβ *Hinf* I polymorphism. This site is situated between the strongly associated Bβ *Mnl* I and Bβ *Bcl* I polymorphisms, and yet exhibits little association with either, as evidenced by D values close to zero. Given the low overall rate of recombination in this region, this data indicates that both Aα *Dde* I and Bβ *Hinf* I occur randomly with respect to background haplotype, and suggests that they may be mutational hot spots. Although the γ intron 9 polymorphism also exhibits linkage equilibrium with the three Aα and three Bβ loci examined, this may simply reflect recombination in the γ/Aα interval.

TABLE 2. Pairwise linkage disequilibrium coefficients between selected polymorphisms[a]

	Aα *Dde* I	Aα *Rsa* I	Aα *Taq* I	Bβ *Bcl* I	Bβ *Hinf* I	Bβ *Mnl* I
γ intron 9	0.0664	−0.0487	−0.0648	−0.0249	−0.0393	−0.0277
Aα *Dde* I		−0.0107	−0.0239	0.0108	−0.0118	0.0090
Aα *Rsa* I			0.1680	−0.0122	−0.0726	−0.0072
Aα *Taq* I				−0.0334	−0.0540	−0.0324
Bβ *Bcl* I					−0.0464	0.1520
Bβ *Hinf* I						0.0445

[a]Estimated in 50 normal individuals. *Boxed values* are significant ($p < 0.001$).

DISCUSSION

Because Bβ chain production is considered rate limiting for assembly and secretion of mature fibrinogen,[12] most studies have focused on polymorphisms within the Bβ gene. Data obtained in the limited study performed here confirms and extends the Bβ haplotype data obtained from the ECTIM study.[1] This revealed strong linkage disequilibrium across the entire Bβ gene, including the Bβ promoter, and defined two ancestral Bβ haplotypes. Association of alleles of the rare ($f = 0.19$) haplotype with increased plasma fibrinogen[13] and risk of peripheral atherosclerosis[14] suggests that one or more of the constituent polymorphisms directly affects fibrinogen expression. Recent studies suggest that polymorphisms −455 and −854 from the transcription start site control the level of Bβ transcript by influencing the binding of transcriptional repressors to the Bβ promoter.[15] These findings support the large number of studies that associate the −455 G/A polymorphism with plasma fibrinogen levels,[16] disease risk,[17] and disease progression.[18] The possibility remains, however, that other polymorphisms of the rare Bβ haplotype also functionally contribute to raised fibrinogen.

Of the Bβ polymorphisms identified in this study, neither Bβ codon 159 nor Bβ *Bsm*A I are likely to function in this manner unless they affect mRNA processing. Currently, this possibility cannot be excluded. The Bβ *Hinf* I polymorphism in the Bβ 3′ untranslated region has the potential to influence mRNA stability by analogy to a similar mutation in the prothrombin gene.[19] However, although its association with fibrinogen levels has not been examined, it is clear from the linkage analysis undertaken here that Bβ *Hinf* I is not strongly associated with other loci linked to high fibrinogen levels.

REFERENCES

1. BEHAGUE, I., *et al.* 1996. β fibrinogen gene polymorphisms are associated with plasma fibrinogen and coronary artery disease in patients with myocardial infarction. The ECTIM Study. Etude Cas-Temoins sur l'Infarctus du Myocarde. Circulation **93**: 440–449.
2. GARDEMANN, A., *et al.* 1997. Positive association of the β fibrinogen H1/H2 gene variation to basal fibrinogen levels and to the increase in fibrinogen concentration during acute phase reaction but not to coronary artery disease and myocardial infarction. Thromb. Hæmost. **77**: 1120–1126.
3. SCHMIDT, H., *et al.* 1998. β-fibrinogen gene polymorphism (−148 C→T) is associated with carotid atherosclerosis—results of the Austrian stroke prevention study. Arterioscler. Thromb. Vasc. Biol. **18**: 487–492.
4. CARTER, A.M., *et al.* 1997. Gender-specific associations of the fibrinogen Bβ 448 polymorphism, fibrinogen levels, and acute cerebrovascular disease. Arterioscler. Thromb. Vasc. Biol. **17**: 589–594.
5. ZITO, F., *et al.* 1997. *Bcl* I polymorphism in the fibrinogen β-chain gene is associated with the risk of familial myocardial infarction by increasing plasma fibrinogen levels: a case-control study in a sample of GISSI-2 patients. Arterioscler. Thromb. Vasc. Biol. **17**: 3489–3494.
6. BRENNAN, S.O., *et al.* 2000. Hypofibrinogenemia in an individual with two coding (γ82 A→G and Bβ235 P→L) and two non-coding mutations. Blood **95**: 1709–1713.
7. BAUMANN, R.E. & A.H. HENSCHEN. 1994. Linkage disequilibrium relationships among four polymorphisms within the human fibrinogen gene cluster. Hum. Genet. **94**: 165–170.

at either of these times. However, neutrophil counts were significantly different between the two groups at day five with the $FG^{-/-}$ mice showing a threefold increase over that of WT (see FIGURE 3). Interestingly, neutrophil levels at day five in BAL from $FG^{-/-}$ mice were unchanged relative to day three. It is unclear as to whether this is due to increased migration, proliferation, or diminished clearance of these cells. This observation is currently under investigation.

CONCLUSIONS

A number of clinical correlates have implicated fibrin(ogen) in facilitating the development of pulmonary fibrosis.[7,11–15] Using a model of bleomycin induced pulmonary fibrosis, the data show that fibrin(ogen) is not required for the formation of fibrotic lesions in acute lung injury/repair. This result indicates that enhanced fibrosis seen in $PG^{-/-}$ mice, and mice that overexpress PAI-1, may not necessarily be a consequence of unresolved pulmonary fibrin deposits, but could be due to other mechanisms, such as the inability to resolve other matrix proteins directly or to activate MMPs as a result of loss of plasmin activity. Alternatively, fibrin persistence may still facilitate the developing fibrotic process seen in the $PG^{-/-}$ and PAI-1 overexpressing mice, but other mechanisms may compensate when there is a loss of fibrin. Experiments with a $FG^{-/-}/PG^{-/-}$ animals should aid in further elucidating the relevance of these proteins in pulmonary fibrotic lesion formation.

REFERENCES

1. BROWN, L.F., N. LANIR, J. MCDONAGH, et al. 1993. Fibroblast migration in fibrin gel matrices. Am. J. Pathol. **142:** 273–283.
2. LANGUINO, L.R., A. DUPERRAY, K.J. JOGANIC, et al. 1995. Regulation of leudocyte-endothelium interaction and leukocyte transendothelial migration by intracellular adhesion molecule 1-fibrinogen recognition. Proc. Natl. Acad. Sci. U.S.A. **92:** 1505–1509.
3. ROKITA, H., R. NETA & J.D. SIPE. 1993. Increased fibrinogen synthesis in mice during the acute phase response: co-operative interaction of interleukin 1, interleukin 6, and interleukin 1 receptor antagonist. Cytokine **5:** 454–458.
4. KOTANI, I., A. SATO, H. HAYAKAWA, et al. 1995. Increased procoagulant and antifibrinolytic activities in the lungs with idiopathic pulmonary fibrosis. Thromb. Res. **77:** 493–504.
5. SIKIC, B.I. 1986. Biochemical and cellular determinants of bleomycin cytotoxicity. Cancer Surv. **5:** 81–91.
6. GUNTHER, A., P. MOSAVI, C. RUPPERT, et al. 2000. Enhanced tissue factor pathway activity and fibrin turnover in the alveolar compartment of patients with interstitial lung disease. Thromb. Hæmost. **83:** 853–860.
7. OLMAN, M.A., W.L. SIMMONS, D.J. POLLMAN, et al. 1996. Polymerization of fibrinogen in murine bleomycin-induced lung injury. Am. J. Pathol. **271:** L519–526.
8. SWAISGOOD, C.M., E.L. FRENCH, C. NOGA, et al. 2000. The development of bleomycin-induced pulmonary fibrosis in mice deficient for components of the fibrinolytic system. Am. J. Pathol. **157:** 177–187.
9. EITZMAN, D.T., R.D. MCCOY, X. ZHENG, et al. 1996. Bleomycin-induced pulmonary fibrosis in transgenic mice that either lack or overexpress the murine plasminogen activator inhibitor-1 gene. J. Clin. Invest. **97:** 232–237.
10. PLOPLIS, V.A., J.A. WILBERDING, L. MCLENNON, et al. 2000. A total fibrinogen deficiency is compatible with the development of pulmonary fibrosis in mice. Am. J. Pathol. **157:** 703–708.

11. CHAPMAN, H.A., C.L. ALLEN & O.L. STONE. 1986. Abnormalities in pathways of alveolar fibrin turnover among patients with interstitial lung disease. Am. Rev. Respir. Dis. **133:** 437–444.
12. CROUCH, E. 1990. Pathobiology of pulmonary fibrosis. Am. J. Physiol. **259:** L159–L184.
13. IDELL, S., K.K. JAMES, C. GILLIES, *et al.* 1989. Abnormalities of pathways of fibrin turnover in lung lavage of rats with oleic acid and bleomycin-induced lung injury support alveolar fibrin deposition. Am. J. Pathol. **135:** 387–399.
14. KUHN, C., III, C.J. BOLDT, T.E. KING JR., *et al.* 1989. An immunohistochemical study of architectural remodeling and connective tissue synthesis in pulmonary fibrosis. Am. Rev. Respir. Dis. **140:** 1693–1703.
15. SITRIN, R.G., P.G. BRUBAKER & J.C. FANZONE. 1987. Tissue fibrin deposition during acute lung injury in rabbits and its relationship to local expression of procoagulant and fibrinolytic activities. Am. Rev. Respir. Dis. **135:** 930–938.

Fibrinogen Polymorphisms and Atherothrombotic Disease

F.R. GREEN

Department of Cardiovascular Medicine, University of Oxford,
Wellcome Trust Centre for Human Genetics,
Roosevelt Drive, Headington, Oxford, United Kingdom

ABSTRACT: Common polymorphisms of the fibrinogen gene cluster are associated with circulating fibrinogen level and with susceptibility to and/or severity of atherothrombotic disease. The frequencies of the polymorphisms vary among different ethnic groups but there is strong linkage disequilibrium at the β-fibrinogen gene locus so that, in caucasian populations, there are only four common β-fibrinogen haplotypes. One of these haplotypes, defined by the β-fibrinogen −455A allele, is associated with elevated fibrinogen level and increased risk of atherothrombotic disease. The molecular mechanism of these associations is currently under investigation.

KEYWORDS: Fibrinogen polymorphism; Atherothrombotic disease.

INTRODUCTION

We know from twin studies that there is a measurable genetic contribution to at least one disease that has atherothrombosis as its cause, that is myocardial infarction.[1] It is now very well accepted that the circulating level of fibrinogen is prospectively associated with risk of almost all manifestations of atherothrombosis, including myocardial infarction, coronary artery disease, peripheral artery disease, and stroke. The precise role of fibrinogen in the disease is far from understood, although many of its normal functions could contribute to the progression of atherosclerosis and the formation of thrombus, if fibrinogen is present to excess or in a variant form. Fibrinogen is an acute phase reactant whose level naturally rises about 2–4-fold in response to trauma and/or inflammation and, as such, is implicated as a marker of preexisting atherothrombotic disease, as well as in the molecular pathology of the disease itself. Fibrinogen genetic variation is, therefore, a candidate for determining interindividual variability in circulating fibrinogen level and/or function and in its response to stimulation by a variety of agents. In turn, this genetic variation may be one of the several genetic factors that determine individual susceptibility to atherothrombotic disease.

Address for correspondence: F.R. Green, Ph.D., Department of Cardiovascular Medicine, University of Oxford, Wellcome Trust Centre for Human Genetics, Roosevelt Drive, Headington, Oxford, OX3 7BN, U.K. Voice: +44 1865 287582; fax: +44 1865 287599.
fionag@well.ox.ac.uk

DESCRIPTION OF FIBRINOGEN POLYMORPHISMS

There are numerous polymorphisms across the fibrinogen locus (see FIGURE 1), with several in or close to the β-fibrinogen gene. This apparent clustering is probably simply because this has been the most widely studied of the three genes, the reason for this being that, in humans, synthesis of the β-fibrinogen polypeptide chain is thought to be rate limiting in the production of the mature molecule.[2,3]

The frequencies of these polymorphisms in various apparently healthy population samples are shown in TABLE 1, together with references to the original papers. For the β-fibrinogen polymorphisms, some non-caucasian population samples (Japanese and Inuit) tend to have a lower rare allele frequency than the European and US caucasian population samples, whereas the Chinese and Korean samples seem rather more like the caucasian samples. The β-fibrinogen polymorphisms are essentially absent from the Peruvian Quechua sample, the low frequency of rare alleles probably accounted for by European admixture.[4] For the α-fibrinogen polymorphisms, the Inuit have a much higher rare allele frequency, as do the Finns (who are a genetic isolate), and the Koreans.

Polymorphisms within the β-fibrinogen locus show strong linkage disequilibrium with each other, but little if any with those at the adjacent α-fibrinogen locus (see TABLE 2). This probably reflects the genetic history of the locus, with population bottlenecks and random genetic drift resulting in only a limited number of the possible haplotypes. The small number of common haplotypes are very unlikely to be caused by classical Darwinian selection since these fibrinogen polymorphisms do not affect survival until long after reproductive age, if at all. The particular combinations of individual polymorphism genotypes that comprise a haplotype may coincidentally be functionally cooperative.

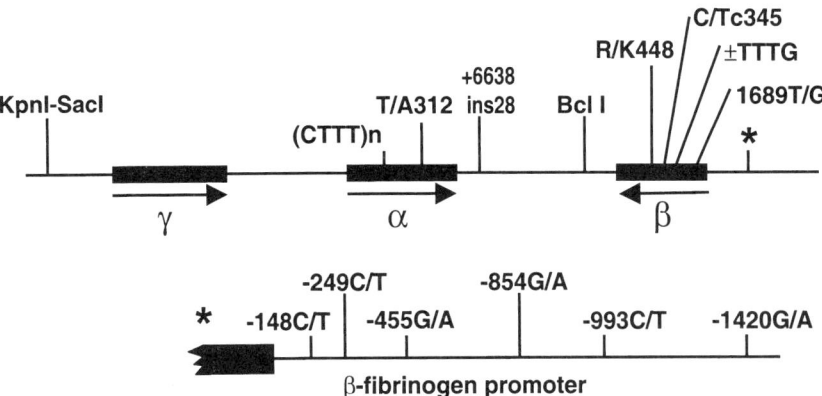

FIGURE 1. The figure shows common polymorphisms of the three fibrinogen genes, with arrows indicating direction of transcription of the genes. The promoter region of the β fibrinogen gene is shown in more detail underneath. The locus is on chromosome 4q28.

TABLE 1. Frequency of fibrinogen rare alleles in control population samples

Gene	Polymorphism	Population Description	Rare Allele Frequency
β	−1420G/A	European caucasian controls[9]	0.21
	−854G/A	European caucasian controls[9,18]	0.15, 0.19
		European caucasian controls[9,10,18–27]	0.17–0.27
		US caucasian controls[20,23]	0.17–0.26
	−455G/A	**US Japanese controls**[23]	**0.12**
	(HaeIII)	**Chinese controls**[28]	**0.27**
		Inuit controls[29]	**0.11**
		Peruvian Quechua controls[4]	**0.02**
	−249C/T	European caucasian controls[9,18]	0.19, 0.20
		European caucasian controls[7,18,21]	0.19, 0.21
	−148C/T	US caucasian controls[20,23]	0.17, 0.26
	(HindIII/AluI)	**US Japanese controls**[23]	**0.12**
		Peruvian Quechua controls[4]	**0.02**
	+1689T/G	US caucasian controls[30]	0.22
	(AvaII)	European caucasian controls[12,21]	0.26, 0.28
		European caucasian controls[9,31]	0.18, 0.19
	R/K448	US caucasian controls[32]	0.15
	(MnlI)	**Korean controls**[33]	**0.14**
		Peruvian Quechua controls[4]	**0.02**
		European caucasian controls[9,12,21,34]	0.17–0.25
	BclI	US caucasian controls[23,30]	0.20, 0.22
		US Japanese controls[23]	**0.13**
		Inuit controls[29]	**0.12**
α	+6638ins28	91 UK controls[9,12,21]	0.25–0.32
	(TaqI)	**Inuit controls**[29]	**0.47**
		US caucasian controls[32]	0.24
	T/A312	European caucasian controls[13–15]	0.23–0.24
	(RsaI)	**Korean controls**[33]	**0.38**
		Finnish controls[35]	**0.33**
γ	KpnI-SacI	US caucasian controls[30]	0.18

Note: Bold type indicates non-caucasian or genetic isolate.

TABLE 2. Linkage disequilibrium relationships among polymorphisms at the fibrinogen locus

Gene		β						α		γ
Gene	Polymorphism	−854G/A	−455G/A	−249C/T	−148C/T	R/K448	BclI	+6638ins28	T/A312	KpnI-SacI
β	−1420G/A	**−0.92**[9]	**0.93**[9]	**−0.92**[9]	**0.93**[9]	**0.97**[9]	**0.98**[9]	**−0.53**[9]	…	…
	−854G/A		**−0.87**[9]	**−0.85**[9]	**−0.87**[9]	**−1.00**[9]	**−1.00**[9]	**−0.87**[9]	…	…
	−455G/A (HaeIII)			**−0.96**[9]	**1.00**[20]	**1.00**[20]	0.46[36]	**−0.17**[36]	−0.58[20]	**−0.19**[36]
					0.97[23]		0.85[21]	−0.09[21]		
					1.00[23]*	**0.96**[9]	**0.96**[9]	−0.53[9]		
							0.91[29]**	**−0.32**[29]**		
							0.85[21]	−0.09[21]		
	−249C/T				**−0.96**[9]	**−1.00**[9]	**−1.00**[9]	**0.82**[9]		
	−148C/T (HindIII/AluI)					**1.00**[20]	**0.96**[9]	**−0.53**[9]	−0.58[20]	
	+1689T/G (AvaII)						0.96[12]	−0.09[12]		0.15[30]
							0.75[30]	0.08[30]		
	R/K448 (MnlI)						**0.95**[9]	**−0.47**[9]	−0.44[20]	…

TABLE 2/continued.

Gene	Polymorphism	β						α		γ
		−854G/A	−455G/A	−249C/T	−148C/T	R/K448	BclI	+6638ins28	T/A312	KpnI-SacI
β	BclI							*−0.16*[12]	...	*0.13*[30]
								−0.07[30]		−0.15[36]
								−0.14[36]		
								0.01[21]		
								−0.37[9]		
								−0.29[29]**		
								0.01[21,30]		
α	+6638ins28 (TaqI)								**0.95**[13]	**−0.88**[30]
										0.70[36]

Note: Some authors quoted Δ[12,21,29,30,36] and some quoted D′[9,13,20,23] for linkage disequilibrium coefficients. Bold type, $p < 0.001$; normal type, $p < 0.01$; italic type, p ns. *Japanese, **Inuit, otherwise US/European caucasian controls.

TABLE 3. Common fibrinogen haplotypes

β-1420	β-854	β-455	β-249	β-148	β+1689	β448	βBclI		
G	G	G	C	C	T	R	1	~40%	(a)
G	A	G	C	C	T	R	1	~20%	(b)
G	G	G	T	C	T	R	1	~20%	(c)
A	G	A	C	T	G	K	2	~20%	(d)

According to the linkage disequilibrium coefficients shown in TABLE 2, there are probably four common fibrinogen haplotypes in the caucasian population (see TABLE 3).

ASSOCIATION OF FIBRINOGEN GENETIC VARIATION WITH FIBRINOGEN LEVEL

There is now considerable evidence that fibrinogen concentration shows a moderate degree of genetic heritability (see TABLE 4). The absolute values for genetic heritability of plasma fibrinogen level vary between about 20 and 50%, but this may be a property of the particular populations tested and perhaps reflects differences in the statistical methods applied.

There are now many studies consistently showing an association between possession of the (d) haplotype of the β fibrinogen gene (TABLE 3) (or the appropriate allele of one of the individual β fibrinogen polymorphisms that make up haplotype (d)) and elevated fibrinogen level, some examples of which are given in TABLE 5. There are also studies where no significant association was observed, two of which also showed low heritability.[5,6] This might indicate a large environmental influence on fibrinogen level in these populations, swamping any genetic effect.

There are data suggesting the existence of gene–environment interactions between the β fibrinogen polymorphisms and factors such as age, sex, smoking habit, and

TABLE 4. Heritability of fibrinogen level

Family structure	Fibrinogen genetic heritability	Reference
Families	51%	Hamsten et al., 1987[37]
Twins	27%	Berg & Kierulf, 1989[6]
Twins	30%	Reed et al., 1994[5]
Families	20–48%	Friedlander et al., 1995[38]
Families	39%	Livshits et al., 1996[39]
Families	34%	Pankow et al., 1998[40]
Twins	44%	de Lange et al., 2001[42]

TABLE 5. Association of fibrinogen polymorphisms with circulating fibrinogen level

β-Fibrinogen Polymorphism	Population Description (all control groups)	Mean (SD/SEM/95%CI) Fibrinogen (g/L) by genotype[a]		
		11	12	22
−854G/A	Swedish ≤ 55 yrs[18] $p < 0.0001$	2.7(0.5)	2.9(0.4)	3.2(0.2)
	Swedish ≤ 45 yrs[19] $p \leq 0.01$	2.9(2.7–3.0)	3.3(3.0–3.5)	
−455G/A	Italian[27] $p < 0.001$	2.8	3.0	
	Scottish[10] $p \leq 50.05$	3.2(0.1)	3.3(0.1)	3.7(1.0)
	Swedish ≤ 50 yrs[18] $p < 0.0001$	2.7(0.4)	2.9(0.5)	3.1(0.3)
−148C/T	US[23] $p = 0.03$	2.9(0.1)	3.1(0.1)	
R/K448	UK[31] $p = 0.01$	3.1(3.0–3.2)	3.4(3.2–3.7)	3.5(2.9–4.2)
BclI	UK[12] $p < 0.25$	2.7(0.5)	3.0(0.6)	3.7(0.7)
	UK[41] $p < 0.004$	2.7(0.6)	3.0(0.8)	3.0(0.6)

[a]11 represents homozygotes for the common allele, 12 represents heterozygotes, and 22 homozygotes for the rare allele for each polymorphism.

acute phase reactions in determining fibrinogen levels, but these data are not consistent. Once again, the statistical detection of such effects may depend on the balance between environmental and genetic impact on fibrinogen level in any given population. This is a disadvantage of population based cross-sectional studies; intervention studies would be more powerful in detecting gene–environment interactions, but these have their own difficulties and no single method offers us the complete answer.

The other major problem with population based association studies is that spurious associations can be generated through population stratification. One way to avoid this is to analyze the relationship between genotype and levels in families, by linkage and by association. We have been able to show, in such family based studies, significant evidence for association between plasma fibrinogen concentration and β fibrinogen −455G/A polymorphism genotype, as well as formally excluding population stratification as a cause for the association (Mayosi, et al., manuscript in preparation). This family study confirms the population based studies that use a very different method, and shows that the associations detected by both approaches are real and not due to population stratification.

ASSOCIATION BETWEEN FIBRINOGEN POLYMORPHISMS AND ATHEROTHROMBOTIC DISEASE

Several studies have shown an association between fibrinogen genotype and severity and/or presence of atherothrombotic disease. The Austrian Stroke Prevention study found that homozygosity for the β-148T allele was associated with a sixfold increased risk for carotid atherosclerosis (odds ratio 6.2, 95%CI 1.7–22.4).[7] The REGRESS study found a higher proportion of patients with three vessel coronary disease at baseline in β-455A homozygotes. After treatment with pravastatin, these patients showed regression of disease, whereas their untreated counterparts showed increased progression, compared to the other genotypes, so that the β-455A homozygotes showed a significantly better response to pravastatin.[8] Similarly in ECTIM, a significantly higher proportion of the β-455A allele carriers had three vessel coronary disease, compared with the β-455G homozygotes.[9] Homozygosity for the β-455A allele is also associated with an approximately twofold increase in risk of peripheral artery disease, as shown in the Edinburgh Artery Study[10] and earlier work on the same study,[11] suggested that homozygosity for the rare allele of the BclI polymorphism might have a similar association. All these alleles reside together on the (d) haplotype, and it is impossible to dissect them genetically unless we can find populations, in which one of the polymorphisms does not segregate. Even the Peruvian Quechua cannot help in this respect, since they do not seem to have this allele at all. The Quechua live at high altitude and need a high hematocrit to obtain sufficient oxygen, with the consequence that blood viscosity is increased. Rupert and colleagues[4] decided to investigate the Quechua fibrinogen genotype on the basis that perhaps under such circumstances, genotypes associated with higher fibrinogen level, leading to higher plasma viscosity, might not be favored. What they found was that the (d) haplotype was essentially absent from the Quechua[4] and it will be very interesting to see what genotype they are at some of the other fibrinogen polymorphic sites, and how this might relate to fibrinogen level. In other ethnic groups studied, there seems to be a trend for those at lower risk of atherothrombotic disease to have a lower frequency of haplotype (d) associated alleles—for example, the Japanese and the Inuit—but the Chinese and Korean samples did not show this trend (TABLE 1).

It had been suggested that the α-fibrinogen polymorphisms may be weakly associated with fibrinogen level,[12] but more recent data from ECTIM refute this.[9,13] On the other hand, Carter and colleagues showed a significant association between the α-fibrinogen A312 allele and increased poststroke mortality following atrial fibrillation and suggested that this might be the result of an increased tendency of the clot formed from A312 fibrinogen to embolise.[14] More recently, the same group found a higher frequency of the A312 allele among patients with pulmonary embolus, compared to those with deep vein thrombosis, and an interaction with the factor XIIIA V/L34 polymorphism, suggesting that differential susceptibility to factor XIIIA mediated crosslinking might be the mechanism.[15] We found an association of the A312 allele with less permeable fibrin clots formed *in vitro*, but this was only in patients who had suffered a myocardial infarction, not in the control group, so it may be that other factors induced as a result of myocardial infarction are required for "expression" of the genotype specific effect (gene–environment interaction).[15a]

FUNCTIONAL STUDIES

Genetics cannot give us the complete answer about which, if any, of the fibrinogen polymorphisms are functional because of the strong linkage disequilibrium at the locus. Consequently, several groups have investigated the potential function of polymorphisms in the promoter of the β fibrinogen gene. We showed that allele specific binding of a liver nuclear protein took place in the region of the β-455G/A polymorphism only in the presence of a G at −455.[16] Brown and Fuller demonstrated preferential binding of a 47-kD nuclear protein to this site in the presence of the −455G allele.[17] They also showed that constructs containing the −455A allele had stronger interleukin 6 stimulated promoter activity on transfection into HepG2 human hepatoma cells, consistent with the epidemiological association of the (d) haplotype with higher circulating fibrinogen. These constructs appropriately differed only at the −455 position; the −455G construct was made from the −455A construct by site directed mutagenesis, which means that the −455/−249/−148 haplotype of the −455G construct is probably G/C/T. Had it been obtained directly from a PCR product, that is, the natural haplotype, it would have been G/C/C, because of the linkage disequilibrium at the locus. It is interesting to speculate about whether the naturally occurring haplotype would give a different result. In our hands, constructs containing a T at −148 tend to have a lower transcriptional activity (Wragg and Green, unpublished data). Another group has performed similar studies, confirming these data under basal conditions and extending them to suggest that the −854G/A polymorphism may also be functional, with the −854A allele associated with enhanced promoter activity, again consistent with the epidemiological data.[18]

CONCLUSIONS

In conclusion, the data show that polymorphism(s) in linkage disequilibrium with the β fibrinogen promoter −455A allele, the (d) haplotype, are associated with higher plasma fibrinogen level and with atherothrombotic disease. However, the biological mechanism(s) involved are as yet not clear, although the −455A and −854A sites are implicated from *in vitro* studies. It may be that other polymorphisms in 5′-flanking region, known or as yet unidentified, contribute to the association, or it may be that polymorphisms in the coding or 3′-flanking regions are implicated. Once the causal relationships between polymorphisms and fibrinogen level and/or function are clarified, it will be possible to gain a better understanding of the observed association of these polymorphisms with atherothrombotic disease itself.

REFERENCES

1. MARENBERG, M.E. *et al.* 1994. Genetic susceptibility to death from coronary heart disease in a study of twins. N. Engl. J. Med. **330:** 1041–1046.
2. ROY, S.N., G. MUKHOPADHYAY & C.M. REDMAN. 1990. Regulation of fibrinogen assembly. Transfection of Hep G2 cells with B beta cDNA specifically enhances synthesis of the three component chains of fibrinogen. J. Biol. Chem. **265:** 6389–6393.
3. YU, S., *et al.* 1983. Intracellular assembly of human fibrinogen. J. Biol. Chem. **258:** 13407–13410.
4. RUPERT, J.L., *et al.* 1999. Beta-fibrinogen allele frequencies in Peruvian Quechua, a high-altitude native population. Am. J. Phys. Anthropol. **109:** 181–186.
5. REED, T., R.P. TRACY & R.R. FABSITZ. 1994. Minimal genetic influences on plasma fibrinogen level in adult males in the NHLBI twin study. Clin. Genet. **45:** 71–77.
6. BERG, K. & P. KIERULF. 1989. DNA polymorphisms at fibrinogen loci and plasma fibrinogen concentration. Clin. Genet. **36:** 229–235.
7. SCHMIDT, H., *et al.* 1998. Beta-fibrinogen gene polymorphism (C148→T) is associated with carotid atherosclerosis: results of the Austrian Stroke Prevention Study. Arterioscler. Thromb. Vasc. Biol. **18:** 487–492.

8. DE MAAT, M.P.M., *et al.* 1998. The −455G/A polymorphism of the beta-fibrinogen gene is associated with the progression of coronary atherosclerosis in symptomatic men. Atheroscl. Thromb. Vasc. Biol. **18:** 265–271.
9. BEHAGUE, I., *et al.* 1996. Beta fibrinogen gene polymorphisms are associated with plasma fibrinogen and coronary artery disease in patients with myocardial infarction. The ECTIM Study. Etude Cas-Temoins sur lInfarctus du Myocarde. Circulation **93:** 440–449.
10. LEE, A.J., *et al.* 1999. Fibrinogen, factor VII and PAI-1 genotypes and the risk of coronary and peripheral atherosclerosis: Edinburgh Artery Study. Thromb. Hæmost. **81:** 553–560.
11. FOWKES, F.G., *et al.* 1992. Fibrinogen genotype and risk of peripheral atherosclerosis. Lancet **339:** 693–696.
12. HUMPHRIES, S.E., *et al.* 1987. Role of genetic variation at the fibrinogen locus in determination of plasma fibrinogen concentrations. Lancet **1:** 1452–1455.
13. CURRAN, J.M., *et al.* 1998. The alpha fibrinogen T/A312 polymorphism in the ECTIM study. Thromb. Hæmost. **79:** 1057–1058.
14. CARTER, A.M., A.J. CATTO & P.J. GRANT. 1999. Association of the alpha-fibrinogen Thr312Ala polymorphism with poststroke mortality in subjects with atrial fibrillation. Circulation **99:** 2423–2426.
15. CARTER, A.M., *et al.* 2000. alpha-fibrinogen Thr312Ala polymorphism and venous thromboembolism. Blood **96:** 1177–1179.
15a. CURRAN, J.M., *et al.* 2001. A hypothesis to explain the reported association of the α-fibrinogen A312 allele with thromboembolic disease. Thromb. Hæmost. In press.
16. GREEN, F.R., A. LANE & S.E. HUMPHRIES. 1995. The effect of genetic variation in the beta fibrinogen 5~-flanking region on nuclear protein binding, transcriptional activity and plasma fibrinogen level. Thromb. Hæmost. **73:** 1224.
17. BROWN, E.T. & G.M. FULLER. 1998. Detection of a complex that associates with the Bbeta fibrinogen G-455- A polymorphism. Blood **92:** 3286–3293.
18. VAN I HOOFT, F.M., *et al.* 1999. Two common, functional polymorphisms in the promoter region of the beta-fibrinogen gene contribute to regulation of plasma fibrinogen concentration. Arterioscler Thromb. Vasc. Biol. **19:** 3063–3070.
19. GREEN, F., *et al.* 1993. The role of beta-fibrinogen genotype in determining plasma fibrinogen levels in young survivors of myocardial infarction and healthy controls from Sweden. Thromb. Hæmost. **70:** 915–920.
20. BAUMANN, R.E. & A.H. HENSCHEN. 1994. Linkage disequilibrium relationships among four polymorphisms within the human fibrinogen gene cluster. Hum. Genet. **94:** 165–170.
21. THOMAS, A., *et al.* 1994. Linkage disequilibrium across the fibrinogen locus as shown by five genetic polymorphisms, G/A-455 (HaeIII), C/T-148 (HindIII/AluI), T/G+1689 (AvaII), and BclI (beta-fibrinogen) and TaqI (alpha-fibrinogen), and their detection by PCR. Hum. Mutat. **3:** 79–81.
22. HUMPHRIES, S.E., *et al.* 1995. European Atherosclerosis Research Study: genotype at the fibrinogen locus (G-455-A beta-gene) is associated with differences in plasma fibrinogen levels in young men and women from different regions in Europe. Evidence for gender-genotype environment interaction. Arterioscler Thromb. Vasc. Biol. **15:** 96–104.
23. ISO, H., *et al.* 1995. Polymorphisms of the beta fibrinogen gene and plasma fibrinogen concentration in Caucasian and Japanese population samples. Thromb. Hæmost. **73:** 106–111.
24. GARDEMANN, A., *et al.* 1997. Positive association of the beta fibrinogen H1/H2 gene variation to basal fibrinogen levels and to the increase in fibrinogen concentration during acute phase reaction but not to coronary artery disease and myocardial infarction. Thromb. Hæmost. **77:** 1120–1126.
25. TYBJAERG HANSEN, A., *et al.* 1997. A common mutation (G-455→ A) in the beta-fibrinogen promoter is an independent predictor of plasma fibrinogen, but not of ischernic heart disease. A study of 9,127 individuals based on the Copenhagen City Heart Study. J. Clin. Invest. **99:** 3034–3039.

26. KESSLER, C., et al. 1997. The apolipoprotein E and beta-fibrinogen G/A-455 gene polymorphisms are associated with ischernic stroke involving large-vessel disease. Arterioscler Thromb. Vasc. Biol. **17:** 2880–2884.
27. MARGAGLIONE, M., et al. 1998. Fibrinogen plasma levels in an apparently healthy general population--relation to environmental and genetic determinants. Thromb. Hæmost. **80:** 805–810.
28. LAM, K.S., et al. 1999. Beta-fibrinogen gene G/A-455 polymorphism in relation to fibrinogen concentrations and ischaernic heart disease in Chinese patients with type 11 diabetes. Diabetologia. **42:** 1250–1253.
29. DE MAAT, M.P., et al. 1995. Gender-related association between beta-fibrinogen genotype and plasma fibrinogen levels and linkage disequilibrium at the fibrinogen locus in Greenland Inuit. Arterioscler Thromb. Vasc. Biol. **15:** 856–860.
30. MURRAY, J.C., et al. 1985. Linkage disequilibrium of RFLPs at the beta fibrinogen (FGB) and gamma fibrinogen (FGG) loci on chromosome 4. Cytogenet Cell Genet. **40:** 707–708.
31. CARTER, A.M., et al. 1997. Association of the platelet PI(A) polymorphism of glycoprotein IIb/111a and the fibrinogen Bbeta 448 polymorphism with myocardial infarction and extent of coronary artery disease [see comments]. Circulation **96:** 1424–1431.
32. BAUMANN, R.E. & A.H. HENSCHEN. 1993. Human fibrinogen polymorphic. site analysis by restriction endonuclease digestion and allele-specific polymerase chain reaction amplification: identification of polymorphisms at positions A alpha 312 and B beta 448. Blood **82:** 2117–2124.
33. LEE, D.E., et al. 1999. Fibrinogen gene polymorphism in a non-Caucasian population. Clin. Biochem. **32:** 113–117.
34. ZITO, F., et al. 1997. Bcl I polymorphisin in the fibrinogen beta-chain gene is associated with the risk of familial myocardial infarction by increasing plasma fibrinogen levels. A case-control study in a sample of GISSI-2 patients. Arterioscler Thromb. Vasc. Biol. **17:** 3489–3494.
35. RAURAMAA, R., et al. 2000. The Rsal polymorphism. in the alpha-fibrinogen gene and response of plasma fibrinogen to physical training—a controlled randomised clinical trial in men. Thromb. Hæmost. **83:** 803–806.
36. CONNOR, J.M., et al. 1992. Genetic variation at fibrinogen loci and plasma fibrinogen levels. J. Med. Genet. **29:** 480–482.
37. HAMSTEN, A., et al. 1987. Genetic and cultural inheritance of plasma fibrinogen concentration. Lancet **2:** 988–991.
38. FRIEDLANDER, Y., et al. 1995. Genetic and environmental sources of fibrinogen variability in Israeli families: the Kibbutzim Family Study. Am. J. Hum. Genet. **56:** 1194–1206.
39. LIVSHITS, G., et al. 1996. Tel Aviv-Heidelberg three-generation offspring study: genetic determinants of plasma fibrinogen level. Am. J. Med. Genet. **63:** 509–517.
40. PANKOW, J.S., et al. 1998. Segregation analysis of plasminogen activator inhibitor- I and fibrinogen levels in the NHLBI family heart study. Arterioscler Thromb. Vasc. Biol. **18:** 1559–1567.
41. THOMAS, A.E., et al. 1991. Variation in the promoter region of the beta fibrinogen gene is associated with plasma fibrinogen levels in smokers and non-smokers. Thromb. Hæmost. **65:** 487–490.
42. DE LANGE, M., et al. 2001. The genetics of hæmostasis: a twin study. Lancet **357:** 101–105.

Fibrinogen and Its Degradation Products as Thrombotic Risk Factors

GORDON D.O. LOWE AND ANN RUMLEY

University Department of Medicine, Royal Infirmary, 10 Alexandra Parade, Glasgow, United Kingdom

ABSTRACT: Recent meta-analyses of prospective studies have shown that plasma levels of both fibrinogen and fibrin D-dimer are independent predictors of ischemic heart disease. Although at present reported studies using different assays do not show heterogeneity, there is a need for prospective comparison of different assays, as well as for development of standards. Collaborative development of the clinical use of these risk predictors is also required. Finally, the causal significance of the associations of fibrinogen and D-dimer with thrombotic events remains to be established by randomized controlled trials of reduction in their plasma levels.

KEYWORDS: Fibrinogen; Fibrin degradation products; Thrombosis

INTRODUCTION

The aim of this review is to estimate the predictive values of (1) plasma fibrinogen, and (2) plasma levels of fibrin degradation products (FDP), commonly measured as fibrin D-dimer, for thrombotic events: myocardial infarction (MI) or fatal ischaemic heart disease (IHD), stroke, and venous thrombosis. The influences of assay methodology are discussed, as are the biological and clinical implications. We concentrate on recent findings, updating our review of studies published up to 1999.[1]

FIBRINOGEN AS A RISK PREDICTOR

A recent systematic review of 18 prospective studies of plasma fibrinogen and IHD events observed that the risk ratio was 1.8 (95% CI 1.6–2.0) for individuals in the top third compared to individuals in the bottom third of baseline assays.[2] In most of these studies, adjustment had been made for potential confounders of the relationship between fibrinogen and IHD (e.g., age, smoking habit, and presence of baseline evidence of cardiovascular disease). The risk ratio was similar between individuals with and without baseline evidence of cardiovascular disease. There was no significant heterogeneity between studies, that used a variety of coagulation, heat precipitation, and immunoprecipitation assays.[2]

Address for correspondence: Professor G.D.O. Lowe, University Department of Medicine, Royal Infirmary, 10 Alexandra Parade, Glasgow G31 2ER, United Kingdom. Voice: 44-141-211-5412; fax: 44-141-211-0414.
gdl1j@clinmed.gla.ac.uk

TABLE 1. Suggested adjustment to the American Heart Association multivariable risk profile formulation for ischaemic heart disease risk (IHD)[a]

Fibrinogen (g/L)	Multiplication Factor for Multivariate IHD Risk	
	Men	Women
<2.35	0.83	0.77
2.35–3.35	1.00	1.00
>3.35	1.20	1.30

[a]To include the independent effect of plasma fibrinogen. Modified from Reference 6, based on Framingham data for fibrinogen thirds. Tertiles should be modified for different fibrinogen assays.

The independent association of fibrinogen and IHD is, therefore, well established, both in prospective studies and cross sectional studies.[2,3] The association is similar to that of conventional, modifiable IHD risk factors, such as smoking habit, blood pressure, and serum cholesterol. Fibrinogen adds to the predictive value of these traditional IHD factors,[4] and is a stronger predictor for all cause mortality.[4,5] For these reasons, it has been suggested that fibrinogen be added to multivariate IHD risk factor profiles[6] (see TABLE 1). Future collaborative analyses should refine the place of fibrinogen in multivariate risk prediction of IHD, especially if they can access individual data rather than grouped data. Readers interested in such collaboration are invited to contact Dr. J. Danesh, Fibrinogen Studies Collaboration, CTSU, Harkness Building, Radcliffe Infirmary, Oxford OX2 6HE, UK.

In other prospective studies, plasma fibrinogen has been associated with increased risks of stroke, peripheral arterial disease, venous thrombosis, atrial fibrillation, heart failure, restenosis following angioplasty or bypass grafting, and progression of arterial narrowing (see Refs. 1 and 3). Additional studies and systematic reviews are required to establish, with confidence, the strength of these other associations of fibrinogen.

THE INFLUENCE OF ASSAY METHODOLOGY ON THE RELATIONSHIP BETWEEN FIBRINOGEN AND IHD

As noted above, a recent systematic review of prospective studies of plasma fibrinogen and IHD events observed no significant heterogeneity between studies that used a variety of fibrinogen assays.[2] However, such overviews are less sensitive than direct comparisons between various assays to detect effects of assay methodology. In the only direct comparison study reported to date, a heat precipitation nephelometric assay was a significantly stronger predictor of IHD events than a semi-automated clotting rate assay.[7] In a further study by the same groups, the former assay also showed a closer relationship to the trans-European gradient of IHD events than the latter assay.[8] These studies suggest that further comparative studies are required in order to establish the relative power of different fibrinogen assays for prediction of IHD and of other thrombotic events. They also suggest that qualitative differences in plasma fibrinogen are related to risk of IHD.

Nieuwenhuizen and colleagues[9] have also demonstrated that qualitative differences in plasma fibrinogen are related to IHD (acute MI), by calculating the ratio of clottable to intact fibrinogen. The latter was assayed using an ELISA for the total of higher molecular weight (HMW) and lower molecular weight (LMW) fibrinogen. Normal donors had a ratio close to 1,whereas patients with acute MI had a mean ratio of 1:6 before thrombolytic therapy, rising to 2:1 at 24–72 hours after thrombolytic therapy. It was suggested that this increased ratio represents "hyperfunctional" fibrinogen (i.e., faster clotting than normal) in acute MI, perhaps because of a shift in the ratio of HMW: LMW: LMW' fibrinogen.[9]

Adopting Nieuwenhuizen's suggestion that this ratio be evaluated in epidemiological studies of cardiovascular disease because this may be a more sophisticated marker of fibrinogen activity than routine assays, such as the von Clauss assay,[9] we performed two case control studies, one of previous myocardial infarction[10] and one of chronic peripheral arterial disease.[11] In both studies, we observed significant elevations of the ratio, that were smaller than the elevations in patients with acute MI. The elevations observed in individuals with previous MI were not due to "reactant" increases in plasma fibrinogen, because they persisted in multivariate analysis, including plasma levels of C-reactive protein and interleukin-6 (unpublished observations).

Collectively, these data suggest that qualitative changes in plasma fibrinogen are related to both acute and chronic IHD. Such observations have implications, not only for standardization of fibrinogen assays, but also for the development of fibrinogen standards.[12] To date, fibrinogen standards have been defined in terms of assays of clottable fibrinogen.[12] In future, we suggest that consideration be given to the inclusion of different fibrinogen assays when defining the "content" of fibrinogen standards, as well as to the development of both "high" and "low" fibrinogen standards, which may well differ in ratio between fibrinogen assays. The choice of both assay and standard is clinically important in selection of tertiles for risk prediction (TABLE 1).

SIGNIFICANCE OF THE FIBRINOGEN–THROMBOSIS RELATIONSHIP

To date, there is little evidence for an association between known fibrinogen gene polymorphisms and IHD or other cardiovascular disorders[13] (see also review by Green, in this volume). Hence, the associations between plasma fibrinogen and these clinical disorders may be posttranslational. Studies of increased ratios of clottable, intact, "native" plasma fibrinogen in individuals with IHD (reviewed in the previous section) are consistent with this hypothesis. However, the posttranslational influences that cause these associations are at present unclear: environmental influences (age, smoking, overt cardiovascular disease, and acute phase reactions) do not appear to explain the association. Additional collaborative studies between fibrinogen biochemists and epidemiologists may shed light on this question.

Although there are several plausible biological mechanisms through which increasing plasma fibrinogen may increase risk of thrombosis and arterial disease,[1] randomized controlled trials of fibrinogen reduction are required to establish whether or not this relationship is causal.[1] At present, agents reducing plasma fibrinogen are limited:

- Fibrates reduce mean plasma levels by only about 10% and also affect blood lipids.
- Ticlopidine also reduces mean plasma levels by only about 10% and also affects blood platelets.
- Thrombolytics reduce mean plasma levels by 50–90% (depending on agent and regimen) and also lyse thrombi and hemostatic plugs.
- Defibrinogenating enzymes such as ancrod selectively reduce mean plasma levels by about 90%, and offer promising initial results in prophylaxis of venous thromboembolism[14] and in treatment of acute ischemic stroke.[15] However, this treatment is limited by the need for parenteral administration and the development of resistance after several weeks, due to antibody formation.

Progress in this field would be greatly facilitated by the development of oral agents with a more potent, long term effect on plasma fibrinogen levels, for example 25% reduction in mean plasma level. In other words, the equivalent of statins in reduction of LDL cholesterol is needed.

FIBRIN D-DIMER AS A RISK PREDICTOR

A recent review of seven prospective studies of plasma D-dimer and IHD events (see Ref. 1) and of an additional large, case control study, has observed that the risk ratio was 1:7 (95% CI 1.3–2.2) for individuals in the top third compared to those in the bottom third of baseline assays.[16] This association is of similar strength to that between plasma fibrinogen and IHD. In most of these studies, adjustment had been made for potential confounders; and the risk ratio was similar in patients with and without cardiovascular disease. There was no significant heterogeneity between studies. Two studies also adjusted for plasma fibrinogen (which correlates weakly with plasma D-dimer).

Two prospective studies, that compared two assays, showed similar associations between assays and IHD events[17,18] Further comparative studies are required. As with fibrinogen, D-dimer showed an association with the trans-European gradient in IHD.[8] Fewer data are available for studies relating D-dimer to stroke, venous thrombosis, or other thrombotic disorders.

SIGNIFICANCE OF THE D-DIMER-THROMBOSIS RELATIONSHIP

The most obvious explanation of the independent association of plasma fibrin D-dimer and IHD (and possibly other thrombotic events) is that D-dimer is a measure of hypercoagulability; that is, the formation and lysis of crosslinked fibrin, which, in individuals without inflammatory or neoplastic disease, is likely to be predominantly intravascular. Support for this hypothesis comes from our studies in patients with atrial fibrillation, in whom elevated plasma D-dimer levels are rapidly normalized by cardioversion[19] or warfarinisation[20,21] each of which normalizes the increased risk of thromboembolic stroke. The time is ripe for randomized controlled trials[1,22] of

(1) selective anticoagulant prophylaxis in patients with elevated plasma D-dimer, and (2) control of anticoagulant prophylaxis by normalization of elevated D-dimer levels, in contrast to the historical use of blood clotting times.

CONCLUSION

We conclude that both plasma fibrinogen and fibrin D-dimer are significantly associated with risk of thrombotic ischaemic heart disease. Additional data from prospective studies (as well as collaborative systematic reviews) are required in order to establish, with confidence, the size of these risks, the influence of assay methodology, the influence of genetic and environmental factors, their associations with thrombotic stroke and venous thrombosis, and finally the causal significance of such associations as shown by controlled trials of reduction in plasma levels.

REFERENCES

1. LOWE, G.D.O. & A. RUMLEY. 1999. Use of fibrinogen and fibrin D-dimer in prediction of arterial thrombotic events. Thromb. Hæmost. **82:** 667–672.
2. DANESH, J., R. COLLINS, P. APPLEBY & R. PETO. 1998. Fibrinogen, C-reactive protein, albumin or white cell count: meta analysis of prospective studies of coronary heart disease. JAMA 279: 1477–1482.
3. MARESCA, G., A. DI BLASIO, R. MARCHIOLI, & G. DI MINNO. 1999. Measuring plasma fibrinogen to predict stroke and myocardial infarction: an update. Arterioscler. Thromb. Vasc. Biol. **19:** 1368–1377.
4. WOODWARD, M., G.D.O. LOWE, A. RUMLEY, & H. TUNSTALL-PEDOE. 1998. Fibrinogen as a risk factor for coronary heart disease and mortality in middle-aged men and women: The Scottish Heart Health Study. Eur. Heart J. **19:** 55–62
5. LOWE, G.D.O., A. RUMLEY, J. NORRIE, et al. 2001. Blood rheology, cardiovascular risk factors, and cardiovascular disease: the West of Scotland Coronary Prevention Study. Thromb. Hæmost. **84:** 553–558.
6. KANNEL, W.B. 1997. Influence of fibrinogen on cardiovascular disease. Drugs **54:** S32–S40.
7. SWEETNAM, P.M., J.W.G. YARNELL, G.D.O. LOWE, et al. 1998. The relative power of heat-precipitation nephelometric and clottable (Clauss) fibrinogen in the prediction of ischaemic heart disease: the Caerphilly and Speedwell studies. Br. J. Hæmatol. **100:** 582–588
8. YARNELL, J.W.G., E. MCCRUM, A. EVANS, et al. 1999. Thrombotic risk factors in the WHO MONICA Project: population levels, determinants and risk of CHD. Blood Coagul. Fibrinolys. **10**(Suppl): 10.
9. NIEUWENHUIZEN, W. 1995. Biochemistry and measurement of fibrinogen. Eur. Heart J. **16**(Suppl. A): 6–10
10. RUMLEY, A., G. LOWE, M. WOODWARD, C. MORRISON, et al. 1999. Comparison of four fibrinogen assays in a case-control study of previous myocardial infarction. Thromb. Hæmost. **83**(Suppl.): 847.
11. LOWE, G.D.O., A. RUMLEY, W. NIEUWENHUIZEN, et al. 1999. Hyperfunctional fibrinogen in chronic peripheral arterial disease: evidence from a case-control study. Thromb. Hæmost. **83**(Suppl.): 274.
12. GAFFNEY, P.J. & M.Y. WONG. 1992. Collaborative Study of proposed international standard for plasma fibrinogen measurement. Thromb. Hæmost. **68:** 428–432.
13. LANE, D. & P.J. GRANT. 2000. Role of haemostatic gene polymorphisms in venous and arterial thrombotic disease. Blood **95:** 1517–1532

14. LOWE, G.D.O., A.F. CAMPBELL, D.R. MEEK, *et al.* 1978. Subcutaneous ancrod in prevention of deep vein thrombosis after operation for fractured neck of femur. Lancet **ii:** 698–700
15. SHERMAN, D.C., R.P. ATKINSON, T. CHIPPENDALE, *et al.* 2000. Intravenous ancrod for treatment of acute ischemic stroke. The STAT Study: a randomized clinical trial. JAMA **283:** 2395–2403.
16. DANESH, J., P. WHINCUP, M. WALKER, *et al.* 2001. Fibrin D-dimer and coronary heart disease: prospective study and meta-analysis. Circulation In press.
17. LOWE, G.D.O., A. RUMLEY, P.M. SWEETNAM, *et al.* 2001. Fibrin D-dimer, markers of coagulation activation, and the risk of major ischaemic heart disease in the Caerphilly Study. Thromb. Hæmost. In press.
18. MARDER, V.J., W. ZAREBA, J.T. HORAN, *et al.* 1999. Automated latex agglutination and ELISA testing yield equivalent D-dimer results in patients with recent myocardial infarction. Thrombo Research Investigation. Thromb. Hæmost. **82:**1412–1416.
19. LIP, G.Y.H., A. RUMLEY, F.G. DUNN & G.D.O.LOWE. 1995. Plasma fibrinogen and fibrin D-dimer in patients with atrial fibrillation: effects of cardioversion to sinus rhythm. Int. J. Cardiol. **51:** 245–251.
20. LIP, G.Y.H., G.D.O. LOWE, A. RUMLEY & F.G. DUNN. 1995. Increased markers of thrombogenesis in chronic atrial fibrillation: effects of warfarin treatment. Br. Heart J. **73:** 527–533.
21. LIP, G.Y.H., J. ZAFIRIS, R.D.S. WATSON, *et al.* 1996. Fibrin D-dimer and B-thromboglobulin as markers of thrombogenesis and platelet activation in atrial fibrillation: effects of introducing ultra-low-dose warfarin and aspirin. Circulation **94:** 425–431.
22. LIP, G.Y.H. & G.D.O. LOWE. 1995. Fibrin D-dimer: a useful clinical marker of thrombogenesis? Clin. Sci. Mol. Med. **89:** 205–214.

Wound Healing

Role of Commercial Fibrin Sealants

DAVID L. AMRANI,[a] JAMES P. DIORIO,[a] AND YVES DELMOTTE[b]

[a]*Baxter Healthcare, CRTS, Round Lake, IL, USA, and*
[b]*Baxter Healthcare, CRTS, Nivelles, Belgium*

ABSTRACT: This paper focuses on the use of commercial fibrin sealant (FS) in specific wound healing applications. This review is not intended to be all inclusive, but to examine *in vitro* and *in vivo* models, as well as select clinical conditions that are representative of specific wound healing applications of FS.

KEYWORDS: Wound healing; Fibrin sealant.

INTRODUCTION

Formation of fibrin occurs during the penultimate step in the coagulation cascade.[1,2] Fibrin is formed in the cascade as a result of the formation of thrombin that cleaves two sets of peptides, FpA and FpB, from fibrinogen resulting in fibrin monomer formation leading to creation of an insoluble polymer.[1–3] This physiologically insoluble polymer is created by polymerization of fibrin monomers, and results from an orderly assembly of these monomers, followed by fibril formation, lateral association, and branching, processes that occur as soon as monomers are formed.[1,2] The rate and extent of polymerization, as well as the structure of this polymer is dependent on many factors, including the concentration of the fibrinogen, thrombin, presence of certain other plasma proteins (e.g., albumin), ionic strength, calcium concentration, and temperature.[4–8] The physiologically insoluble fibrin polymer is further stabilized by the introduction of isopeptide bonds resulting in crosslinking among fibrin α, β, and γ chains.[9–11] Transglutaminases (the plasma form, factor XIII, or tissue forms) catalyze these bonds.[9–11]

Commercial FSs are currently composed of plasma derived complexes of purified, pathogen inactivated human fibrinogen and thrombin.[12–18] Other components, such as factor XIII and bovine aprotinin are currently added or contained in European, Canadian, Japanese, and US formulations.[16–19] These commercial products were originally developed to mimic and rapidly allow the last stage of blood coagulation to occur rapidly at sites of bleeding. The relatively rapid formation of fibrin clots with commercial sealants is primarily due to the relatively high concentration of the major components (fibrinogen, 80–100 mg/ml; thrombin, 4–1000 IU/ml). Commercial fibrin sealant kits of one manufacturer, available in Europe, Canada, and Japan, actually contain two different thrombin preparations to allow for either very rapid

---

Address for correspondence: David L. Amrani, Ph.D., Baxter Healthcare, RLT-12, Route 120 and Wilson Rd., Round Lake, IL 60073, USA. Voice: 847-270-4448; fax 847-270-6360.
David_amrani@baxter.com

(about 5 sec) or slow (about 60–90 sec) setting after mixing the two components. Current commercial FS formulations result in semirigid, water resistant polymers. The quality of these polymers also depends on factors such as ionic strength, delivery methods, and application device used. This polymer interacts non-covalently and covalently with cellular structures. FS adheres and binds to tissues or biomaterials at sites of injury facilitating hemostasis and wound healing.[12–16]

Commercial FSs have been used in a variety of clinical arenas, including cardiovascular operations,[21] and various types of surgery,[21] especially for patients with hemostatic disorders,[22] primarily to prevent bleeding or as an aid in the adhesive sealing of wounds. Although FSs have proven useful and effective in hemostasis and associated wound repair, they are versatile enough to be used to seal lung air leaks. At sites of tissue injury, FSs with primarily human derived components appear to be generally biocompatible without inducing excessive inflammation, foreign body reactions, tissue necrosis, or extensive fibrosis,[14] positively impacting the overall process of wound healing. In turn, these responses reflect resorption kinetics of FSs that depend on tissue location, specific composition of the product, amount applied, and mode of delivery. Generally, fibrin degradation is achieved within a few days to a few weeks, aiding in local angiogenesis, tissue growth, and repair.[24] The time of degradation may vary due to differing amounts of plasminogen activators and their inhibitors that are produced at distinct tissue sites.

HISTORICAL PERSPECTIVE

Fibrin was first used for wound healing as early as 1909. Bergel[25] used dry plasma as a source of fibrinogen with recalcification to produce fibrin films to achieve hemostasis in surgical applications. During World War I, the fibrin made in this manner was used to stop bleeding of lung, liver, and kidney and for dura repair in head injuries.[26,27] In the 1940s with the coming of World War II, a renewed interest in enhancing wound repair led physicians and scientists to evaluate fibrin as a basis for a wound matrix. Young and Medwar[28] applied fibrin as sutures for peripheral nerve repair and studied its impact on regeneration in an animal model, and later in humans.[29] Tarlov and Benjamin[30] reunited nerves with fibrin in the form of plasma clot. In 1944, two groups, Tidrick and Warner[31] and Cronkite et al.,[32] reported on the use of semipurified fibrinogen mixed with bovine thrombin in defined concentrations to produce FS. Both groups reported enhanced healing and sealing of skin grafts resulting from the use of fibrin even in the absence of sutures, which were normally used to attach the grafts. However, due to the relatively low concentration and semipure nature of the fibrinogen and thrombin, the authors in both studies experienced variation in fibrin strength and stability. The credit for modern revival of fibrin as a sealing and wound healing adjunct stems from the studies published by Matras et al.[33,34] In these papers, sealing nerve anastomoses by using fibrinogen at relatively high concentrations from cryoprecipitate clotted with high concentrations of bovine thrombin demonstrated the potential of high protein concentration sealants for wound healing.

GENERAL WOUND HEALING

The formation of fibrin is one of the primary events that occurs following injury.[23,34,35] Prolonged bleeding at the site of injury is prevented by the activation of coagulation, resulting in the formation of a thrombus that incorporates both platelets and fibrin to form a plug. With the release of growth factors, that is, PDGF, from platelets and chemokines from various protein and cellular sources,[24,35] and the onset of the inflammatory process, neutrophils migrate to the site of injury from the peripheral circulation. These cells begin the process of cellular debriding of the area, and are followed by monocytes, that differentiate into macrophages. Both cell types release proteolytic enzymes, that begin to remodel and degrade the fibrin plug. At the same time, macrophages release cytokines, growth factors, and chemotactic agents, that induce fibroblast, epithelial, and endothelial cell migration to the area for tissue remodeling. The fibrin may bind to surrounding biological tissues by crosslinking, catalyzed by transglutaminases such as factors XIII or tissue transglutaminases.

Fibrinogen, fibronectin, plasma factor XIII, plasminogen, and α_2-plasmin inhibitor, PAI-2, on activation of the coagulation system, become a part of the wound filling clot.[24,35,36] This fibrin clot also incorporates platelets, monocytes, and RBCs as well as growth factors and chemotactic agents released by these diverse cell types. These components were rapidly incorporated (10–30 min) into the fibrin clot, which forms in incisional or excisional, wounds.[36,37] The presence of these components and their impact on the fibrin clot structure ultimately affect the considerable remodeling and reabsorption of the fibrin clot during the first week of the wound healing process. This clot serves as a provisional matrix upon which migrating fibroblasts, epithelial cells, and endothelial cells adhere and it induces specific cellular functions associated with wound healing, that is, upregulation of growth factor genes with concomitant secretion, matrix protein synthesis, and integrin expression.

Depending on the tissue site and relative concentration of fibrinolytic activators and enzymes, the rate and modification of the fibrin clot can produce various matrix structures. These matrices provide infiltrating fibroblasts, epithelial cells, and endothelial cells with unique opportunities to interact with key receptors, for example, integrins, for appropriate cellular/tissue assembly. Infiltrating fibroblasts and epithelial cells also deposit collagen into the provisional tissue matrix formed from the fibrin clot and with the further release of plasminogen activators contribute to fibrin lysis, angiogenesis, and cellular/tissue repair. The role of fibrin(ogen) in wound healing is obvious from the impact of defibrination or dysfibrinogenemia leading to delayed healing.[35]

Crosslinking either by plasma transglutaminase, factor XIII, or tissue transglutaminases is critical to the stability of the fibrin gel. Additional components such as fibronectin are covalently incorporated[38] and play major roles in the subsequent remodeling and cellular responses. Fibronectin crosslinked to fibrin(ogen) by transglutaminases was originally shown to provide a much more specific adhesive surface for fibroblasts *in vitro* than either component alone, or without FXIII.[39] FXIII is important for wound healing as evidenced by a defect in wound healing occurring in FXIII deficiency.[40]

FIBRIN SEALANTS IN WOUND HEALING

In Vitro *and Animal Model Systems for Evaluation of Their Use in Selected Clinical Situations*

It is now a well recognized fact that the formation and structure of fibrin is produced by the conversion of fibrinogen by thrombin *in vivo*. It is influenced by many factors including fibrinogen and thrombin concentration, presence of other protein components, such as fibronectin and albumin, transglutaminase type and concentration, and divalent cations.[2,35,41,42] In addition, the incorporation *in vivo* of platelets and RBCs.[2,35,41] also significantly alters the properties of the resulting fibrin clot.

In the past, the understanding of these components both in affecting fibrin clot structure and cellular responsiveness in wound healing has been revealed through the use of *in vitro* (non-cellular test systems and single or multicell systems) systems. Appropriate animal models have been used to examine the response of altered compositions of fibrin in order to determine the role of individual components or conditions. In the late 1940s, Ferry[42] was probably the first to recognize the significance of fibrinogen and thrombin concentration on fibrin clot structure. His studies on physical measurements of fibrin clot elasticity and mechanical strength, with suggestions concerning structure, provided the first clues as to the significance of increasing the concentrations of the fibrinogen and thrombin, as well as the impact of crosslinking, ionic strength, and divalent cations. His descriptions of coarse and white, and fine and clear clots, together with estimations of fibrin fiber thickness, were coupled with observations of potential use of fibrin films for wound healing applications.[43] At the same time, involvement of fibroblast infiltration in wound healing associated with the fibrin clot was clearly observed.[44]

In the early 1960s, Beck *et al.*[45,46] determined that Factor XIII was critical for normal fibroblast growth into the fibrin clot. This observation, a result of earlier studies that examined the role of fibroblast migration and proliferation in wound healing,[44] was reinforced by studies on the role of fibrin(ogen) and thrombin in inducing fibroblast growth.[47] The requirement for crosslinked or non-crosslinked fibrin may depend on the cell type and composition of the fibrin gel. Kasai *et al.*[48] indicated that crosslinked fibrin is essential for adherence of fibroblasts to the substrate and for well oriented cell growth. Recently, Dallabrida *et al.*[49] demonstrated that human endothelial cells could be oriented to migrate and proliferate on a surface coated with Factor XIIIa, but not factor XIII. These studies imply a potential specificity, either conferred on fibrin in the presence of thrombin leading to Factor XIII activation, or to a direct interaction with a specific cellular receptor that leads to dimerization or multimerization of receptors and modulation of endothelial function.

Focussing on the fundamental formation of fibrin and its impact on cellular binding, migration, and infiltration into fibrin matrices, Dvorak and colleagues established *in vitro* cell systems involving macrophages, fibroblasts, and endothelial cells in order to determine the conditions for formation of fibrin matrices that impact on wound healing.[50–54] In the initial studies on macrophage migration *in vitro* systems were development for the evaluation of fibrin impregnated nitrocellulose filters[50] and later[51] three-dimensional fibrin matrices for changes in fibrin structure or components on cellular migration. Fibrin cast on filters were abnormally crosslinked in that only γ chain crosslinking occurred and macrophages did not migrate very well into these fibrins. With three-dimensional fibrin matrices, they observed that at a physiological concentration range of fibrinogen (1–4 mg/ml) macrophage migration was inversely proportional to fibrinogen concentration. There was little or no fibrinolytic activity associated with this migration, rather the cells migrated through the matrices by marked distortions of shape and presumably with specific membrane receptor interactions. The inclusion of fibronectin and/or factor XIII as a function of their respective concentrations resulted in delayed migration of macrophages through fibrin matrices. Fibrin matrices containing fibronectin demonstrated increased macrophage migration on the addition of an inhibitor for integrin receptors (the hexapeptide, GRGDSP) and no effect by GRGESP, a control peptide. The implication is that sites both within the fibrin and the bound fibronectin interact with integrin receptors,[55] binding to and impeding macrophage migration. Incorporation of hyaluronic acid, heparin, and heparan sulfate also resulted in delayed macrophage migration.

Therefore, fibrin together with other macromolecules found either in the circulation, at sites of wound injury, or incorporated into man made FSs can significantly impact cellular migration. This is further evidenced by the fact that fibrin matrices can induce angiogenesis and that inclusion of chemoattractants such as N-formlymethionine can further enhance this process.[52] Since the wound healing process produces locally increased microvascular permeability, the resulting fibrinogen rich exudate fills the wound area, becomes converted to fibrin, and crosslinked by factor XIIIa.[53]

Subsequent cellular response of macrophages and endothelial cells is also seen with fibroblasts and epithelial cells.[56,57] The latter cells migrate into and organize the provisional fibrin clot into granulation tissue, later into mature collagenous connective tissue. Brown et al.[54] developed a quantitative in vitro assay that permitted the study of fibroblast migration in fibrin clots. As with the macrophages, optimal fibrinogen concentration for migration was found in the physiologic range. However, in contrast to macrophages, fibroblast migration was significantly enhanced by the presence of crosslinked fibrin. Complete α and γ chain crosslinking[41] was required for maximal fibroblast migration, as seen when only γ chain crosslinking occurred to result in poor fibroblast migration. Furthermore, in the absence of factor XIII mediated crosslinking, little or no fibroblast migration was observed. This requirement could have been due to the need for fibroblasts to recognize a given matrix structure or the binding of a specific component, fibronectin.[57]

With the advent of high concentration fibrinogen and thrombin kits in the late 1970s, FSs with multiple factors, including fibronectin, plasminogen, and antifibrinolytics were examined for their impact on wound healing.[13,58–62] In the early to mid 1980s, Redl and Schlag[58,59] examined the principle effects of FSs and the impact of its mode of application for wound healing. Emanating from the initial studies of Matras,[33,34] Redl and Schlag[58,59] used FS formulations and determined that both low (4 IU/ml) and high (500 IU/ml) concentrations of thrombin at a constant fibrinogen concentration would provide hemostatic, wound healing, and sealing options. These sealants clot within 2–5 sec. or 30–90 sec depending on thrombin concentration, and developed sufficient adhesive strength as judged from a rat skin in vitro testing system. As noted above and by these authors, factor XIII mediated crosslinking of fibrin was correlated with stimulation of fibroblast growth and migration.

However, given the high concentration of fibrinogen in these preparations, these authors did not examine fibrin formed in the absence of FXIII and did not assess the potential effect of increased fibrin clot rigidity, as opposed to crosslinking on fibroblast migration.

The same authors examined the impact of different preparations of FS on fibroblast responses. The sealants included cryoprecipitate, autologous fibrin, and two commercial FSs, one containing physiological salt concentrations and other having a high salt concentration. The latter was included in the fibrinogen component of the kit to facilitate more rapid fibrinogen reconstitution. They used a fibroblast (MRC5 lung fibroblasts) proliferation *in vitro* system in which equal volumes of the respective FSs were formed for one hour at 37°C using 4 IU/ml thrombin-40 mM $CaCl_2$ solution. The fibrin mixtures (0.5 ml) were placed in tissue culture 24-well plates, some were washed with isotonic saline and others were not. The fibroblasts were placed directly over the fibrin and incubated for 24 hours at 37°C. Quantification of cell migration was determined by histologic and SEM examination. The two commercial FSs had similar rates of fibrin formation and crosslinking. The FS with high salt had a lower intrinsic tensile strength and finer clot structure as assessed by SEM. Fibroblast proliferation, although extensive on the other FS surface, demonstrated rounded and poorly attached fibroblasts on the high salt FS. Washing the high salt clots resulted in somewhat improved fibroblast attachment and growth, but was still less than that observed for the other sealants, suggesting that not only physiological ionic strength but also a physiological fibrin structure *per se* is essential for optimal cell attachment and growth. These studies find that ionic strength and crosslinking of commercial FS clots probably impact significantly on fibroblast attachment and infiltration into these clots *in vivo*. These data also suggest that course, porous, crosslinked clots rather than tight pore, fine fiber clots enhance fibroblast infiltration.

Dinges *et al.*[60] examined the effect of FS, Tisseel™, on the formation of granulation tissue. In these studies, they developed a spongiosa based granulation tissue model in which the spongiosa blocks were subcutaneously implanted into rats. The cavities of the spongiosa were either filled with a substance that influenced local wound healing (e.g., homologous FS) or left empty as a control. Infiltration of fibroblasts into the spongiosa blocks was determined by sample removal at certain time intervals and evaluated morphometrically. The FS induced significant migration of fibroblasts into the surgically implanted spongiosa blocks in comparison with controls (16% vs. 22% fibroblast per volume granulation tissue). The inclusion of a cytostatic agent, the antibiotic (Adriamycin) that inhibited granulation tissue formation could not be overcome by the presence of the FS. Hence, although providing an appropriate matrix for cellular attachment and proliferation, FS could not overcome the presence of small molecular weight inhibitors.

The current FSs like Tisseel™ that contain physiological salt concentrations, varying amounts of FXIII, and fibronectin probably aid the infiltration of fibroblasts into the actual, as well as artificially created, wound models (e.g., implanted spongiosa blocks). The presence of antifibrinolytics agents may also influence the ability of fibroblasts to infiltrate clots made from FS. Use of commercial fibrin sealants with high salt concentrations, but lacking FXIII, most likely lead to inhibition of cellular migration. Clearly, the concentration and type of components as well as the conditions of the FS can impact on wound healing and enhancement of granulation tissue formation.

Other *in vitro* models have also been developed in order to assess FS biomechanical strength for wound healing applications. Two groups, Sierra *et al.*[61] and Vogel *et al.*[62] used an *in vitro* system, Instron 4301, to evaluate tensile strength of adhesives. Stainless steel jigs were made to specification with a 2 cm × 2 cm adhesive surface. Using fresh cut pig skin on both surfaces, FS preparations with different concentration and types of components were placed on the pig tissue and examined for their impact tensile strength on living tissue. The FS was allowed to clot, and the tissue sections placed into the testing system and pulled apart at a constant rate. The energy to failure was recorded for multiple samples to account for intersample variability. Vogel *et al.*[62] found that maximal binding strengths were dependent on fibrinogen concentration between 40–70 mg/ml. An optimal thrombin concentration at these high fibrinogen concentrations contributed to the speed of polymerization. Relatively low levels of factor XIII (1–10 unit/ml) already present in fibrinogen preparations showed little additional impact on increasing adhesive strength. Toriumi[63] also performed adhesive strength testing using shear testing of overlapping flaps of pig subdermal tissue. FS was placed between the two overlapping flaps, sutured, and allowed to heal between 1 and 28 days, after which the tissue flaps were examined for the impact of the FS on wound healing. The sutures were removed and the flaps were fastened to jigs, designed to fit the system, in which the tissue was pulled apart at a constant rate. During the first 48 hours the FS provided increased strength compared with control tissues. By 21 days, the wound strength between control and FS treated tissue was the same. One additional observation was that the amount of FS applied, as well as the concentration of fibrinogen and thrombin, were critical for optimal tissue healing and wound strength. Models, such as these, can provide *in vitro* systems for development and assessment of FS effectiveness for wound healing applications.

POST-SURGICAL ADHESION PREVENTION

Impact on Controlling Cellular Responses with Commercial FS

In normal wound healing, an appropriate cellular response and tissue reorganization results in repair of the injury. In contrast, many wounds are healed but demonstrate a relatively large connective tissue deposition, primarily composed of collagen.[64–66] In large tissue injury sites, such as occur during gastrointestinal, gynecological, or cardiac surgical interventions, thick, extensive, often vascularized connective tissue adhesions form.

With expanded use of FS for surgical procedures, the potential effect on postsurgical adhesion formation has been examined.[67–76] In 1984, Lindenberg *et al.*[67] reported for the first time that FS might be useful in minimizing the formation of these adhesions. Using their rat peritoneal muscular layer suture model,[68] they further determined that the relative thickness of a commercial FS (Tisseel™) layer and the presence of an antiplasmin agent (aprotinin) could affect the length of peritoneal adhesions. The lower antiplasmin (3,000 vs. 20,000 KIU/ml) with a thick layer (0.1 ml) covering a 1 × 3 cm defect resulted in the smallest adhesions.

In the early 1990s, a series of reports differed on the potential usefulness of FS for preventing postsurgical adhesions. Gauwerky et al.[69] reported using a rat uterine horn injury model that a commercial FS (Tissucol®) with or without a patch did not prevent the formation of any type of postsurgical adhesions when compared with controls. In referencing the data of Lindenberg et al.,[67,68] they indicated that the differences observed using FS from the same commercial source might be due to differences in trauma, site of trauma, or scoring of adhesions. They never considered the formulation of the preventive agent used in the two reports. Several reports subsequently showed that, in similar rat abdominal wall suture models, FS was only poorly able to prevent postsurgical adhesions.[70–72] These studies all used a fibrin sealant with high concentrations of an antiplasmin agent, aprotinin.

The studies failed to examine the potential basis for the ability of FS to prevent the postsurgical adhesion formation. For example, it has been noted that physiologically generated fibrin is generally removed from such sites by proteolytic degradation and absorption, but residual amounts of this material can act as sites for fibroblast proliferation and connective tissue formation.[35,47,48,54,56] Use of FS with high concentrations of antiplasmin agents may prolong the usual halflife of the FS (4–7 days) resulting in increased fibroblast proliferation and connective tissue adhesion formation.

Using similar peritoneal muscular layer injury and suture rat models, other studies[73–77] examined the potential of FSs to inhibit intra-abdominal adhesion formation. Virgilio et al.[73] in a carefully controlled study examined the prevention of adhesions after peritoneal muscular wall injury using fibrinogen prepared from human cryoprecipitate (1.77 vs. 23.6 mg/ml) clotted with thrombin (1000 U/ml), and $CaCl_2$ but containing no antiplasmin agent. The FS was allowed to set for four minutes, the laparotomies were closed with suture, and the rats were evaluated on day seven. Adhesions were graded 0–4 with 4 indicating that 75% of the injured area showed adhesions. Control rats or those treated with low fibrinogen containing FS had a similar number of type 3–4 adhesions, and all animals exhibited at least one form of adhesion. In contrast, rats treated with high concentration containing FS showed a significantly lower incidence of adhesions, only 3 of 15 showed grade 3 or 4 adhesions, and 11 no adhesions. A series of complementary studies aimed at understanding the cellular response over the seven day period found that in comparison to a control group, the high fibrinogen group showed a healing response similar to normal healing in the absence of collagenous adhesions. After day one, the fibrin layer covered the peritoneal muscular layer defect and was scattered with polymorphonuclear leukocytes (PMNs); day three showed a similar healing response with some collagen deposition. By day five, the fibrin was mostly resorbed and a marked fibroblastic response was seen. At day seven the fibroblastic response was greater with collagen deposition and the ingrowth of new capillaries. This response was greater in the FS treated animals, with a more consistent, healthy wound repair. The pattern of control animal formation of adhesions versus the FS treated animals is consistent with the pathogenesis of intradominal adhesion formation as described by Milligan and Rafferty.[74] In 1993, Sheppard et al.[75] performed a follow-up long term study in which they confirmed their original observations[73] on FS inhibition of peritoneal adhesions during the first 5–7 days postsurgery. This study also showed no additional adhesion formation after 30 days. These data demonstrated the importance of

TABLE 1. Post-Surgical Adhesion[a] Inhibition by Fibrin Sealant

Group	Number of Animals with Adhesions	Percent Decrease in Adhesion Incidence[b]
Control	30/34	11.8
Fibrin Sealant	11/35	68.9

[a]Rat cecal-peritoneal wall injury model.
[b]Percent decrease of adhesion incidence is the adhesion score divided by the value for the control group. This value is multiplied by 100 and then the resulting value subtracted from 100.

relatively high fibrinogen concentration with no antiplasmin. Furthermore, histological studies confirmed that normal fibroblastic infiltration occurs but without formation of collagenous adhesions. Clearly, the impact of FS in preventing adhesion is within the first week postsurgery. Subsequently, a number of preclinical animal studies in rats and rabbits have demonstrated, with varying degrees of efficacy, the potential for FS sealant prevention of postsurgical adhesions.[76–81]

Delmotte et al.[82] investigated the potential for modulation of an *in vivo* cellular wound healing response to surgical trauma in a rat caecal-parietal abrasion model, by correlating fibrin structure with prevention of postsurgical adhesion formation. Our summary data (see TABLE 1) demonstrates that modulation of fibrin structure by appropriate concentration of fibrinogen and thrombin result in a time dependent reduction in macrophage and neutrophil migration at the wound site. Tightpore fibrin structures resulted in reduced cellular migration and correlated with significant decrease in postsurgical adhesion formations. Our findings suggest that tight pore, rigid fibrin structures formed by rapid polymerization of FS resulted in a decrease in the neutrophil and macrophage migration. Fibrin clot remodeling into provisional matrix is delayed. The FS may not require Factor XIII for prevention of adhesion since in its absence, the FS was also observed to inhibit formation of postsurgical adhesion. It is also possible that the presence of tissue transglutaminases may play a role in modifying regions of applied fibrin. In the wound healing process, fibroblast deposition of collagen is subsequently delayed. We speculate that unregulated and too rapid fibroblast and epithelial cell migration to the wound site prior to proper provisional clot remodeling results in unregulated collagen and other adhesive glycoprotein expression facilitating interactions between tissues.

CONCLUSIONS

Specific Uses of FS for Clinically Relevant Wound Healing Applications

Fibrin sealants have been found useful in a variety of clinical settings for primarily hemostatic and surgical sealing applications.[14–24] Currently, FS is used extensively in cardiovascular surgery, neurosurgery, and thoracic surgery.[22] These applications are primarily intended to provide a structural adjunct to the tissue as well as minimize bleeding at sites of surgery. Thus, the main intent of FS is for a material, that will provide optimal sealing of anastomoses and puncture sites in all types of surgeries. Along with these intended applications, FSs appear to provide improved wound healing in various applications, such as burn injury repair, where FS functions to help seal grafts and promote the wound healing process with little inflammatory reaction resulting in earlier healing patients.[83,84] In cardiovascular applications, the treatment of non-traditional areas such as ventricular septal defects with FS has been achieved.[85] In structural (bone) and plastic surgery applications,[86,87] FSs such as Tissucol have been successfully used in oncologic plastic surgery where reduction of immediate complications, such as inflammation and partial separation of the surgical wound with associated scar hypertrophy, were significantly decreased. Recent uses for peptic ulcer bleeding[88] and repair and pancreatic fistula repair[89] provide further evidence of enhanced wound repair with appearance of normal healing granulation tissue. The use in preventing postsurgical adhesions following excision of large ovarian endometriomas produced a significant decrease in adhesion formation.[90] Although FSs have been clearly documented to be useful for many applications, current commercial formulations may not be as effective in all clinical conditions as they are in their inability to aid in healing perianal fistulas.[91] This latter finding indicates that FS may need to be reformulated to produce structures and contain components that provide an optimal balance for cellular infiltration, provisional matrix remodeling, and resorption in specific tissues. An understanding of tissue specific wound healing responses and the formulation of fibrin sealants to match the tissue response is required for maximal wound healing efficacy.

ACKNOWLEDGMENTS

Critical and thoughtful reading of this manuscript by Drs. T. Seelich and H. Redl is greatly appreciated.

REFERENCES

1. DOOLITTLE, R.F. 1983. The structure and evolution of vertebrate fibrinogen. Ann. N.Y. Acad. Sci. **408:** 13–26.
2. MOSESSON, M.W. 1990. Fibrin Polymerization and its regulatory role in hemostasis. J. Lab. Clin. Med. **116:** 8–17.
3. EVERSE, S.J., G. SPRAGGON & R.F. DOOLITTLE. 1998. A three-dimensional consideration of variant human fibrinogens. Thromb. Hæmostas. **80:** 1–9.
4. HARDY, J.J., N.A. CARRELL & J. MCDONAGH. 1983. Calcium ion function in fibrinogen conversion to fibrin. Ann. N.Y. Acad. Sci. **408:** 279–287.
5. GALANAKIS, D.K., B.P. LANE & S.R. SIMON. 1987. Albumin modulates lateral assembly of fibrin polymers: evidence of enhanced fine fibril formation & of unique synergism with fibrinogen. Biochemistry **26:** 2389–2400.
6. OKADA, M. & B. BLOMBACK. 1983. Factors influencing fibrin gel structure studied by flow measurement. Ann. N.Y. Acad. Sci. **408:** 233–253.
7. CARR, M.E. & J. HERMANS. 1978. Size & density of fibrin fibers from turbidity. Macromol. **11:** 52–56.
8. CARR, M.E. *et al.* 1987. Platelet factor 4 enhances fibrin fiber polymerization. Thromb. Res. **45:** 539–534.
9. PISANO, J.J., J.S. FINLAYSON & M.P. PEYTON. 1968. Crosslink in fibrin polymerized by factor XIII: ε-(γ-glutamyl) lysine. Science **160:** 892–893.

10. CHEN, R. & R.F. DOOLITTLE. 1971. γ-γ Cross-linking sites in human and bovine fibrin. Biochemistry **10:** 4486–4491.
11. GRØN, B. *et al.* 1993. Cross-linked α_s-γ_t-chain hybrids in plasma clots studied by 1D & 2D electrophoresis & western blotting. Thromb. Hæmost. **70:** 438–442.
12. GRINNELL, F., M. FELD & D. MINTER. 1980. Fibroblast adhesion to fibrinogen & fibrin substrata: requirement for cold-insoluble globulin (plasma fibronectin). Cell **19:** 517–525.
13. H. REDL, G. SCHLAG & H.P. DINGES. 1985. The use of biocompatibility of a human fibrin sealant for hemostasis and tissue sealing. *In* Biocompatibility of Natural Tissue and Their Synthetic Analogues. D.F. Williams, Ed.: 135–137. CRC Press Inc., Boca Raton.
14. SCHLAG, G. & H. REDL. 1988. Fibrin sealant in orthopedic surgery. Clin. Ortho. Related Res. **227:** 269–285.
15. THOMPSON, D.F., N.A. LESTASSY & G. D. THOMPSON. 1988. Fibrin glue: a review of its preparation, efficacy, and adverse effects as a topical hemostat. Drug Intell. Clin. Pharm. **22:** 946–952.
16. BRENNAN, M. 1991. Fibrin glue. Blood Rev. **5**(4): 240–244.
17. SIERRA, D.H. 1993. Fibrin sealant adhesive systems: a review of their chemistry, material properties, and clinical applications. J. Biomaterial Appl. **7:** 309–352.
18. ALVING, B.M. *et al.* 1995. Fibrin sealant: summary of a conference on characteristics and clinical uses. Transfusion **35:** 783–790.
19. JACKSON, M.R. & B.M. ALVING. 1999. Fibrin sealant in preclinical and clinical studies. Curr. Opin. Hematol. **6:** 415–419.
20. RADOSEVICH, M., H.A. GOUBRAN & T. BURNOUF. 1997. Fibrin sealant: Scientific rationale, production methods, properties, and current clinical use. Vox Sanguinis **72:** 133–143.
21. BORST, H.G. *et al.* 1982. Fibrin adhesive: an important hemostatic adjunct in cardiovascular operations. J. Thorac. Cardiovasc. Surg. **84:** 548–553.
22. DUNN, C.J. & K.L., GOA. 1999. Fibrin sealant: a review of its use in surgery and endoscopy. Drugs **58**(5): 863–886.
23. MARTINOWITZ, U. *et al.* 1996. Role of fibrin sealants in surgical procedures on patients with hemostatic disorders. Clin. Ortho. Related Res. **328:** 65–75.
24. SCHLAG, G. *et al.* 1986. The importance of fibrin in wound repair. *In* Fibrin Sealant in Operative Medicine Otohinolaryngology. G. Schlag & H. Redl, Eds.: 3–12. Springer-Verlag, Berlin.
25. BERGEL, S. 1909. Uber Wirkungen des fibrins. Deustsch Wochenschr. **35:** 633–665.
26. GREY, E.G. 1915. Fibrin as a hemostatic in cerebral surgery. Surg. Gynecol. Obstet. **21:** 452–454.
27. HARVEY, S.C. 1916. The use of fibrin paper and forms in surgery. Boston Med. Surg. J. **658**–659.
28. YOUNG, J.Z. & P.B. MEDAWAR. 1940. Fibrin suture of peripheral nerves: measurement of the rate of regeneration. Lancet **ii:** 126–128.
29. SEDDON, H.J. & P. B. MEDAWAR. 1942. Fibrin sutures of human nerves. Lancet **ii:** 87–92.
30. TARLOV, I.M. & B. BENJAMIN. 1943. Plasma clot and silk suture of nerves. Surg. Gynecol. Obstet. **76:** 366–374.
31. TIDRICK, R.T. & E.D. WARNER. 1944. Fibrin fixation of skin transplants. Surgery **15:** 90–95.
32. CRONKITE, E.P. E. L. LOZNER & J.M. DEAVER. 1944. Use of thrombin and fibrinogen in skin grafting. JAMA **124:** 976–978.
33. MATRAS, H. *et al.* 1972. Zur nahtlosen interfaszikularen nerven-transplantation im tierexperiment. Wien. Med. Wochenschr. **122:** 517–523.
34. MATRAS, H. 1985. Fibrin seal: the state of the art. J. Oral Maxillofac. Surg. **43:** 605–611.
35. CLARK, R.A.F. & R. B. COLVIN. 1985. Wound Repair. *In* Plasma Fibronectin: Structure and Function. J. McDonagh, Ed: 197–262. Marcel-Dekker, New York.

36. BROWN, L.F., A.M. DVORAK & H.F. DVORAK. 1989. Leaky vessels, fibrin deposition, & fibrosis: a sequence of events common to solid tumors and to many other types of disease. Am. Rev. Resp. Dis. **140**(4): 1104–1107.
37. KAPLAN, J.E. 1976. Comparative disappearance & localization of isotopically labelled opsonic protein and soluble albumin following surgical trauma. J. Reticuloendothelial Soc. **20**: 375–384.
38. MOSHER, D.F. 1976. Action of fibrin-stablizing factor on cold-insoluble globulin and α_2-macroglobulin in clotting plasma. J. Biol. Chem. **251**: 1639–1645.
39. GRINNELL, F., M. FELD & D. MINTER. 1980. Fibroblast adhesion to fibrinogen and fibrin substrata: requirement for cold-insoluble globulin (plasma fibronectin). Cell **19**: 517–525.
40. DUCKERT, F. 1972. Documentation of the plasma factor XIII deficiency in man. Ann. N.Y. Acad. Sci. **202**: 190–199.
41. MOSESSON, M.W., et al. 1987. Studies on the ultrastructure of fibrin lacking fibrinopeptide B (β-fibrin). Blood **69**: 1073–1081.
42. FERRY, J.D. & P.R. MORRISON. 1947. Preparation and properties of serum and plasma proteins. VIII. The conversion of human fibrinogen to fibrin under various conditions. J. Amer. Chem. Soc. **69**: 388–400.
43. FERRY, J.D. & P.R. MORRISON. 1944. Chemical, clinical, and immunological studies on the products of human plasma fractionation. XVI. Fibrin clots, fibrin films, and fibrinogen plastics. J. Clin. Invest. **23**: 566–572.
44. SCHILLING, J.A., B.V. FAVATA & M. RADAKOVICH. 1953. Studies of fibroplasia in wound healing. Surg. Gyn. Obstet. **96**: 143–149.
45. BECK, E., F. DUCKERT & M. ERNST. 1961. The influence of fibrin-stabilizing factor on the growth of fibroblasts *in vitro* and wound healing. Thromb. Diath. Hæmorrh. **6**: 485–491.
46. BECK, E.F. 1962. Der Einfluss des fibrinstabilisierenden facktors (FSF) auf funktion und morphologie von fibroblasten *in vitro*. Z. Zellorsch. **57**: 327–346.
47. POHL, J.H.D. BRUHN & E. CHRISTOPHERS. 1979. Thrombin and fibrin-induced growth of fibroblasts: role in wound repair & thrombus organization. Klin. Wochenschr. **57**: 273–277.
48. KASAI, S., T. KUNIMOTO & K. NITTA. 1983. Cross-linking of fibrin by activated factor XIII stimulates attachment, morphological changes, and proliferation of fibroblasts. Biomed. Res. **4**: 155–160.
49. DALLABRIDA, S.M., L.A. FALLS & D.H. FARRELL. 2000. Factor XIIIa supports microvascular endothelial cell adhesion and inhibits capillary tube formation in fibrin. Blood **95**(8): 2586–92.
50. CIANO, P.S., et al. 1986. Macrophage migration in fibrin gel matrices. Lab. Invest. **54**: 62–70.
51. LANIR, N.P.S., et al. 1987. Macrophage migration in fibrin gel matrices: II. Effects of clotting factor XIII, fibronectin, and glycosaminoglycan content on cell migration. J. Immunol. **140**: 2340–2349.
52. DVORAK, H.F. 1987. Fibrin containing gels induce angiogenesis. Implications for tumor stroma generation and wound healing. Lab. Invest. **57**(6): 673–686.
53. BROWN, L.F., et al. 1988. Fibrinogen influx and accumulation of cross-linked fibrin in healing wounds and in tumor stroma. Am. J. Pathol. **130**(3): 455–465.
54. BROWN, L.F., et al. 1993. Fibroblast migration in fibrin gel matrices. Am. J. Pathol. **142**(1): 273–283.
55. HARRIS, E.S., et al. 2000. The Leukocyte Integrins. J. Biol. Chem. **275**(31): 23409–23412.
56. CLARK, R.A., et al. 1982. Fibronectin and fibrin provide a provisional matrix for epidermal cell migration during wound reepithelialization. J. Invest. Dermatol. **79**: 264–269.
57. CORBETT, S.A., C.L. WILSON & J. E. SCHWARZBAUER. 1996. Changes in cell spreading & cytoskeletal organization are induced by adhesion to a fibronectin-fibrin matrix. Blood **88**(1): 158–166.
58. REDL, H. & G. SCHLAG. 1985. Principles of fibrin sealant and its mode of application. Facial Plastic Surg. **2**(4): 315–321.

59. REDL, H. & G. SCHLAG. 1986. Properties of different tissue sealants with special emphasis on fibrinogen-based preparations. In Fibrin Sealant in Operative Medicine Otorhinolaryngology. G. Schag & H. Redl, Eds.: 29–38. Springer-Verlag, Berlin.
60. DINGES, H.P., et al. 1986. Morphometric studies on wound healing after systemic administration of Adriamycin and local application of fibrin sealant: application of a new wound healing model using spongiosa implants. Path. Res. Pract. **101:** 746–754.
61. SIERRA, D., et al. 1992. A method to determine shear adhesive strength of fibrin sealants. J. Appl. Biomat. **3:** 147–151.
62. VOGEL, A,, et al. 1993. Surgical tissue adhesives in facial plastic and reconstructive surgery. Fac. Plast. Surg. Monogr. **9:** 49–57.
63. TORIUMI, D. 1996. Surgical tissue adhesives: host tissue response, adhesive strength, and clinical performance. In Surgical Adhesives and Sealants. D. Sierra & R.Saltz, Eds.: 61–69. Technomic, Lancaster.
64. CONNOLLY, J.E. & J.W. SMITH. 1960. The prevention and treatment of intestinal adhesions. Int. Abstr. Surg. **110:** 417–431.
65. ELLIS, H. 1971. The cause and prevention of postoperative intraperitoneal adhesions. Surg. Gynec. Obstet. **133:** 497–511.
66. DIZEREGA, G.S. 1994. Contemporary adhesion prevention. Fertil. Steril. **61:** 219–235.
67. LINDENBERG, S. & J.G. LAURITSEN. 1984. Prevention of peritoneal adhesion formation by fibrin sealant. Ann. Chir. Gyn. **73:** 11–13.
68. LINDENBERG, S.P. et al. 1985. Studies on prevention of intra-abdominal adhesion formation by fibrin sealant. Acta Chir. Sci. **151:** 525–527.
69. GAUWERKY, J.F.H., J. MANN & G. BASTERT. 1990. The effect of fibrin glue and peritoneal grafts in the prevention of intraperitoneal adhesions. Arch. Gynecol. Obstet. **247:** 161–166.
70. BOTHIN, C. & D. HALLBERG. 1992. Treatment of postsurgical adhesion formation with fibrin sealant. Surg. Res. Comn. **13:** 233–237.
71. CABALLERO, J. & T. TULANDI. 1992. Effects of Ringer's lactate and fibrin glue on postsurgical adhesions. J. Reprod. Med. **37**(2): 141–143.
72. BYRNE, D.J., et al. 1992. Adverse influence of fibrin sealant on the healing of high-risk sutured colonic anastomoses. J. Rog. Coll. Edinb. **37:** 394–398.
73. VIRGILIO, C., et al. 1990. Fibrin glue inhibits intra-abdominal adhesion formation. Arch. Surg. **125:** 1378–1382.
74. MILLIGAN, D.W. & A.T. RAFTERY. 1974. Observations on the pathogenesis of peritoneal adhesions: a light and electron microscopical study. Br. J. Surg. **61:** 274–280.
75. SHEPPARD, B.B., et al. 1993. Inhibition of Intra-abdominal adhesions: Fibrin glue in a long term model. Am. Surgery **59**(12): 786–790.
76. HARRIS, E.S., R.F. MORGAN & G.T. RODEHEAVER. 1995. Analysis of the kinetics of peritoneal adhesion formation in the rat and evaluation of potential antiadhesive agents. Surgery **117**(6): 663–669.
77. FRYKMAN, E.S. JACOBSSON & B. WIDENFALK. 1993. Fibrin sealant in prevention of flexor tendon adhesions: an experimental study in the rabbit. J. Hand Surg. **18A**(1): 68–75.
78. DEIACO, P., et al. 1994. Fibrin sealant in laparoscopic adhesion prevention in the rabbit uterine horn model. Fert. Steril. **62**(2): 400–404.
79. BORIS, W.J., J GU & L.B. MCGRATH. 1996. Effectiveness of fibrin glue in the reduction of postoperative intrapericardial adhesions. J. Invest. Surg. **9:** 327–333.
80. TAKEUCHI, H., et al. 1997. Effects of fibrin glue on postsurgical adhesions after uterine or ovarian surgery in rabbits. J. Obstet.Gynaecol. Res. **23**(5): 479–484.
81. OZEREN, S., et al. 1998. The effects of human amniotic membrane and fibrin sealant in the prevention of postoperative adhesion formation in the rabbit ovary model. Aust. New Zealand J. Obstet. Gynae. **38**(2): 207–209.
82. DELMOTTE, Y., et al. 2001. Modulation of fibrin structure can regulate post-surgical connective tissue formation. Submitted.
83. ADANT P.J., et al. 1993. Skin grafting with fibrin glue in burns. Europ. J. Plastic Surg. **16**(6): 292–297.
84. VEDUNG, S. & A. HEDLUND. 1993. Fibrin glue: Its use for skin grafting of contaminated burn wounds in areas difficult to immobilize. J. Burn Care Rehab. **14**(3): 356–358.

85. LECA, F., et al. 1994. Surgical treatment of multiple ventricular septal defects using a biologic glue. J. Thoracic Cardiovas. Surg. **107**(1): 96–102.
86. FABRIZIO, T., et al. 1990. Usefulness of fibrin glue in oncologic plastic surgery. Europ. J. Plastic Surg. **13**(5): 201–205.
87. MATSUMOTO, K., et al. 1998. Restoration of small bone defects at craniotomy using autologous bone dust and fibrin glue. Surg. Neurol. **50**(4): 344–346.
88. PESCATORE, P., et al. 1998. Fibrin sealing in peptic ulcer bleeding: the fate of the clot. Endoscopy. **30**(6): 519–523.
89. OHWADA, S., et al. 1998. Fibrin glue sandwich prevents pancreatic fistula following distal pancreatectomy. World J. Surg. **22**(5): 494–498.
90. TAKEUCHI, H., et al. 1996. Reduction of adhesions with fibrin glue after laparoscopic excision of large ovarian endometriomas. J. Amer. Gynec. Lapar. **3**(4): 575–579.
91. AITOLA, P., K.-M. HILTUNEN & M. MATIKAINEN. 1999. Fibrin glue in perianal fistulas—a pilot study. Ann. Chir. Gynaec. **88**(2): 136–138.

Fibrinogen Non-Inherited Heterogeneity and Its Relationship to Function in Health and Disease

AGNES H. HENSCHEN-EDMAN

Department of Molecular Biology and Biochemistry, University of California, Irvine, California, USA

ABSTRACT: In healthy individuals fibrinogen occurs in more than one million non-identical forms because of the many possible combinations of biosynthetically or postbiosynthetically modified or genetically polymorphic sites. The various forms may show considerable differences in their functional properties. Normal variant sites are due to alternative splicing, modification of certain amino acid residues, and proteolysis. Both the Aα and the γ chain occur in two splice forms, and it is known that only the shorter γ chain can interact with platelets, but the longer may bind thrombin and factor XIII. Many types of posttranslationally modified amino acid residues are present in fibrinogen. The Aα chain is partially phosphorylated at two sites, possibly leading to protection against proteolysis. The Bβ chain is N-glycosylated and partially proline hydroxylated, each at one site. The γ chain is N-glycosylated at one site and the longer splice form doubly tyrosine-sulfated. The glycosylations are believed to protect against polymerization and proteolysis. All three chains are partially oxidized at methionine residues and deamidated at asparagine and glutamine residues. The Aα and γ chain are partially carboxy-terminally degraded by proteolysis, the shorter forms causing a decrease in polymerization, crosslinking, and clot stability. Abnormal variants occur in patients with diabetes mellitus, in the form of glycated lysine residues; in patients with certain types of cancer, in the form of crosslinked degradation products; in patients with certain types of autoimmune disease, in the form of complexes with antibodies; in cigarette smokers; and in individuals treated with acetylsalicylic acid, in the form of acetylated lysine residues.

KEYWORDS: Heterogeneity; Alternative splicing; Posttranslational modification; Phosphorylation; Sulfation; Hydroxylation; Oxidation; Glycosylation; Proteolytic degradation; Glycation; Acetylation; Procoagulant; Anticoagulant; Fibrin network structure; Aspirin; Diabetes; Oxidative stress.

INTRODUCTION

The most significant biological role of fibrinogen is related to its ability to form the scaffold of a blood clot and thereby prevent the loss of blood after injury.[1] It is of greatest consequence for the maintenance of health that the capacity of soluble

---

Address for correspondence: Agnes H. Henschen-Edman, M.D., Ph.D., Department of Molecular Biology and Biochemistry, University of California, Irvine, CA 92697-3900, USA. Voice: 949-824-2465; fax: 949-824-8551.
ahensche@uci.edu

fibrinogen to be converted into insoluble fibrin is meticulously regulated, and that stable fibrin is formed only when and where it is needed. Fibrinogen plays also a role in the physiological and pathological processes related to wound healing, tumor growth, and metastasis as well as defense mechanisms. In order to fulfill its numerous functions, fibrinogen interacts in highly specific ways with a large number of other proteins as well as lower molecular weight and cellular components, primarily in the blood stream or blood vessel wall.[1-3] All functions and interactions are mediated by distinct and specific structural elements of the fibrin(ogen) molecule—the functional sites. They correspond to short amino acid sequences or to regional conformations and may be differentially expressed in fibrinogen, fibrin, and their degradation products.

The various functional sites and specific interactions may, in many cases, be classified as being of the procoagulant or the anticoagulant type, respectively. A tentative classification is presented in TABLE 1. However, since the hemostatic system is highly complex, certain interactions can be of both types. Thus, for instance, the interaction between fibrin(ogen) and thrombin is procoagulant in the sense that it results in clot formation, but anticoagulant in the sense that thrombin remains bound to the fibrin clot and thereby is prevented from causing additional clotting. Certain sites have been well identified from a structural as well as a functional point of view; for example, the thrombin and plasmin cleavage sites. Other sites are only partly known; for example, the crosslinking and platelet-binding sites. However, for many of the functional properties, the molecular counterparts in fibrinogen and fibrin have yet to be located.

TABLE 1. Functional sites in the fibrinogen structure, tentative classification

Interaction Type	Procoagulant Effect	Anticoagulant Effect	Uncertain Effect
Intrinsic	thrombin cleavage polymerization crosslinking	plasmin cleavage	
Protein interaction	thrombin Factor XIII α_2-antiplasmin	thrombin plasmin(ogen) plasminogen activators albumin complement protein C1q	fibronectin thrombospondin collagen lipoprotein(a)
Cell interaction	platelets (glycoprotein IIb/IIIa) erythrocytes		monocytes macrophages endothelial cells fibroblasts staphylococci, streptococci
Ion binding	calcium	heparin zinc citrate EDTA	

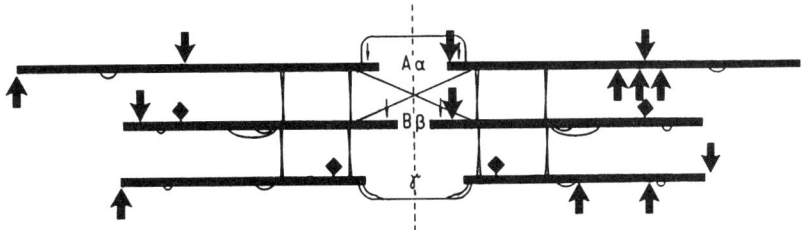

FIGURE 1. Human fibrinogen, a model of the covalent structure including major variant sites. The peptide chains are aligned according to homology with the N-terminal ends in the center. The *thin,* connecting *lines* represent disulfide bonds, the *thin arrows* thrombin cleavage sites, and the *diamonds* carbohydrate side chains. On the left side, the *fat arrows* pointing *upward* indicate the sites for alternative processing during biosynthesis and those pointing *downward* the genetically polymorphic sites. On the right side, the *fat arrows* pointing *upward* indicate the proteolytic cleavage sites and those pointing *downward* the phosphorylation sites in the Aα chain, the hydroxylation site in the Bβ chain, and the sulfation sites in the γ chain.

It is well established that in healthy individuals fibrinogen occurs in more than one million non-identical forms, due to the many possible combinations of alternative, posttranslationally modified, or genetically polymorphic sites.[1–4] The positions of the major variant sites within the human fibrinogen molecule are indicated in FIGURE 1. It has repeatedly been recognized that the various molecular forms may show considerable differences in their functional properties. It may be assumed that the majority of the forms occurring in a healthy individual are functionally beneficial with respect to type and abundance. However, conspicuous alterations in the distributions among preexisting forms, as well as additional forms, have often been observed in conjunction with differences in age of the individual and many types of disease. It is important to consider that it is often unknown whether the characteristic forms may indeed cause or aggravate the disease or they may appear as a result of the disease. The variant sites may be classified as occurring already in the normal, healthy individual, or primarily in conjunction with disease or other external factors.

NORMAL NON-INHERITED REGIONAL VARIANTS

Human fibrinogen is one of the most heterogeneous proteins so far known, since numerous structural elements are modified during or after biosynthesis. The modifications are in most cases incomplete, that is, two alternative forms or an unmodified and a modified form are present. Since these structural elements are combined in various ways in the individual molecules, the number of non-identical molecules becomes very large, in excess of one million.[2,3] Normal variant sites are caused by alternative splicing, modification of certain amino acid residues, and proteolysis as listed in TABLE 2.

Both the Aα and the γ chain occur in two alternative splice forms. Thus, a few percent of the Aα chains contain a 237-amino acid residue C-terminal extension.[5]

The specific functional properties of the corresponding higher molecular weight molecules are not yet known. In about 10% of the γ chains the most C-terminal four residues are replaced by 20 residues.[6] It is known that only the shorter γ chain can interact with platelets, but the longer may specifically bind thrombin and factor XIII.

Many types of posttranslationally modified amino acid residues occur in fibrinogen. The Aα chain is phosphorylated to about 25% at two serine residues, once in fibrinopeptide A and once in the middle of the chain.[7,8] Recent data indicate that the phosphorylation in the fibrinopeptide may lead to increased binding to thrombin.[9] However, fibrinopeptide release rate is independent of this modification.[12] The degree of phosphorylation is increased up to 70% in conjunction with an acute phase reaction,[8] most likely as a result of the increased synthesis and turnover rates characteristic of this reaction. It has been suggested that fibrinogen is synthesized in a highly phosphorylated form and that the phosphate groups are subsequently removed by a phosphatase.[10] The normally occurring 25% of modification would then correspond to the steady state condition in a healthy person. It is of interest that also in the healthy fetus or newborn the degree of phosphorylation is considerably increased.[11] In fibrinopeptide A the phosphate group protects against degradation by aminopeptidases (see below).[12]

Tyrosine sulfation of fibrinopeptide B is found in many animal species, but human fibrinogen lacks tyrosine in its fibrinopeptides and only the longer splice form of the γ chain is sulfated at both tyrosines.[13] The modification is quantitative, as judged by the appearance of a single peak on HPLC of a related fragment. It has been shown to be relevant to the binding of factor XIII.

The N-terminal glutamine encoded in the gene sequence of the Bβ chain is completely cyclized to pyroglutamic acid.[7] This is the normal case with N-terminal glutamine residues. It has been suggested that this serves as a protection against degradation by aminopeptidases. Hydroxylation of proline residues, a modification characteristic of collagen type proteins, occurs to about 20% at a single proline in the N-terminal region of the Bβ chain.[14] The effect on function is unknown.

Both the Bβ and the γ chain are N-glycosylated, the Bβ chain in the C-terminal region,[15] the γ chain in the N-terminal region,[16] and each chain at a single asparagine residue. The carbohydrate side chains have the structure occurring in many plasma glycoproteins and contain N-acetyl-glucosamine, mannose, galactose, and sialic (neuraminic) acid in a biantennary configuration. The glycosylation sites are fully occupied. Only the degree of sialylation at the end of each of the antennæ varies as judged by the appearance of several peaks on HPLC and mass-spectrometric analysis of tryptic glycopeptides.[17] Sialylation has been reported to be increased in very young and old individuals, and in those with liver disease.[11,17,18] The glycosylations are believed to protect against excessive polymerization and proteolysis.

All three chains are partially oxidized at methionine residues, as concluded from the observation that certain, identified methionines are resistant to cyanogen bromide cleavage since they are present in the sulfoxide form.[19] This modification is not an artifact of the fibrinogen purification procedure because it is found also in fibrin freshly isolated from plasma by clotting with thrombin. The modification may serve as an indicator of oxidative stress (see below). All three chains are also partially deamidated at asparagine and glutamine residues, although the positions are not yet known. This modification is believed to be related to the molecular aging process.

TABLE 2. Normal, non-inherited regional variants in the fibrinogen structure

Type	Site	Normal Levels	Function	Change in Distribution
Alternative splicing[5]	Aα C-terminal	99% short, 1% long form	unknown	fetus: 97% short, 3% long form
Alternative splicing[6]	γ C-terminal	90% short, 10% long form	long form binds thrombin and Factor XIII, but unable to bind to platelets	unknown
Phosphorylation[7–12]	Aα 3 Ser, Aα 345 Ser	25% modified form	protection against fibrinolysis and other proteolysis	fetus: 70% modified form, during acute-phase reaction: up to 70% modified form
Sulfation[13]	γ 418 Tyr, γ 422 Tyr, both in long splice form	100% modified form	binding of thrombin and Factor XIII	unknown
Hydroxylation[14]	Bβ 31 Pro	20% modified form	unknown	during acute phase reaction: decreased?
Glutamine-cyclization[7]	Bβ 1 Gln	100% modified form	protection against proteolysis	unknown
Glycosylation[15–18]	Bβ 364 Asn and γ 52 Asn	100% modified form, 42 and 27% disialylated, 58 and 73% monosialylated, 0% asialylated in Bβ and γ	protection against polymerization and proteolysis	fetus, high age and during liver disease: hypersialylation

TABLE 2/continued.

Type	Site	Normal Levels	Function	Change in Distribution
Methionine-oxidation[19]	Many Met in Aα, Bβ, and γ	0–20% modified form	polymerization inhibition	high age, during oxidative stress, smoking: increased?
Deamidation[12]	Several Asn and Gln in Aα, Bβ, and γ	0–10% modified form?	unknown	high age: increased?
Proteolysis[7,12]	Aα 1 Ala lost	10% modified form	unknown	fetus and during acute-phase reaction: decreased
Proteolysis	Aα C-terminal 27 amino acids lost	25% modified form	unknown	unknown
Proteolysis[20–30]	Aα C-terminal 300–340 amino acids lost	20% modified forms	decreased polymerization rate and clot stability	high age: increased; during acute-phase reaction: decreased
Proteolysis[32]	γ long splice form C-terminal 4 amino acids lost		loss of thrombin binding	unknown
Proteolysis[33]	γ C-terminal 100–250 amino acids lost	7% modified forms	loss of crosslinking and platelet binding ability	unknown

The Aα chain is proteolytically degraded from both ends. At the N-terminal end only the first residue, an alanine, is removed from 10% of the chains by an aminopeptidase.[7] N-Terminal sequence analysis showed that fibrinopeptide A phosphorylated in position 3 (see above) never lost this alanine.[12] Evidently the aminopeptidase can only act on the portion of the Aα chains that already has been dephosphorylated. At the C-terminal end the Aα chain may lose up to 340 amino acid residues.[20] The least extensive loss, that is, of the last 27 amino acids, may be caused by plasmic degradation, since it corresponds to the earliest plasmin cleavage site (see TABLE 3). It is found in about 25% of the Aα chains. The loss of the C-terminal 300 to 340 amino acids is highly heterogeneous and occurs in 20% of the chains.[20] The main variants have been reported to end in positions 269, 297, and 309.[20] As a result fibrinogen is found in three different molecular weight forms, 340, 305, and 270 kD, designated HMW, LMW, and LMW'. Thus, the LMW form contains one long and one short Aα chain and the LMW' form two short Aα chains. The three forms occur as 70, 25, and 5% of plasma fibrinogen in a healthy person.[21,22]

There has been considerable interest in the function of the C-terminal region of the Aα chain[22-27] as well as the origin and relevance of the degraded Aα chains.[20,22,28-31] However, to date no enzyme has been found that could cause a proteolytic cleavage of an appropriate kind. N-terminal sequence analyses of the LMW

TABLE 3. Proteolytic degradation of native fibrinogen

Enzyme	Order of Earliest Cleavages	Comment
Thrombin[1,7]	1: Aα16, 2: Bβ14	
Snake venom enzymes[12]	1: Aα16, 2: Aα C-terminus and/or Bβ N-terminus	
Plasmin 1:2000, <30min[1,28]	1: Aα583, 2: Aα206 + 219 + 230, 3: Aα239 + 424 + Bβ42	
Leukocyte elastase 1:2000, <22h[28,30]	1: Aα21 + Bβ46 + 48 + 50, 2: Aα93 + 271 + 450 + 498 + 499 + 503 + 504 + 522 + 568 + Bβ138 + 139 + γ79 + 82	
MMP-1 (Collagenase) 1:4500, 24h[12,20]	Aα475 + 477	incompatible MW of fragments
MMP-2 (Gelatinase)[20]	Aα?	incompatible MW of fragments
MMP-3 (Stromelysin) 1:900, <24h[12,20]	1: Aα520, 2: Aα228+231, 3: Aα480, 4: Aα217	incompatible MW of fragments
MMP-7 (Matrilysin) 1:3000, <24h[12,20]	Aα413 + 477	incompatible MW of fragments
Trypsin 1:6000, 1h[12]	Aα19 + 348 + 424 + 491 + 556 + 602 + Bβ42	
Chymotrypsin, elastase, thermolysin, proteinase K[12]	multiple sites in Aα and Bβ N-terminus and many other regions	

and LMW' forms clearly show that all three chains are N-terminally intact, except for the partial loss of alanine in the first position of the Aα chain.[28] The earliest proteolytic cleavages obtained in native human fibrinogen with a large number of proteases are summarized in TABLE 3.[12] The data are based on N-terminal sequence analyses of the complete digests present after various periods of time.[12,28,30] Most of the enzymes, such as thrombin, snake venom enzymes, leukocyte elastase,[28,30] trypsin, and other common serine proteases, attack the N-terminal regions of the Aα and Bβ chains before virtually any more C-terminal bond is cleaved. Furthermore, those enzymes that indeed attack the C-terminal region of the Aα chain, plasmin,[28] leukocyte elastase,[28,30] and various matrix metalloproteinases (MMPs),[12] split the chain far away from and often considerably closer to the N-terminus than the C-termini of the short Aα chains present in plasma fibrinogen. In addition, the molecular weights of the fragments obtained have been reported to differ from those of LMW and LMW' fibrinogen.[20] An identification of the enzymes involved in the formation of the shorter Aα chains would considerably enhance our understanding of the relevance of the corresponding fibrinogen variants. Their distribution changes during acute phase reactions, when more of the HMW form is present.[22,27] The two LMW forms are characterized by impaired fibrin polymerization.[21,26]

The γ chain is partially degraded by proteolytic removal of C-terminal regions. Thus, the last four amino acids of the longer splice variant may be missing.[32] In about 6% of the γ chains the last approximately 110 amino acids are absent, possibly due to an unusual and premature plasmin cleavage.[33] Furthermore, in 1% of the chains the molecular weight corresponds to a cleavage of as much as 250 residues from the C-terminal.[33] The shorter γ chains have to be devoid of crosslinking and platelet binding ability since the corresponding regions are absent.

ABNORMAL NON-INHERITED REGIONAL VARIANTS

Certain postbiosynthetic protein modifications may be specifically caused by diseases or drugs. Several different modifications of these types have been found to occur in fibrinogen as listed in TABLE 4. The structural alterations typically change the functional properties of the protein.

Diabetes is one of the important risk factors for cardiovascular disease and it has repeatedly been discussed that at least part of the increased risk is related to the high glucose levels found in patients with insufficiently controlled or controllable insulin treatment and a resulting fibrinogen modification.[34,35] Thus, it has been suggested that monitoring fibrinogen glycation may be of value for the identification of short term fluctuations in the blood glucose level.[34] Recent investigations provide strong evidence for the effect of glucose on fibrinogen structure and function.[12] Mass-spectrometric analyses of fragments of fibrinogen that previously had been incubated with glucose showed, in several cases, the addition of 162 mass units to the original mass of the fragment. This mass increase is characteristic of glycation of lysine residues. Additional evidence for the reaction with glucose was obtained by sequence analysis of tryptic fragments retained by the glycation specific boronate affinity column. Certain sites within the fibrinogen structure were preferentially glycated. It is expected that some of these correspond to plasmic cleavage sites and

TABLE 4. Abnormal, non-inherited regional variants in the fibrinogen structure

Type	Sites	Normal Level	Function	Occurrence
Glycation (glucosylation)[12,34,35]	certain amino groups	0%?	protection against fibrinolysis; decreased polymerization rate	diabetes mellitus
Oxidation[12,19,36–40]	Met and certain other amino acids	low	decreased polymerization rate	oxidative stress
Oxidation[12,41,42]	unknown	0%	decreased polymerization rate	cigarette smoking
Additional degradation, crosslinking[12,43]	proteolytic cleavage sites, crosslinking sites	0%	decreased polymerization rate and clot stability?	certain types of cancer
Antibody binding[44,45]	antigenic epitope	0%	polymerization inhibition; increased degradation rate	autoimmune diseases, myeloma
Acetylation[12,46,47]	certain amino groups	0%	increase in fibrin gel porosity and susceptibility to fibrinolysis; decreased polymerization rate and clot stability	acetylsalicylic acid treatment

crosslinking sites, that is, glycation may in the patient interfere both with proteolytic fibrin degradation and crosslinking. The influence of glucose addition on function was tested in a preliminary way by measuring the time dependence and extent of thrombin induced fibrinopeptide release, monitored by HPLC, and by polymerization, monitored as an increase in light absorbance. The effect of glycation on peptide release was marginal, but there was extensive, dose dependent inhibition of fibrin polymerization. Similar evidence for glycation was obtained with samples of fibrinogen from diabetic patients.[12]

Reactive oxygen species of several types have, in recent years, been shown to be involved in a wide range of pathophysiological conditions, such as cardiovascular diseases and inflammation. In a number of model systems, where specific oxidative molecules are generated, the functional properties of fibrinogen have been demonstrated to be damaged. Photo-oxidation with visible light in the presence of dyes like methylene blue leads to a complete loss of clottability due to the irradiation time dependent modification of specific histidine residues.[36,37] In this case, polymerization is impaired even though fibrinopeptides are released. Metal ion catalyzed oxidation and γ-irradiation create carbonyl groups and other modifications in unidentified sites and these alterations inhibit thrombin induced clot formation.[38,39] Chloramine-T oxidation of identified methionine residues in the C-terminal regions of the Bβ and γ chain is connected with a dose dependent decrease in fibrin polymerization.[19,40] The E domain retains its function, but the D domain gradually loses its ability to bind to fibrin attached to a solid support. It is of interest that the methionines oxidized by chloramine-T are the same as those partially present in the sulfoxide form already in normal plasma fibrinogen (see above).[19] Cigarette smoking has been demonstrated to introduce carbonyl functions into plasma proteins.[41] When fibrinogen is subjected to cigarette smoke in a model experiment it gradually loses its ability to clot on thrombin addition.[12,42] In all these studies of the effects of oxidative status or stress on the ability of fibrin to polymerize and thereby cause an increase in light absorption has been used as a first and simple test of fibrinogen function. Obviously, since fibrin(ogen) is involved in so many and quite different molecular interactions (see TABLE 1) a much wider range of tests would have to be employed to fully evaluate the impact of various types of oxidation.

Blood coagulation related problems are often the first symptoms leading to the detection of cancer. The types of fibrinogen modifications sometimes observed in cancer patients may serve as examples of the highly complex structural alterations that can accompany certain diseases.[43] The structural changes may include proteolytic degradation by plasmin and, possibly, tumor specific proteases, as well as factor XIII mediated and, possibly, tumor specific transglutaminases, and the combinations can lead to the appearance of unusual types of fragments.[12] Fibrinogen specific or unrelated antibodies may influence blood coagulation in patients with autoimmune diseases.[44] Recently, patients have been described with stable, virtually stoichiometric complexes between fibrinogen, an immunoglobulin and the complement protein C1q in their blood.[45]

It is well established that the risk of cardiovascular disease can be lowered by aspirin treatment since this drug acetylates a cyclooxygenase of the prostaglandin system and thereby inhibits platelet function. More recently it has been discovered that aspirin treatment will also lead to a looser fibrin clot structure that seems to be less dangerous because it is more accessible to proteolytic enzymes.[46] Model experiments have been performed to find evidence for the effect of aspirin on fibrinogen structure and function.[12,47] Mass-spectrometric analyses of fragments of fibrinogen previously incubated with aspirin, showed in many cases the addition of 42 mass units to the original mass of the fragment. This mass increase is characteristic of acetylation of lysine residues. Certain lysine amino groups in the fibrinogen structure were more readily modified than others and, just as with glucose induced modification, there was evidence that lysine residues normally involved in plasmic degradation and crosslinking had become unavailable. The influence of aspirin modification on function was tested in a preliminary way by measuring the time dependence and extent of thrombin induced fibrinopeptide release, monitored by HPLC, and polymerization, monitored as an increase in light absorbance. There was virtually no effect of acetylation on fibrinopeptide release, but an extensive, dose dependent inhibition of fibrin polymerization was observed.[12,47]

All the fibrinogen modifications mentioned above occurring in conjunction with diseases or medication change the functional properties of fibrin(ogen). The effects are expected to be of pathophysiological relevance and the characterization of the effects will contribute to our understanding of, for example, the increased risk of cardiovascular disease associated with diabetes and the decreased risk associated with aspirin medication. It may at first seem surprising that both the clinically beneficial fibrinogen modification by aspirin and the presumably detrimental modifications by glucose or various oxidizing agents are characterized by fibrin polymerization inhibition. However, fibrin formation and the fibrin network structure have so far only been characterized in terms of time dependent increase in light absorption, but not in terms of change in other, more relevant physical properties, such as gel porosity and mechanical stability. Thus, it has been preliminarily observed that the fibrin clots obtained from glucose treated fibrinogen were rigid, but those obtained from aspirin treated fibrinogen were loose and soft. The critical evaluation of the influence on health and disease of the various abnormal modifications requires a panel of tests corresponding to a large number of the procoagulant and anticoagulant sites and interactions of fibrinogen mentioned in TABLE 1.

CONCLUSIONS

Much work remains to be performed in order to firmly establish both the existence and the functional relevance of variant sites in fibrinogen and fibrin. It is quite likely that several types of modification, both of the kind primarily present in the healthy individual and of the kind present in conjunction with disease or medical treatment, not yet have been detected. It is often difficult to study the effect of each modification separately, even under laboratory conditions, because they always occur together. However, in special cases it has been feasible to fractionate the population of fibrinogen molecules in a variant site specific way. Thus, molecules corresponding to the Aα and γ chain splice variants[5,32,48] and the three molecular weight forms[20,21,28,49] have been successfully isolated from each other. Furthermore, certain modifications may be selectively removed or added, leaving the other modifications untouched. Thus, treatment with phosphatase has been used to produce phosphate free material and treatment with different glycosidases to produce material devoid of sialic acid or the complete carbohydrate side chains.[18,50] The material can also be examined before and after reaction with glucose, oxidizing agents, or aspirin.[12,19,36–40,42,47] In order to evaluate the true relevance to health and disease of the variant sites a large number of individuals have to be screened and for this purpose simplified analytical procedures, suited for the clinical laboratory, have to be developed. Subsequently, it may be of considerable diagnostic and prognostic value to monitor the variant levels that occur in various categories of patients.

ACKNOWLEDGMENTS

The author would like to thank the following for stimulating and fruitful discussions and collaboration as well as assistance: Dr. Birger Blombäck, Dr. Margareta Blombäck, Dr. Dennis Galanakis, Dr. Shu He, Dr. Susan Lord, John Paul Almeda, Vance Cao, Harry Hu, Rakhi Sinha, Ali Razmara, Quin Vu, and Gina Yoo.

REFERENCES

1. HENSCHEN, A. & J. MCDONAGH. 1986. Fibrinogen, fibrin and factor XIII. *In* Blood Coagulation. R.F.A. Zwaal & H.C. Hemker. Eds.: 171–241. Elsevier, Amsterdam.
2. HENSCHEN, A.H. 1993. Human fibrinogen—structural variants and functional sites. Thromb. Hæmost. **70:** 42–47.
3. HENSCHEN-EDMAN, A.H. 1999. On the identification of beneficial and detrimental molecular forms of fibrinogen. Hæmost. **29:** 179–186.
4. BAUMANN, R.E. & A.H. HENSCHEN. 1993. Human fibrinogen polymorphic site analysis by restriction endonuclease digestion and allele-specific polymerase chain reaction amplification: identification of polymorphisms at positions Aα312 and Bβ448. Blood **82:** 2117–2124.
5. FU, Y., J.Z. ZHANG, C.M. REDMAN, *et al.* 1998. Formation of the human fibrinogen subclass Fib$_{420}$: disulfide bonds and glycosylation in its unique ($α_E$ chain) domains. Blood **92:** 3302–3308.
6. FORNACE, A.J., D.E. CUMMINGS, C.M. COMEAU, *et al.* 1984. Structure of the human γ-fibrinogen gene. Alternate mRNA splicing near the 3′ end of the gene produces γA and γB forms of γ-fibrinogen. J. Biol. Chem. **259:** 12826–12830.
7. BLOMBÄCK, B., M. BLOMBÄCK, P. EDMAN, *et al.* 1966. Human fibrinopeptides: isolation, characterization and structure. Biochim. Biophys. Acta **115:** 371–396.
8. SEYDEWITZ, H.H. & I. WITT. 1985. Increased phosphorylation of human fibrinopeptide A under acute phase conditions. Thromb. Res. **40:** 29–39.
9. MAURER, M.C., J.L. PENG, S.S. AN, *et al.* 1998. Structural examination of the influence of phosphorylation on the binding of fibrinopeptide A to bovine thrombin. Biochemistry **37:** 5888–5902.
10. KUDRYK, B., M. OKADA, C.M. REDMAN, *et al.* 1982. Biosynthesis of dog fibrinogen. Characterization of nascent fibrinogen in the rough endoplasmic reticulum. Eur. J. Biochem. **125:** 673–682.
11. GALANAKIS, D.K., J. MARTINEZ, C. MCDEWITT, *et al.* 1983. Human fetal fibrinogen: its characteristics of delayed fibrin formation, high sialic acid and AP peptide content are more marked in the pre-term than in term samples. Ann. N.Y. Acad. Sci. **408:** 640–643.
12. HENSCHEN-EDMAN, A.H., *et al.* unpublished data.
13. HENSCHEN, A.H. 1993. On the occurrence and significance of tyrosine sulfate in fibrinogen. Blood Coagul. Fibrinol. **4:** 822.

14. HENSCHEN, A.H., I. THEODOR & H. PIRKLE. 1991. Hydroxyproline, a posttranslational modification of proline, is a constituent of human fibrinogen. Thromb. Hæmost. **65:** 821.
15. TÖPFER-PETERSEN, E., F. LOTTSPEICH & A. HENSCHEN. 1976. Carbohydrate linkage site in the β-chain of human fibrin. Hoppe-Seyler's Z. Physiol. Chem. **357:** 1509–1513.
16. BLOMBÄCK, B., N.J. GRÖNDAHL, B. HESSEL, et al. 1973. Primary structure of human fibrinogen and fibrin. II. Structural studies on NH_2-terminal part of γ-chain. J. Biol. Chem. **248:** 5806–5820.
17. HENSCHEN-EDMAN, A.H., V.H. CAO & D. GALANAKIS. 1998. Screening of adult and fetal human fibrinogen with matrix-assisted laser-desorption ionization time-of-flight (MALDI-TOF) mass spectrometry. Blood Coagul. Fibrinol. **9:** 698.
18. MARTINEZ, J. & C. BARSIGIAN. 1987. Biology of disease. Carbohydrate abnormalities of N-linked plasma glycoproteins in liver disease. Lab. Invest. **57:** 240–257.
19. CHEN, N. & A. HENSCHEN. 1994. Identification of methionine sulfoxide in native and oxidized fibrinogen. Protein Sci. 3(Suppl. 1): 147.
20. NAKASHIMA, A., S. SASAKI, K. MIYAZAKI, et al. 1992. Human fibrinogen heterogeneity: the COOH-terminal residues of defective Aα chains of fibrinogen II. Blood Coagul. Fibrinol. **3:** 361–370.
21. HOLM, B., F. BROSSTAD, P. KIERULF, et al. 1985. Polymerization properties of two normally circulating fibrinogens, HMW and LMW. Evidence that the COOH-terminal end of the α-chain is of importance for fibrin polymerization. Thromb. Res. **39:** 595–606.
22. NIEUWENHUIZEN, W. 1995. Biochemistry and measurement of fibrinogen. Eur. Heart J. **16**(Suppl. A): 6–10.
23. VEKLICH, Y.I., O.V. GORKUN, L.V. MEDVED', et al. 1993. Carboxyl-terminal portions of the α chains of fibrinogen and fibrin. J. Biol. Chem. **268:** 13577–13585.
24. GORKUN, O.V., Y.I. VEKLICH, L.V. MEDVED', et al. 1994. Role of the αC domains of fibrin in clot formation. Biochemistry **33:** 6986–6997.
25. LITVINOVICH, S.V., A.H. HENSCHEN, K.G. KRIEGLSTEIN, et al. 1995. Structural and functional characterization of proteolytic fragments derived from the C-terminal regions of bovine fibrinogen. Eur. Biochem. J. **229:** 605–614.
26. GORKUN, O.V., A. HENSCHEN-EDMAN, L.F. PING, et al. 1998. Analysis of Aα251 fibrinogen: the αC domain has a role in polymerization, albeit more subtle than anticipated from the analogous proteolytic fragment X. Biochemistry **37:** 15434–15441.
27. REGANON, E., V. VILA, F. FERNANDO, et al. 1999. Elevated high molecular weight fibrinogen in plasma is predictive of coronary ischemic events after acute myocardial infarction. Thromb. Hæmost. **82:** 1403–1405.
28. MÜLLER, E. & A. HENSCHEN. 1988. Isolation and characterization of human plasma fibrinogen molecular-size-variants by high-performance liquid chromatography and amino acid sequence analysis. In Fibrinogen—Biochemistry, Biological Functions, Gene Regulation and Expression. M.W. Mosesson, D.L Amrani, K.R. Siebenlist, et al., Eds.: 279–282. Elsevier, Amsterdam.
29. STERRENBERG, L., H.L. HAAK, E.J.P. BROMMER, et al. 1985. Evidence of fibrinogen breakdown by leukocyte enzymes in a patient with acute promyelocytic leukemia. Hæmost. **15:** 126–133.
30. GORETZKI, L., E. MÜLLER & A. HENSCHEN. 1987. On the cleavage specificity of leukocyte elastase towards fibrinogen. In Fibrinogen—Biochemistry, Physiology and Clinical Relevance. G.D.O. Lowe, J.T. Douglas, C.D. Forbes, et al., Eds.: 293–296. Elsevier, Amsterdam.
31. BACH-GANSMO, E.T., S. HALVORSEN, H.C. GODAL, et al. 1994. Impaired coagulation of fibrinogen due to digestion of the C-terminal end of the Aα-chain by human neutrophil elastase. Thromb. Res. **73:** 61–68.
32. FRANCIS, C.W., E. MÜLLER, A. HENSCHEN, et al. 1988. Carboxy-terminal amino acid sequences of two large variant forms of the human plasma fibrinogen γ chain. Proc. Natl. Acad. Sci. U.S.A. **85:** 3358–3362.

33. HENSCHEN, A. & P. EDMAN. 1972. Large-scale preparation of S-carboxymethylated chains of human fibrin and fibrinogen and the occurrence of γ-chain variants. Biochim. Biophys. Acta **263:** 361–367.
34. HAMMER, M.R., P.N. JOHN, M.D. FLYNN, et al. 1989. Glycated fibrinogen: a new index of short-term diabetic control. Ann. Clin. Biochem. **26:** 58–62.
35. JÖRNESKOG, G., K. FATAH & M. BLOMBÄCK. 1998. Fibrin gel structure in diabetic patients before and during treatment with acetylsalicylic acid: a pilot study. Fibrinol. Proteol. **12:** 360–365.
36. YOSHIMOTO, T., Y. SAITO & Y. INADA. 1987. Non-clottable fibrinogen by photooxidation in blood plasma. Photochem. Photobiol. **45:** 675–676.
37. HENSCHEN-EDMAN, A.H. 1997. Photo-oxidation of histidine as a probe for aminoterminal conformational changes during fibrinogen-fibrin conversion. Cell. Molec. Life Sci. **53:** 28–32.
38. KARPEL, R., G. MARX & M. CHEVION. 1991. Free radical-induced fibrinogen coagulation: modulation of neofibe formation by concentration, pH and temperature. Israel J. Med. Sci. **27:** 61–66.
39. SHACHTER, E., J.A. WILLIAMS & R.L. LEVINE. 1995. Oxidative modification of fibrinogen inhibits thrombin-catalyzed clot formation. Free Radical Biol. Med. **18:** 815–821.
40. STIEF, T.W. 1993. Oxidized fibrin stimulates the activation of pro-urokinase and is the preferential substrate of human plasmin. Blood Coagul. Fibrinol. **4:** 117–121.
41. REZNICK, A.Z., C.E. CROSS, M.L. HU, et al. 1992. Modification of plasma proteins by cigarette smoke as measured by protein carbonyl formation. Biochem. J. **286:** 607–611.
42. GALANAKIS, D.K., P. LAURENT & A. JANOFF. 1982. Cigarette smoke contains anticoagulants against fibrin polymerization and factor XIIIa in plasma. Science **217:** 642–645.
43. WILHELM, O., R. HAFTER, A. HENSCHEN, et al. 1990. Role of plasmin in the degradation of the stroma-derived fibrin in human ovarian carcinoma. Blood **75:** 1673–1678.
44. CARR, M.E., R.M. DENT & S.L. CARR. 1996. Abnormal fibrin structure and inhibition of fibrinolysis in patients with multiple myeloma. J. Lab. Clin. Med. **128:** 83–88.
45. GALANAKIS, D.K., A. HENSCHEN, J. WEISEL, et al. 2001. Circulating immune complexes with anti-fibrinogen IgG in a hypodysfibrinogenemic proband: isolation, stoichiometry, and partial characterization. Ann. N.Y. Acad. Sci. **936:** this volume.
46. WILLIAMS, S., K. FATAH, P. HJEMDAHL, et al. 1998. Better increase in fibrin gel porosity by low dose than by intermediate dose acetylsalicylic acid. Eur. Heart J. **19:** 1666–1672.
47. SHU, H., M. BLOMBÄCK, G. YOO, et al. 2001. Modified clotting properties of fibrinogen in presence of acetylsalicylic acid in a purified system. Ann. N.Y. Acad. Sci. **936:** this volume.
48. WOLFENSTEIN-TODEL, C. & M.W. MOSESSON. 1980. Human plasma fibrinogen heterogeneity: evidence for an extended carboxyl-terminal sequence in a normal γ chain variant (γ'). Proc. Natl. Acad. Sci. U.S.A. **77:** 5069–5073.
49. GALANAKIS, D.K. & M.W. MOSESSON. 1983. Human fibrinogen heterogeneities: determination of the major Aα chain derivatives in blood. Thromb. Res. **31:** 403–413.
50. LANGER, B.G., J.W. WEISEL, P.A. DINAUER, et al. 1988. Deglycosylation of fibrinogen accelerates polymerization and increases lateral aggregation of fibrin fibers. J. Biol. Chem. **263:** 15056–15063.

Fibrin Degradation Products

A Review of Structures Found *in Vitro* and *in Vivo*

PATRICK J. GAFFNEY

Division of Hæmatology, National Institute for Biological Standards and Control, Blanche Lane, South Mimms, Potters Bar, Hertfordshire, United Kingdom

ABSTRACT: This review attempts to relate subunit structures of fibrin degradation products (FnDP) made *in vitro* with structures found *in vivo*. The domainally directed fragmentation *in vitro* of both fibrinogen and fibrin is emphasized, all fragments being various associations of the two core fibrin fragments D and E. The digestion of fibrinogen by plasmin *in vivo* is rare, and the crosslinking of fibrin *in vivo* takes place at a very early stage in the clotting/polymerization process. The notion that as fibrin forms *in vivo* it orchestrates its own destruction is developed. Plasmas from patients suffering from dessiminated intravascular coagulation demonstrate very large crosslinked FnDP fragments in their plasmas which seem to contain not alone fibrinopeptide A but also subunits with intact alpha chains. This is interpreted to mean that many of the large soluble fragments found *in vivo*, and heretofore known as FnDP, are in reality long fibrin polymers, random parts of whose structures have been converted to FnDP by lysis of the carboxy terminal regions of the alpha chains in the polymer. The ratio of intact fibrin to FnDP in these large soluble structures may be a useful clinical marker; however, such data can only be relied upon when blood samples are taken into an anticoagulant mix that contains a fibrinolytic inhibitor. Some of the biological effects of FnDP structures *in vivo* (fibrinogen synthesis, vasoactivity) are still quite ambiguous.

KEYWORDS: Fibrin degradation products; Structures *in vitro*; Structures *in vivo*.

INTRODUCTION

In this review the author exercises his privilege to be selective on many aspects of fibrin degradation product (FnDP) biochemistry and physiology. Since much of our biochemical data have been obtained using purified systems, some of the FnDP structures defined are probably quite irrelevant physiologically. However, this overview includes a brief historical and biochemical outline of the development of our current knowledge of FnDP. The literature is vast on this subject, thus this author will have to be selective with respect to quotation, observing, in as much as is possible, recognition of primary discovery and reporting of strategic firsts, but paying little attention to the cosmetic adjustment of previous discoveries by many authors. Despite the presumption that fibrinolysis is a major defence against thrombotic disease, which is a major cause of mortality in Western society, only rarely are tests for fibrinolysis con-

Address for correspondence: Patrick J. Gaffney, D.Sc., Ph.D., Academic Department of Surgery, St. Thomas' Hospital, Lambeth Palace Road, London SE1 7EH, UK.
pejaygaffney@yahoo.com

ducted against a clinical setting. There must be some explanation of this seeming contradiction. This author feels that an explanation may involve the detailed structural analysis of the various fibrinogen degradation products (FgDP) and FnDP using purified fibrinogen and fibrin[1-6] although there is a paucity of data on the structural nature of FgDP and FnDP found in plasma, both of which are relevant in various clinical pathologies. In this review I try to compensate for this historic deficiency, laying special emphasis on Fg/FnDP in plasma and invoking author's privilege to speculate on the relevance of certain of these fragments found in human plasma. The reader is reminded that although we normally understand degradation products of fibrinogen and fibrin to be the results of plasmin activity, the interaction of thrombin with fibrinogen also generates a product called fibrin. That fibrin orchestrates its own destruction by plasmin is now accepted.[7-10] Thus this overview attempts to intertwine the competing interactions of thrombin and plasmin with fibrinogen in the generation of FnDP and, indeed, to relate these opinions with our experience of the measurement of degradation products in plasma. As is outlined below, investigations of the FDP fraction in plasma have allowed us to place in physiological perspective the structures that have been elucidated by examining the degradation of purified fibrinogen and fibrin. These data have led this author to speculate that degradation products of fibrinogen are rare *in vivo* and their frequent reportage in the literature has been due to the

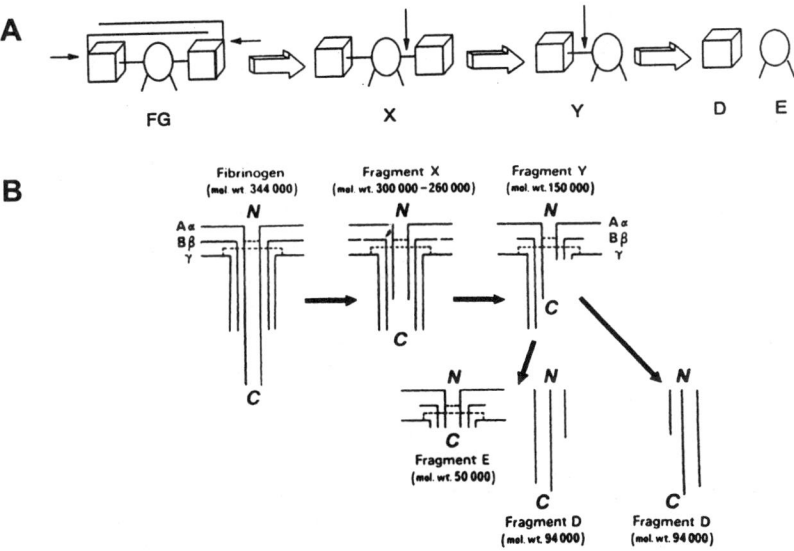

FIGURE 1. A. A domainal perspective of the plasmin mediated conversion of fibrinogen (*FG*) to its fragments *D* and *E* showing the intermediate fragments *X* and *Y*. **B.** The major plasmin digestion takes place at the C-terminals of the Aα chains, the N-terminals of the Bβ chains, and the coiled coil peptide sequences joining the plasmin resistant domains D and E. The various structures shown in the figure indicate the polypeptide compositions of the various fragments. This domainally directed digestion determines the structure of the fragments generated by plasmin from both fibrinogen and fibrin. (Reproduced from Refs. 2 and 69, with permission).

fact that we had thought that fragments containing fibrinopeptide A (FPA), of necessity, originated in fibrinogen, a suggestion that we now know to be most probably erroneous. I try to deal with these physiologically important features of FnDP interpretation in plasma.

SOME EARLY HISTORICAL OBSERVATIONS

Although the presence of fibrin in plasma and the existence of fibrinolytic activity has been known for some time, the earliest structural observations on fibrinogen–plasmin interaction was made by the late Professor Seegers when he described[11] α and β fibrinogen degradation products, which we now call core fragments D and E. Nussenzweig and colleagues[12] made one of the earliest comprehensive studies of the plasmin–fibrinogen interaction and classified the DEAE column chromatographic fractions of this interaction as A, B, C, D, and E in order of their elution from the ion exchange column. We now know that fractions A, B, and C were mainly peptide groups derived from the carboxy termini of the Aα chains and the NH_2 termini of the Bβ chain of fibrinogen, whereas the D and E fractions are plasmin resistant core fragments that make up 70% of the total molecule of fibrinogen (see FIGURE 1).

FURTHER DETAILS OF THE FIBRINOGEN–PLASMIN INTERACTION

This author opines that the interaction of plasmin with fibrinogen is a fairly unlikely physiological event (except during exogenous use of streptokinase or urokinase in thrombolytic therapy). However, much work has been performed on the structural events that take place during this reaction (FIG. 1). This work was obviously performed with a view to understanding the structure of fibrin degradation *in vivo*. Although this has guided us in understanding the domainal fragmentation of fibrinogen and fibrin, it is really of little practical significance since our more recent data has suggested two important facts (1) the interaction of plasmin with intact fibrinogen is unlikely, and (2) the presence of non-crosslinked fibrin (NXL-FN) in plasma is also unlikely in that crosslinking seems to occur between the γ chains of forming fibrin very early, probably at the fibrin dimer or trimer level. Despite these facts (which we develop later) it is worthwhile expressing here the highlights of the development of our knowledge of the fibrinogen–plasmin interaction. Following the observations of Nussenzweig *et al.*[12] in 1961, the next milestone on which there was consensus was the asymmetric scheme of the digestion of fibrinogen.[13,14] This is depicted in the schemes shown in FIGURE 1, one scheme showing the polypeptide chain composition of the major fragments released by plasmin from fibrinogen (FIG. 1 B), and the other scheme showing these fragments as amalgams of the basic two core fragments of fibrinogen, namely D and E (FIG. 1 A). Human fibrinogen is composed of a pair of three dissimilar polypeptide chains named Aα, Bβ and γ.[15] Each chain is disulphide bonded to each of the other two. The two Aα chains are also joined by disulphide bridges, as are the two γ chains near the NH_2 terminus of the molecule.[16,17] The molecular weights of the Aα, Bβ and γ chains have been reported[18] as 67,000, 58,000, and 47,000 daltons, respectively, suggesting a molecular weight for the entire molecule of 344,000 daltons. Additional structural detail of this large molecule can be found elsewhere.[19,20] In this presentation, the domainal structure of human fibrinogen is discussed, since it has such an impact on the structures found in the various fibrin degradation products. Human fibrinogen contains 29 disulphide bonds and, although some are intrachain, the majority are interchain and are concentrated in clusters that endow the molecule with a domainal structure.[16] FIGURE 1 shows the polypeptide chains of fibrinogen with a crude representation of the disulphide clusters that determine the various domains. One cluster links the NH_2 terminals of all three chains and also join the two Aα and γ chains in what has been described as an NH_2 terminal disulphide knot (NDSK).[16] Two other clusters of interchain disulphides are shown on each side of the dimeric molecule, each denoting a disulphide linked D domain.

The structure and shape of fibrinogen seems to dictate the sequence of degradation reactions by which plasmin attacks the molecule. Despite the fact that plasmin has a general hydrolytic affinity for most arginines and lysines, only a limited number are available in fibrinogen because it contains three major disulphide rich inaccessible core or domain regions. The carboxy terminal (40,000 MW) of the two Aα chains of fibrinogen are first digested by plasmin[18,21] followed by segregation of one disulphide rich region of the molecule, known as fragment D. As is shown in FIGURE 1, there is a consensus that the digestion of fibrinogen to its plasmin resistant core fragments D and E takes place through an asymmetric sequence[13] that involves the generation of an intermediate fragment called Y, which subsequently degrades to its two core fragments, D and E.[12] The detail of this sequence of degradation reactions has been reviewed elsewhere[1-6] and FIGURE 1 B is adequate here to demonstrate that the digestion is domainally directed during the degradation sequence.

FIBRIN FORMATION AND LYSIS BY PLASMIN

Prior to considering the interaction of plasmin with fibrin it is useful to present some notion of what fibrin is in relation to fibrinogen. This will also help us to understand at a physiological level what is meant by soluble fibrin (SF), which circulates in plasma during certain pathological states. This is dealt with below when we consider the relationship between fibrin degradation products (FnDP) and circulating soluble fibrin (SF), particularly when related to clinical conditions such as disseminated intravascular coagulation (DIC).

FIBRIN FORMATION

The interaction of thrombin with fibrinogen has been extensively studied (for a review see Ref. 19), and it seems that a series of delicately modulated interdependent reactions take place during the transition of fibrinogen into crosslinked fibrin, which is the form of fibrin found in most thrombi.[22,23] Fibrin formation is initiated by the thrombin-mediated release of two sets (two of each) of small peptides known as fibrinopeptides A and B (FPA and FPB) from the N-terminal ends of the dimeric

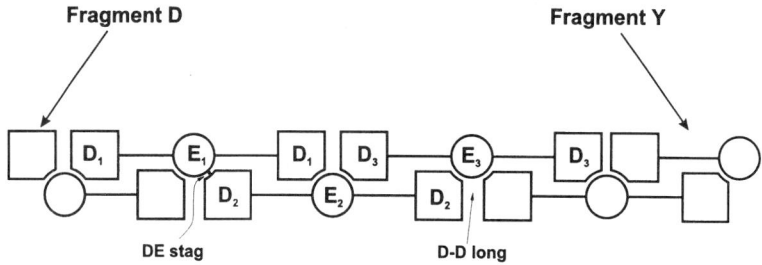

FIGURE 2. An arrangement of fibrin molecules (shown here as two square D domains, linked by a central circular E domain) during polymerisation with a suggested mechanism as to how fibrin fragments Y and D inhibit the polymerisation of fibrin subunits and thus inhibit the clotting of fibrinogen. The half stagger arrangement of the fibrin molecules is dictated by D-E stag polymerisation sites and supplemented by D-D long polymerisation sites. The straight lines used to join the various domains can be considered as regions of plasmin susceptibility. Various sized cross-linked FDP can be generated from structures such as those outlined here. (Reproduced with permission from Ref. 20).

fibrinogen Aα and Bβ chains.[15] Indeed, only the release of FPA seems to be required for the exposure of a polymerization site near the N-terminus of fibrinogen such that des-AA-fibrin (known also as fibrin 1, with an empirical formula $\alpha_2 B\beta_2 \gamma_2$) can polymerize with its complementary site on any nearby fibrinogen or fibrin molecule; this complementary site is located in the D domain of fibrinogen or fibrin.[24] The release of FPB[25] follows that of FPA, and there is evidence that the former exposes other sites of polymerization that may be associated with a lateral growth of fibrin fibers, whereas end-to-end aggregation of fibrin seems to be associated with FPA release.[20] However, at a structural level, the role of FPB release is quite unclear and it may be stated that FPB still seeks a role in the overall scheme of fibrin aggregation.[26]

The domainal locations (see FIG. 1 for domainal nomenclature) of polymerization sites that are relevant during the initial stages of fibrin organization are as follows: Polymerization sites are exposed on the E domain by the release of FPA peptides, allowing it to form a D-E half-stagger polymerization arrangement with the D domains of two adjacent fibrinogen or fibrin molecules; (e.g., fibrin 2 binds fibrin 1 and fibrin 3) (see FIGURE 2). A set of polymerization sites (D-D long) mediate the association between adjacent fibrin subunits (between D_2 and D_3). This explains the strong association between [D_1-D_3]E in a D dimer-E complex (discussed below). The half-stagger of fibrin molecules allows each fibrin chain to stabilize the other, using the dimeric E-located D-E stag sites. Additional stabilization is introduced into each chain by the D-D long polymerization sites. Factor XIII-mediated crosslinks form between these two D domains near the D-D long site.

There are many other features associated with fibrin organization described by Hermans and McDonagh,[20] but those shown in FIGURE 2, together with information on the factor XIII-induced crosslinking of the fibrin subunits, are adequate to explain the fragments of fibrin found in blood during *in vivo* expression of fibrinolysis. Active factor XIII rapidly introduces glutamyl-lysyl[27,28] crosslinks between the γ chains of adjacent fibrin molecules, forming γ-γ dimers (probably near the D-D *long* polymerization sites shown in FIG. 2). Similar chemical crosslinks are catalyzed more slowly between the α chains of fibrin, giving rise to large (greater than 400,000 MW) α chain associations that adversely affect the lysis of fibrin by plasmin.[29] The role of these crosslinks in fibrin stability is paramount.

FIBRIN DIGESTION

This author is here expressing an opinion that it is essentially only crosslinked fibrin (XL-FN) digestion that is of consequence physiologically, and, thus, herein only the plasmin mediated lysis of XL-FnDP is considered.

There is a strong similarity between fragments obtained from fibrin whose γ chains only are crosslinked and those whose γ chains and α chains are crosslinked (TXL-FN).[30] The crosslinking of α chains endows fibrin with resistance to lysis by plasmin,[29,31] whereas the γ chain crosslinks are instrumental in defining the structure of fibrin degradation products.[30,32,33] Whenever fibrin contains crosslinked α chains, the γ chains are also crosslinked; the converse does not apply since α chain crosslinking takes place more slowly than, and seems to depend on the presence of,

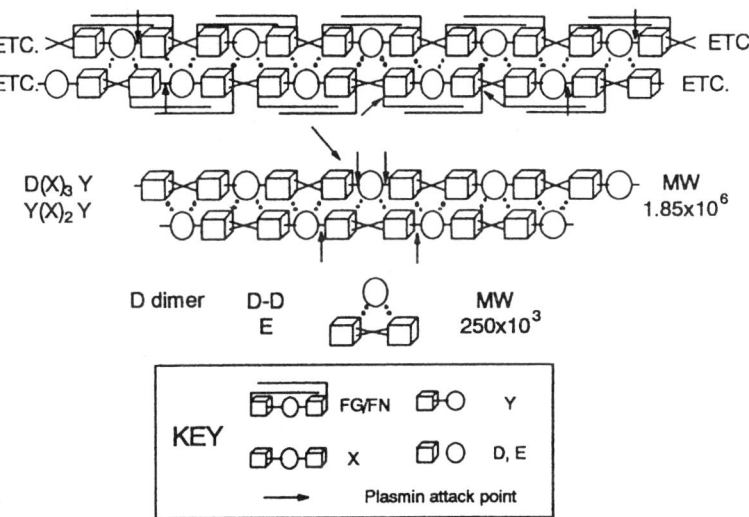

FIGURE 3. Domainally directed fragmentation of cross linked fibrin (XL-FN): an example of typical cross linked fragments. The *top* of the figure depicts the half-stagger two-chain structure of fibrin, showing crosslinks between the D domains of adjacent fibrin subunits. The action of plasmin generates various molecular weight structures, all having various X,Y,D,E domains associated by the γ-chain crosslinks in the originating fibrin. Only a typical intermediate fraction (MW = 2×10^6) and the terminal cross linked fragment D dimer-E are shown. Lysis by plasmin of only COOH-Aα regions and the coiled coil sequences between the D and E domains dominates the sequence of lytic events. (Reproduced from Ref. 69, with permission).

γ chain crosslinks. In 1973, a unique fragment was described in crosslinked fibrin digests. This was named D dimer[30] and was shown to contain two D domains from adjacent fibrin molecules, held together by the crosslinked γ chain remnants of the originating fibrin (see FIG. 2). At about the same time other workers[34] described a similar fragment from bovine fibrin, which they called "double D". It was presumed that D dimer would be the largest soluble fragment derived from crosslinked insoluble fibrin. However, subsequently soluble domainal aggregates of about 2×10^6 molecular weight have been generated from fibrin by plasmin under controlled conditions of lysis,[35] and similar fragments have been found in blood during a variety of clinical states.[36–39] These variously sized fragments of fibrin are held together by both the non-covalent fibrin polymerization bonds and the γ chain crosslinks that were present in the originating fibrin (see reviews in Refs. 1 and 3).

FIGURE 3 shows the domainal structures of some of the fragments found in digestion mixtures of XL-FN. It is convenient to use the domainal nomenclature developed during studies on plasmin-fibrinogen in naming the fragments shown in this figure. Only a few of the various possible XL-FDP that could arise from the digestion of crosslinked fibrin are shown in FIGURE 3. Large aggregates (X-oligomers) held together by both γ chain crosslinks and fibrin polymerization sites can have molecular weights as large as or exceeding 2×10^6 daltons. The group of fragments indicated here as X-oligomers[37,38] has been characterized into various sized fractions,[3] one of the major features of which is the number of X-like structures in each of the subclassifications. Although FIGURE 3 shows only two types of XL-FDP (one large and one small) the general formula for these fragments can be given as $(Y/D-X_n-Y/D)_2$. It is also evident that the D dimer will be a recurring structural feature in all fragments having this general formula. More important than the large variety of the crosslinked fragments is both the presence and relevance of these individual fragments in patient plasma.[37–39] Besides various X-oligomers, the crosslinked Y-D fragments are prominent in all XL-FN digests, as is the D dimer fragment, all these having structures that depend on the γ chain crosslinks derived from the originating fibrin. Under controlled conditions of fibrin digestion, it was also found that the D dimer is always strongly but non-covalently associated with an E domain, forming an electrophoretically stable D dimer-E complex[40] (see FIG. 3). However, it was subsequently indicated that only XL-fibrin lacking FPA and FPB yields the D dimer-E complex, whereas XL-fibrin lacking only FPA yields free D dimer.[41] This was taken to indicate that FPB release exposed an extra set of polymerization sites on forming fibrin, allowing what was previously alluded to as lateral polymerization[26] to take place; these latter sites remained intact in the FDP fraction and gave rise to the stable D dimer-E complex. A similar argument can be applied to the Y-D and X-oligomer groups of fragments, suggesting that two-stranded forms of these crosslinked fragments can occur and that their presence depends on the release of FPB from the originating fibrin (i.e., on the formation of fibrin II). The relevance of this to the interpretation of the clinical impact of episodes of intravascular coagulation requires further investigation. However, crosslinked fibrin fragments (XL-FDP) in blood (notably D dimer) have come under close scrutiny. Their measurement in plasma, using monoclonal antibodies, may lead to a better understanding of the role of fibrinolysis in health and disease.

PLASMA FNDP: A PHYSIOLOGICAL INTERPRETATION

Although DIC may be caused by a variety of underlying primary disease processes, its biochemical expression in blood involves the generation of thrombin and plasmin. As we indicated elsewhere in this review, and as has been clearly expressed elsewhere in the literature, the expression of plasmin seems to require the prior formation of thrombin and its subsequent product *in vivo*, fibrin. Both thrombin and plasmin perform a number of physiological functions. Thrombin, among its varied activities, converts fibrinogen to fibrin, activates platelets and coagulation factors, and modulates endothelial cell protein release, whereas plasmin degrades fibrin, activates platelets, and degrades coagulation factors.

For several years there has been little further information on the structure of FgDP and FnDP in purified systems, and the literature on FDPs mainly involves routine assays for fragments, such as D dimer in a variety of clinical settings. However, some workers[36–39,42] have reported on the nature of the FDP structures observed in plasma during various pathologies and much of the rest of this overview deals with these reports. It would seem that the occurrence of large complex structures predominate in plasma (see FIGURE 4) whereas very little of the well described domainal structures such as D and E occur in plasma. The first comprehensive characterization of *in vivo* generated fibrin fragments was conducted by the group of Professor Graeff[37,38] in the plasma of obstetric patients and these workers first described large crosslinked fragments that they named X-oligomers, since the basic structural feature of this heterogeneous group of fragments was the X domain described above.

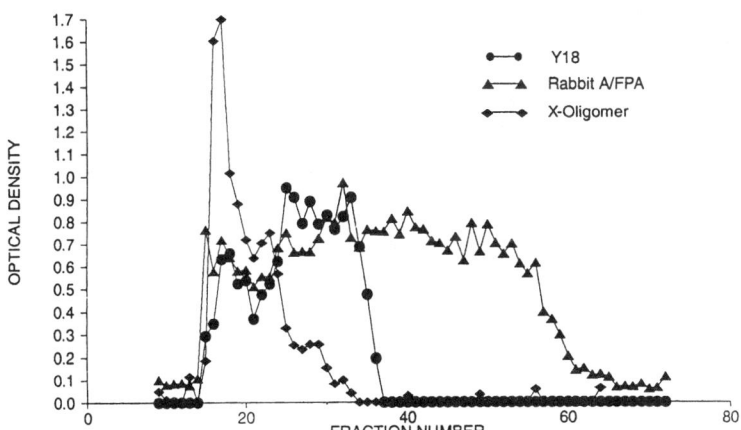

FIGURE 4. FPA and X-oligomer in plasma from patients with DIC. The elution profile shown is from a gel exclusion column Biogel A-15. The elution fractions were monitored using three specific antibodies. The major FDP peak eluted at the void volume (2×10^6 D) and crossreacted with the X-oligomer mab (NIBn-123); this same fraction crossreacted with antibodies to FPA (A/FPA) and with an antibody (Y18) to fragments that contained FPA. A variety of other fractions reacted with Y18 and crossreactivity with A/FPA persisted to near the total exclusion volume of the column. (Reproduced from Ref. 43, with permission).

Another early study of plasma fibrin fragments observed during the DIC associated with amniotic fluid embolism (AFE) showed that D dimer and crosslinked Y-D were the predominant fragments present.[36] This may be due to the hyperfibrinolytic response in the plasma of AFE patients, which evinces rapid digestion of the large crosslinked fragments normally seen in DIC plasma to the smallest crosslinked fragment, D dimer.

More recently a number of groups have looked at the structures of the fibrin fragments found in plasma during a variety of clinical conditions. During DIC it was established[43] by column chromatography (FIG. 4) that plasma contained very large crosslinked fragments, probably quite similar to the X-oligomers described above.[37,38] Similar fragments have been identified by Gron and colleagues[42] in the plasma of fibrinaemic patients. These latter fragments, as well as, being crosslinked, contained significant quantities of FPA suggesting that possibly large amounts of fibrinogen were incorporated into the fibrin and subsequently digested with it. The high proportion of fragments containing FPA in the degradation fraction found in resting plasma, seems to suggest that fibrin polymers with one FPA intact are the source of the FDP fraction, and casts doubt on the interpretation of Nieuwenhuizen and colleagues[44,45] when they define fibrinogen degradation products as those degradation products that contain FPA. Thus, this author favors the view that primary fibrinogenolysis is an extremely rare phenomenon, being restricted to exogenously induced hyperfibrinolysis during thrombolytic therapy and endogenously induced

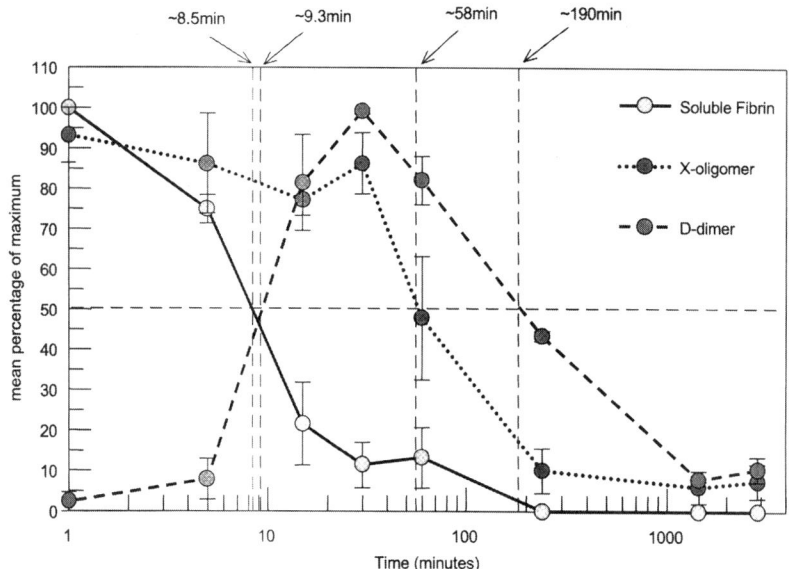

FIGURE 5. Clearance of soluble fibrin (SF) from the rabbit circulation. The rapid clearance of SF as measured by specific SF assay was accompanied by an increase in the level of FnDP (e.g., D dimer), which in turn was cleared more slowly from the circulation. The clearance of X-oligomer was more difficult to interpret.

hyperfibrinolysis associated with rare episodes of extreme shock (e.g., amniotic fluid embolism during pregnancy). This is, of course, at variance with the recently published conclusion of others.[46]

There has been much speculation on the value of measuring what is referred to as soluble fibrin (SF) in plasma. An assay for SF has been demonstrated to be of value in identifying a variety of thrombotic conditions.[47] SF may be regarded as a degradation product of fibrinogen mediated by thrombin and, theoretically, this should be a marker of the prethrombotic state. However, it seems that there is great difficulty in structurally defining SF and we mostly regard it as a polymer of fibrin with possibly some fibrinogen attached. The major problem with proposing a SF fibrin assay as a marker of the prethrombotic state is that the conversion of this soluble fibrin to fibrin degradation products is very rapid. The rapid digestion of soluble fibrin *in vivo* presents us with a slight theoretical enigma. There is solid data[48] that the major plasmin inhibitor PI (also known as α_2-antiplasmin) is crosslinked during fibrin formation to the α chain and it is also known (Gaffney, unpublished data) that this crosslinkage reaction takes place prior to crosslinking of α chains themselves, and at the same pace as the γ-crosslinking of fibrin in the soluble state. There is, however, more recent data suggesting that soluble fibrin with crosslinked α chains may be present *in vivo*.[49] This suggests that, if Aoki and colleagues are correct[48] in proposing PI binding to fibrin as a structural advantage in neutralizing the plasmin bound

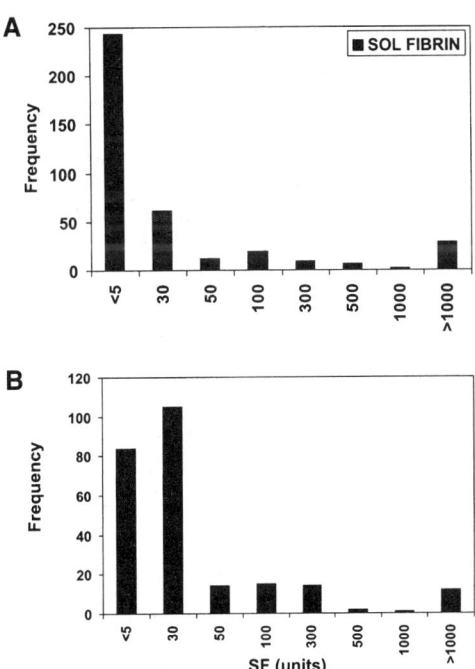

FIGURE 6. Histograms of SF levels in the plasma of (**A**) age matched controls and (**B**) stroke patients.

to fibrin, the maintenance of the earliest fibrin structures is a primary feature in generating adequate hæmostatic plugs at a hemorrhagic site. Coagulant competence is essential for species survival, hemorrhage being a major hazard to such survival. An alternative interpretation may be the utilization of the respective lysine binding sites of plasminogen and PI in binding to fibrin, suggesting that plasmin–PI interaction on the surface of fibrin may be greatly reduced. The enigma, referred to above, resides in this very rapid incorporation of PI into fibrin in the soluble state and the reality that, under normal circumstances of fibrin formation *in vivo*, fibrin is rapidly digested by the fibrinolytic system. Indeed, this rapid clearance of fibrin is demonstrated in FIGURE 5 which shows the levels of human SF and human FnDP in an *in vivo* rabbit model following iv injection of a human SF preparation. The clearance of SF has a halflife of 6.5 minutes and, thereafter, a level of FnDP is observed with a far longer halflife.[50] Thus, the orchestration of its own destruction by fibrin, especially long fibrin polymers, is very rapid. Indeed this may be the major natural substrate for the fibrinolytic enzyme, plasmin, and it is no surprise that the level of so called SF found in patients during some presumed hypercoagulable states is highly variable. It appears that the SF level may depend on when the sample is taken from the patient during a hypercoagulable state. This is amply demonstrated in the histograms in FIGURE 6, showing the levels of SF found in both a group of clinical controls and a group of stroke patients to be highly heterogeneous with little ability of the SF assay to distinguish the two groups.

Thus, it seems that a heterogeneous group of large crosslinked fragments containing varying amounts of FPA are the most likely structural identity of FDP in plasma. The origin of such a group of fragments appears to be the large crosslinked soluble fibrin polymers generated in blood during the hypercoagulable state. The question arises as to whether there is a clear cut distinction between these two groups of large fragments in blood, one being an indication of the hypercoagulable state, the other a reflection of fibrinolytic activity. Structurally it seems that the major difference

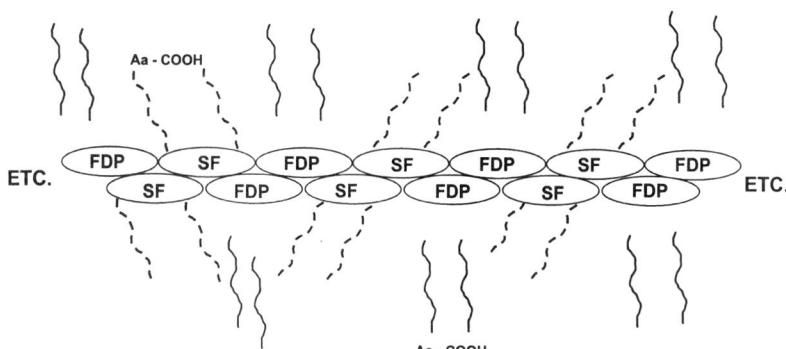

FIGURE 7. Scheme showing fibrin polymers (probably cross linked and containing some FPA as suggested in FIG. 4) part of which have their COOH-Aα chains removed by enzyme (probably plasmin) activity and can be defined as FDP. Some of the same moiety can be defined as SF containing intact COOH-Aα peptide regions (taken from Ref. 50).

between these two groups of fragments is that one contains intact COOH terminal Aα-chains and the other does not (see FIG. 3). An interesting experiment was performed[50] in that a number of DIC plasmas were charged onto 96-well plates coated with an antibody that reacted with crosslinked FnDP, such as D dimer. These same antibodies were demonstrated not to react with soluble fibrin. However, when a tag antibody that reacted with the COOH terminal α chain of soluble fibrin was used, the bound FnDP was shown to contain varying amounts of material that we can define as SF. The reasonable conclusion from these data is that when soluble fibrin forms it begins to orchestrate its own destruction by generating plasmin on its surface. This plasmin attacks the fibrin at various random parts of its structure and mixed FDP/SF structures (see FIGURE 7) are formed *in vivo*. This suggests a dynamic and rapid conversion of SF to FnDP with a variety of SF/FnDP intermediates. The ratio of SF/FnDP in the plasma intermediates may provide a useful marker of the clinical progress during any episode of overt or transient DIC. Thus, the original hemostatic balance notion proposed by Astrup[51] in 1956 seems to be played out on the surface of forming fibrin. It is proposed here that a monoclonal antibody with a specificity for both soluble fibrin and FnDP can be used to monitor this hemostatic balance by using second antibodies individually to soluble fibrin and to the FnDP.

EFFECT OF FIBRIN DEGRADATION PRODUCTS (FNDP) ON THE SYNTHESIS OF PLASMA FIBRINOGEN

Since plasma fibrinogen levels have been related to the risk of thrombotic disease,[52] mechanisms that regulate fibrinogen levels *in vivo* need to be fully understood. Although other comments on the effect of FnDP on plasma fibrinogen synthesis may exist in the literature, my earliest observations have been two papers by Bocci and Pacini[53] and Olis and Rapaport[54] the former indicating an enhancing effect of fibrinogen fragments on plasma fibrinogen levels in the rabbit, whereas the latter authors showed no effect in the same rabbit model. This direct contradiction between the data from these two laboratories has been mirrored during the subsequent 20 years. There has been a sequence of totally contradicting data on this subject, mostly in reputable journals and the role for Fg/FnDP in fibrinogen synthesis has not been as yet clarified.

Ittyerah *et al.*[55] indicated that homologous fibrin fragments X, Y, D, and E did not affect fibrinogen synthesis in a rabbit model. Kessler and Bell[56] demonstrated by infusing rabbits with urokinase and ancrod that fibrinogen degradation products (FgDP) enhanced fibrinogen synthesis, whereas fibrin degradation products (FnDP) did not. Increased levels of fibrinogen mRNA and elevated plasma fibrinogen concentrations were observed in rats after administration of homologous fibrinogen degradation products X, Y, D, and somewhat less so with fragment E.[57] Following these contradictory findings a number of groups looked at the effect of purified fragments of fibrinogen and fibrin on synthesis. Franks *et al.*[58] indicated that only homologous fibrinogen fragment D, and not fragment E, enhanced fibrinogen synthesis in a rat model, whereas others[59] showed that both fibrinogen fragments D and E enhanced synthesis in a rabbit model. Richie et al.[60] using cultured monolayers of rat hepatocytes showed no direct effect of fragments D and E, but an indirect enhancement of synthesis via a leukocyte mediated pathway. Contradicting this data using the same experimental system of cultured monolayers of rat hepatocytes, Qureshi et al.[61] showed that only fibrinogen fragment E directly enhanced the rate of fibrinogen biosynthesis, and others[62] presented evidence that only homologous fibrinogen fragment D enhanced fibrinogen synthesis in the same experimental system.

It is difficult to imagine a greater degree of disparity in data both *in vivo* and *in vitro* from a number of laboratories, and these data leave the question unresolved as to whether the products of fibrinogen/fibrin degradation *in vivo* affect the subsequent level of circulating fibrinogen.

CONCLUDING REMARKS

This overview has covered structural features of fibrin degradation products generated *in vitro* and summarized the limited information in the literature about the structures of the FnDP fraction found in plasma *in vivo*.

The evolution of our structural knowledge about fibrin degradation products (FnDP) has been slow and indeed may not have reached the final stage of relating structural possibilities resulting from *in vitro* experimentation with structural realities derived from *in vivo* data. The large body of literature dealing with *in vitro* derived FnDP structures has been summarized briefly above and an attempt has been made to relate these structures to those found *in vivo*. Although the major distinction between fibrinogen and fibrin degradation products was the description of γ crosslinked D dimer,[30,34] the next finding related to the presence of a non-covalently bound D dimer-E complex,[40] whereas in 1979, Graeff and Hafter suggested that large crosslinked FnDP called X-oligomer were the prominent FnDP fraction *in vivo*.[35–39] This was confirmed by this author[35] and further data from *in vivo* tests indicated that the FnDP fraction contained high levels of fibrinopeptide A.[42,43] This was explained by the fact that crosslinked fibrin containing one fibrinopeptide A was the source of this FnDP.[43] More recent data using specific monoclonal antibodies for soluble fibrin and FnDP allowed us to propose that soluble fibrin and FnDP may not be distinct entities but may both be part of a common backbone of fibrins undergoing dynamic if somewhat spatially erratic degradation. Thus, it seems that continuing work on the structural nature of FnDP in plasma under a variety of clinical conditions will allow a more complete understanding of such structures.

One of the clinical expressions of DIC accompanied by high levels of FnDP is hemorrhage, which is partly due to consumption of fibrinogen during its conversion to fibrin and subsequently to FnDP. The high level of FnDP further inhibits the clotting of the residual fibrinogen. This latter biological activity of FnDP is the only one that has been established with surety.[14] Although it has been suggested that the products of thrombin/plasmin interactions with fibrinogen have other biological activities, the details of these activities are rather vague. The effect of fibrin-fibrinogen degradation products on fibrinogen synthesis has been reviewed herein and, as mentioned above, the contradictory aspect of these conclusions are close to being humorous. A somewhat similar situation exists with respect to the vasoactivity of the degradation products in that some authors have reported positive data,[63] whereas others report a negative effect.[64] However, this latter ambivalence is somewhat more excusable than the experimental contradictions on fibrinogen synthesis in that the animal models used differ and these models are quite difficult to set up with reliability. The extent of experimental data about the various structures derived from the interaction of plasmin, thrombin, and other enzymes with fibrinogen is considerable, but the amount of work on the nature and structure of the FnDP found in clinical material is comparatively limited. This is indeed a pity since ultimately the utility of such studies lies in the development of assays that can help us to understand the pathophysiology of fibrinogen/fibrin digestion. Although the level of fibrinopeptide A (FPA) in plasma was popular for a short time as a marker of the prethrombotic state, the difficulty in taking samples and the short halflife of FPA has led to its demise as a test. The only robust test in this area seems to be that for D dimer, which was suggested as far back as 1972 by this author[65] as a marker of hemostatic imbalance. Currently, there are a number of such assays based on monoclonal antibodies to various preparations of D dimer that make the standardization of these assays quite difficult.[66,67] Despite this the only sure clinical advantage of a D dimer assay seems to be the exclusion of DVT/PE when the level of D dimer is normal.[68] It has been insinuated that an elevated D dimer may indicate a hypercoagulable state, which may augur forms of thrombosis such as myocardial infarction.[69] As use of the FPA assay has declined, there is an increasing interest in the measurement of soluble fibrin (SF). However, as mentioned above, the short halflife of SF and its rapid conversion to FnDP make this an unreliable approach. The information above indeed suggests that SF and FnDP may be part of a common structural backbone, and assays that define a ratio of SF/FnDP on this shared structure may provide a useful marker of the progression for any individual hypercoagulable episode. Our data (FIG. 7) suggests that it is essential to take blood samples into an anticoagulant containing a fibrinolytic inhibitor in order to arrest the conversion of SF→FnDP after blood sampling. Thus, in conclusion, it is well for us to think in terms of a dynamic state of fibrinogen→fibrin→FnDP, with no distinct demarcation line separating one type of fragment from the other and where assays targeting specific amounts of any one group of fragments can be constructed. Thus, the relative amounts of each group of structures in the complex dynamically changing mixture of structures residing on the common fibrin-like backbone can suggest a clinical interpretation of a patient's condition.

REFERENCES

1. GAFFNEY, P.J. 1977. The biochemistry of fibrinogen and fibrin degradation products. *In* The Biochemistry, Physiology and Pathology of Hæmostatis. D. Ogston & B. Bennett, Eds.: 105–108. Wiley, New York.
2. GAFFNEY, P.J. 1977. Structure of fibrinogen and degradation products on fibrinogen and fibrin. Br. Med. Bull. **33:** 245–252.
3. FRANCIS, C.W. & V.J. MARDER. 1982. A molecular model of plasmic degradation of crosslinked fibrin. Semin. Thromb. Hæmost. **8:** 25–35.
4. GAFFNEY, P.J. 1987. Fibrinolysis. In Hæmostasis and Thrombosis, 2nd edit. A.L. Bloom & D.P. Thomas, Eds.: 223–244. Churchill Livingstone, Edinburgh.
5. MARGUERIE, G., G. HUDRY-CLERGEON & M. SUSCILLON. 1975. On the structure of fibrinogen related to its plasmic degradation products D and E. Thromb. Diath. Hæmorrh. **34**(3): 664–670.

6. GAFFNEY, P.J. 1993. D dimer. History of the discovery, characterisation and utility of this and other fibrin fragments. Fibrinolysis (Suppl.) **2:** 2–8.
7. WIMAN, B. & D. COLLEN. 1978. Molecular mechanisms of physiological fibrinolysis. Nature **272:** 549–550.
8. COLLEN, D. 1980. On the regulation and control of fibrinolysis. Thromb. Hæmost. **43:** 77–89.
9. COLLEN, D. & H.R. LIJNEN. 1990. Molecular mechanisms of thrombolysis. Implications for therapy. Biochem. Pharmacol. **40:** 177–186.
10. MARSH, N.A. & P.J. GAFFNEY. 1982. Exercise-induced fibrinolysis—fact or fiction? Thromb. Hæmost. **48:** 201–203.
11. SEEGERS, W.H., M.L. NIEFT & J.M. VANDENBELT. 1945. Decomposition products of fibrinogen and fibrin. Arch. Biochem. **7:** 15–19.
12. NUSSENZWEIG, V., M. SELIGMAN & J. PELMONT, et al. 1961. Les produits de degradation du fibrinogene humain par la plasmine. I: Separation et proprietes physicochemiques. Ann. Inst. Pasteur. **100:** 377–387.
13. MARDER, V.J., N.R. SHULMAN & W.R. CARROLL. 1969. High molecular weight derivatives of human fibrinogen produced by plasmin. I: Physiochemical and immunological characterization. J. Biol. Chem. **244:** 2111–2119.
14. MARDER, V.J. & N.R. SHULMAN. 1969. High molecular weight derivatives of human fibrinogen produced by plasmin. J. Biol. Chem. **244:** 2120–2124.
15. BETTELHEIM, F.R. & K. BAILEY. 1952. The products of the action of thrombin on fibrinogen. Biochim. Biophys. Acta **9:** 578–579.
16. BLOMBACK, B., M. BLOMBACK & A. HENSCHEN, et al. 1968. N-terminal disulphide knot of human fibrinogen. Nature **218:** 130–134.
17. HESSEL, B. 1975. On the structure of the COOH-terminal part of the alpha chain of human fibrinogen. Thromb. Res. **7:** 75–87.
18. GAFFNEY, P.J. & P. DOBOS. 1971. A structural aspect of human fibrinogen suggested by its plasmin degradation. FEBS Letters **15:** 13–16.
19. DOOLITTLE, R.F. 1973. Structural aspects of the fibrinogen-fibrin conversion. Adv. Protein. Chem. **27:** 1–109.
20. HERMANS, J. & J. MCDONAGH. 1982. Fibrin: structure and interactions. Semin. Thromb. Hæmost. **8:** 11–24.
21. MOSESSON, M.W., J.S. FINLAYSON & D.K. GALANAKIS. 1973. The essential covalent structure of human fibrinogen evinced by analysis of derivatives formed during plasmic hydrolysis. J. Biol. Chem. **248:** 7913–7929.
22. FINLAYSON, J.S. & D.L. ARONSON. 1974. Crosslinking of rabbit fibrin in vivo. Thromb. Diath. Hæmorrh. **31:** 435–438.
23. GAFFNEY, P.J., M. BRASHER, K. LORD, et al. 1976. Fibrin subunits in venous and arterial thromboembolism. Cardiovasc. Res. **10:** 421–426.
24. KUDRYK, B.J., D. COLLEN, K.R. WOODS, et al. 1994. Evidence for localisation of polymerisation sites in fibrinogen. J. Biol. Chem. **249:** 3322–3325.
25. LAHIRI, B. & J.R. SHAINOFF. 1973. Fate of ribrinopeptides in the reaction between human plasmin and fibrinogen. Biochim. Biophys. Acta **303:** 161–170.
26. LAURENT, T.C. & B. BLOMBACK. 1958. On the significance of the release of two different peptides from fibrinogen during clotting. Acta Chem. Scand. **12:** 1875–1877.
27. MATACIC, S. & A.G. LOEWY. 1968. The identification of isopeptide crosslinks in insoluble fibrin. Biochem. Biophys. Res. Commun. **30:** 356–362.
28. PISANO, J.J., J.S. FINLAYSON & M.P. PEYTON. 1968. Crosslink in fibrin polymerized by factor XIII: ε-(γ-glutamyl) lysine. Science **160:** 892–893.
29. GAFFNEY, P.J. & A.N. WHITAKER. 1979. Fibrin crosslinks and lysis rates. Thromb. Res. **14:** 85–94.
30. GAFFNEY, P.J. 1973. Subunit relationships between fibrinogen and fibrin degradation products. Thromb. Res. **2:** 201–218.
31. SCHWARTZ, M.L., S.V. PIZZO, R.C. HILL, et al. 1973. Human factor XIII from plasma and platelets. Molecular weights, subunit structure, proteolytic activation and crosslinking of fibrinogen and fibrin. J. Biol. Chem. **248:** 1395–1407.
32. PIZZO, S.V., M.L. SCHWARTZ, R.L. HILL, et al. 1973. The effect of plasmin on the subunit structure of human fibrin. J. Biol. Chem. **248:** 4574–4583.

33. PIZZO, S.V., L.M. TAYLOR, M.L. SCHWARTZ, et al. 1973. Subunit structure of fragment D from ribrinogen and crosslinked fibrin. J. Biol. Chem. **248:** 4584–4590.
34. KOPEC, M., E. TEISSEYRE, G. DUDEK-WOJCIECHOUSKA et al. 1973. Studies on the "double D" fragment from stabilised bovine fibrin. Thromb. Res. **2:** 283–292.
35. GAFFNEY, P.J., F. JOE & M. MAHMOUD. 1980. Giant fibrin fragments derived from crosslinked fibrin: structural and clinical implication. Thromb. Res. **20:** 647–662.
36. GAFFNEY, P.J. 1975. Distinction between fibrinogen and fibrin degradation products in plasma. Clin. Chim. Acta **65:** 109–115.
37. GRAEFF, H. & R. HAFTER. 1982. Detection and relevance of crosslinked fibrin derivatives in blood. Semin. Thromb. Hæmost. **8:** 57–68.
38. GRAEFF, H., R. HAFTER & L. BACHMANN. 1979. Subunit and macromolecular structure of circulating fibrin from obstetric patients with intravascular coagulation. Thromb. Res. **16:** 313–328.
39. GAFFNEY, P.J. 1983. The occurrence and clinical relevance of fibrin fragments in blood. Ann. N.Y. Acad. Sci. **408:** 407–423.
40. GAFFNEY, P.J., D.A. LANE & M. BRASHER. 1975. Soluble high molecular weight E fragments in the plasmin-induced degradation products of cross-linked human fibrin. Clin. Sci. Mol. Med. **49:** 149–156.
41. OLEXA, S.A. & A.Z. BUDZYNSKI. 1980. Effects of fibrinopeptide cleavage on the plasmic degradation pathways of human crosslinked fibrin. Biochem. **19:** 647–651.
42. GRON, B., A. BENNICK, W. NIEUWENHUIZEN, et al. 1990. Normal and fibrinaemic patient plasma contain high molecular weight cross-linked fibrin(ogen) derivatives with intact fibrinopeptide A. Thromb. Res. **57:** 259–270.
43. GAFFNEY, P.J., T. EDGELL & L.J. CREIGHTON-KEMPSFORD. 1995. Fibrin degradation products (FnDP) assays. Analysis of standardisation issues and target antigens in plasma. Br. J. Hæmatol. **90:** 187–194.
44. NIEUWENHUIZEN, W., B. HOEGEE-DENOBEL & R. LATERVEER. 1992. A rapid monoclonal antibody-based enzyme immunoassay (EIA) for the quantitative determination of soluble fibrin in plasma. Thromb. Hæmost. **68:** 273–277.
45. KOPPERT, P.W., W. KUIPERS, B. HOEGEE-DENOBEL, et al. 1987. A quantitative enzyme immunoassay (EIA) for primary fibrinogenolysis products in plasma. Thromb. Hæmost. **57:** 25–28.
46. EL-SAYED, M.S. & W. NIEUWENHUIZEN. 2000. The effect of alcohol ingestion on the exercise-induced changes in fibrin and fibrinogen degradation products in man. Blood Coagul. Fibrinolysis **11:** 359–365.
47. BOS, R., G.H. LATERVEER-VREESWIJK, D. LOCKWOOD, et al. 1999. A new enzyme immunoassay for soluble fibrin in plasma, with a high discriminating power for thrombotic disorders. Thromb. Hæmost. **81:** 54–59.
48. SAKATA, Y. & N. AOKI. 1982. Significance of cross-linking of α2-plasmin inhibitor to fibrin in inhibition of fibrinolysis and in hemostasis. J. Clin. Invest. **69:** 536–542.
49. SANDBAEK, G.S., S. BJORNSEN, J.H. SOBEL, et al. 2000. Soluble fibrin species in arterial thrombi. Blood Coagul. Fibrinolysis **11:** 1–5.
50. GAFFNEY, P.J. & T.A. EDGELL. 1997. Fibrinolysis and the haemostatic balance: Harmonisation of some old and new concepts. In Recent Progress in Blood Coagulation and Fibrinolysis. T. Takada, D. Collen & P.J. Gaffney, Eds.: 127–141. Elsevier, Amsterdam.
51. ASTRUP, T. 1956. The biological significance of fibrinolysis. Lancet **ii:** 565–568.
52. MEADE, T.W. 1987. Hypercoagulability and ischaemic heart disease. Blood Rev. **1:** 2–8.
53. BOCCI, V. & A. PACINI. 1973. Factors regulatory plasma protein synthesis II: Influence of fibrinogenolytic products on plasma fibrinogen concentration. Thromb. Diath. Hæmorrh. (Stuttg.) **29:** 63–75.
54. OTIS, P.T. & S.I. RAPAPORT. 1973. Failure of fibrinogen degradation products to increase plasma fibrinogen in rabbits. Proc. Soc. Exp. Biol. Med. **144:** 124–129.
55. ITTYERAH, T.R., N. WEIDNER, R.D. WOCHNER, et al. 1979. Effect of fibrin degradation products and thrombin on fibrinogen synthesis. Br. J. Hæmatol. **43:** 661–668.
56. KESSLER, C.M. & W.R. BELL. 1980. Stimulation of fibrinogen synthesis: a possible functional role of fibrinogen degradation products. Blood **55:** 40–47.

57. PRINCER, H.M.G., H.J. MOSHAGE, J.J. EMEIS, et al. 1985. Fibrinogen fragments X,Y,D & E increase levels of plasma fibrinogen and liver mRNAs coding for fibrinogen polypeptides in rats. Thromb. Hæmost. **53:** 212–215.
58. FRANKS, J.J., R.E. KIRSCH, L.O. FRITH, et al. 1981. Effect of fibrinogenolytic products D & E on fibrinogen and albumin synthesis in the rat. J. Clin. Invest. **67:** 575–580.
59. BELL, W.R., C.M. KESSLER & R.R. TOWNSEND. 1983. Stimulation of fibrinogen biosynthesis by fibrinogen fragments D & E. Br. J. Hæmatol. **54:** 599–610.
60. RICHIE, D.G., B.A. LEVY, M.A. ADAMS, et al. 1982. Regulation of fibrinogen synthesis by plasmin-derived fragments of fibrinogen and fibrin: an indirect feedback pathway. Proc. Natl. Acad. Sci. U.S.A. **79:** 1530–1534.
61. QURESHI, G.D., P.S. GUZELIAN, R.M. VENNART, et al. 1985. Stimulation of fibrinogen synthesis in cultured rat hepatolytes by fibrinogen fragment E. Biochem. Biophys. Acta **844:** 288–295.
62. LADUCA, F.M., L.A. TINSLEY, C.V. DANG, et al. 1989. Stimulation of fibrinogen synthesis in cultured rat hepatocytes by fibrinogen degradation products fragment D. Proc. Natl. Acad. Sci. U.S.A. **86:** 8788–8792.
63. SALDEEN, T. 1983. Vasoactive peptides derived from degradation of fibrinogen and fibrin. Ann. N.Y. Acad. Sci. **408:** 424–437.
64. BOUTCHER, P.A., P.J. GAFFNEY, S. RAUT, et al. 1996. Effects of early plasmin digests of fibrinogen on isometric tension development in isolated rings of rat pulmonary artery. Thromb. Res. **81:** 231–239.
65. GAFFNEY, P.J. 1972. FDP. Lancet **ii:** 1422.
66. NIEUWENHUIZEN, W. 1997. A reference material for haemonisation of D dimer assays. Thromb. Hæmost. **77:** 1031–1033.
67. EISENBERG, P.R., D.B. RYLATT, F. RUSTICALI, et al. 1997. The importance of antibody specificity in measuring crosslinked from degradation products by ELISA. Blood Coagul. Fibrinolysis **8:** 105–113.
68. BOUNAMEAUX, H., D. SLOSMAN, P.H. DE MOERLOOSE, et al. 1988. Diagnostic value of plasma D dimer in suspected pulmonary embolism. Lancet **ii:** 628–629.
69. GAFFNEY, P.J., L.J. CREIGHTON-KEMPSFORD & P.M. TYMKEWYCZ. 1990. Detection of fibrin and its fragments using monoclonal antibodies. Clinical implications. *In* Thrombosis and Haemorrhagic Disorders. I. Nagy, H. Losenczy & H. Vinazzer, Eds.: 121–134. Schmitt and Meyer GmbH., Wurzburg.

Antifibrinogen IgG, Fibrinogen, and C1q Complexes Circulating in a Hypodysfibrinogenemic Proband

Isolation, Stoichiometry, and Partial Characterization

DENNIS K. GALANAKIS,[a] AGNES HENSCHEN-EDMAN,[b] JOHN WEISEL,[c] AND SILVIA SPITZER[a]

[a]*Departments of Pathology and Medicine, School of Medicine, State University of New York, Stony Brook, New York 11794, USA*

[b]*Department of Molecular Biology and Biochemistry, University of California at Irvine, Irvine, California 92717, USA*

[c]*Department of Cell Biology and Biochemistry, University of Pennsylvania, Philadelphia, Pennsylvania 19104, USA*

ABSTRACT: Circulating antifibrinogen antibodies have been reported in rare afibrinogenemic propositi, apparently occurring following fibrinogen replacement therapy, but immune complexes have not been described. In this report we describe circulating immune complexes formed by a monoclonal antifibrinogen IgG in a heterozygous hypodysfibrinogenemic (Aα 16 Arg → Cys) proband. Estimated by partial protein sequence and by other analyses, each immune complex consisted of one fibrin(ogen), one C1q, and 3–4 IgG molecules. The complexes were cryoprecipitable, a property also displayed by mixtures of proband IgG and normal fibrinogen. Indicating that both D and E domains were necessary for this behavior, cryoprecipitability was abolished by preincubation of the isolated IgG with either isolated normal fibrinogen fragment D100 or E. Consistent with the crossreactivity of the IgG with normal and mutant fibrinogen, the results suggest that the primary epitope resides on a D–E locus on the fibrin polymer formed by normal and abnormal molecules containing the uncleaved (or mutant) peptide A.

KEYWORDS: Fibrinogen; Antifibrinogen IgG; Immune complex; Cryoprecipitating antifibrinogen IgG; Hypodysfibrinogenemia; Dysfibrinogenemia.

Circulating immune complexes (IC) containing fibrinogen, C1q and antifibrinogen antibodies have not been described among reports of circulating immunoglobulins with antifibrinogen properties.[1–10] Moreover, such antibodies have been described in afibrinogenemic,[8–10] but not in dysfibrinogenemic probands. The investigations reported here describe the isolation of IC from an Aα 16Arg→Cys, heterozygous hypodysfibrinogenemic proband's plasma, and the fibrinogen precipitating properties of the IgG. Evaluations of the dysfibrinogen (fibrinogen Northspring), and more

---

Address for correspondence: Dennis Galanakis, M.D., SUNY, Stony Brook, NY 11794, USA. Voice: 631-444-2625; fax: 631-444-3137.

dgalanak@path.som.sunysb.edu

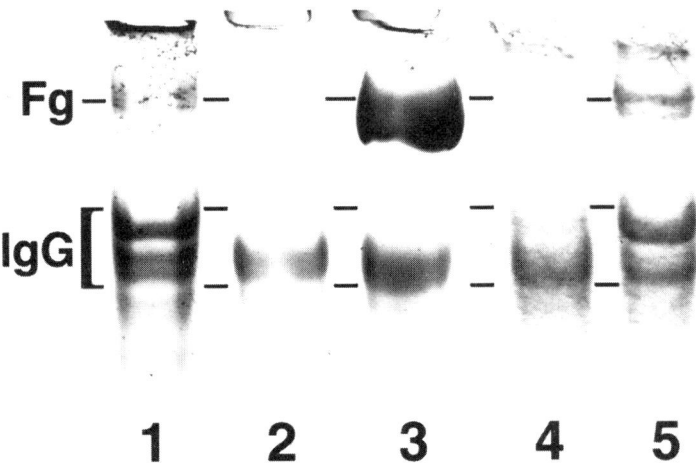

FIGURE 1. PAGE-SDS electrophoretograms of various preparations showing fibrinogen/ IgG complexes, *lanes 1, 3,* and *5*. *Lane 1*, eluate from protein A beads following exposure to proband plasma. *Lane 2*, proband IgG isolated by DEACc chromatography. *Lane 3,* proband fibrinogen isolate containing IgG, obtained by the ammonium sulfate/glycine precipitation procedure, and used to obtain IgG shown in *lane 2*. *Lanes 4* and *5* are from separate runs. *Lane 4*, proband IgG isolated from a fibrin affinity chromatogram. (see text) of a proband serum IgG fraction.[14] *Lane 5*, eluate from protein A beads exposed to dialyzed and unfractionated polyethylene glycol (PEG) precipitate from proband plasma.

extensive characterization of this antibody and the IC are reported elsewhere.[11] In order to enable IC isolation from plasma, modifications were made to procedures available for isolation of IC from serum[12] and others available for plasma fractionation. The resulting isolates obtained from citrated, heparinized, or ACD (by plasmapheresis) plasma treated with aprotinin (100 U/ml) and hirudin (5 U/ml) were routinely examined electrophoretically in sodium dodecyl sulfate polyacrylamide gels,[13] SDS-PAGE. To assess for the possible presence of IC, whole plasma or its fractions were mixed in equal aliquants in 0.1 M Borate buffer, pH 8.3, with suspensions of microbead agarose immobilized Protein A (from either Pierce, Rockford, IL or Gamma Biologics, Houston, TX) and slowly mixed by rocking motion (for at least 2 h at 23°). Following buffer washes to remove unbound proteins, eluates obtained with urea–SDS buffer were examined, FIGURE 1, lane 1. Alternatively, proband plasma or its fractions were subjected to standard Sepharose–fibrin affinity chromatography or bead suspensions, and bound proteins were eluted with 1 M KBr at pH 5.5. When proband serum or its IgG fraction[14] was used, IgG bound to the immobilized fibrin was desorbed with urea–SDS–phosphate, pH 7 (FIGURE 1, lane 4). Although both yielded similar results, the protein A procedure was used routinely because of its ease of use. These results indicated the presence of immune complexes in proband plasma and led to efforts to isolate them by non-detergent means for further studies. In one procedure, isolates were obtained by fractionation of plasma precipitates induced by dissolving glycine to 2.1 M or $(NH_4)_2SO_4$ to 1 M and allowing the mixture to stand at 4° at least four hours, or overnight. The precipitate was dissolved in phosphate saline buffer (pH 6.4 or 7) and reprecipitated three times with the same initial salt precipitant. $(NH_4)_2SO_4$ was always used for the final precipitation. Isolates were dialyzed versus 0.3 M NaCl, and used or stored at −80° until use. In a second procedure, plasma proteins precipitable with 2.5% polyethylene glycol (PEG, MW 6,000) at similarly low temperatures were either reprecipitated as detailed above, or the precipitates were washed repeatedly with an excess of cold buffer containing PEG, to extract soluble proteins. In a third procedure, isolates were obtained by plasma cryoprecipitation. After standard freezing (−20°) and thawing (4°), the harvested cryoprecipitates were exhaustively washed with ice cold buffer (pH 6.4 or 8.3) to extract soluble proteins. The resulting normal plasma cryoprecipitates displayed absence or trace amounts of non-fibrinogen proteins, relative to the fibrinogen/fibrin electrophoretic bands. In additional experiments, triplicate mixtures of isolated normal fibrinogen (1.2 mg/ml, pH 6.4) and normal (from a donor pool larger than 40) or proband IgG (3 mg/ml) were frozen and thawed at 4°. The resulting cryoprecipitates contained 68% (range 64–79%) of the total protein in the proband IgG/normal fibrinogen mixture and displayed major amounts of both proteins. In contrast, isolated patient or normal IgG and normal fibrinogen/normal IgG mixtures failed to yield cryoprecipitates, underscoring the fact that cryoprecipitation required the presence of both fibrinogen and proband IgG. To ascertain if this effect could be demonstrated in the proband plasma environment, three aliquots of proband plasma from which cryoprecipitate had been harvested were treated by addition of normal fibrinogen (0, 1, and 2 mg/ml, respectively) and subjected to the freeze–thawing procedure. Although the latter two aliquots yielded cryoprecipitates containing major amounts of fibrinogen and IgG, the first aliquot (lacking fibrinogen) yielded no cryoprecipitate. In addition to fibrinogen and IgG electrophoretic bands of proband isolates, faintly staining bands were apparent in the MW about 50–60 kDa positions in isolates from all three procedures, but could only be demonstrated with heavy protein loads. Examination of two isolates by Western blot confirmed that these were Clq bands. In fact, partial amino acid sequence determination of two isolates from proband plasma disclosed estimated fibrinogen:IgG:Clq molar ratios of 1:4:1 and 1:3.5:1. This also disclosed monoclonal $IgG_1\kappa$, consistent with immunoelectrophoresis results.

Preliminary observations disclosing that proband IgG/fibrinogen complexes in solution displayed a marked increase in opacity when exposed to subambient temperatures were explored further. For this purpose, both normal IgG isolates[14] and proband isolates (*vide supra*) were subjected to DEAEc gradient elution chromatography.[15] This yielded a proband IgG (early) peak free from fibrinogen, although it represented approximately 30% of the IgG in the mixture and the remainder coeluted with the fibrinogen in a late peak. Since proband fibrinogen could not be obtained free from IgG by non-detergent procedures, and normal fibrinogen was cryoprecipitable by the IgG, normal fibrinogen from fibrin depleted plasma[16] was used for this purpose. A major increase in turbidity (350 nm) was readily demonstrated when solution mixtures of the two proteins were subjected to subambient temperatures, FIGURE 2, and it was both fibrinogen and IgG concentration dependent.

From these experiments (FIG. 2, insert) it was calculated that cryoprecipitability was appreciable at an IgG:fibrinogen molar ratio of about 1:1 and reached maximal levels at ratios of at least 4:1. Characteristically, the maximum turbidity of the solution mixtures reached at 4° decreased to that of controls within 10 minutes or so on rewarming to 37°, not shown. The possible effect of plasmic fibrin(ogen) core fragments[16,17] on this cryoprecipitation was also explored, and this disclosed that

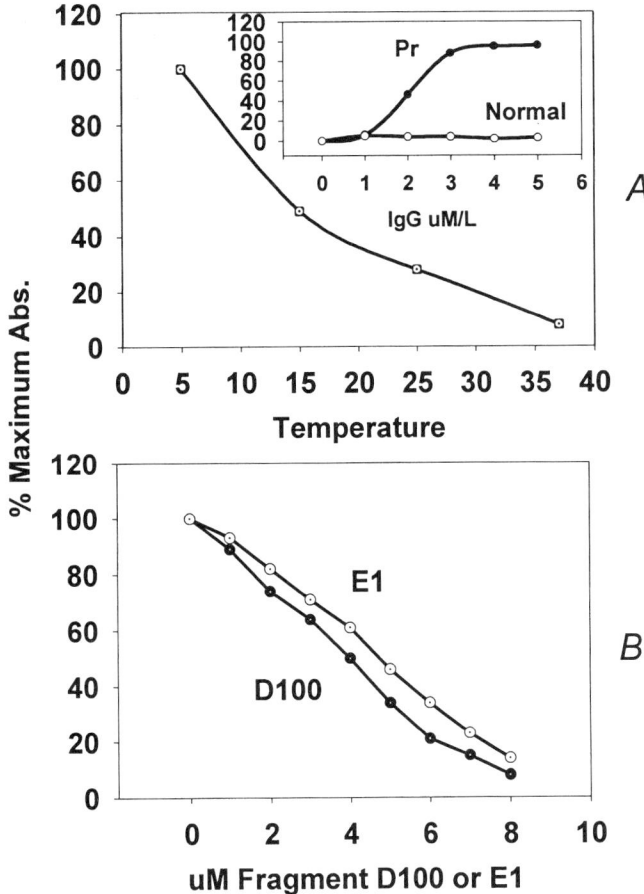

FIGURE 2. Characteristics of fibrinogen precipitation by Pr IgG. *Ordinate* reflects maximum absorbance at 350 nm for 30 minute, and each point plotted represents a separate experiment. **Panel A**, effect of temperature (□) and IgG concentration (*inset*, ○). Pr, proband; N, normal. Conditions: fibrinogen 1 µM, pH 7, µ = 0.15, 5°, IgG 6 µM, except for *inset* where IgG concentrations were varied as shown. **Panel B**, inhibition by fibrinogen fragments E1 (○) and D100 (●) on the cryoprecipitation capacity of the IgG. Each fragment, prepared as described in References 17 and 18, was preincubated with the IgG for at least 30 minutes prior to each experiment.

isolated fragments D100, DD, or E1 abolished the capacity of the IgG to cryoprecipitate fibrinogen. That is to say, appreciable inhibition was demonstrable at equimolar IgG:fragment ratios and progressively increased at higher ratios. At ratios of 1:8 or greater, for example, no cryoprecipitation occurred irrespective of the fragment. Not shown were similar results obtained by using fragments DD and fibrinogen E, and also when normal plasma (5–10% in buffer) was used. The foregoing investigations, demonstrated that: (1) the IC consisted of fibrinogen, IgG, and C1q, (2) the IC were cryoprecipitable, (3) both D and E fibrinogen domains were necessary for IC cryoprecipitation, and (4) the target epitope(s) is present on both normal and proband fibrinogen.

Evidence of interaction between both D and E domains and the IgG is further supported by electronmicroscopy and binding data detailed elsewhere.[11] In the absence of a second mutation, the data are consistent with a primary epitope on the D domain and a mimicking epitotope on the E domain. Alternatively, the primary epitope may be on a D–E locus on the fibrin/fibrin(ogen) polymer.

REFERENCES

1. MARNMEN, E.F., K.P. SCHMIDT & M. BARNHARD. 1967. Thrombophlebitis migrans associated with circulating antibodies against fibrinogen: a case report. Thromb. Diath. Hæmorrh. **18:** 605–611.
2. GALANAKIS, D.K., E.M. GINZLER & S FIKRIG. 1978. Monoclonal IgG anti-coagulant delaying fibrin aggregation in two patients with systemic lupus erythematosus (SLE). Blood **52:** 1037–1045.
3. GALANAKIS, D.K. 1984. Dysfibrinogenemia: a current perspective. Clin. Lab. Clin. Med. **4:** 395–418.
4. GHOSH, S., P. MCEVOY & P.A. MCVERRY. 1983. Idiopathic antoantibody that inhibits fibrin monomer polymerization. Br. J. Hæmatol. **53:** 65–72.
5. HOOTS, W.K., N.A. CARRELL, R.H. WAGNER, et al. 1981. A naturally occurring antibody that inhibits fibrin polymerization. New Eng. J. Med. **304:** 857–861.
6. MARCINIAC, E. & M.F. GREENWOOD. 1979. Acquired coagulation inhibitor delaying fibrinopeptide release. Blood **53:** 81–92.
7. ROSEMBERG, R.D., R.W. COLMAN & L. LORAND. 1974. A new hemorrhagic disorder with defective fibrin stabilization and cryofibrinogenemia. Br. J. Hæmatol. **26:** 269–284.
8. RA'RANI, P., Y. LEVI, D. VARON, et al. 1991. Congenital afibrinogenernia with bleeding, bone cysts and antibodies to fibrinogen. Harfuah **121:** 291–293.
9. BRONNIMANN, R. 1954. Kongenitale afibrinogenemie: Mitteilung eines Falles mit multiplen Knochenzystern und Bildung eines Spezifischen Antikorpers (antifibrinogen) nach Bluttransfusionen. Acta Hæmatol. **11:** 40–51.
10. DEVRIES, A., T. ROSENBERG, S. KOCHWA & J.H. BOSS. 1961. Precipitating antifibrinogen antibody appearing after fibrinogen infusion in a patient with congenital afibrinogenemia. Amer. J. Med. **30:** 486–494.
11. GALANAKIS, D.K., et al. 2001. Manuscript in preparation.
12. GHEBREHIWET, B. 1983. The role of complement in immune complex-mediated pathological disorders. Plasma Ther. **4:** 129–142.
13. WEBER, K., & M. OSBORN. 1969. The reliability of molecular weight determinations by dodecyl sulfatepolyacrylamide gel electrophoresis. J. Biol. Chem. **244:** 4406.
14. STEINBUCH, M. & R. AUDRAN. 1969. Isolation of IgG from mammalian sera with the aid of caprylic acid. Arch. Bioch. Biophys. **134:** 279–284.
15. FINLAYSON, J.S. & M.W. MOSESSON. 1963. Heterogeneity of human fibrinogen. Biochemistry **2:** 42–46.

16. GALANAKIS, D.K. 1995. Studies of plasma cryoprecipitation: major increase in fibrinogen yield by albumin enrichment of plasma. Thromb. Res. **78:** 303–313.
17. OLEXA, A.S. & A.B.Z. BUDZYNSKI. 1979. Primary soluble plasmic degradation product of human cross-linked fibrin. Isolation and stoichiometry of the (DD)E complex. Biochemistry **18:** 991–995.
18. HAVERKATE, F. & G. TIMAN. 1977. Protective effect of calcium in the plasmin degradation of fibrinogen and fibrin fragments D. Thromb. Res. **10:** 803–812.

Factor XIII Prevents Development of Myocardial Edema in Children Undergoing Surgery for Congenital Heart Disease

GERNOLD WOZNIAK,[a] THOMAS NOLL,[b] HAKAN AKINTÜRK,[a] JOSEF THUL,[c] AND MATTHIAS MÜLLER[d]

[a]*Department of Cardiovascular Surgery, Justus-Liebig-University, Giessen, Germany*

[b]*Institute of Physiology, Justus-Liebig-University, Giessen, Germany*

[c]*Department of Pediatric Cardiology, Justus-Liebig-University, Giessen, Germany*

[d]*Department of Anæsthesiology, Justus-Liebig-University, Giessen, Germany*

ABSTRACT: In a prospective investigation of perioperative cardiac edema formation requiring a delayed sternal closure, we identified thrombin increase combined with a simultaneous decrease of factor XIII as a probable cause. After experimental studies additionally revealed that factor XIII could protect endothelial barrier function, we did another prospective randomized trial in which factor XIII or placebo was preoperatively substituted. The substitution finally showed distinct effects minimizing the incidence of myocardial swelling. Therefore, the clinical application of factor XIII may have a valuable therapeutic benefit in cases of leakage syndrome during extracorporeal circulation in congenital heart surgery.

KEYWORDS: Factor XIII; Myocardial edema; Congenital heart disease; Endothelial barrier function.

Myocardial edema (ME) requiring delayed sternal closure (DSC) is a frequent complication of surgical intervention in newborns and young children suffering from congenital heart disease (CHD).[1] Myocardial hypoxia, reperfusion injury, congestive heart failure, or pulmonary hypertension are considered to be causal for the development of myocardial edema.[2,3] In neonates, the edema incidence is significantly enhanced in young patients with low body weight who have had to undergo a prolonged period of extracorporal circulation (ECC).[4] At present, the mechanism of myocardial edema formation is not fully understood. Myocardial swelling may develop due to elevated perfusion pressure and/or decreased lymphatic drain.[5] However, interstitial edema may also occur when the endothelial barrier function fails.[6] Failure of endothelial barrier function causes extravasation of water, solutes, and macromolecules and it can be induced by a variety of mediators circulating in the blood stream under ECC conditions. Thrombin, bradykinin, or anaphylatoxins originating from the complement cascade,[7] can activate endothelial cells. As a result,

---

Address for correspondence: Gernold Wozniak, M.D., Department of Cardiovascular Surgery, Justus-Liebig-University Giessen, Rudolf-Buchheim-Str. 7, D-35392 Giessen, Germany. Voice: 49 641 99 44301; fax: 49 641 99 44309.

gernold.wozniak@chiru.med.uni-giessen.de

cell–cell contacts are loosened, causing leaks to appear between adjacent cells, and thus increasing paracellular permeability across the endothelium.[8,9] Depending on the extent, capillary leakage can cause locally confined or generalized edeme. In practice, no proven prophylactic or therapeutic means are available at present to protect endothelial barrier against pathological disrupture.

Our clinical observations showed that the plasma level of the coagulation factor XIII (FXIII, plasmatransglutaminase) is significantly reduced in those newborns who develop myocardial edema during congenital heart surgery with ECC.[10,11] Based on this observation, we tested to see whether there is a causal relationship between FXIII and endothelial barrier function in a culture model of endothelial cells. In these studies, the flux of albumin across an endothelial monolayer was applied as a measure of paracellular permeability.[6] It was found that thrombin activated FXIII led to an immediate reduction of permeability by $30 \pm 11\%$ (compared with control) within 20 min. This was maintained at that low level during a subsequent 60-min observation period (see FIGURE 1). Non-activated FXIII had no effect on permeability of aortic endothelial monolayers. Activated FXIII also protects endothelial cells against failure of barrier function provoked by metabolic stress; that is, if mitochondrial and glycolytic energy production are inhibited by sodium cyanide and 2-deoxy-D-glucose, respectively.[9] Immunofluorescence microscopy revealed that the activated FXIII was deposited at the sites of the interendothelial clefts and that the deposition of FXIII correlates with the reduction of

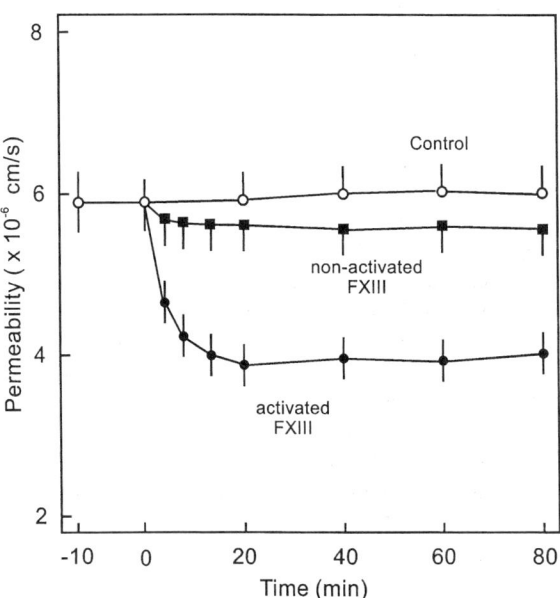

FIGURE 1. Effect of FXIII on paracellular albumin permeability of porcine aortic endothelial monolayers: non-activated FXIII (■, 1 U/mL); activated FXIII (●, 1 U/mL).

paracellular permeability.[6] The FXIII effect on endothelial barrier function was confirmed in the intact coronary vascular system. In a model of isolated perfused hearts, activated FXIII protects against edema formation provoked by ischemia followed by reperfusion. These experimental studies demonstrate an as yet undocumented function of FXIII: it stabilizes the endothelial barrier and can protect against an imminent failure of barrier function.

A prospective randomized trial was performed to test whether preoperative substitution of plasmatic Factor XIII can protect against myocardial edema formation during open heart surgery in neonates and young children suffering from cyanotic cardiac malformation. The incidence of a distinct myocardial edema requiring delayed sternal closure was used as main clinical parameter for evaluation of the effect of preoperative FXIII substitution. Generalized edema formation was estimated by circumference measurement of arms and legs and sonographic measurement of tissue thickness.

Group I ($n = 12$) received preoperatively 250 U FXIII (Fibrogammin HS; Aventis Behring, Marburg, Germany); Group II ($n = 18$) placebo (4 mL 0.9% NaCl solution). The preoperative (prior to substitution) FXIII concentrations were not significantly different ($62 \pm 13\%$ [Group II] versus $75 \pm 15\%$ [Group I]). After substitution the average activity of FXIII was $143 \pm 19\%$ in Group I (FXIII) and $66 \pm 14\%$ in Group II (placebo). In Group I only one patient (1/12, 8%) developed ME, whereas ME was seen in six patients (6/18, 33.3%) in Group II (see TABLE 1). Preoperative substitution also distinctly reduced the extension of generalized edema formation, as could be observed from the results of circumference and sonographic measurement.

Formation of thrombin (measured by prothrombin fragment 1+2) increased in both groups during operative treatment with extracorporeal circulation, despite complete anticoagulation averaging 6.3 ± 1.9 nmol/L.

TABLE 1. Demographic data and measurement results from prospective randomized clinical trails (intermediate results)[a]

	Substitution ($n = 12$) [4 mL/250 U FXIII]	Substitution ($n = 18$) [4 mL/NaCl]
Body weight (gr)	$5,060 \pm 2,060$	$6,970 \pm 1,820$
ECC time (min)	164.3 ± 67.8	190.1 ± 74.2
Ischemic time (min)	94.9 ± 27.5	107.5 ± 24.9
C.M. right arm difference pre–post-OP (mm)	0.29 ± 0.45	0.65 ± 0.83
Sonography tissue thickness 5. rib difference pre–post-OP (mm)	1.4 ± 0.99	3.6 ± 3.51
Distinct ME as delayed sternal closure	1/12 (8.3%)	6/18 (33.3%)

[a]ABBREVIATIONS: ECC, extracorporeal circulation; C.M., circumference measurement; M.E., myocardial edema.

In young children undergoing open heart surgery for congenital cyanotic malformation, myocardial edema formation is closely associated with low factor XIII levels and elevated generation of mediators like thrombin. Taken together with our recent finding that activated FXIII has an endothelial protective effect, we speculate that a deficiency of FXIII can become clinically relevant if the endothelial barrier function is compromised. Preoperative substitution of FXIII can significantly reduce the incidence of myocardial edema and can protect against an imminent generalized capillary leakage syndrome.

ACKNOWLEDGMENT

This work was in part supported by the Deutsche Forschungsgesellschaft, Grant A3 of SFB 547.

REFERENCE

1. IYER, R.S., *et al.* 1997. Outcomes after delayed sternal closure in pediatric heart operations: a 10-year experience. Ann. Thorac. Surg. **63:** 489–491.
2. NAVAS, J.P., *et al.* 1993. Pathophysiology of edema in congestive heart failure. Heart Dis. Stroke **2:** 325–329.
3. DAVIS, K.L., *et al.* 1995. Myocardial edema, left ventricular function, and pulmonary hypertension. J. Appl. Physiol. **78:** 132–137.
4. MAEHARA, T., *et al.* 1991. Perioperative monitoring of total body water by bio-electrical impedance in children undergoing open heart surgery. Eur. J. Cardiothorac. Surg. **5:** 258–265.
5. MEHLHORN, U., *et al.* 1996. Cardiac surgical conditions induced by β-blockade: effect on myocardial fluid balance. Ann. Thorac. Surg. **62:** 143–150.
6. NOLL, T., *et al.* 1999. Effect of Factor XIII on endothelial barrier function. J. Exp. Med. **189:** 1373–1382.
7. GARCIA, J.G.N., *et al.* 1996. Regulation of thrombin-mediated endothelial cell contraction and permeability. Sem. Thromb. Hemost. **22:** 309–315.
8. WOZNIAK, G. 1998 Beeinflussung der endothelialen Schrankenfunktion durch die Plasma-Transglutaminase (Faktor XIII). Shaker-Verlag, Aachen.
9. NOLL, T., *et al.* 1995. Initiation of hyperpermeability in energy-depleted coronary endothelial monolayers. Am. J. Physiol. **268:** H1462–H1470.
10. WOZNIAK, G., *et al.* 1997. Loss of endothelial barrier function during congenital heart surgery in children and newborns: factors causing edema formation (Abst.). Cardiovasc. Surg. **5:** 48.
11. WOZNIAK, G., *et al.* 1999. Einfluß von faktor XIII auf die endotheliale barriere—klinik und experiment. *In* Klinische Aspekte des Faktor-XIII-Mangels. R. Egbring, R. Seitz & G. Wozniak, Eds.: 16–25. Karger, Basel.

Fibrinogen: A Vascular Risk Factor, Why?

Contributing Effect of Oncostatin M on Both Fibrinogen Biosynthesis by Hepatocytes and Participation in Atherothrombotic Risk Related to Modifications of Endothelial Cells

F. MIRSHAHI, M. VASSE, L. VINCENT, V. TROCHON,
J. POURTAU, J.P. VANNIER, H. LI, J. SORIA, AND C. SORIA

*Diféma, Faculté de Médecine et de Pharmacie de Rouen. 76000 Rouen,
Laboratoire de Biochimie A and INSERM EMI E9912, Hôtel Dieu. 75004 Paris, France*

ABSTRACT: Fibrinogen is a vascular risk factor. We suggest that it is a marker of cytokine secretion that simultaneously stimulates fibrinogen biosynthesis and vascular modifications responsible for atherothrombosis. Among these cytokines, oncostatin M (OSM) is the most potent cytokine for inducing fibrinogen biosynthesis by hepatocytes, and it could contribute to endothelial cell anomalies involved in the atherothrombotic process. Here we show that OSM acts (1) by inducing the secretion involved in invasion of the vessel wall by monocytes; (2) by inducing angiogenesis it promotes plaque destabilization, rupture, and consequently thrombosis; and (3) by decreasing fibrinolysis on macrovascular endothelial cells.

KEYWORDS: Fibrinogen; Vascular risk factor; Oncostatin M; Fibrinogen biosynthesis; Hepatocytes; Atherothrombic risk; Endothelial cell.

INTRODUCTION

In the past several years, many epidemiological studies have shown that a high fibrinogen level is a predictor of ischemic heart disease and stroke. However, although highly significant ($p < 0.001$), the difference between the level of fibrinogen in the group that did not develop a vascular event during the five year follow-up and the group that did develop an ischemic disorder was rather low (2.59 [2.17–3.01 g/L] versus 2.8 [2.35–3/30 g/L]).[1] Therefore, we suggest that the increase in fibrinogen is not directly responsible for the vascular disease, but represents, rather, a marker of vascular risk: the increase in fibrinogen could represent the secretion of cytokines that simultaneously stimulate fibrinogen biosynthesis and induce vascular modifications leading to atherothrombosis. The aim of our study was to explore the potential role played by oncostatin M (OSM) in vascular risk. OSM, a member of the IL-6 family of cytokines secreted by monocytes, was considered because, when

Address for correspondence: Claudine Soria, Pr. Faculté de Médecine et de Pharmacie de Rouen, Laboratoire Diféma, 22 boulevard Gambetta, 76000 Rouen, France.
claudine.soria@lrb.ap-hop-paris.fr

compared to other cytokines, it was shown to induce the major effect on fibrinogen biosynthesis by hepatocytes.[2]

Several sources of evidence indicate that endothelial cells may be primary targets of OSM, because these cells have 10- to 20-fold greater expression of OSM receptors in comparison with other cell types.[3] Thus, we were prompted to test the effect of OSM on endothelial cells and its contribution to atherothrombosis. Our study demonstrates that OSM can contribute to endothelial cell anomalies that subsequently contribute to the atherothrombotic process by the various means described in this paper.

OSM INDUCES THE SECRETION OF MCP-1 BY ENDOTHELIAL CELLS

The migration of monocytes across the vascular endothelium plays an important role during atherogenesis. Monocyte chemoattractant protein (MCP-1) is believed to play an important role in this process. MCP-1 is expressed by a variety of cell types, including vascular endothelial cells, in response to various stimuli.[4]

We showed that incubation of microvascular endothelial cells with OSM induced a marked and dose dependent increase in MCP-1 secretion (five-fold increase for 112 pmol/L OSM). This was due to increased synthesis because it was associated with an increase in mRNA expression and it could represent one of the pathogenic roles of OSM in vascular disease.

OSM INDUCES ANGIOGENESIS

Neovascularization of atherosclerotic plaque is responsible for plaque weakening and consequently for its rupture.[5,6] Plaque rupture leads to thrombus formation and is responsible for an acute ischemic event (myocardial infarction or stroke). We tested the effect of OSM on angiogenesis because, in the plaque, macrophages that secrete OSM reside at the site of angiogenesis.

On human microvascular endothelial cells (HMEC-1s), OSM (22 to 112 pmol/L) induced a dose dependent increase in cell proliferation that is greater than that induced by the well known angiogenic factors, vascular endothelial growth factor (VEGF) and basic fibroblast growth factor (bFGF), whereas IL-6 had no effect. Furthermore, in a modified Boyden chamber model, OSM was chemoattractant for HMEC-1, and in a tridimensional gel of fibrin, it increased tube formation and tube length. This effect of OSM appears to depend on endothelial cell origin, since OSM (up to 112.5 pmol/L) did not induce human umbilical vein endothelial cell proliferation. The angiogenic effect of OSM was also demonstrated *in vivo* in a rabbit corneal model.[7] Therefore, OSM could contribute to atherothrombotic risk by favoring plaque rupture related to angiogenesis.

OSM DECREASES FIBRINOLYSIS ON MACROVASCULAR ENDOTHELIAL CELLS

Hypofibrinolysis is considered to be a major cause of vascular occlusion. Hypofibrinolysis can be caused either by an increase in plasminogen activator inhibitor (PAI-1) or a decrease in tissue type plasminogen activator (tPA) by endothelial cells. Our study demonstrates[8] that, when incubated with human umbilical vein endothelial cells, OSM induced an increase in PAI-1 (172% of control cells with 112 pmol/L OSM) and a decreased secretion of t-PA (50% reduction with 112 pmol/L OSM), indicating a hypofibrinolytic stage, which could participate to the thrombogenicity of OSM on large vessels. In contrast, on a human microvasculature endothelial cell line, OSM did not modify PAI-1, inducing a weak but significant increase in tPA secretion, similar to that observed with IL-1. This could explain the paradoxical increase in t-PA level in patients with atheroslerotic lesions.

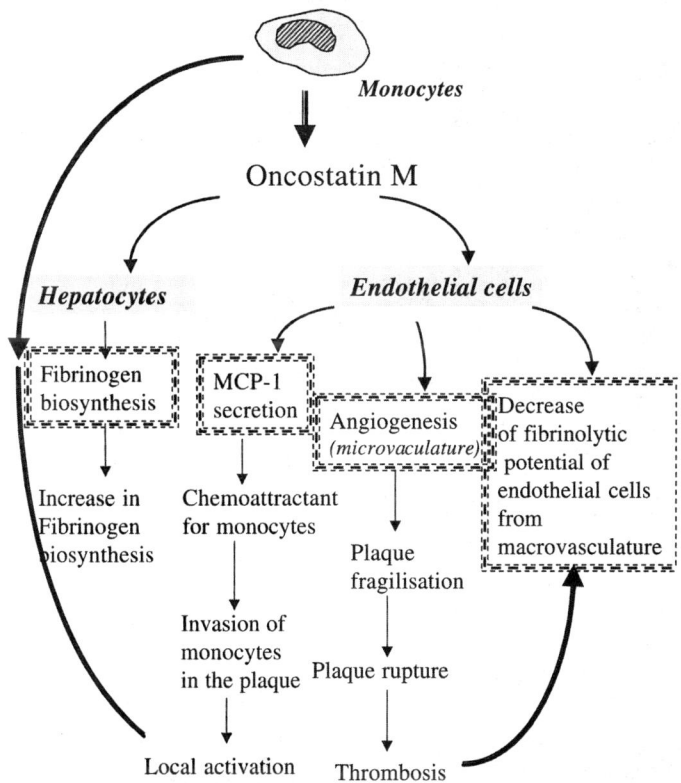

FIGURE 1. Oncostatin is responsible for both fibrinogen biosynthesis and for endothelial cell contribution in atherothrombosis.

CONCLUSION

As is shown in FIGURE 1, our results indicate that the increased fibrinogen level in the patients with high thrombotic risk could be related to the secretion of OSM because it is the strongest cytokine in inducing fibrinogen synthesis and it induces changes in endothelial cells that may contribute to atherothrombotic risk.

ACKNOWLEDGMENT

The authors thank Richard Medeiros for his advice in editing the manuscript.

REFERENCES

1. SMITH, F.B., A.J. LEE, F.G. FOWKES, et al. 1997. Hemostatic factors as predictors of ischemic heart disease and stroke in the Edinburgh Artery Study. Arterioscler. Thromb. Vasc. Biol. **17:** 3321–3325.
2. VASSE, M., J.P. COLLET, J. SORIA, et al. 1995. Fibrinogen, a vascular risk factor: a simple marker or a real cause of vascular disease? Thromb. Res. **75:** 349–352.
3. LINSLEY, P.S., M. BOLTON-HANSON, D. HORN, et al. 1989. Identification and characterization of cellular receptors for the growth regulator, Oncostatin M. J. Biol. Chem. **264:** 4282–4289.
4. GOEBELER, M., T. YOSHIMURA, A. TOKSOY, et al. 1997. The chemokine repertoire of human dermal microvascular endothelial cells and its regulation by inflammatory cytokines. J. Invest. Dermatol. **108:** 445–451.
5. THOMPSON, M.M., L. JONES, A. NASIM, et al. 1996. Angiogenesis in abdominal aortic aneurysms. Eur. J. Vasc. Endovasc. Surg. **11:** 464–469.
6. WARE, J.A. & M. SIMONS. 1997. Angiogenesis in ischemic heart disease. Nat. Med. **3:** 158–164.
7. VASSE, M., J. POURTAU, V. TROCHON, et al. 1999. Oncostatin M induces angiogenesis *in vitro* and *in vivo*. Arterioscler. Thromb. Vasc. Biol. **19:** 1835–1842.
8. POURTAU, J., C. SORIA, J. PAYSANT, et al. 1998. *In vitro* effect of oncostatin M on the release by endothelial cells of von Willebrand factor, tissue-type plasminogen activator and plasminogen activator inhibitor-1. Blood Coagul. Fibrinol. **9:** 609–615.

Purification of Fibrinogen and Virus Removal Using Preparative Electrophoresis

ANDREW GILBERT, MICHAEL EVTUSHENKO, AND HARI NAIR

Gradipore Ltd, Frenchs Forest, New South Wales, Australia

ABSTRACT: The Gradiflow is a novel, scalable preparative electrophoresis technique that uses the dual characteristics of size and charge to isolate target macro- and micromolecules from complex biological solutions. It does this with high resolution and in rapid time. The mild buffers are used to assist in retaining biological activity of the isolated protein. Gradiflow technology employs a sandwich of three polyacrylamide membranes configured to allow passage of macromolecules ranging in size from 10 kDa to 1,500 kDa. Fibrinogen was isolated from cryoprecipitate 1 using a single phase process. This separation was achieved within three hours with yields of 85%. Purified fibrinogen was then characterized using biophysical characterization of fibrin clot structure and compared with clots derived from a commercially available product and human plasma. Significantly, clots developed from Gradiflow fibrinogen had characteristics closer to human plasma. Viral removal characteristics of the Gradiflow were investigated by spiking the source material (cryoprecipitate 1) with canine parvovirus and testing for its presence in the isolated fibrinogen using PCR. Parvo removal was found to be greater than 4 logs and was achieved during the purification process. The Gradiflow offers the advantage of large-scale separation of macromolecules and provides a new approach to fibrinogen separation that is quite distinct from other present-day technologies. The technology is capable of isolating protein with high purity, recovery, and functionality in combination with the removal of viruses during the purification. Furthermore, it is capable of integrating into present production systems, significantly improving yield and functionality of target molecules.

KEYWORDS: Fibrinogen; Fibrin glue; Preparative electrophoresis; Gradiflow; Plasma fractionation; Viruses.

INTRODUCTION

The processing of complex biological solutions is a major bottleneck in the biotechnology industry. As a result, there is a demand for a low-cost alternative to conventional precipitation and chromatographic techniques, which are not only expensive but often denature the desired product. Preparative electrophoresis represents a novel method for the purification of native and functional biological molecules.[1] Collectively, however, the resultant product needs to be pathogen-free and any method that incorporates rapid protein isolation and also a viral inactivation step is highly desirable. A major source of concern in fibrinogen preparations is contamination with non-enveloped viruses such as Parvo B19.

Address for correspondence: Hari Nair, Ph.D., Gradipore Ltd, P.O. Box 6126, Frenchs Forest NSW 2086, Australia. Voice: +61 289779042; fax: +61 289779099.
hnair@gradipore.com

The conversion of fibrinogen to fibrin forms the infrastructure on which other components of blood interact in hemostasis. Fibrin also has other functional roles in a myriad of physiological processes, including wound healing, tumour growth, and bone fracture repair.[2] Purified fibrinogen is used in surgery as a hemostatic adjuvant called fibrin glue and it has found significant roles in cardiovascular and neurosurgery. Currently, available methods of fibrinogen purification can take up to three days, with yields ranging from 6% to 40%, depending on the method used.[3]

This paper establishes a methodology for the rapid purification of fibrinogen, with high yield, using the Gradiflow technology and the use of this technology for the clearance of Parvo B19—a non-enveloped virus.

METHODS

Purification of Fibrinogen

Cryoprecipitate 1, produced by thawing frozen plasma at 4°C overnight was removed from plasma by centrifugation at $10,000g$ at 4°C for five minutes. The precipitate was redissolved in citrate saline, pH 7.4 and placed in the upstream of a BF200 (Gradipore Ltd, Sydney, Australia). A potential of 250 V was applied across a 700–1,500–700-kDa cutoff cartridge configuration, and run for two hours at pH 8.0. The product was dialysed against phosphate buffered saline, pH 7.2 (PBS) and analysed for clotting activity by the addition of calcium and thrombin (final concentrations 10 mM and 10 NIHunit/mL respectively).

The purity of the sample was established using reduced SDS PAGE (Gradipore Ltd, Sydney, Australia) and the presence of fibrinogen confirmed by Western analysis. Western blots were stained with rabbit antihuman fibrinogen conjugated to HRP (DAKO, CA, USA) and developed with 4-chloro-naphthol (4CN).

Fibrinogen estimation was performed using an in-house ELISA. Fibrinogen standards and samples were captured using polyclonal rabbit antihuman fibrinogen (DAKO) and tagged with rabbit antihuman fibrinogen HRP conjugate (DAKO). The ELISA was developed with o-tolidene. Fibrinogen fractions were compared with the starting sample.

Fibrin fiber thickness was measured using mass to length,[4,5] and compaction was used to determine tensile properties of the fibrin fibers.[6,7] In each case fibrinogen purified using the Gradiflow was compared to grade L fibrinogen (Kabi, Stockholm, Sweden) and fibrinogen in plasma. Fibrinogen concentration was standardized at 1.7 mg/mL.

Decontamination of Canine Parvo B-19

Cryoprecipitate 1 was spiked with 1×10^8 of canine parvo. Fibrinogen was isolated using size exclusion as the major strategy for removal of the virus.

The presence of canine parvo was investigated using PCR. Canine parvo DNA was extracted from the starting material and purified fibrinogen fractions according to standard procedure.[8] PCR was adapted from Schwartz et al.,[9] using Primer 1 (GCAGTTAACGGAAACATGGC) and Primer 2 (TCTCCTTCTGGATATCTTCC).

PCR products were characterized using PAGE and quantified using serial dilutions.

DISCUSSION

The Gradiflow technology, a preparative electrophoresis system, employs a combination of charge- and size-based separation strategies with the use of mild buffers. The mild purification conditions used in preparative electrophoresis are in contradiction to conventional precipitation techniques and column chromatography, where "harsh" chemical and physical separation methods are used. Loss of fibrinogen functionality in purified preparations is partially attributed to these isolation conditions.

In addition to mild conditions, the electrophoretic procedure allows high recovery of protein. This is important in large-scale purification schemes, particularly in view of the fact that in the plasma fractionation industry, plasma sources are limited. Fibrinogen recovery was estimated to exceed 75%, compared to the 30–40% achieved in most modern industrial schemes (see TABLE 1). Furthermore, biophysical analysis of the resultant fibrin clot showed that Gradiflow fibrin had characteristics similar to clots developed from plasma, but significantly different from clots developed from purified fibrinogen using modified Cohn separation techniques.

These results point to the potential of the Gradiflow to produce a fibrinogen, that in therapeutic regimes is less degraded and similar to that found in plasma. The resultant fibrin is perhaps more functional in arresting blood loss and promoting cell proliferation.

Significantly, these results show that Parvo B19, a 16-nm non-enveloped virus, can be effectively removed by at least four logs (see FIGURE 1). When four logs of virus has been removed, this reduction is considered substantial and the step may, in fact, remove far greater quantities than can be quantified or claimed. Current alcohol precipitation schemes exclude parvo by additional nanofiltration steps. If used together with solvent detergents or other pasteurization steps, the Gradiflow procedure results in a high yielding, highly pure and viral-free product, alluding to a new generation of separation technology ultimately capable of replacing Cohn fractionation.

TABLE 1. A comparison of fibrinogen purified using conventional schemes with that isolated using preparative electrolysis

Fibrinogen Characteristic	Fibrinogen Purification Technique	
	Conventional Scheme	Preparative Electrophoresis
Purity	> 90%	> 90%
Recovery	30%	75%
Separation time	3 days	2 hours
Clottability	90%	90%
Fibrin nativity	dissimilar to plasma	similar to plasma
Parvo removal	additional steps	in process
Cost	high	low

NOTE: Conventional schemes produce a preparation that requires further processing to remove parvo contamination as opposed to Gradiflow technology, which removes virus during the purification.

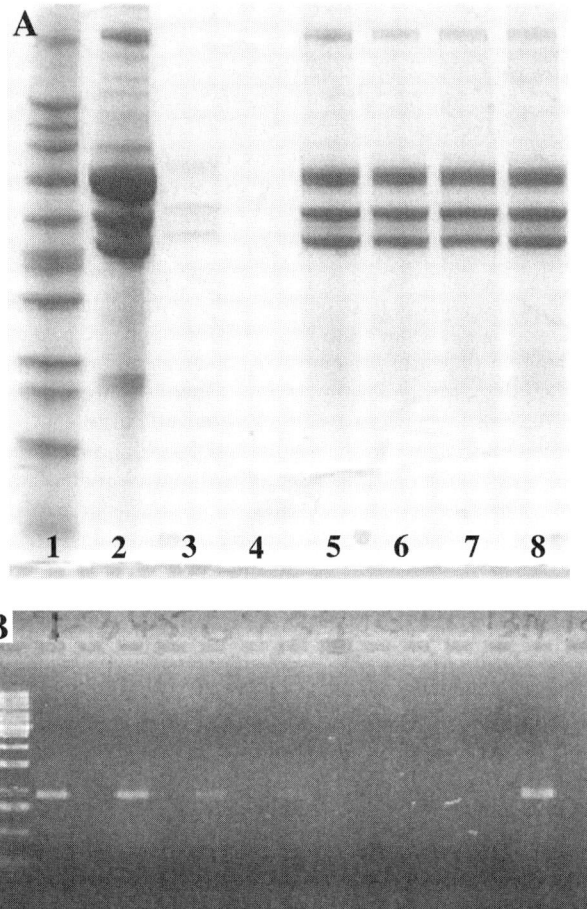

FIGURE 1. Fibrinogen (*lanes 5–8*) was purified from cryoprecipitate (*lane 2*) using preparative electrophoresis and analyzed using reduced SDS PAGE. The characteristic Aα, Bβ, and χ bands are evident. Fibrinogen samples were analyzed for viral clearance using PCR. Pure fibrinogen samples were found to have at least a four log parvo decrease.

REFERENCES

1. HORVATH, S.Z., G.L. CORTHALS, C.W. WRIGLEY & J. MARGOLIS. 1994. Multifunctional apparatus for electrokinetic processing of proteins. Electrophoresis **15:** 968.
2. MARTINOWITZ, U. & R. SALTZ. 1996. Fibrin sealant. Curr. Opin. Hematol. **3:** 395.
3. FURLAN, M. 1984. Purification of fibrinogen. *In* Variants of Human Fibrinogen. E.A. Beck & M. Furlan, Eds.: 133–145. Hans Huber Publishers, Berne.

4. CARR, JR., M.E. & J. HERMANS. 1978. Size and density of fibrin from turbidity. Macromolecules **11:** 46.
5. NAIR, C.H., A. AZHAR & D.P. DHALL. 1991. Studies of fibrin network structure in human plasma. Part one: methods for clinical application. Thromb. Res. **64:** 455–476.
6. DHALL, T.Z., W.A.J. BRYCE & D.P. DHALL. 1976. Effects of dextran on the molecular structure and tensile behaviour of human fibrinogen. Thromb. Diath. Hæmorrh **35:** 737.
7. NAIR, C.H. & E.A. SHATS. 1997. Compaction as a method to characterise fibrin network structure: kinetic studies and relationship to crosslinking. Thromb. Res. **88:** 381.
8. SAMBROOK, J., E.F. FRITSCH & T. MANIATIS. 1989. Molecular Cloning—A Laboratory Manual, 2nd edit. 9.38–9.42. Cold Spring Harbour Laboratory Press, Cold Spring Harbour.
9. SCHWARTZ, T.F., J. JAGER, W. HOLZGREVE & M. ROGGENDORF. 1992. Diagnosis of human parvovirus B19 infections by polymerase chain reaction. Scand. J. Infect. Dis. **24:** 691–696.

Effect of Moderate Alcohol Consumption on Fibrinogen Levels in Healthy Volunteers Is Discordant with Effects on C-Reactive Protein

AAFJE SIERKSMA,[a,b] MARTIJN S. VAN DER GAAG,[a,*] CORNELIS KLUFT,[c] AND HENK F.J. HENDRIKS[a]

[a]*TNO Nutrition and Food Research, Zeist, the Netherlands*

[b]*University Medical Center Utrecht, Julius Center for Patient Oriented Research, Utrecht, the Netherlands*

[c]*TNO Prevention and Health, Leiden, the Netherlands*

> ABSTRACT: In a diet-controlled, crossover trial with 10 middle-aged men and 9 postmenopausal women, baseline concentrations of fibrinogen influenced the magnitude of decrease of fibrinogen after moderate alcohol consumption. The mechanism of reduction is specific for fibrinogen and unrelated to a reduction in C-reactive protein.
>
> KEYWORDS: Alcohol; Fibrinogen; C-reactive protein.

INTRODUCTION

Epidemiological[1] and experimental[2,3] studies have shown that moderate alcohol consumption decreases fibrinogen level, which is associated with a decreased risk for cardiovascular disease. The mechanisms of this decrease in fibrinogen are not known. The aim of the study reported here was to evaluate whether baseline concentrations of fibrinogen influence the magnitude of this decrease and whether or not an antiinflammatory mechanism is involved, monitored via C-reactive protein (CRP).

SUBJECTS AND METHODS

In a diet-controlled, crossover trial, ten middle aged men (aged 45–64 years) and nine postmenopausal women (aged 49–62 years), all apparently healthy, non-smoking, and moderate alcohol drinkers, consumed, respectively, four and three glasses of beer or non-alcoholic beer (control) with dinner. Beverages were switched after three weeks in a random order. To exclude carryover effects, there was a one week wash-out period (no alcohol) before each treatment. All food and drink was supplied during the six weeks of treatment. The diet, which was essentially the same

---

*Martijn S. van der Gaag died in January 2000.
Address for correspondence: Aafje Sierksma, M.Sc., TNO Nutrition and Food Research, Department of Nutritional Physiology, P.O. Box 360, 3700 AJ Zeist, the Netherlands. Voice: +31 30 6944236; fax: +31 30 6944928.
sierksma@voeding.tno.nl

during the two periods, contained adequate amounts of macro- and micronutrients. During the beer period, alcohol intake was equal to 40 and 30 grams a day, respectively, for men and women. Fibrinogen was analyzed by the Clauss method.[4] CRP was measured in one batch with a sensitive in-house sandwich enzyme–immunoassay (intra CV 5.2%). Treatment effects were assessed by analysis of variance, by use of a general linear model, in which CRP measurements were log-transformed. No carryover effects were seen.

RESULTS

Plasma fibrinogen levels significantly decreased by 12.4% after three weeks of beer consumption, as compared with non-alcoholic beer consumption (see FIGURE 1). Fibrinogen levels after beer consumption showed a significant positive correlation with fibrinogen levels after non-alcoholic beer consumption (Pearson; $r = 0.90$ and $p = 0.0001$). In terms of percentage, the decrease in fibrinogen was greater for subjects with higher baseline concentrations of fibrinogen (the non-alcoholic beer values) (see examples in FIGURE 2). A potential mechanism by which alcohol may reduce fibrinogen is the antiinflammatory effect.

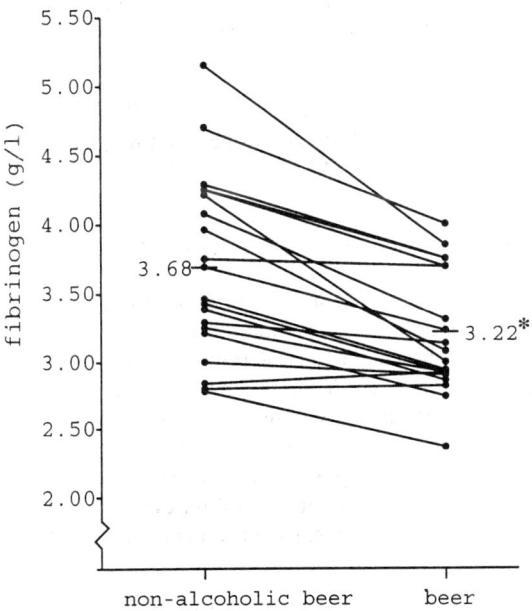

FIGURE 1. Individual and mean plasma fibrinogen levels in middle aged men and postmenopausal women after three weeks of beer and non-alcoholic beer consumption. *$p = 0.0001$.

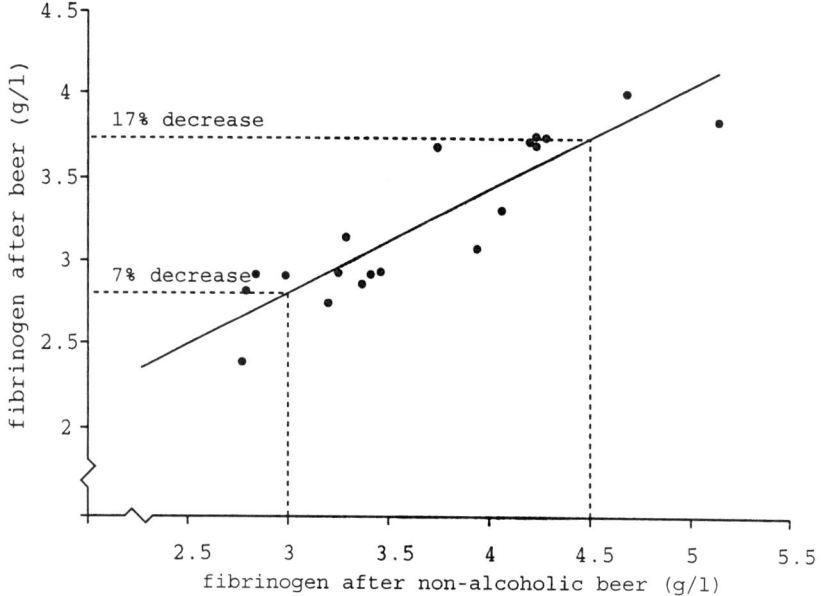

FIGURE 2. Correlation between plasma fibrinogen levels in middle aged men and postmenopausal women after three weeks of beer and non-alcoholic beer consumption. *Regression line*: beer = 0.944042 + 0.619257×non-alcoholic beer (Pearson; $r = 0.90$, $p = 0.0001$).

There was a good correlation between CRP and fibrinogen for both beverages (Pearson; $r = 0.61–0.67$, $p < 0.006$), but not between relative changes in fibrinogen and CRP (Pearson; $r = 0.23$, $p = 0.35$).

DISCUSSION AND CONCLUSION

Baseline concentrations of fibrinogen influence the magnitude of decrease in fibrinogen after moderate alcohol consumption. The mechanism of reduction is specific for fibrinogen and unrelated to a reduction in CRP. Effects on fibrinogen, unrelated to CRP changes, have been shown previously for the effect of Ticlopidine.[5] In that study the change in fibrinogen was related to the method of detection, suggesting a change in properties of fibrinogen. This and other types of fibrinogen related mechanisms should be further explored with various methods for detection. It also suggests better defining the effects of alcohol on cytokines and vascular adhesion molecules in order to delineate the exact antiinflammatory mechanism.

REFERENCES

1. HENDRIKS, H.F.J. & M.S. VAN DER GAAG. 1998. Alcohol, coagulation and fibrinolysis. *In* Alcohol and Cardiovascular Diseases. (Novartis Foundation Symposium 216): 111–124. Wiley, Chichester.

2. DIMMITT, S.B., V. RAKIC, I.B. PUDDEY, et al. 1998. The effects of alcohol on coagulation and fibrinolytic factors: a controlled trial. Blood Coagul. Fibrinol. **76:** 39–45.
3. PELLEGRINI, N., F.I. PARETI, F. STABILE, et al. 1996. Effects of moderate consumption of red wine on platelet aggregation and hæmostatic variables in healthy volunteers. Eur. J. Clin. Nutr. **50:** 209–213.
4. VON CLAUSS, A. 1957. Gerinnungsphysiologische Schnellmethode zur Bestimmung des Fibrinogens. Acta Hæmatol. **17:** 231–237.
5. DE MAAT, M.P., A.E. ARNOLD, S. VAN BUUREN, et al. 1996. Modulation of plasma fibrinogen levels by ticlopidine in healthy volunteers and patients with stable agina pectoris. Thromb. Hæmost. **76:** 166–170.

Fibrin Network Structure

Changes in Characteristics in Response to Physical Activity in Combination with a Pre-exercise Meal

S.J. MOSS,[a] M.M. MALAN,[b] L.I. DREYER,[a] AND H.H. VORSTER[c]

[a]*Institute for Biokinetics,* [b]*School of Pharmacy,*
[c]*School of Physiology, Nutrition and Family Ecology,*
Potchefstroom University for Christian Higher Education, Potchefstroom, South Africa

ABSTRACT: The metabolic environment determines the characteristics of fibrin network structure (FNS) in plasma. Physical activity or changes in the diet (e.g., high and low glycemic index meals) can initiate these changes. The FNS were measured by means of mass–length ratio, turbidity, and compaction of the plasma obtained from male subjects. Samples were taken before and after the introduction of the different diets and physical activity. The results indicate that changes in the FNS do occur, but further investigation is required to obtain conclusive results.

KEYWORDS: Physical activity; Males; Fibrin network structure; Glycemic index; Pre-exercise meal.

INTRODUCTION

Regular physical activity is associated with a lower risk of CVD. It is hypothesized that part of this association may be mediated by the quality of fibrin network structures. The plasma metabolic environment determines the characteristics of the fibrin network structure. Regular physical activity changes the metabolic environment, which is expected to influence formation of fibrin network structures. Changes in diet also change the metabolic environment. A high glycemic index (GI) diet results in large variations in blood glucose and insulin responses, whereas the ingestion of low GI meals results in the slow release of glucose and insulin with less variation.[1] The effects of physical activity and a combination of activity with low or high GI pre-exercise meals on the formation of the fibrin network structures are not known. The purpose of this study was to determine the influence of a maximal exercise session on fibrin network characteristics. The combinations of a low or high GI pre-exercise meal and physical activity on the network structure characteristics were also investigated.

Address for correspondence: Dr. S.J. Moss, Ph.D., 148 Kamp Street, Potchefstroom 2531, South Africa. Voice: +27 18 297 6906; fax: +27 18 297 6906.
   profile@intekom.co.za

METHODS

Fifteen active and 14 sedentary males were recruited to participate voluntarily in the study. The respondents were subjected to a maximal exercise, blood samples taken while fasting, during maximal activity, and after 30 minutes of recovery. Maximal activity was achieved by exercising the respondents to exhaustion on a Monark bicycle ergometer.

The permeability coefficient (K_s),[2] mass–length ratio (MLR),[3] and compaction[4] at the various times were determined. The same study design was followed with a high and low pre-exercise GI meal given in a random order. Blood samples were taken while fasting and at one hour post meal, maximal exercise, and after 30 minutes of recovery. K_s, MLR, compaction, and selected biochemical variables were determined at each of the times.

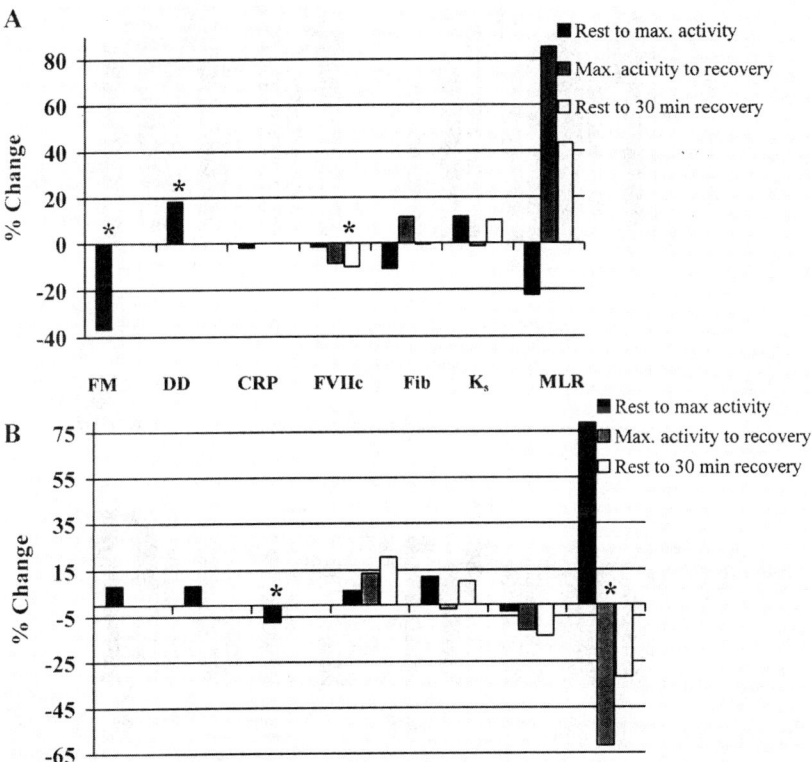

FIGURE 1. Changes in the fibrin network structure characteristics and hemostatic variables of active (**A**) and sedentary (**B**) males as a percentage change from rest to maximum activity, maximum activity to 30 min of recovery, and rest to 30 minutes of recovery. *Significantly different ($p = 0.05$). FM, fibrin monomer; DD, D-dimer; CRP, C-reactive protein; FVIIc, factor VIIc; Fib, fibrinogen; K_s, permeability; MLR, mass–length ratio.

RESULTS

Maximal exercise resulted in an increase in K_s values from plasma of active males, whereas the MLR values decreased when compared to those at rest. After 30 minutes of recovery the K_s values decreased slightly, but remained higher than the K_s values obtained at rest. The MLR values increased after 30 minutes of recovery, significantly higher than at rest. It seems that the active males formed fibrin network structures that were less resistant to lysis[3] after 30 minutes of recovery than at the start of activity (see FIGURE 1). In the sedentary males, the K_s values decreased and the MLR values increased in response to maximal activity. After 30 minutes of

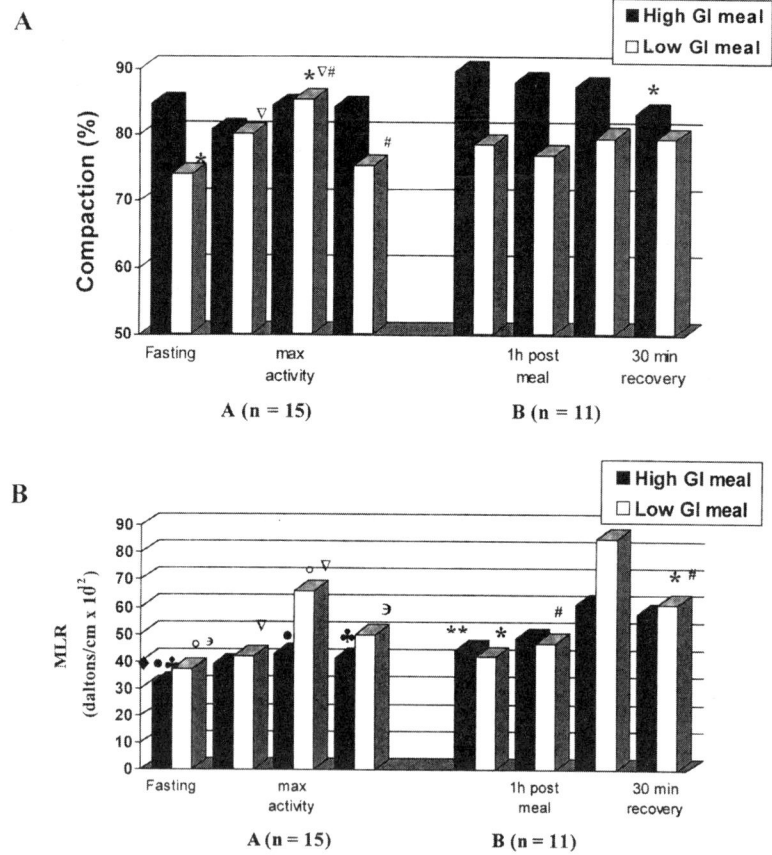

FIGURE 2. Changes in compaction (**A**) and mass–length ratio (MLR) (**B**) of the active (*A*) and sedentary (*B*) males obtained at fasting, one hour post-glycemic index preexercise meals, maximal activity, and after 30 minutes of recovery. *Similar symbols* signify significant differences within groups ($p = 0.05$). **Value was 75.60 ± 67.36 as a result of two outlying values.

recovery, K_s values for sedentary males decreased even more, to result in less permeable fibrin network structures than prior to activity. MLR values increased with maximal activity, but after 30 minutes of recovery decreased to values lower than prior to activity. It seems that the sedentary males formed fibrin network structures after maximal activity that were less permeable and more resistant to lysis, possibly due to longer and thinner fibers.

The low GI meal resulted in a smaller insulin response together with an increase in compaction when compared with the high GI meal. In the sedentary group the low GI meal also exhibited a smaller insulin response, with an increase in MLR. This increased MLR suggests that shorter and thicker fibrin network structures were formed with the ingestion of a low GI meal when compared to a high GI meal (see FIGURE 2). The advantage of the low GI meal compared to the high GI meal was seen in the increased MLR values and compaction in response to maximal activity in the active males. A slight increase was also seen in K_s values. This suggests that the active males formed more permeable fibrin network structures, that are more readily dissolved, than those formed with the high GI meal. The low GI meal also increased compaction. The results found in the sedentary males were not clear. The characteristics of the fibrin networks indicated a trend toward a decrease in K_s values, whereas the high GI meal decreased compaction and the low GI meal increased compaction.

Recovery from the maximal exercise resulted in a decrease in K_s values for the active males with both the high and low GI meals. The low GI meal resulted in values higher than initially measured at fasting. The MLR values showed similar changes. MLR values after activity were greater in combination with the low GI meal than with the high GI meal. The same trend was found in sedentary males, but the MLR and K_s values after recovery, in combination with the high GI meal, were lower than at fasting. Values returned to the fasting values with the low GI meal. An increase in K_s value suggests an increase in permeability of the fibrin network structure, whereas an increase in MLR value suggests shorter and thicker fibrin fibers.[5]

CONCLUSION

It is concluded that the effects of physical fitness and acute sessions of intensive exercise on fibrin networks formed from plasma of healthy young males may differ. Fibrin networks formed from plasma of healthy young males are influenced by the glycemic properties of the preexercise meal. Changes in the metabolic environment occurred that possibly affected the characteristics of the fibrin network structures during exercise and after meals. More research is needed for a better understanding of underlying mechanisms, and to relate these differences to health outcomes.

REFERENCES

1. WOLEVER, T.M.S. 1990. The glycaemic index. World Rev. Nutrit. Diet. **62:** 120–185.
2. BLOMBÄCK, B. & M. OKADA. 1982. Fibrin gel structure and clotting time. Thromb. Res. **25:** 51–70.
3. CARR, M.E. & J. HERMANS. 1978. Size and density of fibrin fibres from turbidity. Macromolecules **11:** 46–50.

4. DHALL, T.Z., *et al.* 1976. Effect of dextran on the molecular structure and tensile strength. Thromb. Hæmostas. **35:** 737–745.
5. BLOMBÄCK, B., *et al.* 1990. Native Fibrin Gel Networks and Factors Influencing Their Formation in Health and Disease. 1–23. Plenum Press, New York.

Influence of Plasma Triglyceride and Plasma Cholesterol Levels on the Clearance Rate of Fibrinogen

MAARTJE VERSCHUUR, MARIAN BEKKERS, MONIQUE G.M. VAN ERCK, JEF J. EMEIS, AND MONIEK P.M. DE MAAT

Gaubius Laboratory, TNO Prevention and Health, Leiden, the Netherlands

ABSTRACT: Increased plasma levels of fibrinogen are associated with a higher risk of cardiovascular disease. It has been suggested that lipid levels may influence the fibrinogen levels by a mechanism other than the synthesis rate, for example a decreased clearance rate. We performed a pilot study to explore this possibility. Twelve male Wistar rats were fed for four weeks with a control low fat/low cholesterol diet, a high fat/high cholesterol diet, and a high fat/high cholesterol diet with an additional 0.5% cholic acid. Labeled ^{125}I fibrinogen was injected, and blood was sampled repeatedly. From the plasma radioactivity of the samples, fibrinogen halflife time was calculated for each animal. Our results suggest that plasma lipids lengthen the fibrinogen halflife times, although the differences were not statistically significant in this small study. Our final conclusion from this study is that lipids may have an effect on the turnover rate of fibrinogen and possibly affect fibrinogen levels through this mechanism.

KEYWORDS: Fibrinogen clearance rate; Plasma triglyceride; Plasma cholesterol.

INTRODUCTION

It is well established that increased plasma levels of fibrinogen are associated with a higher risk of cardiovascular disease. Several factors are known to influence fibrinogen levels, for example, genetic background, acute phase events,[1] and the clearance rate of fibrinogen.[2] Subjects with high lipid levels often also have increased plasma fibrinogen levels, but no underlying mechanism has yet been elucidated. In a recent study by Rezaee *et al.*,[3] mice receiving a high fat diet exhibited increased fibrinogen plasma levels, but no increased mRNA levels of any of the three fibrinogen chains. This suggests that the lipid levels may influence fibrinogen levels by a mechanism other than the synthesis rate, possibly a decreased clearance rate of fibrinogen. We performed a pilot study to explore this possibility.

Address for correspondence: Maartje Verschuur, M.Sc., Gaubius Laboratory, TNO Prevention and Health, P.O. Box 2215, 2301 CE Leiden, the Netherlands. Voice: 31-71-5181486; fax: 31-71-5181904.

m.verschuur@pg.tno.nl

MATERIALS AND METHODS

Materials and Animals

Twelve male Wistar rats were separated into three groups. The first group was fed a control low fat, low cholesterol diet containing basically sucrose and nutrients (purchased from Hope Farms, Woerden, the Netherlands). The second group received a high fat, high cholesterol diet (HFC/0) containing 1% (w/v) cholesterol, to increase the plasma triglyceride levels. The third group was fed the same high fat, high cholesterol diet with an additional 0.5% (w/v) cholic acid (HFC/0.5), to increase plasma cholesterol levels.[4] These diets were given for a period of four weeks prior to the start of the experiment.

Preparation of Labeled ^{125}I Fibrinogen

Human fibrinogen was prepared as described elsewhere,[5] and was more than 95% clottable. This fibrinogen was labeled with $Na^{125}I$ using the Iodo Beads® method. The clottability of the labeled fibrinogen was 85%. 0.44 ml of the ^{125}I-fibrinogen solution (0.5 mg/ml in PBS, 1.35×10^5 cpm/µl) was injected into a male Wistar rat. After eight hours, blood was collected by aortic puncture and citrated plasma was prepared. The clottability of this screened fibrinogen was 92%.

In Vivo Experiment

Into each of the 12 rats (420–490 grams) screened plasma (±800.000 cpm) was injected via the dorsal vein of the penis (under Nembutal anesthesia). Five minutes after injection, a blood sample was collected from each animal by aspiration into a heparinized glass capillary after making a small incision in one of the tail veins. Blood sampling was repeated at 1, 2, 4, 6, 8, 24, and 48 hours after injection, plasma was prepared and the radioactivity of the plasma was determined.

Treatment of the Experimental Data

The plasma radioactivity at five minutes was taken as 100%, and the activities found at other times were expressed as fractions of this first sample. For each animal, the tracer data were analyzed using a two-component mathematical model.[2] The halflife time of fibrinogen was calculated by using a computer fit of the ^{125}I count data to the following equation:

$$ae^{-\alpha t} + be^{-\beta t}$$

RESULTS

The plasma lipid levels after the dietary period and the calculated fibrinogen β-halflife times are shown in TABLE 1. There was no significant relation between plasma lipid levels and fibrinogen halflife times: for cholesterol $\rho = 0.40$ with $p = 0.20$, and for triglycerides $\rho = 0.13$ with $p = 0.70$ (see FIGURE 1).

TABLE 1. Plasma triglyceride and cholesterol levels after the dietary period, and the calculated halflife times of fibrinogen[a]

	LFC ($n = 4$)	HFC/0 ($n = 4$)	HFC/0.5 ($n = 4$)
Triglyceride level (mM)	3.2 ± 0.6	8.3 ± 1.3	3.2 ± 2.0
Cholesterol level (mM)	2.7 ± 0.3	4.5 ± 0.4	5.3 ± 0.6
Fibrinogen halflife time (h)	24.9 ± 3.9	31.3 ± 11.8	28.8 ± 7.0

[a]Data are presented as mean ± SD. LFC, low fat, low cholesterol diet; HFC/0, high fat, high cholesterol diet; HFC/0.5, high fat, high cholesterol diet with an additional 0.5% cholic acid.

DISCUSSION

Our study suggests that plasma lipids lengthen the fibrinogen halflife times. Although the average differences in halflife times between the LFC group and HFC/0 and HFC/0.5 groups were not significant in this pilot study, it was striking that large variations were found within the HFC groups. In some rats in the two HFC groups, halflife times were doubled compared to other animals in the same dietary group. These findings suggest that response to a stimulus (e.g., the lipid composition of the diet) may be determined individually.

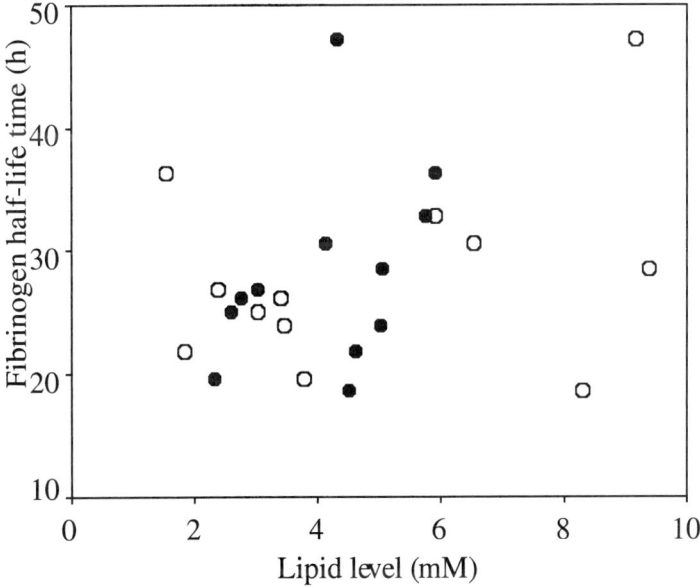

FIGURE 1. Plasma fibrinogen halflife times in relation to lipid levels: ○, triglyceride level (mM); ●, cholesterol level (mM).

Our final conclusion from this pilot study is that lipids may have an effect on the turnover rate of fibrinogen and, thus, possibly affect fibrinogen levels through this mechanism.

REFERENCES

1. GABAY, C. & I. KUSHNER. 1999. Acute-phase protein and other systemic responses to inflammation. New. Engl. J. Med. **340:** 448–454.
2. NIEUWENHUIZEN, W., J.J. EMEIS, A. VERMOND, et al. 1980. Studies on the catabolism and distribution of fibrinogen in rats. Application of the Iodogen labelling technique. Biochem. Biophys. Res. Commun. **97:** 49–55.
3. REZAEE, F., A. MAAS, J.H. VERHEIJEN, et al. 2000. Effect of genetic background and diet on plasma fibrinogen in mice. Possible relation with susceptibility for atherosclerosis (Abstr.). Atherosclerosis **151:** 65.
4. POST, S.M., J.P. ZOETEWEIJ, M.H.A. BOS, et al. 1999. Acyl-Coenzyme A: cholesterol acyltransferase inhibitor, Avasimibe, stimulates bile acid synthesis and cholesterol 7a-hydroxylase in cultured rat hepatocytes and in vivo in rat. Hepatology **30:** 491–500.
5. RUIVEN-VERMEER, VAN, I.A., W. NIEUWENHUIZEN, F. HAVERKATE, et al. 1979. A novel method for the rapid purification of human and rat fibrin(ogen) degradation products in high yields. Hoppe-Seyler's Z. Physiol. Chem. **360:** 633–637.

Index of Contributors

Akintürk, H., 617–620
Alesci, S., 210–214
Amrani, D.L., 566–579
Anglés-Cano, E., 125–128, **261–275**
Argiriou, S., 210–214
Arocha-Piñango, C.L., **223–225**
Arosio, D., 456–458
Ayala, Y.M., 456–458

Bach, T.L., 386–405
Bajzar, L., 247–260
Bekkers, M., 639–642
Bennett, J.S., 340–354
Bernocco, S., 167–184
Bezerra, J.A., 276–290
Biot, F., 89–90
Blombäck, B., 1–10
Blombäck, M., 531–535
Boffa, M., 247–260
Bolliger, B, 219–222
Bonnefoy, A., 459–463
Booth, N.A., 215–218
Brennan, S.O., 91–100, 522–**525**, 536–541

Callea, F., 522–525
Cantwell, A.M., 133–146
Castellino, F.J., 466–468, **542–548**
Chase, B., 147–166
Christiansen, V.J., 335–339
Clark, J.A.F., 355–367
Coates, A.I., 129–132
Collen, A., 426–437
Coller, B.S., 464–465
Cornelissen, I., 542–548
Cuniberti, C., 167–184

Danton, M.J.S., 276–290
Daub, M., 449–455
de Groot, P.G., 444–448
de la Peña Díaz, A., 261–275

de Maat, M.P.M., 509–521, 639–642
DeFord, M.E., 542–548
Degen, J.L., 276–290
Delmotte, Y., 566–579
Dempfle, C.-E., 210–214
Di Cera, E., 133–146, 456–458
DiBello, P.M., 147–166
Dierichs, R., 449–455
DiOrio, J.P., 566–579
Doolittle, R.F., 31–43
Drew, A.F., 276–290
Dreyer, L.I., 634–638

Emeis, J.J., 639–642
Endo, H., 223–225
Evtushenko, M., 625–629

Fabbretti, G., 522–525
Feldman, M., 542–548
Fellowes, A.P., 91–100, 536–541
Ferber, A., 386–405
Ferri, F., 167–184
Frojmovic, M.M., 459–463
Fuller, G.M., 469–479

Gaffney, P.J., 594–610
Galanakis, D.K., 444–448, 611–616
George, P.M., 91–100, 522–525, 536–541
Gilbert, A., 625–629
Gorkun, O.V., 101–116
Green, F.R., 549–559
Grieninger, G., 44–64

Hamada, E., 526–530
Hanss, M., 89–90
Hantgan, R.R., 129–132
He, S., 531–535
Heene, D.L., 210–214
Hendriks, H.F.J., 630–633

Henschen-Edman, A.H, 129–132, 531–535, 580–593, 611–616
Hogan, K.A., 117–121, 205–209, 219–222, 444–448
Holmbäck, K., 276–290
Horrevoets, A., 247–260

Jackson, K.W., 335–339
Jerome, W.G., 459–463
Jiroušková, M., 464–465

Kaneko, H., 526–530
Kluckman, K.D., 117–121
Kluft, C., 630–633
Kombrinck, K.W., 276–290
Koolwijk, P., 426–437
Kucher, K., 210–214
Kudryk, B., 522–525

Lawrie, L.C., 215–218
Lee, C.S., 335–339
Lee, K.N., 335–339
Lefkowitz, J.B., 129–132
Legrand, C., 459–463
Li, H., 621–624
Liang, Z., 542–548
Lijnen, H.R., 226–236
Lill, H., 147–166
Liu, Q., 459–463
Lorand, L., 291–311
Lord, S.T., 101–116, 117–121, 129–132, 205–209, 219–222, 331–334, 444–448
Loukinov, D., 122–124
Lounes, K.C., 129–132, 205–209, 444–448
Lowe, G.D.O., 560–565
Loyau, S., 125–128, 261–275

Maeda, N., 117–121
Malan, M.M., 634–638
Marchi, R., 125–128
Martinez, J., 386–405
Matsuda, M., 65–88, 223–225

McKee, P.A., 335–339
McLennan, L., 542–548
Medicina, D., 522–525
Medved, L., 122–124, 185–205, 312–327, 328–330
Meh, D.A., 11–30
Mirshahi, F., 621–624
Mitkevich, O.V., 147–166
Mochalkin, I., 31–43
Morgenstern, E., 449–455
Mosesson, M.W., xi–xii, 11–30, 215–218
Moss, S.J., 634–638
Müller, M., 617–620
Müller-Peltzer, H., 210–214
Mullin, J.L., 331–334

Nair, H., 625–629
Nakamikawa, C., 223–225
Neerman-Arbez, M., 496–508
Nesheim, M., 247–260
Nieuwenhuizen, W., 237–246
Noll, T., 617–620
Norfolk, S.E., 331–334

Okumura, N, 219–222, 205–209, 526–530

Palumbo, J.S., 276–290
Pereira, M., 438–443
Ping, L., 205–209
Ploplis, V.A., 466–468, 542–548
Pourtau, J., 621–624
Profumo, A., 167–184

Redman, C.M., 480–495
Remijn, J.A., 444–448
Ritchie, H., 215–218
Rocco, M., 167–184
Rosen, E.D., 542–548
Rübsamen, K., 210–214
Rumley, A., 560–565
Rybarczyk, B., 406–425

INDEX OF CONTRIBUTORS

Sekine, O., 223–225
Shainoff, J.R., 147–166
Siebenlist, K.R., 11–30
Sierksma, A., 630–633
Simpson-Haidaris, P.J., 406–425, 438–443
Sinha, R., 531–535
Sixma, J.J., 444–448
Smejkal, G.B., 147–166
Soria, C., 621–624
Soria, J., 621–624
Spitzer, S., 611–616
Sugo, T., 65–88, 223–225
Suh, T.T., 276–290

Tae, W.-C., 335–339
Terasawa, F., 526–530
Thul, J., 617–620
Trochon, V., 621–624
Tsurupa, G., 185–205, 328–330
Turci, M., 167–184

Ugarova, T.P., 368–385

Van der Gaag, M.S., 630–633
van Erck, M.G.M., 639–642

van Hinsbergh, V.W.M., 426–437
Vannier, J.P., 621–624
Vasse, M., 621–624
Verschuur, M., 639–642
Vincent, L., 621–624
Vorster, H.H., 634–638

Walker, J., 247–260
Wang, W., 247–260
Weisel, J.W., 125–128, 312–327, 331–334, 611–616
Wilberding, J.A., 542–548
Wozniak, G., 617–620

Xia, H., 480–495

Yaen, C.H., 386–405
Yakovlev, S., 122–124, 185–205
Yakubenko, V.P., 368–385
Yang, Z., 31–43
Yonekawa, O., 526–530
Yoo, G., 531–535

Zhang, Z., 469–479